CMOS Analog Integrated Circuits

Amplifiers, Comparators, Multipliers, Filters, and Oscillators

CMOS Analog Integrated Circuits

Amplifiers, Comparators, Multipliers, Filters, and Oscillators

Tertulien Ndjountche

CRC Press
Taylor & Francis Group
Boca Raton London New York

CRC Press is an imprint of the
Taylor & Francis Group, an **informa** business

CRC Press
Taylor & Francis Group
6000 Broken Sound Parkway NW, Suite 300
Boca Raton, FL 33487-2742

First issued in paperback 2020

ISBN-13: 978-1-138-59972-7 (hbk)
ISBN-13: 978-0-367-73310-0 (pbk)

Visit the Taylor & Francis Web site at
http://www.taylorandfrancis.com

and the CRC Press Web site at
http://www.crcpress.com

Contents

Preface

Since the publication of the first edition of the book, *CMOS Analog Integrated Circuits*, the complementary metal-oxide semiconductor (CMOS) process used for the fabrication of integrated circuits has been constantly scaled down. Current nanometer CMOS processes have the advantages of increasing the chip density and circuit speed, and reducing the power supply voltage. However, significant challenges (leakage currents, variability of technological parameters) for the analog circuit design can be related to the use of the nanometer CMOS process, especially below 65 nm. They will only be tackled by using appropriate analog synthesis techniques and computer-aided design tools at the circuit and physical levels.

Hardware developments have been a major vehicle in popularizing the applications of signal processing theory in both science and engineering. The book describes the important trends of designing high-speed and power-efficient front-end analog circuits, which can be used alone or to interface modern digital signal processors and micro-controllers in various applications such as multimedia, communication, instrumentation, and control systems.

The book contains resources to allow the reader to design CMOS analog integrated circuits with improved electrical performance. It offers a complete understanding of architectural- and transistor-level design issues of analog integrated circuits. It provides a comprehensive, self-contained, up-to-date, and in-depth treatment of design techniques, with an emphasis on practical aspects relevant to integrated circuit implementations.

Starting from an understanding of the basic physical behavior and modeling of MOS transistors, we review design techniques for more complex components such as amplifiers, comparators, and multipliers. The book details all aspects from specifications to the final chip related to the development and implementation process of filters, analog-to-digital converters (ADCs) and digital-to-analog converters (DACs), phase-locked loops (PLLs) and delay locked loops (DLLs). It provides analyses of architectures and performance limitation issues affecting the circuit operation. The focus is on designing and verifying analog integrated circuits.

The book is intended to serve as a text for the core courses in analog integrated circuits and as a valuable guide and reference resource for analog circuit designers and graduate students in electrical engineering programs. It provides balanced coverage of both theoretical and practical issues in hierarchically organized format. With easy-to-follow mathematical derivations of all equations and formulas, the book also contains graphical plots, and a number of open-ended design problems to help determine the most suitable circuit architecture satisfying a given set of performance specifications. To appreciate the material in this book, it is expected that the reader has a rudimentary understanding of semiconductor physics, electronics, and signal processing.

New to this edition

Every chapter in the second edition has been revised to reflect the evolution of modern CMOS process technology. Furthermore, the text emphasizes paradigms that needed to be mastered and covers new material such as:

1. MOS transistor short channel effects and capacitor modeling
2. Transistor sizing based on the g_m/I_D methodology
3. Temperature compensation of voltage and current references
4. Frequency compensation of three-stage operational amplifiers
5. Comparator design based on settling time specification
6. Analysis of track-and-hold circuits

Content overview

The book contains seven chapters and three appendices.

Chapter 1
MOS Transistors
Descriptions of CMOS technology and transistors are provided. Different equivalent transistor models, including SPICE representations, are covered. The shrinking process of transistors results in an increase in the device physics complexity. Thus, advanced models, which can accurately describe the different electrical characteristics, are required to meet the circuit design specifications.

Chapter 2
Physical Design of MOS Integrated Circuits
Physical design and fabrication considerations for high-density integrated-circuits (ICs) in deep-submicrometer processes are reviewed. In addition to considering RLC models for the interconnect and package parasitic components, it appears also necessary to take into account the coupling through the

common substrate during the IC design. Advances in packaging technology will be required to support high-performance ICs.

Chapter 3
Bias and Current Reference Circuits
Circuit structures for the design of current mirrors (current sources and sinks) and voltage references are reviewed. Generally, the effects of the IC process and power supply variations on these basic blocks should be minimized to improve the overall performance of a device. Current mirrors are used in a variety of circuit building blocks to copy or scale a reference current, while voltage references are required to set an accurate and stable voltage for biasing circuits irrespective of fluctuations of the supply voltage and changes in operating temperature.

Chapter 4
CMOS Amplifiers
Topologies of amplifiers, which are suitable for the design of analog circuits, are described. The factors determining the nonideal behavior of an amplifier circuit are considered. To be tailored for a given application, an architecture has to meet the trade-off requirement among the different specifications, such as gain, bandwidth, phase margin, signal swing, noise, and slew rate. Design methods, which result in the optimization of specific performance characteristics, are summarized.

Chapter 5
Nonlinear Analog Components
Circuit architectures for comparators and multipliers are reviewed. Theoretical analysis is carried out for design and optimization purposes. The performances of comparators are essentially limited by the switching speed and mismatches of transistor characteristics, resulting in voltage offsets, while the main limitations affecting the operation of multipliers are nonlinear distortions. The design challenge is to meet the requirements of low-voltage and low-power circuits.

Chapter 6
Continuous-Time Circuits
Continuous-time circuits are required to interface digital signal processors to real-world signals. They are based on components such as transistors, resistors, capacitors, and inductors. The choice of an architecture and design technique depends on the performance parameters and application frequency range. The use of inductors, which can only be integrated with moderate efficiency (low quality factor, parasitic elements) in CMOS processes, is restricted to high-frequency building blocks with tuned characteristics. Using active components (transistor and operational amplifier) and capacitors, MOSFET-C and G_m-C structures have proven reliable for the design of integrated circuits in the video

frequency (or MHz) range.

Chapter 7
Switched-Capacitor Circuits
Switched-capacitor circuits are used in the design of large-scale integrated systems. They are based on basic building blocks such as sample-and-hold, integrator, and gain stage, which can be optimized to meet the requirements of low power consumption and chip area. Design techniques, which result in the minimization of the circuit sensitivity to component imperfections, are described. Accurate switched-capacitor filters are obtained by performing the synthesis in the z-domain, and using stray insensitive circuits for the implementation of the resulting signal-flow graph.

Appendices
Three appendices cover the following topics:
Appendix A: Transistor Sizing in Building Blocks
Appendix B: Signal Flow Graph
Appendix C: Notes on Track-and-Hold Circuit Analysis

Acknowledgments

Many of the changes in this edition were made in response to feedback received from some readers of the first edition. I would like to thank all those who took the time to send me messages.

I am grateful for the support of colleagues and students whose remarks helped refine the content of this book.

I would like to thank Prof. Dr.-Ing. h.c. R. Unbehauen (Erlangen-Nuremberg University, Germany). His continuing support, the discussions I had with him, and the comments he made have been very useful.

I express my sincere gratitude for all the support and spontaneous help I received from Dr. Fa-Long Luo (Element CXI, USA).

I wish to acknowledge the suggestions and comments provided by Prof. Avebe Zibi (UY-I, CM) and Prof. Emmanuel Tonye (ENSP, CM) during the early phase of this project.

While doing this work, I received much spontaneous help from some international experts: Prof. Ramesh Harjani (University of Minnesota, Minneapolis, Minnesota), Prof. Antonio Petraglia (Universidade Federal do Rio de Janeiro, Brazil), Dr. Schmid Hanspeter (Institute of Microelectronics, Windisch, Switzerland), Prof. Sanjit K. Mitra (University of California, Santa Babara, California), and Prof. August Kaelin (Siemens Schweiz AG, Zurich, Switzerland). I would like to express my thanks to all of them.

I am also indebted to the publisher, Nora Konopka, the project coordinator, Kyra Lindholm, the production editor, Michele Dimont, and the CRC Press editorial team of the previous edition, Jessica Vakili, Karen Simon, Brittany Gilbert, Stephany Wilken, Christian Munoz, and Shashi Kumar, for their valuable comments and reviews at various stages of the manuscript preparation, and their quality production of the book.

Finally, I would like to truly thank all members of my family and friends for the continual love and support they have given during the writing of this book.

1

MOS Transistors

CONTENTS

In almost all modern electronic circuits, transistors are the key active element. By reducing the dimensions of MOS transistors and the wires connecting them in integrated circuits (ICs), it has been possible to increase the density and complexity of integrated systems. Figure 1.1 illustrates the reduction in feature size over time. It is expected that a chip designed in a 35-nm IC process will include more than 10^{11} transistors in a few years [1]. But up to now, the scaling progress was essentially attributed to the improvements in manufacturing technology. However, as the physical limits are being met, some changes to the device structures and new materials will be necessary. Next, we will describe the transistor structure and the different equivalent models that are generally used for simulations.

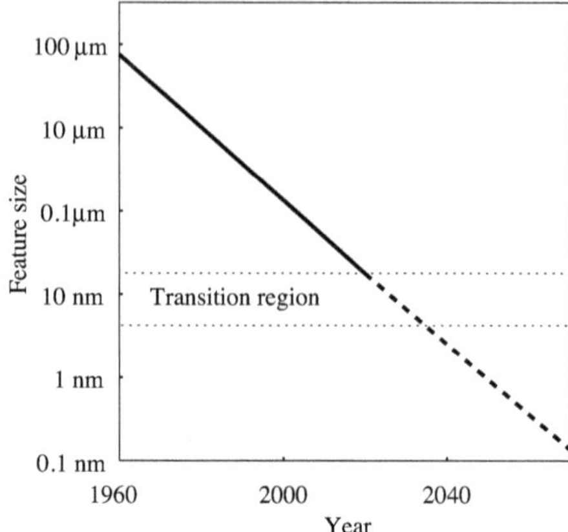

FIGURE 1.1
Transistor scaling advance: A plot of the feature size versus time.

1.1 Transistor structure

Structures of MOS transistors are shown in Figure 1.2. The drain and source of the nMOS transistor (see Figure 1.2(a)) are realized by two heavily doped n-type semiconductor regions, which are implanted into a lightly doped p-type substrate or bulk. A thin layer of silicon dioxide (SiO_2) is thermally grown over the region between the source and drain, and is covered by a polycrystalline silicon (also called polysilicon or poly), which forms the gate of the transistor. The thickness t_{ox} of the oxide layer is on the order of a few angstroms. The useful charge transfer takes place in the induced channel of the transistor, which is the substrate region under the gate oxide. The length L and the width W of the gate are estimated along and perpendicularly to the drain-source path, respectively. The substrate connection is provided by a doped p^+ region. Generally, the substrate is connected to the most negative supply voltage of the circuit so that the source-drain junction diodes are reverse-biased.

In the case of the pMOS transistor (see Figure 1.2(b)), the drain and source are formed by p^+ diffusions in the n-type substrate. A doped n^+ region is required for the realization of the substrate connection. Otherwise, the cross-sections for the two transistor types are similar. Here, the substrate is to be connected to the most positive supply voltage.

The gate and substrate of an nMOS transistor can be assumed to form the plates of a capacitor using silicon dioxide as the dielectric. As a positive voltage is applied to the gate, there is a movement of charges resulting in an augmentation of holes at the gate side and electrons at the substrate edge. Initially, the mobile holes are pushed away from the substrate surface, leaving

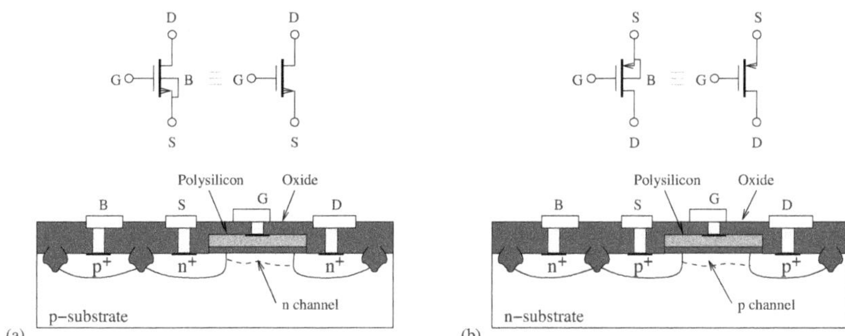

FIGURE 1.2
Model and structure of (a) nMOS transistor and (b) pMOS transistor.

behind a *depletion region* below the gate as shown in Figure 1.3, and the electron enhancement is first attributed to a *weak inversion*. By increasing the gate voltage, the concentration of electrons (minority carriers) can become larger than the one of holes (majority carriers) at the surface of the p-type substrate. This is known as *strong inversion*, which occurs for voltages greater than two times the Fermi potential. When the gate voltage is negative, the concentration of holes will increase at the surface and an *accumulation region* is formed.

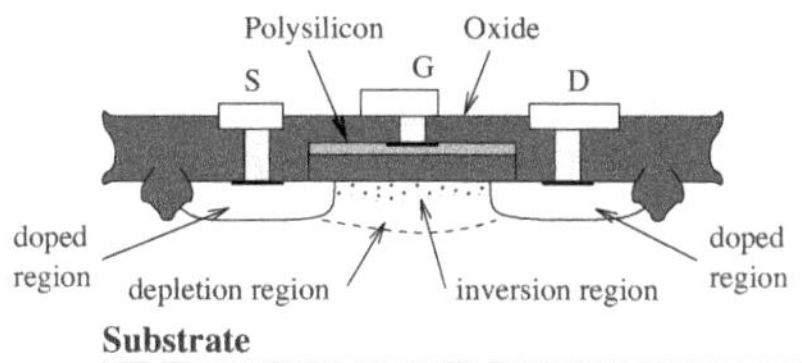

FIGURE 1.3
Localization of the inversion and depletion regions in a MOS transistor.

For a pMOS transistor, the current is carried by holes instead of electrons. The source should be the node biased at the most positive voltage. Similar results can then be obtained for the pMOS device connected to a gate voltage with inverse polarity. Generally, the electrical characteristics of pMOS transistors can be obtained from the ones of nMOS transistors by reversing the sign of all currents and voltages.

1.1.1 I-V characteristics of MOS transistors

The static characteristics of a MOS device can be determined by solving numerically a set of differential equations governing the movement of electrical charges, given the relevant boundary conditions. The behavior of the internal electrostatic potential ϕ is given by the Poisson law,

$$\nabla^2\phi = -\frac{q}{\epsilon}(N_d - N_a + p - n) \tag{1.1}$$

where q is the charge of the electron, ϵ is the dielectric constant of the semiconductor, and N_a and N_d are the concentrations of the acceptors (n-type dopant) and donors (p-type dopant), respectively. The electron and hole concentrations, n and p, can be derived, respectively, from the following conservation equations,

$$\frac{\partial n}{\partial t} = \frac{1}{q}\nabla \cdot \mathbf{J}_n - R_r + R_g \tag{1.2}$$

$$\frac{\partial p}{\partial t} = -\frac{1}{q}\nabla \cdot \mathbf{J}_p - R_r + R_g \tag{1.3}$$

where R_r and R_g denote, respectively, the recombination and generation rate of electrons and holes. The electron and hole current densities, $\mathbf{J}_n$ and $\mathbf{J}_p$, are, respectively, given by

$$\mathbf{J}_n = q\mu_n\left(-n\nabla\phi + \frac{kT}{q}\nabla n\right) \tag{1.4}$$

$$\mathbf{J}_p = q\mu_p\left(-p\nabla\phi - \frac{kT}{q}\nabla p\right) \tag{1.5}$$

where μ_n and μ_p are the electron and hole mobilities, respectively, k is Boltzmann's constant ($k = 1.38 \times 10^{-23}$ J/K), and T represents the absolute temperature (in K). It should be noted that simple and compact transistor models are generally used for the analysis and design at the circuit level.

1.1.2 Drain current in the strong inversion approximation

Based on Boltzmann's distribution [2, 3], the concentration of electrons and holes can be respectively computed as

$$n = n_{dp}\exp\left[\frac{q}{kT}(\phi - V - V_{SB})\right] \tag{1.6}$$

$$p = p_{dp}\exp\left(-\frac{q\phi}{kT}\right) \tag{1.7}$$

where n_{dp} is the electron concentration within the diffusion region on the p side, p_{dp} is the corresponding concentration for holes, ϕ represents the potential of the field $\mathbf{E} = -\nabla\phi$, and V is the applied voltage. Given the relation, $p_{dp} \simeq N_a$, we can write $n_{dp} \simeq n_i^2/N_A$, where n_i is the intrinsic charge density.

With the assumption that the current flow is essentially one-dimensional from the source to drain and the mobility is constant throughout the channel, the current density becomes

$$J_n(x,y) = q\mu_n n(x)\frac{\partial V}{\partial y} \tag{1.8}$$

and using a spatial integration, the drain-source current for long channel transistors can be written as

$$I_{DS}\int_0^L dy = \int_0^W dz \int_0^{V_{DS}} dV \int_0^{x_W} q\mu_n n(x)dx \tag{1.9}$$

and

$$I_{DS} = \mu_n\left(\frac{W}{L}\right)\int_0^{V_{DS}} Q_n dV \tag{1.10}$$

where

$$Q_n = q\int_0^{x_W} n(x)dx \tag{1.11}$$

The mobile charge per unit area in the inversion region, Q_n, is related to the charge in the semiconductor, Q_s, and the charge in the depletion region, Q_d, that is,

$$Q_n = Q_s - Q_d \tag{1.12}$$

To proceed further, we assume that

$$V_{GS} - V_{FB} = V_{GB} = V_{ox} + \phi_s \tag{1.13}$$

where V_{FB} is the flat-band voltage, V_{ox} is the voltage drop across the oxide, and ϕ_s is the potential at the silicon-oxide interface referenced to the bulk. Furthermore,

$$\phi_s = \phi_s(0) + V(y) = -2\psi_F + V(y) \tag{1.14}$$

with ψ_F being the bulk Fermi potential, which is negative for a *p*-type substrate.

Figure 1.4 shows the variation of the total charge per unit area, Q_s, as a function of the surface potential. The following relation can be written:

$$Q_s = -C_{ox}V_{ox} = -C_{ox}[V_{GB} + 2\psi_F - V(y)] \tag{1.15}$$

where $C_{ox} = \epsilon_{ox}/t_{ox}$ is the gate oxide capacitance per unit area, ϵ_{ox} is the oxide permittivity, and t_{ox} denotes the oxide thickness. As a small positive voltage is applied to the gate, holes are lessened from the vicinity of the oxide-silicon interface, and a space-charge region consisting of stationary acceptor ions is established. The depletion charge is then given by

$$Q_d = -qN_AW_d = -\sqrt{2q\epsilon N_A[V(y) - 2\psi_F]} \tag{1.16}$$

where $W_d = \sqrt{2\epsilon[V(y) - 2\psi_F]/qN_A}$ is the width of the depletion region. In the

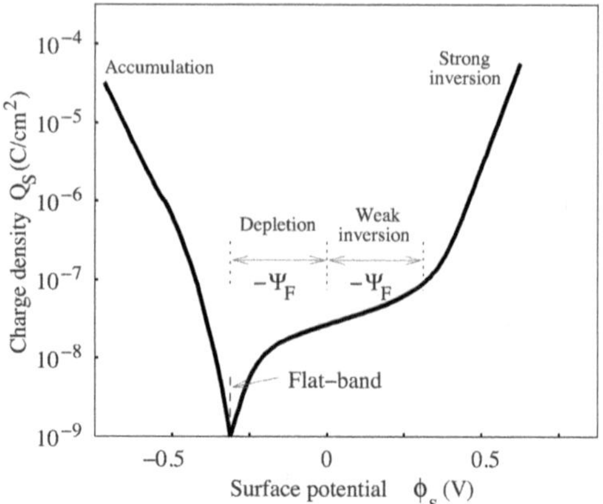

FIGURE 1.4
Plot of the charge density versus the surface potential.

small-signal analysis, the depletion capacitance is obtained as $C_d = \partial Q_d/\partial V$. The expression of the drain current reads

$$I_{DS} = \mu_n C_{ox}\left(\frac{W}{L}\right) \times \int_0^{V_{DS}} \left[V_{GB} + 2\psi_F - V(y) - \frac{1}{C_{ox}} 2q\epsilon N_A [V(y) - 2\psi_F]^{1/2}\right] dV \tag{1.17}$$

Hence,

$$I_{DS} = \mu_n C_{ox}\left(\frac{W}{L}\right)\left[\left(V_{GS} - V_{FB} + 2\psi_F - \frac{V_{DS}}{2}\right)V_{DS} - \frac{2}{3}\gamma\left((V_{DS} - 2\psi_F)^{3/2} - (-2\psi_F)^{3/2}\right)\right] \tag{1.18}$$

where

$$\gamma = \frac{\sqrt{2\epsilon q N_A}}{C_{ox}} \tag{1.19}$$

It is generally assumed that $V_{DS} \ll -2\psi_F$ and the current I_{DS} in the triode region can be reduced to

$$I_{DS} = \mu_n C_{ox}\left(\frac{W}{L}\right)(V_{GS} - V_T - V_{DS}/2)V_{DS} \tag{1.20}$$

where

$$V_T = V_{FB} - 2\psi_F + \gamma\sqrt{-2\psi_F} \tag{1.21}$$

The current I_{DS} reaches its maximum at $V_{DS(sat)} = V_{GS} - V_T$, which can be obtained by solving the equation $\partial I_{DS}/\partial V_{DS} = 0$. Based on this result, the drain-source current in the saturation region is deduced from Equation (1.20) as follows:

$$I_{DS} = \frac{1}{2}\mu_n C_{ox}\left(\frac{W}{L}\right)(V_{GS} - V_T)^2 \tag{1.22}$$

Without the above simplifying assumption, the equation of the drain current can be derived as

$$\begin{aligned} I_{DS} = \mu_n C_{ox}\left(\frac{W}{L}\right)\Bigg[&\left(V_{GS}-V_T + \gamma\sqrt{-2\psi_F} - \frac{V_{DS}}{2}\right)V_{DS} \\ &- \frac{2}{3}\gamma\left((V_{DS} - 2\psi_F)^{3/2} - (-2\psi_F)^{3/2}\right)\Bigg] \end{aligned} \tag{1.23}$$

In the saturation region, the drain-source voltage can be obtained by solving $Q_n(L) = 0$ with $V(L) = V_{DS}$. Hence,

$$V_{DS(sat)} = V_{GS} - V_{FB} + 2\psi_F + \frac{\gamma^2}{2}\left[1 - \sqrt{1 + \frac{4(V_{GS} - V_{FB})}{\gamma^2}}\right] \tag{1.24}$$

The triode and saturation regions are illustrated on the I-V characteristics shown in Figure 1.5. For a low drain-source voltage, V_{DS}, the charges in the inversion layer are uniformly induced along the channel, resulting in a current flowing from the source to the drain. The current, I_{DS}, grows proportionally to V_{DS} and the channel behaves as a voltage-controlled resistor in the *triode or linear region*. By increasing the drain voltage, there is a reduction of the charges at the drain boundary and the channel is *pinched off*. That is, for $V_{DS} \geq V_{DS(sat)} = V_{GS} - V_T$, where V_{GS} denotes the gate-source voltage, the channel ceases to further conduct electricity and its conductivity is significantly reduced. The current is now due to the charge drift and I_{DS} remains practically constant. In this case, the transistor is considered to operate in the saturation region.

A plot of the square root of the normalized drain current versus the gate-source voltage is shown in Figure 1.6 for several values of the bulk-source voltage, V_{BS}. Note that the threshold voltage changes with V_{BS}, provided that this latter is different from zero.

1.1.3 Drain current in the subthreshold region

For a gate voltage, V_G, greater than 0 and less than V_T, the drain-source current still exhibits a magnitude different from zero, which decreases exponentially. This corresponds to the subthreshold region. Here, the current I_{DS} is due to the diffusion [4] instead of the drift process, as is the case in the strong inversion region. Thus,

$$I_{DS} = -\mu_n q A \frac{kT}{q}\frac{\triangle n}{\triangle y} \tag{1.25}$$

$$= \mu_n q A \frac{kT}{q}\frac{n(0) - n(L)}{L} \tag{1.26}$$

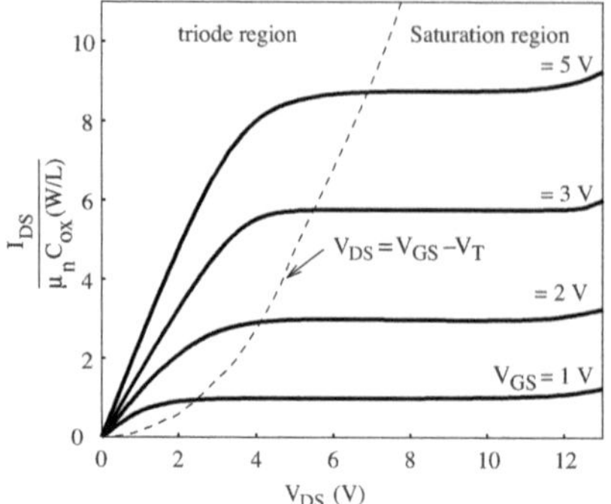

FIGURE 1.5
Plot of the drain current versus the drain-source voltage.

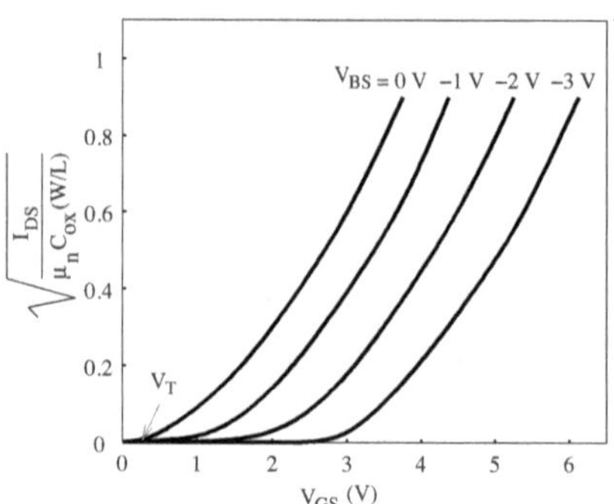

FIGURE 1.6
Plot of the drain current versus the gate-source voltage.

where $A = W \cdot \triangle d$ is the cross-section area of the current flow; $\triangle d$ is the channel depth, which is defined as the distance from the silicon-oxide interface at which the potential is lowered by kT/q, i.e., $\triangle d = kT/qE_s$; and E_s is the surface field given by

$$E_s = -\frac{Q_d}{\epsilon} = \sqrt{\frac{2qN_A\phi_s}{\epsilon}} \tag{1.27}$$

The electron densities near the source, $n(0)$, and drain, $n(L)$, are, respectively, given by

$$n(0) = n_{dp} \exp\left[\frac{q}{kT}(\phi_s + 2\psi_F - V_{SB})\right] \tag{1.28}$$

$$n(L) = n_{dp} \exp\left[\frac{q}{kT}(\phi_s + 2\psi_F - V_{DB})\right] \tag{1.29}$$

where $n_{dp} \simeq N_a$, as the concentration of acceptors is the most significant. The drain-source current can be written as

$$\begin{aligned} I_{DS} = {} & \mu_n q \left(\frac{W}{L}\right)\left(\frac{kT}{q}\right)^2 \\ & \times \sqrt{\frac{\epsilon N_a}{2q\phi_s}} \exp\left[\frac{q}{kT}(\phi_s + 2\psi_F - V_{SB})\right]\left[1 - \exp\left(-\frac{qV_{DS}}{kT}\right)\right] \end{aligned} \tag{1.30}$$

From Gauss's law of charge balance applied to the silicon-oxide interface [5], the gate-source voltage for a given surface potential, ϕ_s, is of the form

$$V_{GS} = V_{FB} + \phi_s - V_{SB} - \frac{Q_d + Q_i}{C_{ox}} \tag{1.31}$$

where V_{FB} represents the flat-band voltage, V_{SB} is the source-substrate bias voltage, Q_d denotes the depletion charge, Q_i is the inversion charge at the silicon-oxide interface, and C_{ox} is the oxide capacitance per unit area. Due to the nonlinear dependence of the charge on ϕ_s, an expression of ϕ_s is easily obtained from the next first-order Taylor series of V_{GS} around $\phi_{so} + V_{SB}$ [6],

$$V_{GS} = V_{GS}^* + \eta(\phi_s - \phi_{so} - V_{SB}) \tag{1.32}$$

where

$$V_{GS}^* = V_{GS}|_{\phi_s=\phi_{so}+V_{SB}} \tag{1.33}$$

and

$$\eta = \left.\frac{dV_{GS}}{d\phi_s}\right|_{\phi_s=\phi_{so}+V_{SB}} \tag{1.34}$$

Thus,

$$\phi_s = \frac{V_{GS} - V_{GS}^*}{\eta} + \phi_{so} + V_{SB} \tag{1.35}$$

and the subthreshold current becomes

$$\begin{aligned} I_{DS} =& \mu_n \left(\frac{W}{L}\right)\left(\frac{kT}{q}\right)^2 C_d \\ & \times \exp\left[\frac{q}{kT}\left(\frac{V_{GS} - V_{GS}^*}{\eta} + \phi_{so} + 2\psi_F\right)\right]\left[1 - \exp\left(-\frac{qV_{DS}}{kT}\right)\right] \end{aligned} \tag{1.36}$$

where C_d denotes the depletion capacitance given by

$$C_d = \sqrt{\frac{\epsilon q N_a}{2\phi_s}} \tag{1.37}$$

In the weak inversion region, $-\psi_F + V_{SB} < \phi_s < -2\psi_F + V_{SB}$, and great accuracy can be ensured by choosing the reference point $\phi_{so} = -3\psi_F/2$. However, with $\phi_{so} = -2\psi_F$, the voltage V_{GS}^* is reduced to V_T. The current I_{DS} depends on the gate-source and drain-source voltages. But, its dependence on the voltage, V_{DS}, is considerably reduced as V_{DS} becomes greater than a few kT/q.

The I-V characteristics of an nMOS transistor are shown in Figure 1.7. For $V_{DS} > 0.1$ V, the current I_{DS} is almost independent of V_{DS}, and we have

$$I_{DS} \simeq \mu_n \left(\frac{W}{L}\right)\left(\frac{kT}{q}\right)^2 C_d \exp\left[\frac{q}{kT}\left(\frac{V_{GS} - V_{GS}^*}{\eta} + \phi_{so} + 2\psi_F\right)\right] \tag{1.38}$$

It should be noted that nMOS and pMOS devices feature the characteristics that are the mirror of each other. In addition, pMOS transistors generally feature the lower holes mobility, $\mu_p \simeq \mu_n/4$. This can result in a lower current drive, transconductance and output resistance.

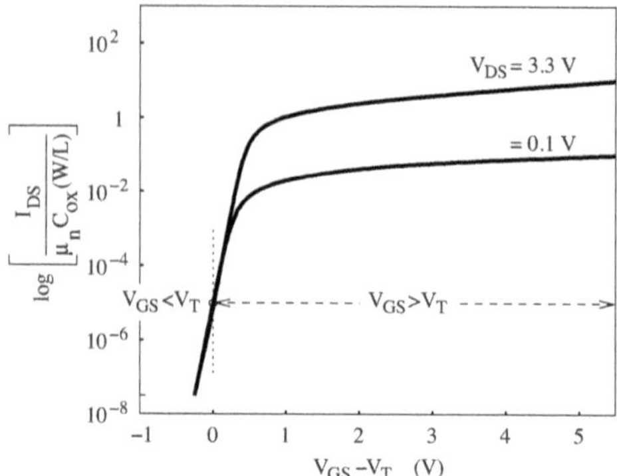

FIGURE 1.7
I-V characteristics of an nMOS transistor.

1.1.4 MOS transistor capacitances

A model for MOS transistor capacitances is required to accurately predict the *ac* behavior of circuits. The silicon oxide, which provides the isolation of the gate from the channel, can be considered as the dielectric of a capacitor with the value C_{ox}. Due to the lateral diffusion, the effective channel length, L_{eff}, is shorter than the drawn length, L, as shown in Figure 1.8, and overlap capacitors are formed between the gate and drain/source. The expression of the overlap capacitance can be given by

$$C_{gso} = C_{gdo} = C_{ox} W \triangle L \tag{1.39}$$

where $\triangle L = (L - L_{eff})/2$. Note that the accurate determination of the overlap capacitance per unit width, C_{ov}, can require more precise calculations. The capacitors related to the silicon oxide, $C_g = C_{ox}WL$, and the depletion region, C_d, exist between the gate and channel and between the channel and substrate. In addition, junction capacitors are present between the source/drain and substrate. They consist of two geometry-dependent components related, respectively, to the bottom-plate and side-wall of the junction.

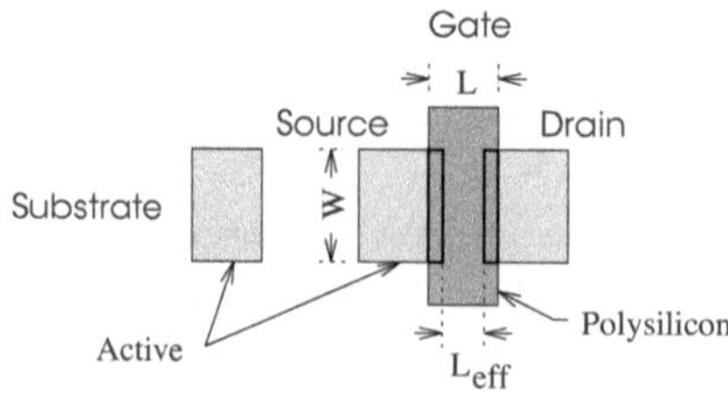

FIGURE 1.8
MOS transistor layout.

A transistor can then be represented as shown in Figure 1.9 with parasitic capacitors between every two of the four output nodes [7]. It is assumed that the capacitance between the source and drain is negligible.

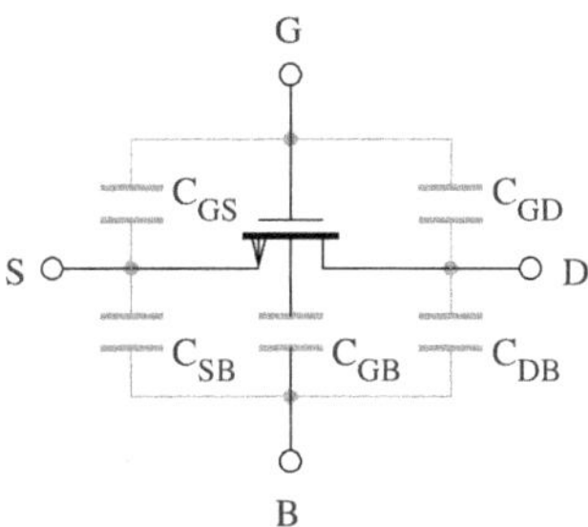

FIGURE 1.9
MOS transistor capacitance model.

Let Q_G, Q_B, Q_S, and Q_D be the gate, bulk, source, and drain charges, respectively. The charge conservation principle results in the following relation:

$$Q_G + Q_B + Q_S + Q_D = 0 \tag{1.40}$$

The charges Q_G and Q_B can be computed from the Poisson equation. The sum of Q_D and Q_S is equal to the channel charge, and the drain and source charge partition changes uniformly from the Q_D/Q_S ratio of 50/50 in the triode region to 40/60 in the saturation region [8]. However, the charge partition will become closer to the ratio 0/100 when the transistor is switched at a speed greater than the channel charging time. Based on the charge model, the capacitances can be defined as

$$C_{GS} = \frac{\partial Q_G}{\partial V_S} \tag{1.41}$$

$$C_{GD} = \frac{\partial Q_G}{\partial V_D} \tag{1.42}$$

and

$$C_{GB} = \frac{\partial Q_G}{\partial V_B} \tag{1.43}$$

There are normally two capacitors between a couple of nodes. Here, the capacitors are nonlinear and the condition of reciprocity is not fulfilled. That is, $C_{SG} = \partial Q_S/\partial V_G$, $C_{DG} = \partial Q_D/\partial V_G$, and $C_{BG} = \partial Q_B/\partial V_G$. The three capacitances associated with the gate are represented in Figures 1.10 and 1.11.

Capacitors between the source/drain and substrate, C_{SB} and C_{DB}, are caused by the charge in the depletion region of the *pn* regions. They can be considered passive, but dependent on the reverse voltage across the junction. Note that the gate-source, gate-drain, and parasitic source/drain-bulk capacitances per unit gate width of pMOS and nMOS transistors, which are on

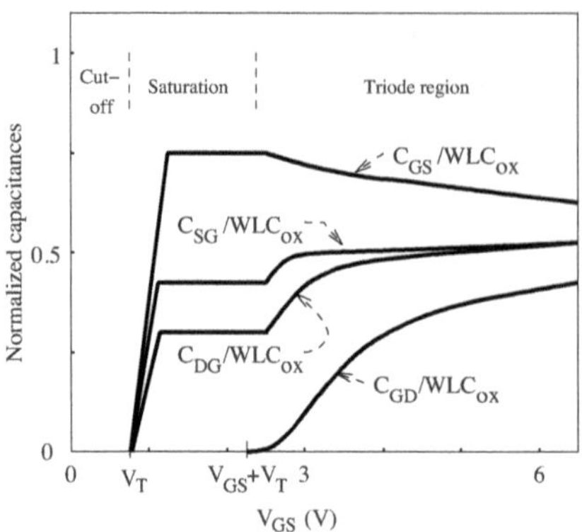

FIGURE 1.10
Plot of normalized capacitances (C_{DG}, C_{GD}, C_{GS}, and C_{SG}) versus the gate-source voltage. (Adapted from [7], ©1978 IEEE.)

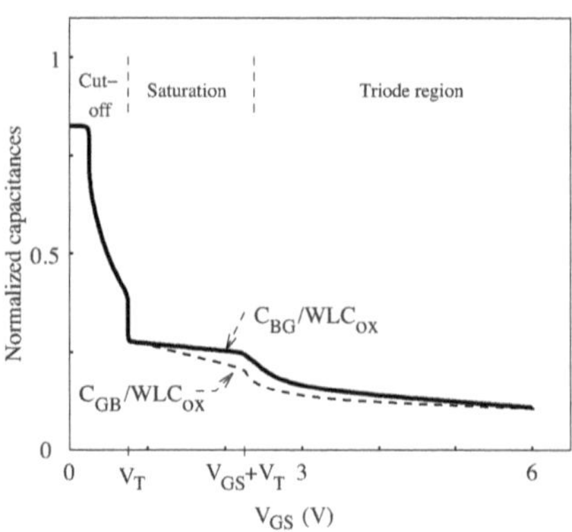

FIGURE 1.11
Plot of normalized capacitances (C_{GB} and C_{BG}) versus the gate-source voltage. (Adapted from [7], ©1978 IEEE.)

the order of 1 fF/µm, 0.5 fF/µm, and 1.5 fF/µm, respectively, remain almost unchanged across technology nodes [9].

1.1.5 Scaling effects on MOS transistors

Typical values of the threshold voltages and transconductance characteristics are provided in Table 1.1 for the 0.25-µm, 0.18-µm, and 0.13-µm CMOS processes. Note that the transconductance is defined as half of the product of the charge carrier mobility and oxide capacitance.

TABLE 1.1
CMOS Process Characteristics

	0.25 µm $V_{DD} = 2.5$ V		0.18 µm $V_{DD} = 1.8$ V		0.13 µm $V_{DD} = 1.5$ V	
	nMOS	pMOS	nMOS	pMOS	nMOS	pMOS
V_T (V)	0.65	−0.51	0.49	−0.43	0.44	−0.42
K' (µA/V^2)	114.9	25.5	154.0	33.2	283.1	49.2

While resulting in the improvement of the circuit performance (area, speed, power dissipation), the reduction of the transistor size is affected by limitations associated with the thickness and electrical characteristics of the gate dielectric. With the constant-voltage scaling, the magnitude of the electric field in the channel can increase considerably, placing some limitations in the transistor miniaturization.

- Small-geometry effects
  When the effective channel length is on the order of the source/drain junction depletion width, the potential distribution in the channel becomes dependent on the lateral electric field component in addition to the normal one and features a two-dimensional representation. As a result, the threshold voltage is now dependent on the drain bias, channel length, and channel width. The threshold voltage is decreased as the channel length is reduced and also as the drain bias is raised. By reducing excessively the channel length, the depletion region of the drain junction can punch through the one of the source junction and the drain-source current ceases to be controlled by the gate-source voltage. The variation of the threshold voltage with the drain bias is caused by the drain-induced barrier lowering effect at the source junction.

- Hot carrier effects
  The lateral electric field increases quickly as the transistor is scaled down. During the displacement from the source to the drain, carriers can be accelerated when they can acquire the energy related to the high electric field existing in the channel. They may collide with fixed atoms to generate additional electron/hole pairs. This process is repetitive and results in an abnormal increase in the drain current. The high-energy electrons may also be trapped at the silicon-oxide interface, and in turn cause the degradation of the device characteristics, such as the voltage threshold and transconductance. For the long-term operation, the transistor can exhibit reliability problems, due to the variation of the I-V characteristics.

- Gate-induced drain leakage current
  The gate-induced drain leakage current is observed in a transistor biased in the off-state. Due to the tunneling effect through the gate insulator, it is dependent on the drain-gate voltage and becomes important as the gate-dielectric thickness is decreased.

1.2 Transistor SPICE models

1.2.1 Electrical characteristics

The first transistor SPICE models are known as level 1, 2, and 3 [10]. They were developed based on the equations characterizing the physical behavior of MOS devices. (Note that SPICE stands for Simulation Program with Integrated Circuit Emphasis).

The accuracy of the level 1 model is adequate for transistors with channel length and width greater than 4 µm. The drain-source current, I_{DS}, in the cutoff, triode, and saturation regions is given by the next equation [11],

$$I_{DS} = \begin{cases} 0 & \text{for } V_{GS} \leq V_T \\ K\left[2(V_{GS} - V_T)V_{DS} - V_{DS}^2\right] & \text{for } V_{GS} > V_T \text{ and } V_{DS} < V_{GS} - V_T \\ K(V_{GS} - V_T)^2(1 + \lambda V_{DS}) & \text{for } V_{GS} > V_T \text{ and } V_{DS} \geq V_{GS} - V_T \end{cases} \tag{1.44}$$

where

$$V_T = V_{T0} + \gamma(\sqrt{\psi_B + V_{SB}} - \sqrt{\psi_B}) \tag{1.45}$$

$$V_{T0} = V_{FB} + \psi_B + \gamma\sqrt{\psi_B} \tag{1.46}$$

and

$$\psi_B \simeq -2\psi_F \tag{1.47}$$

The transconductance parameter K is defined as

$$\beta = 2K = \mu_n C_{ox} \frac{W}{L_{eff}} \tag{1.48}$$

where L_{eff} is the effective channel length, and the saturation voltage is given by $V_{DS(sat)} = V_{GS} - V_T$. The parameter γ represents the body-effect coefficient, ψ_B is the surface potential in the strong inversion for zero back-gate bias, ψ_F represents the bulk Fermi potential, and λ is the channel-length modulation coefficient. By introducing the parameter λ, the slight increase in the drain current in the saturation region is taken into account. However, the drain current in the triode region can be multiplied by the term $1 + \lambda V_{DS}$ to provide a continuous transition to the saturation current during the simulation. To model the drain-source resistance near the transistor bias point, the channel-length modulation coefficient can be expressed as,

$$\lambda = 1/V_A \tag{1.49}$$

where V_A is the Early voltage. The drain-bulk and source-bulk currents, I_{DB} and I_{SB}, can be written as

$$I_{DB} = I_{D0}\left[\exp\left(\frac{qV_{DB}}{kT}\right) - 1\right] \tag{1.50}$$

and

$$I_{SB} = I_{S0}\left[\exp\left(\frac{qV_{SB}}{kT}\right) - 1\right] \tag{1.51}$$

where I_{D0} and I_{S0} are the saturation currents of the drain and source junctions, respectively.

A small-signal equivalent model of a MOS transistor [12] is shown in Figure 1.12. The gate-source and source-bulk (or simply bulk) transconductances,

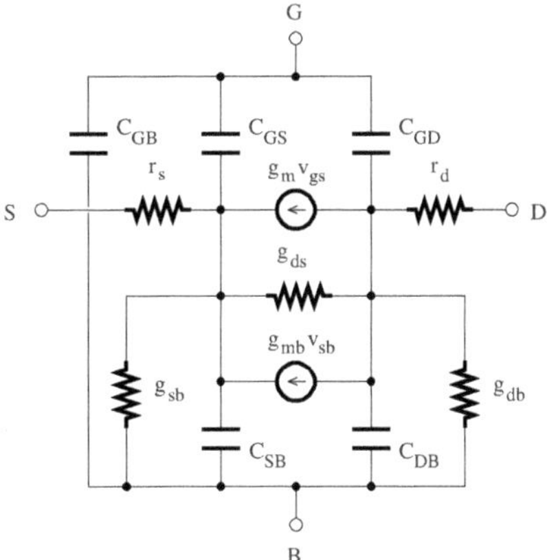

FIGURE 1.12
Small-signal equivalent model of a MOS transistor.

g_m and g_{mb}, and the drain-source conductance, g_{ds}, can be related to the I-V characteristic as follows:

$$g_m = \left.\frac{\partial I_{DS}}{\partial V_{GS}}\right|_{V_{DS},V_{BS}} \tag{1.52}$$

$$g_{mb} = \left.\frac{\partial I_{DS}}{\partial V_{SB}}\right|_{V_{GS},V_{DS}} \tag{1.53}$$

and

$$g_{ds} = \left.\frac{\partial I_{DS}}{\partial V_{DS}}\right|_{V_{GS},V_{BS}} \tag{1.54}$$

The five capacitances, C_{GD}, C_{GS}, C_{GB}, C_{DB}, and C_{SB}, are determined by the corresponding charges. The ohmic resistances, r_d and r_s, are not affected by the bias condition and are lower than the other transistor resistances. Due to the fact that the drain-bulk and source-bulk junctions are reverse-biased, the effect of the conductances $g_{db} = \partial I_{DB}/\partial V_{DB}$ and $g_{sb} = \partial I_{SB}/\partial V_{SB}$ on the device behavior is limited. It should be noted that additional substrate-coupling resistors and the gate resistor are to be included in the compact model of Figure 1.12 for an accurate description of the radio-frequency response.

Find the transconductances, g_m and g_{mb}, and the conductance, g_{ds}, for a transistor operating in the saturation region.

From the drain-source current in the saturation region,

$$I_{DS} = I_D = K(V_{GS} - V_T)^2(1 + \lambda V_{DS}) \tag{1.55}$$

the threshold voltage,

$$V_T = V_{T0} + \gamma(\sqrt{\psi_B + V_{SB}} - \sqrt{\psi_B}) \tag{1.56}$$

and with the assumption that $V_{BS} = -V_{SB}$ and $\psi_B = 2\phi_F$, we can obtain the next relations

$$g_m = \left.\frac{\partial I_D}{\partial V_{GS}}\right|_{V_{DS},V_{BS}} = \frac{2I_D}{V_{GS} - V_T} \tag{1.57}$$

$$g_{ds} = \left.\frac{\partial I_D}{\partial V_{DS}}\right|_{V_{GS},V_{BS}} = \lambda\frac{I_D}{1 + \lambda V_{DS}} \tag{1.58}$$

and

$$\begin{aligned} g_{mb} &= \left.\frac{\partial I_D}{\partial V_{BS}}\right|_{V_{GS},V_{DS}} \\ &= \left.\frac{\partial I_D}{\partial V_T}\left(-\frac{\partial V_T}{\partial V_{SB}}\right)\right|_{V_{GS},V_{DS}} \\ &= g_m \left.\frac{\partial V_T}{\partial V_{SB}}\right|_{V_{GS},V_{DS}} \\ &= g_m\frac{\gamma}{\sqrt{2\phi_F + V_{SB}}} = (n-1)g_m \end{aligned} \tag{1.59}$$

where n is the substrate factor. Note that $g_{ds} \simeq \lambda I_D = I_D/V_A$ if $V_{DS} > V_{DS(sat)}$.

The level 1 model is adequate for initial design calculations and does not support the transistor operation in the subthreshold region.

The level 2 model includes the bulk-source voltage, V_{BS}, and the body-effect parameter, γ, in the I-V characteristics. By defining the ON voltage as $V_{ON} = V_T + NkT/q$, the drain-source current in the subthreshold region can be written as

$$I_{DS} = I_{DS(ON)} \exp\left(q\frac{V_{GS} - V_{ON}}{NkT}\right) \tag{1.60}$$

where $N = 1 + qNFS/C_{ox} + C_d/C_{ox}$, NFS is an empirical parameter, and C_d is the depletion capacitance. During the normal operation of the transistor, the current I_{DS} takes the next form

$$\begin{aligned} I_{DS} = 2K_{eff}\Big[&(V_{GS} - V_{FB} - \psi_B - V_{DS}/2)V_{DS} \\ &- \frac{2}{3}\gamma[(V_{DS} - V_{BS} + \psi_B)^{3/2} - (-V_{BS} + \psi_B)^{3/2}]\Big] \end{aligned} \tag{1.61}$$

where

$$K_{eff} = K\left(\frac{1}{1-\lambda V_{DS}}\right)\left(\frac{E_{CRIT}\epsilon/C_{ox}}{V_{GS}-V_{ON}}\right)^{U_{EXP}} \tag{1.62}$$

and

$$\gamma = \frac{\sqrt{2\epsilon q N_A}}{C_{ox}} \tag{1.63}$$

Note that E_{CRIT} and U_{EXP} are empirical parameters, V_{ON} is the ON voltage, V_{FB} is the flat-band voltage, and N_A denotes the substrate doping concentration. The parameter $I_{DS(ON)}$ is the value of the above drain-source current at the subthreshold boundary, that is, for $V_{GS} = V_{ON}$. Here, the drain-source voltage in the saturation region becomes

$$V_{DS(sat)} = V_{GS} - V_{FB} - \psi_B + \frac{\gamma^2}{2}\left[1 - \sqrt{1 + 4\frac{(V_{GS} - V_{FB} - V_{BS})}{\gamma^2}}\right] \tag{1.64}$$

In the level 2 representation, the slope of I_{DS} features a discontinuity between the subthreshold and strong inversion regions.

To improve the efficiency of the simulation results for transistors with channel length on the order of 1 µm, the level 3 model relies on empirical equations of the effective mobility and a correction factor, F_B, in order to take into account short-channel effects. The drain-source current is given by

$$I_{DS} = 2K\left[V_{GS} - V_T - \left(\frac{1+F_B}{2}\right)V_{DS}\right]V_{DS} \tag{1.65}$$

where

$$K = \mu_{eff} C_{ox}\frac{W_{eff}}{L_{eff}} \tag{1.66}$$

W_{eff} and L_{eff} are the effective width and length, respectively. Let

$$V_C = \frac{V_{MAX} L_{eff}}{\mu_s} \tag{1.67}$$

where μ_s is the surface mobility and V_{MAX} is used to steer the carrier velocity saturation. The expression of the effective mobility takes into account the degradation due to the lateral field and the carrier velocity saturation. Thus,

$$\mu_{eff} = \begin{cases} \dfrac{\mu_s}{1 + V_{DS}/V_C} & \text{for} \quad V_{MAX} > 0 \\ \mu_s & \text{otherwise} \end{cases} \tag{1.68}$$

and

$$\mu_s = \frac{U0}{1 + \theta(V_{GS} - V_T)} \tag{1.69}$$

where $U0$ denotes the low-field mobility constant, θ is the mobility degradation coefficient, V_{DS} is the drain-source voltage, V_{GS} is the gate-source voltage, and V_T is the threshold voltage.

In general, the carrier mobility depends on many process parameters (gate oxide thickness, substrate doping concentration) and bias conditions (threshold voltage, gate and substrate voltages). The effective mobility can be modeled as follows

$$\mu_{eff} = \frac{\mu_0}{1 + (E_{eff}/E_0)^{\nu}} \tag{1.70}$$

where μ_0 is the low-field mobility, E_{eff} is the effective electrical field, E_0 is the electrical field at the nominal temperature, and ν is a constant. By considering E_{eff} as the average electrical field experienced by carriers in the inversion layer, it can be shown that

$$E_{eff} = -\frac{Q_b + \eta_E Q_i}{\epsilon_{si}} \tag{1.71}$$

where η_E is approximately equal to 1/2 for electrons and 1/3 for holes, Q_b is the bulk depletion charge per unit area, Q_i is the inversion layer charge per unit area, and ϵ_{si} (1.0359×10^{-10} F/m) is the dielectric constant of silicon. Typical values of the mobility parameters are given in Table 1.2.

TABLE 1.2
Typical Values of Mobility Parameters

Parameter	electron	hole
μ_0 $(cm^2/V \cdot s)$	670	250
E_0 (MV/cm)	0.67	0.7
ν	1.6	1.0

To reduce the computing time, empirical expressions or Taylor expansions of the mobility, whose coefficients are determined to fit experimental data, are commonly used in computer-aided design programs such as SPICE and BSIM MOSFET models. Hence, the effective mobility can be estimated using expressions of the following form:

$$\mu_{eff} = \frac{\mu_0}{1 + U_a\left(\frac{V_{GS} - V_T}{t_{ox}}\right) + U_b\left(\frac{V_{GS} - V_T}{t_{ox}}\right)^2 + U_c V_{SB}} \tag{1.72}$$

where V_{GS} is the gate-source voltage, V_{SB} represents the source-body voltage, V_T is the threshold voltage, t_{ox} denotes the gate-oxide thickness, and U_a, U_b, and U_c are fitting coefficients.

By including the coefficients K_1, K_2, and η in the threshold voltage, which is now given by

$$V_T = V_{FB} + \psi_B + K_1\sqrt{\psi_B - V_{BS}} - K_2(\psi_B - V_{BS}) - \eta V_{DS} \tag{1.73}$$

the contributions due to the body, short-channel, narrow width, and drain-induced barrier lowering effects are considered.

The drain-source voltage in the saturation region, that is, where the displacement of carriers is governed by a constant mobility, is obtained as

$$V_{DS(sat)} = V_C + \frac{V_{GS} - V_T}{1 + F_B} - \sqrt{V_C^2 + \left(\frac{V_{GS} - V_T}{1 + F_B}\right)^2} \tag{1.74}$$

The level 3 model exhibits some discontinuities (output conductance at $V_{DS} = V_{DS(sat)}$, transconductance at $V_{GS} = V_T$), which can result in the nonconvergence of simulations. The Berkeley short-channel IGFET models (BSIMs) were proposed to achieve an accurate description of transistors with submicrometer sizes [13] (IGFET stands for Insulated Gate Field Effect Transistor). They rely on empirical parameters obtained by data fitting to reduce the complexity of the equations of the device characteristics. In the BSIM4 transistor model, the continuity of the drain-source current and conductances is maintained using a single current equation for the different operating regions of the transistor.

It should be noted that BSIM-common-multi-gate, or BSIM-CMG, was proposed for circuit simulation of multi-gate transistors (FinFETs). In comparison with other versions, it uses a different device structure and maintains the source/drain symmetry (that is, the second derivative of the drain current and charge are continuous for the drain-source voltage equal to zero).

1.2.2 Temperature effects

Due to the fact that a circuit can operate at a temperature different from the nominal one ($T_0 = 300\ K$) at which the model parameters are extracted, the temperature effects must be taken into account in the transistor representation. The mobility, μ, and the threshold voltage, V_T, are related to the absolute temperature, T, according to

$$\mu(T) = \mu(T_0)\left(\frac{T}{T_0}\right)^{3/2} \tag{1.75}$$

and

$$V_T(T) = V_T(T_0) + \text{VTC}\left(\frac{T}{T_0} - 1\right) \tag{1.76}$$

where VTC is a voltage temperature coefficient. The saturation current of the junction diodes (at the drain and source side) can be written as

$$J_s(T) = J_s(T_0)\exp\left\{\frac{1}{N}\left[\frac{q}{k}\left(\frac{E_g(T_0)}{T_0} - \frac{E_g(T)}{T}\right) + XT \cdot \ln\left(\frac{T}{T_0}\right)\right]\right\} \quad (1.77)$$

where N and XT are two constant parameters, and the bandgap (or energy gap) of the silicon, E_g, is given by

$$E_g(T) = 1.16 - \frac{7.02 \times 10^{-4}T^2}{T + 1108} \quad (1.78)$$

The next equations can be used to express the temperature dependence of a resistor and capacitor:

$$R(T) = R(T_0) + \text{RTC}\left(\frac{T}{T_0} - 1\right) \quad (1.79)$$

and

$$C(T) = C(T_0) + \text{CTC}(T - T_0) \quad (1.80)$$

where RTC and CTC are the resistor and capacitor temperature coefficients, respectively.

1.2.3 Noise models

Different noise sources caused by fluctuations of the device characteristics affect the operation of MOS transistors. A simplified noise model of a transistor is depicted in Figure 1.13(a). It includes the flicker (or $1/f$ noise) and thermal channel noises, whose spectrum density is represented in Figure 1.13(b). Note that $\overline{i^2_{1/f}} = g^2_m \overline{v^2_{1/f}}$, where g_m is the transistor transconductance. The corner frequency, f_c, denotes the point at which both noise types intersect.

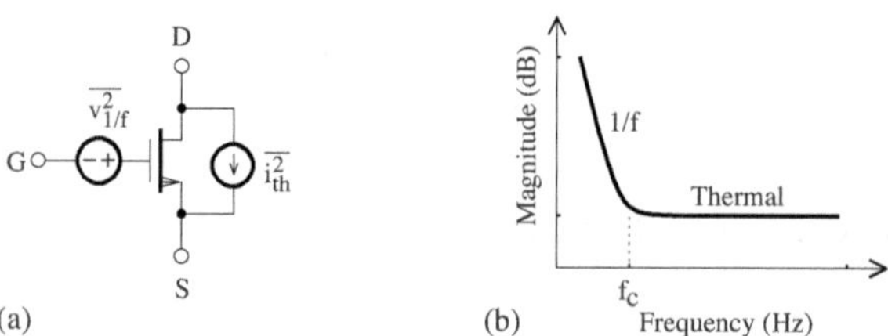

FIGURE 1.13
(a) Simplified noise model of a transistor; (b) plot of the noise spectrum density versus the frequency.

Specifically, the flicker noise, the spectral density of which is inversely proportional to the frequency, is given by an empirical equation of the form

$$\overline{i^2_d} = \frac{KF \cdot I^{AF}_{DS}}{C_{ox}L^2_{eff}}\frac{1}{f^{EF}} \quad (1.81)$$

where KF is the flicker noise coefficient, I_{DS} is the drain-source current, AF is the flicker noise current exponent, C_{ox} is the oxide capacitance per unit area, L_{eff} is the effective length of the transistor, f is the operating frequency, and EF is the flicker noise frequency exponent. The other expressions are

$$\overline{i_d^2} = \frac{KF \cdot I_{DS}^{AF}}{C_{ox} W_{eff} L_{eff}} \frac{1}{f} \tag{1.82}$$

and

$$\overline{i_d^2} = \frac{KF \cdot g_m^2}{C_{ox} W_{eff} L_{eff}} \frac{1}{f^{EF}} \tag{1.83}$$

where W_{eff} and g_m are the effective width and transconductance of the transistor, respectively. In the BSIM model used for submicrometer transistors, the flicker noise equation includes more parameters because it takes into account the fluctuations of the carrier number and surface mobility.

Let k represent the Boltzmann constant and T denote the absolute temperature. The drain channel thermal noise can be written as [14]

$$\frac{\overline{i_d^2}}{\triangle f} = 4kT\theta g_{do} \tag{1.84}$$

where $\triangle f$ is the noise bandwidth, g_{do} is the channel conductance at $V_{DS} = 0$, i.e., $g_{do} = (\partial I_{DS}/\partial V_{DS})|_{V_{DS}=0}$, and θ is a bias-dependent noise coefficient, which, for long channel transistors, is unity at zero drain bias and about 2/3 in the saturation region. However, the channel thermal noise can be accurately predicted for submicrometer transistors by the next model,

$$\frac{\overline{i_d^2}}{\triangle f} = \text{FP} \frac{4kT}{r_{ds} + \dfrac{L_{eff}^2}{\mu_{eff}|Q_{inv}|}} \tag{1.85}$$

where r_{ds} is the drain-source resistance, μ_{eff} is the effective mobility, Q_{inv} is the inversion layer charge, and FP is a fitting parameter.

The induced gate current noise, $\overline{i_g^2}$, which is generally negligible, can become important for submicrometer devices operating at high frequencies or close to the transition frequency of the transistor. It is partially correlated with the drain current noise and is expressed as [15]

$$\frac{\overline{i_g^2}}{\triangle f} = 4kT\delta g_g(1 - |c|^2) + 4kT\delta g_g|c|^2 \tag{1.86}$$

where c is a correlation coefficient given by

$$c = \frac{\overline{i_g i_d^*}}{\sqrt{\overline{i_g^2}\,\overline{i_d^2}}} \tag{1.87}$$

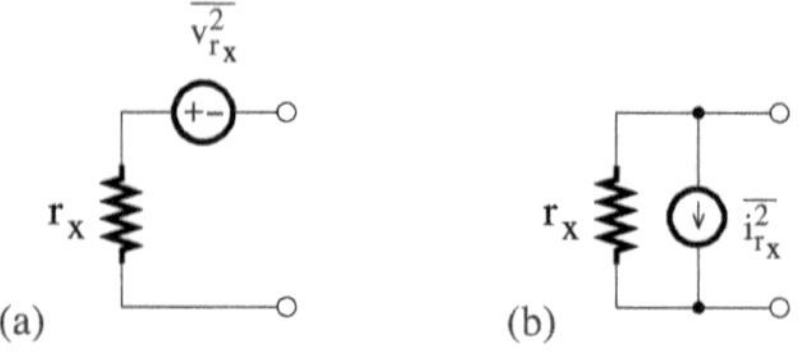

FIGURE 1.14
Representation of a resistor thermal noise: (a) voltage source, (b) current source.

and δ is the gate noise coefficient. Because the conductance g_g increases with the square of the frequency, the power spectral density of the gate noise is different from that of a white noise.

Other noise sources include the thermal noise, which is related to various terminal resistances, and shot noise. For a resistor r_x, the thermal noise can be modeled either by a series voltage source with the power spectral density, $\overline{v^2_{r_x}}$, as shown in Figure 1.14(a), and given by

$$\frac{\overline{v^2_{r_x}}}{\triangle f} = 4kTr_x \tag{1.88}$$

or by a shunt current source with the power spectral density, $\overline{i^2_{r_x}}$, as illustrated in Figure 1.14(b), and of the form

$$\frac{\overline{i^2_{r_x}}}{\triangle f} = \frac{4kT}{r_x} \tag{1.89}$$

where k is Boltzmann's constant, T is the absolute temperature, and $\triangle f$ denotes the measurement bandwidth in hertz (Hz).

Considering a resistor $r_x = 1$ kΩ, the power spectral densities of the thermal noise voltage and current at room temperature (i.e., $T = 300$ K) are

$$\overline{v^2_{r_x}}/\triangle f \simeq 16 \times 10^{-18}\ \mathrm{V}^2/\mathrm{Hz} \quad \text{and} \quad \overline{i^2_{r_x}}/\triangle f \simeq 16 \times 10^{-22}\ \mathrm{A}^2/\mathrm{Hz}$$

or equivalently in terms of root-mean-square (rms) units,

$$\sqrt{\overline{v^2_{r_x}}/\triangle f} \simeq 4\ \mathrm{nV}/\sqrt{Hz} \quad \text{and} \quad \sqrt{\overline{i^2_{r_x}}/\triangle f} \simeq 4\ \mathrm{pA}/\sqrt{Hz},$$

respectively. Assuming a bandwidth of 1 kHz, we can obtain

$$\sqrt{\overline{v^2_{r_x}}} \simeq 4\ \mu\mathrm{V}_{rms} \quad \text{and} \quad \sqrt{\overline{i^2_{r_x}}} \simeq 4\ \mathrm{nA}_{rms}$$

Note that the contributions of independent noise sources should be added using mean squared quantities, instead of rms quantities.

The shot noise is related to the tunneling currents and can be written as

$$\frac{\overline{i_{sh}^2}}{\triangle f} = 2MqI \tag{1.90}$$

where M is a multiplication factor, q is the electron charge, and I is the forward junction current. Note that the thermal noise is due to the random motion of charge carriers caused by an increase in the temperature, while the shot noise depends on the energy of carriers near a potential barrier or junction.

Using the transistor transconductance, g_m, the drain current noise, $\overline{i_d^2}$, can be related to the gate voltage noise, $\overline{v_{gs}^2}$, or the input referred noise. That is, $\overline{i_d^2} = g_m^2 \overline{v_{gs}^2}$. Note that the thermal noise generated by the transistor substrate is transmitted through the bulk transconductance, g_{mb}, as a noise current source at the drain. Its contribution to the input-referred voltage noise is $4kTr_b(g_{mb}/g_m)^2$, where r_b is the bulk resistance.

Typical values of the equivalent input voltage and current noises are generally expressed in pA/$\sqrt{\text{Hz}}$ and nV/$\sqrt{\text{Hz}}$ units, respectively.

1.3 Drain-source current valid in all regions of operation

In the weak inversion region, where $V_{GS} < V_T$, the drain-source current of an n-channel transistor can be written as

$$I_D = 2n\mu_n C_{ox} U_T^2 \left(\frac{W}{L}\right) \exp\left(\frac{V_{GS} - V_T}{nU_T}\right)\left[1 - \exp\left(-\frac{V_{DS}}{U_T}\right)\right] \tag{1.91}$$

where n is the substrate factor and is approximately comprised between 1.1 and 1.5 for typical CMOS process, μ_n is the channel carrier mobility, C_{ox} is the gate-oxide capacitance per unit area, V_T is the threshold voltage, and $U_T = kT/q$ is the thermal voltage ($U_T = 25.9$ mV at room temperature, $T = 300$ K).

When $V_{DS} \gg V_T$, we have $\exp(-V_{DS}/U_T) \ll 1$, and the last term of I_D is approximately equal to one and can then be suppressed. The accuracy of this approximation is in the order of 2% for $V_{DS} \geq 4U_T$, because $\exp(-4) \simeq 0.018$. The expression of the drain-source current is then reduced to

$$I_D \simeq 2n\mu_n C_{ox} U_T^2 \left(\frac{W}{L}\right) \exp\left(\frac{V_{GS} - V_T}{nU_T}\right) \tag{1.92}$$

A transistor operating in the weak inversion region can be kept in the saturation mode, where the current flowing through the transistor is less affected by changes in the drain-source voltage, by setting the value of V_{DS} to about $4U_T$. Here, the fact that the drain-source saturation voltage is independent of the gate-source voltage can be advantageously exploited in low-voltage circuit designs.

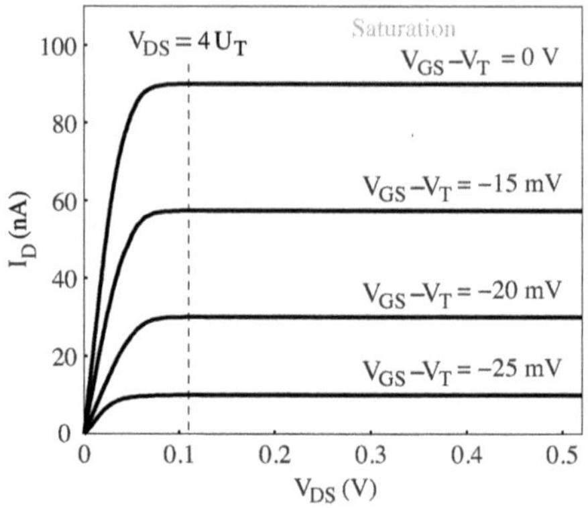

FIGURE 1.15
Plot of I_D versus V_{DS} in the weak inversion region.

For a MOS transistor operating in the weak inversion region, and assuming that $V_{BS} = 0$ and $n = 1.3$, Figure 1.15 shows the plot of I_D versus V_{DS} with $V_{GS} - V_T$ as the family variable. It can be observed that the drain current drops off considerably as the gate-source voltage is reduced.

In the weak inversion region, the transconductance can be obtained as:

$$g_m = \frac{\partial I_D}{\partial V_{GS}} \tag{1.93}$$

$$= \frac{I_D}{nU_T} \tag{1.94}$$

MOS transistors operating in the weak inversion region are essentially used for the design of low-voltage circuits dedicated to the processing of low-frequency signals.

In a strong inversion region, where $V_{GS} > V_T$, the transistor is turned on and can operate either in the linear mode or in the saturation mode.

As the value of V_{DS} is increased starting from zero, a channel is formed, so that a current can flow between the drain and the source. The transistor then behaves like a resistor controlled by the gate voltage relative to the drain voltage. The drain-source current is modeled as:

$$I_D = \frac{1}{2}\mu_n C_{ox}\left(\frac{W}{L}\right)\left[2(V_{GS} - V_T)V_{DS} - V_{DS}^2\right] \tag{1.95}$$

By differentiating I_D with respect to V_{DS}, it can be shown that $\partial I_D/\partial V_{DS}$ becomes zero at $V_{DS} = V_{DS(sat)}$. That is,

$$\frac{\partial I_D}{\partial V_{DS}} = \mu_n C_{ox}\left(\frac{W}{L}\right)(V_{GS} - V_T - V_{DS}) = 0 \tag{1.96}$$

and

$$V_{DS} = V_{DS(sat)} = V_{GS} - V_T \tag{1.97}$$

where $V_{DS(sat)}$ is the drain-source saturation (or overdrive) voltage. The transistor operates in linear mode provided $V_{DS} < V_{DS(sat)} = V_{GS} - V_T$.

When V_{DS} approaches $V_{DS(sat)}$, the conducting channel narrows at the drain terminal. Due to the conducting channel pinch-off, the drain-source current is now almost independent of V_{DS} and is said to be saturated. The transistor then operates as a voltage-controlled current source. The drain-source current in the saturation mode can be obtained by substituting $V_{DS(sat)}$ for V_{DS} into Equation (1.95). It follows a square law of the form,

$$I_D \simeq \frac{1}{2}\mu_n C_{ox}\left(\frac{W}{L}\right)\left(V_{GS} - V_T\right)^2 \tag{1.98}$$

where $V_{DS} \geq V_{DS(sat)} = V_{GS} - V_T$. In practice, the drain-source current varies slightly with changes of the drain-source voltage due to the effect of the channel-length modulation. This effect can be accounted for by rewriting the drain-source current as follows,

$$I_D = \frac{1}{2}\mu_n C_{ox}\left(\frac{W}{L}\right)\left(V_{GS} - V_T\right)^2\left(1 + \frac{V_{DS}}{V_A}\right) \tag{1.99}$$

where V_A is the Early voltage.

Assuming that $V_{DS} \ll V_A$, the transconductance is given by

$$g_m = \frac{\partial I_D}{\partial V_{GS}} \tag{1.100}$$

$$= \mu_n C_{ox}\left(\frac{W}{L}\right)\left(V_{GS} - V_T\right) = \sqrt{2\mu_n C_{ox}\left(\frac{W}{L}\right)I_D} = \frac{2I_D}{V_{DS(sat)}} \tag{1.101}$$

where $V_{DS(sat)} = V_{GS} - V_T$ is the saturation voltage.

The plot of I_D versus V_{GS} is shown in Figure 1.16.

The transconductance efficiency, g_m/I_D, can be expressed as the derivative of the logarithmic of I_D with respect to V_{GS}.

$$\frac{g_m}{I_D} = \frac{1}{I_D}\frac{\partial I_D}{\partial V_{GS}} = \frac{\partial[\ln(I_D)]}{\partial V_{GS}} \tag{1.102}$$

Furthermore, using the fact that $\partial \ln(W/L)/\partial V_{GS}$ is equal to 0, it can be shown that

$$\frac{g_m}{I_D} = \frac{\partial[\ln(I_D) - \ln(W/L)]}{\partial V_{GS}} = \frac{\partial[\ln(I_D/(W/L)]}{\partial V_{GS}} \tag{1.103}$$

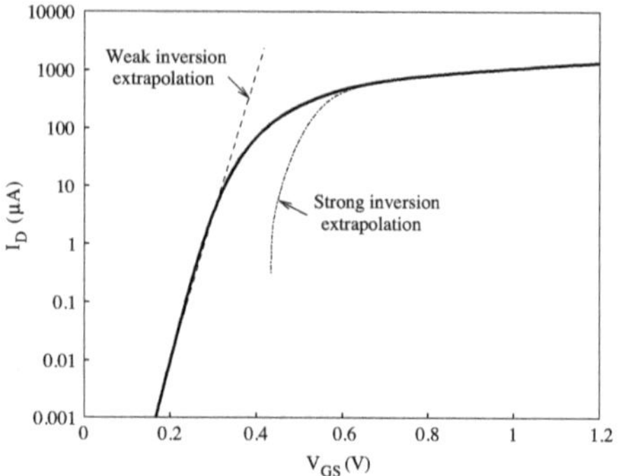

FIGURE 1.16
Plot of I_D versus V_{GS}.

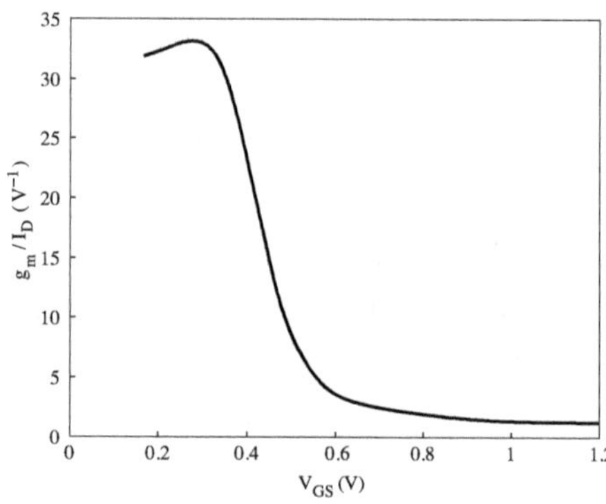

FIGURE 1.17
Plot of g_m/I_D versus V_{GS}.

The normalized current $I_D/(W/L)$ is, by definition, independent of the transistor size. As a consequence, the ratio g_m/I_D is also independent of the transistor size. It can then be considered as a universal characteristic during the design phase, when the transistor aspect ratios (W/L) aren't already known.

The derivative is maximum in the weak inversion region where I_D is an exponential function of V_{GS} and almost linear in the strong inversion region where I_D is a quadratic function of V_{GS}. It can then be deduced that the ratio g_m/I_D is related to the transistor operating region [16]. Figure 1.17 shows the plot of g_m/I_D versus V_{GS}.

In addition to the weak inversion (or subthreshold) region (V_{GS} less than V_T) and the strong inversion region (V_{GS} greater than V_T), a MOS transistor can also operate in the moderate inversion region (V_{GS} around V_T).

A transistor operating in the weak inversion region exhibits a large g_m/I_D ratio but a low speed, and in the strong inversion region, a high speed but a small g_m/I_D ratio, while the g_m/I_D ratio versus speed trade-off governs the operation in the moderate inversion region.

The drain current is normalized to yield the inversion coefficient, which is a measure of the channel inversion. The inversion coefficient is defined as

$$IC = \frac{I_D}{I_S} \tag{1.104}$$

where $I_S = I_T(W/L)$ is the specific (or normalization) current, I_T is the technology current, W is the effective channel width of the transistor, and L the effective channel length. Based on (1.92), we can obtain

$$IC = \frac{I_D}{2n\mu_n C_{ox} U_T^2 \left(\frac{W}{L}\right)} \simeq \frac{I_D}{2n\mu_0 C_{ox} U_T^2 \left(\frac{W}{L}\right)} \tag{1.105}$$

where μ_0 is the low-field carrier mobility. The inversion coefficient, IC, provides a numerical representation of the inversion level and is then directly related to the transistor operating regions:

- weak inversion region if $IC \leq 0.1$;
- moderate inversion region if $0.1 < IC \leq 10$;
- strong inversion region if $IC > 10$.

For a transistor in the off-state, where $V_{GS} = V_T$, the inversion coefficient is equal to about 0.48.

Note that the substrate factor n is slightly reduced with the increasing inversion level. Its value can be on the order of $1.4 - 1.5$ in the weak inversion region, 1.35 in the moderate inversion region, and 1.3 in the strong inversion region.

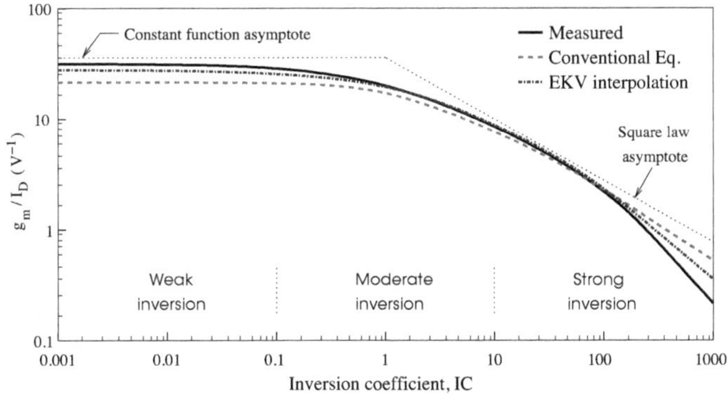

FIGURE 1.18
Plots of g_m/I_D versus IC.

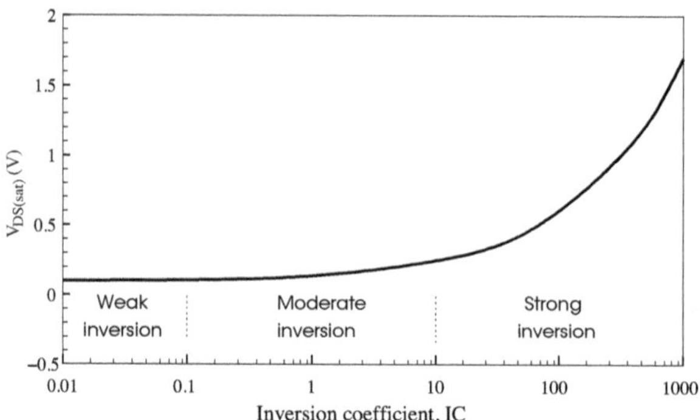

FIGURE 1.19
Plot of $V_{DS(sat)}$ versus IC.

A single-piece expression of the drain-source current that is valid in all inversion regions is obtained as follows:

$$I_D \simeq 2n\mu_n C_{ox} U_T^2 \left(\frac{W}{L}\right) \ln^2\left(1 + \exp\left(\frac{V_{GS} - V_T}{2nU_T}\right)\right) \tag{1.106}$$

The drain-source current given by (1.106) can be rewritten in the form,

$$I_D \simeq 2n\mu_n C_{ox} U_T^2 \left(\frac{W}{L}\right) IC \tag{1.107}$$

where the inversion coefficient IC is now expressed in terms of transistor voltages as,

$$IC = \ln^2\left(1 + \exp\left(\frac{V_{GS} - V_T}{2nU_T}\right)\right) \tag{1.108}$$

By solving (1.108) for $V_{GS} - V_T$, it can be shown that

$$V_{DS(sat)} = V_{GS} - V_T = 2nU_T \ln(\exp(\sqrt{IC}) - 1) \tag{1.109}$$

The operation in the saturation mode, where the transistor acts as a voltage-controlled current source with an optimal value of the transconductance, g_m, and a drain-source conductance, g_{ds}, requires that $V_{DS} > V_{DS(sat)}$. Figure 1.19 shows the plot of $V_{DS(sat)}$ versus the inversion coefficient, IC. The voltage $V_{DS(sat)}$ exhibits a minimum and constant value in the weak inversion region and increases as the square root of IC in the strong inversion region.

Based on the drain-source current given by (1.106), the transconductance efficiency, g_m/I_D, can be written as

$$\frac{g_m}{I_D} = \frac{1}{I_D}\frac{\partial I_D}{\partial V_{GS}} \tag{1.110}$$

$$= \frac{1}{nU_T \ln\left(1+\exp\left(\frac{V_{GS}-V_T}{2nU_T}\right)\right)} \cdot \frac{\exp\left(\frac{V_{GS}-V_T}{2nU_T}\right)}{1+\exp\left(\frac{V_{GS}-V_T}{2nU_T}\right)} \tag{1.111}$$

Combining (1.108) and (1.111), we obtain

$$\frac{g_m}{I_D} = \frac{1-\exp(-\sqrt{IC})}{nU_T\sqrt{IC}} \tag{1.112}$$

Although the expression (1.112), that is based on the conventional current equation, gives a sufficient accuracy, it is not adequate because it cannot be solved for the inversion coefficient as a function of the transconductance efficiency. To improve the accuracy in the weak inversion region, a simple interpolation function is used in the EKV MOSFET model and is of the form:

$$\frac{g_m}{I_D} = \frac{1}{nU_T\sqrt{IC + (1/2)\sqrt{IC} + 1}} \tag{1.113}$$

The inversion coefficient can readily be expressed in term of the transconductance efficiency by solving (1.113). Hence,

$$IC = \left(-\frac{1}{4} + \sqrt{\frac{1}{n^2U_T^2(g_m/I_D)^2} - \frac{15}{16}}\right)^2 \tag{1.114}$$

Based on measured data, the conventional equation, and the EKV interpolation, plots of g_m/I_D versus IC are sketched in Figure 1.18. The ratio g_m/I_D is such that $g_m/I_D \leq 1/(nU_T)$. The curve associated with measured data follows the predicted $1/\sqrt{IC}$ shape of other g_m/I_D curves until the effects of carrier velocity saturation become significant (i.e., the slope of the curve drops below the theoretical $-1/2$) in the strong inversion region of the response, especially for transistors with small channel lengths.

In contrast to the plot of (g_m/I_D) versus IC that has almost the same the appearance for transistors of the same type and manufacturing batch, the plot of g_{ds}/I_D or $(g_{ds}/I_D)^{-1}$ versus IC is a function of the transistor length through the dependence on the Early voltage, as can be observed in Figures 1.20 and 1.21, where L_j $(j = 1, 2, 3, 4)$ denotes the transistor length. Note that the Early effect is better analyzed by representing the inverse of the ratio g_{ds}/I_D. It is not actually linear. In Figure 1.20, it can be observed that the function $(g_{ds}/I_D)^{-1}$ decreases significantly around $IC = 5$, or when $V_{DS(sat)}$ is drawing near to V_{DS}, while in Figure 1.21, where V_{DS} is always

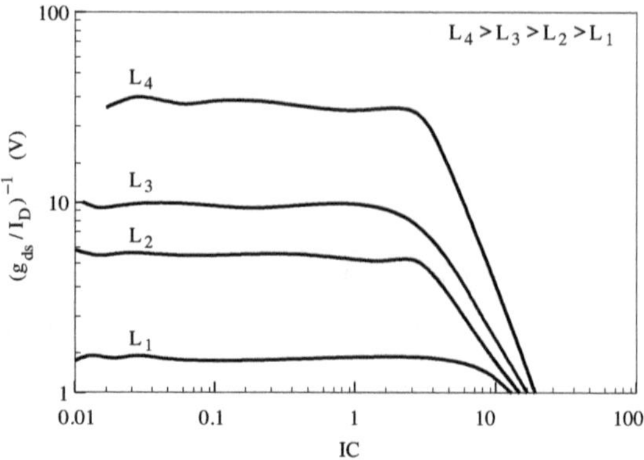

FIGURE 1.20
Plot of $(g_{ds}/I_D)^{-1}$ versus IC at $V_{DS} = 0.25$ V.

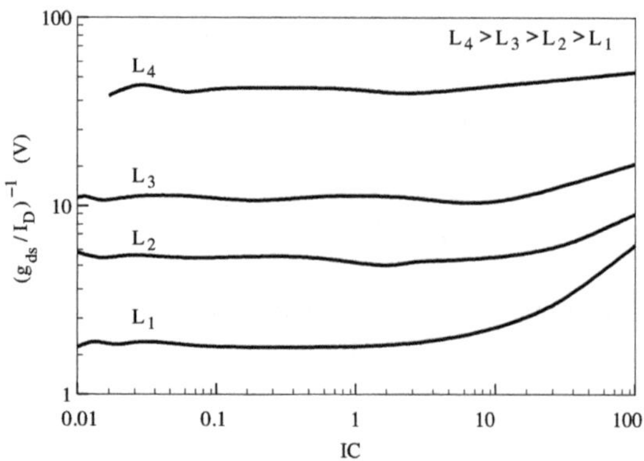

FIGURE 1.21
Plot of $(g_{ds}/I_D)^{-1}$ versus IC at $V_{DS} = V_{GS}$.

greater than $V_{DS(sat)}$, the function $(g_{ds}/I_D)^{-1}$ appears to steadily increase from the weak inversion region to the strong inversion region.

Typically, the Early voltage is far more important in the small-signal analysis than in the large-signal analysis of transistors operating in the saturation region.

Example

Determine the parameters g_m, g_{mb}, and g_{ds}, implicit to the MOSFET I-V relationship,

$$I_D = I_S \exp\left(\frac{V_{GS} - V_T}{nU_T}\right)\left[1 - \exp\left(-\frac{V_{DS}}{U_T}\right)\right], \quad V_{GS} < V_T \qquad (1.115)$$

where

$$V_T = V_{FB} + 2\phi_F + \gamma\sqrt{2\phi_F + V_{SB}} \tag{1.116}$$

$$n = 1 + \frac{\gamma}{2\sqrt{2\phi_F + V_{SB}}} \tag{1.117}$$

Deduce the ratio g_{mb}/g_m.

It can be shown that the transconductance is

$$g_m = \left.\frac{\partial I_D}{\partial V_{GS}}\right|_{V_{DS},V_{BS}} \tag{1.118}$$

$$= \frac{1}{n}\frac{I_D}{U_T} \tag{1.119}$$

In the case of the bulk transconductance, we have

$$g_{mb} = \left.\frac{\partial I_D}{\partial V_{BS}}\right|_{V_{DS},V_{GS}} \tag{1.120}$$

$$= \left.\frac{\partial I_D}{\partial V_T}\right|_{V_{DS},V_{GS}} \left(\frac{\partial V_T}{\partial V_{BS}}\right) \tag{1.121}$$

$$= -\frac{I_D}{nU_T}\left(\frac{\partial V_T}{\partial V_{BS}}\right) = \frac{I_D}{nU_T}\left(\frac{\partial V_T}{\partial V_{SB}}\right) \tag{1.122}$$

$$= \frac{I_D}{nU_T}\frac{\gamma}{2\sqrt{2\phi_F + V_{SB}}} = \frac{n-1}{n}\frac{I_D}{U_T} \tag{1.123}$$

The conductance is given by

$$g_{ds} = \left.\frac{\partial I_D}{\partial V_{DS}}\right|_{V_{BS},V_{GS}} \tag{1.124}$$

$$\simeq \frac{\exp\left(-\frac{V_{DS}}{U_T}\right)}{1 - \exp\left(-\frac{V_{DS}}{U_T}\right)}\frac{I_D}{U_T} \tag{1.125}$$

The ratio g_{mb}/g_m is of the form,

$$\frac{g_{mb}}{g_m} = n - 1 \tag{1.126}$$

The value of the thermal voltage U_T is 25.9 mV at room temperature (300 K) and n is generally greater than 1 and typically ranges from 1.2 to 1.5 for nanometer CMOS technologies.

1.4 Small-geometry effects

As the transistor channel length L is scaled down to increase the operation speed and to decrease the supply voltage, the circuit performance becomes increasingly limited by short-channel effects. The physical mechanisms responsible for short-channel effects are reflected in the penetration of junction fields into the channel region. In contrast to long-channel transistors, short-channel transistors feature a two-dimensional potential distribution and high horizontal and vertical electric fields in the channel region.

At high electric fields normally encountered in small channel length transistors, the average velocity of carriers is no longer proportional to the field and saturates at approximately 10^5 m/s for both electrons and holes due to scattering effects. As a consequence, the drain current is lower than that predicted by the mobility model and increases linearly with $V_{GS} - V_T$ rather than quadratically in the saturation region.

The velocity saturation of the horizontal electric field, E, can be modeled as follows:

$$v_d = \begin{cases} \dfrac{\mu E}{1 + E/E_C} & \text{if} \quad E < E_C \\ v_{sat} & \text{if} \quad E \geq E_C \end{cases} \tag{1.127}$$

where v_d is the drift velocity, μ denotes the carrier mobility, E_C is the critical electrical field at which the carrier velocity becomes saturated, and v_{sat} represents the saturation velocity. The velocity model is continuous at $E = E_C$ provided $E_C = 2v_{sat}/\mu$.

Taking into account the velocity saturation (VS), the drain current in the strong inversion region, $I_{D,VS}$, can be written as

$$I_{D,VS} = \frac{I_D}{1 + \dfrac{V_{DS}}{E_C L}} \tag{1.128}$$

where E_C is the critical electrical field.

A short-channel transistor is characterized by a width and a length that are small enough so that the edge effects cannot be neglected. The condition $E_C L \gg V_{GS} - V_T$, where E_C is the vertical critical electrical field, can be satisfied when L is large or when V_{GS} is close to V_T, as is the case in analog circuits where the gate is biased around V_T. Assuming that $E_C = 5.6 \times 10^4$ V/cm and $V_{GS} - V_T = 0.1$ V, a channel length as low as 0.1 μm would still satisfy the aforementioned inequality. As a consequence, the transistor would exhibit long-channel characteristics, such as the square-law I-V relationship and the constant mobility.

The channel length modulation (CLM) in a MOS transistor is caused by the extension of the depletion layer width at the drain as the drain voltage is increased. This leads to a reduction of the channel length, and consequently, an increase of the drain current. In order to model the CLM effect, the drain current becomes

$$I_{D,CLM} = \frac{I_D}{1 - l_p/L} \tag{1.129}$$

$$\simeq I_D(1 + l_p/L) \quad \text{if} \quad l_p/L \ll 1 \tag{1.130}$$

where l_p is the length of the velocity saturation region. In a pseudo two-dimensional formulation, the length l_p can determined as,

$$l_p = \begin{cases} l_c \ln\left(1 + \dfrac{V_{DS} - V_{DS(sat)}}{V_E}\right) & \text{if} \quad V_{DS} \geq V_{DS(sat)} \\ 0 & \text{if} \quad V_{DS} < V_{DS(sat)} \end{cases} \tag{1.131}$$

where l_c is the characteristic length and V_E is considered as a fitting parameter, although it can be expressed in terms of physical quantities.

The drain-induced barrier lowering (DIBL) effect originally refers to a reduction of the threshold voltage when the drain is biased at a high potential. This is related to the diminution of the potential barrier between the source and the channel, leading to an increase of the number of carriers injected into the channel from the source. Due to the DIBL effect, the threshold voltage is lowered by an amount $\triangle V_T$ and can be written as,

$$V_{T,DIBL} = V_T - \triangle V_T \tag{1.132}$$

Expressions of $\triangle V_T$ are often determined empirically as a function of L and V_{DS}.

The DIBL effect can also show up as the lateral shift of the I-V characteristics to lower V_{GS} values, resulting in an increase in the drain current at a given drain bias, and consequently, an increase of the transistor output conductance. This increase is additional to one induced by the channel length modulation, and cannot be accounted for by only the threshold voltage reduction.

Taking into account some small-geometry effects, the equation of the drain current can be summarized as follows:

If $V_{DS} < V_{DS(sat)}$, the drain current can be written as

$$I_D = \frac{\mu C_{ox}}{2\left(1 + \dfrac{V_{DS}}{E_C L}\right)} \frac{W}{L} [2(V_{GS} - V_T)V_{DS} - V_{DS}^2] \tag{1.133}$$

where $V_{DS(sat)}$ is the drain-source saturation voltage. At $V_{DS} = V_{DS(sat)}$, the drain current is given by,

$$I_{D(sat)} = \mu C_{ox} W (V_{GS} - V_T - V_{DS(sat)}) E_C \tag{1.134}$$

By solving $I_D = I_{D(sat)}$ with the assumption that $V_{DS} = V_{DS(sat)}$, we obtain

$$V_{DS(sat)} = \frac{(V_{GS} - V_T)E_C L}{(V_{GS} - V_T) + E_C L} \tag{1.135}$$

If $V_{DS} \geq V_{DS(sat)}$, the drain current can be put into the form,

$$I_D \simeq I_{D(sat)} \left[1 + \frac{1}{C_{clm}} \ln\left(\frac{V_A}{V_{A(sat)}}\right)\right]\left(1 + \frac{V_{DS} - V_{DS(sat)}}{V_{A,DIBL}}\right) \quad (1.136)$$

where $V_A = V_{A(sat)} + V_{A,CLM}$, $V_{A(sat)}$ is the Early voltage at $V_{DS} = V_{DS(sat)}$, $V_{A,CLM}$ is the Early voltage due to the CLM effect, and $V_{A,DIBL}$ is the Early voltage due to the DIBL effect.

The drain current is a function of the gate and drain voltages. But the dependence on the drain voltage is weak in the saturation region and is exploited in the BSIM transistor model to put I_D into the form,

$$I_D = I_{D(sat)} + \int_{V_{DS(sat)}}^{V_{DS}} \frac{\partial I_D}{\partial V_D} \cdot dV_D \quad (1.137)$$

$$= I_{D(sat)} \left[1 + \int_{V_{DS(sat)}}^{V_{DS}} \frac{1}{V_A} \cdot dV_D\right] \quad (1.138)$$

The Early voltage can be defined as,

$$V_A = I_{D(sat)} \left[\frac{\partial I_D}{\partial V_D}\right]^{-1} \quad (1.139)$$

The Early voltage due to the channel length modulation effect can be estimated as,

$$V_{A,CLM} = I_{D(sat)} \left[\frac{\partial I_D}{\partial L} \cdot \frac{\partial L}{\partial V_D}\right]^{-1} \quad (1.140)$$

The Early voltage caused by the drain-induced barrier lowering effect is given by,

$$V_{A,DIBL} = I_{D(sat)} \left[\frac{\partial I_D}{\partial V_T} \cdot \frac{\partial V_T}{\partial V_D}\right]^{-1} \quad (1.141)$$

Note that V_A, $V_{A,CLM}$, and $V_{A,DIBL}$ are dependent on the transistor length and bias.

Each of the physical mechanisms that affects the Early voltage (or the output conductance) in the saturation region dominates in a specific range of the drain-source voltage. The range for the CLM effect starts from the beginning of the saturation region and is followed by the one for the DIBL effect.

1.5 Design-oriented MOSFET models

Piece-wise regional-based models (or models valid for each region of operation) are extensively used due to the ease with which new effects can be taken into account. However, they may be plagued by the ever increasing number of parameters required to encompass effects associated with the downscaling of the transistor channel length into the sub-100-nm range. Semi-empirical (EKV, advanced compact MOSFET (ACM)) formulations seem to be suitable for the description of transistor characteristics in analog circuit design applications.

The ACM model hereinafter is based on the charge-sheet formulation and the linear relationship between the variations of the inversion charge density and the surface potential. It is characterized by a drain-source current that is formulated in terms of forward and reverse components and is valid in weak, moderate, and strong inversion regions.

Let V_C be the channel potential that is such that $V_S \leq V_C \leq V_D$, where V_S and V_D are the source and drain voltages, respectively. The drain-source current can be written as,

$$I_D = -\mu_n W Q_i \frac{dV_C}{dy} \tag{1.142}$$

where μ_n is the electron mobility, W is the channel width, Q_i is the inversion charge, and y is the longitudinal distance along the channel. By integrating with respect to y, we obtain

$$I_D = -\mu_n \frac{W}{L} \int_{Q_{IS}}^{Q_{ID}} Q_i \frac{dV_C}{dQ_i} dQ_i \tag{1.143}$$

To proceed further, the variable V_C may be changed to Q_i using the expression

$$dQ_i \left(\frac{1}{nC_{ox}} - \frac{U_T}{Q_i} \right) = dV_C \tag{1.144}$$

Then,

$$I_D = -\mu_n \frac{W}{L} \int_{Q_{IS}}^{Q_{ID}} Q_i \left(\frac{1}{nC_{ox}} - \frac{U_T}{Q_i} \right) dQ_i \tag{1.145}$$

so that the integration from the source to drain results in

$$I_D = \mu_n \frac{W}{L} \left[\frac{Q_{IS}^2 - Q_{ID}^2}{2nC_{ox}} - U_T (Q_{IS} - Q_{ID}) \right] \tag{1.146}$$

where

$$n = 1 + \frac{\gamma}{2\sqrt{\phi_{sa} - U_T}} \tag{1.147}$$

with ϕ_{sa} corresponding to the surface potential calculated with the assumption that the inversion charge is negligible, and γ is the body effect factor. The integration of (1.144) from a channel potential V_C to the pinch-off voltage V_P yields

$$V_P - V_C = U_T \left[\frac{Q_{IP} - Q_I}{nC_{ox} U_T} + \ln \left(\frac{Q_I}{Q_{IP}} \right) \right] \tag{1.148}$$

where Q_{IP} is the value of Q_I at the pinch-off voltage V_P.

The drain-source current can be rewritten as,

$$I_D = I_F - I_R = I_S(i_f - i_r) \tag{1.149}$$

where I_F and I_R are the forward and reverse components of the drain-source current, respectively, $I_S = \mu_n C_{ox} n(U_T^2/2)(W/L)$ is the specific current, and i_f and i_r are the forward and reverse normalized currents, respectively. It can be shown that

$$I_F = I_S\left(q_{IS}^2 + 2q_{IS}\right) \tag{1.150}$$

$$I_R = I_S\left(q_{ID}^2 + 2q_{ID}\right) \tag{1.151}$$

where the normalized inversion charge density at the source and the drain are, respectively, given by

$$q_{IS} = -\frac{Q_{IS}}{nC_{ox}U_T} = \sqrt{1+i_f} - 1 \tag{1.152}$$

$$q_{ID} = -\frac{Q_{ID}}{nC_{ox}U_T} = \sqrt{1+i_r} - 1 \tag{1.153}$$

Note that $IC = \max(i_f, i_r)$, where IC is the inversion coefficient that indicates whether the transistor is in weak ($IC < 0.1$), moderate ($0.1 < IC < 10$), or strong inversion ($IC > 10$).

The drain-source voltage is of the form,

$$\frac{V_{DS}}{U_T} = q_{IS} - q_{ID} + \ln\left(\frac{q_{IS}}{q_{ID}}\right) \tag{1.154}$$

or equivalently

$$\frac{V_{DS}}{U_T} = \sqrt{1+i_f} - \sqrt{1+i_r} + \ln\left(\frac{\sqrt{1+i_f}-1}{\sqrt{1+i_r}-1}\right) \tag{1.155}$$

The drain-source saturation voltage, $V_{DS(sat)}$ can be defined as the value of V_{DS} at which the ratio $\xi = q_{ID}/q_{IS}$ is a number much smaller than unity. For values of ξ close 1, the transistor operates in the linear region, and for values of xi close to 0 the operating point of the transistor is in the saturation region. The voltage $V_{DS(sat)}$ can be derived from (1.154) as

$$V_{DS(sat)} = U_T\left[\ln\left(\frac{1}{\xi}\right) + (1-\xi)(\sqrt{1+i_f} - 1)\right] \tag{1.156}$$

This expression of $V_{DS(sat)}$ defines the boundary between the linear and saturation regions in terms of the inversion level.

1.5.1 Small-signal transconductances

For small variations of the gate, source, drain, and bulk voltages, the change of the drain-source current can be obtained as follows:

$$\triangle I_D = g_{mg}\triangle V_G - g_{ms}\triangle V_S + g_{md}\triangle V_D + g_{mb}\triangle V_B \tag{1.157}$$

where

$$g_{mg} = \frac{\partial I_D}{\partial V_G} \quad g_{ms} = -\frac{\partial I_D}{\partial V_S} \quad g_{md} = \frac{\partial I_D}{\partial V_D} \quad \text{and} \quad g_{mb} = \frac{\partial I_D}{\partial V_B} \tag{1.158}$$

represent the gate, source, drain, and bulk transconductances, respectively.

In cases where the variations of the gate, source, drain, and bulk voltages are identical, $\triangle I_D = 0$. As a consequence, it can be deduced that

$$g_{mg} + g_{md} + g_{mb} = g_{ms} \tag{1.159}$$

At low frequencies, the small-signal behavior of the transistor can be characterized using only three of the four transconductances.

The drain and source transconductances can be derived from the drain-source current given by

$$I_D = -\frac{\mu_n W}{L}\int_{V_S}^{V_D} Q_I dV_C \tag{1.160}$$

as[1]

$$g_{ms} = -\mu_n \frac{W}{L} Q_{IS} = \frac{2I_S}{U_T}(\sqrt{1+i_f} - 1) \tag{1.161}$$

$$g_{md} = -\mu_n \frac{W}{L} Q_{ID} = \frac{2I_S}{U_T}(\sqrt{1+i_r} - 1) \tag{1.162}$$

Using (1.159), it can be shown that

$$g_{mg} = \frac{g_{ms} - g_{md}}{n} \tag{1.163}$$

and

$$g_{mb} = (n-1)g_{mg} \tag{1.164}$$

where n represents the slope factor.

1.5.2 Transistor parameters in various CMOS technologies

Table 1.3 [19] gives the magnitude orders of some transistor parameters in various CMOS technologies, where t_{ox} is the gate oxide thickness, the peak

---

[1]Let $f(x)$ be a continuous function of the variable x, such that $a \leq x \leq b$. We have

$$\frac{\partial}{\partial b}\int_a^b f(x)dx = f(b) \quad \text{and} \quad \frac{\partial}{\partial a}\int_a^b f(x)dx = -f(a)$$

TABLE 1.3
Transistor Parameters in Various CMOS Technologies

Technology (nm)	250	180	130	90	65
V_{DD} (V)	2.5	1.8	1.5	1.2	1
V_T (V)	0.44	0.43	0.34	0.36	0.24
g_m (S/m)	335	500	720	1060	1400
g_{ds} (S/m)	22	40	65	100	230
F_t (GHz)	35	53	94	140	210
$1/t_{ox}$ (nm)	6.2	4.45	3.12	2.2	1.8

value is chosen for the transconductance, g_m, and the transit frequency, $F_t = g_m/(2\pi C_G)$, with C_G being the gate capacitance.

1.5.3 Capacitances

MOSFET parasitic capacitances can be subdivided into extrinsic and intrinsic capacitances.

Extrinsic capacitances are related to overlap and junction capacitances that can be modeled by adding lumped capacitances between each pair of transistor terminals and between the well and the bulk if the transistor is fabricated in a well.

Intrinsic capacitances are associated with the region between the metallurgical source and drain junctions (or the channel region) and can be derived directly from the terminal charges.

The high-frequency capability of MOS transistors can be assessed using the transit frequency that is defined as the frequency at which the extrapolated small-signal current gain of the transistor in common-source configuration is reduced to unity and that can be expressed as

$$F_t = \frac{g_m}{2\pi C_G} \tag{1.165}$$

where g_m is the transconductance and C_G represents the total gate capacitance that is composed of intrinsic and extrinsic capacitances. The transit frequency, F_t, gives an estimate of the transistor delay because it can be approximately related to the transit time from the source to the drain, $\tau_t \simeq 1/(2\pi F_t)$.
For long-channel transistors, the drain current in the saturation

region can be written as

$$I_D = \frac{1}{2}\mu C_{ox}\frac{W}{L}(V_{GS} - V_T)^2 \tag{1.166}$$

The small-signal transconductance is given by

$$g_m = \frac{\partial I_D}{\partial V_{GS}} = \mu C_{ox}\frac{W}{L}(V_{GS} - V_T) \tag{1.167}$$

Assuming that the gate capacitor C_G is dominated by the gate-source capacitor given by $C_{GS} = 2WLC_{ox}/3$, we arrive at

$$F_t = \frac{g_m}{2\pi C_G} = \frac{3}{4\pi}\frac{\mu(V_{GS} - V_T)}{L^2} \tag{1.168}$$

where $V_{ov} = V_{GS} - V_T$ represents the overdrive voltage. The transit frequency is proportional to the overdrive voltage and inversely proportional to the square of the transistor length.
For short-channel transistors, the drain current that includes the effect of the velocity saturation can take the following form:

$$I_{D,VS} = \frac{I_D}{1 + V_{DS}/E_C L} \tag{1.169}$$

where I_D is the drain current in the linear region, $E_C = 2v_{sat}/\mu$ is the critical field, and v_{sat} is the saturation velocity. When the values of L are very small, $V_{DS}/E_C L \gg 1$ and the drain current in the saturation region then approaches the following limit:

$$I_{D,VS} \simeq \mu C_{ox} W (V_{GS} - V_T) E_C \tag{1.170}$$

The small-signal transconductance is obtained as

$$g_m = \frac{\partial I_{D,vs}}{\partial V_{GS}} \simeq \mu C_{ox} W E_C \tag{1.171}$$

The transit frequency becomes

$$F_t = \frac{g_m}{2\pi C_G} \simeq \frac{3}{4\pi}\frac{\mu E_C}{L} \tag{1.172}$$

Now, it depends on the critical field (or the saturation velocity) and is inversely proportional to the transistor length.
Due to the dependence of g_m on the inversion coefficient (IC), the transit frequency, F_t, can be considered as a function of IC. In radio-frequency applications, it can be related to parameters, such as the gain and figure of merit of a transistor stage loaded by another transistor stage. Typical values of F_t are in the range

from 50 to 250 GHz. Note that nMOS transistors exhibit F_t values that are higher than the ones of pMOS transistors.
Note that the maximum frequency f_{max}, that is, the frequency at which the maximum power gain is equal to unity, is another metric used to characterize the high-frequency performance of a transistor. In addition to the gate capacitor, it takes into account the gate resistor.

Assuming the quasi-static approximation, the charging current can be expressed as,

$$\frac{dQ_k}{dt} = \frac{\partial Q_k}{\partial V_G}\frac{dV_G}{dt} + \frac{\partial Q_k}{\partial V_S}\frac{dV_S}{dt} + \frac{\partial Q_k}{\partial V_D}\frac{dV_D}{dt} + \frac{\partial Q_k}{\partial V_B}\frac{dV_B}{dt} \tag{1.173}$$

Let the intrinsic transcapacitances and self-capacitances be defined by

$$C_{kl} = -\frac{\partial Q_k}{\partial V_l} \quad k \neq l \tag{1.174}$$

and

$$C_{kk} = \frac{\partial Q_k}{\partial V_k} \tag{1.175}$$

where the derivatives are taken around the bias point, and k and l can represent any of the transistor nodes (G, S, D, B). Note that the capacitances are non-reciprocal, that is, $C_{kl} \neq C_{lk}$.

The capacitance model of a MOSFET can be described as follows [20],

$$\begin{bmatrix} dQ_G/dt \\ dQ_S/dt \\ dQ_D/dt \\ dQ_B/dt \end{bmatrix} = \begin{bmatrix} C_{GG} & -C_{GS} & -C_{GD} & -C_{GB} \\ -C_{SG} & C_{SS} & -C_{SD} & -C_{SB} \\ -C_{DG} & -C_{DS} & C_{DD} & -C_{DB} \\ -C_{BG} & -C_{BS} & -C_{BD} & C_{BB} \end{bmatrix} \begin{bmatrix} dV_G/dt \\ dV_S/dt \\ dV_D/dt \\ dV_B/dt \end{bmatrix} \tag{1.176}$$

With the assumption that $V_G = V_S = V_D = V_B = V$, we obtain

$$\frac{dQ_G}{dt} = (C_{GG} - C_{GS} - C_{GD} - C_{GB})\frac{dV}{dt} = 0 \tag{1.177}$$

and

$$C_{GG} = C_{GS} + C_{GD} + C_{GB} \tag{1.178}$$

Furthermore, by assuming that

$$\frac{dV_G}{dt} = \frac{dV_S}{dt} = \frac{dV_D}{dt} = 0 \tag{1.179}$$

we have

$$\begin{aligned}\frac{dQ_G}{dt} &= C_{GG}\frac{dV_G}{dt} & \frac{dQ_S}{dt} &= -C_{CG}\frac{dV_G}{dt} \\ \frac{dQ_D}{dt} &= -C_{DG}\frac{dV_G}{dt} & \frac{dQ_B}{dt} &= -C_{BG}\frac{dV_G}{dt}\end{aligned} \tag{1.180}$$

and

$$\frac{dQ_G}{dt} + \frac{dQ_S}{dt} + \frac{dQ_D}{dt} + \frac{dQ_B}{dt} = (C_{GG} - C_{SG} - C_{DG} - C_{BG})\frac{dV_G}{dt} \tag{1.181}$$

If we exploit the charge conservation law, $d(Q_G + Q_S + Q_D + Q_B)/dt = 0$, we will then arrive at

$$C_{GG} = C_{SG} + C_{DG} + C_{BG} \tag{1.182}$$

By proceeding in the same way for the remaining nodes, other self-capacitance expressions can be derived. Hence,

$$C_{GG} = C_{GS} + C_{GD} + C_{GB} = C_{SG} + C_{DG} + C_{BG} \tag{1.183}$$

$$C_{SS} = C_{SG} + C_{SD} + C_{SB} = C_{GS} + C_{DS} + C_{BS} \tag{1.184}$$

$$C_{DD} = C_{DG} + C_{DS} + C_{DB} = C_{GD} + C_{SD} + C_{BD} \tag{1.185}$$

$$C_{BB} = C_{BG} + C_{BS} + C_{BD} = C_{GB} + C_{SB} + C_{DB} \tag{1.186}$$

As a result of the charge conservation and the fact that only three voltage differences out of four can be chosen independently, the MOSFET capacitive model is characterized by only nine capacitances. Therefore, we can choose to determine C_{GS}, C_{GD}, C_{GB}, C_{BG}, C_{BS}, C_{BD}, C_{DS}, C_{SD}, and C_{DG}. That is, we can combine (1.176) and (1.183)–(1.186) to show that

$$\frac{dQ_G}{dt} = C_{GS}\frac{dV_{GS}}{dt} + C_{GD}\frac{dV_{GD}}{dt} + C_{GB}\frac{dV_{GB}}{dt} \tag{1.187}$$

$$\frac{dQ_B}{dt} = C_{GB}\frac{dV_{BG}}{dt} + C_Q\frac{dV_{BG}}{dt} + C_{BS}\frac{dV_{BS}}{dt} + C_{BD}\frac{dV_{BD}}{dt} \tag{1.188}$$

$$\begin{aligned}\frac{dQ_D}{dt} = C_{GD}\frac{dV_{DG}}{dt} + C_{BD}\frac{dV_{DB}}{dt} - \\ C_P\frac{dV_{GB}}{dt} + C_{SD}\frac{dV_{DB}}{dt} - C_{DS}\frac{dV_{SB}}{dt}\end{aligned} \tag{1.189}$$

where $C_P = C_{DG} - C_{GD}$ and $C_Q = C_{BG} - C_{GB}$.

The combination of the following equations

$$I_D = -\mu_n W Q_i \frac{dV_C}{dy} \tag{1.190}$$

and

$$dV_C = d\phi_s - U_T\frac{dQ_i}{Q_i} \tag{1.191}$$

leads to the charge-sheet formulation of the drain-source current as the sum of the drift and diffusion components:

$$I_D = I_{drift} + I_{diff} = -\mu_n W Q_i \frac{d\phi_s}{dy} + \mu_n W U_T \frac{dQ_i}{dy} \tag{1.192}$$

where μ_n is the electron mobility, W is the channel width, U_T is the thermal voltage, Q_i is the inversion charge density, ϕ_s is the surface potential, and y is the longitudinal distance along the channel. From the charge-sheet approximation, the inversion charge per unit area, Q_i, can be related to the bulk charge per unit area, Q_b, as stated in

$$\begin{aligned} Q_i &= -C_{ox}(V_G - V_{FB} - \phi_s - \gamma\sqrt{\phi_s - U_T}) \\ &= -C_{ox}(V_G - V_{FB} - \phi_s) - Q_b \end{aligned} \tag{1.193}$$

where C_{ox} is the oxide capacitance per unit area, V_{FB} is the flat-band voltage, ϕ_s is the surface potential, and γ is the body factor. At constant gate voltage V_G, it can be shown that

$$dQ_i = \left(C_{ox} - \frac{dQ_b}{d\phi_s}\right) d\phi_s = (C_{ox} + C_b) d\phi_s = nC_{ox} d\phi_s \tag{1.194}$$

Combining (1.192) and (1.194), the drain-source current can take the next form,

$$I_D = -\frac{\mu_n W}{nC_{ox}} (Q_i - nU_T C_{ox}) \frac{dQ_i}{dy} \tag{1.195}$$

To proceed further, it is convenient to define the new variable

$$Q_{it} = Q_i - nC_{ox} U_T \tag{1.196}$$

Due to the fact that the parameter n is not a function of the distance y, it can be shown that

$$dQ_i = dQ_{it} \tag{1.197}$$

and

$$I_D = -\frac{\mu_n W}{nC_{ox}} Q_{it} \frac{dQ_{it}}{dy} \tag{1.198}$$

The integration along the channel extending from the source to the drain leads to

$$I_D \int_0^L dy = -\frac{\mu_n W}{nC_{ox}} \int_{Q_F}^{Q_R} Q_{it} \, dQ_{it} \tag{1.199}$$

The drain-source current is then given by

$$I_D = -\frac{\mu_n}{2nC_{ox}} \frac{W}{L} (Q_F^2 - Q_R^2) \tag{1.200}$$

where Q_F and Q_R represent the values of Q_{it} estimated at the source and drain of the transistor, respectively. Hence,

$$Q_F = Q_{IS} - nC_{ox}U_T \tag{1.201}$$

$$Q_R = Q_{ID} - nC_{ox}U_T \tag{1.202}$$

Let the total inversion charge stored in the channel be of the form

$$Q_I = W\int_0^L Q_i dy \tag{1.203}$$

Using (1.196) and (1.198) to change the variable of (1.203) gives

$$Q_I = -\frac{\mu_n W^2}{nC_{ox}I_D}\int_{Q_F}^{Q_R}(Q_{it} + nC_{ox}U_T)Q_{it}\, dQ_{it} \tag{1.204}$$

The integration is then carried out with respect to Q_{it}, resulting in

$$Q_I = -\frac{\mu_n W^2}{nC_{ox}I_D}\left[\frac{Q_R^3 - Q_F^3}{3} + nC_{ox}U_T\frac{Q_R^2 - Q_F^2}{2}\right] \tag{1.205}$$

Substituting (1.200) into (1.205), we obtain

$$Q_I = WL\left(\frac{2}{3}\frac{Q_F^2 + Q_F Q_R + Q_R^2}{Q_F + Q_R} + nC_{ox}U_T\right) \tag{1.206}$$

Both drain and source terminals are in contact with the channel region, whose charge Q_I is empirically assumed to be partitioned into a charge Q_D associated with the drain and a charge Q_S associated with the source, such that

$$Q_I = Q_D + Q_S \tag{1.207}$$

Approaches used to partition Q_I into Q_D and Q_S can vary from an equal division of Q_I across both terminals ($Q_D = Q_S = Q_I/2$) to Q_I multiplied by linear weighting factors. The drain and source charges can be computed based on a linear charge partition scheme as [20],

$$Q_D = W\int_0^L \frac{y}{L}Q_i dy \tag{1.208}$$

and

$$Q_S = W\int_0^L\left(1 - \frac{y}{L}\right)Q_i dy \tag{1.209}$$

By substituting (1.196) and (1.198) into (1.208), we obtain

$$Q_D = -\frac{\mu_n^2 W^3}{2L(nC_{ox}I_D)^2}\int_{Q_F}^{Q_R}(Q_F^2 - Q_{it}^2)(Q_{it} + nC_{ox}U_T)Q_{it}\, dQ_{it} \tag{1.210}$$

By carrying out the calculation of the integral, we arrive at

$$Q_D = \frac{\mu_n^2 W^3}{2L(nC_{ox}I_D)^2}\times \left[\frac{3Q_R^3 + 6Q_F Q_R^2 + 4Q_F^2 Q_R + 2Q_F^3}{15}(Q_F - Q_R)^2 + \frac{n}{4}C_{ox}U_T(Q_F^2 - Q_R^2)^2\right] \tag{1.211}$$

The substitution of the drain-source current given by (1.200) into (1.211) yields,

$$Q_D = WL\left(\frac{6Q_R^3 + 12Q_F Q_R^2 + 8Q_F^2 Q_R + 4Q_F^3}{15(Q_F + Q_R)^2} + \frac{n}{2}C_{ox}U_T\right) \tag{1.212}$$

The source charge can be determined using either (1.207) or (1.209), or simply by exploiting the source and drain symmetry. Hence,

$$Q_S = WL\left(\frac{6Q_F^3 + 12Q_R Q_F^2 + 8Q_R^2 Q_F + 4Q_R^3}{15(Q_F + Q_R)^2} + \frac{n}{2}C_{ox}U_T\right) \tag{1.213}$$

According to the charge-sheet approximation, the inversion and bulk charges per unit area can be written as,

$$Q_i = -C_{ox}\left(V_G - V_{FG} - \phi_s - \gamma\sqrt{\phi_s - U_T}\right) \tag{1.214}$$

and

$$Q_b = -C_{ox}\gamma\sqrt{\phi_s - U_T} \tag{1.215}$$

To find simple charge equations, it is necessary to establish a linear relationship between Q_b and ϕ_s. For this purpose, the square root in (1.215) is approximated as a Taylor series about the reference surface potential ϕ_{sa}:

$$\sqrt{\phi_s - U_T} \simeq \sqrt{\phi_{sa} - U_T} + \frac{1}{2\sqrt{\phi_{sa} - U_T}}(\phi_s - \phi_{sa}) \tag{1.216}$$

where

$$V_G - V_{FG} = \phi_{sa} + \gamma\sqrt{\phi_{sa} - U_T} \tag{1.217}$$

and ϕ_{sa} is the surface potential estimated with the assumption that the inversion charge is negligible. Using (1.216) and (1.217), the expressions of Q_i and Q_b given by (1.214) and (1.215), respectively, become

$$Q_i \simeq nC_{ox}(\phi_s - \phi_{sa}) \tag{1.218}$$

and

$$Q_b \simeq -\frac{n-1}{n}Q_i + Q_{ba} \tag{1.219}$$

where

$$n = 1 + \frac{\gamma}{2\sqrt{\phi_{sa} - U_T}} \tag{1.220}$$

and

$$Q_{ba} = -\gamma C_{ox}\sqrt{\phi_{sa} - U_T} = -C_{ox}\left(V_G - V_{FG} - \phi_{sa}\right) \tag{1.221}$$

Note that Q_{ba} can be considered as the depletion charge per unit area for a negligible carrier charge density.

The total gate charge is determined from the charge conservation condition

$$Q_G = -Q_B - Q_I - Q_O \tag{1.222}$$

or equivalently,

$$Q_G \simeq -\frac{Q_I}{n} - Q_{ba}WL - Q_O \tag{1.223}$$

where Q_O is the effective oxide charge, assumed to be independent of the transistor terminal voltages.

The application of the gate-source and gate-drain capacitance definitions to the gate charge expression given by (1.223) leads to

$$C_{GS} = -\frac{\partial Q_G}{\partial V_S} = \frac{1}{n}\frac{\partial Q_I}{\partial V_S} \tag{1.224}$$

$$C_{GD} = -\frac{\partial Q_G}{\partial V_D} = \frac{1}{n}\frac{\partial Q_I}{\partial V_D} \tag{1.225}$$

The inversion charge is a function of Q_F and Q_R that are defined by (1.201) and (1.202), respectively. Hence,

$$C_{GS} = \frac{1}{n}\frac{\partial Q_I}{\partial Q_F}\frac{\partial Q_F}{\partial V_S} = \frac{2WL}{3n}\left[\frac{Q_F^2 + 2Q_FQ_R}{(Q_F+Q_R)^2}\right]\frac{\partial Q_{IS}}{\partial V_S} \tag{1.226}$$

$$C_{GD} = \frac{1}{n}\frac{\partial Q_I}{\partial Q_R}\frac{\partial Q_R}{\partial V_D} = \frac{2WL}{3n}\left[\frac{Q_R^2 + 2Q_RQ_F}{(Q_F+Q_R)^2}\right]\frac{\partial Q_{ID}}{\partial V_S} \tag{1.227}$$

where

$$\frac{\partial Q_{IS}}{\partial V_S} = \left(\frac{1}{nC_{ox}} - \frac{U_T}{Q_{IS}}\right)^{-1} = nC_{ox}\frac{Q_{IS}}{Q_{IS} - nC_{ox}U_T} \tag{1.228}$$

$$\frac{\partial Q_{ID}}{\partial V_S} = \left(\frac{1}{nC_{ox}} - \frac{U_T}{Q_{ID}}\right)^{-1} = nC_{ox}\frac{Q_{ID}}{Q_{ID} - nC_{ox}U_T} \tag{1.229}$$

Finally, we have

$$C_{GS} = \frac{2}{3}C_{ox}WL\left[\frac{Q_F^2 + 2Q_FQ_R}{(Q_F+Q_R)^2}\right]\left(1 + \frac{nC_{ox}U_T}{Q_F}\right) \tag{1.230}$$

$$C_{GD} = \frac{2}{3}C_{ox}WL\left[\frac{Q_R^2 + 2Q_RQ_F}{(Q_F+Q_R)^2}\right]\left(1 + \frac{nC_{ox}U_T}{Q_R}\right) \tag{1.231}$$

Using (1.213) and (1.212), the source-drain and drain-source capacitances can, respectively, be obtained as follows:

$$C_{SD} = -\frac{\partial Q_S}{\partial V_D} \tag{1.232}$$

$$= -\frac{\partial Q_S}{\partial Q_R}\frac{\partial Q_R}{\partial V_D} \tag{1.233}$$

$$= -\frac{4}{15}nC_{ox}WL\frac{Q_R^3 + 3Q_R^2 Q_F + Q_R Q_F^2}{(Q_F + Q_R)^3}\left(1 + \frac{nC_{ox}U_T}{Q_R}\right) \tag{1.234}$$

and

$$C_{DS} = -\frac{\partial Q_D}{\partial V_S} \tag{1.235}$$

$$= -\frac{\partial Q_D}{\partial Q_F}\frac{\partial Q_F}{\partial V_S} \tag{1.236}$$

$$= -\frac{4}{15}nC_{ox}WL\frac{Q_F^3 + 3Q_F^2 Q_R + Q_F Q_R^2}{(Q_F + Q_R)^3}\left(1 + \frac{nC_{ox}U_T}{Q_F}\right) \tag{1.237}$$

For the bulk-source and bulk-drain capacitances, using (1.219), it can be shown that

$$C_{BS} = -\frac{\partial Q_B}{\partial V_S} = \frac{n-1}{n}\frac{\partial Q_I}{\partial V_S} = (n-1)C_{GS} \tag{1.238}$$

$$C_{BD} = -\frac{\partial Q_B}{\partial V_D} = \frac{n-1}{n}\frac{\partial Q_I}{\partial V_D} = (n-1)C_{GD} \tag{1.239}$$

From (1.223) and (1.219), we can respectively put the gate-bulk and bulk-gate capacitances into the following forms:

$$C_{GB} = -\frac{\partial Q_G}{\partial V_B} = \frac{1}{n}\frac{\partial Q_I}{\partial V_B} + \frac{n-1}{n}C_{ox}WL \tag{1.240}$$

$$C_{BG} = -\frac{\partial Q_B}{\partial V_G} = \frac{n-1}{n}\frac{\partial Q_I}{\partial V_G} + \frac{n-1}{n}C_{ox}WL \tag{1.241}$$

where it was assumed that

$$\frac{\partial Q_{Ba}}{\partial V_B} = -\frac{\partial Q_{Ba}}{\partial V_G} = \frac{n-1}{n}C_{ox}WL \tag{1.242}$$

Recalling the fact that the partial derivatives of Q_I with respect to V_B and V_G can be expressed as,

$$\frac{\partial Q_I}{\partial V_B} = -\frac{n-1}{n}\frac{\partial Q_I}{\partial Q_F}\frac{\partial Q_{IS}}{\partial V_S} - \frac{n-1}{n}\frac{\partial Q_I}{\partial Q_R}\frac{\partial Q_{ID}}{\partial V_D} \tag{1.243}$$

$$= -(n-1)(C_{GS} + C_{GD}) \tag{1.244}$$

and

$$\frac{\partial Q_I}{\partial V_G} = -\frac{1}{n}\frac{\partial Q_I}{\partial Q_F}\frac{\partial Q_{IS}}{\partial V_S} - \frac{1}{n}\frac{\partial Q_I}{\partial Q_R}\frac{\partial Q_{ID}}{\partial V_S} \tag{1.245}$$

$$= -(C_{GS} + C_{GD}) \tag{1.246}$$

we then have

$$C_{GB} = C_{BG} = \frac{n-1}{n}(C_{ox}WL - C_{GS} - C_{GD}) \tag{1.247}$$

The equality $C_{GB} = C_{BG}$ is obtained as a result of the approximations that were made, but accurate calculations based on the charge sheet model show that $C_{GB} < C_{BG}$. The source-gate and gate-source capacitances are given by

$$C_{SG} = -\frac{\partial Q_S}{\partial V_G} = (C_{SS} - C_{SD})/n \tag{1.248}$$

and

$$C_{GS} = -\frac{\partial Q_G}{\partial V_S} = (C_{SS} - C_{DS})/n \tag{1.249}$$

Assuming that $C_{GB} = C_{BG}$, it can be deduced that

$$C_P = C_{DG} - C_{GD} = C_{GS} - C_{SG} = (C_{SD} - C_{DS})/n \tag{1.250}$$

Let q_{ID} and q_{IS} be the normalized inversion charges given by

$$q_{ID} = -\frac{Q_{ID}}{nC_{ox}U_T} \tag{1.251}$$

$$q_{IS} = -\frac{Q_{IS}}{nC_{ox}U_T} \tag{1.252}$$

The use of the saturation coefficient computed as

$$\alpha = \frac{Q_R}{Q_F} = \frac{Q_{ID} - nC_{ox}U_T}{Q_{IS} - nC_{ox}U_T} = \frac{q_{ID}+1}{q_{IS}+1} \tag{1.253}$$

leads to the following capacitance expressions:

$$C_{GS} = \frac{2}{3}C_{ox}WL\frac{1+2\alpha}{(1+\alpha)^2}\frac{q_{IS}}{1+q_{IS}} \tag{1.254}$$

$$C_{GD} = \frac{2}{3}C_{ox}WL\frac{\alpha^2+2\alpha}{(1+\alpha)^2}\frac{q_{ID}}{1+q_{ID}} \tag{1.255}$$

$$C_{BS} = (n-1)C_{GS} \tag{1.256}$$

$$C_{BD} = (n-1)C_{GD} \tag{1.257}$$

$$C_{GB} = C_{BG} = \frac{n-1}{n}(C_{ox}WL - C_{GS} - C_{GD}) \tag{1.258}$$

$$C_{SD} = -\frac{4}{15}nC_{ox}WL\frac{\alpha+3\alpha^2+\alpha^3}{(1+\alpha)^3}\frac{q_{ID}}{1+q_{ID}} \tag{1.259}$$

$$C_{DS} = -\frac{4}{15}nC_{ox}WL\frac{1+3\alpha+\alpha^2}{(1+\alpha)^3}\frac{q_{IS}}{1+q_{IS}} \tag{1.260}$$

$$C_{DG} - C_{GD} = (C_{SD} - C_{DS})/n \tag{1.261}$$

Generally, the saturation coefficient α is in the range $0 < \alpha \leq 1$. For $V_{DS} = 0$, $q_{IS} = q_{ID}$ and $\alpha = 1$. In the weak inversion, $\alpha \simeq 1$ from the linear region to the saturation region. In the strong inversion, α decreases from 1 initially to $\alpha \simeq 0$ in the saturation region, where $q_{IS} \gg 1$ and q_{ID} tends to 0.

To obtain a capacitance per unit area, each of the above total capacitances should be divided by WL.

It should be noted that, in addition to intrinsic capacitances, a complete transistor model should include extrinsic capacitances which consist of the inner and outer fringing capacitances between the polysilicon gate and the source/drain regions, the overlap capacitances between the gate and the heavily doped source/drain regions, the overlap capacitances between the gate and the lightly doped source/drain regions, and the bulk-source and bulk-drain junction capacitances.

In some applications, the gate capacitive effect can be modeled by the gate-source and gate-drain overlapping capacitances, and the capacitances, C_{GS}, C_{GD}, and C_{GB}, that are dependent on the biasing condition, while the equivalent junction parasitic capacitances can be combined into C_{SB} and C_{DB}. By reducing the number of capacitances, the symmetry of the transistor model is still maintained. But, the charge conservation principle may not be satisfied.

1.6 Summary

Miniaturization of transistors has influenced almost all levels of the circuit design. While being subject to laws of physics, device technology development, and economic factors, the limits to this trend should also be application dependent due to power consumption issues. Accurate device models are essential for circuit design and analysis. They should exhibit continuous and scalable electrical characteristics to meet the requirements of mixed-signal circuits based on submicrometer technologies. Recent modeling approaches rely on device physics and a suitable choice of empirical parameters to account for the different short-channel effects of transistors.

1.7 Circuit design assessment

1. **Capacitance due to the depletion charge**
   Consider a *pn* junction. The junction diffusion potential, ψ_0, is given

by

$$\psi_0 = \psi_n - \psi_p = \frac{kT}{q} \ln \frac{N_a N_d}{n_i^2} \tag{1.262}$$

where $\psi_n = (kT/q)\ln(N_d/n_i)$ and $\psi_p = -(kT/q)\ln(N_a/n_i)$ are the potentials across the n-type and p-type regions, respectively; n_i is the intrinsic carrier concentration; and N_a and N_d are the acceptor and donor concentrations, respectively. The depletion width on the n and p regions, x_n and x_p, respectively, are related to the overall width of the depletion region, W_d, according to

$$x_n = \frac{N_a}{N_a + N_d} W_d \tag{1.263}$$

and

$$x_n = \frac{N_d}{N_a + N_d} W_d \tag{1.264}$$

The Poisson equation can then be written as

$$\frac{d^2\phi(x)}{dx^2} = \begin{cases} \dfrac{qN_a}{\epsilon} & \text{for} \quad -x_p \leq x < 0 \\ -\dfrac{qN_d}{\epsilon} & \text{for} \quad 0 < x \leq x_n \end{cases} \tag{1.265}$$

Show that the junction voltage is given by

$$\phi(x) = \begin{cases} \dfrac{qN_a}{\epsilon}\left(\dfrac{x^2}{2} + x_p x + \dfrac{x_p^2}{2}\right) & \text{for} \quad -x_p \leq x < 0 \\ \\ \dfrac{qN_d}{\epsilon}\left(-\dfrac{x^2}{2} + x_n x + \dfrac{x_n x_p}{2}\right) & \text{for} \quad 0 < x \leq x_n \end{cases} \tag{1.266}$$

Based on the relation, $\psi_0 + V_r = \phi(x_n)$, determine x_n and x_p.

The depletion charge per unit area, Q_d, for a junction under reverse bias (i.e., the applied voltage V_r is negative) reads

$$Q_d = qN_d x_n = qN_a x_p = \sqrt{2q\epsilon \frac{N_a N_d}{N_a + N_d}(\psi_0 + V_r)} \tag{1.267}$$

where q is the electron charge and ϵ is the silicon dielectric.

Verify that the junction capacitance per unit area can be expressed as

$$C_j = \frac{dQ_d}{dV_r} = \frac{C_{j0}}{\sqrt{1 + \dfrac{V_r}{\psi_0}}} \tag{1.268}$$

where the capacitance under the zero-bias condition, C_{j0}, is given by

$$C_{j0} = \sqrt{\frac{q\epsilon}{2}\left(\frac{N_a N_d}{N_a + N_d}\right)\frac{1}{\psi_0}}. \quad (1.269)$$

Find the expression of C_{j0} in the case where $N_a \gg N_d$.
Use the following parameters, $C_{j0} = 0.47$ fF/µm^2, $\psi_0 = 0.65$ V, and $V_r = -3.3$ V to determine C_j.

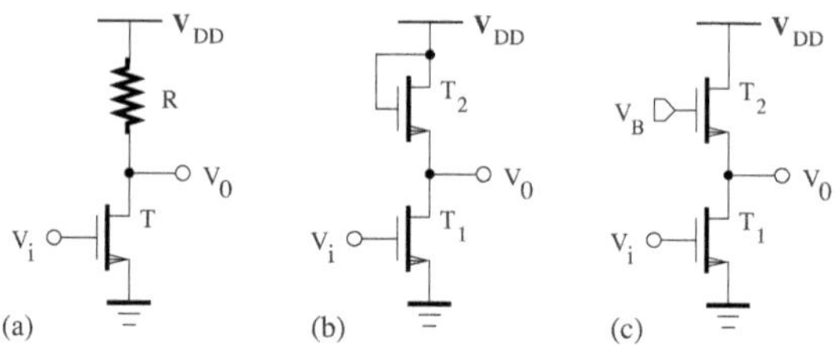

FIGURE 1.22
Circuit diagram of common-source amplifier stages.

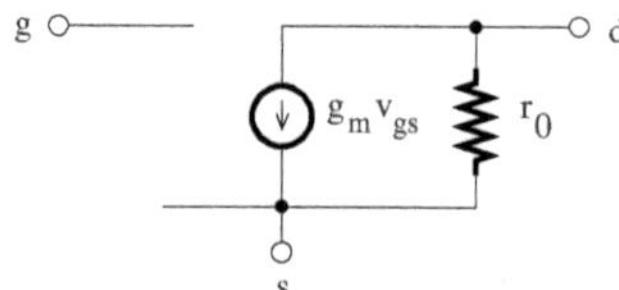

FIGURE 1.23
Small-signal model of the transistor.

2. **Analysis of common-source amplifier stages**
   Consider the inverting amplifier stages shown in Figure 1.22. Figure 1.23 shows the small-signal model of the transistor, where $g_m = \partial I_{DS}/\partial V_{GS}|_{V_{DS}}$ and $r_0 = \partial V_{DS}/\partial I_D$. Only the width, W, and length, L, of the transistors are assumed to be variable.

   Determine the small-signal voltage gain $A_v = v_0/v_i$.
   Compare the analytical analysis and SPICE simulations, and justify the dissimilarity between both results.

3. **Analysis of source-degeneration amplifier stages**
   Repeat the previous exercise using the source-degeneration amplifier stages of Figure 1.24.

4. **Cascode amplifier**
   Using the transistor equivalent model of Figure 1.23, estimate the small-signal gain $G_m = v_0/i_i$ and the output resistance, r_{out}, of the cascode amplifier stage shown in Figure 1.25.

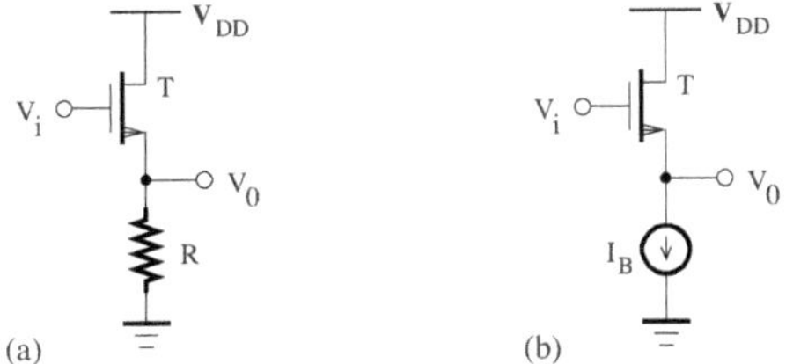

FIGURE 1.24
Circuit diagram of source-degeneration amplifier stages.

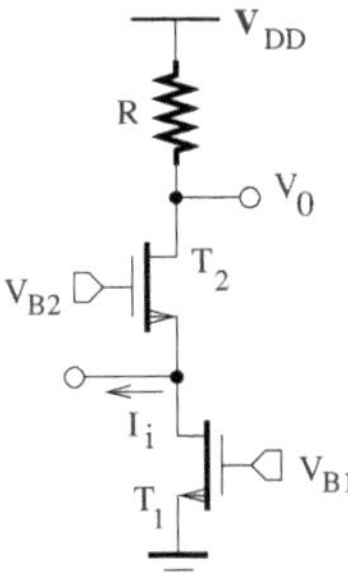

FIGURE 1.25
Circuit diagram of a cascode amplifier stage.

5. **CMOS inverter**
   The circuit diagram of a CMOS inverter is depicted in Figure 1.26. When the input voltage is equal to V_{DD}, the transistor T_1 is on, while T_2 is off. On the other hand, the transistors T_1 and T_2 are, respectively, off and on when the input voltage is equal to zero.

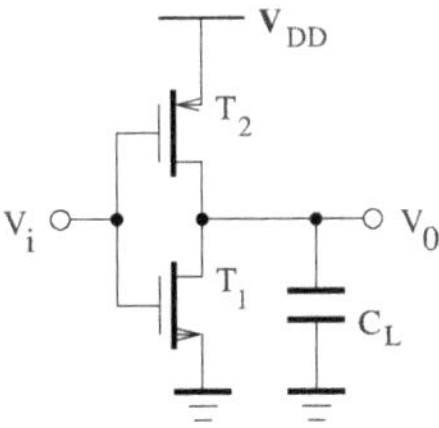

FIGURE 1.26
Circuit diagram of a CMOS inverter.

Determine the small-signal voltage gain $A_v = v_0/v_i$.
Let the charge stored on the capacitor C_L be

$$Q = C_L V_{DD} = I_0 T \tag{1.270}$$

where I_O is the output current and $T = 1/f$ is the signal period. Show that

$$P = V_{DD}I_{DDQ} + C_L V_{DD}^2 f \tag{1.271}$$

where P is the power dissipated over a single signal cycle and I_{DDQ} denotes the quiescent leakage current, which flows through the transistor when it is off (V_{SB} or V_{DB} is different from zero).

6. **Transistor arrays**

Consider the transistor array shown in Figure 1.27(a). The transistors T_i ($i = 1, 2, 3$) are designed to have the same length but different widths. Verify that the structure of Figure 1.27(a) is equivalent to a single transistor (see Figure 1.27(b)) with the width equal to the sum of T_i widths and the same length as T_i.

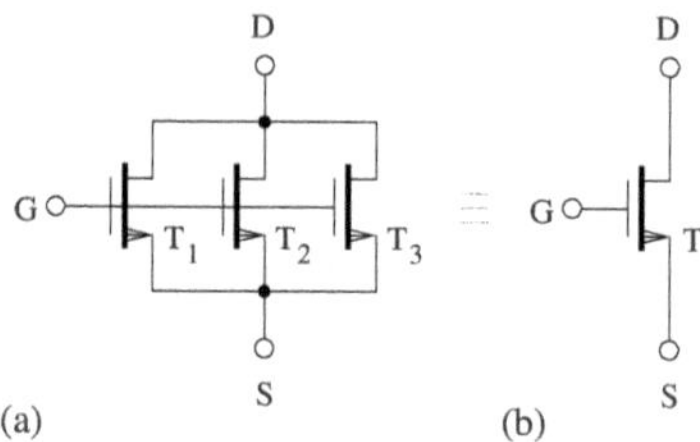

FIGURE 1.27
Circuit diagram of a transistor array.

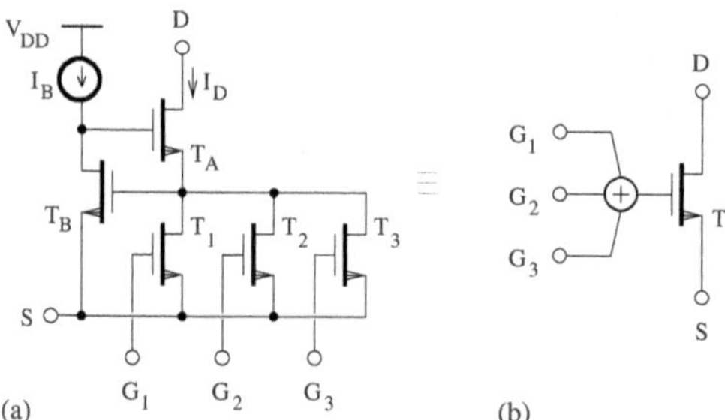

FIGURE 1.28
Circuit diagram of an array of weighted transistors.

For the circuit of Figure 1.28(a), the transistors T_i operate in the triode region and their drain-source voltage is fixed by the loop consisting of T_A, T_B, and the bias current I_B [21].
Find the current I_D and establish the equivalence between the structures of Figure 1.28.
Show that the transconductance of a transistor T_i is given by

$$g_{mi} = \frac{dI_{Di}}{dV_{GSi}} = K_i V_{DS} \tag{1.272}$$

where K_i is the transconductance parameter.

7. **Simplified noise analysis of a single-stage amplifier**
Verify that the output noise per unit bandwidth of the amplifier stage shown in Figure 1.22(a) is given by

$$\overline{v_{n,o}^2} = (\overline{i_{th}^2} + \overline{i_{1/f}^2})R^2 + 4kTR \tag{1.273}$$

where $\overline{i_{th}^2}$ and $\overline{i_{1/f}^2}$ denote the thermal and flicker noises of the transistor, respectively. It should be assumed that the thermal noise of the resistor R is equivalent to a current source with the value $4kT/R$ and $\overline{i_{1/f}^2} = g_m^2 \overline{v_{1/f}^2}$, where g_m is the transistor transconductance.

Bibliography

[1] Semiconductor Industry Association (SIA), *International Technology Roadmap for Semiconductor*, San Jose, CA: SIA, 1999.

[2] S. M. Sze, *Physics of Semiconductor Devices*, New York, NY: John Wiley & Sons, 1981.

[3] D. P. Foty, *MOSFET Modeling with SPICE: Principles and Practice*, Upper Saddle River, NJ: Prentice-Hall, 1996.

[4] R. J. Van Overstraeten, G. Declerck, and G. L. Broux, "Inadequacy of the classical theory of the MOS transistor operating in weak inversion," *IEEE Trans. on Electron Devices*, vol. 20, pp. 1150–1153, Dec. 1973.

[5] K. N. Ratnakumar and J. D. Meindl, "Short-channel MOST threshold voltage model," *IEEE J. of Solid-State Circuits*, vol. 17, pp. 937–948, Oct. 1982.

[6] T. Grotjohn and B. Hoefflinger, "A parametric short-channel MOS transistor model for subthreshold and strong inversion current," *IEEE Trans. on Electron Devices*, vol. 31, pp. 234–246, Feb. 1984.

[7] D. E. Ward and R. W. Dutton, "A charge-oriented model for MOS transistor capacitances," *IEEE J. of Solid-State Circuits*, vol. 17, pp. 703–708, Oct. 1978.

[8] B. J. Sheu, D. L. Scharfetter, C. Hu, and D. O. Pederson, "A compact IGFET charge model," *IEEE Trans. on Circuits and Syst.*, vol. 31, pp. 745–748, Aug. 1984.

[9] N. Weste and D. Harris, *CMOS VLSI Design*, Boston, MA: Addison-Wesley, 2005.

[10] T. Quarles, A. R. Newton, D. O. Pederson, and A. Sangiovanni-Vincentelli, *SPICE 3 Version 3F5 User's Manual*, University of California, Berkeley, CA, 1994.

[11] H. Shichman and D. A. Hodges, "Modeling and simulation of insulated field effect transistor switching circuits," *IEEE J. of Solid-State Circuits*, vol. 3, pp. 285–289, Sept. 1968.

[12] S. Liu and L. W. Nagel, "Small-signal MOSFET models for analog circuit design," *IEEE J. of Solid-State Circuits*, vol. 17, pp. 983–998, Oct. 1982.

[13] Y. Chen, M.-C. Jeng, Z. Liu, J. Huang, M. Chan, K. Chen, P. K. Ko, and C. Hu, "A physical and scalable *I*-*V* model in BSIM3v3 for analog/digital circuit simulation," *IEEE Trans. on Electron Dev.*, vol. 44, pp. 277–287, Feb. 1997.

[14] S. Tedja, J. Van der Spiegel, and H. H. Williams, "Analytical and experimental studies of thermal noise in MOSFET's," *IEEE Trans. on Electron Dev.*, vol. 41, pp. 2069–2075, Nov. 1994.

[15] A. van der Ziel, *Noise in Solid State Devices and Circuits*, New York, NY: John Wiley & Sons, 1986.

[16] F. Silveira, D. Flandre, and P. G. A. Jespers, "A g_m/I_D based methodology for the design of CMOS analog circuits and its application to the synthesis of a silicon-on-insulator micropower OTA," *IEEE J. of Solid-State Circuits*, vol. 31, pp. 1314–1319, Sept. 1996.

[17] C. C. Enz, F. Krummenacher, and E. A. Vittoz, "An analytical MOS transistor model valid in all regions of operation and dedicated to low-voltage and low-current applications," *Analog Integrated Circuits and Signal Processing*, vol. 8, no. 1, pp. 83–114, 1995.

[18] A. I. A. Cunha, M. C. Schneider, and C. Galup-Montoro, "An MOS transistor model for analog circuit design," *IEEE J. Solid-State Circuits*, vol. 33, no. 10, pp. 1510–1519, Oct. 1998.

[19] J. Pekarik, et al., "RFCMOS Technology from 0.25 μm to 65 nm: The state of the art," in the *Proceedings of the 2004 IEEE CICC*, pp. 217–224.

[20] D. E. Ward and R. W. Dutton, "A charge-oriented model for MOS transistor capacitances," *IEEE J. Solid-State Circuits*, vol. SC-13, no. 5, pp. 703–708, Oct. 1978.

[21] Z. Czarnul, T. Iida, and K. Tsuji, "A low-voltage highly linear multiple weighted CMOS transconductor," *IEEE Trans. on Circuits and Systems*, vol. 42, pp. 362–364, May 1995.

2

Physical Design of MOS Integrated Circuits

CONTENTS

Modern lithography systems used in integrated circuit (IC) fabrication employ an optical projection printing that operates almost at the diffraction limit. The mask of each IC layout layer is projected onto the wafer substrate, which has been coated with a photoresist material, also called resist. The structure of the resist is altered by exposure to light so that, after the development, a silicon pattern can emerge. The next step can consist of implanting the dopant ions. Note that the small feature, which can be printed in this way, is about the wavelength of the light used.

Even with the great capabilities of lithography, it is impossible to produce completely identical devices over an entire wafer, or a much smaller die. As the size features are scaled down, the likelihood of process variations increases and the actual performance of a chip becomes more unpredictable. It is then necessary to take care of the process variation effects during the IC design phase, so that even worst-case fabrication conditions may provide a working circuit. The suitable design technique can consist of using worst-case (slow), nominal (typical), and best-case (fast) transistor models based on device measurements for SPICE simulations. However, to meet the demand for greater bandwidths and higher levels of integration, interconnects have to undergo a considerable reduction in size and currently represent an important limitation to the development of IC technology. To address this challenge, CAD and testing tools are required to link the interconnect characteristics to the circuit performance. Efficient interconnect models rely on simplifying assumptions, the validity of which is to be confirmed by on-chip measurements.

2.1 MOS transistors

Two MOS transistors are considered: a MOS field-effect transistor (MOSFET) that has a planar structure and a fin field-effect transistor (FinFET) that exhibits a three-dimensional structure.

2.1.1 MOS field-effect transistor

The structures of MOSFETs [1, 2] are shown in Figure 2.1. The drain and source of the nMOSFET (see Figure 2.1(a)) are realized by two heavily doped n-type semiconductor regions, which are implanted into a lightly doped p-type substrate or bulk. A thin layer of silicon dioxide (SiO_2) is thermally grown over the region between the source and drain, and is covered by polycrystalline silicon (also called polysilicon or poly), which forms the gate of the transistor. The thickness t_{ox} of the oxide layer is on the order of a few angstroms. The length L and the width W of the gate are estimated along and perpendicularly to the drain-source path, respectively. Due to the lateral diffusion of the doped regions, the actual length is slightly smaller than L. The overlap region is symmetrical and is extended on the distance L_D on both sides. As a result, the effective length between the drain and source is given by, $L_{eff} = L_{drawn} - 2L_D$, where L_{drawn} is the dimension drawn in the layout. The width of the gate is also less than the drawn value due to a reduction in the active area by the field oxide growth. The substrate connection is provided by a doped p^+ region.

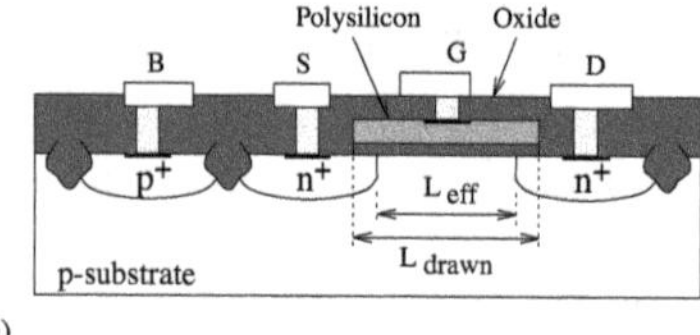

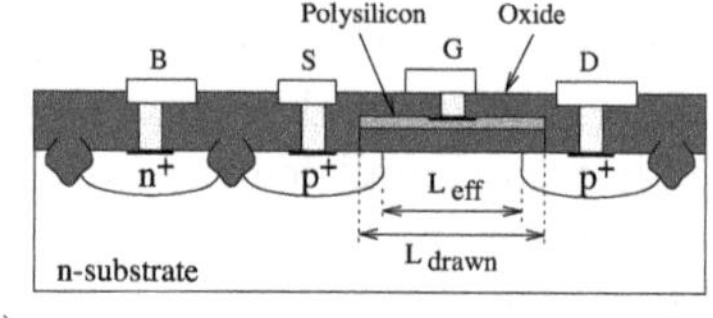

FIGURE 2.1
(a) nMOSFET and (b) pMOSFET structures.

Generally, the substrate is connected to the most negative supply voltage of the circuit so that the source-drain junction diodes are reverse-biased.

In the case of the pMOSFET (see Figure 2.1(b)), the drain and source are formed by p^+ diffusions in the n-type substrate. A doped n^+ region is required for the realization of the substrate connection. Otherwise, the cross-sections for the two transistor types are similar. Here, the substrate is to be connected to the most positive supply voltage.

The layout of a MOSFET is shown in Figure 2.2. In addition to the active region representing the substrate, it is formed by the overlap of the active and polysilicon layers.

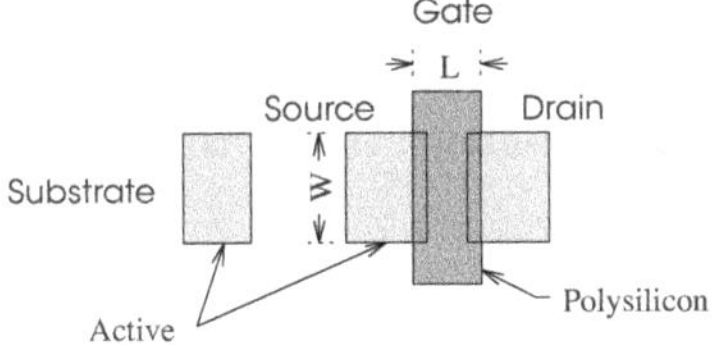

FIGURE 2.2
MOSFET layout.

In practice, nMOS and pMOS transistors are fabricated on a wafer based on only one substrate, say of the p-type. In this case, the pMOS device is realized in an n-well as shown in Figure 2.3. Two MOSFET layouts are shown

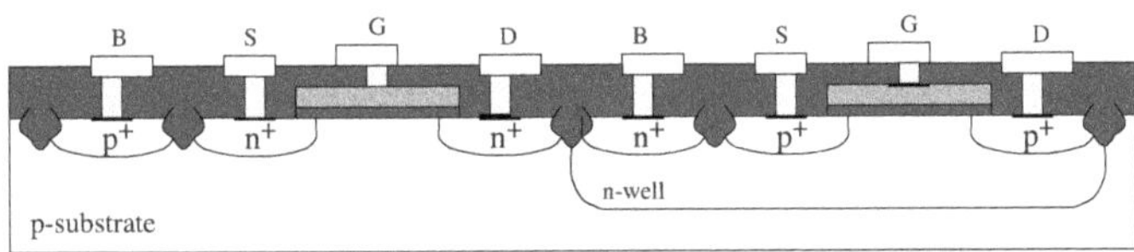

FIGURE 2.3
MOSFET structure in a CMOS process with an n-well.

in Figures 2.4(a) and (b). The transistor can be fabricated in a well as is the case in Figure 2.4(b). The active contact is a cut in the oxide, which allows the connection from the active and polysilicon regions to the first layer of metal.

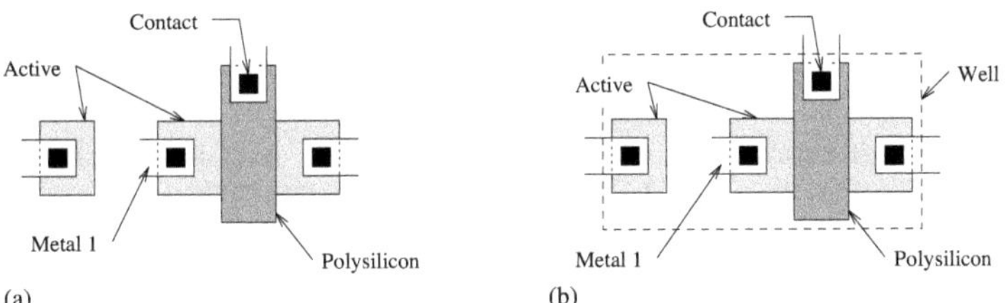

FIGURE 2.4
(a) Layout of a MOSFET realized on the substrate; (b) layout of a MOSFET fabricated in a well.

2.1.2 Fin field-effect transistor

As the MOSFET gate length is scaled down to 22 nanometers in order to increase the switching speed, the undesirable effect of the quantum tunneling current from the gate into the channel becomes significant, and further reduction of the gate oxide thickness can result in high static leakage current and high electric power consumption even when the transistor is turned off. An approach that can allow future scaling down relies on multi-gate transistor structures, such as the fin field-effect transistor (FinFET).

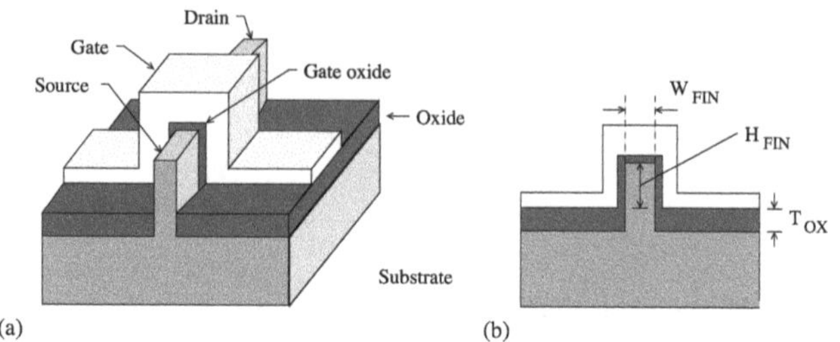

FIGURE 2.5
FinFET (a) structure and (b) cross section.

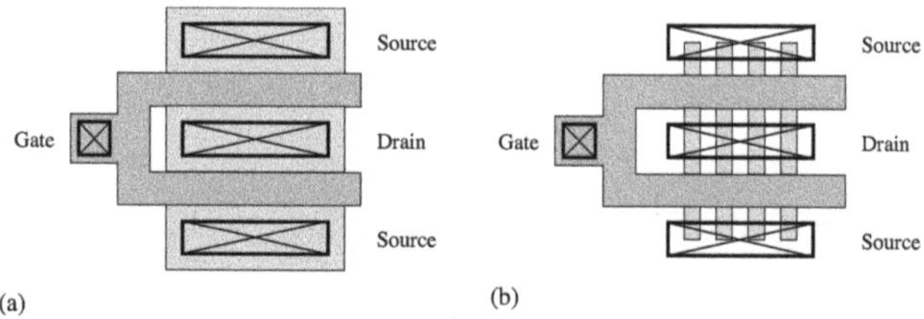

FIGURE 2.6
(a) MOSFET layout; (b) FinFET layout.

In a FinFET, the source-drain channel is a vertical fin that is surrounded by the gate. Due to the fact that it can then be controlled electrostatically by the gate from more than one side, the FinFET is considered as a multi-gate transistor. The improved gate control of the FinFET has the advantage of attenuating short channel effects. Because the output conductance is also reduced, the voltage gain, that is useful in analog circuits, is increased. An operational amplifier designed using FinFETs instead of MOSFETs exhibits an open-loop *dc* gain higher by about 20 dB, but the gain is lower at higher frequencies due to high source and drain resistances.

In comparison with MOSFETS, FinFETs have other advantages such as a higher integration density due to the vertical orientation of the channel and a lower statistical variability caused by (gate and fin) geometry and process fluctuations, but the inconvenience of exhibiting higher gate capacitances.

A three-dimensional structure and cross-section of a FinFET are depicted in Figures 2.5(a) and (b), respectively. Figures 2.6(a) and (b) show the MOSFET and FinFET layouts, respectively.

The channel of a MOSFET is laid horizontally, while the one of a FinFET is vertical with a height, H_{FIN}, and a thickness, T_{FIN}. Typically, the fin height is kept below four times the fin thickness. The effective width of a FinFET can then be increased by using multiple fins. Hence, arbitrary FinFET widths are not possible. This is known as the width quantization of FinFETs.

2.2 Passive components

The integrated-circuit (IC) performance is determined by the characteristics of basic components [1, 2] or elements, which can be fabricated in the MOS technology.

2.2.1 Capacitors

Let us consider a structure with a dominant parallel-plate capacitor having the length, L, width, W, and thickness, t_{ox}. The capacitance can be obtained as,

$$C = \frac{\epsilon_0 \epsilon_{ox} LW}{t_{ox}} \tag{2.1}$$

where ϵ_0 is the vacuum dielectric constant and ϵ_{ox} represents the relative dielectric constant of the insulator (silicon dioxide). Accurate capacitances are the result of the reduction in the errors associated with the dielectric constants, also called oxide effects, and the variations due to geometrical parameters, known as edge effects.

A capacitor can be realized using the structures of Figure 2.7. The silicon dioxide dielectric layer can be deposited between a heavily doped crystalline silicon and polycrystalline silicon (see Figure 2.7(a)), two polycrystalline silicon layers (see Figure 2.7(b)), or polycrystalline silicon and metal layers (see Figure 2.7(c)). The capacitor value depends on the dielectric thickness, which

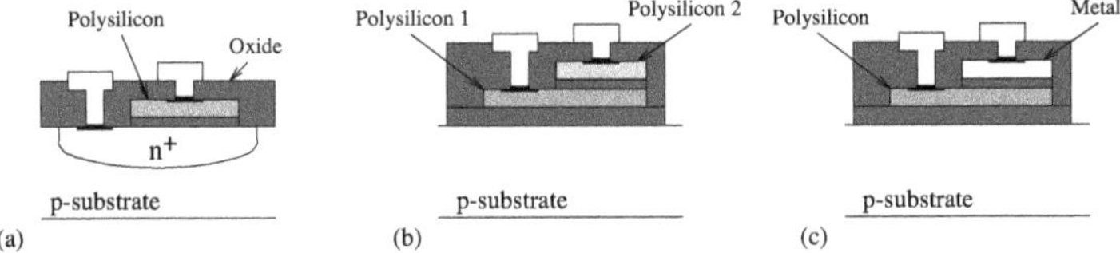

FIGURE 2.7
(a) Poly-diffusion capacitor; (b) poly-poly capacitor; (c) metal-poly capacitor.

can exhibit large variations in the case of the silicon dioxide. As a result, capacitance values can fluctuate by a few percent and the achievable device matching is limited. Specifically, the width of the depletion regions formed at the oxide contact-surface of the capacitor shown in Figure 2.7(a) depends on the applied voltage. As a result, the effective dielectric thickness is not constant and the capacitance is affected by the voltage variation and the bottom-plate parasitic capacitance. In the structures of Figure 2.7(b)–(c), the polysilicon region is isolated from the substrate by an oxide layer, which forms a parasitic capacitance to be included in the equivalent circuit model.

The layout of a capacitor based on two polysilicon layers is depicted in Figure 2.8. To minimize the contact resistance, contacts can be made everywhere it is possible on the polysilicon layer. It should be noted that a given

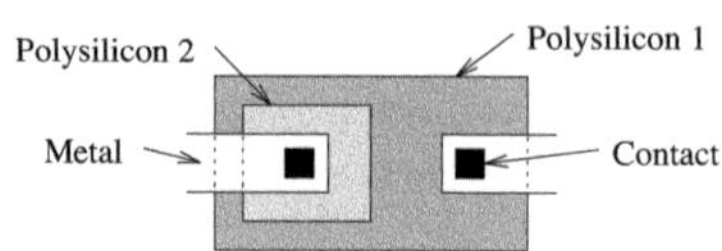

FIGURE 2.8
Layout of a poly-poly capacitor.

capacitor consisting of two conducting planes separated by a thin insulator can occupy more area than the one implemented as a combination of several layers of conductors and insulators.

Metal-oxide-metal (MOM) capacitors are realized as an inter-digitated multi-finger structure formed by metal layers separated by a dielectric (oxide).

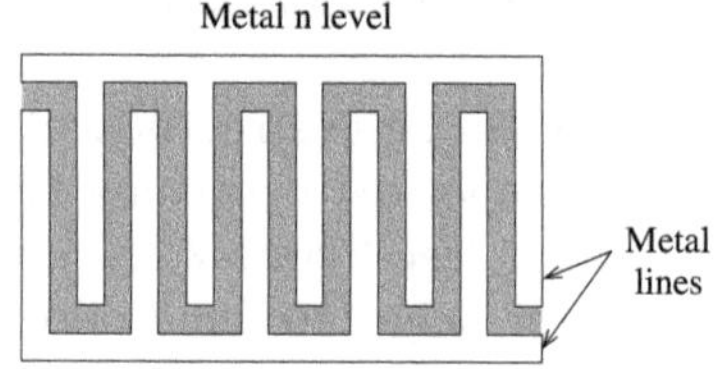

FIGURE 2.9
MOM capacitor layout.

Figure 2.9 shows the layout of an inter-digitated MOM capacitor, that is realized in a given metallization layer.

Ideally, every pair of two metal lines separated by a dielectric material can be used to form a MOM capacitor. In advanced CMOS processes, the lateral and vertical intervals between metal layers are close enough to guarantee a significant capacitance density.

MOM capacitors are available in various (two-dimensional and three-dimensional) configurations. Especially, several metal layers can be connected in parallel through vias to increase the capacitance density (up to about 2 fF/μm^2) of MOM capacitors.

MOM capacitors can have a high capacitance density, use standard CMOS process and scalability, and feature good matching characteristics and low parasitics.

Metal-insulator-metal (MIM) capacitors are realized by depositing a dielectric between the last top two metal layers of the wafer so as to reduce the substrate coupling effect. They are characterized by depletion-free and high-conductance electrodes and minimized capacitance loss to the silicon substrate.

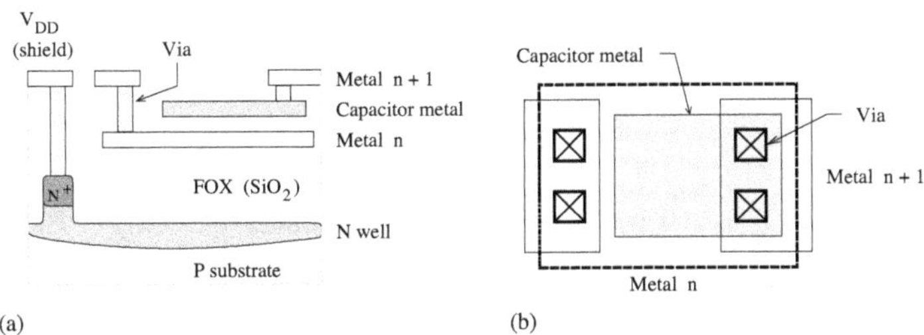

FIGURE 2.10
MIM capacitor (a) structure and (b) layout.

The structure and layout of a MIM capacitor are represented in Figures 2.10(a) and (b), respectively.

The capacitance variation with respect to the applied voltage can be expressed as,

$$C(V) = C_0(\alpha V^2 + \beta V + 1) \tag{2.2}$$

where V is the voltage applied between the capacitor electrodes, C_0 represents the capacitance at zero voltage, and α and β are the quadratic and linear voltage coefficients, respectively. Note that the linear coefficient β can be cancelled out by adopting a cross-coupled circuit configuration.

The key parameters of a capacitor for IC applications are the capacitance density (or area occupied in a chip), voltage linearity, leakage current density, and Q factor (or resistive losses). For MIM capacitors with a capacitance density of 5 fF/μm^2, the quadratic and linear coefficients are in the order of 238 ppm/V^2 and 206 ppm/V, respectively, at the frequency of 1 MHz. The leakage current can be smaller than 10^{-8} A/cm^2 at room temperature and at the maximum supply voltage V_{DD}, and the Q factor can be greater than 50 for a 1-pF capacitor at 5 GHz.

Generally, a MIM capacitor can exhibit a capacitance density in the order of 1 fF/μm^2, and has more accurate and stable characteristics than a MOM capacitor. However, the occupied silicon area of a MIM capacitor can be about two times larger than the one of the equivalent MOM capacitor (based on 65-nm CMOS process). Furthermore, MIM capacitors have the inconvenience of requiring extra mask and process steps.

The transistor configuration of Figure 2.11(a) can also be used as a capacitor. The resulting capacitance, as shown in Figure 2.11(b), where V_{FB} and V_T are the flatband and threshold voltages, respectively, is a function of the voltage, $V = V_{GS} = V_{GD}$, which should be chosen sufficiently greater than V_T, to take advantage of the linear capacitance-voltage characteristic of the transistor operating in the strong inversion region. The total capacitance, C, is given by the series connection of the gate-oxide capacitance, C_{ox}, and the depletion capacitance, C_d, between the channel induced under the oxide by increasing V_{GS}. That is,

$$C = \left(\frac{1}{C_{ox}} + \frac{1}{C_d}\right)^{-1} \tag{2.3}$$

The capacitance C_d is a function of the gate-substrate voltage. Between the gate and drain/source, we simply have overlap capacitances. Note that a MOSFET capacitor is generally designed with a minimum length to reduce the influence of the channel resistance.

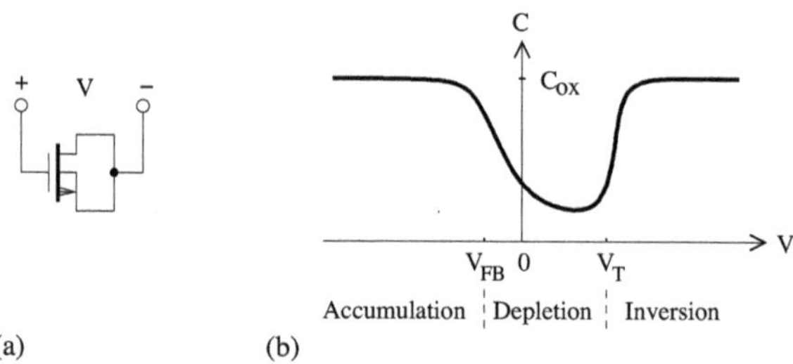

FIGURE 2.11
(a) A capacitor based on a MOS transistor and (b) its capacitance-voltage characteristic.

Note that in the accumulation region, there is a build-up of electrons at the oxide-semiconductor interface; in the depletion region, electrons are repelled from the oxide-semiconductor interface, and in the inversion region, charges at the oxide-semiconductor interface become positive.

2.2.2 Resistors

The resistance of a uniformly doped structure of the length, L, width, W, and thickness, t, can be computed as

$$R = \frac{\rho}{t}\frac{L}{W} = R_{\square}\frac{L}{W} \tag{2.4}$$

where ρ is the resistivity of the sample and $R_{\square}$ denotes the sheet resistance. The resistivity is determined by the type and concentration of impurity atoms.

The temperature variation of the resistance can be modeled as

$$R = R(T_0[1 + (T - T_0)TC(R)] \tag{2.5}$$

where the first-order temperature coefficient is given by

$$TC(R) = \frac{1}{R}\frac{dR}{dT} \tag{2.6}$$

The cross-section diagram of a polysilicon resistor is shown in Figure 2.12(a).

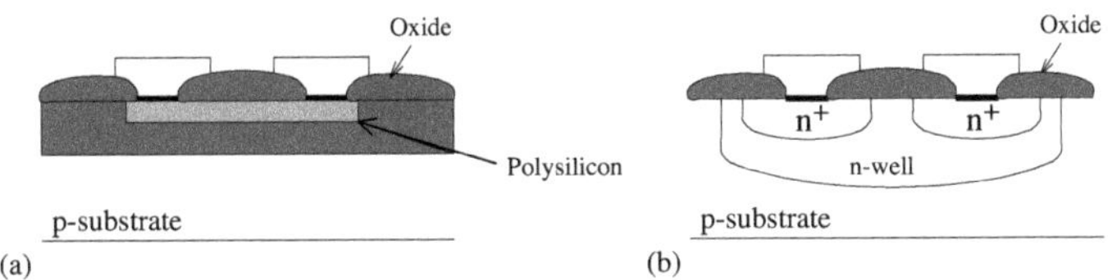

FIGURE 2.12
Structures of (a) polysilicon and (b) n-well resistors.

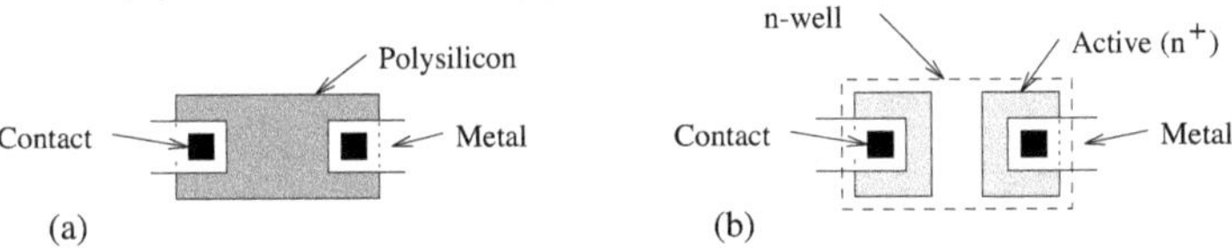

FIGURE 2.13
Layouts of (a) polysilicon and (b) n-well resistors.

The corresponding layouts are depicted in Figure 2.13. The resistor consists of the silicon dioxide thermally grown on a p-type substrate, undoped polysilicon settled by a low-pressure chemical-vapor deposition, and ohmic contact realized with aluminum electrodes. Experimental results show that the resistances vary depending on the polysilicon thickness. The resistor can feature a good linearity. However, for an accurate characterization of the high-frequency performance, the parallel-plate capacitance formed between the polysilicon and substrate must be taken into account.

The cross-section diagram of an n^+ diffusion resistor is shown in Figure 2.12(b). The diffusion layer is located within the n-well and isolated using the silicon dioxide. The sheet resistance of the device can be on the order of a kiloohm with a typical absolute value tolerance of a few percent.

2.2.3 Inductors

Although inductors are generally realized off-chip, there is an increasing interest in the integration of all circuits elements in a single structure. To this end, MOS processes made up of lightly doped substrate and wells can be used for the inductor design. Due to the mutual magnetic coupling, the inductance

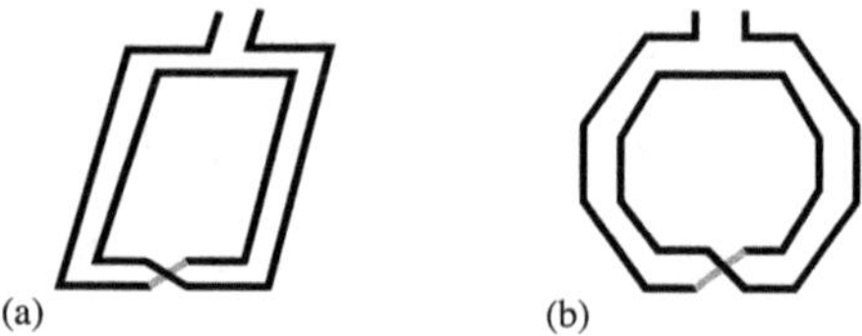

FIGURE 2.14
Layout of integrated inductors.

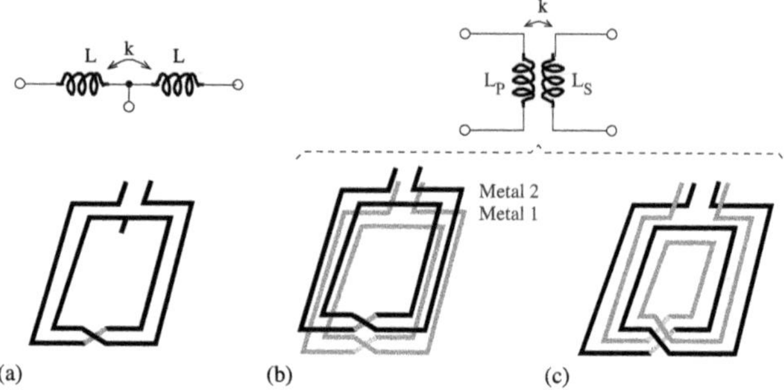

FIGURE 2.15
Layout of coupled inductors or transformers.

value can be increased using a series connection of N wire segments in the metal layer. This results in the inductance given by

$$L = \sum_{i=1}^{N} L_i + 2\sum_{i=1}^{N}\sum_{i \neq j} M_{ij} \tag{2.7}$$

where L_i is the inductance of the metal segment i, and M_{ij} is the mutual magnetic coupling between the segment i and j. The mutual inductance between two segments is determined by their length, separation distance, and intersection angle. The coupling between two perpendicular wires is negligible. However, the behavior of on-chip inductors is perturbed by dissipative mechanisms beyond the ones of a simple conductor loss. The planar spiral inductor has parasitic capacitors between the metal layer and substrate due to the insulating oxide. At high frequencies, a current can flow through these capacitors into the substrate, which can be modeled as RC networks. The achievable Q-factor is then about 5–20 for inductances in the range of a few nanohenries.

The inductors shown in Figure 2.14 consist of windings, which are fabricated with a metal layer patterned on the field oxide. The total self-inductance increases with the number of layers. Note that the difficulty in the realization of high-Q IC inductors is related to the creation of local insulating regions on the wafer for the device isolation.

A coupled inductor can be implemented using a center tapped inductor, as shown in Figure 2.15(a). The magnetic coupling coefficient is determined by the line width, spacing between conductors, and the substrate thickness.

A monolithic transformer can be designed using conductors overlaid as stacked metals or interwound in the same plane [3], as shown in Figures 2.15(b) and (c). The mutual inductance (and capacitance) is related to the common periphery between conductors. In practice, the achievable coupling coefficient can be as high as 0.9. Transformers using stacked conductors provide a slightly higher coupling coefficient, but their operating frequency can be limited by the lower self-resonance frequency associated with the large parallel-plate capacitance available between the windings.

2.3 Integrated-circuit (IC) interconnects

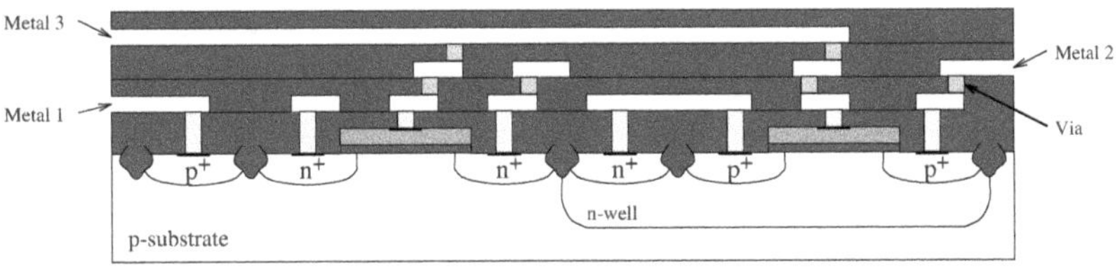

FIGURE 2.16
An IC cross-section with three metal layers.

The fabrication of an IC starts with the production of silicon wafers, which are then processed using various deposition and etching techniques, such as photolithography. The transistors are formed in wafer regions protected from oxidation and often referred to as moat regions. The area between adjacent transistors, which is generally covered with a thick-field oxide to prevent the formation of parasitic channel components, is called the field region. Figure 2.16 shows the cross-section of an IC with three metal layers.

Unlike transistors, the interconnect performance is not improved through miniaturization. By scaling down the interconnect size, the crosstalk and latency effects are increased, while the inductance influence on the signal transients becomes dominant for large interconnect geometries. This is the case for power wires, which are connected to a large number of devices. Basically, a signal propagating on an interconnect is influenced by effects (delay, attenuation, reflection, and crosstalk) similar to the ones known in transmission lines. Depending on the IC process and signal operating frequency, the interconnects can be described by a lumped, or distributed (frequency independent/dependent parameters) model. All wires have a resistance, which is a function of the material resistivity. In the case of thin-film aluminum (Al) and copper (Cu), the resistivity is 3.3 mΩ·cm and 2.2 mΩ·cm, respectively. Interconnections between metal layers, also called plugs or vias (see Figure 2.16), and made of tungsten (W) for aluminum wires, appear to be somewhat resistive (a few ohms). The current density in a metal wire is limited by the *electromigration* effect. At high current densities, the aluminum ions tend to migrate from one wire end, leaving voids, which can grow to a discontinuity after some time. At the other end, an accumulation of atoms in microscopic structures called hillocks can be observed. This failure is prevented by using via arrays for long wires and specifying design rules, which includes the minimum sizes of the wire to keep the current density less than 0.5-1 mA/µm. At the process level, alloying elements such as copper can be added to aluminum to avoid the ion displacement. Copper has also been introduced as a substitute for aluminum. In this case, the effect of electromigration is reduced because the IC process also permits vias made of copper.

The schematic of the equivalent model of a two-wire interconnect with the active line linked to a signal generator [4] is illustrated in Figure 2.17. To reduce the resistance, wires are generally designed to be taller than they

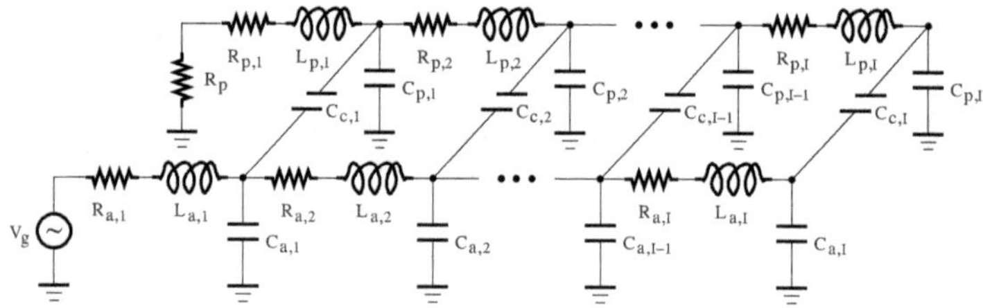

FIGURE 2.17
Equivalent circuit model of the interconnect.

are wide. Coupling noise can be important due to the high values of parasitic capacitance and inductance. The top and bottom capacitors are typically assumed to be grounded because they are associated with a set of orthogonally routed conductors that, averaged over the wire length, maintain a constant voltage. Due to the fact that a given wire is coupled to the neighbors by significant capacitances rather than ground, the signal integrity becomes dependent on the neighboring wire activity. At high frequencies, inductive effects become significant and the inductance can dominate the impedance of interconnect wires. Specifically, the coupling related to inductors is a non-local effect because the mutual inductance does not sink rapidly, as does the coupling capacitance. In fact, the inductance can be directly determined in cases where a time-varying current is supposed to flow in a known closed loop and the flux of the resulting magnetic field is proportional to the loop area. But, the ambiguity in the determination of the current return path, which may include the ground, power lines, and neighboring wires, complicates the estimation of the interconnect inductance.

The computation of the interconnect resistance, inductance, and capacitance involves the solution of quasi-static electromagnetic systems. Due to the high amount of parasitic data to be extracted, CAD tools are required for high-density designs in a deep-submicrometer technology. The matching of impedances, insertion of repeaters (or inverters with large transistors), and the use of three-dimensional structures consisting of multiple levels of devices and layers can offer prospects for relief from the parasitic effects of interconnects.

2.4 Physical design considerations

Generally, the physical design involves steps of floor planning, timing optimization, placement, routing, and layout generation. The design process is increasingly interlaced with the verification, which consists of electrical rule checking (ERC), design rule checking (DRC), layout-versus-schematic (LVS) comparison, as well as resistance, inductance and capacitance parasitic extraction, and interconnect and signal integrity characterization.

Lay out the inverter shown in Figure 2.18(a) using the design

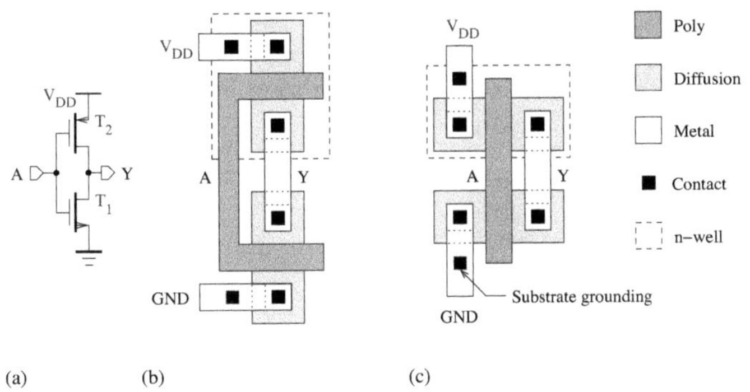

FIGURE 2.18
CMOS inverter: (a) Circuit diagram, (b) and (c) two possible layouts.

rules of a given CMOS process with an n-well.
Two possible layouts can be derived, as illustrated in Figures 2.18(b) and (c). A metal layer is required between the n and p diffusion areas, as they should not be directly in contact. Note that contact openings can only be realized between metal and any other layer. To avoid the forward-biasing of pn junc-

tions, the substrate and n-well are tied to the ground and the voltage V_{DD}, respectively.

ERC involves checking a circuit design for proper geometry and connectivity specifications. Examples of electrical rule violations include floating nodes, open circuits, improper supply voltage/ground and well/substrate connections, and limited power-carrying capacity on wires. The main objective of DRC is to provide a high overall die yield and reliability for a given design. Design rules specify certain geometric and connectivity restrictions (active-to-active spacing, metal-to-metal spacing, well-to-well spacing, minimum metal width, minimum channel length of transistors, etc.) to ensure sufficient margins to account for common variations, and are specific to an integrated-circuit fabrication process. They can be based on the λ rule or micron rule. In the first case, all geometric specifications in a design are specified as integer multiples of the scalable parameter, λ, which is chosen such that the minimum gate length of a transistor is equal to 2λ, and micrometer units (or absolute measurements) are used in the second case. With the λ rule, the migration from one fabrication process to another is simplified. On the other hand, micron rules have the advantage of minimizing the resulting layout size. Computer-aided tools for DRC and ERC enable the verification of a circuit design prior to the fabrication. The ERC performs the syntax analysis on the circuit network, while the DRC verifies that the circuit layout does not have any specific errors.

Physical design results in a layout or geometric patterns, which are used for the IC manufacturing. Each fabrication step is based on a different layer mask extracted from the layout.

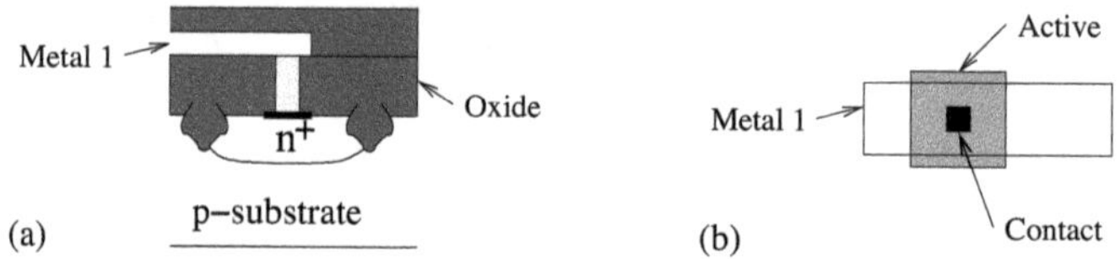

FIGURE 2.19
(a) Cross-section of an n^+ region and (b) the corresponding layout.

The fabrication of basic MOS structures requires several masking sequences, such as n-well, active, polysilicon, select, contact, metal, via, and overglass. The geometric size of features on every layer is defined by relevant design rules. Let us consider the n^+-doped region shown in Figure 2.19. It is realized by implanting ions into the substrate through the area described by the active mask. The minimum width of the active area is specified by the design rule. The active contact is shaped in the oxide to allow a connection between the first layer of metal and the active region. Here, the minimum spacing between the active and contact area, and the vertical and horizontal sizes of the contact have to be defined. The metal, which is deposited after the oxide layer, is used as an interconnect. It is subject to rules such as metal-to-active contact minimum spacing and the minimum width of the metal line.

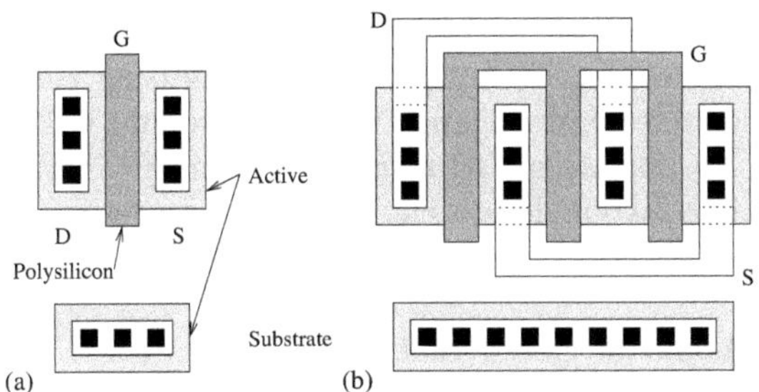

FIGURE 2.20
(a) Layout of a single transistor; (b) layout of a large transistor with multiple (three) fingers.

Different layout techniques are often adopted to minimize the parasitic effect and improve the device matching. As shown in the transistor layout of Figure 2.20(a), the overall contact resistance can be reduced by using as many contacts as the design rules allow. With N contacts, the resulting resistance is $1/N$ times the value of a single contact, as is the case in a parallel configuration of resistors.

The design hierarchy permits the use of library cells to construct more complex circuits. The layout of a large transistor, which consists of a parallel connection of three minimum-size structures with the same length, is shown in Figure 2.20(b). The final transistor features the length of a single device and the sum of the individual widths. Note that by using a transistor with parallel fingers, the gate resistance can be reduced while the capacitance related to the source-to-drain areas increases. For this reason, excessively long geometries should preferably be avoided in the design of wide transistors.

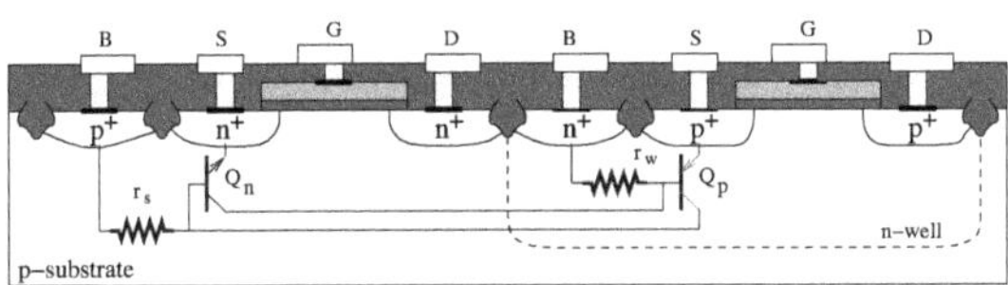

FIGURE 2.21
Parasitic resistors and bipolar transistors in a CMOS circuit.

In CMOS circuits in which a thyristor consisting of parasitic npn and pnp bipolar transistors, Q_n and Q_p (see Figure 2.21), is formed, a *latch-up* effect can occur [5]. The parasitic circuit includes the substrate and well resistors,

r_s and r_w, and a positive feedback is formed around Q_n and Q_p. Due to transient noises, one of the two transistors can become forward biased and feeds the base of the other transistor. As a result, a current flows between the supply voltage lines and the circuit is unable to deliver a response to an input signal. Provided the feedback gain is greater than or equal to unity, this current will increase until the circuit burns out.

The latch-up effect can be mitigated by minimizing the current gains of the parasitic bipolar transistors, and the substrate and well resistances. This can be achieved by placing guard rings or n^+ and p^+ regions around MOS transistors. Suitable layout design rules, and appropriate selection of doping concentrations and profiles, can contribute to the reduction of latch-up susceptibility.

However, it should be noted that a substrate with a low resistivity can exhibit various parasitic paths between on-chip devices, thereby coupling the substrate noise to the signal of interest. The problem of *substrate coupling* is remarkable in mixed-signal circuits, where the analog components can be perturbed by the switching noise generated during the transitions of the clock signal used to control the switched devices. It is solved at the circuit level using a differential configuration, which is less sensitive to the common-mode noise. Another solution can consist of isolating sensitive circuit sections from the substrate noise.

Device matching can be enhanced by adopting layout techniques, which can reduce the statistical variations of the IC process. To this end, symmetry must be applied to the device layout as well as its nearby environment. In a set of transistors, for instance, a better matching is achieved by adopting the same orientation to place the transistors in the layout. This is due to the fact that the transconductances of MOS transistors depend on carrier mobilities, which are known to be sensitive to orientation-dependent stress.

Matched pairs of devices can be laid out with interdigitated or common-centroid structure [6]. Two layout examples of two capacitors X and Y are shown in Figure 2.22. In Figure 2.22(a), the construction is realized simply by cross-coupling the connection between the different cells. A common-centroid layout (see Figure 2.22(b)) features more symmetry and can be obtained by arranging elements as $XYYXXYYX$, or $XYXYYXYX$. It can also be obtained by using a two-dimensional configuration of the following form:

$$\begin{array}{c} XYYX \\ YXXY \\ XYYX \end{array}$$

As a result, the linear gradient in the oxide thickness, for instance, is equally distributed to the capacitor pairs and its effect is attenuated. Note that patterns of the form

$$XYXY$$

and

$$XYYXXY$$

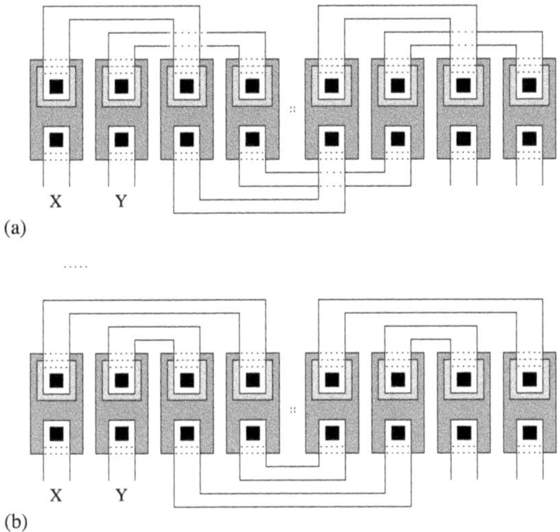

FIGURE 2.22
Poly-poly capacitor: (a) Interdigitated and (b) common-centroid layouts.

are often avoided because they exhibit a symmetry for each set of elements rather than for the whole pattern or are not uniformly distributed. The layout of Figure 2.23 can be used for the implementation of two capacitors C_1 and C_2. The unit cell located in the layout center is required for C_1, while C_2 is formed using eight unit capacitors. Dummy capacitors without an electrical role are added to similarly provide the same adjacent environment to both devices.

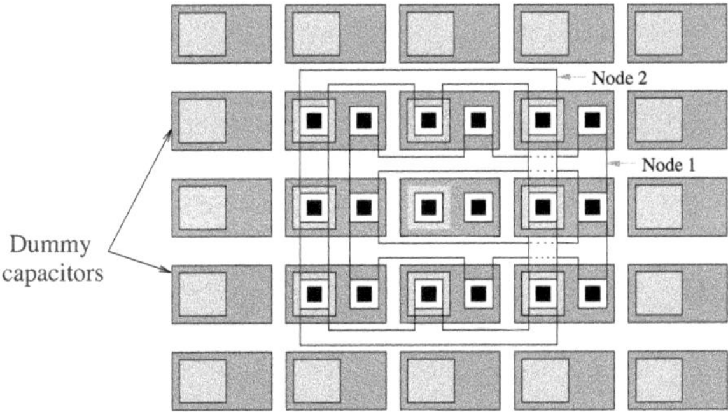

FIGURE 2.23
Layouts with dummy capacitors.

Generally, the performance of a circuit is determined by the matching characteristics of basic building blocks such as current mirrors and differential transistor pairs. In the case of current mirrors, the dominant mismatch effects are usually related to threshold voltage variations.

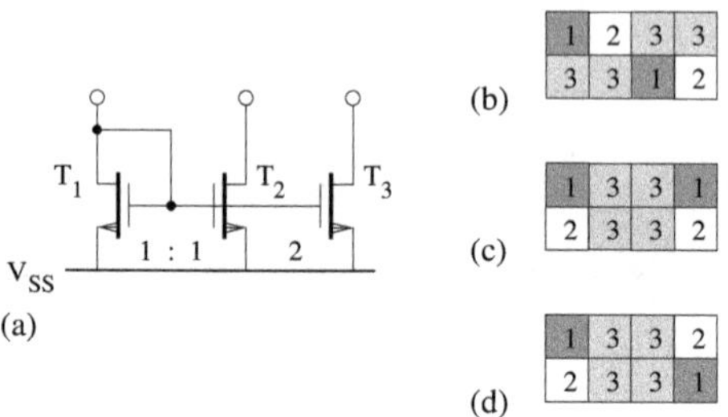

FIGURE 2.24
Two-output current mirror (a) with possible floorplans (b), (c), and (d).

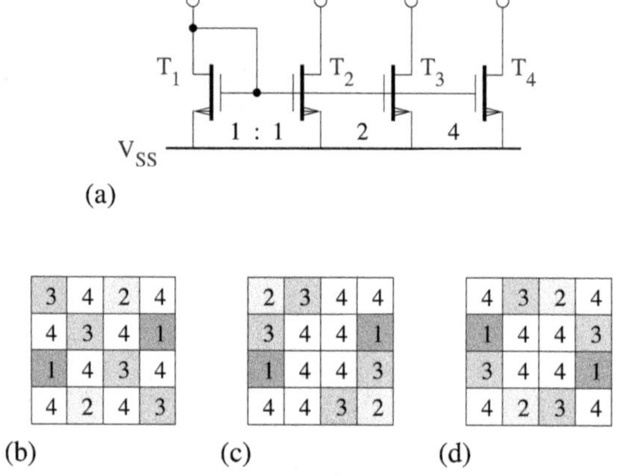

FIGURE 2.25
Three-output current mirror (a) with possible floorplans (b), (c), and (d).

Consider the current mirror shown in Figure 2.24(a), where the transistors T_1 and T_2 have the same ratio W/L, while the ratio of T_3 is $2W/L$. It can be verified that among the three possible floorplans of Figures 2.24(b), (c) and (d), the one of Figure 2.24(d) provides the best matching between transistors.
Figure 2.25(a) shows a current mirror, where the ratios of transistors T_1 and T_2 are W/L, the ratio of T_3 is $2W/L$, and the ratio

of T_4 is $4W/L$. Of the three possible floorplans that are represented in Figures 2.25(b), (c) and (d), the one of Figure 2.25(d) provides the best matching between transistors.
This is due to the fact that the transistors are symmetrically placed and with the same orientations. Note that, in each floorplan, a given transistor is represented by square boxes with the corresponding number.
In practice, manually placing devices to form common-centroid structures can be tedious. By programming in a scripting language (SKILL, PERL, TCL, C-shell), parameterized cells can be generated and used to automatically place devices in a common-centroid pattern. However, the manual routing of connections between placed devices can be time consuming, and the performance of automatic routing tools may not always be optimal.

2.5 IC packaging

The IC package should provide a protection from external environments and facilitate the mounting of components in a test fixture without the risk of damage. In most demanding applications, multi-chip modules seem to be one of the few moderate-to-high lead count structures that can provide adequate electrical and thermal performance. The trend is toward thinner, smaller, and lighter packages with a high pin count. Advances in electronic packaging are based on improvements in the IC fabrication facilities, as well as in the deposited and laminated multi-chip processes. Generally, the interconnect nonideal components together with the resistor, inductor, and capacitor of the package, pads, and bond wires form various parasitic RLC circuits that can add substantial noises to the signal of interest. The quality factor determines the amplitude and sharpness of the resonance. It can be tuned by steering the value of on-chip and off-chip passive devices. An accurate circuit model for packages is then indispensable for high-performance designs.

2.6 Summary

The design of high-density ICs with deep-submicrometer processes faces increasingly difficult challenges. Due to the scaling, transistors have become smaller and faster, while chip power dissipation has increased. The size reduction of interconnects and packaging has led to an increase of parasitic effects. It is then necessary to adopt innovative approaches for modeling, design, simulation, fabrication, and testing. Design technology, which includes software and hardware tools, and methodology, is thus the mean in approaching and realizing the limits imposed at the different levels of IC development. Its quality determines the design time, performance, cost, and reliability of the final chip.

2.7 Circuit design assessment

1. **High-dynamic-range MOS capacitor**
   The main disadvantage of capacitors implemented using a MOS transistor is the nonlinear capacitance-voltage characteristic due to different charge distributions in the accumulation, depletion, and inversion regions. Two gate-coupled transistors, $T_1 - T_2$, can be configured as shown in Figure 2.26 to compensate for the voltage dependence and provide a linear capacitor over a wide dynamic range [7,8].

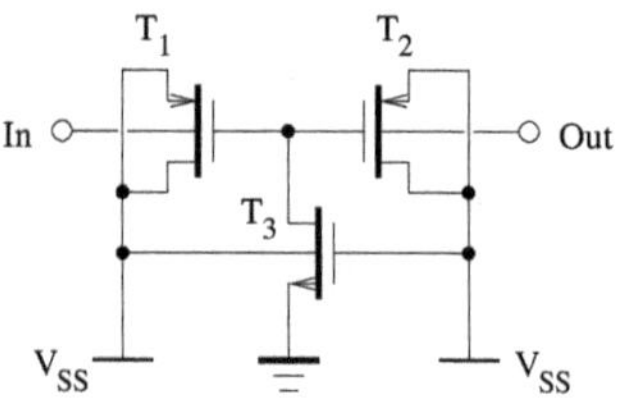

FIGURE 2.26
Circuit diagram of a high-dynamic-range MOS capacitor.

   Analyze the capacitance-voltage characteristics of the circuit using SPICE simulations.

2. **Inductor analysis**
   Consider the equivalent circuit model of the MOS spiral inductor shown in Figure 2.27 [9]. Estimate the Q-factor defined as

$$Q = 2\pi \frac{E_m - E_e}{\triangle E} \tag{2.8}$$

   where E_m and E_e are the peak magnetic and electric energies, respectively, and $\triangle E$ denotes the energy loss in one oscillation cycle. Verify that the maximum value of Q, $Q_{max} = \omega L/R_p$, is obtained at low frequencies.

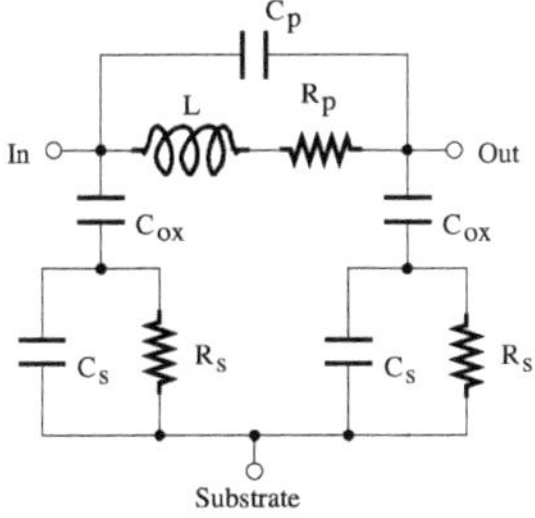

FIGURE 2.27
Equivalent circuit model of a MOS inductor.

3. **NAND gate and current mirror layouts**
   • The circuit diagram and layout of a NAND gate are shown in Figures 2.28(a) and (b), respectively.

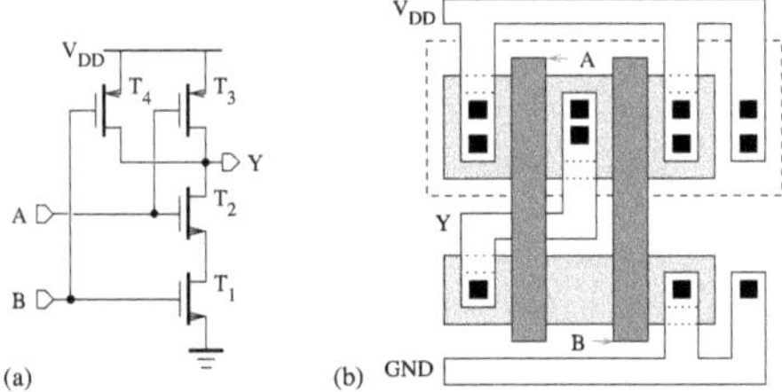

FIGURE 2.28
NAND gate: (a) Circuit diagram, (b) layout.

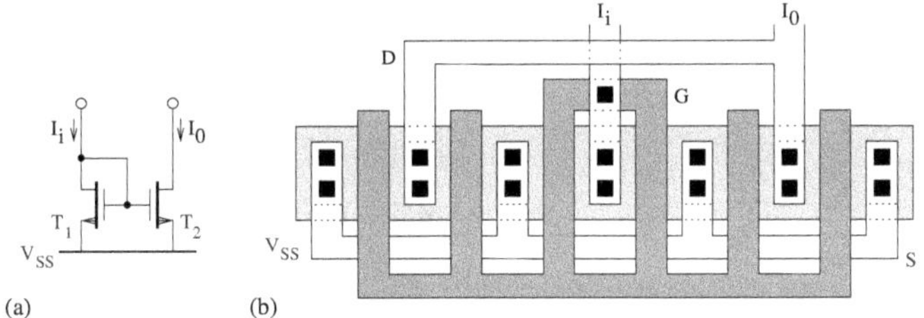

FIGURE 2.29
Current mirror: (a) Circuit diagram, (b) layout.

   Verify the NAND gate layout and estimate the aspect ratios of transistors.

   • Consider the circuit diagram and layout of a current mirror depicted in Figures 2.29(a) and (b), respectively.

   Determine the aspect ratios of the transistors from the layout.

Lay out the current mirror assuming that each transistor is now realized with a single gate structure.

Determine the errors introduced in the current ratio in each of the layouts.

4. **Differential amplifier layout**

To obtain a rectangular layout of a circuit, a transistor with the aspect ratio W/L can be realized as a single or a series stack structure.

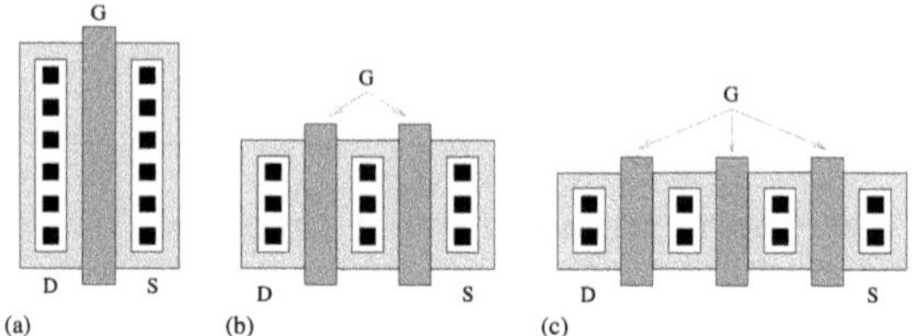

FIGURE 2.30
Layouts of a transistor and the equivalent series stack transistors.

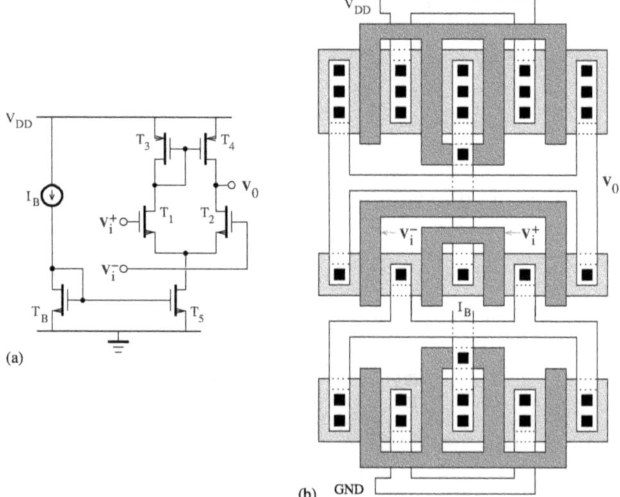

FIGURE 2.31
Differential amplifier layout.

Establish the equivalence between the characteristics (e.g., drain-substrate and source-substrate capacitances) of the transistor described by the layouts depicted in Figure 2.30, where the transistor widths are (a) W, (b) $W/2$, and (c) $W/3$.

The differential amplifier of Figure 2.31(a) can be laid out as shown in Figure 2.31(b). The aspect ratios of transistors are as follows:

$$T_1\text{-}T_2\ : W/L = 3 \quad T_3\text{-}T_4\ : W/L = 9 \quad T_5\text{-}T_6'\ : W/L = 6$$

Assuming that the differential amplifier should be realized using a CMOS process with an n-well, complete the layout by placing all pMOS transistors in the same well to be tied to the voltage V_{DD}, and by connecting the substrate to the ground.

Use a layout tool to perform the design rule check.

Bibliography

[1] D. J. Allstot and W. C. Black, Jr., "Technological design considerations for monolithic MOS switched-capacitor filtering systems," *Proc. of the IEEE*, vol. 71, pp. 967–986, Aug. 1983.

[2] B. Razavi, *Design of Analog CMOS Integrated Circuits*, New York, NY: McGraw-Hill, 2001.

[3] J. R. Long, "Monolithic transformers for silicon RF IC design," *IEEE J. of Solid-State Circuits*, vol. 35, pp. 1368–1382, Sept. 2000.

[4] Y. Ismail, E. Friedman, and J. Neves, "Equivalent Elmore delay for RLC trees," *IEEE Trans. on Computer-Aided Design*, vol. 19, pp. 83–97, Jan. 2000.

[5] T. Cabara, "Reduced ground bounce and improved latch-up suppression through substrate conduction," *IEEE J. of Solid-State Circuits*, vol. 23, pp. 1224–1232, Oct. 1988.

[6] R. Jacob Baker, Harry W. Li, David E. Boyce, *CMOS Circuit Design, Layout, and Simulation*, Piscataway, NJ: IEEE Press, 1998.

[7] P. Larsson, "A 2-1600-MHz CMOS clock recovery PLL with low-V_{dd} capability," *IEEE J. of Solid-State Circuits*, vol. 34, pp. 1951–1960, Dec. 1999.

[8] T. Tille, J. Sauerbrey, and D. Schmitt-Landsiedel, "A 1.8-V MOSFET-only $\Sigma\Delta$ modulator using substrate biased depletion-mode MOS capacitors in series compensation," *IEEE J. of Solid-State Circuits*, vol. 36, pp. 1041–1047, July 2001.

[9] T. H. Lee and S. S. Wong, "CMOS RF integrated circuits at 5 GHz and beyond," *Proc. of the IEEE*, vol. 88, pp. 1560–1571, Oct. 2000.

3

Bias and Current Reference Circuits

CONTENTS

As the power supply is scaled down, the operation of analog circuits becomes difficult because the threshold voltage of MOS transistors and *dc* operating points are not scaling proportionally. The bias and current reference circuits should be designed to feature good stability over the IC process, and supply voltage and temperature variations. A review of different techniques used to achieve this goal is then necessary.

Current mirrors should desirably feature high-output resistance, which is required for an accurate replication of currents, and high-output swing, which is essential for a low-voltage operation. There are many current mirror architectures, each with its advantages and inconveniences.

A voltage reference is commonly used in IC design for providing a constant voltage in spite of the IC process, supply voltage, and temperature variations. Various voltage reference structures are available with varying degrees of initial accuracy and drift over the temperature range of operation. Some voltage reference circuits provide an output voltage that is determined by a resistor or transistor. However, the resulting reference voltage is limited by IC process deviations. Band-gap voltage reference circuits generate an output voltage that is less sensitive to variations of the operating temperature. But often, external voltage references, which exhibit a much higher accuracy and lower drift than on-chip voltage references, are required in high-precision applications.

3.1 Current mirrors

Current mirrors find applications in analog ICs as biasing elements, which can set transistor bias levels such that the circuit characteristics are less affected by power supply and temperature variations. They are also used in the amplifier design as load devices, whose high impedance is exploited to increase the resulting gain. The operation of current mirrors relies on the principle that a reference current in the input stage is either sinked or sourced in such a way as to be reproduced with a given scaling factor in the output stage. Generally, the generated output current or bias current is used to drive a given load.

3.1.1 Simple current mirror

The circuit diagram of a simple current mirror is shown in Figure 3.1(a). Because the transistor T_1 is diode-connected, it operates in the saturation region, and T_2 is also assumed to operate in the saturation region.

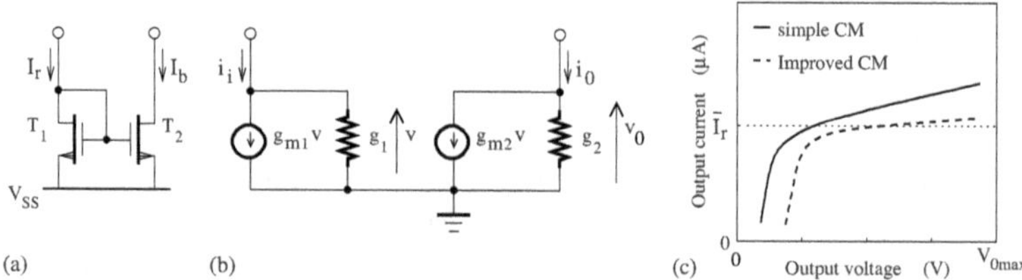

FIGURE 3.1
(a) Simple current mirror (CM); (b) small-signal equivalent model; (c) output characteristics of a simple CM and an improved CM.

The minimum voltage at the output node of the current mirror is

$$V_{bmin} = V_{DS(sat)} = V_{GS} - V_T \tag{3.1}$$

where $V_{DS(sat)}$ is the drain-source saturation voltage. The drain currents I_{D1} and I_{D2} are given by

$$I_{D_1} = K_1' \left(\frac{W_1}{L_1}\right)(V_{GS_1} - V_T)^2(1 + \lambda V_{DS_1}) \tag{3.2}$$

$$I_{D_2} = K_2' \left(\frac{W_2}{L_2}\right)(V_{GS_2} - V_T)^2(1 + \lambda V_{DS_2}) \tag{3.3}$$

where $K_1' = K_2' = \mu C_{ox}/2$ is the transconductance parameter, μ denotes the electron mobility, C_{ox} is the gate oxide capacitance per unit area, and λ is the channel-length modulation coefficient. With $V_{GS_1} = V_{GS_2} = V_{GS}$, the current ratio can be obtained as

$$\frac{I_{D_2}}{I_{D_1}} = \frac{W_2/L_2}{W_1/L_1} \cdot \frac{1+\lambda V_{DS_2}}{1+\lambda V_{DS_1}} \tag{3.4}$$

Neglecting the effect due to the channel-length modulation and assuming that $I_{D_1} = I_r$ and $I_{D_2} = I_b$, we can write

$$\frac{I_{D_2}}{I_{D_1}} = \frac{I_b}{I_r} \simeq \frac{W_2/L_2}{W_1/L_1} \tag{3.5}$$

With reference to the small-signal equivalent model of Figure 3.1(b), where $i_i = 0$ and the transistor body transconductances are assumed to be negligible, it can be deduced that $v = 0$ and the output resistance is obtained as

$$r_0 = \frac{v_0}{i_0} = \frac{1}{g_2} = r_{DS_2} = \frac{1+\lambda V_{DS_2}}{\lambda I_{D_2}} \tag{3.6}$$

For typical values of the transistor parameters, the resistance r_0, which is about a few hundred kiloohms, can appear to be too small in high-accuracy applications. Figure 3.1(c) shows the output characteristics of current mirrors. The effect of the higher output resistance exhibited by an improved current mirror can be clearly observed in these plots.

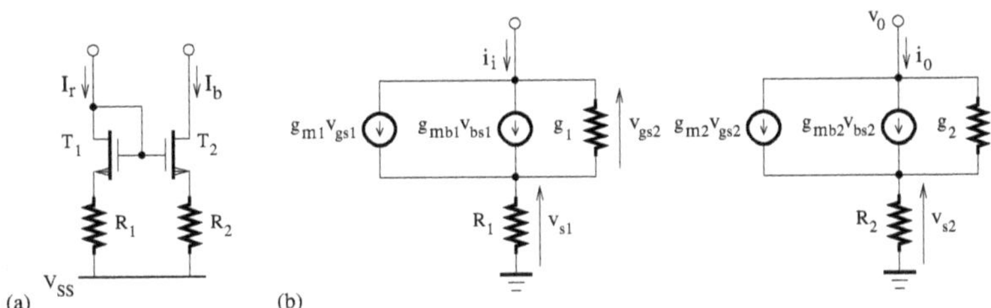

FIGURE 3.2
(a) Current mirror with source degeneration resistors and (b) its small-signal equivalent model.

A simple approach to increase the current mirror resistance can consist of using source degeneration resistors, as illustrated in Figure 3.2. Both transistors operate in the saturation region and it can be shown that

$$R_2 I_b + V_{GS_2} = R_1 I_r + V_{GS_1} \tag{3.7}$$

where $V_{GS_1} = V_{DS_1(sat)} + V_{T_1}$ and $V_{GS_2} = V_{DS_2(sat)} + V_{T_2}$. Because the transistor threshold voltages are matched and $V_{DS_1(sat)} = \sqrt{I_b/[K_2'(W_2/L_2)]}$, we have

$$R_2 I_b + \sqrt{\frac{I_b}{K_2'(W_2/L_2)}} - (R_1 I_r + V_{DS_1(sat)}) = 0 \tag{3.8}$$

The square of the only positive root of this quadratic equation is given by

$$I_b = \frac{1}{4R_2^2}\left(-\sqrt{\frac{1}{K_2'(W_2/L_2)}}+\sqrt{\frac{1}{K_2'(W_2/L_2)}+4R_2\left(R_1 I_r + V_{DS_1(sat)}\right)}\right)^2 \tag{3.9}$$

where $V_{DS_2(sat)} = \sqrt{I_r/[K_1'(W_1/L_1)]}$. In the specific case of a Widlar MOS current mirror, $R_1 = 0$ and the source of T_1 is directly connected either to a supply voltage terminal or ground.

With reference to the small-signal equivalent model of Figure 3.2(b), we have

$$v_O = [i_O - (g_{m2}v_{gs2} + g_{mb2}v_{bs2})]/g_2 + v_{s2} \tag{3.10}$$

where $v_{s2} = R_2 i_O$. For the determination of the output resistance, the voltage at the input node is set to zero, that is, $v_{d1} = v_{g1} = v_{g2} = 0$. As a result, $v_{gs2} = -v_{s2}$. With $v_{bs2} \simeq -v_{s2}$, the output resistance can be computed as

$$r_O = \frac{v_O}{i_O} = \frac{1}{g_2} + R_2\left(1 + \frac{g_{m2}+g_{mb2}}{g_2}\right) \tag{3.11}$$

Hence, the use of source degeneration resistors leads to an increase in the current mirror output resistance. By reducing the sensitivity of the output current to variations of the voltage at the current mirror output, a high output resistance helps provide an accurate replication of currents.

3.1.2 Cascode current mirror

In order to obtain high resistance, current mirrors can be designed using cascode transistor structures. However, adequate transistor biasing is required to achieve a high-output swing, which is desired for the low-voltage operation.

The cascode current mirror represented in Figure 3.3(a) has the advantage of reducing the effect of the channel-length modulation and increasing the output resistance. It is assumed that the effects due to the bulk of the transistors T_1 and T_2 are negligible, and all transistors are matched. To ensure the operation of all the transistors in the saturation region, the drain-source voltages of the diode-connected transistors, T_1 and T_3, should be at least $V_{DS_1} = V_{DS_3} = V_{DS(sat)} + V_T$, yielding the minimum voltages of the value

$$V_{rmin} = V_{DS_1} + V_{DS_3} = V_{G_2} = 2(V_{DS(sat)} + V_T) \tag{3.12}$$

and

$$V_{bmin} = V_{G_2} - V_T = 2(V_{DS(sat)} + V_T) - V_T = 2V_{DS(sat)} + V_T \tag{3.13}$$

respectively, at the input and output nodes. Note that V_{bmin} represents the

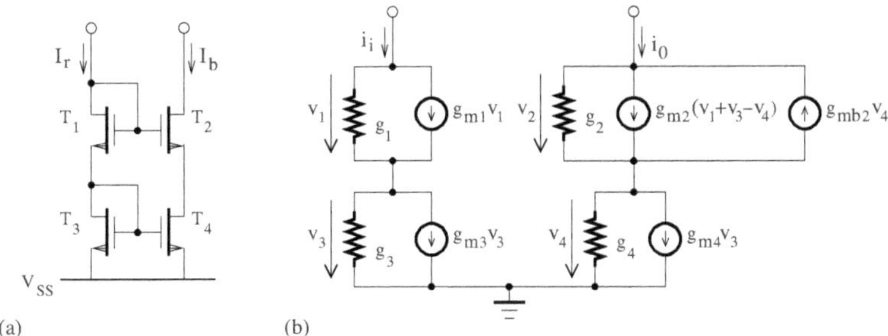

FIGURE 3.3
(a) Cascode current mirror and (b) its small-signal equivalent model.

sum of two overdrive voltages and one threshold voltage. The drain-source voltages of the transistors T_4 and T_2 are then given by

$$\begin{aligned} V_{DS_4} &= V_{G_2} - V_{GS_2} \\ &= 2(V_{DS(sat)} + V_T) - (V_{DS(sat)} + V_T) = V_{DS(sat)} + V_T \end{aligned} \tag{3.14}$$

and

$$V_{DS_2} = V_{bmin} - V_{DS_4} = V_{DS(sat)} \tag{3.15}$$

In general, the relation $V_{GS_1} + V_{DS_3} = V_{GS_2} + V_{DS_4}$ holds for the circuit of Figure 3.3(a). Thus, $V_{GS_1} = V_{GS_2}$ and $V_{DS_3} = V_{DS_4}$ provided $(W_2/L_2)(W_3/L_3) = (W_1/L_1)(W_4/L_4)$. In this case, the current I_b has a value close to the one of I_r.

The output resistance of the cascode current mirror can be derived using the small-signal equivalent circuit shown in Figure 3.3(b). Then, we can write

$$v_2 = [i_0 + g_{mb2}v_4 - g_{m2}(v_1 + v_3 - v_4)]/g_2 \tag{3.16}$$

$$v_4 = (i_0 - g_{m4}v_3)/g_4 \tag{3.17}$$

Assuming that $i_r = 0$, the values of the voltage v_1 and v_3 are reduced to zero and the output resistance is given by

$$r_0 = \frac{v_0}{i_0} = \frac{1}{g_2} + \frac{1}{g_4} + \frac{g_{m2} + g_{mb2}}{g_2 g_4} \tag{3.18}$$

where $v_0 = v_2 + v_4$. Note that $v_{sb2} \neq 0$ in contrast to the source-substrate voltage of the other transistors. The body effect due to the g_{mb2} contribution tends to increase the output resistance.

In the cascode current mirror shown in Figure 3.4(a), the output current will follow I_r if V_B is chosen such that $V_{G2} = V_{D2}$. Applying Kirchhoff's current and voltage laws to the small-signal equivalent model of Figure 3.4(b),

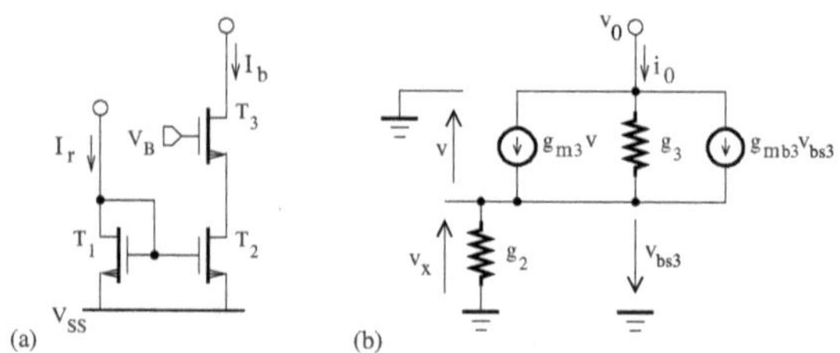

FIGURE 3.4
(a) Cascode current mirror with a single input transistor and (b) its small-signal equivalent model.

we obtain

$$i_0 = g_{m3}v + g_{mb3}v_{bs3} + g_3(v_0 - v_x) \tag{3.19}$$

$$v_x = i_0/g_2 \tag{3.20}$$

$$v_{bs3} = v \ = -v_x \tag{3.21}$$

where g_{mb3} denotes the transconductance due to the body of the transistor T_3. The output resistance r_0 is then given by

$$r_0 = \frac{v_0}{i_0} = \frac{1}{g_2} + \frac{1}{g_3} + \frac{g_{m3} + g_{mb3}}{g_2 g_3} \tag{3.22}$$

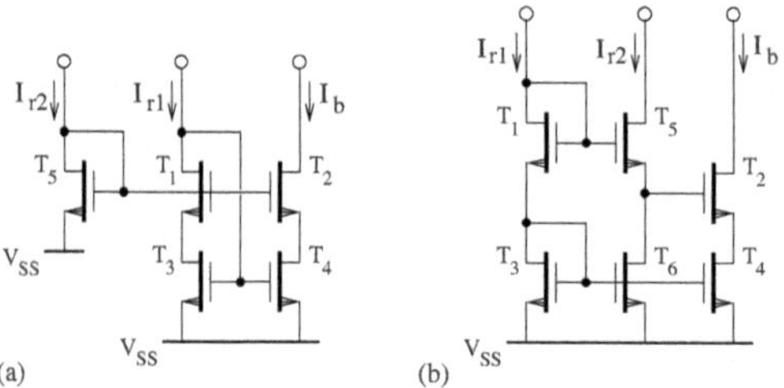

FIGURE 3.5
(a) High-swing cascode current mirror; (b) high-swing cascode current mirror using a source follower output stage.

A high-swing cascode current mirror is illustrated in Figure 3.5(a), where $I_{r1} = I_{r2} = I_r$. All transistors operate in the saturation region where the I-V characteristic is reduced to a square law for $V_{DS} \geq V_{DS(sat)} = V_{GS} - V_T$. The currents flowing through transistors T_5, T_4, and T_2 can be respectively expressed as

$$I_r = K_5'\left(\frac{W_5}{L_5}\right)(V_{GS_5} - V_{T_5})^2 \quad \text{or} \quad V_{GS_5} = \sqrt{\frac{I_r}{K_5'(W_5/L_5)}} + V_{T_5} \tag{3.23}$$

$$I_r = K_4' \left(\frac{W_4}{L_4}\right) (V_{GS_4} - V_{T_4})^2 \quad \text{or} \quad V_{GS_4} = \sqrt{\frac{I_r}{K_4'(W_4/L_4)}} + V_{T_2} \tag{3.24}$$

and

$$I_r = K_2' \left(\frac{W_2}{L_2}\right) (V_{GS_2} - V_{T_2})^2 \quad \text{or} \quad V_{GS_2} = \sqrt{\frac{I_r}{K_2'(W_2/L_2)}} + V_{T_2} \tag{3.25}$$

With the assumption that the size of transistors T_1 and T_3 is identical to the one of T_2 and T_4, the current I_b should be a replica of I_r. The drain-source voltage of T_4 is given by

$$V_{DS_4} = V_{GS_5} - V_{GS_2} \tag{3.26}$$

$$= V_{DS_4(sat)} = V_{GS_4} - V_{T_4} \tag{3.27}$$

Upon substitution of V_{GS_5}, V_{GS_2}, and V_{GS_4} into Equations (3.26) and (3.27), we obtain

$$\left(\sqrt{\frac{I_r}{K_5'(W_5/L_5)}} + V_{T_5}\right) - \left(\sqrt{\frac{I_r}{K_2'(W_2/L_2)}} + V_{T_2}\right) = \sqrt{\frac{I_r}{K_4'(W_4/L_4)}} \tag{3.28}$$

or, equivalently,

$$\sqrt{\frac{1}{K_4'(W_4/L_4)}} + \sqrt{\frac{1}{K_2'(W_2/L_2)}} - \sqrt{\frac{1}{K_5'(W_5/L_5)}} = \frac{V_{T_5} - V_{T_2}}{\sqrt{I_r}} \tag{3.29}$$

Assuming that transistors are designed with identical parameters K_i' and V_{T_i} $(i = 1, 2, 3, 4, 5)$, the requirement, $V_{DS_4} = V_{DS_4(sat)}$, is met provided

$$\sqrt{\frac{1}{(W_5/L_5)}} = \sqrt{\frac{1}{(W_4/L_4)}} + \sqrt{\frac{1}{(W_2/L_2)}} \tag{3.30}$$

A proper operation of the current mirror then relies on the insensitivity of the achievable transistor matching to the effects of process variations. Furthermore, the use of long channel transistors whose substrates are connected to the corresponding sources may be necessary to reduce current variations related to the effects of the channel length modulation.

In the particular cases where the transistors $T_1 - T_4$ are designed with the same W/L ratio while the ratio of T_5 is $W/(4L)$, the minimum voltages needed at the input and output nodes can be expressed as

$$V_{r1min} = V_{GS_3} = V_{GS_4} = V_{GS_2} = \sqrt{\frac{I_r}{K'(W/L)}} + V_T \tag{3.31}$$

$$= V_{DS(sat)} + V_T$$

and

$$V_{r2min} = V_{GS_5} = \sqrt{\frac{I_r}{K'(W/4L)}} + V_T = 2V_{DS(sat)} + V_T \tag{3.32}$$

where

$$V_{bmin} = V_{DS_4} + V_{DS_2} = (V_{GS_4} - V_T) + (V_{GS_2} - V_T) = 2V_{DS(sat)} \tag{3.33}$$

and $K' = \mu C_{ox}/2$. Here, the minimum output voltage or compliance voltage is reduced to the sum of two overdrive voltages.

Consider the high-swing cascode current mirror using a source follower output stage, as shown in Figure 3.5(b). It is assumed that $I_{r1} = I_{r2} = I_b$ and all transistors operate in the saturation region. The use of Kirchhoff's voltage law can yield

$$V_{DS_4} = V_{GS_3} + V_{GS_1} - V_{GS_5} - V_{GS_2} \tag{3.34}$$

and

$$V_{GS_3} = V_{GS_6} = V_{GS_4} \tag{3.35}$$

Because $V_{DS_4} = V_{DS_4(sat)} = V_{GS_4} - V_{T_4}$ and $V_{GS_4} = V_{GS_3}$, Equation (3.34) becomes

$$-V_{T_4} = V_{GS_1} - V_{GS_5} - V_{GS_2} \tag{3.36}$$

Based on the simplified $I-V$ characteristic for MOS transistors operating in the saturation region, we obtain

$$I_r = K_5'\left(\frac{W_5}{L_5}\right)(V_{GS_5} - V_{T_5})^2 \quad \text{or} \quad V_{GS_5} = \sqrt{\frac{I_r}{K_5'(W_5/L_5)}} + V_{T_5} \tag{3.37}$$

$$I_r = K_2'\left(\frac{W_2}{L_2}\right)(V_{GS_2} - V_{T_2})^2 \quad \text{or} \quad V_{GS_2} = \sqrt{\frac{I_r}{K_2'(W_2/L_2)}} + V_{T_2} \tag{3.38}$$

and

$$I_r = K_1'\left(\frac{W_1}{L_1}\right)(V_{GS_1} - V_{T_1})^2 \quad \text{or} \quad V_{GS_1} = \sqrt{\frac{I_r}{K_1'(W_1/L_1)}} + V_{T_1} \tag{3.39}$$

Upon substitution of V_{GS_5}, V_{GS_2}, and V_{GS_1} into Equation (3.36), we arrive at

$$\sqrt{\frac{1}{K_4'(W_4/L_4)}} - \sqrt{\frac{1}{K_2'(W_2/L_2)}} - \sqrt{\frac{1}{K_5'(W_5/L_5)}} = \frac{V_{T_2} + V_{T_5} - (V_{T_1} + V_{T_4})}{\sqrt{I_r}} \tag{3.40}$$

In the case where the parameters K_i' and V_{T_i} $(i = 1, 2, 3, 4, 5, 6)$ are identical for all transistors, it can be deduced that

$$\sqrt{\frac{1}{K_4'(W_4/L_4)}} = \sqrt{\frac{1}{K_2'(W_2/L_2)}} + \sqrt{\frac{1}{K_5'(W_5/L_5)}} \tag{3.41}$$

The choice of the same W/L ratio for transistors $T_2 - T_6$ results in a ratio of $W/(4L)$ for T_1. Hence, the gate-source voltages for transistors T_5, T_3, T_2, and T_1 are given by

$$V_{GS_3} = V_{GS_5} = V_{GS_2} = \sqrt{\frac{I_r}{K'(W/L)}} + V_T = V_{DS(sat)} + V_T \tag{3.42}$$

$$V_{GS_1} = \sqrt{\frac{I_r}{K'(W/(4L))}} + V_T = 2V_{DS(sat)} + V_T \tag{3.43}$$

where $K' = \mu C_{ox}/2$. Because the transistors T_6 and T_2 are biased at the boundary of the saturation region, we have

$$V_{DS_6} = V_{DS_2} = V_{DS(sat)} \tag{3.44}$$

The minimum voltages required at the input and output nodes can then be obtained as

$$V_{r1min} = V_{GS_1} + V_{GS_3} = 3V_{DS(sat)} + 2V_T \tag{3.45}$$

$$V_{r2min} = V_{DS_5} + V_{DS_6} = (V_{r1min} - V_{GS_5}) + V_{DS_6} = 3V_{DS(sat)} + V_T \tag{3.46}$$

and

$$V_{bmin} = V_{DS_4} + V_{DS_2} = (V_{DS_6} - V_{GS_2}) + V_{DS_2} = 2V_{DS(sat)} \tag{3.47}$$

Although the aforementioned cascode current mirror requires a minimum output voltage of only two overdrive voltages, its accuracy can be limited by the difference in the body effects of transistors T_2 and T_5, and the discrepancy between the drain-source voltages of transistors T_4 and T_6.

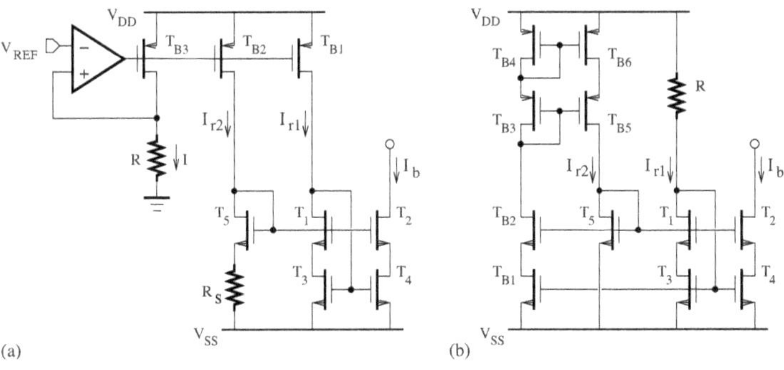

FIGURE 3.6
(a) High-swing cascode current mirror with voltage-controlled reference currents; (b) high-swing cascode current mirror with a reference current duplicated by a complementary circuit.

In general, variations in the IC process and supply voltage can lead to changes in the saturation voltage margin so that the output resistance and swing of the current mirror are reduced. The use of an adequate bias circuit may then be required to maintain the cascode connected transistors in the saturation region regardless of changes in the current and voltage levels due to nonideal effects.

The circuit diagram of a high-swing cascode current mirror with a bias circuit driven by a reference voltage is shown in Figure 3.6(a). An accurate control of the bias voltage applied between the drain and source of the transistor T_3 is achieved by inserting the resistor R_S in series with the source of T_5 [1]. With the assumption that the operational amplifier is ideal, the reference current I is of the form $I = V_{REF}/R$. For transistors T_{B3}, T_{B2}, and T_{B1} designed with identical threshold voltage and process transconductance parameters, and operating with the same biasing condition, the drain current is proportional to the ratio of the width to the length, and we have

$$\frac{I}{W_{B3}/L_{B3}} = \frac{I_{r2}}{W_{B2}/L_{B2}} = \frac{I_{r1}}{W_{B1}/L_{B1}} \tag{3.48}$$

Because $V_{GS_5} = V_{GS_1}$, the drain-source voltage of T_3 is given by

$$V_{DS_3} = V_{S_5} = R_S I_{r2} = \frac{(W_{B2}/L_{B2})}{(W_{B3}/L_{B3})} \frac{R_S}{R} V_{REF} \tag{3.49}$$

In order to reduce the dependence of the current mirror performance to variations in the IC process, supply voltage, and temperature, the voltage V_{REF} should be generated by a bandgap reference source and the resistor ratio should be accurately matched.

Another cascode bias circuit is illustrated in Figure 3.6(b) [2]. The transistors T_{B1} and T_{B2} of the bias circuit dynamically detect the actual gate voltages of the current mirror transistors $T_1 - T_4$ and this information is used to appropriately set the value of the current I_{r2}. As a result, the current mirror output exhibits a low swing regardless of variations in the IC process and supply voltage, and the operating range or headroom available to the output load is maximized.

Note that degeneration resistors may be included between the transistor sources and V_{SS} to help maintain a constant current flowing through transistors over a wider range of the supply voltages.

The regulated cascode structure [8], which is based on a transistor arrangement featuring a very high output resistance, can be used to improve the performance of current mirrors, as shown in Figure 3.7. In the circuit of Figure 3.7(a), the input and output currents are denoted by I_{r1} and I_b, respectively, while I_{r2} is used to bias the transistors T_1 and T_2. With reference to the practical implementation shown in Figure 3.7(b), the input and biasing currents delivered, respectively, by transistors T_{B1} and T_{B2} should nominally be identical to achieve $V_{GS_1} = V_{GS_4}$ for any level of the input current, which is mirrored through T_3 [3]. If T_1 and T_3 have the same size, $V_{GS_1} = V_{DS_3} = V_{GS_3}$

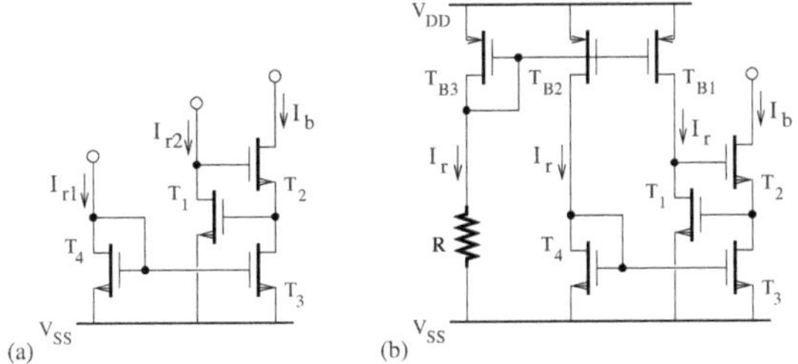

FIGURE 3.7
(a) Active-feedback cascode current mirror; (b) active-feedback cascode current mirror with a duplicated reference current.

and T_3 will be saturated because $V_{DS_3} \geq V_{GS_3} - V_T$. The minimum output voltage is then primarily determined by the drain-source saturation voltage required to keep T_2 in the saturation region. However, the accuracy of the equality between the input and output currents can be limited by the achievable matching between the gate-source voltages of T_1 and T_4.

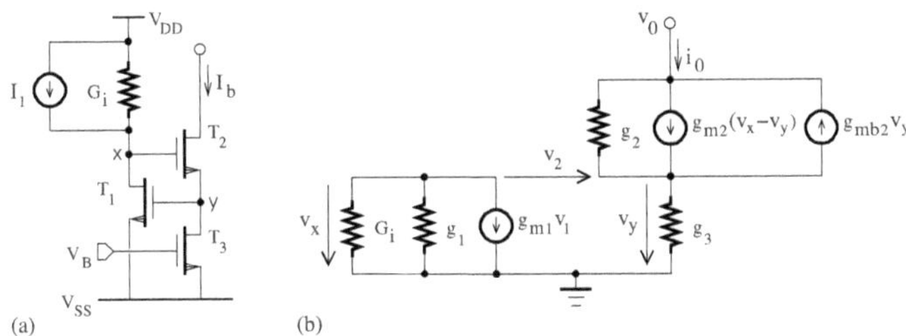

FIGURE 3.8
(a) Equivalent circuit of the active-feedback cascode current mirror and (b) its small-signal equivalent model.

Consider the equivalent circuit of the active-feedback cascode current mirror shown in Figure 3.8(a) [8], where V_B is a bias voltage, and G_i is the conductance of the current source I_1. From its small-signal equivalent model depicted in Figure 3.8(a), the output current can be computed as

$$i_0 = g_{m2}(v_x - v_y) - g_{mb2}v_y + g_2(v_0 - v_y) \tag{3.50}$$

Because

$$v_y = v_1 = i_0/g_3 \tag{3.51}$$

and

$$v_x - v_y = -\frac{g_{m1}v_1}{g_1 + G_i} - v_1 = -\left(1 + \frac{g_{m1}}{g_1 + G_i}\right)\frac{i_0}{g_3} \tag{3.52}$$

we obtain

$$i_0 = -\left[\frac{g_{m2}}{g_3}\left(1 + \frac{g_{m1}}{g_1 + G_i}\right) + \frac{g_{mb2}}{g_3} + \frac{g_2}{g_3}\right] i_0 + g_2 v_0 \tag{3.53}$$

Hence, the output resistance is given by

$$r_0 = \frac{v_0}{i_0} = \frac{1}{g_2}\left[1 + \frac{g_{m2}}{g_3}\left(1 + \frac{g_{m1}}{g_1 + G_i}\right) + \frac{g_{mb2}}{g_3} + \frac{g_2}{g_3}\right] \tag{3.54}$$

Assuming that the drain-source voltage of the transistor T_3 is kept constant by the structure consisting of $T_1 - T_2$ and the current source I_1 to reduce the effect of the body transconductance, g_{mb2}, and that the ratio g_2/g_3 is relatively negligible, we can write

$$r_0 = \frac{v_0}{i_0} \simeq \frac{g_{m1} g_{m2}}{g_2 g_3 (g_1 + G_i)} \tag{3.55}$$

With identical transconductances and all conductances being equal to the inverse of the drain-source resistance, the output resistance of the active-feedback cascode current mirror is approximately $g_m^2 r_{ds}^3/2$. On the other hand, the transistors should operate in the saturation region to ensure proper operation of the current mirror, leading to the minimum output voltage of

$$V_{bmin} = V_{GS_1} + V_{DS_2} = 2V_{DS(sat)} + V_T \tag{3.56}$$

It was assumed that the transistors feature the same drain-source saturation voltage and threshold voltage.

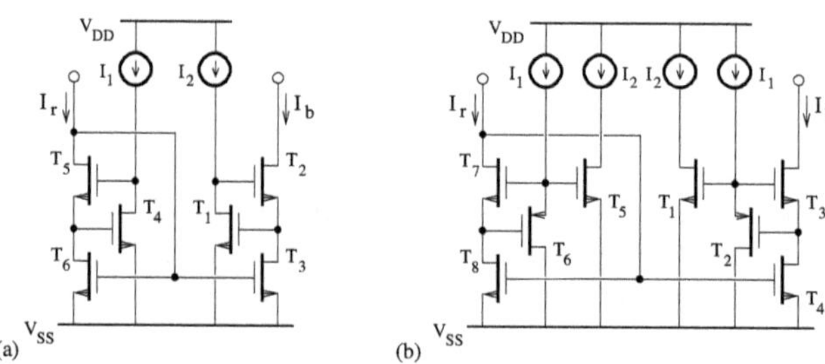

FIGURE 3.9
(a) Regulated cascode current mirror; (b) regulated cascode current mirror with the optimum output swing.

The regulated cascode current mirror of Figure 3.9(a) consists of the same transistor structure in the input and output paths and the biasing current sources I_1 and I_2. Its overall symmetric structure results in the reduction of the systematic matching error in the current ratio at *dc*.

For low-voltage applications, the minimum output voltage of the regulated cascode current mirror can be reduced using source followers operating as level shifters to lower the transistor bias voltages [4], as shown in Figure 3.9(b). It is assumed that the pMOS and nMOS transistors are designed with the process transconductance and threshold voltage of the form $K'_n = K'_p = K'$ and $V_{T_n} = -V_{T_p} = V_T$, and that the value of the bias current I_2 is four times that of I_1. Under these conditions, we can have

$$V_{G_3} = V_{GS_1} = \sqrt{\frac{I_2}{K'(W/L)}} + V_T = 2V_{DS(sat)} + V_T \tag{3.57}$$

where $I_2 = 4I_1$ and $V_{DS(sat)} = \sqrt{I_1/[K'(W/L)]}$. The drain-source voltage of the transistor T_4 is computed as

$$V_{DS_4} = V_{GS_1} + V_{GS_2} = V_{DS(sat)} \tag{3.58}$$

where $V_{GS_2} = -V_{DS(sat)} - V_T$, and the gate-source voltage of the transistor T_3 is given by

$$V_{GS_3} = V_{GS_1} - V_{DS_4} = V_{DS(sat)} + V_T \tag{3.59}$$

The minimum output voltage is reduced to

$$V_{bmin} = V_{DS_3} + V_{DS_4} = 2V_{DS(sat)} \tag{3.60}$$

where $V_{DS_3} = V_{GS_3} - V_T = V_{DS(sat)}$. The current mirror then features a high output impedance over a wide output swing, but this performance can be affected by mismatches between the parameters of the nMOS and pMOS transistors.

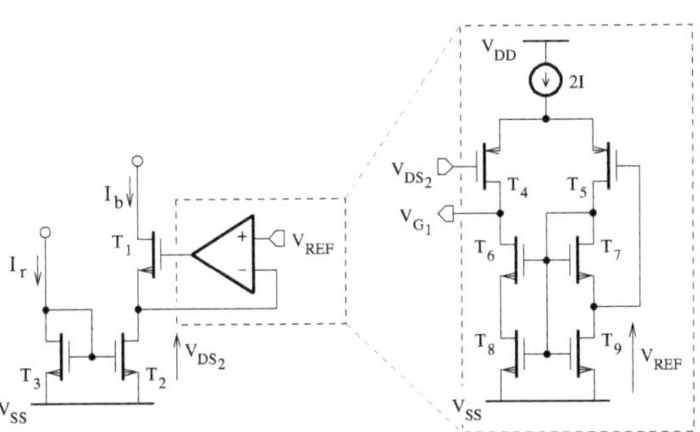

FIGURE 3.10
Active cascode current mirror with a wide output swing.

Another architecture for a cascode current mirror with a high output resistance and a wide output swing is shown in Figure 3.10 [5]. In the ideal case, the amplifier should have a high *dc* gain and a zero offset voltage. This makes the voltages at both input terminals of the amplifier equal, that is, $V_{DS_2} = V_{REF}$.

The structure of the amplifier is chosen such that the transistor is biased to operate in the saturation region with $V_{DS_1} = V_{DS_1(sat)}$ and the reference voltage, V_{REF}, is determined by the drain-source voltage of T_9. Hence, the minimum output voltage is given by

$$V_{omin} = V_{DS_1(sat)} + V_{REF} \tag{3.61}$$

The amplifier is designed such that the level of V_{REF} is appropriate for the biasing of the transistor T_2 in the saturation region, while maintaining the minimum output voltage as low as possible. With reference to the amplifier of Figure 3.10, we have

$$V_{GS_9} = V_{DS_7} + V_{DS_9} \tag{3.62}$$

where $V_{REF} = V_{DS_9}$. The transistors $T_6 - T_9$ all have equal threshold voltages. Because the transistor T_7 operates in the saturation region, its I-V characteristic leads to

$$V_{DS_7} = V_{GS_7} = \sqrt{\frac{I}{K_7'(W_7/L_7)}} + V_T \tag{3.63}$$

In the case of the transistor T_9, which operates in the triode region, we can write

$$V_{GS_9} = \frac{1}{2V_{DS_9}}\left(\frac{I}{K_9'(W_9/L_9)} + V_{DS_9}^2\right) + V_T \tag{3.64}$$

Substituting Equations (3.63) and (3.64) into Equation (3.62) gives

$$V_{DS_9}^2 + 2V_{DS_7}V_{DS_9} - \frac{I}{K_9'(W_9/L_9)} = 0. \tag{3.65}$$

The only positive solution of this quadratic equation is of the form

$$V_{DS_9} = \sqrt{\frac{I}{K_7'(W_7/L_7)}}\left(\sqrt{1 + \frac{K_7'(W_7/L_7)}{K_9'(W_9/L_9)}} - 1\right) \tag{3.66}$$

The transistor parameters can then be sized such that the reference voltage is on the order of the drain-source saturation voltage of T_2.

The small-signal equivalent model depicted in Figure 3.11 can be considered for the determination of the output resistance. With the assumption that the input voltage is equal to zero, $v_{gs3} = v_{gs2} = 0$, we have

$$v_o = [i_o - (g_{m1}v_{gs1} + g_{mb1}v_{bs1})]/g_1 + i_o/g_2 \tag{3.67}$$

where $v_{bs1} = -v_{ds2}$. The effect of the body transconductance of T_2 is considered negligible because $v_{bs2} = 0$. Noting that the gate voltage of the transistor T_1 is $v_{g1} = -Av_{ds2}$, we can write

$$v_{gs1} = v_{g1} - v_{s1} = -(A+1)v_{ds2} \tag{3.68}$$

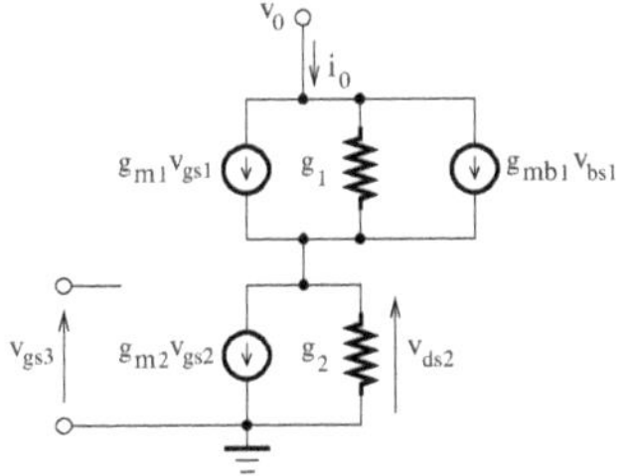

FIGURE 3.11
Equivalent model of the active cascode current mirror.

where $v_{ds2} = i_0/g_2$ and A denotes the amplifier gain. The output resistance can then be derived as

$$r_0 = \frac{v_0}{i_0} = \frac{1}{g_1} + \frac{1}{g_2} + \frac{1}{g_1 g_2}[(A+1)g_{m1} + g_{mb1}] \tag{3.69}$$

In addition to a low output compliance, the active cascode current mirror can also feature a high output resistance.

The aforementioned cascode current mirrors do provide a high output resistance, but at the cost of a large power consumption due to the additional circuit branches or components required for adequate biasing. Self-biased cascode current mirrors, which require only one reference current source, may be used to reduce power consumption.

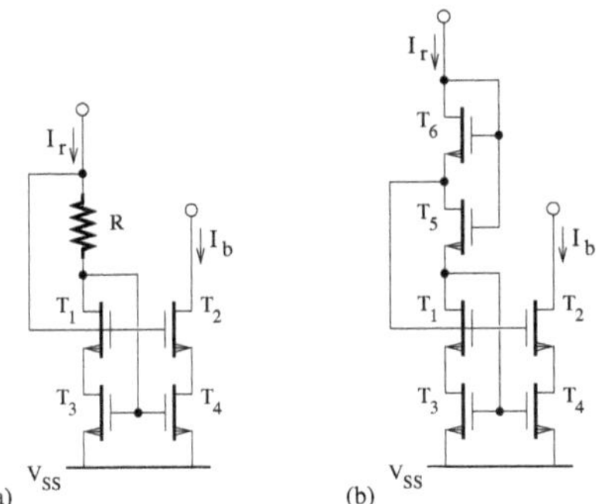

FIGURE 3.12
(a) Self-biased high-swing cascode current mirror with the bias voltage fixed by a resistor; (b) self-biased high-swing cascode current mirror with the bias voltage fixed by an active load.

With reference to Figure 3.12(a), a self-biased high-swing cascode current mirror with the bias voltage fixed by a resistor [6] is shown in schematic form. Assuming that all transistors are matched and operate in the saturation region, we have $V_{GS_3} = V_{GS_4}$ and it follows that $I_b = I_r$. The transistors T_3

and T_1, which are in series, are traversed by the same current. Their gate-source voltages are then equal, that is,

$$V_{GS_3} = V_{GS_1} = V_T + V_{DS(sat)} \tag{3.70}$$

The current I_r flowing through the resistor R establishes a bias voltage at the gate of T_2, which is by an amount RI_r greater than the one at the gate of T_4. With an appropriate selection of R, $V_{DS(sat)} = RI_r$ and we obtain

$$V_{G_1} = V_{G_2} = V_T + 2V_{DS(sat)} \tag{3.71}$$

Hence, a minimum output voltage of $V_{bmin} = 2V_{DS(sat)}$ is required to maintain a high resistance at the output of the self-biased cascode current mirror.

An alternative self-biased cascode current mirror is depicted in Figure 3.12(b) [7], where the difference between the gate voltages of transistors T_4 and T_2 is equal to the drain-source voltage of the transistor T_5. It is assumed that all transistors are designed with the same process transconductance parameter, K', and threshold voltage, V_T. With $V_{DS_6} \geq V_{DS_6(sat)} = V_{GS_6} - V_T$, the transistor T_6 operates in the saturation region and we can write

$$I_r = K'(W_6/L_6)(V_{GS_6} - V_T)^2 \tag{3.72}$$

where $K' = \mu C_{ox}/2$. The characteristic of T_5, which is assumed to operate in the triode region, is given by

$$I_r = K'(W_5/L_5)[2(V_{GS_5} - V_T)V_{DS_5} - V_{DS_5}^2] \tag{3.73}$$

when $V_{DS_5} \leq V_{DS_5(sat)} = V_{GS_5} - V_T$. For the optimum biasing condition,

$$V_{DS_5} = V_{DS(sat)} \tag{3.74}$$

and

$$V_{GS_6} = V_{DS(sat)} + V_T \tag{3.75}$$

Applying Kirchhoff's voltage law around the loop including transistors T_6 and T_5, we obtain

$$V_{GS_5} = V_{GS_6} + V_{DS_5} = 2V_{DS(sat)} + V_T \tag{3.76}$$

Because the same current flows through transistors T_6 and T_5, it can be deduced that

$$K'(W_6/L_6)V_{DS(sat)}^2 = K'(W_5/L_5)[2(2V_{DS(sat)})V_{DS(sat)} - V_{DS(sat)}^2] \tag{3.77}$$

Hence,

$$W_5/L_5 = (1/3)(W_6/L_6) \tag{3.78}$$

All the transistors except T_5 operate in the saturation region and can be designed with the same width-to-length ratio as the transistor T_6. The currents

I_r and I_b are matched provided the gate-source voltages of transistors T_3 and T_4 are identical, that is,

$$V_{GS_3} = V_{GS_4} = V_{DS(sat)} + V_T \tag{3.79}$$

The voltage at the gates of transistors T_1 and T_2 is

$$V_{G_1} = V_{G_2} = 2V_{DS(sat)} + V_T \tag{3.80}$$

and the minimum output voltage for which a high resistance is ensured is $2V_{DS(sat)}$.

3.1.3 Low-voltage active current mirror

Cascode-based current mirrors generally feature a high output resistance with a minimum output voltage of about $2V_{DS(sat)}$. This, however, may still too large for low-voltage applications. In the current mirror of Figure 3.13(a), a sensing amplifier is used to force the similar transistors T_1 and T_2 to have the same drain-source voltage, and hence the same drain current. The minimum output voltage is reduced to $V_{DS(sat)}$ when the transistor operates in the saturation region.

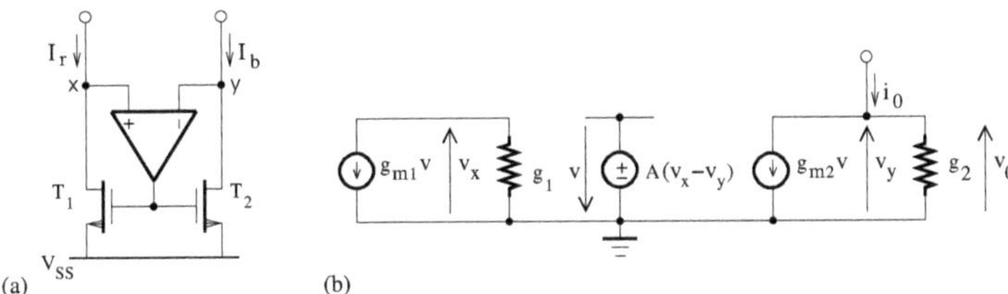

FIGURE 3.13
(a) Current mirror using a sensing amplifier and (b) its small-signal equivalent circuit.

Consider the small-signal equivalent circuit shown in Figure 3.13(b). Applying the current and voltage laws, we have

$$i_0 = g_{m2}v + g_2 v_y \tag{3.81}$$

$$v_x = -g_{m1} g_1 v \tag{3.82}$$

$$v = A(v_x - v_y) \tag{3.83}$$

The output resistance, r_0, of the current mirror can then be computed as

$$r_0 = \frac{v_0}{i_0} = \frac{1 + A g_{m1}/g_1}{g_2(1 + A g_{m1}/g_1 - A g_{m2}/g_2)} \tag{3.84}$$

where $v_0 = v_y$ and A denotes the amplifier gain. It should be noted that a more accurate analysis taking into account the parasitic impedances may be necessary in order to predict the stability and high-frequency behavior of the circuit.

3.2 Current and voltage references

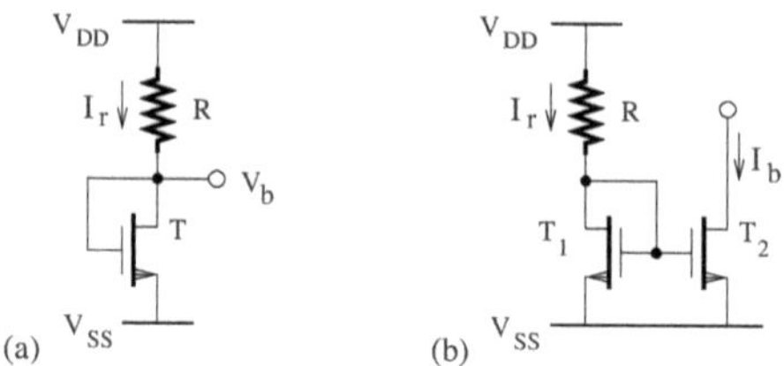

FIGURE 3.14
Simple (a) voltage and (b) current references.

A simple voltage reference is shown in Figure 3.14(a). It operates with a current level set by the resistor R. The transistor T, whose gate and drain are linked, operates in the saturation region provided that $V_{GS} \geq V_T$. That is, we have

$$I_r = K(V_b - V_{SS} - V_T)^2 \tag{3.85}$$

$$= \frac{V_{DD} - V_b}{R} \tag{3.86}$$

where $K = (1/2)\mu C_{ox}(W/L)$ is the transconductance parameter. The equation for the output voltage can be written as

$$KRV_b^2 + [1 - 2KR(V_{SS} + V_T)]V_b + KR(V_{SS} + V_T)^2 - V_{DD} = 0 \tag{3.87}$$

Because $V_{GS} = V_b - V_{SS} \geq V_T$, the only valid root of this quadratic equation is given by

$$V_b = V_{SS} + V_T - \frac{1}{2KR} + \sqrt{\frac{1}{KR}\left[V_{DD} - (V_{SS} + V_T) + \frac{1}{4KR}\right]} \tag{3.88}$$

The bias voltage, V_b, is a function of the resistance; the threshold voltage, V_T, of the transistor; and the supply voltages. Thus, the accuracy of the current reference will be affected by the supply-voltage variations and the changes in V_T due to the temperature and IC process fluctuations.

A current reference based on a simple current is shown in Figure 3.14(b). The reference current I_r is defined by the resistor R. Let

$$I_{D1} = I_r = \frac{V_{DD} - V_{GS} - V_{SS}}{R} \tag{3.89}$$

The current I_{D2} reads

$$I_{D2} = I_b = \frac{(W_2/L_2)}{(W_1/L_1)} \frac{V_{DD} - V_{GS} - V_{SS}}{R} \tag{3.90}$$

Thus, the bias current I_b is affected by the variations in the supply voltages.

The sensitivity of I_b to the supply voltage, $V_{sup} = V_{DD} - V_{SS}$, can be defined by

$$S_{V_{sup}}^{I_b} = \lim_{\triangle V_{sup} \to 0} \frac{\triangle I_b / I_b}{\triangle V_{sup} / V_{sup}} = \frac{V_{sup}}{I_b} \frac{\partial I_b}{\partial V_{sup}} \tag{3.91}$$

Assuming that V_{GS} is constant, we can write

$$S_{V_{sup}}^{I_b} = \frac{1}{1 - \dfrac{V_{GS}}{V_{sup}}} \tag{3.92}$$

Given the percentage variation in V_{sup}, the change in the bias current can be computed as

$$\frac{\triangle I_b}{I_b} = S_{V_{sup}}^{I_b} \frac{\triangle V_{sup}}{V_{sup}} \tag{3.93}$$

Generally, the reference circuit must be designed to feature a sensitivity to the supply-voltage change that is less than 1%.

In the case of the temperature dependence of I_b, the fractional temperature coefficient, which is given by

$$TC(I_b) = \frac{1}{I_b} \frac{\partial I_b}{\partial T} \tag{3.94}$$

appears to be useful. Using the expression of the bias current, we obtain

$$TC(I_b) = -\frac{1}{I_b} \left[\frac{(W_2/L_2)}{(W_1/L_1)} \frac{1}{R} \frac{\partial V_{GS}}{\partial T} + \frac{I_b}{R} \frac{\partial R}{\partial T} \right] \tag{3.95}$$

The voltage V_{GS} is related to the parameters, such as the transistor transconductance parameter and threshold voltage, and the resistor R, whose values depend on the temperature. Note that the fractional temperature coefficient is a function of the temperature, that is, it is specified only at a given temperature. Ideally, a reference circuit can be temperature-independent due to the cancellation of the temperature coefficients of the individual components.

A proportional-to-absolute temperature (PTAT) voltage (current) reference generates a voltage (current) that gets larger as the temperature is increased. A complementary-to-absolute temperature (CTAT) voltage (current) reference generates a voltage (current) that decreases as the temperature is reduced. PTAT voltage and current references can be designed by relying on the fact that the difference of the gate-source voltage of two MOS transistors or the base-emitter voltage of two bipolar transistors is proportional to the temperature. The negative temperature coefficient of the threshold voltage of a MOS transistor or the base-emitter voltage of a bipolar transistor can be exploited in the design of CTAT voltage and current references. Temperature-independent voltage and current references can be realized by appropriately combining the outputs of PTAT and CTAT voltage and current sources.

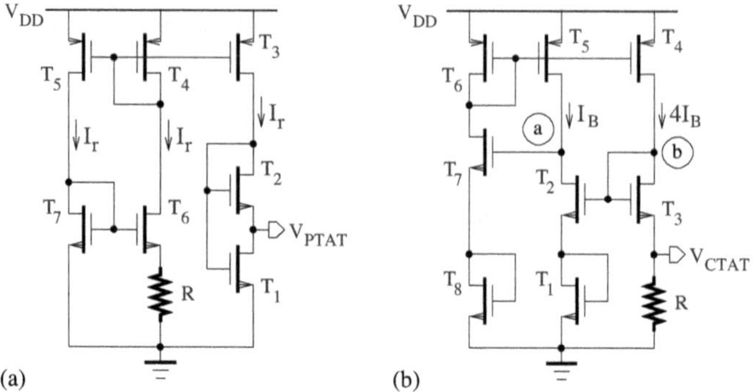

FIGURE 3.15
(a) PTAT voltage reference; (b) CTAT voltage reference.

A PTAT voltage generator, as shown in Figure 3.15(a), can be used to compensate the thermal dependence of current and voltage references. Transistors T_1 and T_2 are assumed to operate in the subthreshold region, and the voltage V_{PTAT} can be obtained as,

$$V_{PTAT} = V_{GS_1} - V_{GS_2} = nU_T \ln\left(\frac{W_2/L_2}{W_1/L_1}\right) \tag{3.96}$$

where $U_T = kT/q$ is the thermal voltage, T is the absolute temperature in Kelvin, k is Boltzmann's constant, q is the electron charge, and n is the subthreshold slope factor.

A CTAT voltage reference is depicted in Figure 3.15(b). All transistors operate in the saturation region and the threshold voltages are matched. With $W_1/L_1 = W_2/L_2 = W_3/L_3$ and $W_4/L_4 = 4(W_5/L_5) = 4(W_6/L_6)$, it can be shown that

$$V_{CTAT} = V_{GS_1} + V_{GS_2} - V_{GS_3} = V_{T_n} \tag{3.97}$$

where V_{T_n} is the threshold voltage whose relationship with temperature is given by

$$V_{T_n}(T) = V_{T_n}(T_0) - \alpha_{V_{T_n}}(T - T_0) \tag{3.98}$$

and $V_{T_n}(T_0)$ is the threshold voltage at the temperature T_0 at which the proportionality constant (or temperature coefficient) $\alpha_{V_{T_n}}$ has been estimated. The sensibility of V_{CTAT} to non-idealities, such as the supply-voltage noise and transistor channel-length modulation [9], is reduced by sizing the transistors T_7 and T_8 to minimize the voltage difference between the nodes a and b.

3.2.1 Supply-voltage independent current and voltage references

A supply-voltage independent current reference is depicted in Figure 3.16(a) [10]. It is based on two current mirrors interconnected into a closed loop. A resistor R is linked to the source of the transistor T_1. The requirement, $I_1 = I_2 = I_r$, is met because the p-channel transistors are designed to have the same size. The next equations can be written as

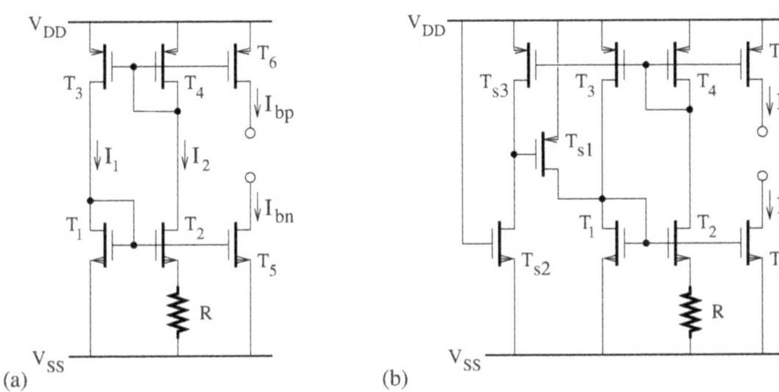

FIGURE 3.16
(a) Supply-voltage independent current reference; (b) current reference including a start-up circuit.

$$V_{GS1} = V_{GS2} + RI_r \tag{3.99}$$

$$V_{GS1} = \sqrt{\frac{2I_r}{\mu_n C_{ox}(W_1/L_1)}} + V_{T1} \tag{3.100}$$

$$V_{GS2} = \sqrt{\frac{2I_r}{\mu_n C_{ox}(W_2/L_2)}} + V_{T2} \tag{3.101}$$

With the assumption that $V_{T1} = V_{T2}$ and $(W_1/L_1) = \kappa(W_2/L_2)$, the current I_r is given by

$$I_r = \frac{2}{\mu_n C_{ox}(W_1/L_1)}\left[\frac{1}{R}\left(1 - \frac{1}{\sqrt{\kappa}}\right)\right]^2 \tag{3.102}$$

The versions of the reference current, I_{bn} and I_{bp}, which are respectively suitable for the biasing of n- and p-channel transistors, are generated by T_5 and T_6.

The start-up circuit $T_{s1} - T_{s3}$ of the reference circuit of Figure 3.16(b) injects a current into one branch during the initial power-on to prevent the occurrence of the zero current state. It is designed so that T_{s1} will turn off (i.e., $V_{SG} < V_T$) during normal operation of the current reference.

Given the currents I_1 and I_2, both operating points with the desired cur-

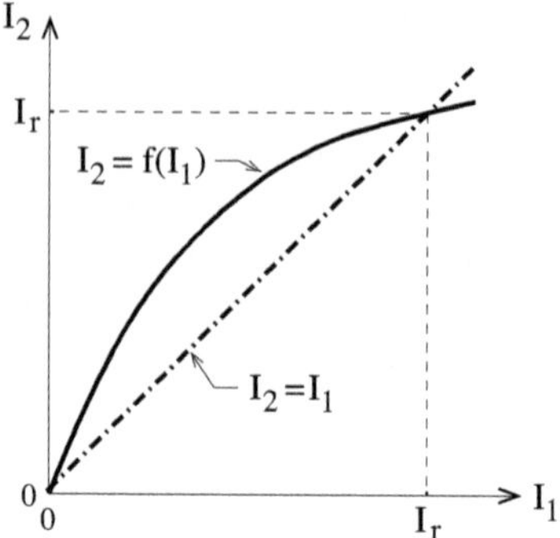

FIGURE 3.17
Characteristic of the current reference.

rent of I_r and with no current flowing through the current reference are illustrated in Figure 3.17. This is due to the fact that the operation of the current reference is governed by two equations, $V_{GS_1} = V_{GS_2} + RI_2$ and $V_{GS_3} = V_{GS_4}$, that can lead to relationships of the form, $I_2 = f(I_1)$, where f is a nonlinear function, and $I_2 = I_1$, respectively.

In general, the desired operating point of self-biased current or voltage references is quickly set using a start-up circuit.

The circuit of Figure 3.16(a) can also serve as a PTAT current source. Its operation requires that transistors T_1 and T_2 operate in the weak inversion (or subthreshold) region and the temperature coefficient of the resistor R is compensated.

In the subthreshold region, the drain-source current of an n-channel transistor can be reduced to

$$I_D \simeq I_T \left(\frac{W}{L}\right) \exp\left(\frac{V_{GS} - V_{T_n}}{nU_T}\right) \quad V_{DS} \geq 4U_T \tag{3.103}$$

where $I_T = 2n\mu_n C_{ox} U_T^2$, n $(1 < n < 3)$ is the subthreshold slope factor, U_T is the thermal voltage, and V_{T_n} is the threshold voltage. Typical values of the technology current, I_T, range from 100 nA to 500 nA.

Assuming that $I_1 = I_2 = I_r$, it can be shown that

$$I_r = \frac{V_{GS_1} - V_{GS_2}}{R} \tag{3.104}$$

The threshold voltages and drain currents of transistors T_1 and T_2 are considered to be equal. Hence,

$$I_r = \frac{nU_T}{R} \ln\left(\frac{W_2/L_2}{W_1/L_1}\right) \tag{3.105}$$

where $U_T = kT/q$ is the thermal voltage, T is the absolute temperature in

Kelvin, k is Boltzmann's constant, and q is the electron charge. The current source can exhibit a high power-supply rejection ratio because the current I_r is independent of the supply voltage.

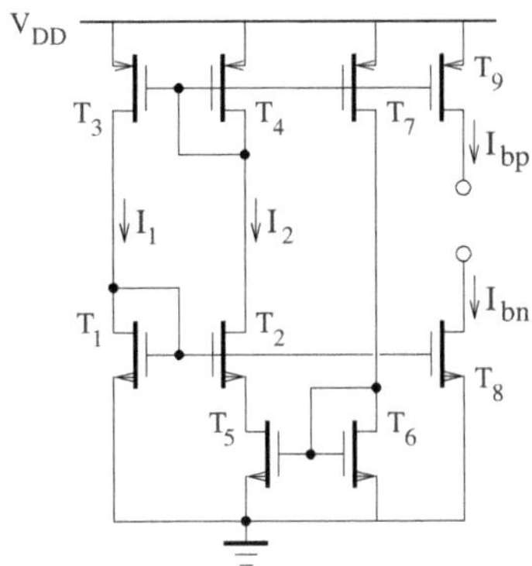

FIGURE 3.18
Current reference without resistor.

For applications requiring a low current, the value of the resistor R used in the current reference may be too high to be practical. Figure 3.18 shows the circuit diagram of the current reference without resistor [11]. The transistor T_5 operates in the triode region, while T_6 and T_7 are biased in the saturation region. The drain currents of transistors T_5 and T_6 can be written as

$$I_{D_5} = K_5[2(V_{GS_5} - V_{T_n})V_{DS_5} - V_{DS_5}^2] \tag{3.106}$$

and

$$I_{D_6} = K_6(V_{GS_6} - V_{T_n})^2 \tag{3.107}$$

The expression of the current I_r is of the form,

$$I_r = \frac{V_{GS_1} - V_{GS_2}}{R_{DS_5}} = \frac{nU_T}{R_{DS_5}} \ln\left(\frac{W_2/L_2}{W_1/L_1}\right) \tag{3.108}$$

where

$$R_{DS_5} = \frac{V_{DS_5}}{I_{D_5}} \simeq \frac{1}{2K_5(V_{GS_5} - V_{T_n})} \tag{3.109}$$

and $I_{D_1} = I_{D_2} = I_r = I_{D_6}$. Using the fact that

$$V_{GS_5} = V_{GS_6} = V_{T_n} + \sqrt{\frac{I_r}{K_6}} \tag{3.110}$$

we can then obtain

$$I_r = 4n^2U_T^2\left(\frac{K_5^2}{K_6}\right)\ln^2\left(\frac{W_2/L_2}{W_1/L_1}\right) \tag{3.111}$$

In the case where the carrier mobilities remain constant over the temperature range, the current I_r is proportional to the square of the temperature.

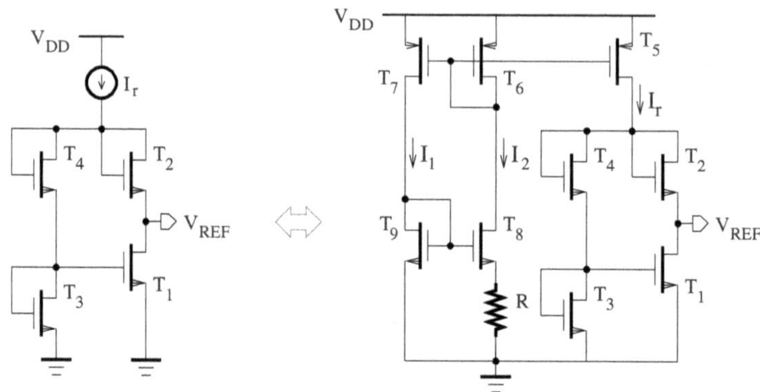

FIGURE 3.19
Voltage reference with compensated temperature coefficient.

A voltage reference with compensated temperature coefficient [12] is depicted in Figure 3.19. All transistors operate in the saturation region. The drain current and the gate-source voltage can be written as

$$I_{D_j} = K_j(V_{GS_j} - V_{T_n})^2 \quad j = 1, 2, 3, 4 \tag{3.112}$$

and

$$V_{GS_j} = V_{T_n} + \sqrt{\frac{I_{D_j}}{K_j}} \tag{3.113}$$

respectively, where $K_j = (\mu_n C_{ox}/2)(W_j/L_j)$. Using Kirchhoff's voltage law, it can be shown that

$$V_{REF} = V_{GS_4} + V_{GS_3} - V_{GS_2} \tag{3.114}$$

and

$$V_{GS_3} = V_{GS_1} \tag{3.115}$$

Assuming that the transistor threshold voltages are matched and substituting (3.113) for $j = 1, 3$ into (3.115) yields

$$I_{D_3} = \frac{W_3/L_3}{W_1/L_1} I_{D_1} \tag{3.116}$$

Because $I_{D_4} = I_{D_3}$ and $I_{D_2} = I_{D_1}$, the current I_r can be expressed as,

$$I_r = I_{D_4} + I_{D_2} = I_{D_3} + I_{D_1} \tag{3.117}$$

Combining (3.114), (3.113) for $j = 2, 3, 4$, (3.116), and (3.117) gives

$$V_{REF} = V_{T_n} + \sqrt{\frac{I_r}{1 + \dfrac{W_3/L_3}{W_1/L_1}}} \left[\frac{1}{\sqrt{K_1}} \left(1 + \sqrt{\frac{W_3/L_3}{W_4/L_4}} \right) - \frac{1}{\sqrt{K_2}} \right] \tag{3.118}$$

Variations of the threshold voltage with respect to the temperature can be computed as,

$$V_{T_n} = V_{T_n}(T_0) - \alpha_{V_{T_n}}(T - T_0) \tag{3.119}$$

where $V_{T_n}(T_0)$ is the threshold voltage at the temperature T_0 at which the proportionality constant (or temperature coefficient) $\alpha_{V_{T_n}}$ has been estimated. The reference voltage, V_{REF}, exhibits a zero temperature coefficient provided the current I_r is proportional to $\mu_n(T)T^2$, where μ_n is the electron mobility. By setting V_{REF} equal to $2V_{T_n}$, the channel length modulation effect can also be cancelled out so that $\partial V_{REF}/\partial T = 0$.

3.2.2 Bandgap references

The bandgap reference is designed to provide a voltage that is independent of supply voltage and temperature variations. The temperature insensitivity is practically realized by adding a voltage with a positive temperature coefficient to a voltage with an equal but negative temperature coefficient. The first type of voltage can be obtained by amplifying the difference between the base-emitter voltages of two forward-biased bipolar transistors operating at different current densities, while the second one is proportional to the base-emitter voltage of a forward-biased bipolar transistor. Numerous variations can be introduced in the bandgap reference circuitry to improve some performance characteristics, such as the temperature coefficient, initial accuracy, noise, and stability.

The collector current of a bipolar transistor operating in the forward-active region, that is, $V_{BC} \leq 0$, may be expressed as

$$I_c = I_s(e^{V_{BE}/\eta U_T} - 1) \simeq I_s e^{V_{BE}/\eta U_T} \tag{3.120}$$

where I_s is the saturation current, $U_T = kT/q$ is the thermal voltage, T is the absolute temperature in Kelvin (K), $k = 1.38 \cdot 10^{-23}$ J/K is Boltzmann's constant, $q = 1.6 \cdot 10^{-19}$ C is the electronic charge, and η is an empirical scaling constant, which depends on the geometry, material, and doping levels. The current I_s can be defined by

$$I_s = CT^{1/2} e^{-E_g/kT} \tag{3.121}$$

where $E_g \simeq 1.12$ eV is the silicon bandgap (or energy gap) and C is a proportionality constant. With

$$V_{BE} = \eta U_T \ln\left(\frac{I_c}{I_s}\right) \tag{3.122}$$

we can compute

$$\frac{\partial V_{BE}}{\partial T} = \eta\left[\frac{\partial U_T}{\partial T}\ln\left(\frac{I_c}{I_s}\right) - \frac{\partial I_s}{\partial T}\frac{U_T}{I_s}\right] \tag{3.123}$$

Therefore,

$$\frac{\partial V_{BE}}{\partial T} = \eta\frac{V_{BE} - U_T/2 - E_g/q}{T} \tag{3.124}$$

Let $\eta = 1$, $T = 300$K or 27°C (room temperature), and $V_{BE} \simeq 0.75$ V, $\partial V_{BE}/\partial T$ is on the order of -2.2 mV/^{o}C, while U_T, which has a PTAT characteristic, exhibits a temperature coefficient of $+0.086$ mV/^{o}C. The base-emitter voltage of a bipolar transistor possesses a negative temperature coefficient and has a CTAT dependency. In general, the bandgap voltage, V_{REF}, is generated as a linear combination of U_T and V_{BE}, that is,

$$V_{REF} = V_{BE} + \alpha U_T \tag{3.125}$$

where the constant parameter α is set by the circuit component such that V_{REF} is almost independent of the temperature variations.

In the bandgap reference of Figure 3.20(a), the voltage drop across R_3 is equal to the difference, $\triangle V_{BE}$, between the V_{BE}'s of the bipolar transistors. The reference voltage is then a combination of the V_{BE} of one transistor and a version of $\triangle V_{BE}$ scaled by a factor determined by the resistances, which may be chosen so that the opposite temperature coefficients of V_{BE} and $\triangle V_{BE}$ counterbalance. We have

$$V_{REF} = V_{BE1} + (R_1 + R_3)I_1 \tag{3.126}$$

For a high-gain amplifier, the voltage drop between the input nodes is reduced to zero, and the next relationships can be written:

$$I_1 = \frac{V_{BE2} - V_{BE1}}{R_3} \tag{3.127}$$

$$R_1 I_1 = R_2 I_2 \tag{3.128}$$

The reference voltage V_{REF} is then given by

$$V_{REF} = V_{BE1} + \left(1 + \frac{R_1}{R_3}\right) U_T \ln\left(\frac{R_1}{R_2}\frac{I_{S1}}{I_{S2}}\right) \tag{3.129}$$

where I_{S1} and I_{S2} are the saturation currents of transistors Q_1 and Q_2. The temperature coefficient $\partial U_T/\partial T$ is in the order of $+0.087$ mV/K. For

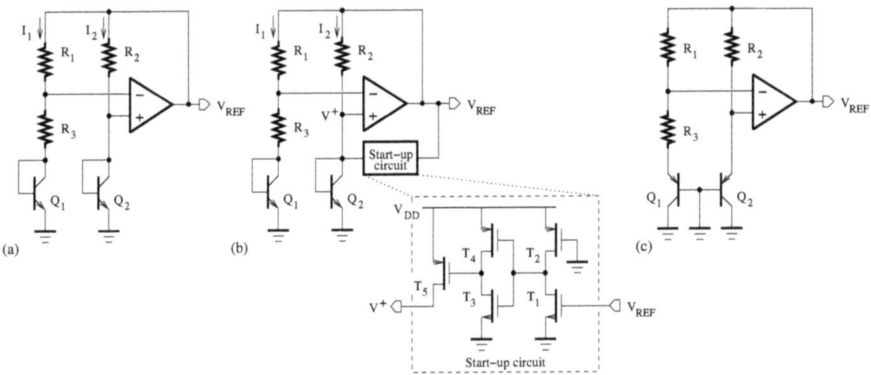

FIGURE 3.20
(a) Bandgap reference using *npn* transistors; (b) bandgap reference with a start-up circuit; (c) modified bandgap reference with grounded collector *pnp* transistors suitable for *n*-well fabrication process.

$(1 + R_1/R_3)\ln(R_1 I_{S1}/R_2 I_{S2}) \simeq 17.2$, the reference voltage will feature a zero temperature coefficient. Furthermore, by using the relationship $\partial V_{REF}/\partial T = 0$, the dependence of V_{REF} to the bandgap voltage, E_g, can be established.

In order to provide an output voltage that is independent of the variations in the supply voltage, the bandgap reference circuit is designed to be self-biased. As a consequence, the operating point obtained when all currents in the circuit are equal to zero can also be stable. A start-up circuit is required to force the bandgap reference circuit to operate with nonzero bias currents. Figure 3.20(b) shows a bandgap reference with a start-up circuit [13] connected between the output and the noninverting input of the amplifier. In normal operation, the output level of the bandgap reference circuit is greater than the threshold voltage of the transistor T_1, assumed to be the start-up voltage threshold, and the start-up circuit is disabled so as not to interfere with the normal operation of the band-gap reference circuit. On the contrary, if the bandgap reference circuit is in the zero-current state, the output of the inverter including transistors T_3 and T_4 will generate a logic high voltage and an initial current will be injected into the bandgap reference circuit by the start-up circuit. Note that the *p*-channel transistor T_2 is always activated because its gate is connected to ground.

In practice, the accurate prediction of the temperature coefficient should rely on simulations due to the nonideal effects such as:

- The temperature dependence of the collector current
- The amplifier offset voltage and output impedance

Furthermore, note that the temperature compensation achieved in a bandgap reference is limited to first order.

Due to the fact that the transistor collectors are not connected to the most negative supply voltage, the circuit of Figure 3.20(a) can be incompatible with implementations using the CMOS process where the p-type substrate has to serve as the collector. This problem is solved in the bandgap reference shown in Figure 3.20(c).

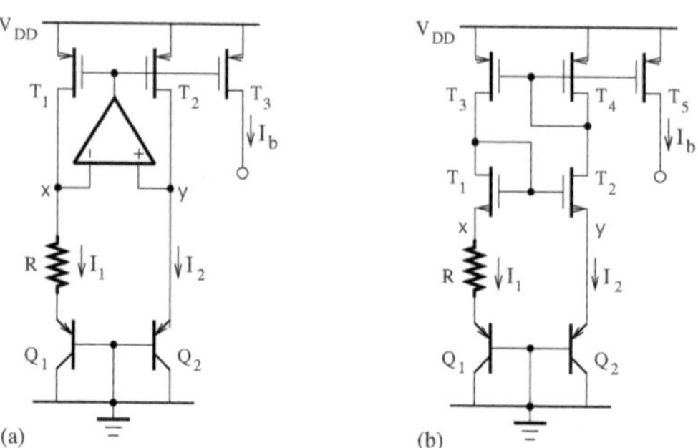

FIGURE 3.21
Bandgap reference generating a bias current: Implementations based on (a) amplifier and (b) transistor loads.

Bandgap reference implementations using amplifier and transistor loads are shown in Figures 3.21(a) and (b), respectively. These structures generate a bias current, I_b, which is proportional-to-absolute temperature. With the assumption that all MOS transistors are identical, I_b is similar to the currents I_1 and I_2, which are given by $I_1 = I_2 = (U_T/R)\ln(I_{S1}/I_{S2})$, provided $V_x = V_y$. A supply voltage dependence can be observed in the circuit of Figure 3.21(b) due to the channel-length modulation of the MOS transistors. A solution to this problem can consist of using cascode structures.

3.2.2.1 Low-voltage bandgap voltage reference

Generally, the normal operation of conventional bandgap references requires the use of a power supply voltage higher than the resulting reference voltage. For instance, a 2.5-V supply voltage is needed for the generation of an output voltage of about 1.25 V. To satisfy the requirements for low-voltage applications, the current-mode operation principle, which relies on the combination of current sources with positive and negative temperature coefficients to create a temperature-independent current, can be exploited, as illustrated in Figure 3.22(a) [14]. The resulting current is transferred by a current mirror to a network operating as a current divider, and the output voltage level can be scaled to a given value of the form

$$V_{REF} = \frac{R_3}{R_2 + R_3}(V_{EB3} + R_2 I) \tag{3.130}$$

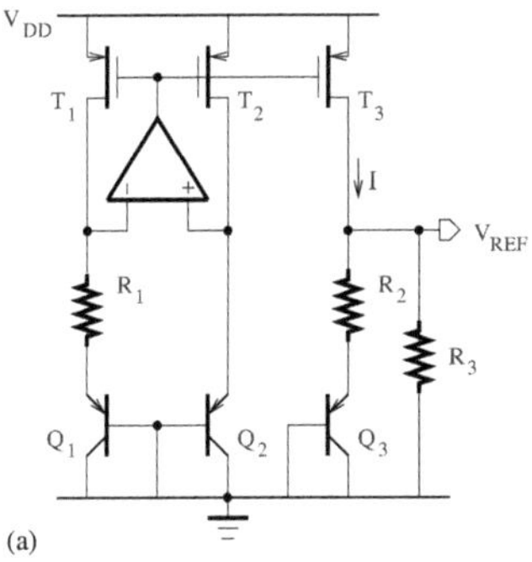

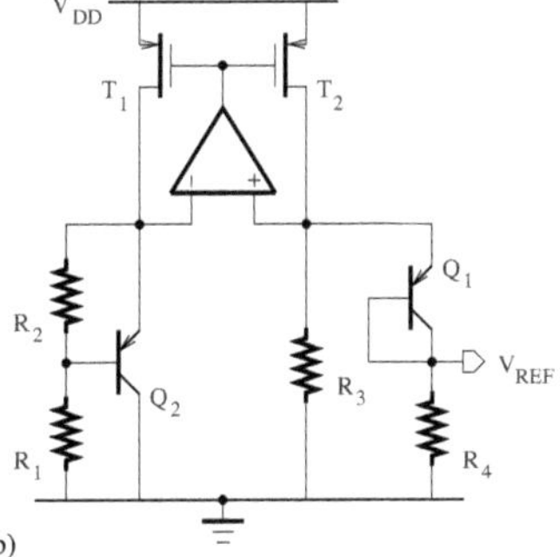

FIGURE 3.22
(a) Current mode low-supply voltage bandgap reference; (b) low-supply voltage bandgap reference based on the reverse bandgap voltage principle.

Hence,

$$V_{REF} = \frac{R_3}{R_2 + R_3}\left[V_{EB3} + \frac{R_2}{R_1}U_T \ln\left(\frac{I_{S1}}{I_{S2}}\right)\right] \tag{3.131}$$

It should be noted that a low-voltage amplifier structure may be required to reduce the influence of the input common-mode range on the minimum supply voltage of the bandgap reference. Furthermore, the accuracy of the reference voltage is limited by the nonideal characteristics of the current mirror, resistor mismatches, and noises.

To reduce the nonideal effect introduced by the active current mirror, the reversed bandgap voltage principle can be exploited for the voltage reference design, as illustrated in Figure 3.22(b) [17]. Assuming that the transistors T_1 and T_2 are identical and $R_3 = R_1 + R_2$, the collector currents of transistors Q_1 and Q_2 should be equal, provided the base current of Q_2 is negligible. According to the voltage divider principle, the voltage applied to the inverting node of the amplifier is the sum of V_{EB_2} and $V_{EB_2}(R_1/R_2)$. Because the voltages at the input nodes of an ideal amplifier are equal, we have

$$V_{EB_2}\left(1 + \frac{R_1}{R_2}\right) = V_{REF} + V_{EB_1} \tag{3.132}$$

Hence,

$$V_{REF} = \frac{R_1}{R_2}V_{EB_2} + \triangle V_{EB} \tag{3.133}$$

where $\triangle V_{EB} = V_{EB_2} - V_{EB_1}$. For bipolar transistors, the collector currents are of the form

$$I_{c_1} = I_{s_1}\mathrm{e}^{V_{EB_1}/U_T} \tag{3.134}$$

$$I_{c_2} = I_{s_2}\mathrm{e}^{V_{EB_2}/U_T} \tag{3.135}$$

Here $I_{c_1} = I_{c_2}$, and we can obtain

$$\Delta V_{EB} = V_{EB_2} - V_{EB_1} = U_T \ln\left(I_{s_1}/I_{s_2}\right) \tag{3.136}$$

The reference voltage can then be expressed as

$$V_{REF} = \frac{R_1}{R_2} V_{EB_2} + U_T \ln\left(\frac{I_{s_1}}{I_{s_2}}\right) \tag{3.137}$$

With the voltage reference architecture based on the reversed bandgap voltage, the supply voltage can be on the order of 1 V while the value of V_{REF} is set as low as 200 mV.

3.2.2.2 Curvature-compensated bandgap voltage reference

In practice, the base-emitter voltage of a bipolar transistor exhibits a nonlinear temperature relationship. However, the temperature compensation of the reference voltage in classical structures of the bandgap voltage reference is realized only in a vicinity of the reference temperature value and is limited to the first order. Due to this nonlinearity, the reference voltage is generated with a temperature curvature error, which is proportional to $T\ln(T)$, where T is the absolute temperature. Uncompensated bandgap references have a temperature coefficient[1] greater than 20 ppm/^{o}C (parts-per-million per degree Celsius) over a typical industrial temperature range from -40^oC to 85^oC.

The bandgap voltage reference structure shown in Figure 3.23(a) is designed to exhibit an output voltage that is almost independent of the temperature. This is achieved by compensating for the curvature introduced by the nonlinear temperature dependence of bipolar transistor characteristics on the output voltage.

The base-emitter voltage can be expressed as [18]

$$V_{BE} = V_{G0} - [V_{G0} - V_{BE}(T_R)]\frac{T}{T_R} - (\eta - x)U_T \ln\left(\frac{T}{T_R}\right) \tag{3.138}$$

where V_{G0} is the energy-bandgap voltage at zero degree Kelvin, T is the absolute temperature in degrees Kelvin, T_R is the reference temperature, $V_{BE}(T_R)$ is the base-emitter voltage at the temperature T_R, n denotes a temperature

[1]The temperature coefficient TC can be defined using the difference in the maximum and minimum values of the reference voltage over the entire temperature range, i.e.,

$$TC\ (\text{ppm}/^o\text{C}) = \frac{10^6}{T_{max} - T_{min}}\left(\frac{V_{REF,max} - V_{REF,min}}{V_{REF,25^o\text{C}}}\right)$$

where $V_{REF,25^o\text{C}}$ is the value of the reference voltage at the room temperature, and T_{max} and T_{min} are, respectively, the maximum and minimum temperatures.

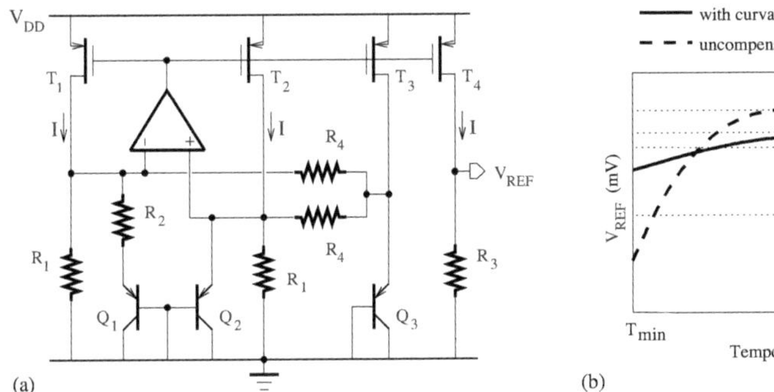

FIGURE 3.23
(a) Low-supply voltage bandgap reference with the temperature curvature compensation; (b) variations of V_{REF} as a function of the temperature.

constant depending on the IC process, η represents a process-dependent constant, x is the exponential order of the temperature dependence of the collector current (i.e., $I_C \propto T^x$), and U_T denotes the thermal voltage. Typical values for these parameters are $V_{G0} = 1.17$ V, $\eta = 3.5$, and at $T_R = 300$K, $V_{BE}(T_R) = 0.65$ V. A high-order compensation of the bandgap voltage reference is required to reduce the effect of the nonlinearity related to the logarithmic temperature ratio. Its principle can be based on a proper combination of the base-emitter voltage across a junction with a temperature-independent current ($x = 0$) and the one across a junction with a PTAT current ($x = 1$). The transistor Q_2 is biased by a PTAT current so that

$$V_{EB_2} = V_{G0} - [V_{G0} - V_{EB_2}(T_R)]\frac{T}{T_R} - (\eta - 1)U_T \ln\left(\frac{T}{T_R}\right) \tag{3.139}$$

while Q_3 is biased by a temperature-independent current delivered by a p-channel MOS transistor, and

$$V_{EB_3} = V_{G0} - [V_{G0} - V_{EB_3}(T_R)]\frac{T}{T_R} - \eta U_T \ln\left(\frac{T}{T_R}\right) \tag{3.140}$$

Because the voltage across the resistor R_4 is equal to the difference of the base-emitter voltages of Q_2 and Q_3, the curvature correction current is given by

$$I' = \frac{V_{EB_2} - V_{EB_3}}{R_4} = \frac{U_T}{R_4}\ln\left(\frac{T}{T_R}\right) \tag{3.141}$$

The reference voltage can then be computed as

$$V_{REF} = R_3 I \tag{3.142}$$

where

$$I = \frac{V_{EB_2} - V_{EB_1}}{R_2} + \frac{V_{EB_2}}{R_1} + I'$$
$$= \frac{U_T}{R_2} \ln\left(\frac{I_{s_1}}{I_{s_2}}\right) + \frac{V_{EB_2}}{R_1} + \frac{U_T}{R_4} \ln\left(\frac{T}{T_R}\right) \tag{3.143}$$

Hence,

$$V_{REF} = \frac{R_3}{R_1}\left[\frac{R_1}{R_2} U_T \ln\left(\frac{I_{s_1}}{I_{s_2}}\right) + V_{EB_2} + \frac{R_1}{R_4} U_T \ln\left(\frac{T}{T_R}\right)\right] \tag{3.144}$$

The curvature compensation will be realized if the following requirement [19] is met:

$$R_4 = R_1/(\eta - 1) \tag{3.145}$$

and the value of the ratio R_1/R_2 should be determined to minimize the drift of V_{REF} due to the linear variation of the temperature. Ideally, the resulting voltage reference is reduced to

$$V_{REF} = R_3 V_{G0}/R_1 \tag{3.146}$$

However, the output voltage reference, as illustrated in Figure 3.23(b), may not remain constant with respect to the temperature due to component mismatches and the ignored temperature coefficients of resistors.

3.2.3 Floating-gate voltage reference

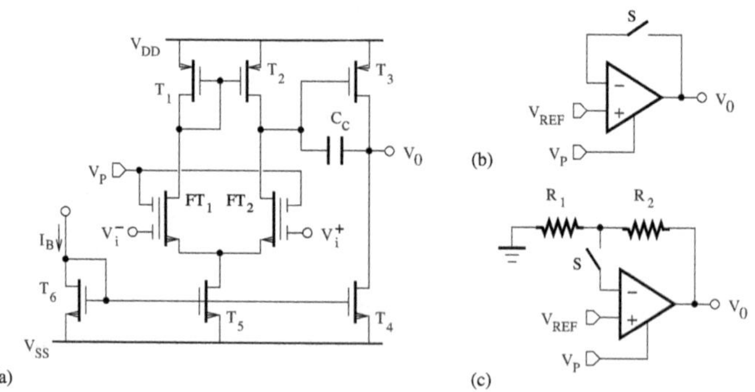

FIGURE 3.24
Floating-gate voltage reference.

A typical bandgap voltage reference, which consists of bipolar transistors combined with either the resistive feedback network of an amplifier or MOS transistor load, can exhibit a number of shortcomings. This is the case, for instance, for high circuit complexity, large silicon area, and high power consumption. Furthermore, the reference voltage precision depends on bipolar transistor matching and may be affected by the offset voltage introduced by the amplification circuit required to appropriately scale the base-emitter voltage difference. For applications requiring a reference voltage that is less than 1.25 V, the reference voltage must be scaled to values other than the one set by the silicon bandgap by using additional circuit components, which not only increase the circuit size and power consumption but also can constitute a source of distortions affecting the output accuracy.

One method to overcome the problems related to the implementation of bandgap voltage references using CMOS processes with parasitic bipolar devices and integrated resistors is to design the voltage reference using floating-gate transistors. Generally employed for its capability to store an electrical charge for extended periods of time even without a connection to a power supply, a floating-gate transistor can be fabricated by inserting a layer of oxide to electrically isolate the gate of a conventional MOS transistor and depositing extra secondary gates or input nodes. The control signal applied to each input is capacitively coupled to the floating gate, and then appears to be attenuated by a factor C_i/C_T, where C_i is the coupling capacitance for the *i-th* input and C_T is the total load capacitance seen from the gate. Programming involves adding charges to the floating gate to raise the threshold voltage, while erasing is achieved by removing charge from the floating gate to lower the threshold voltage. This is generally realized under high applied voltages (> 10 V) that can create electric fields with sufficient strength to allow electrons to gain kinetic energy to overcome the potential barrier, or to tunnel through the potential barrier. Essentially, the performance characteristics of physical mechanisms that may be used to vary the amount of charge on the floating gate, such as hot electron injection and Fowler-Nordheim tunneling, are affected by the programming voltage and time.

The design of precise and stable voltage references in a standard CMOS process without using parasitic bipolar devices and integrated resistors makes use of floating-gate transistors, whose charge storage capability can be exploited for post-fabrication programmability. An amplifier based on two floating-gate transistors in the differential configuration, as shown in Figure 3.24(a), can constitute the main building block required for the voltage reference design [20]. This is needed to provide a reproduction of the programmed voltage level at a low-impedance node. Figure 3.24(b) illustrates the operation as a unity-gain voltage follower, where $V_0 = V_{REF}$. In the non-inverting amplifier configuration in Figure 3.24(c), the output voltage is of the form $V_0 = (1 + R_2/R_1)V_{REF}$. Here, the nature of the feedback forces V_0 to be dependent on V_{REF}, which is a function of the difference in the amount of charge stored on the floating gates of the transistor differential pairs. The switch S is in open state during the reference voltage programming, which may require many iterations to obtain the target, depending on the adopted technique.

3.3 Summary

Basic building blocks necessary for the design of active components such as amplifiers, comparators, and multipliers, were presented. They include current mirrors, voltage, and current references. The accuracy of a current mirror is improved by increasing the output resistance. This objective can be met by using cascode or feedback structures. On the other hand, the techniques that can be used to reduce the dependence of the supply voltage and temperature on the output signal provided by the reference circuit were also reviewed.

3.4 Circuit design assessment

1. **Improved cascode current mirror**
   Analyze each of the improved cascode current mirrors of Figure 3.25 with its respective transistor sizes given in Table 3.1 and show that the minimum output voltage is about $2V_{DS(sat)}$.

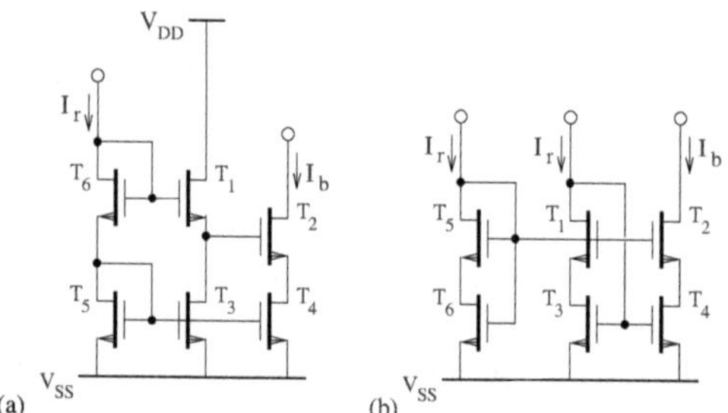

FIGURE 3.25
Circuit diagram of improved cascode current mirrors.

2. **Low-voltage current mirror**
   Determine the small-signal transfer function, $a_i = i_O/i_i$, of the current mirror shown in Figure 3.26.

3. **Self-biasing current reference**
   The self-biasing current reference of Figure 3.27 is designed with $(W_2/L_2) = \kappa(W_1/L_1)$ and $(W_4/L_4) = (W_3/L_3)$. Here, the body

TABLE 3.1
Transistor Sizes

	$T_1 - T_5$	T_6
Figure 3.25(a):	(W/L)	$(1/4)(W/L)$
Figure 3.25(b):	(W/L)	$(1/3)(W/L)$

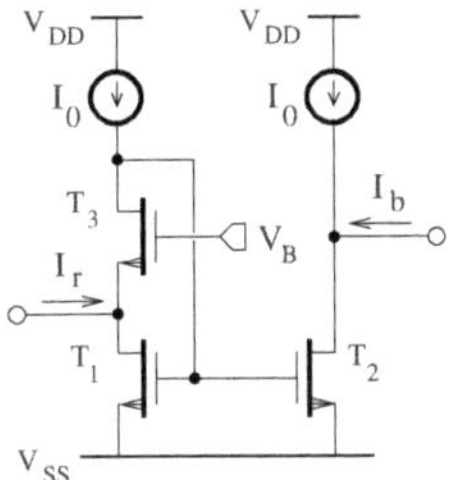

FIGURE 3.26
Circuit diagram of a low-voltage current mirror.

effect is eliminated because the transistors, T_1 and T_2, have the same source potential.

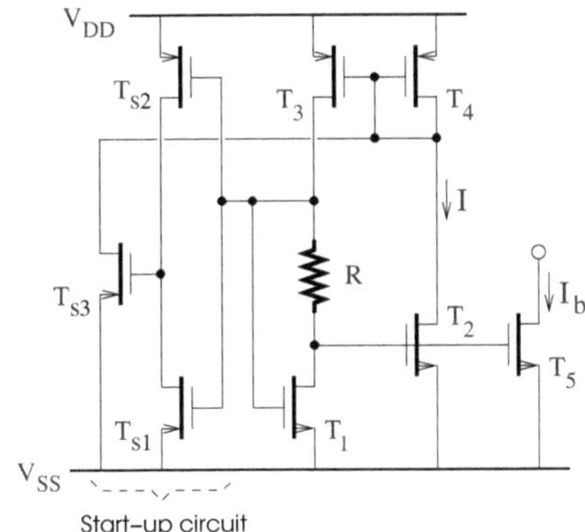

FIGURE 3.27
Circuit diagram of a self-biasing current reference.

Use Kirchhoff's voltage law equation,

$$V_{GS_1} - RI - V_{GS_2} = 0 \tag{3.147}$$

to show that:

– When transistors T_1 and T_2 operate in the strong inversion region,

$$I = \frac{1}{K_1 R^2}\left(1 - \frac{1}{\sqrt{\kappa}}\right)^2 \tag{3.148}$$

where $K_1 = (\mu_n C_{ox}/2)(W_1/L_1)$ is the transconductance of the transistor T_1.

– When transistors T_1 and T_2 operate in the weak inversion region,

$$I = \frac{nU_T}{R}\ln(\kappa) \tag{3.149}$$

where $U_T = kT/q$ is the thermal voltage and n is the subthreshold factor.

Verify that the start-up circuit composed of transistors $T_{s1} - T_{s3}$ delivers a small initialization current only if the current I remains equal to zero.

4. **Voltage reference**
A voltage reference using a diode-connected transistor biased by a PTAT current source, I_{D_5}, is shown in Figure 3.28. All transistors, except T_7, which is biased in the triode region, are assumed to operate in the saturation region.

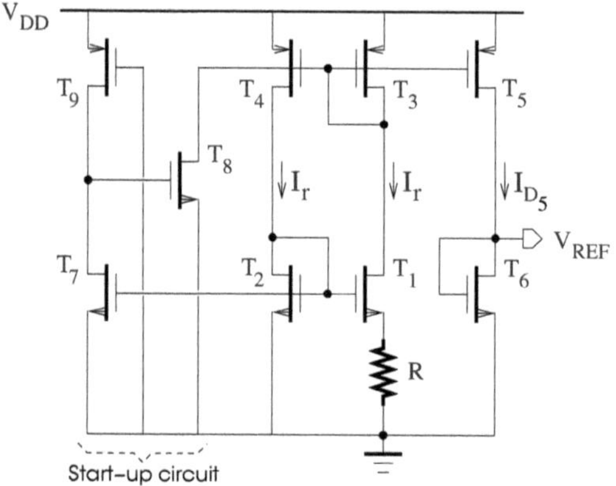

FIGURE 3.28
Voltage reference.

The start-up circuit (formed by transistors $T_7 - T_9$) initially injects a small current into the current reference to prevent the entry to the undesirable state and is then turned off so as not to interfere with the normal operation.

The reference current, I_r, is generated by the circuit stage composed of transistors $T_1 - T_4$ and resistor R, and the current mirror consisting of T_3 and T_5 is used to provide the bias current to the transistor T_6.

Verify that the value of the current I_r is given by

$$I_r = \frac{2}{\mu_n C_{ox}(W_1/L_1)} \frac{1}{R^2} \left(1 - \frac{1}{\sqrt{\kappa}}\right)^2 \tag{3.150}$$

where $W_1/L_1 = \kappa(W_2/L_2)$.

Assuming that

$$R(T) = R_0[1 + \alpha_R(T - T_0)] \tag{3.151}$$

$$\mu_n(T) = \mu_n(T_0)\left(\frac{T}{T_0}\right)^{-m} = \mu_n(T_0)\left(1 + \frac{T - T_0}{T_0}\right)^{-m} \tag{3.152}$$

where α_R is the temperature coefficient of the resistor, m is a constant ($1 \leq m \leq 2.5$), and $T = T_0 + \triangle T$, derive the following first-order approximation:

$$I_r(T) \simeq I_{r0}\Big[1 + \alpha_{I_r}(T - T_0)\Big] \tag{3.153}$$

where

$$\alpha_{I_r} = \frac{m}{T_0} - 2\alpha_R \tag{3.154}$$

$$I_{r0} = \frac{2}{\mu_n(T_0)C_{ox}(W_1/L_1)} \frac{1}{R_0^2} \left(1 - \frac{1}{\sqrt{\kappa}}\right)^2 \tag{3.155}$$

Show that the drain current I_{D_5} can be put into the form,

$$I_{D_5} \simeq \frac{W_5/L_5}{W_3/L_3} I_{r0}\Big[1 + \alpha_{I_r}(T - T_0)\Big] \tag{3.156}$$

Use the expression of the reference voltage,

$$V_{REF} = V_{GS_6} = V_{T_n} + \sqrt{\frac{I_{D_6}}{K_6}} \tag{3.157}$$

where $I_{D_6} = I_{D_5}$, the temperature dependencies of the mobility, $\mu_n(T)$, and the threshold voltage,

$$V_{T_n}(T) = V_{T_n}(T_0) - \alpha_{V_{T_n}}(T - T_0) \tag{3.158}$$

where $\alpha_{V_{T_n}}$ is the temperature coefficient of the threshold voltage, to show that

$$\frac{\delta V_{REF}}{\delta T} = \frac{V_{REF}(T_0 + \triangle T) - V_{REF}(T_0)}{\delta T} \tag{3.159}$$

$$\simeq -\alpha_{V_{T_n}} + \frac{K_6}{2}\sqrt{\frac{W_5/L_5}{W_3/L_3} I_{r0}}\left(\frac{m}{T_0} + \alpha_{I_r}\right) \tag{3.160}$$

Show that the compensation of the temperature effect on the voltage reference, by setting $\delta V_{REF}/\delta T = 0$, yields

$$\sqrt{\frac{W_5/L_5}{W_3/L_3}} = \frac{\alpha_{V_{T_n}}}{K_6\sqrt{I_{r0}}\left(\frac{m}{T_0} - \alpha_R\right)} \tag{3.161}$$

Verify that the minimum required power supply voltage is given by $V_{T_n} + V_{SD_5(sat)}$, where $V_{SD_5(sat)}$ is the source-drain saturation voltage.

5. **Analysis of an active cascode current mirror**
Consider the current mirror shown in Figure 3.29 [21]. The difference between the drain-source voltages of transistors T_2 and T_3 is detected by the amplifier and used to control the gate voltage of T_1. Assuming that all transistors feature the same drain-source saturation voltage, $V_{DS(sat)}$, determine the resistance R as a function of I_r.

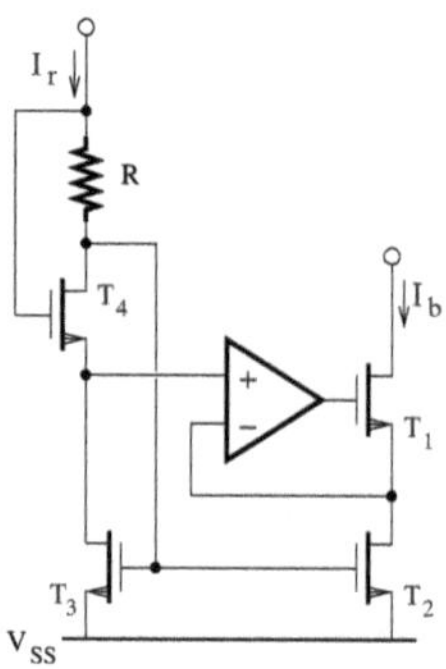

FIGURE 3.29
Active cascode current mirror.

Estimate the minimum output voltage of the current mirror.

Derive an expression for the output resistance r_0.

6. **Voltage-controlled current source/sink**
Voltage-controlled current source and sink can be realized as shown in Figure 3.30. In each case, the operational transconductance amplifier drives a transistor and a resistor in a negative feedback loop. The transistors, T_{B_2} and T_{B_3}, of the current mirror operate in the saturation region.

Verify that:

– the value of the reference current I_r is kV_{REF}/R, where the current mirror gain $k = (W_{B_3}/L_{B_3})/(W_{B_2}/L_{B_2})$, and the operating

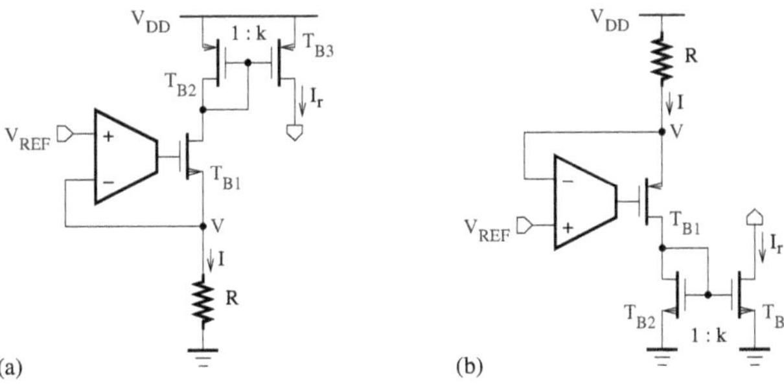

FIGURE 3.30
Voltage-controlled current (a) source and (b) sink.

range stretches from 0 to $V_{DD} - (|V_{T_p}| + V_{DS(sat)})$ for the circuit of Figure 3.30(a);

– the reference current I_r is equal to $k(V_{DD} - V_{REF})/R$, and the operating range is from $V_{T_n} + V_{DS(sat)}$ to V_{DD} for the circuit of Figure 3.30(b).

For the voltage-controlled current source/sink of Figure 3.31, determine the current I as a function of the resistor R.

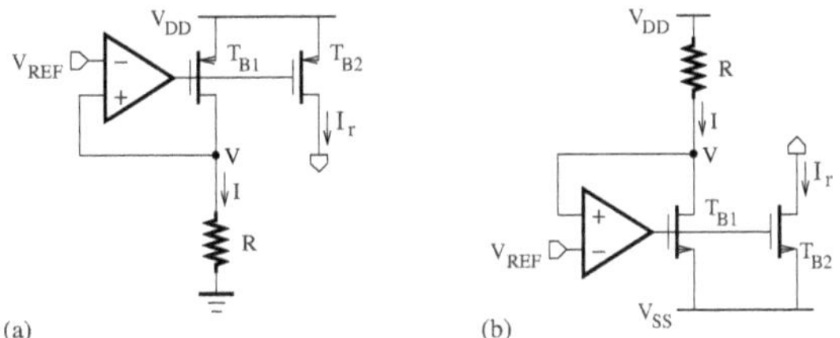

FIGURE 3.31
Wide-swing voltage-controlled current (a) source and (b) sink.

Show that $I_r = I$, provided the transistors T_{B1} and T_{B2} are matched.

Consider the circuit structures shown in Figure 3.32 [15]. The current mirror transistors T_{B3} and T_{B4} are used to improve the accuracy of the relation between the current I and the output current, I_r.

Assuming that the width-to-length ratio of T_{B1} and T_{B3} is equal to W_1/L_1, and that for T_{B2} and T_{B4} is equal to W_2/L_2, determine the current I_r as a function of I.

With the sum $V_{DS_{B1}} + V_{DS_{B3}}$ being equal either to $V_{DD} - V_{REF}$ or

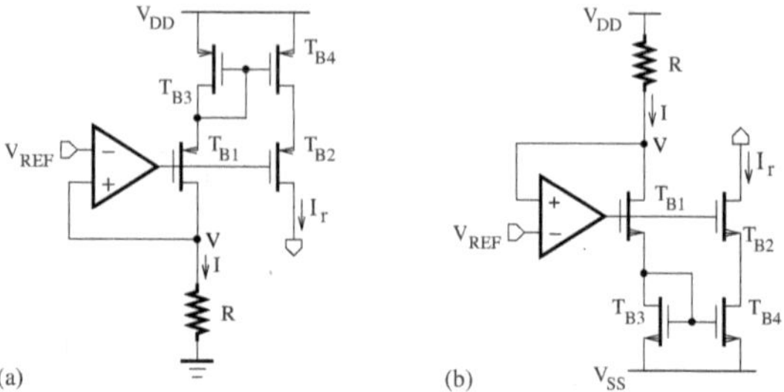

FIGURE 3.32
Improved voltage-controlled current (a) source and (b) sink.

to $V_{REF} - V_{SS}$, find the minimum value of V_{REF} to maintain the transistors in the saturation region.

7. **Bandgap reference for low supply-voltage applications**
The implementation of a low-voltage bandgap reference [16] is depicted in Figure 3.33(a). To reduce the minimum supply voltage to about 1 V, MOS transistors with a low threshold voltage are required in the amplifier implementation. The transistors are assumed to be identical and $R_1 = R_3$.

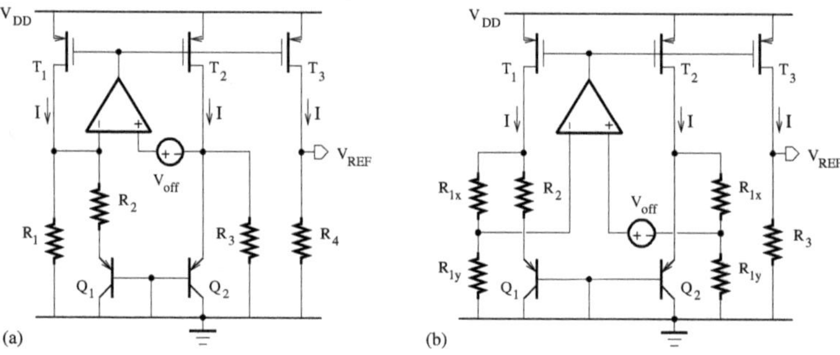

FIGURE 3.33
Low supply-voltage bandgap references.

Taking into account the amplifier offset voltage, V_{off}, show that

$$V_{REF} = \frac{R_4}{R_1}\left[V_{EB2} + \frac{R_1}{R_2}U_T \ln\left(\frac{I_{S1}}{I_{S2}}\right) - \left(1 + \frac{R_1}{R_2}\right)V_{off}\right] \quad (3.162)$$

Estimate the contribution of the offset dispersion, ΔV_{off}, on the variation of V_{REF}.

The requirement of using a CMOS fabrication process with low threshold voltages can be avoided in the case of the low-voltage bandgap reference shown in Figure 3.33(b) [22]. To maintain the same voltage at the drain of transistors T_1 and T_2, it is necessary to have $R_{1x} = R_{1y}$ and $R_1 = R_{1x} + R_{1y}$.

Verify that

$$V_{REF} = \frac{R_3}{R_1}\left[V_{EB2} + \frac{R_1}{R_2}\left(U_T \ln\left(\frac{I_{S1}}{I_{S2}}\right) + 2V_{off}\right)\right] \quad (3.163)$$

Compare these circuits for use as bandgap voltage references with a low supply-voltage.

Bibliography

[1] C. Abel, "Bias circuit for high-swing cascode current mirrors," U.S. Patent 7,208,998, filed April 12, 2005; issued April 24, 2007.

[2] J. B. Hughes and I. C. MacBeth, "Circuit arrangement for processing sampled analogue electrical signals," U.S. Patent 4,897,596, filed December 16, 1988; issued January 30, 1990.

[3] P. C. Kwong, "Cascode current mirror with amplifier," U.S. Patent 6,124,705, filed August 20, 1999; issued September 26, 2000.

[4] S. K. Hoon and J. Chen, "Regulated cascode current source with wide output swing," U.S. Patent 6,903,539, filed November 19, 2003; issued June 7, 2005.

[5] B. R. Gregoire, Jr., "Low voltage enhanced output impedance current mirror," U.S. Patent 6,707,286, filed February 24, 2003; issued March 16, 2004.

[6] T. L. Brooks and M. A. Rybicki, "Self-biased cascode current mirror having high voltage swing and low power consumption," U.S. Patent 5,359,296, filed September 10, 1993; issued October 25, 1994.

[7] N. S. Sooch, "MOS Cascode current mirror," U.S. Patent 4,550,284, filed May 16, 1984; issued October 29, 1985.

[8] E. Säckinger and W. Guggenbühl, "A high-swing, high-impedance MOS cascode circuit," *IEEE J. of Solid-State Circuits*, vol. 25, pp. 289–298, Feb. 1990.

[9] Z.-k. Zhou, P.-s. Zhu, Y. Shi, H.-y. Wang, Y.-q. Ma, X.-z. Xu, L. Tan, X. Ming, and B. Zhang, "A CMOS voltage reference based on mutual compensation of Vtn and Vtp," *IEEE Trans. on Circuits and Systems–II*, vol. 59, no. 6, pp. 341–345, June 2012.

[10] E. Vittoz and J. Fellrath, "CMOS analog circuits based on weak inversion operation," *IEEE J. of Solid-State Circuits*, vol. 12, pp. 224–231, June 1977.

[11] H. J. Oguey and D. Aebischer, "CMOS current reference without resistance," *IEEE Journal of Solid-State Circuits*, vol. 32, no. 7, pp. 1132–1135, Jul. 1997.

[12] G. De Vita and G. Iannaccone, "A sub-1-V, 10 ppm/^oC, nanopower voltage reference generator," *IEEE J. of Solid-State Circuits*, vol. 42, no. 7, pp. 1536–1542, July 2007.

[13] D. Y. Yu, "Startup circuit for band-gap reference circuit," U.S. Patent 5,867,013, filed November 20, 1997; issued February 2, 1999.

[14] H. Neuteboom, B. M. J. Kup, and M. Janssens, "A DSP-based hearing instrument IC," *IEEE J. of Solid-State Circuits*, vol. 32, pp. 1790–1806, Nov. 1997.

[15] C.-J. Yen and C.-H. Cheng, "Voltage-controlled current source," U.S. Patent 7,417,415, filed June 10, 2005; issued August 26, 2008.

[16] H. Banba, H. Shiga, A. Umezawa, T. Miyaba, T. Tanzawa, S. Atsumi, and K. Sakui, "A CMOS bandgap reference circuit with sub-1-V operation," *IEEE J. of Solid-State Circuits*, vol. 34, pp. 670–674, May. 1999.

[17] V. V. Ivanov and K. E. Sanborn, "Precision reversed bandgap voltage reference circuits and method," U.S. Patent 7,411,443, filed June 22, 2006; issued August 12, 2008.

[18] Y. P. Tsividis, "Accurate analysis of temperature effects in I_C-V_{BE} characteristics with application to bandgap reference sources," *IEEE J. Solid-State Circuits*, vol. 15, pp. 1076–1084, Dec. 1980.

[19] M. Gunawan, G. C. M. Meijer, J. Fonderie, and J. H. Huijsing, "A curvature-corrected low-voltage bandgap reference," *IEEE J. Solid-State Circuits*, vol. 28, pp. 667–670, June 1993.

[20] K. C. Adkins, "Structure and method for programmable and non-volatile analog signal storage for a precision voltage reference," U.S. Patent 6,791,879, filed September 23, 2002; issued September 14, 2004.

[21] T. Itakura and Z. Czarnul, "Current mirror circuit," U.S. Patent 5,986,507, filed September 12, 1996; issued November 16, 1999.

[22] K. N. Leung and P. K. T. Mok, "A sub-1-V 15-ppm/^{o}C CMOS bandgap voltage reference without requiring low threshold Voltage Device," *IEEE J. of Solid-State Circuits*, vol. 37, pp. 526–530, April 2002.

4

CMOS Amplifiers

CONTENTS

The performance of mixed-signal circuits are generally determined by the characteristics of active components. Analog cells used inside the chip have to drive well-defined loads that are often purely capacitive, while the output buffer may demonstrate the capability of driving variable resistive and capacitive loads with low distortion. For this reason, the choice of the architecture and the design approach are mainly determined by the function to be realized. Furthermore, the realization of an amplifier, which features both high dc gain and bandwidth, can lead to conflicting demands. The high gain requirement is met by multistage designs with long-channel transistors biased at a low current level, while the high bandwidth specification is achieved in single-stage structures using short-channel transistors biased by a high-level current source.

Many architectures are available for the design of amplifiers [1]. However, these integrated circuits (ICs) would only benefit marginally from the decrease in transistor feature sizes and power supply voltages due to noise and offset requirements. Efficient solutions should be able to associate the low-voltage operation with high power efficiency and result in simple structures, that is, with a low die area.

Various performance specifications can be taken into consideration for the choice of an amplifier structure. They include dc characteristics such as input common-mode range, output swing, and offset voltage. In addition, there are specifications that can be described in the frequency domain (gain, bandwidth, phase margin) and in the time domain (slew rate, settling time).

4.1 Differential amplifier

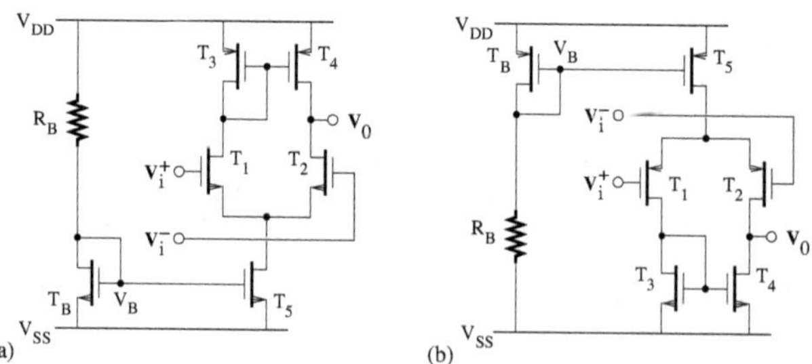

FIGURE 4.1
Circuit diagram of the single-stage differential amplifier based on (a) n-channel and (b) p-channel transistor pair.

The circuit diagram of a basic differential amplifier is shown in Figure 4.1. Note that, to achieve an accurate biasing of the amplifier in the presence of process, voltage, and temperature variations, the resistor R_B is generally implemented off-chip. The input differential transistor pair is biased by a constant current source and loaded by a current mirror. In the n-well process, a better threshold voltage matching can be achieved with the differential stage using n-channel transistors (see Figure 4.1(a)). On the other hand, the body effect can be removed by implementing the amplifier input stage with p-channel transistors as shown in Figure 4.1(b). Furthermore, the output noise is minimized because a p-channel transistor is less affected by $1/f$ noise than an n-channel transistor. In general, a large transconductance is obtained by designing the input transistors with a high W/L ratio. The parameter L is chosen to keep the channel length modulation effects low and W is sized for a given gate-source voltage.

4.1.1 Dynamic range

In general, the minimum supply voltage is determined based on the input common-mode range and output swing specifications. The input common-mode range is the range of the dc input voltage for which the specified gain of the differential stage is maintained unchanged. The output swing is the range over which the output voltage can be driven without being distorted. These specifications are estimated by assuming that the input voltage applied to both inputs is reduced to its common-mode dc component.

A differential stage is designed to amplify the voltage difference between both inputs and reject the common-mode input voltage. Let us consider the differential stage shown in Figure 4.1(a). In the worst case, the common-mode input voltage can be given by

$$V_{ICM} = V_{DD} - V_{SG_3} - V_{DS_1} + V_{GS_1} \tag{4.1}$$

or

$$V_{ICM} = V_{SS} + V_{DS_5} + V_{GS_1} \tag{4.2}$$

Because $V_{DS_1} = V_{DS_1(sat)} = V_{GS_1} - V_{T_n}$, $V_{SG_3} = V_{SD_3} = V_{SD_3(sat)} - V_{T_p}$, and $V_{DS_5} = V_{DS_5(sat)} = V_{GS_5} - V_{T_n}$, the maximum and minimum values of the common-mode input voltage can be found as

$$V_{ICM} = V_{DD} - V_{SD_3(sat)} + V_{T_p} + V_{T_n} = V_{ICM,max} \tag{4.3}$$

and

$$V_{ICM} = V_{SS} + V_{DS_5(sat)} + V_{DS_1(sat)} + V_{T_n} = V_{ICM,min} \tag{4.4}$$

Thus,

$$V_{SS} + V_{DS_5(sat)} + V_{DS_1(sat)} + V_{T_n} \leq V_{ICM} \leq V_{DD} - V_{SD_3(sat)} + V_{T_p} + V_{T_n} \tag{4.5}$$

The limits of the output voltage swing are set by the requirement of maintaining the transistors T_2 and T_4 in the saturation region. For the transistor T_2, this results in the condition

$$V_{DS_2} \geq V_{DS_2(sat)} = V_{GS_2} - V_{T_n} \tag{4.6}$$

Because $V_{DS_2} = V_O - V_{S_2}$ and $V_{GS_2} = V_{ICM} - V_{S_2}$, we can obtain

$$V_O \geq V_{ICM} - V_{T_n} \tag{4.7}$$

In the case of the transistor T_4, we have

$$V_{SD_4} \geq V_{SD_4(sat)} = V_{SG_4} + V_{T_p} \tag{4.8}$$

where $V_{SD_4} = V_{DD} - V_O$. It can be shown that

$$V_O \leq V_{DD} - V_{SG_4} - V_{T_p} = V_{DD} - V_{SD_4(sat)} \tag{4.9}$$

Hence,

$$V_{ICM} - V_{T_n} \leq V_O \leq V_{DD} - V_{SD_4(sat)} \tag{4.10}$$

An analogous analysis can be performed for the differential stage with p-channel input transistors of Figure 4.1(b). The input common-mode voltage can be expressed as

$$V_{ICM} = -V_{SG_1} + V_{SD_1} + V_{GS_3} + V_{SS} \tag{4.11}$$

and

$$V_{ICM} = V_{DD} - V_{SD_5} - V_{SG_1} \tag{4.12}$$

where $V_{GS_3} = V_{DS_3}$. Assuming that $V_{GS_3} = V_{DS_3(sat)} + V_{T_n}$, the input common-mode range is then given by

$$V_{T_p} + V_{T_n} + V_{DS_3(sat)} + V_{SS} \leq V_{ICM} \leq V_{DD} - V_{SD_5(sat)} - V_{SD_1(sat)} + V_{T_p} \tag{4.13}$$

The transistors T_4 and T_2 will remain in the saturation region if the output swing is

$$V_{SS} + V_{DS_4(sat)} \leq V_O \leq V_{ICM} + V_{T_p} \tag{4.14}$$

It was assumed that $V_{T_n} > 0$, $V_{T_p} < 0$, $V_{DD} > 0$, and $V_{SS} \leq 0$. With the difference between the threshold voltages of n-channel and p-channel transistors not exceeding the drain-source saturation voltage, whose typical value is on the order of 0.3 V, it can be noted that the n-channel input stage exhibits a wide positive input common-mode swing, while the p-channel structure has a wide negative input common-mode swing. This suggests the use of a parallel combination of n-channel and p-channel differential pairs to achieve an almost rail-rail input range. On the other hand, the basic differential stage can only provide a limited output swing.

4.1.2 Source-coupled differential transistor pair

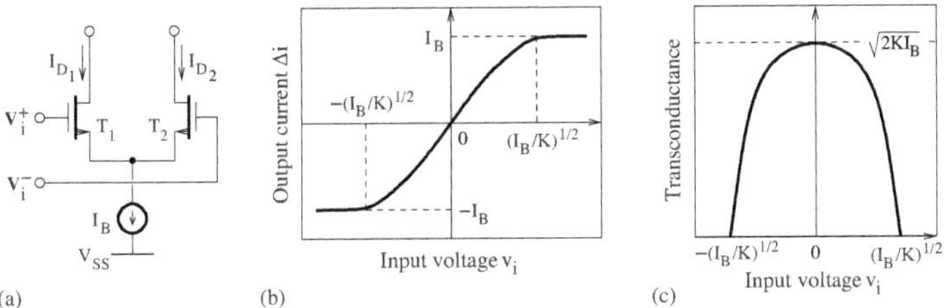

FIGURE 4.2
Source-coupled differential transistor pair: (a) Circuit diagram, (b) I-V transfer characteristic, (c) transconductance plot.

Let the n-channel transistor be described by the next equation,

$$I_D = \begin{cases} K(V_{GS} - V_{T_n})^2 & \text{if} \quad V_{GS} \geq V_{T_n} \quad \text{saturation region} \\ 0 & \text{if} \quad V_{GS} < V_{T_n} \quad \text{cutoff region,} \end{cases} \tag{4.15}$$

where $K = \mu_n(C_{ox}/2)(W/L)$ is the transconductance parameter; μ_n is the effective surface carrier mobility; C_{ox} is the gate-oxide capacitance per unit area; W and L are the channel width and length, respectively; and V_{T_n} denotes the threshold voltage. The bias current, I_B, of the source-coupled transistor structure of Figure 4.2(a) can be expressed as

$$I_{D_1} + I_{D_2} = I_B \tag{4.16}$$

Applying Kirchhoff's voltage law to the loop involving the noninverting and inverting input nodes, the differential signal, V_i, is derived as

$$V_i = V_i^+ - V_i^- = V_{GS_1} - V_{GS_2} \tag{4.17}$$

Because the transistors T_1 and T_2 are identical and operate in the saturation region, we have

$$V_{GS_1} = \sqrt{\frac{I_{D_1}}{K}} + V_{T_n} \tag{4.18}$$

and

$$V_{GS_2} = \sqrt{\frac{I_{D_2}}{K}} + V_{T_n} \tag{4.19}$$

Hence, Equation (4.17) becomes

$$\sqrt{I_{D_1}} - \sqrt{I_{D_2}} = \sqrt{K} V_i \tag{4.20}$$

Solving the system of Equations (4.16) and (4.20) gives

$$I_{D_1} = \frac{I_B}{2} \pm \sqrt{2KI_B}\,\frac{V_i}{2}\sqrt{1 - \frac{V_i^2}{2I_B/K}} \tag{4.21}$$

Because the drain current should be greater than $I_B/2$ when the input voltage increases, the only acceptable solution is given by

$$I_{D_1} = \frac{I_B}{2} + \sqrt{2KI_B}\,\frac{V_i}{2}\sqrt{1 - \frac{V_i^2}{2I_B/K}} \tag{4.22}$$

Using Equation (4.16), the drain current of T_2 can then be computed as

$$I_{D_2} = \frac{I_B}{2} - \sqrt{2KI_B}\,\frac{V_i}{2}\sqrt{1 - \frac{V_i^2}{2I_B/K}} \tag{4.23}$$

For input voltages within the linear range, the current flowing through one transistor will increase while the current in the other transistor will decrease. The maximum value of the input voltage is reached when T_1 operates in the saturation region and T_2 turns off, resulting in $I_{D_1} = I_B$ and $I_{D_2} = 0$, while the minimum value of the input voltage is attained when T_1 turns off and T_2 operates in the saturation region, yielding $I_{D_1} = 0$ and $I_{D_2} = I_B$. Using these values together with Equation (4.20), the input linear range can be expressed as

$$|V_i| \leq \sqrt{\frac{I_B}{K}} \tag{4.24}$$

The differential output current, Δi, is derived as follows:

$$\Delta i = I_{D_1} - I_{D_2} = \begin{cases} \sqrt{2KI_B}\,V_i\sqrt{1 - \dfrac{V_i^2}{2I_B/K}} & \text{if} \quad |V_i| \leq \sqrt{\dfrac{I_B}{K}} \\ I_B\,\mathrm{sign}(V_i) & \text{if} \quad |V_i| > \sqrt{\dfrac{I_B}{K}} \end{cases} \tag{4.25}$$

This last expression is valid provided the transistors do not turn off. The linear region can be made large by increasing the bias current or transistor lengths. The transconductance of the differential transistor pair represents the slope of the I-V characteristic depicted in Figure 4.2(b). Taking the derivative of Δi with respect to V_i gives

$$g_m = \frac{d\Delta i}{dV_i} = \begin{cases} \sqrt{2KI_B}\left(\sqrt{1 - \dfrac{V_i^2}{2I_B/K}} - \dfrac{\dfrac{V_i^2}{2I_B/K}}{\sqrt{1 - \dfrac{V_i^2}{2I_B/K}}}\right) & \text{if} \quad |V_i| \leq \sqrt{\dfrac{I_B}{K}} \\ 0 & \text{if} \quad |V_i| > \sqrt{\dfrac{I_B}{K}} \end{cases} \tag{4.26}$$

Figure 4.2(c) shows the plot of the transconductance versus the input voltage. At $V_i = 0$, the transconductance is reduced to $\sqrt{2KI_B}$. The square root term is associated with nonlinear distortions that can affect the transconductance for high values of the input voltages.

4.1.3 Current mirror

The circuit diagram of a current mirror using p-channel transistors is depicted in Figure 4.3(a). The currents flowing through the transistors T_1 and T_2 are given by

$$I_i = K_1(V_{SG_1} + V_{T_p})^2 \tag{4.27}$$

and

$$I_0 = K_2(V_{SG_2} + V_{T_p})^2 \tag{4.28}$$

respectively, where $K_1 = \mu_p(C_{ox}/2)(W_1/L_1)$ and $K_2 = \mu_p(C_{ox}/2)(W_2/L_2)$. Assuming that $V_{SG_1} = V_{SG_2}$, the current ratio can be expressed as

$$\frac{I_0}{I_i} = \frac{W_2/L_2}{W_1/L_1} \tag{4.29}$$

The output current is represented in Figure 4.3(b) in the case where the input

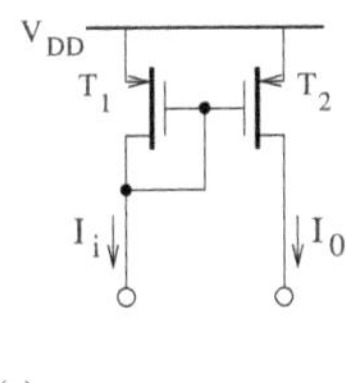

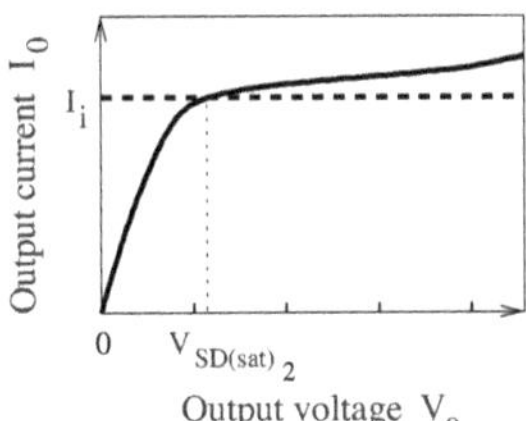

FIGURE 4.3
(a) Circuit diagram and (b) output characteristic of a simple current mirror.

current is constant. The output signal swing is limited by the voltage required to maintain the transistor in the saturation region, that is,

$$V_{SD(sat)} = V_{SG} + V_{T_p} \tag{4.30}$$

Ideally, the input and output current should have the same value provided that the transistors are matched. However, the actual output resistance of the current mirror is limited instead of being infinite, and the input and output currents are equal only when $V_{SG_1} = V_0 = V_{SD_2}$. As a result, the achievable linearity range of a simple current mirror is reduced.

4.1.4 Slew-rate limitation

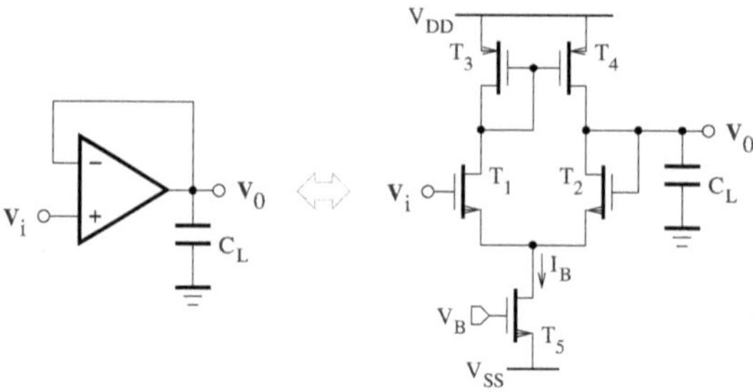

FIGURE 4.4
Differential stage in unity-gain configuration with a capacitive load.

The output slew rate of a differential stage is defined as the maximum rate of change of the output voltage. The current required to charge the output capacitor is given by

$$i = C_L \frac{dV_0}{dt} \tag{4.31}$$

When the amplifier input stage is driven out of the linear region by a large input signal, one of the transistors, T_1 and T_2, turns off, while the other conducts the whole bias current, I_B. Because the maximum current available to charge the load capacitor is I_B, the slew rate of the differential stage shown in Figure 4.4 can be expressed as

$$SR = \max\left(\left|\frac{dV_0}{dt}\right|\right) = \frac{I_B}{C_L} \tag{4.32}$$

The slew rate is generally expressed in units of V/µs. For example, the slew rate of a differential stage with $I_B = 20$ µA and $C_L = 2$ pF is 10 V/µs. In the case of a differential stage configured, as shown in Figure 4.4, to operate as a unity-gain amplifier with a sinusoidal input signal of the form

$$V_i = V_{max} \sin(2\pi f t) \tag{4.33}$$

the derivative of the output voltage is

$$\frac{dV_0}{dt} = 2\pi f V_{max} \cos(2\pi f t) \tag{4.34}$$

where V_{max} is the peak voltage and f denotes the input frequency. To avoid distortions due to slewing, the slew rate must satisfy the following constraint:

$$SR \geq 2\pi f_{max} V_{max} \tag{4.35}$$

where f_{max} is the maximum input frequency.

4.1.5 Small-signal characteristics

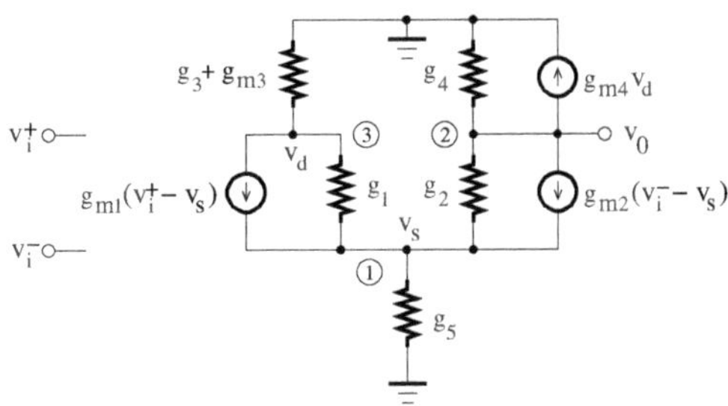

FIGURE 4.5
Small-signal equivalent model of the differential stage.

The small-signal equivalent model of the differential stage of Figure 4.1(a) is depicted in Figure 4.5. The body effects of transistors are assumed to be negligible. The transistor T_5, which generates the bias current, is represented by its output resistance. The diode-connected transistor, T_3, is modeled by a resistor with the conductance $g_3 + g_{m3}$. Applying Kirchhoff's current law to the nodes 1, 2, and 3, we obtain

$$g_{m1}(v_i^+ - v_s) + g_1(v_d - v_s) + g_{m2}(v_i^- - v_s) + g_2(v_0 - v_s) - g_5 v_s = 0 \quad (4.36)$$

$$g_{m2}(v_i^- - v_s) + g_2(v_0 - v_s) + g_{m4}v_d + g_4 v_0 = 0 \quad (4.37)$$

and

$$(g_3 + g_{m3})v_d + g_{m1}(v_i^+ - v_s) + g_1(v_d - v_s) = 0 \quad (4.38)$$

respectively. Under the assumption that the transistors, T_1 and T_2, and T_3 and T_4, are matched, that is, $g_{m1} = g_{m2}$, $g_1 = g_2$, $g_{m3} = g_{m4}$, and $g_3 = g_4$, replacing Equation (4.37) by the difference of Equations (4.38) and (4.37), and Equation (4.38) by the sum of Equations (4.38) and (4.37) gives

$$2g_{m1}v_{ic} + g_1 v_0 + g_1 v_d - [g_5 + 2(g_1 + g_{m1})]v_s = 0 \quad (4.39)$$

$$g_{m1}v_{id} - (g_1 + g_3)v_0 + (g_1 + g_3)v_d = 0 \quad (4.40)$$

and

$$2g_{m1}v_{ic} + (g_1 + g_3)v_0 + (g_1 + g_3 + 2g_{m3})v_d - 2(g_1 + g_{m1})v_s = 0 \quad (4.41)$$

where

$$v_{id} = v_i^+ - v_i^- \quad (4.42)$$

and

$$v_{ic} = \frac{v_i^+ + v_i^-}{2} \quad (4.43)$$

Solving this last system of three equations gives the small-signal output voltage of the form

$$v_O = A_d v_{id} + A_c v_{ic} \tag{4.44}$$

where

$$A_d = \frac{g_{m1}\{g_1 g_5 + (g_3 + 2g_{m3})[g_5 + 2(g_1 + g_{m1})]\}}{2(g_1 + g_3)\{g_1 g_5 + (g_3 + g_{m3})[g_5 + 2(g_1 + g_{m1})]\}} \tag{4.45}$$

and

$$A_c = \frac{-\, g_5 g_{m1}}{g_1 g_5 + (g_3 + g_{m3})[g_5 + 2(g_1 + g_{m1})]} \tag{4.46}$$

The common-mode rejection ratio (CMRR) is defined by

$$CMRR = \left|\frac{A_d}{A_c}\right| \tag{4.47}$$

where A_d and A_c are the differential and common-mode gains, respectively.

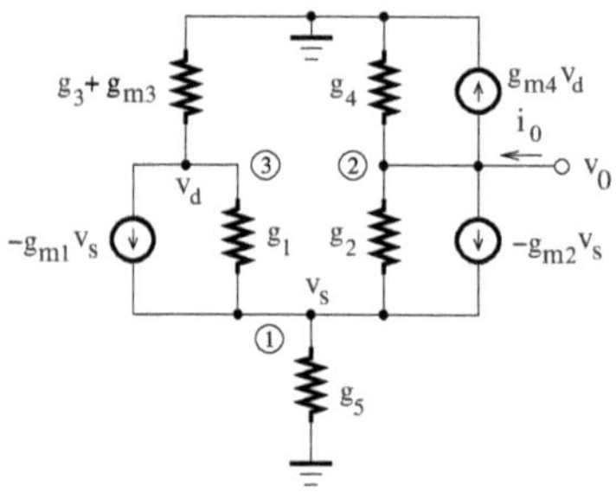

FIGURE 4.6
Small-signal equivalent model of the differential stage for the output resistance determination.

To determine the output resistance, the input nodes of the differential stage are short-circuited to the ground and a test generator is connected to the output node. The corresponding small-signal equivalent model is illustrated in Figure 4.6. Applying Kirchhoff's current law to the nodes 1, 2, and 3, we get

$$-g_{m1}v_s + g_1(v_d - v_s) - g_{m2}v_s + g_2(v_O - v_s) - g_5 v_s = 0 \tag{4.48}$$

$$-g_{m2}v_s + g_2(v_O - v_s) + g_{m4}v_d + g_4 v_O = i_O \tag{4.49}$$

and

$$(g_3 + g_{m3})v_d - g_{m1}v_s + g_1(v_d - v_s) = 0 \tag{4.50}$$

where $g_{m1} = g_{m2}$, $g_{m3} = g_{m4}$, $g_1 = g_2$, and $g_3 = g_4$. These equations can be solved for the output resistance given by

$$r_O = \frac{v_O}{i_O} = \frac{g_1 g_5 + g_1(g_1 + g_{m1}) + (g_3 + g_{m3})[g_5 + 2(g_1 + g_{m1})]}{(g_1 + g_3)\{g_1 g_5 + (g_3 + g_{m3})[g_5 + 2(g_1 + g_{m1})]\}} \tag{4.51}$$

In practice, g_{m1}, $g_{m3} \gg g_1$, g_3, g_5, and it can then be shown that

$$A_d \simeq \frac{g_{m1}}{g_1 + g_3} \tag{4.52}$$

$$A_c \simeq \frac{-g_5}{2g_{m3}} \tag{4.53}$$

$$r_0 \simeq \frac{1}{g_1 + g_3} \tag{4.54}$$

and

$$CMRR \simeq 2\frac{g_{m1}g_{m3}}{g_5(g_1 + g_3)} \tag{4.55}$$

where $g_1 = g_{ds_1}$, $g_3 = g_{ds_3}$, and $g_5 = g_{ds_5}$. The transconductance and conductance are proportional to the W/L ratio of the transistor, and a large differential gain is realized for a given bias current provided the aspect ratio of differential transistors is much greater than the one of load transistors. To obtain a high CMRR, the value of g_5 must be small, as is the case when a bias current source with a high output resistance is used.

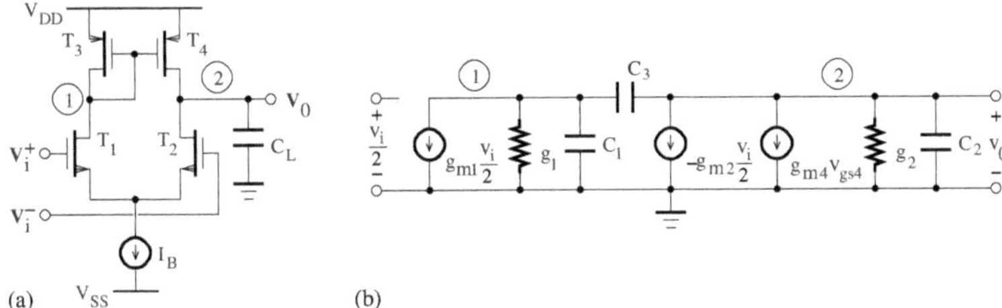

FIGURE 4.7
(a) Single-ended transconductance amplifier; (b) small-signal equivalent model of the transconductance amplifier including parasitic capacitances.

At high frequencies, the effect of parasitic capacitances can no longer be neglected. The amplifier gain and impedances then become dependent on the signal frequency. The circuit diagram of a single-ended transconductance amplifier and its small-signal equivalent model are shown in Figures 4.7(a) and (b), respectively. Using Kirchhoff's current law, the equations for node 1 and node 2 can be written as

$$g_{m1}V_i/2 + (g_1 + SC_1)V_1 - SC_3(V_0 - V_1) = 0 \tag{4.56}$$

and

$$-g_{m2}V_i/2 + SC_3(V_0 - V_1) + g_{m4}V_{gs4} + (g_2 + SC_2)V_0 = 0 \tag{4.57}$$

respectively. The system of Equations (4.56) and (4.57) can be solved for the transfer function given by

$$A(s) = \frac{V_0(s)}{V_i(s)} = \frac{g_{m1}(g_1 + g_{m4} + sC_1)/2}{(g_1 + sC_1)(g_2 + sC_2) + sC_3[g_1 + g_2 + g_{m4} + s(C_1 + C_2)]} \tag{4.58}$$

where $C_1 = C_{gd1} + C_{db1} + C_{db3} + C_{gs3} + C_{gs4}$, $C_2 = C_{gd2} + C_{db2} + C_{db4} + C_L$, $C_3 = C_{gd4}$, $g_1 = g_{ds1} + g_{ds3} + g_{m3}$, and $g_2 = g_{ds2} + g_{ds4}$. Due to the relative low value of C_{gd4}, the capacitance C_3 can be considered negligible. Furthermore, because $g_{ds1} + g_{ds3} + g_{m3} \simeq g_{m3}$, and the transistors T_3 and T_4 are matched ($g_{m3} = g_{m4}$), we obtain

$$A(s) = \frac{V_0(s)}{V_i(s)} \simeq A_0 \frac{1 - s/\omega_{z_1}}{(1 - s/\omega_{p_1})(1 - s/\omega_{p_2})} \tag{4.59}$$

where $A_0 = g_{m1}/(g_{ds2} + g_{ds4})$, $\omega_{z_1} = -2g_{m3}/C_1$, $\omega_{p_1} = -(g_{ds2} + g_{ds4})/C_2$ and $\omega_{p_2} = -g_{m3}/C_1$. The main contribution to the value of C_1 can be attributed to the sum of equal capacitances C_{gs3} and C_{gs4}. The second pole is then located at approximately half the transition frequency of the transistor T_3, while the zero occurs almost at the transition frequency, which is generally in the range of few hundred megahertz. Therefore, the zero and second pole are rejected at very high frequencies, and the amplifier frequency response can be described by a transfer function with a single dominant pole.

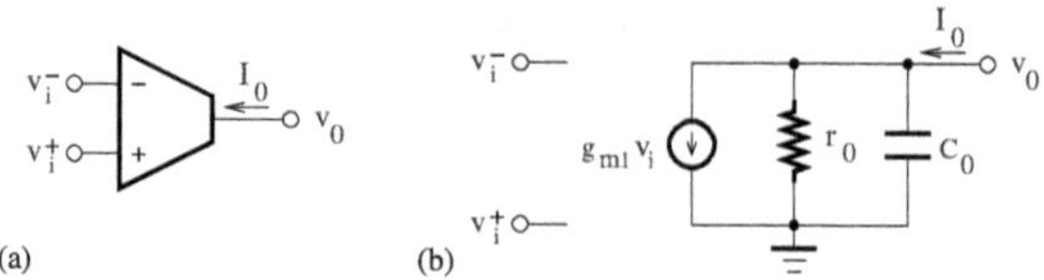

FIGURE 4.8
(a) Symbol of a single-ended transconductance amplifier; (b) small-signal equivalent model of a single-ended transconductance amplifier.

When the differential stage is used with a capacitive load, it operates as an operational transconductance amplifier (OTA), which is a differential voltage-controlled current source. Figure 4.8(a) shows the symbol of a single-ended OTA. A single-stage transconductance amplifier essentially features a dominant-pole behavior and can be modeled as a first-order system. Its small-signal equivalent model is depicted in Figure 4.8(b), where the current delivered to the overall output load, or the parallel combination of the output resistor and capacitor, is directly related to the input differential voltage. The proportionality factor, g_m, between the output current and input differential voltage is known as the transconductance. In the frequency domain, the voltage-gain transfer function can be written as

$$A(s) = \frac{V_0(s)}{V_i(s)} = g_{m1} Z_0 \tag{4.60}$$

where $Z_0 = r_0 \parallel 1/(sC_0)$, $s = j\omega$ and $\omega = 2\pi f$. It can then be found that

$$A(s) = \frac{V_0(s)}{V_i(s)} = \frac{A_0}{1 + s/\omega_c} \tag{4.61}$$

where the dc gain is of the form, $A_0 = g_{m1} r_0$, and $\omega_c = 1/r_0 C_0$ is the 3-dB cutoff frequency. The *gain-bandwidth product* is defined as

$$\text{GBW} = A_0 \omega_c = g_{m1}/C_0 \tag{4.62}$$

Note that GBW is measured in rad/s, while the unit of $GBW/(2\pi)$ is Hz. In practice, ω_c is generally very small and $\omega \gg \omega_c$. Hence,

$$A(s) = \frac{V_0(s)}{V_i(s)} \simeq \frac{\omega_t}{s} \tag{4.63}$$

where ω_t = GBW. Because $|A(j\omega_t)| = 1$, the parameter ω_t is called the *unity-gain frequency* or *transition frequency.* The basic differential stage can exhibit a gain-bandwidth product in the megahertz range, but its gain is generally less than 40 dB.

Hand analysis of the differential stage in the frequency domain using a high-frequency equivalent model of transistors is quite tedious when no simplifying assumption is made. Thus, an insight into the frequency response is usually gained using computer-aided design programs, such as SPICE and Spectre.

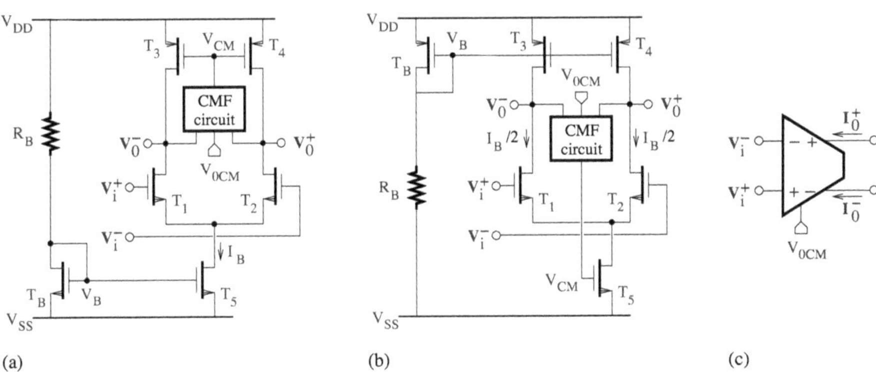

FIGURE 4.9
Circuit diagram of fully differential amplifiers with (a) output and (b) input common-mode control; (c) symbol of a fully differential amplifier.

The output swing and linearity can be further improved using fully differential amplifier structures, as shown in Figure 4.8(a) and (b), which feature differential input and output voltages. The transistors T_1 and T_2, and T_3 and T_4, are assumed to be matched. The common-mode gain of the differential input stage is approximately $A_c \simeq -g_5/(2g_l)$, where g_l is the conductance of the load transistor T_3 or T_4. In general, the conductance, g_5, of the transistor T_5, or say the output conductance of the bias current source, is small and A_c has a very low value. Therefore, only limited control can be exerted by the common-mode input voltage on the common-mode output voltage through the local feedback for common-mode signals. A common-mode feedback (CMF) circuit is then required to regulate the output common-mode voltage to the desired value, V_{OCM}. Furthermore, the CMF circuit can suppress the common-mode variations at the output, thereby preventing a change in the operation region for the load transistors due to IC process variations and loading conditions. Ideally, the differential output voltage is not affected by common-mode signals and the small-signal differential gain is approximately given by

$$A_d = \frac{v_0}{v_i} \simeq g_{m1} r_0 \tag{4.64}$$

where $r_0 \simeq 1/(g_1 + g_3)$, $v_i = v_i^+ - v_i^-$, and $v_0 = v_0^+ - v_0^-$. On the other hand, the voltage at each output node can swing symmetrically around the output common-mode level set by the CMF circuit. The symbol of a fully differential amplifier is shown in Figure 4.8(c). For simplification, the V_{0CM} node is generally omitted. A fully differential amplifier has the advantage over its single-ended counterpart of rejecting the common-mode noise. It should also be noted that even-order nonlinearities are cancelled at the outputs of a differential circuit that is balanced or whose both sides are electrically similar and symmetrical with respect to the ground.

4.1.6 Offset voltage

Ideally, if both inputs of a differential amplifier stage are connected to the same voltage, the differential output voltage should be zero. In practice, due to device mismatches, a small voltage that is known as the input offset voltage must be applied in series with one of the inputs to bring the differential output to zero.

Let us consider the differential amplifier stage shown in Figure 4.10. The offset voltage is primarily caused by mismatches of input transistors, T_1 and T_2, and current mirror transistors, T_3 and T_4. In the offset voltage analysis, it is assumed that all transistors operate in the saturation region.

Using the drain current expressions of transistors T_1 and T_2,

$$I_{D_1} = K_1(V_{GS_1} - V_{T_{n_1}})^2 \tag{4.65}$$

$$I_{D_2} = K_2(V_{GS_2} - V_{T_{n_2}})^2 \tag{4.66}$$

we can obtain

$$V_{GS_1} = V_{T_{n_1}} + \sqrt{\frac{I_{D_1}}{K_1}} \tag{4.67}$$

$$V_{GS_2} = V_{T_{n_2}} + \sqrt{\frac{I_{D_2}}{K_2}} \tag{4.68}$$

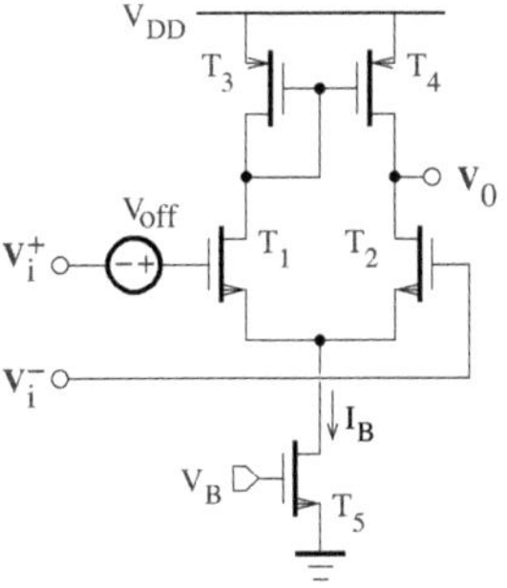

FIGURE 4.10
Circuit diagram of a differential amplifier stage with the input-referred offset voltage.

Let

$$\triangle I_D = I_{D_1} - I_{D_2} \tag{4.69}$$

and

$$I_B = I_{D_1} + I_{D_2} \tag{4.70}$$

where I_B is the bias current of the input differential transistor pair. Combining (4.67)–(4.70), the input-referred offset voltage is given by

$$V_{off} = V_{GS_1} - V_{GS_2} \tag{4.71}$$

$$= \triangle V_{T_n} + \sqrt{\frac{I_B}{2K_1}\left(1 + \frac{\triangle I_D}{I_B}\right)} - \sqrt{\frac{I_B}{2K_2}\left(1 - \frac{\triangle I_D}{I_B}\right)} \tag{4.72}$$

where $\triangle V_{T_n} = V_{T_{n_1}} - V_{T_{n_2}}$. For small variations of the drain currents, $\triangle I_D/I_B \ll 1$, we can exploit $(1 + \triangle I_D/I_B)^{1/2} \simeq 1 + \triangle I_D/(2I_B)$ and obtain

$$V_{off} \simeq \triangle V_{T_n} + \left(\sqrt{\frac{I_B}{2K_1}} + \sqrt{\frac{I_B}{2K_2}}\right)\frac{\triangle I_D}{2I_B} + \sqrt{\frac{I_B}{2K_1}} - \sqrt{\frac{I_B}{2K_2}} \tag{4.73}$$

where $K_2 = K_1 + \triangle K_1$. Using $(1 + \triangle K_1/K_1)^{-1/2} \simeq 1 - \triangle K_1/(2K_1)$ when $\triangle K_1/K_1 \ll 1$ gives the first-order approximation

$$V_{off} \simeq \triangle V_{T_n} + \frac{\triangle I_D}{I_B}\sqrt{\frac{I_B}{2K_1}} + \frac{\triangle K_1}{2K_1}\sqrt{\frac{I_B}{2K_1}} \tag{4.74}$$

Considering the current mirror transistors T_3 and T_4, we have

$$I_{D_3} = K_3(V_{SG_3} - |V_{T_{p3}}|)^2 \tag{4.75}$$

$$I_{D_4} = K_4(V_{SG_4} - |V_{T_{p4}}|)^2 \tag{4.76}$$

and

$$V_{SG_3} = V_{SG_4} \tag{4.77}$$

Substituting (4.75) and (4.76) into (4.77), it can be shown that

$$\sqrt{\frac{I_{D_3}}{K_3}} - \sqrt{\frac{I_{D_4}}{K_4}} = \triangle|V_{T_p}| \tag{4.78}$$

where $\triangle|V_{T_p}| = |V_{T_{p4}}| - |V_{T_{p3}}|$ and $K_4 = K_3 + \triangle K_3$. For small variations of the transconductance parameters, we have $\triangle K_3/K_3 \ll 1$, and $(1 + \triangle K_3/K_3)^{-1/2} \simeq 1 - \triangle K_3/(2K_3)$ so that (4.78) becomes

$$\sqrt{I_{D_3}} - \sqrt{I_{D_4}} \simeq \sqrt{K_3}\left(\triangle|V_{T_p}| - \frac{1}{2}\frac{\triangle K_3}{K_3}\sqrt{\frac{I_{D_4}}{K_3}}\right) \tag{4.79}$$

Due to the fact that $I_{D_3} = I_{D_1}$ and $I_{D_4} = I_{D_2}$, the drain currents of transistors T_3 and T_4 can be expressed as,

$$I_{D_3} = \frac{I_B}{2}\left(1 + \frac{\triangle I_D}{I_B}\right) \tag{4.80}$$

$$I_{D_4} = \frac{I_B}{2}\left(1 - \frac{\triangle I_D}{I_B}\right) \tag{4.81}$$

Recalling that $(1 \pm \triangle I_D/I_B)^{1/2} \simeq 1 \pm \triangle I_D/(2I_B)$ when the drain current variations remain small, namely $\triangle I_D/I_B \ll 1$, (4.79) can be put into the form,

$$\frac{\triangle I_D}{I_B} \simeq \sqrt{\frac{2K_3}{I_B}}\left(\triangle|V_{T_p}| - \frac{\triangle K_3}{2K_3}\sqrt{\frac{I_B}{2K_3}}\right) \tag{4.82}$$

The input-referred offset voltage due to transistor mismatches is then given by

$$V_{off} \simeq \triangle V_{T_n} + \triangle|V_{T_p}|\sqrt{\frac{K_3}{K_1}} + \left(\frac{\triangle K_1}{2K_1} - \frac{\triangle K_3}{2K_3}\right)\sqrt{\frac{I_B}{2K_1}} \tag{4.83}$$

Its variance can be written as

$$\sigma^2_{V_{off}} \simeq \sigma^2_{\triangle V_{T_n}} + \sigma^2_{\triangle|V_{T_p}|}\left(\frac{K_3}{K_1}\right) + \left(\frac{\sigma^2_{\triangle K_1^2}}{K_1^2} + \frac{\sigma^2_{\triangle K_3^2}}{K_3^2}\right)\frac{I_B}{8K_1} \tag{4.84}$$

The standard deviations of the threshold voltages and transconductance parameter between two identically drawn transistors can be predicted using the next formula [2],

$$\sigma^2_{V_{T_n}} = \frac{A^2_{V_{T_n}}}{W \cdot L} + S^2_{V_{T_n}} D^2 \tag{4.85}$$

$$\sigma^2_{|V_{T_p}|} = \frac{A^2_{|V_{T_p}|}}{W \cdot L} + S^2_{|V_{T_p}|} D^2 \tag{4.86}$$

and

$$\frac{\sigma^2_{\Delta K_j}}{K_j^2} = \frac{A^2_{K_j}}{W \cdot L} + S^2_{K_j} D^2 \quad \text{for} \quad j = 1, 3 \tag{4.87}$$

where W and L are the width and length of the transistor pair, D is the on-chip distance between the matching transistor, and $A_{V_{T_n}}$, $A_{|V_{T_p}|}$ and A_{K_j} are the area proportionality constants for the (n-channel and p-channel) threshold voltages and transconductance parameter, respectively; $S_{V_{T_n}}$, $S_{|V_{T_p}|}$, and S_{K_j} represent the variations of the (n-channel and p-channel) threshold voltages and transconductance parameter with the spacing, respectively.

Typical values of the amplifier offset voltage, V_{off}, range from one mV up to a few tens of mV.

4.1.7 Noise

The amplifier noise is a random signal characterized by its average mean-square value. It determines the lower bound of the dynamic range and is primarily due to thermal and Flicker noises of transistors. Thermal noise is associated with fluctuations of transistor charge carriers (electrons or holes) due to temperature. Flicker noise, which is also called $1/f$ noise, is related to the transistor *dc* current. Figure 4.11 shows equivalent representations of MOS transistor noise sources. The $1/f$ noise voltage at the gate is transformed into an equivalent current noise source that is added in parallel between the drain and source nodes.

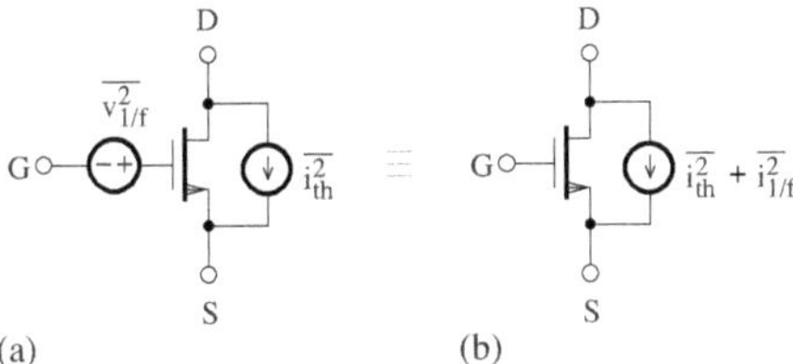

FIGURE 4.11
Equivalent representations of MOS transistor noise sources.

The circuit diagram of the differential amplifier stage with transistor noise sources is depicted in Figure 4.12(a). The noise current source connected between the drain and source terminals of the transistor T_j is of the form,

$$\overline{i^2_{n_j}} = \overline{i^2_{th,j}} + \overline{i^2_{1/f,j}} \tag{4.88}$$

where the power spectral densities of the thermal and Flicker noises are given by

$$\overline{i^2_{th,j}} = 4kT\gamma g_{m_j} \tag{4.89}$$

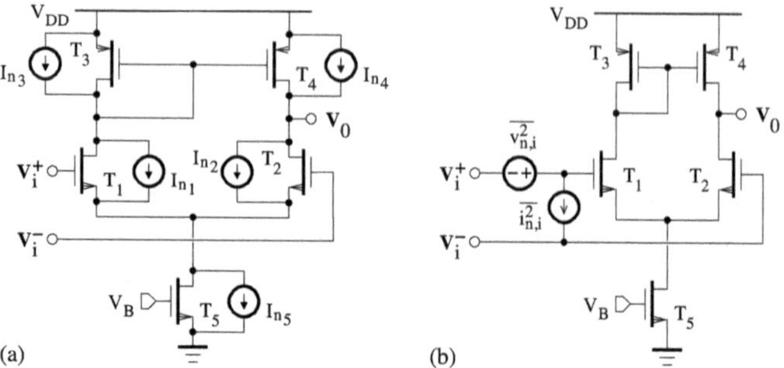

FIGURE 4.12
Circuit diagrams of differential amplifier stages with (a) transistor noise sources and (b) the input-referred noise sources.

and

$$\overline{i^2_{1/f,j}} = \frac{K_f}{C_{ox}L^2}\frac{I_{D_j}}{f} \tag{4.90}$$

respectively. Note that k is Boltzmann's constant, T is the temperature in Kelvin, g_{m_j} denotes the transistor transconductance, γ is the bias-dependent noise parameter, I_{D_j} is the drain current, f is the frequency, K_f is a process-dependent parameter, L is the transistor length, and C_{ox} represents the oxide capacitance per unit area.

The noise corner frequency is defined as the frequency, f_{co}, at which the thermal noise density equals the Flicker noise density, that is,

$$4kT\gamma g_{m_j} = \frac{K_f}{C_{ox}L^2}\frac{I_{D_j}}{f_{co}} \tag{4.91}$$

and

$$f_{co} = \frac{K_f}{4kT\gamma}\frac{1}{L^2}\frac{1}{(g_{m_j}/I_{D_j})} \tag{4.92}$$

The contribution of the transistor T_1 (and T_2) to the input-referred noise voltage can be expressed as, $\overline{i^2_{n_1}}/g^2_{m_1}$. For the transistor T_3 (and T_4), we have $\overline{i^2_{n_3}}/g^2_{m_1}$. Assuming that the amplifier paths are well-matched, the noise contribution of the tail current source T_5 can be neglected. The total input-referred noise is then given by

$$\overline{v^2_{n,i}} = 2(\overline{i^2_{n_1}} + \overline{i^2_{n_3}})/g^2_{m_1} \tag{4.93}$$

At frequencies beyond f_{co}, the contribution of Flicker noise can be ignored, while the thermal noise becomes dominant. The total input-referred noise is then reduced to

$$\overline{v^2_{n,i}} \simeq 2\frac{4kT\gamma}{g_{m_1}}\left(1 + \frac{g_{m_3}}{g_{m_1}}\right) \tag{4.94}$$

At high frequencies, the noise is white (or flat) because its power spectral density does not vary with frequency anymore.

In general, the circuit diagram of the differential amplifier stage with input-referred noise sources can be represented as shown in Figure 4.12(b). The units of the noise voltage and current power spectral densities are V^2/Hz and A^2/Hz, respectively.

The input-referred noise current of the differential MOS transistor stage can be determined by opening the input nodes and expressing the ratio of the spectral density of the total noise voltage at the output node and the square of the transimpedance gain in terms of the circuit noise contributions. Typically, it is primarily dependent on the shot noise of the input bias current and can be assumed to be negligible due to its value, which is generally in the $fA/\sqrt{Hz}$ range.

4.1.8 Operational amplifier

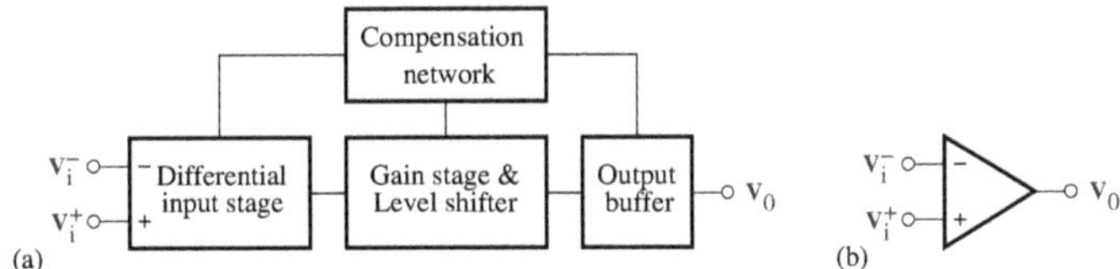

FIGURE 4.13
(a) Block diagram and (b) symbol of a single-ended operational amplifier.

Operational transconductance amplifiers, which are equivalent to a voltage-controlled current source, are generally based on single-stage structures, while additional gain stages and an output buffer are required for the design of operational amplifiers operating as a voltage-controlled voltage source. A high-performance operational amplifier should provide a high input impedance, a high open-loop gain, a large CMRR, a low dc offset voltage, a low noise, and a low output impedance. Figure 4.13 shows the block diagram and symbol of a single-ended operational amplifier. The output of the differential stage is supplied to additional gain stages in order to meet the high gain requirement. A level shifter is needed whenever the *dc* voltage difference introduced between the input and output voltages of a stage should be cancelled. An output buffer with unity gain will be necessary if the amplifier is supposed to drive a resistive load. The use of a compensation network is necessary to avoid the conditions leading to instability when the amplifier operates with a feedback.

A differential-to-single-ended converter can be implemented, as depicted in Figure 4.14. Using the superposition theorem and voltage division principle, it can be shown that

$$V^+ = \frac{R_4}{R_3 + R_4} V_i^+ \tag{4.95}$$

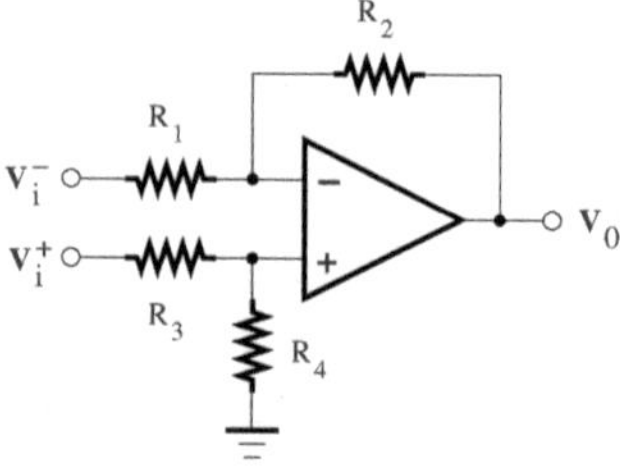

FIGURE 4.14
Circuit diagram of a differential-to-single-ended converter.

and

$$V^- = \frac{R_2}{R_1 + R_2} V_i^- + \frac{R_1}{R_1 + R_2} V_0 \tag{4.96}$$

where $V_0 = A(V^+ - V^-)$. Solving the system of Equations (4.95) and (4.96) gives

$$V_0 = \frac{\dfrac{R_1 + R_2}{R_1} \dfrac{R_4}{R_3 + R_4} V_i^+ - \dfrac{R_2}{R_1} V_i^-}{1 + \dfrac{1}{A} \dfrac{R_1 + R_2}{R_1}} \tag{4.97}$$

Assuming that $R_2/R_1 = R_4/R_3 = k$, we obtain

$$V_0 = \frac{k(V_i^+ - V_i^-)}{1 + \dfrac{1+k}{A}} \tag{4.98}$$

In practice, the amplifier gain, A, is very high and does not affect the output voltage. However, the achievable common-mode rejection ratio may be limited by the resistor mismatches.

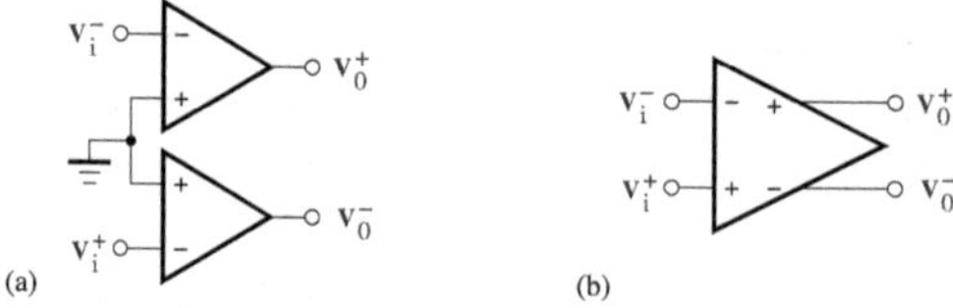

FIGURE 4.15
Symbols of (a) pseudo-differential and (b) fully differential operational amplifiers.

For low-voltage applications, the signal swing may be increased using a pseudo-differential structure based on two single-ended amplifiers with their noninverting inputs connected to the ground, as illustrated in Figure 4.15(a).

However, due to the lack of a common-mode feedback path, the amplifier linearity may be affected by any imbalance between the signal paths. Another solution is to design the operational amplifier with a differential configuration. Figure 4.15(b) shows the symbol of a fully differential operational amplifier.

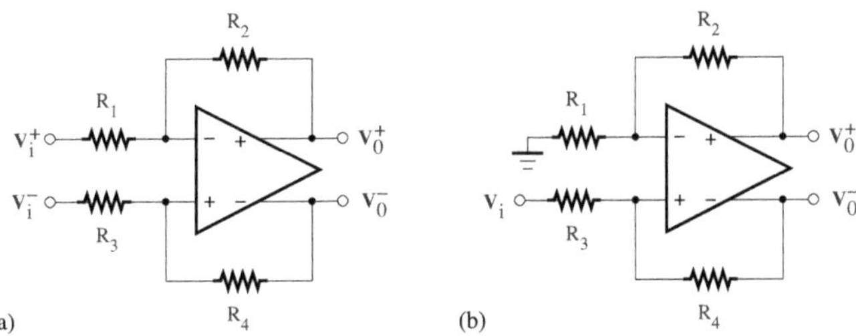

FIGURE 4.16
(a) Differential amplifier with negative feedback; (b) single-ended-to-differential converter.

More insight into the operation of a fully differential amplifier is gained by analyzing the structure shown in Figure 4.16(a). Here, the amplifier output voltage can be expressed as

$$V_0^+ - V_0^- = A(V^+ - V^-) \tag{4.99}$$

where A is the amplifier gain, and V^+ and V^- are the voltage at the noninverting and inverting input nodes, respectively. Using the principles of superposition and voltage division, we find

$$V^- = \frac{R_2}{R_1 + R_2} V_i^+ + \frac{R_1}{R_1 + R_2} V_0^+ \tag{4.100}$$

and

$$V^+ = \frac{R_4}{R_3 + R_4} V_i^- + \frac{R_3}{R_3 + R_4} V_0^- \tag{4.101}$$

Let $\alpha = R_3/(R_3 + R_4)$ and $\beta = R_1/(R_1 + R_2)$. Combining Equations (4.100) and (4.101) with Equation (4.99), we obtain

$$(1 + \beta A)V_0^+ - (1 + \alpha A)V_0^- = A[(1 - \alpha)V_i^- - (1 - \beta)V_i^+] \tag{4.102}$$

The voltage at each output node is independently defined by setting the output common-mode voltage to a given level, V_{0CM}, that is,

$$\frac{V_0^+ + V_0^-}{2} = V_{0CM} \tag{4.103}$$

Solving Equations (4.102) and (4.103) gives

$$V_0^- = \frac{-(1 - \alpha)V_i^- + (1 - \beta)V_i^+ + 2(\beta + 1/A)V_{0CM}}{(\alpha + \beta)\{1 + 2/[(\alpha + \beta)A]\}} \tag{4.104}$$

and

$$V_0^+ = \frac{(1-\alpha)V_i^- - (1-\beta)V_i^+ + 2(\alpha + 1/A)V_{0CM}}{(\alpha+\beta)\{1 + 2/[(\alpha+\beta)A]\}} \tag{4.105}$$

The differential output voltage is then given by

$$V_0 = \frac{2[(1-\alpha)V_i^- - (1-\beta)V_i^+ + (\alpha-\beta)V_{0CM}]}{(\alpha+\beta)\{1 + 2/[(\alpha+\beta)A]\}} \tag{4.106}$$

where $V_0 = V_0^+ - V_0^-$. The differential amplifier is generally designed with identical resistors R_1 and R_3, and R_2 and R_4. Hence, $\alpha = \beta$ and the differential output voltage is reduced to

$$V_0 = \frac{-(1-\alpha)V_i}{\alpha[1 + 1/(\alpha A)]} \tag{4.107}$$

where $V_i = V_i^+ - V_i^-$. In the ideal case, the gain of the amplifier is very high so that $1 + 1/(\alpha A) \simeq 1$. The differential input voltage is then amplified by the factor $(1-\alpha)/\alpha = R_2/R_1$.

Singled-ended signals can be converted to differential signals using the circuit shown in Figure 4.16(b). It can be shown that

$$V_0 = \frac{2[(1-\alpha)V_i + (\alpha-\beta)V_{0CM}]}{(\alpha+\beta)\{1 + 2/[(\alpha+\beta)A]\}} \tag{4.108}$$

where $V_i^- = V_i$, $V_i^+ = 0$, and $V_0^+ - V_0^- = V_0$. To prevent the output common-mode voltage, V_{0CM}, from affecting the output differential voltage, the resistors R_1 and R_3, and R_2 and R_4, should be matched so that $\alpha = \beta$.

4.2 Linearization techniques for transconductors

In order to increase the input dynamic range of the basic differential stage shown in Figure 4.17(a), various linearization techniques can be used. Consider a differential stage consisting of two source-connected transistors operating in the saturation region, where the lowest internal channel resistance can be achieved. Assuming that the transistors are matched, the drain currents are given by

$$I_{D_1} = K(V_{GS_1} - V_T)^2 \tag{4.109}$$

and

$$I_{D_2} = K(V_{GS_2} - V_T)^2 \tag{4.110}$$

The output current of the differential stage can be expressed as

$$\triangle i = I_{D_1} - I_{D_2} = K(V_{GS_1} + V_{GS_2} - 2V_T)(V_{GS_1} - V_{GS_2}) \tag{4.111}$$

where the differential input voltage is equal to the difference of the gate-source voltages, while the linearity of the transconductance is achieved by maintaining the term $V_{GS_1} + V_{GS_2} - 2V_T$, or the sum of the gate-source voltages, constant. Approaches used to linearize the transconductance can then be implemented using a source degeneration resistor or floating dc voltages. In the case of a transconductance realized by transistors operating in the triode region, where the drain current is of the form

$$I_D = K[2(V_{GS} - V_T)V_{DS} - V_{DS}^2] \tag{4.112}$$

the drain-source voltage should be maintained constant to improve the linear range of the differential stage.

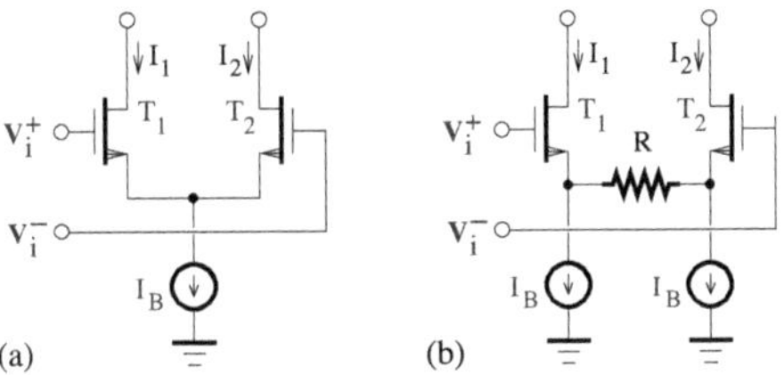

FIGURE 4.17
(a) Basic differential stage; (b) differential stage with resistor source degeneration.

A differential stage with resistor source degeneration is depicted in Figure 4.17(b). The currents flowing through the transistors T_1 and T_2 can be written as

$$I_{D_1} = I_B + \triangle i = K(V_{GS_1} - V_T)^2 \tag{4.113}$$

and

$$I_{D_2} = I_B - \triangle i = K(V_{GS_2} - V_T)^2 \tag{4.114}$$

Applying Kirchhoff's voltage law to the input loop involving T_1, R, and T_2 gives

$$V_i^+ - V_{GS_1} - R\triangle i + V_{GS_2} - V_i^- = 0 \tag{4.115}$$

where

$$V_{GS_1} = V_T + \sqrt{\frac{I_B + \triangle i}{K}} \tag{4.116}$$

and

$$V_{GS_2} = V_T + \sqrt{\frac{I_B - \triangle i}{K}} \tag{4.117}$$

Hence,

$$V_i - R\triangle i = \sqrt{\frac{I_B}{K}}\left[\left(1 + \frac{\triangle i}{I_B}\right)^{1/2} - \left(1 - \frac{\triangle i}{I_B}\right)^{1/2}\right] \tag{4.118}$$

where $V_i = V_i^+ - V_i^-$. Assuming that $\Delta i/I_B \ll 1$, $(1+\Delta i/I_B)^{1/2} \simeq 1+\Delta i/2I_B$ and $(1-\Delta i/I_B)^{1/2} \simeq 1-\Delta i/2I_B$. Thus,

$$\Delta i \simeq \frac{g_m}{1+g_m R} V_i \tag{4.119}$$

where $g_m = \sqrt{KI_B}$.

The linearity can be further improved using amplifiers to sense the input voltages, as shown in Figure 4.18(a). Ideally, the difference between the voltages at the noninverting and inverting nodes of the amplifier is zero, and the input voltages are directly applied across the resistor R due to the feedback path. Hence,

$$\Delta i = V_i/R \tag{4.120}$$

Note that it is generally required to optimize the bandwidth of these extra amplifiers so as not to reduce the speed of the overall circuit. Furthermore, the accuracy of the transconductance, which is inversely proportional to the resistance R, may be affected by process variations.

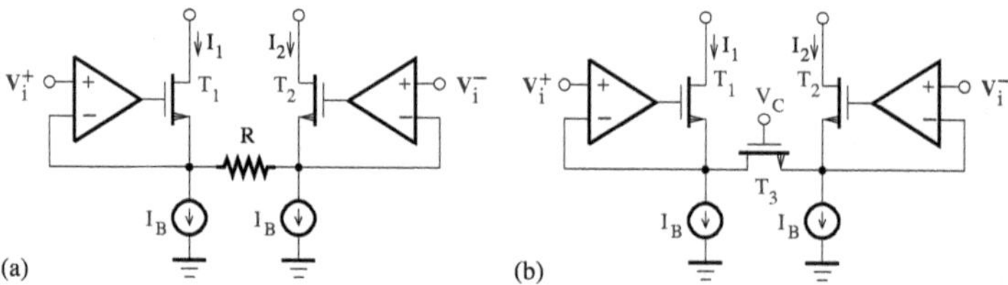

FIGURE 4.18
(a) Improved differential stage with a resistor source degeneration; (b) differential stage with a single transistor source degeneration.

In the case of the differential stage shown in Figure 4.18(b), a transistor operating in the triode region is used to realize the source degeneration resistor [3], whose value can be controlled by means of the drain voltage. The transistors T_1 and T_2 are assumed to operate in the saturation region. They play the role of source followers driving T_3, which is biased in the triode region. Let V_{ICM} and V_C be the dc component of the input voltages and the gate voltage of T_3, respectively. Assuming that $V_i^+ = V_{ICM} + V_i/2$ and $V_i^- = V_{ICM} - V_i/2$, the current flowing through T_3 can be expressed as

$$\Delta i = K[2(V_{GS_3} - V_T)V_{DS_3} - V_{DS_3}^2] \tag{4.121}$$

where

$$V_{GS_3} = V_C - V_{ICM} - \frac{V_i}{2} \tag{4.122}$$

$$V_{DS_3} = V_i \tag{4.123}$$

and $V_i = V_i^+ - V_i^-$ is the differential input voltage. Hence,

$$\triangle i = 2K[(V_C - V_{ICM} - V_T)V_i - V_i^2] \tag{4.124}$$

For large values of the differential input voltage, the transconductance linearity can be degraded due to the significant contribution associated with the V_i^2 term.

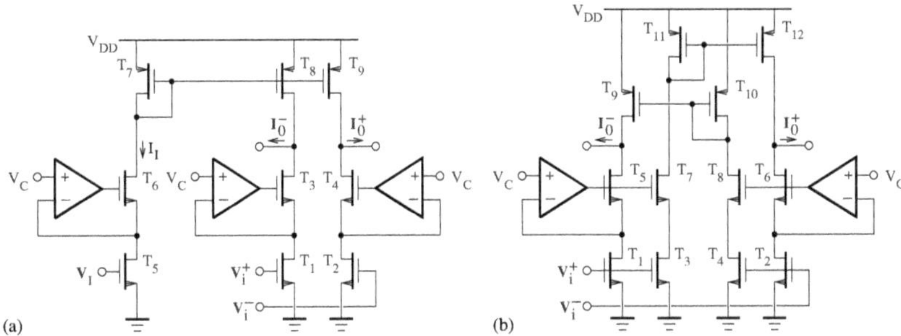

FIGURE 4.19
Differential stages with an active cascode configuration: (a) Basic structure and (b) implementation with improved linearity.

A linear transconductance characteristic can also be provided by transistors operating in the triode region with constant drain-source voltages. Figure 4.19(a) shows the basic structure of a differential stage with an active cascode configuration [4]. The amplifier feedback helps maintain constant the drain-source voltage of transistors T_1 and T_2, thereby guaranteeing the accuracy of the resulting transconductance. Assuming that $V_i^+ = V_{ICM} + V_i/2$ and $V_i^- = V_{ICM} - V_i/2$, the drain currents of the identical transistors T_1, T_2, and T_5 in the triode region can be expressed as

$$I_{D_1} = K[2(V_{GS_1} - V_T)V_{DS_1} - V_{DS_1}^2] \tag{4.125}$$

$$I_{D_2} = K[2(V_{GS_2} - V_T)V_{DS_2} - V_{DS_2}^2] \tag{4.126}$$

and

$$I_I = K[2(V_{GS_5} - V_T)V_{DS_5} - V_{DS_5}^2] \tag{4.127}$$

where $K = (1/2)\mu C_{ox}(W/L)$, $V_{GS_1} = V_{ICM} + V_i/2$, $V_{GS_2} = V_{ICM} - V_i/2$, $V_{GS_5} = V_{ICM}$, and $V_{DS_1} = V_{DS_2} = V_{DS_5} = V_C$. The output currents are, respectively, given by

$$I_0^- = I_I - I_{D_1} = -KV_CV_i \tag{4.128}$$

and

$$I_0^+ = I_I - I_{D_2} = KV_CV_i \tag{4.129}$$

The resulting transconductance is of the form

$$g_m = \triangle i/V_i = 2KV_C \tag{4.130}$$

where $\triangle i = I_0^+ - I_0^-$, and it can be tuned by varying the level of the control voltage V_C. For the transistor operation in the triode region, it is required that $V_{DS} \leq V_{GS} - V_T$, and in the worst case,

$$V_C \leq V_{ICM} - V_T - V_i/2 \tag{4.131}$$

In practice, the transconductance linearity can be degraded due to variations in the mobility, μ, which exhibits a nonlinear dependence with respect to the gate-source voltage, that is, $\mu = f(V_{GS})$, where f is a nonlinear function. To overcome this drawback, the input transistor pairs can be duplicated as shown in Figure 4.19(b) [5]. Assuming that the transistors $T_1 - T_4$ are matched and operate in the triode region, we have

$$I_{D_1} = I_{D_3} = K[2(V_{ICM} + V_i/2 - V_T)V_C - V_C^2] \tag{4.132}$$

and

$$I_{D_2} = I_{D_4} = K[2(V_{ICM} - V_i/2 - V_T)V_C - V_C^2] \tag{4.133}$$

The current mirrors $T_9 - T_{10}$ and $T_{11} - T_{12}$ are required to respectively combine the currents I_{D_4} and I_{D_1}, and I_{D_3} and I_{D_2}, resulting in

$$I_0^- = I_{D_4} - I_{D_1} = -KV_CV_i \tag{4.134}$$

and

$$I_0^+ = I_{D_3} - I_{D_2} = KV_CV_i \tag{4.135}$$

Here, the solution adopted for the common-mode rejection at the output nodes relies on using an extra feedforward transconductor, which can only provide a dc signal equal in magnitude to the common-mode component to be subtracted. Note that the effect of IC process variations affecting the drain currents of the input transistors can also be cancelled in the same way.

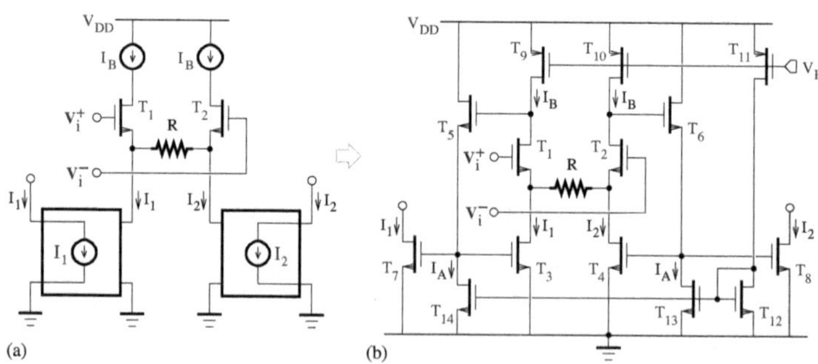

FIGURE 4.20
Differential stage with a resistor source degeneration: (a) Operation principle, (b) implementation.

An alternative structure for the realization of a differential stage with resistor source degeneration is derived by maintaining the same voltage between the gate and source of the input transistors, irrespective of the input voltage magnitude [6]. The principle and implementation of this approach are illustrated in Figures 4.20(a) and (b). The transistors operate in the saturation region. The biasing of each of the differential input transistors, T_1 and T_2, with the identical current source of the value, I_B, and the feedback path realized by T_5 and T_6, force the gate-source voltages of transistors T_1 and T_2 to be constant and equal, that is,

$$V_{GS_1} = V_{GS_2} \tag{4.136}$$

The transistors $T_{11} - T_{14}$ are required to set the bias current, I_A, for the transistors T_5 and T_6. The input voltage is level-shifted and applied across the resistor R, and the current flowing through the resistor R is then given by

$$\triangle i = V_i/R \tag{4.137}$$

where $V_i = V_i^+ - V_i^-$. The drain currents of the transistors T_3 and T_4 are, respectively, of the form

$$I_{D_3} = I_1 = I_B - \triangle i \tag{4.138}$$

and

$$I_{D_4} = I_2 = I_B + \triangle i \tag{4.139}$$

By connecting the gates of transistors T_7 and T_8 to the gates of transistors T_3 and T_4, respectively, we get $I_1 = I_{D_3}$ and $I_2 = I_{D_4}$.

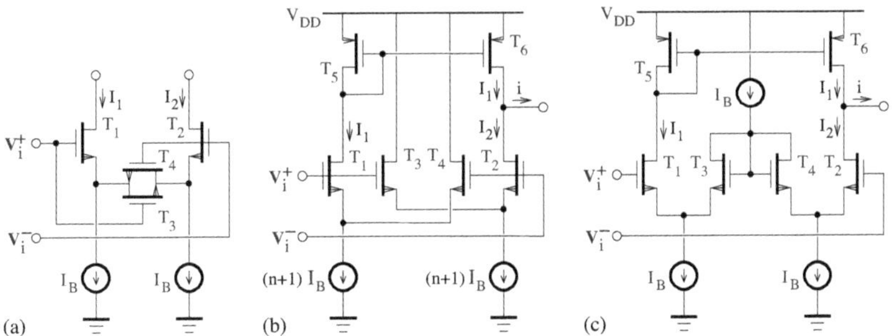

FIGURE 4.21
Differential stage with active source degeneration: (a) Transistor in the triode region, (b) cross-coupled configuration, (c) transistor in the saturation region.

A differential stage with active source degeneration is shown in Figure 4.21(a) [33]. Let V_{ICM} and V_C be the dc component of the input voltages and the gate voltage of T_3, respectively; $V_i^+ = V_{ICM} + V_i/2$; and $V_i^- = V_{ICM} - V_i/2$. The transistors T_3 and T_4, which are identical and biased in the triode region, are equivalent to a parallel configuration of resistors, whose conductances are of the form

$$g_{ds_3} = I_{D_3}/V_{DS_3} = K[2(V_{GS_3} - V_T) - V_{DS_3}] \tag{4.140}$$

and

$$g_{ds_4} = I_{D_4}/V_{DS_4} = K[2(V_{GS_4} - V_T) - V_{DS_4}] \tag{4.141}$$

where

$$V_{GS_3} = V_i + V_{GS_2} \tag{4.142}$$
$$V_{GS_4} = -V_i + V_{GS_1} \tag{4.143}$$
$$V_{DS_3} = V_i - V_{GS_1} + V_{GS_2} = -V_{DS_4} \tag{4.144}$$

and $V_i = V_i^+ - V_i^-$ is the differential input voltage. The transconductance of the differential stage can then be written as

$$g_m = \frac{g_{m_1}}{1 + g_{m_1}/(g_{ds_3} + g_{ds_3})} \tag{4.145}$$

where

$$g_{m_1} = \sqrt{K_1 I_B} = K_1\sqrt{\frac{I_B}{K_1}} \tag{4.146}$$

and

$$g_{ds_3} + g_{ds_4} = 2K_3(V_{GS_1} + V_{GS_2} - 2V_T) \tag{4.147}$$

Assuming that the transistors T_1 and T_2 are matched and operate in the saturation region, and that $\triangle i \ll I_B$, we have

$$V_{GS_1} = V_T + \sqrt{\frac{I_B + \triangle i}{K_1}} \simeq V_T + \sqrt{\frac{I_B}{K_1}}\left(1 + \frac{\triangle i}{2I_B}\right) \tag{4.148}$$

$$V_{GS_2} = V_T + \sqrt{\frac{I_B - \triangle i}{K_1}} \simeq V_T + \sqrt{\frac{I_B}{K_1}}\left(1 - \frac{\triangle i}{2I_B}\right) \tag{4.149}$$

and Equation (4.147) becomes

$$g_{ds_3} + g_{ds_4} \simeq 4K_3\sqrt{\frac{I_B}{K_1}} \tag{4.150}$$

Hence,

$$g_m \simeq \frac{g_{m_1}}{1 + \dfrac{K_1}{4K_3}} = \frac{g_{m_1}}{1 + \dfrac{(W_1/L_1)}{4(W_3/L_3)}} \tag{4.151}$$

where W_1/L_1 and W_3/L_3 denote the gate width-to-length ratios of the transistors T_1 and T_3, respectively.

The input dynamic range of a differential stage can also be improved using two floating dc voltage sources to keep constant the sum of gate-source voltages of input transistors. The differential stage implementation shown in Figure 4.21(b) is based on cross-coupled transistors [8]. Using Kirchhoff's voltage law, the input voltage can be related to the gate-source voltages of transistors $T_1 - T_4$ as

$$V_i = V_{GS_1} - V_{GS_4} = V_{GS_3} - V_{GS_2} \tag{4.152}$$

where

$$V_{GS_j} = V_T + \sqrt{\frac{I_{D_j}}{K_j}} \tag{4.153}$$

and $V_i = V_i^+ - V_i^-$ is the differential input voltage. The transistor pairs T_1 and T_4, and T_3 and T_2, behave as differential stages. Assuming that $K_1 = K_2 = K$ and $K_1 = K_2 = K/n$, and taking into account the fact that

$$I_{D_1} = I_1 = I_B + i_1 \quad I_{D_2} = I_2 = I_B - i_2 \quad I_{D_3} = nI_B + i_2 \quad I_{D_4} = nI_B - i_1 \tag{4.154}$$

Equations (4.152) and (4.153) can be solved for i_1 and i_2. That is,

$$i_1 = \gamma K V_i^2 + \frac{\alpha}{2}\sqrt{KI_B}V_i\sqrt{1 - \eta K V_i^2/I_B} \tag{4.155}$$

and

$$i_2 = -\gamma K V_i^2 + \frac{\alpha}{2}\sqrt{KI_B}V_i\sqrt{1 - \eta K V_i^2/I_B} \tag{4.156}$$

The output current i is then given by

$$i = i_1 + i_2 = \alpha\sqrt{KI_B}V_i\sqrt{1 - \eta K V_i^2/I_B} \tag{4.157}$$

where $\alpha = 4n/(n+1)$, $\eta = n/(n+1)^2$, and $\gamma = n(n-1)/(n+1)^2$. The input range is characterized by $|V_i| \leq \sqrt{I_B/(\eta K)}$. Provided the transistors are accurately matched, the linearity is improved as the value of η is reduced by increasing n, or correspondingly, the bias current.

In the case of the differential stage depicted in Figure 4.21(c), the transconductance linearization is achieved using two differential transistor pairs, T_1 and T_3, and T_2 and T_4, connected in series [9]. The voltages V_i^+ and V_i^- are applied to the gates of transistors T_1 and T_2, respectively. Because each transistor pair is driven by the bias current I_B and the diode-connected transistors T_3 and T_4 are wired to the current source I_B, we can write

$$I_B = I_{D_1} + I_{D_3} \tag{4.158}$$

$$= I_{D_2} + I_{D_4} \tag{4.159}$$

$$= I_{D_3} + I_{D_4} \tag{4.160}$$

Thus,

$$I_{D_1} = I_{D_4} \quad \text{and} \quad I_{D_2} = I_{D_3} \tag{4.161}$$

Assuming that the transistors $T_1 - T_4$ are identical and operate in the saturation region, the voltage applied to each transistor pair is $V_i/2$, where $V_i = V_i^+ - V_i^-$. On the basis of Kirchhoff's voltage law, it can be easily shown that

$$V_i/2 = V_{GS_1} - V_{GS_3} \tag{4.162}$$

$$= V_{GS_4} - V_{GS_2} \tag{4.163}$$

The drain currents of transistors T_1 and T_2 can then be obtained as

$$I_{D_1} = \frac{1}{2}\left[I_B + K\frac{V_i}{2}\sqrt{\frac{2I_B}{K} - \frac{V_i^2}{4}}\right] \tag{4.164}$$

and

$$I_{D_2} = \frac{1}{2}\left[I_B - K\frac{V_i}{2}\sqrt{\frac{2I_B}{K} - \frac{V_i^2}{4}}\right] \tag{4.165}$$

where $|V_i| \leq 2\sqrt{2I_B/K}$. Note that $I_1 = I_{D_1}$ and $I_2 = I_{D_2}$. Therefore, the output current i is given by

$$i = I_{D_1} - I_{D_2} = K\frac{V_i}{2}\sqrt{\frac{2I_B}{K} - \frac{V_i^2}{4}} \tag{4.166}$$

In comparison with the case of the basic differential stage, the input dynamic range is extended by a factor of 2 while the transconductance is decreased by a factor of 2.

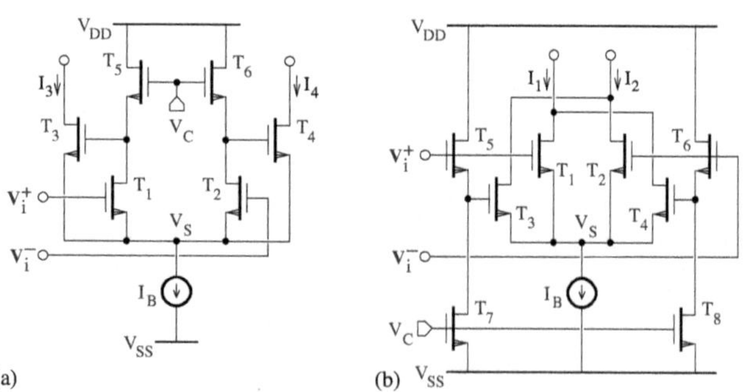

FIGURE 4.22
Circuit diagram of differential stages using voltage-controlled floating dc sources and featuring an output current (a) dependent on and (b) independent of the threshold voltage.

Differential stages can also be designed using voltage-controlled floating dc sources to maintain constant the sum of gate-source voltages of input transistors, as illustrated in Figure 4.22. This approach is known as the bias offset technique. The transistors are assumed to be identical and operate in the saturation region. In the case of the circuit depicted in Figure 4.22(a) [10], the drain currents of transistors T_1 and T_2 can be written as

$$I_{D_1} = K(V_{ICM} + V_i/2 - V_S - V_T)^2 \tag{4.167}$$

and

$$I_{D_2} = K(V_{ICM} - V_i/2 - V_S - V_T)^2 \tag{4.168}$$

where $V_i^+ = V_{ICM} + V_i/2$ and $V_i^- = V_{ICM} - V_i/2$. Because $I_{D_1} = I_{D_5}$ and $I_{D_2} = I_{D_6}$, we have $V_{GS_1} = V_{GS_5}$ and $V_{GS_2} = V_{GS_6}$. The gate voltages of transistors T_3 and T_4 are then of the form

$$V_{G_3} = V_C - V_{GS_5} = V_C - V_{GS_1} = V_C - V_{ICM} - V_i/2 \tag{4.169}$$

and

$$V_{G_4} = V_C - V_{GS_6} = V_C - V_{GS_2} = V_C - V_{ICM} + V_i/2 \tag{4.170}$$

Hence,

$$I_{D_3} = K(V_C - V_{ICM} - V_i/2 - V_S - V_T)^2 \tag{4.171}$$

and

$$I_{D_4} = K(V_C - V_{ICM} + V_i/2 - V_S - V_T)^2 \tag{4.172}$$

The difference in the output currents is given by

$$\triangle i = I_3 - I_4 = I_{D_3} - I_{D_4} = 2K(V_{ICM} - V_C - V_T)V_i \tag{4.173}$$

To remove the effect of the threshold voltage on the output voltage, the differential stage of Figure 4.22(b) [11] can be used. The transistors are identical and biased in the saturation region. Assuming that $V_i^+ = V_{ICM} + V_i/2$ and $V_i^- = V_{ICM} - V_i/2$, the drain currents of transistors T_1 and T_2 can be expressed as

$$I_{D_1} = K(V_{ICM} + V_i/2 - V_S - V_T)^2 \tag{4.174}$$

and

$$I_{D_2} = K(V_{ICM} - V_i/2 - V_S - V_T)^2 \tag{4.175}$$

With the same current flowing through T_5 and T_7, and T_6 and T_8, it can be shown that $V_{GS_5} = V_{GS_7}$ and $V_{GS_6} = V_{GS_8}$. The gate voltages of transistors T_3 and T_4 are then given by

$$V_{G_3} = V_i^+ - V_{GS_5} = V_{ICM} + V_i/2 - V_{GS_7} = V_{ICM} + V_i/2 - V_B \quad (4.176)$$

and

$$V_{G_4} = V_i^- - V_{GS_6} = V_{ICM} - V_i/2 - V_{GS_8} = V_{ICM} - V_i/2 - V_B \quad (4.177)$$

where $V_B = V_C - V_{SS}$. Hence,

$$I_{D_3} = K(V_{ICM} + V_i/2 - V_B - V_S - V_T)^2 \quad (4.178)$$

and

$$I_{D_4} = K(V_{ICM} - V_i/2 - V_B - V_S - V_T)^2 \quad (4.179)$$

The difference in the output currents is of the form

$$\triangle i = I_1 - I_2 = (I_{D_1} + I_{D_4}) - (I_{D_2} + I_{D_3}) = 2KV_BV_i \quad (4.180)$$

Applying Kirchhoff's current law at the common source node of transistors $T_1 - T_4$, we have

$$(I_{D_1} + I_{D_4}) + (I_{D_2} + I_{D_3}) = I_1 + I_2 = I_B \quad (4.181)$$

Solving Equations (4.180) and (4.181) gives

$$I_1 = I_{D_1} + I_{D_4} = I_B/2 + KV_BV_i \quad (4.182)$$

and

$$I_2 = I_{D_2} + I_{D_3} = I_B/2 - KV_BV_i \quad (4.183)$$

The range of input voltages over which the transistors still operate in the saturation region can be determined by the next worst-case requirement,

$$V_{GS_4} = V_{ICM} - V_i/2 - V_B - V_S \geq V_T \quad (4.184)$$

Substituting Equations (4.174) and (4.179) into Equation (4.182), we obtain

$$(V_{ICM} - V_S - V_T)^2 - V_B(V_{ICM} - V_S - V_T) + (V_i^2 + 2V_B^2 - I_B/K)/4 = 0 \quad (4.185)$$

This quadratic equation can be solved for $V_{ICM} - V_S - V_T$. That is,

$$V_{ICM} - V_S - V_T = \frac{V_B}{2} \pm \frac{1}{2}\sqrt{\frac{I_B}{K} - V_B^2 - V_i^2} \quad (4.186)$$

where $I_B/K \geq V_B^2 + V_i^2$. Because the condition set by Equation (4.184) can only be met by the solution with the sign + between both terms, combining Equations (4.186) and (4.184) gives

$$V_i^2 + V_B V_i + V_B^2 - \frac{I_B}{2K} \leq 0 \tag{4.187}$$

or equivalently,

$$|V_i| \leq -\frac{V_B}{2} + \sqrt{\frac{I_B}{2K} - \frac{3V_B^2}{4}} \quad |V_B| \leq \sqrt{\frac{2I_B}{3K}} \tag{4.188}$$

Note that the other possible solution of Equation (4.187) provides a negative bound for the magnitude of the differential input voltage and is not suitable.

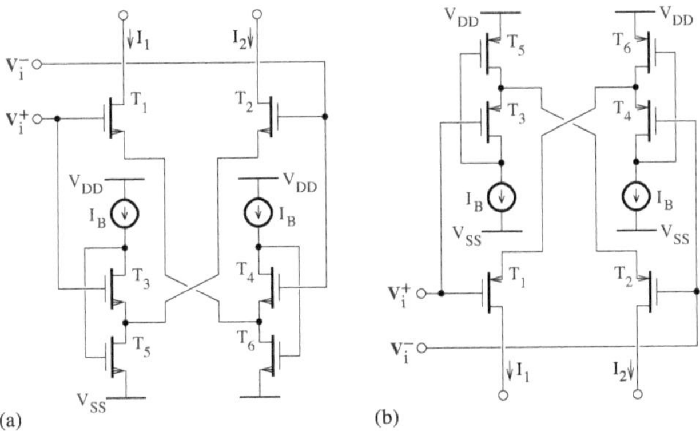

FIGURE 4.23
Circuit diagram of differential stages using floating voltage follower sources: (a) n-channel and (b) p-channel input transistors.

An alternative approach that can be adopted to increase the linearity range consists of using the differential stages shown in Figure 4.23 [12, 13]. Here, two floating dc voltage sources consisting of two transistors biased by a constant current are used to maintain constant the sum of the gate-source voltages associated with the differential transistor pair. One terminal of the input voltage is connected to the gate of the transistors T_1 and T_3, while the other is connected to the gate of transistors T_2 and T_4. The dc voltages available at the source-drain junctions of transistors T_3 and T_5, and T_4 and T_6, are applied to the sources of the input transistors T_2 and T_1, respectively.

Let us consider the differential stage with n-channel input transistors. Assuming that the transistors have the same threshold voltage and operate in the saturation region, we can write

$$I_1 = K_1(V_{GS_1} - V_T)^2 \tag{4.189}$$

$$I_2 = K_2(V_{GS_2} - V_T)^2 \tag{4.190}$$

and

$$I_B = K_3(V_{GS_3} - V_T)^2 \tag{4.191}$$
$$= K_4(V_{GS_4} - V_T)^2 \tag{4.192}$$

Because $V_{G_1} = V_{G_3} = V_i^+$ and $V_{G_2} = V_{G_4} = V_i^-$, it can be shown that

$$V_{GS_1} = V_{G_1} - V_{S_1} = V_i^+ - V_{S_4} = V_i + \sqrt{\frac{I_B}{K_4}} + V_T \tag{4.193}$$

and

$$V_{GS_2} = V_{G_2} - V_{S_2} = V_i^- - V_{S_3} = -V_i + \sqrt{\frac{I_B}{K_3}} + V_T \tag{4.194}$$

where $V_i = V_i^+ - V_i^-$. Hence,

$$I_1 = K_1\left(\sqrt{\frac{I_B}{K_4}} + V_i\right)^2 \tag{4.195}$$

and

$$I_2 = K_2\left(\sqrt{\frac{I_B}{K_3}} - V_i\right)^2 \tag{4.196}$$

With $K_1 = K_2$ and $K_3 = K_4$, the difference in the output currents can be computed as

$$i = I_1 - I_2 = 4K_1\sqrt{\frac{I_B}{K_3}}V_i \tag{4.197}$$

To maintain the input transistors in the saturation region, the currents I_1 and I_2 should not be equal to zero. That is,

$$-\sqrt{\frac{I_B}{K_3}} < V_i < \sqrt{\frac{I_B}{K_3}} \tag{4.198}$$

This differential input stage structure is capable of a class AB operation and is suitable for low-voltage applications.

Another approach for the transconductor linearization consists of using two current-controlled floating dc voltage sources to keep constant the sum of gate-source voltages of input transistors, as illustrated in Figures 4.24(a) and (b) for n-channel and p-channel input transistors, respectively [14]. All transistors operate in the saturation region and it is assumed that the transistors with the same channel type are matched. Let us consider the structure of Figure 4.24(a). Applying Kirchhoff's voltage law to both input loops gives

$$V_i = V_{GS_1} + V_{GS_5} - (V_{GS_7} + V_{GS_4}) \tag{4.199}$$

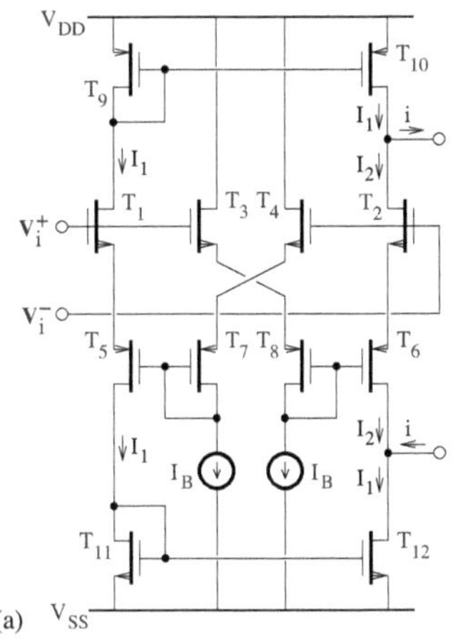

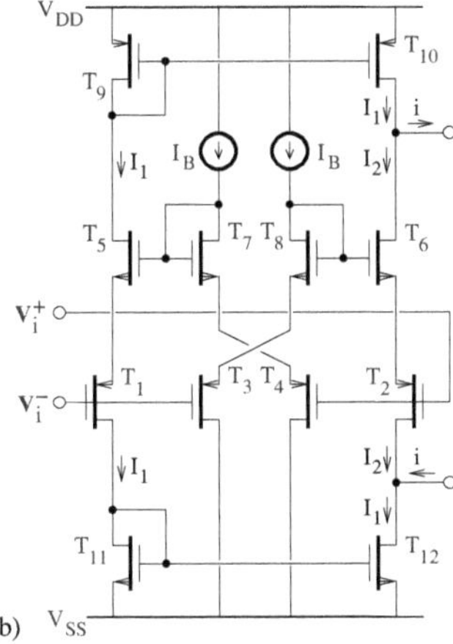

FIGURE 4.24
Circuit diagram of differential stages using current-controlled floating voltage sources: (a) n-channel and (b) p-channel input transistors.

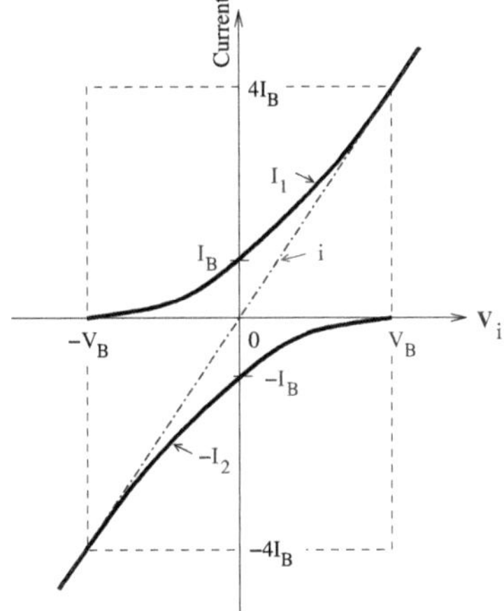

FIGURE 4.25
I-V transfer characteristic of a differential stage using current-controlled floating voltage sources.

and

$$-V_i = V_{GS_2} + V_{GS_6} - (V_{GS_8} + V_{GS_3}) \tag{4.200}$$

where

$$V_{GS_1} = \sqrt{\frac{I_1}{K_n}} + V_{T_n} \tag{4.201}$$

$$V_{GS_2} = \sqrt{\frac{I_2}{K_n}} + V_{T_n} \tag{4.202}$$

$$V_{GS_3} = V_{GS_4} = \sqrt{\frac{I_B}{K_n}} + V_{T_n} \tag{4.203}$$

$$V_{GS_5} = \sqrt{\frac{I_1}{K_p}} + V_{T_p} \tag{4.204}$$

$$V_{GS_6} = \sqrt{\frac{I_2}{K_p}} + V_{T_p} \tag{4.205}$$

and

$$V_{GS_7} = V_{GS_8} = \sqrt{\frac{I_B}{K_p}} + V_{T_p} \tag{4.206}$$

To proceed further, it can be shown that

$$V_i = \sqrt{\frac{I_1}{K_{eq}}} - V_B \tag{4.207}$$

and

$$-V_i = \sqrt{\frac{I_2}{K_{eq}}} - V_B \tag{4.208}$$

where

$$V_B = \sqrt{\frac{I_B}{K_{eq}}} \tag{4.209}$$

and

$$K_{eq} = \frac{K_n K_p}{(\sqrt{K_n} + \sqrt{K_p})^2} \tag{4.210}$$

The currents I_1 and I_2 are, respectively, given by

$$I_1 = K_{eq}(V_B + V_i)^2 \tag{4.211}$$

and

$$I_2 = K_{eq}(V_B - V_i)^2 \tag{4.212}$$

The output current is then obtained as

$$i = I_1 - I_2 = 4K_{eq}V_B V_i = 4\sqrt{K_{eq} I_B}\, V_i \tag{4.213}$$

As illustrated in Figure 4.25, the differential stage remains in the linear range provided the currents I_1 and I_2 are different from zero. From Equations (4.211) and (4.212), it can be deduced that

$$-\sqrt{I_B/K_{eq}} < V_i < \sqrt{I_B/K_{eq}} \tag{4.214}$$

or equivalently,

$$-4I_B < i < 4I_B \tag{4.215}$$

This differential stage design has the advantages of featuring a source node and a sink node for the output current.

4.3 Transconductor operating in the subthreshold region

Transistors operating in the weak inversion (or subthreshold) region are known to provide the highest transconductance efficiency, g_m/I_D, and then appear as the best alternative in low-power applications, especially when a high bandwidth is not an important requirement.

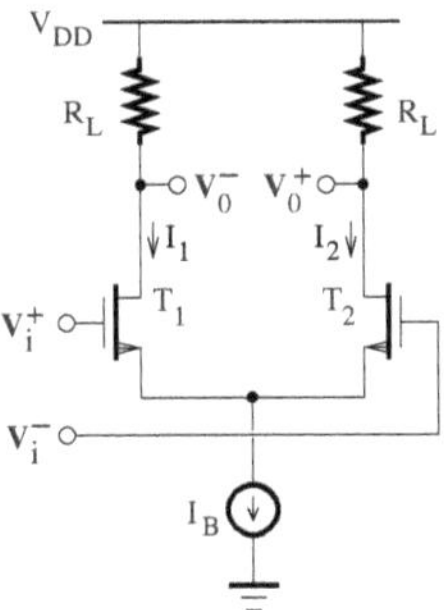

FIGURE 4.26
Transconductor stage.

In the subthreshold region, the drain-source current of an n-channel transistor can be reduced to

$$I_D \simeq I_T\left(\frac{W}{L}\right)\exp\left(\frac{V_{GS}-V_{T_n}}{nU_T}\right) \quad V_{DS} \geq 4U_T \tag{4.216}$$

where $I_T = 2n\mu_n C_{ox}U_T^2$, n $(1 < n < 3)$ is the subthreshold slope factor, U_T is the thermal voltage, and V_{T_n} is the threshold voltage.

Consider the transconductor stage of Figure 4.26. The drain currents of transistors T_1 and T_2 can be expressed as

$$I_{D_1} = I_1 = I_T\left(\frac{W_1}{L_1}\right)\exp\left(\frac{V_i^+ - V_S - V_{T_n}}{nU_T}\right) \tag{4.217}$$

$$I_{D_2} = I_2 = I_T\left(\frac{W_2}{L_2}\right)\exp\left(\frac{V_i^- - V_S - V_{T_n}}{nU_T}\right) \tag{4.218}$$

where $W_1/L_1 = W_2/L_2$. The bias current is given by

$$I_B = I_1 + I_2 \tag{4.219}$$

Substituting (4.217) and (4.218) into (4.219) yields

$$\exp\left(-\frac{V_S + V_{T_n}}{nU_T}\right) = \frac{I_B}{I_T \dfrac{W_1}{L_1}\left[\exp\left(\dfrac{V_i^+}{nU_T}\right) + \exp\left(\dfrac{V_i^-}{nU_T}\right)\right]} \tag{4.220}$$

Combining (4.217), (4.218), and (4.220) gives

$$I_1 = I_B \frac{\exp\left(\dfrac{V_i^+}{nU_T}\right)}{\exp\left(\dfrac{V_i^+}{nU_T}\right) + \exp\left(\dfrac{V_i^-}{nU_T}\right)} \tag{4.221}$$

$$I_2 = I_B \frac{\exp\left(\dfrac{V_i^-}{nU_T}\right)}{\exp\left(\dfrac{V_i^+}{nU_T}\right) + \exp\left(\dfrac{V_i^-}{nU_T}\right)} \tag{4.222}$$

Assuming that $V_i^+ = V_{CM} + V_i/2$ and $V_i^+ = V_{CM} - V_i/2$, where V_{CM} is the common-mode voltage and V_i is the differential input voltage, it can be shown that

$$\triangle i = I_1 - I_2 = I_B \frac{\exp\left(\dfrac{V_i}{2nU_T}\right) - \exp\left(-\dfrac{V_i}{2nU_T}\right)}{\exp\left(\dfrac{V_i}{2nU_T}\right) + \exp\left(-\dfrac{V_i}{2nU_T}\right)} \tag{4.223}$$

$$= I_B \tanh\left(\frac{V_i}{2nU_T}\right) \tag{4.224}$$

and

$$V_0 = R_L \triangle i \tag{4.225}$$

where $V_0 = V_0^+ - V_0^-$ is the differential output voltage. The transconductance is derived as,

$$g_m = \frac{\partial \triangle i}{\partial V_i} = \frac{I_B}{2nU_T} \frac{1}{\cosh^2\left(\dfrac{V_i}{2nU_T}\right)} \tag{4.226}$$

Its maximum value, $I_B/(2nU_T)$, is obtained for V_i equal to zero. Note that the transconductance linear range to achieve a maximum signal distortion of 1% is limited to about ± 7.5 mV for typical transistor characteristics.

In the subthreshold region, the effect of threshold voltage variations on the drain current results in

$$I_D = I_D(V_{T_n} + \triangle V_{T_n}) = I_D(V_{T_n}) \exp\left(-\frac{\triangle V_{T_n}}{nU_T}\right) \tag{4.227}$$

The drain current error can be obtained as[1]

$$\triangle I_D = \frac{I_D(V_{T_n})}{nU_T}\triangle V_{T_n} \tag{4.228}$$

Using the following typical values, $n = 1.35$ and $U_T = 25.9$ mV, it can be found that $\triangle I_D = 28.6 I_D(V_{T_n})\triangle V_{T_n}$.
In the strong inversion region, due to variations of the threshold voltage, the drain current becomes

$$I_D = I_D(V_{T_n} + \triangle V_{T_n}) = K(V_{GS} - V_{T_n} - \triangle V_{T_n})^2 \tag{4.229}$$

By expanding the expression of I_D and keeping only first-order terms, it can be shown that

$$I_D = I_D(V_{T_n})\left(1 - \frac{2\triangle V_{T_n}}{V_{GS} - V_{T_n}}\right) \tag{4.230}$$

The drain current error is then given by

$$\triangle I_D = \frac{2I_D(V_{T_n})}{V_{GS} - V_{T_n}}\triangle V_{T_n} = \frac{2I_D(V_{T_n})}{V_{DS(sat)}}\triangle V_{T_n} \tag{4.231}$$

Typically, $V_{DS(sat)} = 0.25$ V, so that $\triangle I_D = 8I_D(V_{T_n})\triangle V_{T_n}$.

Circuits based on transistors operating in the subthreshold region seem to be limited by the achievable precision on the threshold voltage, V_T. As a result, they essentially find applications in the implementation of current references and current-mode logic digital gates. Another significant advantage of the operation in the subthreshold region is the ability to generate low-level currents between the transistor drain and source, thereby realizing resistances up to giga-ohms.

4.4 Single-stage amplifier

The circuit diagram of a single-stage amplifier [15] is shown in Figure 4.27. The differential transistor pair, $T_1 - T_2$, converts the input voltage into currents, which are directed to the output stage by stacked current mirrors. The tail

[1]Given a function f of a variable x, the variations δf due to small variations δx can be expressed as,

$$\delta f = f(x + \delta x) - f(x) = f'(x)\delta x$$

where f' represents the derivative of f with respect to x.

current of the input stage, the transconductance of which is assumed to be g_m, is set by $T_{15} - T_{16}$. Using Kirchhoff's voltage law, we can obtain

$$V_i^+ = V_{DD} - V_{SG_5} - V_{SG_3} - V_{DS_1} + V_{GS_1} \tag{4.232}$$

$$= V_{GS_1} + V_{DS_{15}} + V_{SS} \tag{4.233}$$

Assuming that the dc voltage applied to both amplifier inputs is V_{ICM}, it can be shown that

$$V_{DS_1(sat)} + V_{DS_{15}(sat)} + V_{T_n} + V_{SS} \leq V_{ICM} \leq V_{DD} - V_{SD_5(sat)} - V_{SD_3(sat)} + 2V_{T_p} + V_{T_n} \tag{4.234}$$

where V_{ICM} denotes the input common-mode voltage. To maintain the transistors T_8, T_{10}, T_{12}, and T_{14} in the saturation region, the signal swing at the output node should be bounded as follows:

$$V_{DS_{12}(sat)} + V_{DS_{14}(sat)} + V_{T_n} + V_{SS} \leq V_0 \leq V_{DD} - V_{SD_{10}(sat)} - V_{SD_8(sat)} + V_{T_p} \tag{4.235}$$

Note that the available output swing can be increased using high-swing cascode current mirrors, which are able to cancel the voltage threshold term.

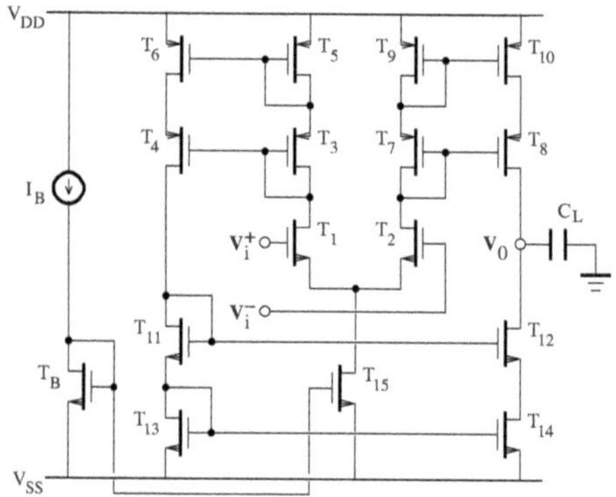

FIGURE 4.27
Circuit diagram of a single-stage amplifier.

Let a_i be the amplification factor provided by the load devices, T_3-T_6 and T_7-T_{10}, and r_0 be the output resistance. The dc voltage gain of the amplifier can be written as

$$A_0 = g_m a_i r_0 \tag{4.236}$$

where g_m is the transconductance of each of the input transistor T_1 or T_2. Assuming that $W_3/L_3 = W_5/L_5$, $W_4/L_4 = W_6/L_6$, $W_7/L_7 = W_9/L_9$, and $W_8/L_8 = W_{10}/L_{10}$, it can be shown that

$$a_i = (W_6/L_6)/(W_5/L_5) = (W_{10}/L_{10})/(W_9/L_9) \tag{4.237}$$

The amplifier is compensated by output capacitive loads. It can be modeled by a single-stage, small-signal equivalent circuit, provided that C_L is sufficiently large so that the pole associated with the output node is dominant. The gain-bandwidth product, GBW, and slew rate, SR, are then given by

$$GBW = \frac{g_m a_i}{C_0} \simeq \frac{g_m a_i}{C_L} \tag{4.238}$$

and

$$SR = \frac{a_i I_B}{C_0} \simeq \frac{a_i I_B}{C_L} \tag{4.239}$$

where I_B is the bias current applied to the input differential stage, C_0 is the total output capacitance, and C_L denotes the external load capacitance connected to the output node.

4.5 Folded-cascode amplifier

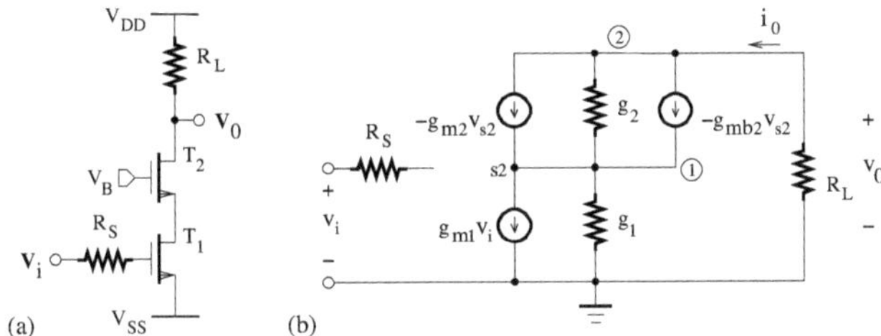

FIGURE 4.28
(a) Circuit diagram and (b) small-signal equivalent model of a cascode amplifier.

An amplifier with a cascode configuration has the advantage of exhibiting a higher bandwidth due its improved isolation between the input and output nodes. Figure 4.28(a) shows the circuit diagram of a cascode amplifier, whose input stage is a common-source transistor driving a common-gate transistor loaded by a resistor. By replacing each transistor with its equivalent model, the small-signal circuit shown in Figure 4.28(b) can be derived. It is assumed that $v_{gs2} = -v_{s2}$ and $v_{bs2} = -v_{s2}$. Using Kirchhoff's current law for the nodes 1 and 2, we may write

$$g_{m_1} v_i + g_1 v_{s2} + (g_{m_2} + g_{mb_2}) v_{s2} + g_2 (v_{s2} - v_O) = 0 \tag{4.240}$$

$$-(g_{m_2} + g_{mb_2}) v_{s2} - g_2 (v_{s2} - v_O) + v_O / R_L = 0 \tag{4.241}$$

Solving the system of Equations (4.240) and (4.241), we get

$$A_v = \frac{v_O}{v_i} = \frac{-g_{m_1}(g_{m_2} + g_{mb_2} + g_2)}{(g_1 + g_2)/R_L + g_2(g_1 + g_{m_2} + g_{mb_2})} \tag{4.242}$$

For the determination of the output resistance, the input voltage is set to zero and a voltage source is applied at the amplifier output node. The output node equation is

$$i_O = -(g_{m_2} + g_{mb_2})v_{s2} + g_2(v_O - v_{s2}) \tag{4.243}$$

where

$$v_{s2} = i_O/g_1 \tag{4.244}$$

It can then be shown that

$$r_O = \frac{v_O}{i_O} = 1/g_1 + 1/g_2 + (g_{m_2} + g_{mb_2})/g_1 g_2 \tag{4.245}$$

where $g_1 = g_{ds_1}$ and $g_2 = g_{ds_2}$. The output resistance is derived as

$$R_O = r_O \parallel R_L \tag{4.246}$$

When R_L is implemented by a passive resistor, it is much lower than r_O, so that $R_O \simeq R_L$ and the voltage gain is reduced to $A_v \simeq -g_{m_1} R_L$. On the other hand, an active load is used for the implementation of R_L in the folded-cascode amplifier in order to meet the high gain specification.

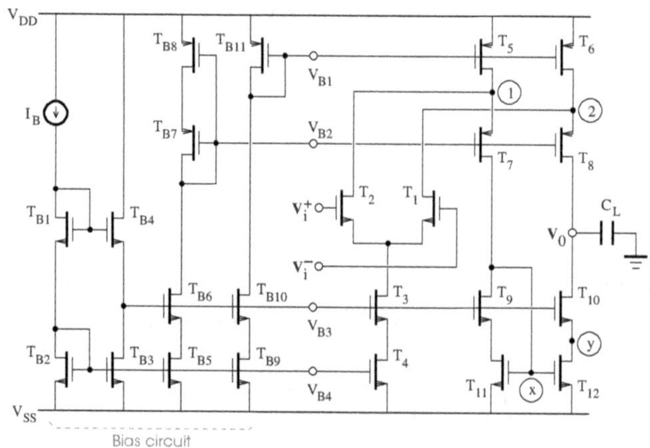

FIGURE 4.29
Circuit diagram of a folded-cascode amplifier.

The folded-cascode amplifier [16,17] schematic with a single-ended output is shown in Figure 4.29. The voltage-to-current converter based on transistors

$T_1 - T_2$ is connected to an output stage with the folded-cascode configuration consisting of $T_5 - T_{12}$. Using Kirchhoff's voltage law, it can be found that

$$V_i^- = V_{DD} - V_{DS_6} - V_{DS_1} + V_{GS_1} \tag{4.247}$$

$$= V_{GS_1} + V_{DS_3} + V_{DS_4} + V_{SS} \tag{4.248}$$

and

$$V_0^- = V_{DD} - V_{SD_6} - V_{SD_8} \tag{4.249}$$

$$= V_{DS_{10}} + V_{DS_{12}} + V_{SS} \tag{4.250}$$

To ensure normal operation of the amplifier, the transistors should be biased slightly above the saturation region. In the worst case, the input common-mode range is given by

$$V_{DS_1(sat)} + V_{T_n} + V_{DS_3(sat)} + V_{DS_4(sat)} + V_{SS} \leq V_{ICM} \leq V_{DD} - V_{SD_6(sat)} + V_{T_n} \tag{4.251}$$

where V_{ICM} is the dc voltage that can be applied to both inputs. To keep the transistors in the saturation region, the output voltage swing should be bounded as follows:

$$V_{DS_{12}(sat)} + V_{DS_{10}(sat)} + V_{SS} \leq V_0 \leq V_{DD} - V_{SD_6(sat)} - V_{SD_8(sat)} \tag{4.252}$$

The biasing circuit [7] consisting of transistors $T_{B1} - T_{B11}$ should be designed to set the quiescent points of the amplifier transistors, such that they can operate in the saturation region. Note that the stability of transistor quiescent points may be affected by variations of the current I_B and fluctuations of the IC process.

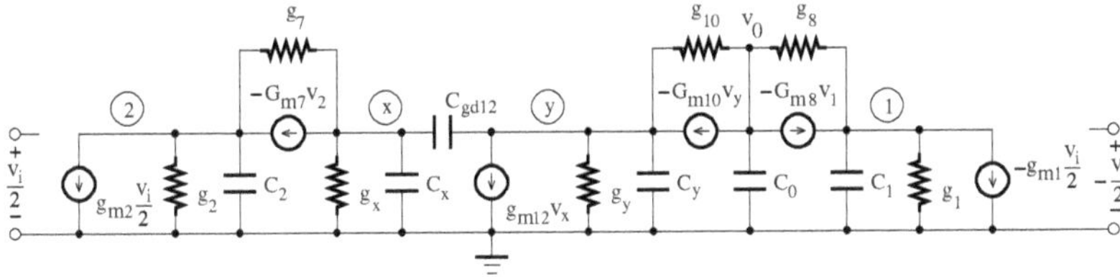

FIGURE 4.30
Equivalent circuit model of a folded-cascode amplifier.

The small-signal equivalent model of the folded-cascode amplifier is depicted in Figure 4.30. Applying Kirchhoff's current law, the equations for node 1, node 2, node x, node y, and output node can be written as

$$-g_{m_1} V_i/2 + (G_{m_8} + g_1 + sC_1)V_1 + g_8(V_1 - V_0) = 0 \tag{4.253}$$

$$g_{m_2} V_i/2 + (g_2 + sC_2 + G_{m_7})V_2 + g_7(V_2 - V_x) = 0 \tag{4.254}$$

$$-G_{m_7}V_2 - g_7(V_2 - V_x) + (g_x + sC_x)V_x = 0 \tag{4.255}$$

$$g_{m_{12}}V_x + (G_{m10} + g_y + sC_y)V_y + g_{10}(V_y - V_O) = 0 \tag{4.256}$$

and

$$-G_{m_{10}}V_y - g_{10}(V_y - V_O) - G_{m_8}V_1 - g_8(V_1 - V_O) + sC_OV_O = 0 \tag{4.257}$$

respectively, where $G_{m_k} = g_{m_k} + g_{mb_k}$, $(k = 7, 8, 10)$, $g_x = g_{m11}$, $g_y = g_{ds_{12}}$, $g_1 = g_{ds_1} + g_{ds_6}$, and $g_2 = g_{ds_2} + g_{ds_5}$. Note that g_{ml} and g_{ds_l} denote the transconductance and the drain-source conductance of the transistor T_l (l is an integer), respectively. For the determination of node capacitances, it can be assumed that the scaling factor of the capacitor $C_{gd_{12}}$ provided by the Miller effect is approximately equal to unity due to the low amplification gain available at the source of the transistor T_{10}. Hence,

$$C_1 = C_{gd_1} + C_{db_1} + C_{gd_6} + C_{db_6} + C_{gs_8} + C_{sb_8} \tag{4.258}$$

$$C_2 = C_{gd_2} + C_{db_2} + C_{gd_5} + C_{db_5} + C_{gs_7} + C_{sb_7} \tag{4.259}$$

$$\begin{aligned} C_x = {} & C_{gd_7} + C_{db_7} + C_{gd_9} + C_{db_9} \\ & + C_{gs_{11}} + C_{gd_{11}} + C_{gb_{11}} + C_{gs_{12}} + C_{gd_{12}} + C_{gb_{12}} \end{aligned} \tag{4.260}$$

$$C_y = C_{gs_{10}} + C_{sb_{10}} + C_{gd_{12}} + C_{db_{12}} \tag{4.261}$$

and

$$C_O = C_L + C_{gd_8} + C_{db_8} + C_{gd_{10}} + C_{db_{10}} \tag{4.262}$$

Assuming that the transconductances are much greater than the conductances, the system of Equations (4.254)–(4.257) can be solved for a transfer function of the form

$$A(s) = \frac{V_O(s)}{V_i(s)} = A_O \frac{(1 - s/\omega_{z_1})(1 - s/\omega_{z_2})}{(1 - s/\omega_{p_1})(1 - s/\omega_{p_2})(1 - s/\omega_{p_3})(1 - s/\omega_{p_4})} \tag{4.263}$$

For practical component values, the first pole is dominant, that is, we have $|\omega_{p_1}| \ll |\omega_{p_2}|, |\omega_{p_3}|, |\omega_{p_4}|$. The amplifier transfer function can then be approximated as

$$A(s) \simeq A_O \frac{(1 - s/\omega_{z_1})(1 - s/\omega_{z_2})}{D(s)} \tag{4.264}$$

where

$$\begin{aligned} D(s) = {} & 1 - \frac{s}{\omega_{p_1}} + \frac{s^2}{\omega_{p_1}} \left(\frac{1}{\omega_{p_2}} + \frac{1}{\omega_{p_3}} + \frac{1}{\omega_{p_4}} \right) \\ & - \frac{s^3}{\omega_{p_1}} \left(\frac{1}{\omega_{p_2}\omega_{p_3}} + \frac{1}{\omega_{p_2}\omega_{p_4}} + \frac{1}{\omega_{p_3}\omega_{p_4}} \right) + \frac{s^4}{\omega_{p_1}\omega_{p_2}\omega_{p_3}\omega_{p_4}} \end{aligned} \tag{4.265}$$

Assuming that the transistor T_1 and T_2 are matched, the small-signal dc gain of the amplifier is written as

$$A_0 \simeq \frac{g_{m_1}}{\frac{(g_{ds_1}+g_{ds_6})g_{ds_8}}{G_{m_8}}+\frac{g_{ds_{10}}g_{ds_{12}}}{G_{m_{10}}}} \tag{4.266}$$

The poles of the transfer function are given by

$$\omega_{p_1} = -\left(\frac{(g_{ds_1}+g_{ds_6})g_{ds_8}}{G_{m_8}}+\frac{g_{ds_{10}}g_{ds_{12}}}{G_{m_{10}}}\right)\frac{1}{C_0} = \frac{g_{m1}}{A_0 C_0} \tag{4.267}$$

$$\omega_{p_2} = -\frac{g_{m_{11}}}{C_x} \tag{4.268}$$

$$\omega_{p_3} = -\frac{G_{m_{10}}}{C_y} \tag{4.269}$$

and

$$\omega_{p_4} = -\frac{G_{m_8}}{C_1} \tag{4.270}$$

while for the zeros, we have

$$\omega_{z_1}, \omega_{z_2} = -\frac{\omega_{p_2}+\omega_{p_3}}{2} \pm \sqrt{\frac{\omega_{p_2}^2+\omega_{p_3}^2}{4}-\frac{3\omega_{p_2}\omega_{p_3}}{2}} \tag{4.271}$$

Note that the dc gain is of the form, $A_0 \simeq g_{m1}R_0$, where R_0 is the output resistance. The frequency response shows that the folded-cascode amplifier has two left-half plane zeros, a dominant pole associated with the output node, and nondominant poles introduced by the current mirror, and n-channel and p-channel cascode transistors.

The zeros, the second and third poles, which are closely located, form two doublets. Generally, pole-zero doublets have less influence on the frequency response, but can degrade the settling response. They should be located at frequencies greater than the unity-gain frequency of the amplifier in order to achieve the optimum settling time.

That is, the frequency behavior of the folded-cascode amplifier can be described by a two-pole system with the next transfer function

$$A(s) = \frac{V_0(s)}{V_i(s)} \simeq A_0 \frac{1}{(1-s/\omega_{p_1})(1-s/\omega_{p_4})} \tag{4.272}$$

The phase margin can be obtained as

$$\begin{aligned} \phi_M &= 180^\circ - \angle A[j(GBW)] & (4.273)\\ &= 180^\circ - \arctan(GBW/\omega_{p_1}) - \arctan(GBW/\omega_{p_4}) & (4.274) \end{aligned}$$

where GBW denotes the gain-bandwidth product or unity-gain frequency. Because $GBW = A_0\omega_{p_1}$ and the dc gain, A_0, is very high, the first arctan term tends to 90°. The expression of the phase margin then becomes

$$\phi_M = 90^\circ - \arctan(GBW/\omega_{p_4}) \tag{4.275}$$

In practice, the nondominant pole ω_{p_4} is located at a high frequency and the frequency response of the folded-cascode amplifier is primarily determined by the single dominant pole.[2] The frequency compensation, which should affect only the output node pole, is implemented by the load capacitor C_L.

4.6 Fully differential amplifier architectures

Differential amplifier architectures offer many design advantages (e.g., improved dynamic range, availability of inverting and noninverting functions on the same structure).

4.6.1 Fully differential folded-cascode amplifier

Depending on the trade-off to be achieved between the gain and speed specifications in a given application, fully differential folded-cascode amplifiers can be designed using either a basic or gain-enhanced structure.

4.6.1.1 Basic structure

The single-stage amplifier with cascode structure [16, 18], as shown in Figure 4.31, can provide an acceptable gain without degrading the high-frequency performance in most applications. It consists of a differential input gain stage ($T_1 - T_4$) followed by a cascode loading structure ($T_5 - T_{12}$). The common-mode feedback (CMF) circuit is used in order to constrain the common-mode (CM) output voltage to a desired dc operating point so that the output voltage swing can be maximized. Assuming that the transistors are biased to exhibit identical saturation voltages, the differential output swing is about $2V_{sup} - 4V_{DS(sat)} - 4V_{SD(sat)}$, where $V_{sup} = V_{DD} - V_{SS}$ represents the total supply voltage.

The small-signal equivalent circuit of the differential amplifier half circuit, as shown in Figure 4.32, is used for the frequency domain analysis. Because both circuit sections around the axis of symmetry are matched, the axis of symmetry can be considered the ac ground. For the node 1, node x, and output

[2]Assuming a dominant pole model for the amplifier, the gain-bandwidth product is approximately equal to the unity-gain or transition frequency.

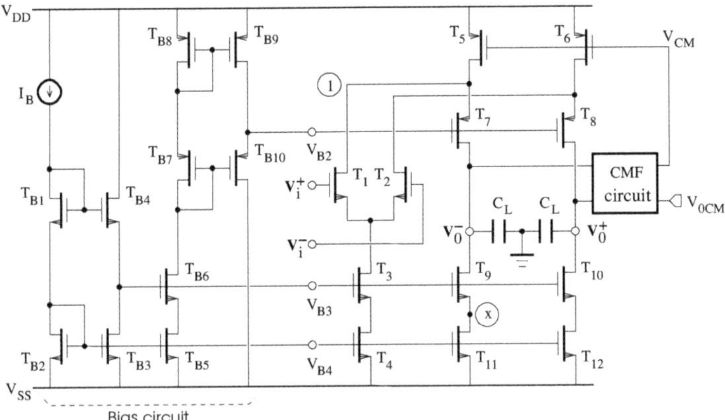

FIGURE 4.31
Fully differential folded-cascode amplifier.

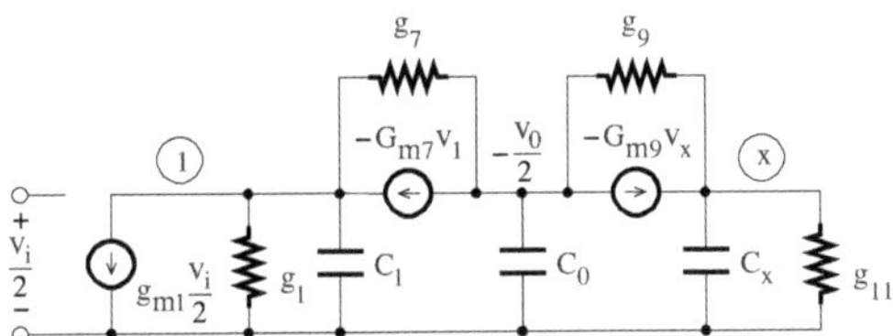

FIGURE 4.32
Equivalent circuit model of the differential amplifier half circuit.

node, the equations obtained using Kirchhoff's current law can be written as

$$g_{m1}V_i/2 + (g_1 + sC_1 + G_{m7})V_1 + g_7(V_1 + V_0/2) = 0 \quad (4.276)$$

$$G_{m9}V_x + g_9(V_x + V_0/2) + (g_{11} + sC_x)V_x = 0 \quad (4.277)$$

and

$$G_{m7}V_1 + g_7(V_1 + V_0/2) + G_{m9}V_x + g_9(V_x + V_0/2) + sC_0V_0/2 = 0 \quad (4.278)$$

respectively, where $G_{mk} = g_{mk} + g_{mbk}$, $(k = 7, 9)$, $g_1 = g_{ds_1} + g_{ds_5}$, $g_7 = g_{ds_7}$, $g_9 = g_{ds_9}$, and $g_{11} = g_{ds_{11}}$. Solving the system of Equations (4.276)–(4.278), the transfer function of the amplifier can be derived in the form

$$A(s) = \frac{V_0(s)}{V_i(s)} = A_0 \frac{1 - s/\omega_{z_1}}{(1 - s/\omega_{p_1})(1 - s/\omega_{p_2})(1 - s/\omega_{p_3})} \quad (4.279)$$

where ω_{z_1} represents the frequency of the zero z_1, and ω_{p_1}, ω_{p_2}, and ω_{p_3} are the frequencies of the poles p_1, p_2, and p_3, respectively. Generally, the amplifier is

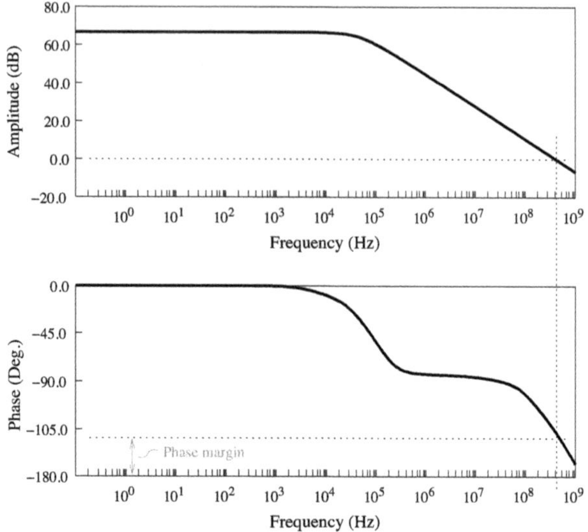

FIGURE 4.33
(a) Magnitude and (b) phase response (output v_0^-) of a folded-cascode amplifier.

designed to feature the behavior of a dominant pole system, whose poles are widely spaced, that is, $|\omega_{p_1}| \ll |\omega_{p_2}|, |\omega_{p_3}|$ and $|\omega_{p_2}| \ll |\omega_{p_3}|$. Hence,

$$A(s) = \frac{V_0(s)}{V_i(s)} \simeq A_0 \frac{1 - s/\omega_{z_1}}{1 - s/\omega_{p_1} + s^2/\omega_{p_1}\omega_{p_2} - s^3/\omega_{p_1}\omega_{p_2}\omega_{p_3}} \tag{4.280}$$

Assuming that the transconductances are much higher than the conductances, we arrive at

$$A_0 \simeq \frac{g_{m1}}{\dfrac{(g_{ds_1} + g_{ds_5})g_{ds_7}}{G_{m7}} + \dfrac{g_{ds_9} g_{ds_{11}}}{G_{m9}}} \tag{4.281}$$

As the zero and third pole, which are located at about $-G_{m9}/C_x$, cancel each other out, the amplifier transfer function is reduced to the one of a second-order system, whose poles are given by

$$\omega_{p_1} = -g_{m1}/A_0 C_0 \tag{4.282}$$

and

$$\omega_{p_2} = -G_{m7}/C_1 \tag{4.283}$$

where $C_0 = C_i' + C_{gd_7} + C_{db_7} + C_{gd_9} + C_{db_9} + C_L$, C_i' is the input capacitance of the CMF circuit, and $C_1 = C_{gd_1} + C_{db_1} + C_{gd_5} + C_{db_5} + C_{gs_7} + C_{sb_7}$. The compensation of the folded-cascode amplifier is achieved by the load capacitor

C_L (see Figure 4.31), which primarily determines the frequency location of the dominant pole. In this case, the gain-bandwidth product, ω_{GBW}, can be expressed as

$$\omega_{GBW} \simeq \frac{g_{m1}}{C_L} \tag{4.284}$$

The slew rate is given by

$$SR \simeq \frac{I_B}{C_L} \tag{4.285}$$

where I_B is the bias current of the differential input stage. But, due to the amplifier stability condition, which imposes that all nondominant poles occur at frequencies past the unity-gain frequency, the amplifier speed is limited by the position of the first nondominant pole. This pole comes from the G_{m_7}/C_1 time constant of the cascode transistor (here T_7) and is specified by a pole frequency approximately at the transition frequency of this transistor. Note that the capacitor C_1 represents the total capacitive load at the source of the related transistor (or node 1). Of practical importance is the parameter K, expressed as

$$K = \tan(\phi_M) \simeq \frac{\omega_{p_2}}{\omega_{GBW}} \tag{4.286}$$

where ω_{p_2} characterizes the first nondominant pole of the amplifier. The phase margin ϕ_M is commonly used for the definition of the stability. It should be noted that the minimum settling time at 0.1% is obtained for a value of ϕ_M around 76°.

Frequency responses of the fully differential amplifier are shown in Figures 4.33(a) and (b), where a phase difference of 180° exists between the two output voltages v_0^+ and v_0^-. For these plots, $V_{DD} = 3.3$ V, $V_{SS} = 0$ V, $C_L = 0.75$ pF, and a common-mode dc voltage of 1.5 V was added to the input voltages.

4.6.1.2 Gain-enhanced structure

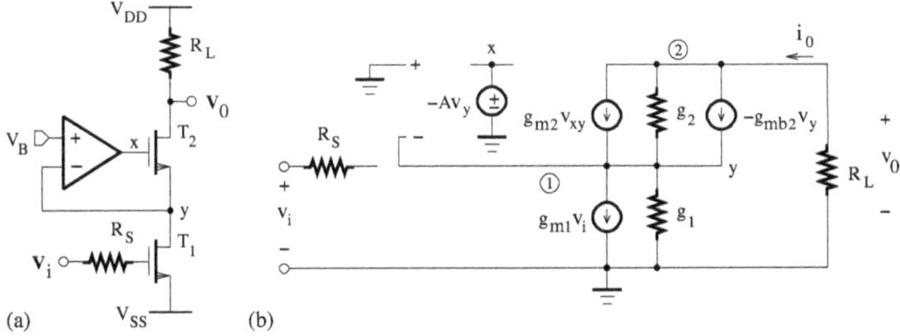

FIGURE 4.34
(a) Circuit diagram and (b) small-signal equivalent model of a gain-enhanced cascode amplifier.

Typically, amplifiers should exhibit a high open-loop gain and a high bandwidth to minimize errors in the output voltage. The high gain requirement can be met by cascading gain stages, the number of which is limited by the need for frequency compensation to enable stable feedback. The gain enhancement technique can also be implemented by inserting each cascode transistor in an amplifier feedback path to increase the overall output resistance.

The circuit diagram of a gain-enhanced cascode amplifier is shown in Figure 4.34(a). The feedback amplifier has a voltage gain, A, and V_B is a constant voltage. A small-signal equivalent model for this amplifier is illustrated in Figure 4.34(b), where $v_{gs_2} = v_{xy}$ and $v_{bs_2} = -v_{ds_1} = -v_y$. Applying Kirchhoff's current law at the nodes 1 and 2 gives

$$g_{m_1} v_i + g_1 v_y - g_{m_2} v_{xy} + g_{mb_2} v_y + g_2(v_y - v_0) = 0 \quad (4.287)$$

$$g_{m_2} v_{xy} - g_{mb_2} v_y - g_2(v_y - v_0) + v_0/R_L = 0 \quad (4.288)$$

where

$$v_{xy} = v_x - v_y = -(A+1)v_y \quad (4.289)$$

Combining Equations (4.287) and (4.288), the voltage gain is obtained as

$$A_v = \frac{v_0}{v_i} = \frac{-g_{m_1}[(1+A)g_{m_2} + g_{mb_2} + g_2]}{[(1+A)g_{m_2} + g_{mb_2} + g_1 + g_2]/R_L + g_1 g_2} \quad (4.290)$$

To find the output resistance, a test generator is connected to the amplifier output and the input node is short-circuited to ground. For the transistor T_1, $g_{m_1} v_i = 0$. The output node current equation can be written as

$$i_0 = g_{m_2} v_{xy} - g_{mb_2} v_y + g_2(v_0 - v_y) \quad (4.291)$$

where

$$v_y = i_0/g_1 \quad (4.292)$$

By solving Equations (4.291) and (4.292), we get

$$r_0 = \frac{v_0}{i_0} = 1/g_1 + 1/g_2 + [(1+A)g_{m_2} + g_{mb_2}]/g_1 g_2 \quad (4.293)$$

Therefore, the overall output resistance is given by

$$R_0 = r_0 \parallel R_L \quad (4.294)$$

Provided the resistance R_L is sufficiently high, the output resistance is enhanced by a factor on the order of the gain of the feedback amplifier or auxiliary amplifier. This principle can be exploited to meet the high gain and fast settling requirements in the design of amplifiers.

A single-stage amplifier based on an active cascode output stage is shown in Figure 4.35. The feedback created around the output transistors helps increase the output impedance of the main amplifier. As a result, its gain is

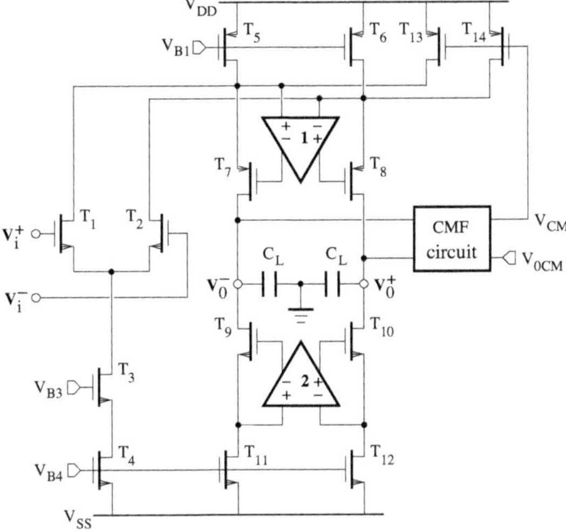

FIGURE 4.35
Fully differential folded-cascode amplifier with gain enhancement.

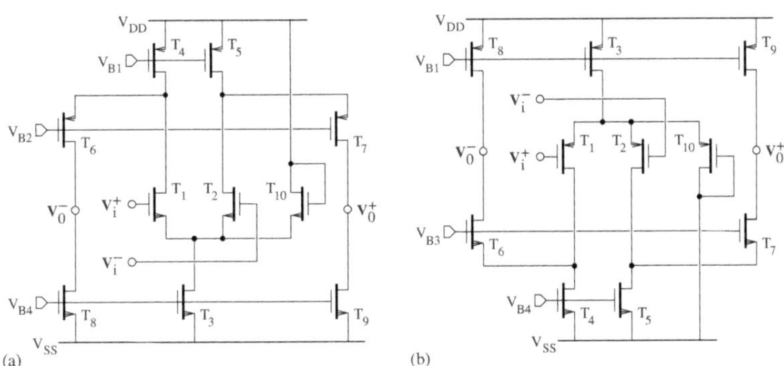

FIGURE 4.36
(a) Differential gain enhancement amplifier 1 and (b) its complementary version for the gain enhancement amplifier 2.

enhanced without altering the bandwidth due to the scaling effect provided by the auxiliary amplifier gain [19–21]. The resulting dc gain, A_0, of the active cascode amplifier is given approximately by

$$A_0 \simeq A_0'(1 + A_0'') \tag{4.295}$$

where A_0' denotes the dc gain of the main amplifier and A_0'' is the dc gain of the auxiliary amplifier. The auxiliary amplifiers, as shown in Figure 4.36, are fully differential and use a single transistor CMF circuit. In this way, only two auxiliary amplifiers are required, resulting in a reduction in area and power consumption in comparison with an architecture based on four single-ended auxiliary amplifiers. To provide a high open-loop gain, the output stage of the auxiliary amplifier must be implemented using transistors with minimum channel length.

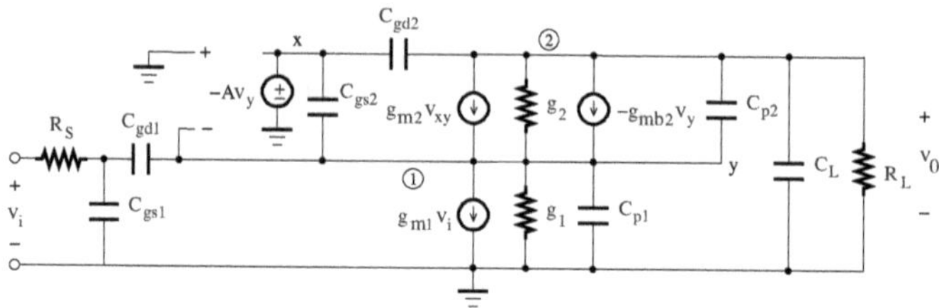

FIGURE 4.37
Small-signal equivalent model of the gain-enhanced cascode amplifier with parasitic and output capacitors.

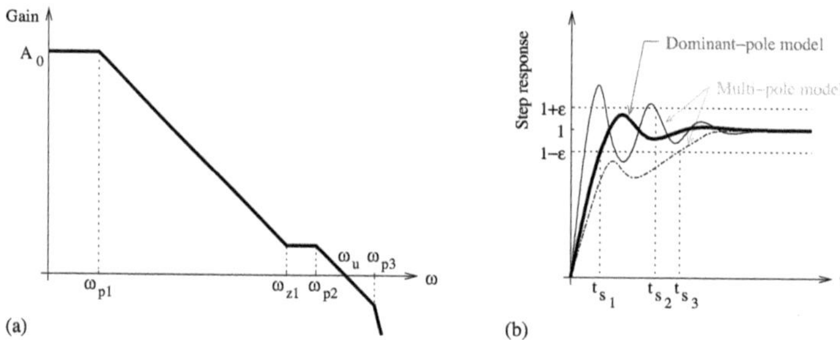

FIGURE 4.38
(a) Bode plot of the amplifier gain magnitude in open-loop configuration; (b) amplifier step response.

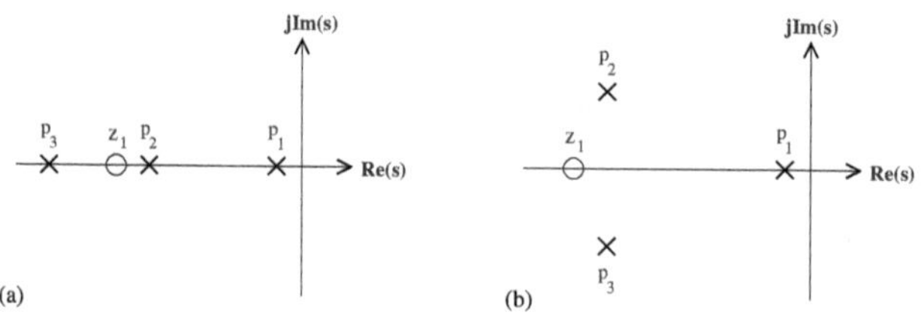

FIGURE 4.39
(a) Pole-zero locations of a nonoptimized gain-enhanced amplifier; (b) pole-zero locations of an optimized gain-enhanced amplifier.

In order to analyze the high-frequency response of the gain-enhanced amplifier, the effects of parasitic and output capacitors are taken into account in the amplifier of Figure 4.34(a) to derive the small-signal equivalent model depicted in Figure 4.37, where C_L is the output load capacitor, and C_{p_1} and C_{p_2} represent all parasitic capacitors seen at the output nodes of transistors T_1 and T_2, respectively. An intuitive analysis can show that the behavior of the gain-enhanced amplifier is determined by the first three poles and the first zero [21–23]. Figure 4.38(a) shows the Bode plot of the gain-enhanced amplifier gain. The first pole, which is located at the lowest frequency and is well separated from the second and third poles, is dominant. Its frequency is on the order of $g_1 g_2 / A_0'' g_{m1} g_{m2} C_L$ and is related to the gain-bandwidth product. When the closely spaced pole and zero (or doublet) characterized by ω_{p2} and ω_{z1} are located near the unity-gain frequency of the auxiliary amplifier, the settling time is increased due to slow components associated with a multi-pole system [24]. Because this is achieved even though the 60° phase margin criterion for the amplifier stability is met, a closed-loop configuration should be considered for the requirement specification to ensure a fast settling response similar to the one of an amplifier with a single dominant pole. In Figure 4.38(b), the settling responses to a voltage step input are shown as a function of time. Note that each small-signal settling time, t_{s_j} ($j = 1, 2, 3$), is defined as the minimum time required for the amplifier output voltage to settle to within an error tolerance, ϵ, of its final steady-state value.

For high-speed applications, the amplifier settling time should be optimized. It can be related to the characteristics of the closed-loop frequency response. For instance, an increase of the 3-dB bandwidth leads to a reduction in the settling time, while any oscillation or ringing in the closed-loop response may increase the settling time. By including the gain-enhanced amplifier in a closed loop with a constant feedback factor β, the closed-loop gain is given by

$$A_{CL}(s) = \frac{V_0(s)}{V_i(s)} = \frac{A(s)}{1 + \beta A(s)} \tag{4.296}$$

where $A(s)$ is the open-loop gain of the amplifier. With the assumption that the unity-gain frequency, ω_u'', of the auxiliary amplifier is greater than the first pole frequency, ω_{p_1}', of the main amplifier and is lower than the unity-gain frequency, ω_u', of the main amplifier, that is,

$$\omega_{p_1}' < \omega_u'' < \omega_u' \tag{4.297}$$

the first pole of the gain-boosted amplifier is moved at a lower frequency than the remaining poles and zero and can then be considered dominant. Hence, the transfer function of the gain-boosted amplifier is approximately given by

$$A(s) \simeq \frac{A_0}{1 + s/\omega_{p_1}} \tag{4.298}$$

This implies that the dominant-pole frequency response is obtained by designing the auxiliary amplifier to be slower than the main amplifier. Note that the gain-boosted amplifier and the main amplifier exhibit the same unity-gain frequency. Combining Equations (4.298) and (4.296) gives

$$A_{CL}(s) \simeq \frac{A_0/(1+\beta A_0)}{1+s/\omega_{p_1}(1+\beta A_0)} \tag{4.299}$$

In the feedback configuration, the first pole is moved at the frequency position $\omega_{p_1}(1+\beta A_0)$, which can be approximated by $\beta A_0 \omega_{p_1}$, or equivalently, $\beta \omega_u$, where ω_u is the unity-gain frequency.

To reduce the effect of slow-settling components on the transient response, the pole-zero doublet should be moved to a higher frequency. This can be achieved by keeping the unity-gain frequency, ω_u'', of the auxiliary amplifier greater than $\beta\omega_u$. On the other hand, the stability requirement is met provided ω_u'' remains lower than the second-pole frequency, ω_{p2}', of the main amplifier. Therefore,

$$\beta\omega_u < \omega_u'' < \omega_{p2}' \tag{4.300}$$

This is satisfied in practical implementations, where the load or compensation capacitor of the auxiliary amplifier is generally chosen to be much smaller than the one of the main amplifier. By increasing ω_u'', while satisfying the condition defined by Equation (4.300), the pole-zero doublet is pushed to higher frequencies until it merges with the third pole to generate a complex-conjugate pole pair and a real zero. Because the real part of this complex-conjugate pole pair determines the decrease rate in the output response, its optimum value can be found by further increasing ω_u'' to obtain a fast settling response. As a result, the phase margin of the auxiliary amplifier can be greater than 60°. Figure 4.39 shows the pole-zero locations of the nonoptimized and optimized gain-enhanced amplifiers.

4.6.2 Telescopic amplifier

The circuit diagram of a telescopic amplifier is depicted in Figure 4.40. A common-mode feedback stage sets the bias voltage V_{CM}. The different bias currents are derived from a master current source using mirrors. An internal bias circuit is adopted for transistors $T_3 - T_4$. The proper common-mode rejection is obtained by maintaining at least a voltage of $V_{DS(sat)}$ on transistors $T_1 - T_2$, $T_7 - T_8$, and T_9. From the amplifier circuit, we can obtain the following relations:

$$V_i^+ = V_{DD} - V_{SD_7} - V_{SD_5} - V_{DS_3} - V_{DS_1} + V_{GS_1} \tag{4.301}$$

$$= V_{GS_1} + V_{DS_9} + V_{SS} \tag{4.302}$$

Due to the amplifier symmetry, the transistors T_1 and T_2, T_3 and T_4, T_5 and T_6, and T_7 and T_8 are matched. Assuming that both input nodes are connected to the same dc voltage V_{ICM}, it can be shown that

$$\begin{aligned} &V_{DS_1(sat)} + V_{DS_9(sat)} + V_{T_n} + V_{SS} \\ &\quad \le V_{ICM} \le V_{DD} - V_{SD_7(sat)} - V_{SD_5(sat)} - V_{DS_3(sat)} + V_{T_n} \end{aligned} \tag{4.303}$$

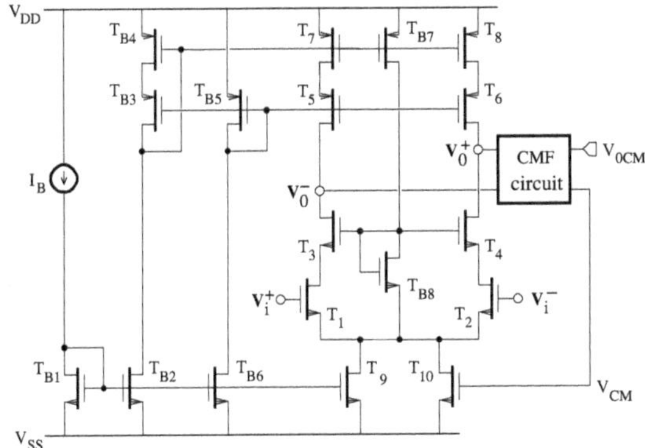

FIGURE 4.40
Circuit diagram of a telescopic amplifier.

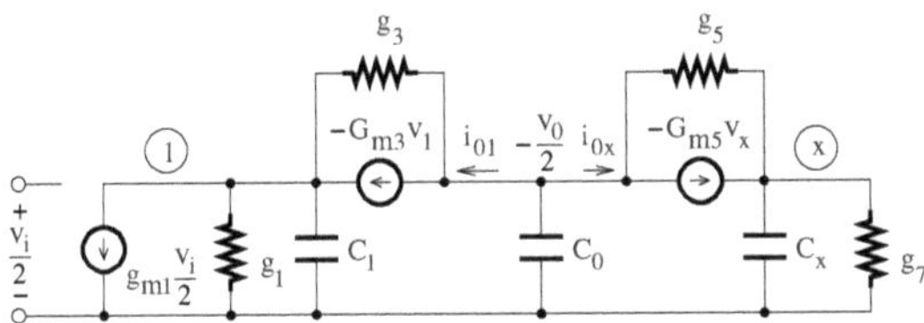

FIGURE 4.41
Equivalent circuit model of the telescopic amplifier.

On the other hand, the expression of V_0^- can be written as

$$V_0^- = V_{DD} - V_{SD_7} - V_{SD_5} \tag{4.304}$$

$$= V_{DS_3} + V_{DS_1} + V_{DS_9} + V_{SS} \tag{4.305}$$

To maintain the transistors in the saturation region, the voltage swing at the inverting output node should be of the form

$$V_{DS_3(sat)} + V_{DS_1(sat)} + V_{DS_9(sat)} + V_{SS} \leq V_0^- \leq V_{DD} - V_{SD_7(sat)} - V_{SD_5(sat)} \tag{4.306}$$

Similar equations can be derived for the voltage swing, V_0^+, at the noninverting output node. To proceed further, we exploit the fact that the voltage V_0^- can also be expressed as

$$V_0^- = V_{DS_3} + V_{DS_1} - V_{GS_1} + V_i^+ \tag{4.307}$$

For a given input common-mode voltage, V_{ICM}, the transistors T_1 and T_3 should be biased slightly above the saturation region to optimize the output swing. That is,

$$V_0^- \leq V_{ICM} + V_{DS_3(sat)} - V_{T_n} \tag{4.308}$$

Therefore, we can conclude that the maximum output voltage is dependent on the input common-mode voltage, whose variations may cause a premature clipping of the output voltage.

The telescopic amplifier, whose number of current paths between the supply voltages is two in comparison to four in the folded-cascode architecture, should consume the smaller static power. It can provide the superior speed but features the smaller differential output swing of about $2[V_{sup} - (3V_{DS(sat)} + 2V_{SD(sat)})]$, where $V_{sup} = V_{DD} - V_{SS}$ and the transistors of the same type are assumed to operate with identical saturation voltages.

The equivalent circuit model of the telescopic amplifier is depicted in Figure 4.41. Based on a dominant-pole model, the frequency response of the amplifier is primarily determined by the dc gain, the first and the second pole frequencies. Because the parasitic and output capacitors act like open circuits at dc, the dc gain is given by

$$A_0 = v_0/v_i \simeq -g_{m1}(r_{01} \parallel r_{0x}) \tag{4.309}$$

where

$$r_{01} = -\left.\frac{v_0}{2i_{01}}\right|_{v_i=0} = \frac{1}{g_1} + \frac{1}{g_3}\left(1 + \frac{G_{m3}}{g_1}\right) \tag{4.310}$$

and

$$r_{0x} = -\frac{v_0}{2i_{0x}} = \frac{1}{g_7} + \frac{1}{g_5}\left(1 + \frac{G_{m5}}{g_7}\right) \tag{4.311}$$

The first pole of the amplifier transfer function can be obtained as

$$\omega_{p_1} = -g_{m1}/A_0C_0 \tag{4.312}$$

where $C_0 = C_L + C_{gd_3} + C_{db_3} + C_{gd_5} + C_{db_5} + C_i'$. Here, C_0 is the total load capacitance at the output node, C_L denotes the external load capacitance connected to the output node, and C_i' represents the input capacitance of the CMF circuit. The second pole of the amplifier is due to the parasitic capacitance at the source of the n-channel cascode transistor, T_3 or T_4, and is of the form

$$\omega_{p_2} \simeq -G_{m3}/C_1 \tag{4.313}$$

where $C_1 = C_{gd_1} + C_{db_1} + C_{gs_3} + C_{sb_3}$ and $G_{m3} = g_{m3} + g_{mb3}$. The slew rate is given by

$$SR \simeq \frac{I_{D_9}}{C_L} \tag{4.314}$$

where $I_{D_9} = I_B$, and I_B is the bias current.

Note that the gain-boosting technique can also be used with the transistors $T_3 - T_4$ and $T_5 - T_6$, to further increase the achievable amplification gain.

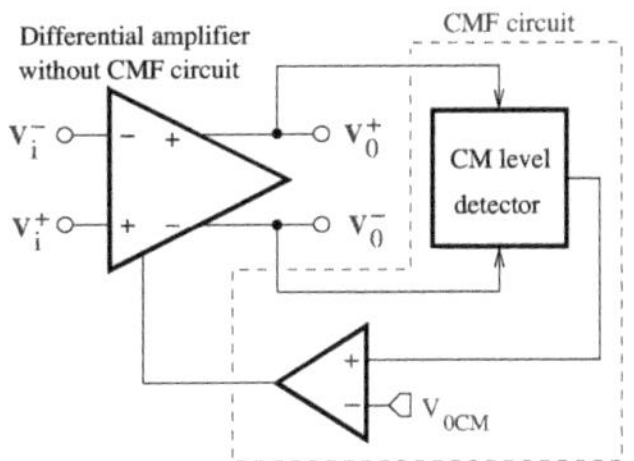

FIGURE 4.42
Block diagram of a differential amplifier.

4.6.3 Common-mode feedback circuits

In fully differential amplifiers, a common-mode circuit is used to control the common-mode voltage level and cancel the undesirable common-mode components of signals. Figure 4.42 shows the block diagram of a differential amplifier. The amplification is achieved by a fully differential stage. The common-mode circuit includes a common-mode level detector and a sense amplifier. Due to the negative feedback loop including the common-mode circuit, the output common-mode voltage is forced to be equal to V_{0CM} in the steady state. In the case where $V_{0CM} = (V_0^+ + V_0^-)/2$, the amplifier should exhibit a large and symmetric output dynamic range. To meet the stability requirement, all poles introduced in the common-mode loop should be well above the ones of the differential loop.

The output common-mode voltage of a single-stage differential amplifier can be defined using two resistors connected as shown in Figure 4.43(a).

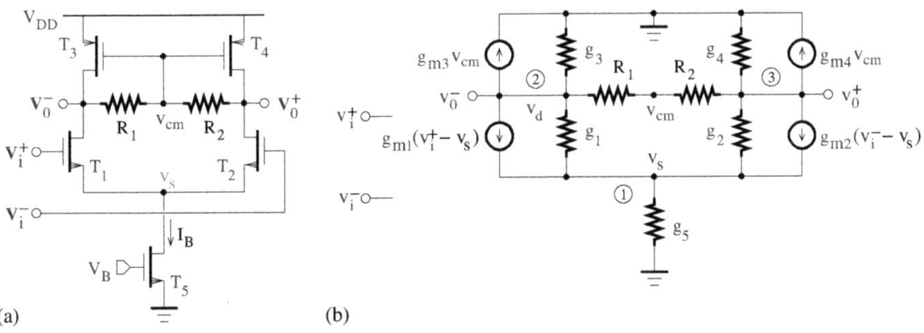

FIGURE 4.43
Differential amplifier (a) and its small-signal equivalent circuit (b).

Determine the voltage gain $A_v = v_0/v_i$, where $v_0 = v_0^+ - v_0^-$ and $v_i = v_i^+ - v_i^-$.
What is the main limitation of this approach?

Kirchhoff's current law equations for nodes 1, 2, and 3, can be written as,

$$g_{m_1}(v_i^+ - v_s) + g_{m_2}(v_i^- - v_s) + g_1(v_0^- - v_s) + g_2(v_0^+ - v_s) = g_5 v_s \quad (4.315)$$

$$g_{m_1}(v_i^+ - v_s) + g_1(v_0^- - v_s) + g_{m_3} v_{cm} + g_3 v_0^- + \frac{v_0^- - v_{cm}}{R_1} = 0 \quad (4.316)$$

and

$$g_{m_2}(v_i^- - v_s) + g_2(v_0^+ - v_s) + g_{m_4} v_{cm} + g_4 v_0^+ + \frac{v_0^+ - v_{cm}}{R_2} = 0 \quad (4.317)$$

where $g_{m_1} = g_{m_2}$, $g_{m_3} = g_{m_4}$, $g_1 = g_2$, and $g_3 = g_4$. Solving (4.315) for v_s, we obtain

$$v_s = \frac{g_{m_1}(v_i^+ + v_i^-)}{2(g_{m_1} + g_1) + g_5} + \frac{g_1(v_0^+ + v_0^-)}{2(g_{m_1} + g_1) + g_5} \quad (4.318)$$

The voltage at the common-source node is related to the input and output common-mode voltages. Subtracting (4.317) from (4.316) with the assumption that $R_1 = R_2 = R$, it can be shown that

$$A_v = \frac{v_0}{v_i} = \frac{g_{m_1}}{g_1 + g_3 + 1/R} \quad (4.319)$$

Hence, the differential output voltage is not affected by the common-mode voltage that is set as, $v_{cm} = (v_0^+ + v_0^-)/2$. In general, subtracting (4.317) from (4.316) yields

$$g_{m_1} v_i = (g_1 + g_3)(v_0^+ - v_0^-) + \frac{R_1 v_0^+ - R_2 v_0^-}{R_1 R_2} + \frac{R_2 - R_1}{R_1 R_2} v_{cm} \quad (4.320)$$

Assuming that $R_1 = R$ and $R_2 = R + \triangle R$, and using the relation $(1 + \triangle R/R)^{-1} \simeq 1 - \triangle R/R$, the next first-order approximation can be obtained:

$$g_{m_1} v_i \simeq \left(g_1 + g_3 + \frac{1}{R}\right)(v_0^+ - v_0^-) - \frac{\triangle R}{R^2}(v_0^+ - v_{cm}) \quad (4.321)$$

The output common-mode voltage will be well defined only if the resistors R_1 and R_2 are matched with a high precision. Furthermore, the use of resistors with a high value helps reduce mismatch errors.

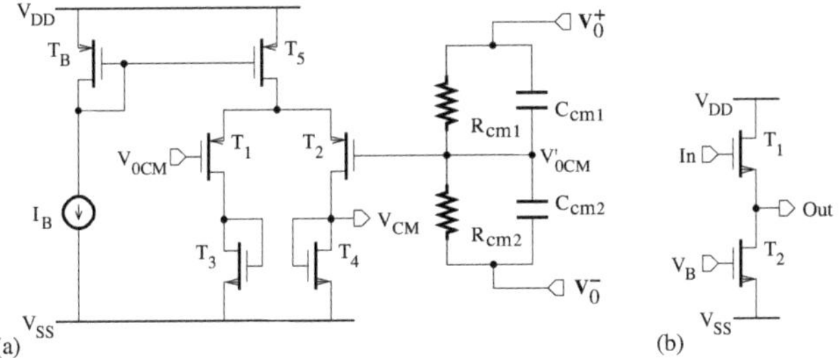

FIGURE 4.44
(a) Circuit diagram of a CMF using an RC network; (b) circuit diagram of a source follower.

4.6.3.1 Continuous-time common-mode feedback circuit

A continuous-time common-mode feedback circuit can be realized as shown in Figure 4.44(a) [25]. The CM detection is performed by an RC network with two equal-valued resistors and capacitors. The role of the capacitors is to stabilize the common-mode feedback loop. Using the principle of superposition and voltage division, we obtain

$$V'_{0CM} = \frac{V_0^+ + V_0^-}{2} \tag{4.322}$$

The voltages V'_{0CM} and V_{0CM} are applied at the inputs of a sense amplifier and the CM error detection and amplification are accomplished by the differential pairs consisting of $T_1 - T_2$. The CM level can be held at a reference potential by increasing or decreasing V_{CM}, which is assumed to be connected to a high resistance node. The implementation of the CMF loop then allows the voltage V_{CM} to be adjusted until $V_{0CM} = V'_{0CM}$. The gain of the CMF circuit should not be excessively high to eliminate any undesirable waveform oscillation at the amplifier output.

Because the resistors used in the CMF circuit can severely degrade the resulting dc gain of the differential amplifier, a source follower, as shown in Figure 4.44(b), is often inserted between each of the V_0^+ and V_0^- terminals and the corresponding node of the RC network. But, source followers can limit the signal swing and increase the noise level. It should also be noted that the use of resistors with a high value is limited by the required large silicon area.

Another structure of the common-mode feedback circuit is shown in Figure 4.45 [16,26]. The input transistors ($T_1 - T_4$) are assumed to operate in the saturation region. Ideally, they have to sense and amplify only the common-mode signal. Through the use of a feedback between the CMF circuit and the amplifier, the CM voltage and V_{0CM} are made equal. This is done by applying the CMF circuit output voltage, V_{CM}, to a suitable internal node of the

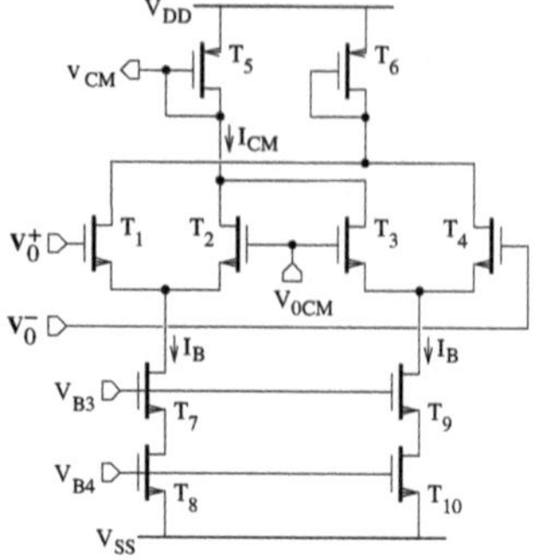

FIGURE 4.45
Circuit diagram of a CMF based on two differential transistor pairs.

amplifier. Let the current I_{CM} be expressed as

$$I_{CM} = I_{D_2} + I_{D_3} \tag{4.323}$$

Because each of the transistor pairs T_1 and T_2, and T_3 and T_4, operate as a differential stage biased by the tail current I_B, it can be found that the drain currents of T_2 and T_3 are, respectively, given by

$$I_{D_2} = \frac{I_B}{2} - \sqrt{2KI_B}\,\frac{V_0^+ - V_{OCM}}{2}\sqrt{1 - \frac{(V_0^+ - V_{OCM})^2}{2I_B/K}} \tag{4.324}$$

and

$$I_{D_3} = \frac{I_B}{2} - \sqrt{2KI_B}\,\frac{V_0^- - V_{OCM}}{2}\sqrt{1 - \frac{(V_0^- - V_{OCM})^2}{2I_B/K}} \tag{4.325}$$

Assuming that

$$V_0^+ = V_{OCM}' + V_0/2 \tag{4.326}$$

$$V_0^- = V_{OCM}' - V_0/2 \tag{4.327}$$

and

$$|V_{OCM}' - V_{OCM}| \ll |V_0| \tag{4.328}$$

we can obtain

$$I_{CM} \simeq I_B - \sqrt{2KI_B}\,\frac{V_{OCM}' - V_{OCM}}{2}\sqrt{1 - \frac{(V_0/2)^2}{2I_B/K}} \tag{4.329}$$

For the normal operation, it is required to have $V_{OCM}' = V_{OCM}$. As a result, the current I_{CM} becomes equal to I_B, which is identical to the constant tail current of the input differential stage of the amplifier requiring the CMF circuit.

The aforementioned CMF circuit has the advantage of exhibiting a high input impedance, which can be useful in the design of operational transconductance amplifiers with a reduced sensitivity to the loading effect.

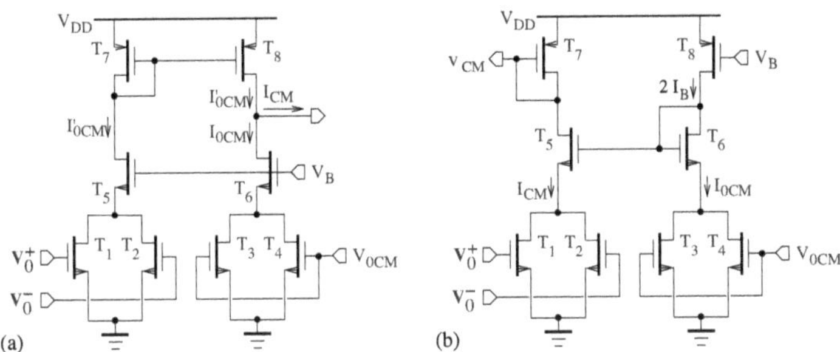

FIGURE 4.46
Circuit diagram of a CMF using two differential transistor pairs operating in the triode region: (a) Voltage output, (b) current output.

For differential amplifiers based on transistors operating in the triode region, the CMF circuit can be designed as shown in Figure 4.46(a) [27, 28]. Here, the transistors $T_1 - T_4$ operate in the triode region, while the remaining transistors are biased in the saturation region. The control current I_{CM} is generated by comparing the currents I'_{0CM} and I_{0CM}, which are related to the output CM voltage, V'_{0CM}, and the reference CM voltage, V_{0CM}, respectively. We can write

$$I_{CM} = I'_{0CM} - I_{0CM} \tag{4.330}$$

where

$$I'_{0CM} = I_{D_1} + I_{D_2} \tag{4.331}$$

and

$$I_{0CM} = I_{D_3} + I_{D_4} \tag{4.332}$$

Assuming that the transistors $T_1 - T_4$ are matched and the same voltage is maintained at the source of T_5 and T_6, we have

$$I_{D_1} = K[2(V'_{0CM} + V_0/2 - V_T)V_D - V_D^2] \tag{4.333}$$

$$I_{D_2} = K[2(V'_{0CM} - V_0/2 - V_T)V_D - V_D^2] \tag{4.334}$$

and

$$I_{D_3} = I_{D_4} = K[2(V_{0CM} - V_T)V_D - V_D^2] \tag{4.335}$$

where $V_D = V_{DS_1} = V_{DS_2} = V_{DS_3} = V_{DS_4}$. Hence,

$$I_{CM} = 4K(V'_{0CM} - V_{0CM}) \tag{4.336}$$

The variations of I_{CM} are exploited to adjust the CM level. When we have $I'_{0CM} = I_{0CM}$, the voltage V_{0CM} is reduced to

$$V_{0CM} = \frac{V_0^+ + V_0^-}{2} \tag{4.337}$$

In the case where the CM control signal should be applied to a high resistance node, a modified version of the CMF circuit, as shown in Figure 4.46(b) [29], may be suitable. The transistors $T_1 - T_6$ are assumed to be identical. Let the drain current of transistors T_1 and T_2, which operate in the triode region, be given by

$$I_{D_1} = K[2(V'_{0CM} + V_0/2 - V_T)V_D - V_D^2] \tag{4.338}$$

and

$$I_{D_2} = K[2(V'_{0CM} - V_0/2 - V_T)V_D - V_D^2] \tag{4.339}$$

where $V_D = V_{DS_1} = V_{DS_2}$. The current I_{CM} is of the form

$$I_{CM} = I_{D_1} + I_{D_2} = 2K[2(V'_{0CM} - V_T)V_D - V_D^2] \tag{4.340}$$

On the other hand, the voltage equation of the loop including T_1, T_5, T_6, and T_2 can be written as

$$V_{DS_1} + V_{GS_5} = V_{DS_3} + V_{GS_6} \tag{4.341}$$

Because the currents flowing through T_5 and T_6 are almost equal, it can be assumed that $V_{GS_5} \simeq V_{GS_6}$ and $V_{DS_1} \simeq V_{DS_3}$. The transistors T_3 and T_4 operate in the triode region, and their drain currents are given by

$$I_{D_3} = I_{D_4} = K[2(V_{0CM} - V_T)V_D - V_D^2] = I_B \tag{4.342}$$

where $V_D = V_{DS_3} = V_{DS_4}$. Taking into account the fact that the triode region is characterized by $0 < V_{DS} < V_{GS} - V_T$, or equivalently $0 < V_D < V_{0CM} - V_T$, the second-order equation (4.342) can be solved for V_D. Therefore,

$$V_D = V_{0CM} - V_T - \sqrt{(V_{0CM} - V_T)^2 - I_B/K} \tag{4.343}$$

where $(V_{0CM} - V_T)^2 - I_B/K \geq 0$. Substituting Equation (4.343) into (4.340) gives

$$I_{CM} = 2K\left[2(V'_{0CM} - V_{0CM})\left(V_{0CM} - V_T - \sqrt{(V_{0CM} - V_T)^2 - I_B/K}\right) + I_B/K\right] \tag{4.344}$$

When $V'_{0CM} = V_{0CM}$, the current I_{CM} is reduced to $2I_B$. The variations in I_{CM} take place around $2I_B$, depending on the difference $V'_{0CM} - V_{0CM}$.

The drawbacks of this approach are essentially due to the fact that the input transistors are required to operate in the triode region, instead of the saturation region. As a consequence, the amplifier output swing and the small-signal gain of the CMF circuit can be reduced.

4.6.3.2 Switched-capacitor common-mode feedback circuit

Switched-capacitor (SC) common-mode feedback (CMF) circuits have several advantages over continuous-time counterparts, including improved output range and linearity, low power dissipation, and silicon area. The circuit diagram and clock waveforms of an SC CMF circuit are shown in Figure 4.47 [30].

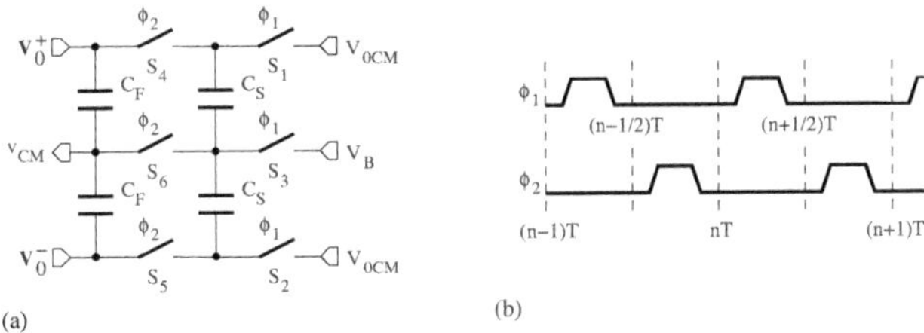

FIGURE 4.47
Circuit diagram of a switched-capacitor CMF circuit.

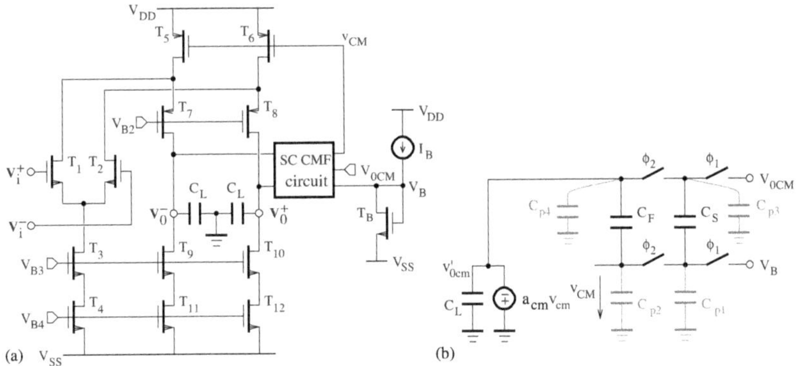

FIGURE 4.48
(a) Folded-cascode fully differential amplifier; (b) equivalent circuit model of the SC CMF loop ($v'_{0cm} = (v_0^+ + v_0^-)/2$ and $v_{CM} = V_{CM} + v_{cm}$).

Capacitors C_F establish an ac feedback path between the amplifier output and CM nodes. To set the dc level of the CM voltage, the capacitors C_S are charged to the difference between the bias voltage, V_B, and the desired CM voltage, V_{OCM}, during the first phase, ϕ_1, and then connected across the capacitors C_F during the clock phase ϕ_2. Because both terminals of the capacitors C_S are periodically switched, a nonoverlapping two-phase clock signal is required to control the switches so as to minimize errors due to charge injection and clock feedthrough. The detection of the actual output CM level is performed using the ac voltage divider composed of capacitors C_F to determine the average of both amplifier outputs, which is periodically compared to the desired CM output voltage available across the switched capacitors C_S. The CM control voltage supplied to the amplifier is then adjusted based on the difference between the actual and desired CM output voltages. When the steady state is reached, there is no longer a charge transfer between the capacitors C_F and C_S. Ideally, the difference between the amplifier output CM voltage and the CM control voltage should then be equal to the difference between the desired CM voltage and the bias voltage.

With reference to the fully differential amplifier depicted in Figure 4.48(a), the equivalent model of the common-mode feedback loop [31] can be derived as shown in Figure 4.48(b). During the clock phase, ϕ_2, or say, for $(n-1/2)T < t \leq nT$, where T is the clock signal period, the equation for the charge conservation at the v_{CM} node is of the form

$$\begin{aligned} &C_S(V_B - V_{OCM}) + C_{p1}V_B \\ &\quad +C_F[v_{CM}(n-1/2) - v'_{0cm}(n-1/2)] + C_{p2}v_{CM}(n-1/2) \\ &\qquad = (C_{p1}+C_{p2})v_{CM}(n) + (C_S+C_F)[v_{CM}(n) - v'_{0cm}(n)] \end{aligned} \tag{4.345}$$

where $v'_{0cm} = (v_0^+ + v_0^-)/2$, and C_{p1} and C_{p2} are parasitic capacitances. When the clock phase ϕ_1 is high, that is, for $(n-1)T \leq t < (n-1/2)T$, we have

$$v_{CM}(n-1/2) = v_{CM}(n-1) \tag{4.346}$$

and

$$v'_{0cm}(n-1/2) = v'_{0cm}(n-1) \tag{4.347}$$

Combining Equations (4.347), (4.346), and (4.345) gives

$$\begin{aligned} &C_S(V_B - V_{OCM}) + C_{p1}V_B \\ &\quad +C_F[v_{CM}(n-1) - v'_{0cm}(n-1)] + C_{p2}v_{CM}(n-1) \\ &\qquad = (C_{p1}+C_{p2})v_{CM}(n) + (C_S+C_F)[v_{CM}(n) - v'_{0cm}(n)] \end{aligned} \tag{4.348}$$

Because $v_{CM} = V_{CM} + v_{cm}$, it can be deduced that

$$v'_{0cm}(n) = -a_{cm}v_{cm}(n) = a_{cm}(V_{CM} - v_{CM}) \tag{4.349}$$

where V_{CM} and v_{cm} denote the dc and ac components of the CM control voltage, respectively. The system of Equations (4.348) and (4.349) can be solved for v_{CM} and v'_{0cm}. Hence,

$$v_{CM}(n) = r v_{CM}(n-1) + p \tag{4.350}$$

and

$$v'_{0cm}(n) = r v'_{0cm}(n-1) + q \tag{4.351}$$

where

$$r = \frac{(1 + a_{cm})C_F + C_{p2}}{(1 + a_{cm})(C_S + C_F) + C_{p1} + C_{p2}} \quad (4.352)$$

$$p = \frac{C_S(a_{cm}V_{CM} - V_{0CM}) + (C_S + C_{p1})V_B}{(1 + a_{cm})(C_S + C_F) + C_{p1} + C_{p2}} \quad (4.353)$$

and

$$q = a_{cm}[(1 - r)V_{CM} - p]. \quad (4.354)$$

Using the iterative method, the solutions of the first-order linear autonomous difference equation, (4.350), can be computed as,

$$v_{CM}(n) = p\sum_{k=0}^{n-1} r^k + r^n v_{CM}(0) \quad (4.355)$$

$$= p\Big(\frac{1 - r^n}{1 - r}\Big) + r^n v_{CM}(0) \quad (4.356)$$

where $r \neq 1$. Similarly, for Equation (4.351), it can be shown that

$$v'_{0CM}(n) = q\Big(\frac{1 - r^n}{1 - r}\Big) + r^n v'_{0CM}(0) \quad (4.357)$$

The sequences $v_{CM}(n)$ and $v'_{0CM}(n)$ can then be expressed as

$$v_{CM}(n) = \overline{v}_{CM} + r^n[v_{CM}(0) - \overline{v}_{CM}] \quad (4.358)$$

and

$$v'_{0CM}(n) = \overline{v'}_{0CM} + r^n[v'_{0CM}(0) - \overline{v'}_{0CM}] \quad (4.359)$$

where $v_{CM}(0)$ and $v'_{0CM}(0)$ represent initial values, and the steady-state values are given by

$$\overline{v}_{CM} = \lim_{n\to\infty} v_{CM}(n) = \frac{p}{1 - r} = \frac{V_{CM} + \left[\left(1 + \dfrac{C_{p1}}{C_S}\right)V_B - V_{0CM}\right]\dfrac{1}{a_{cm}}}{1 + \left(1 + \dfrac{C_{p1}}{C_S}\right)\dfrac{1}{a_{cm}}} \quad (4.360)$$

and

$$\overline{v'}_{0CM} = \lim_{n\to\infty} v'_{0CM}(n) = \frac{q}{1 - r} = \frac{V_{0CM} + \left(1 + \dfrac{C_{p1}}{C_S}\right)(V_{CM} - V_B)}{1 + \left(1 + \dfrac{C_{p1}}{C_S}\right)\dfrac{1}{a_{cm}}} \quad (4.361)$$

Because $r < 1$, it appears that $v_{CM}(n)$ and $v'_{0CM}(n)$ will converge to steady state. To proceed further, we can write

$$\overline{v'}_{0CM} - \overline{v}_{CM} = \frac{(V_{0CM} - V_B)\left(1 + \dfrac{1}{a_{cm}}\right) + \left(V_{CM} - V_B\right)\dfrac{C_{p1}}{C_S} - V_B\dfrac{C_{p1}}{a_{cm}C_S}}{1 + \left(1 + \dfrac{C_{p1}}{C_S}\right)\dfrac{1}{a_{cm}}} \tag{4.362}$$

Assuming that the CM gain a_{cm} is very high, we arrive at

$$\overline{v'}_{0CM} - \overline{v}_{CM} \simeq V_{0CM} - V_B + (V_{CM} - V_B)C_{p1}/C_S \tag{4.363}$$

For accurate control of the output CM, the values of C_S and V_B should be chosen to make the effect of the last term negligible.

Ideally, the SC CMF circuit should exhibit a high gain and bandwidth. For a fast and accurate response, the value of C_S is chosen to be greater than the one of C_F, which can be determined by making the bandwidth of the CM loop at least equal to the one of the differential loop. Note that the amplifier load due to the aforementioned SC CMF circuit is different from one clock phase to another, and may affect the resulting settling time and bandwidth.

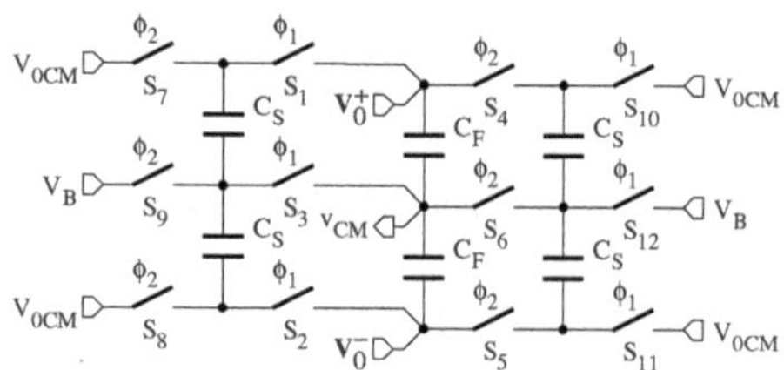

FIGURE 4.49
Circuit diagram of a switched-capacitor CMF circuit with symmetric loading.

For applications where the output voltage is valid during the whole clock period, the SC CMF circuit structure shown in Figure 4.49 [32] may be suitable. Additional capacitors and switches are used to improve the circuit symmetry. The capacitors are switched such that the effect of the SC CMF circuit on the amplifier loading remains the same for both clock phases.

4.6.4 Pseudo fully differential amplifier

In the cases where the variations of the common-mode voltage at the amplifier output remain small, fully differential amplifiers can also be implemented without an extra CMF circuit, which has the drawback of increasing noise and limiting the output swing and speed.

In the single-stage pseudo fully differential amplifier depicted in Figure 4.50 [33, 34], the solution adopted for the stabilization of the common-mode output voltage, $V_{0CM} = (V_0^+ + V_0^-)/2$, defined by the biasing circuit is

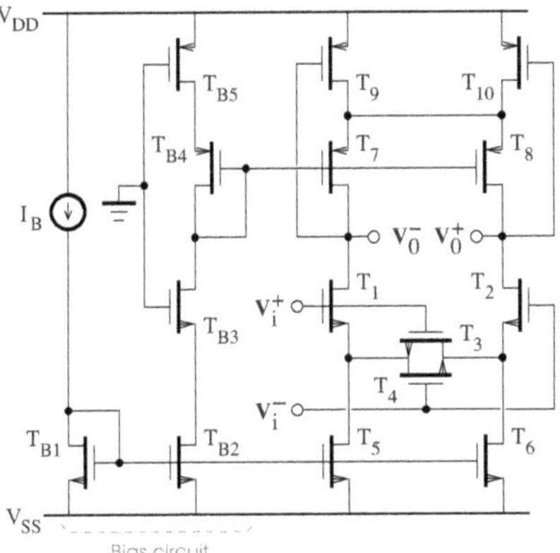

FIGURE 4.50
Circuit diagram of a pseudo fully differential transconductance amplifier using a voltage-controlled output load.

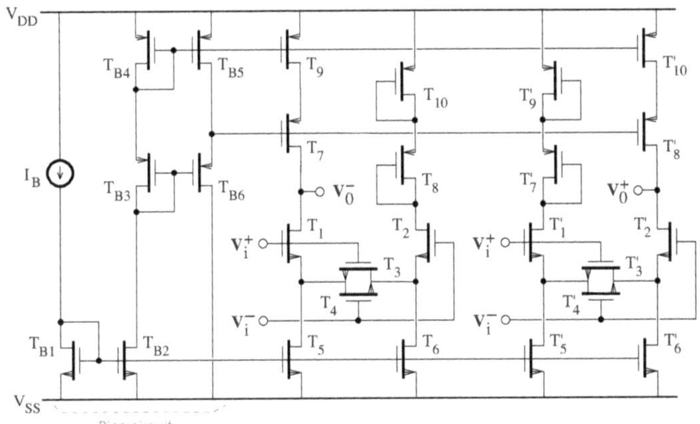

FIGURE 4.51
Circuit diagram of a pseudo fully differential transconductance amplifier with duplicated differential stages.

based on a voltage-controlled output load. The input signal is applied to the gates of transistors $T_1 - T_4$. The bias currents are provided by transistors T_5 and T_6, and the output loads consist of $T_7 - T_{10}$. The linearity of the amplifier transconductance, g_m, is improved by the source degeneration provided by $T_3 - T_4$, which should operate in the triode region. The resulting transfer characteristic is similar to the one of a source-coupled transistor pair, but with a different biasing condition.

An identical dc current flows through T_7 and T_8, which are biased by the same gate voltage chosen such that T_9 and T_{10} operate in the triode region with source-drain voltages set to a stable value. Any variation in the common-mode voltage first affects the resistance of T_9 and T_{10} and then the source-gate voltages of T_7 and T_8. Because the dc currents flowing through the transistors remain constant, this induces a modification of the source-drain voltages of T_7 and T_8, thereby forcing the compensation of the common-mode voltage variations.

The common-mode output voltage can also be defined by duplicating the differential stage [35], as shown in Figure 4.51. In this approach, each differential stage operates as a single-ended circuit and the dc components of the output voltage can be set to a desired level, making the use of a CMF circuit unnecessary. The currents available at the noninverting and inverting output nodes can be expressed as $I_0^+ \simeq I_B/2 + \triangle i$ and $I_0^- \simeq I_B/2 - \triangle i$, respectively. By taking the difference of output currents, the common-mode component is cancelled. Hence, $I_0^+ - I_0^- \simeq 2\triangle i$, where $\triangle i$ represents the output contribution associated with the differential mode. But here, the cancellation of the common-mode component can be limited by the achievable transistor matching. It should also be noted that the variations of the common-mode signals must remain low enough to prevent the transistors from moving outside their operating region.

In general, transconductance amplifiers are designed to have a sufficient g_m tuning range for the correction of IC process variations. The aforementioned transconductance amplifiers are suitable for high-frequency applications due to their single-stage structure. However, they are limited by the achievable gain-bandwidth product.

4.7 Multistage amplifier structures

In the multistage amplifier design approach, the requirement of a high gain is met by combining a differential stage and extra output stages. Various architectures are available for the implementation of output stages, which should be designed to provide an adequate level of the signal power to the amplifier load. The output stage should exhibit a lower output resistance to drive the load as if it were an ideal voltage source.

Output stages are amplification circuits, which can be described in terms of the conduction angle, or the portion of the input signal cycle for which there is an output signal. The output stage can be of the class A, class B, or class AB type, depending on the transistor's conduction angle which typically varies between 0° and 360°.

For class A amplifiers, the conduction angle is 360°. Because the transistors are biased to never reach the cutoff region during the normal operation, a linear amplification of the input signal is performed. However, class A amplifiers typically have a low power efficiency. This is due to the fact that an important part of the supply power is required to bias the transistors during the whole cycle of the input signal such that the ratio of the output signal power to the total input power is relatively small.

Class B amplifiers are generally based on push-pull configuration, consisting of complementary transistors, each of which conducts on alternating half-cycles of the input signal. Each transistor has a conduction angle of about 180°. It is then biased to operate in the linear range during approximately one-half of the input signal cycle and is turned off for the other half cycle. Due to the fact that a gate-source voltage greater than the threshold voltage is required to initiate the transistor conduction, there is a transition region where both transistors of the push-pull amplifier are turned off, thereby producing crossover distortions in the output waveform. Note that in a push-pull configuration, the p-channel transistor sources (or pushes) current to the output load while the n-channel transistor sinks (or pulls) current from the output load.

Class AB amplifiers have the same configuration as class B gain stages, but both transistors are biased slightly above the cutoff region when the value of the input signal is around zero. The conduction angle of each transistor is slightly greater than 180° to reduce the crossover distortions.

4.7.1 Output stage

A common-source amplifier with an n channel input transistor is shown in Figure 4.52(a). Assuming that the transistors are biased in the saturation

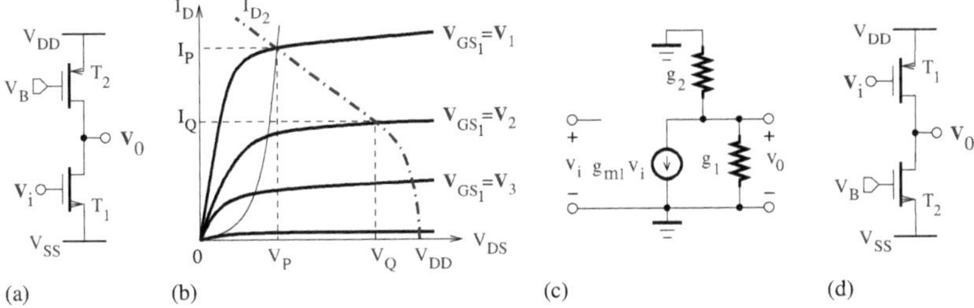

FIGURE 4.52
(a) Circuit diagram of a common-source gain stage with n-channel input transistor; (b) transistor I-V characteristics; (c) small-signal equivalent model; (d) circuit diagram of a common-source gain stage with p-channel input transistor.

region, we have

$$V_{DS_1} \geq V_{DS_1(sat)} = V_{GS_1} - V_{T_n} \tag{4.364}$$

and

$$V_{SD_2} \geq V_{SD_2(sat)} = V_{SG_2} + V_{T_p} \tag{4.365}$$

where $V_{DS_1} = V_0 - V_{SS}$, $V_{GS_1} = V_i - V_{SS}$, $V_{SD_2} = V_{DD} - V_0$, and finally $V_{SG_2} = V_{DD} - V_B$. The output voltage should then remain in the range

$$V_{SS} + V_{DS_1(sat)} \leq V_0 \leq V_{DD} - V_{SD_2(sat)} \tag{4.366}$$

For the normal operation of the output stage, it is required that

$$V_i - V_{T_n} \leq V_0 \leq V_B - V_{T_p} \tag{4.367}$$

The I-V characteristics of the input and load transistors are depicted in Figure 4.52(b). The input voltage, which is directly related to the gate-source voltage, V_{GS_1}, determines the curves of the characteristics to be considered in the case of T_1 for the determination of the operating point. Because the voltage V_B is constant, only the curve of the I-V characteristic associated with $V_{SG_2} = V_{DD} - V_B$ is retained for the load transistor T_2. To achieve a class A amplification, the operating point should be set halfway between the points (I_P, V_P) and (I_Q, V_Q). The small-signal equivalent model of the amplifier is shown in Figure 4.52(c). Applying Kirchhoff's current law gives

$$g_{m_1} v_i + g_1 v_0 + g_2 v_0 = 0 \tag{4.368}$$

The voltage gain can then be computed as

$$A = \frac{v_0}{v_i} = -g_{m_1} r_0 \tag{4.369}$$

where

$$r_0 = 1/(g_1 + g_2) \tag{4.370}$$

Here, $g_1 = g_{ds_1}$, $g_2 = g_{ds_2}$, and r_0 represents the output resistance.

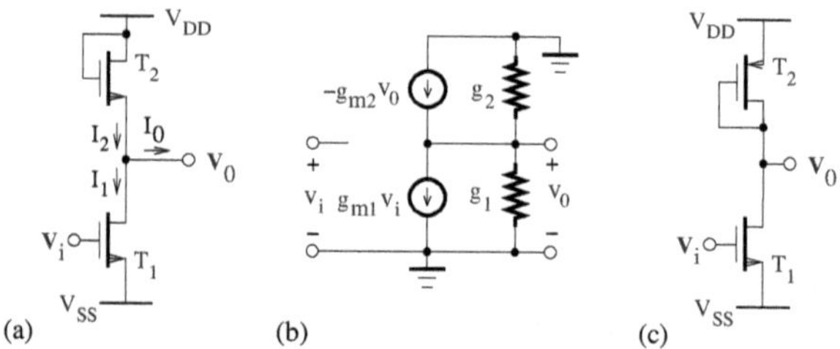

FIGURE 4.53
(a) Common-source gain stage with n-channel input transistor and diode-connected n-channel transistor load; (b) small-signal equivalent model; (c) common-source gain stage with n-channel input transistor and diode-connected p-channel transistor load.

A diode-connected transistor can also be used as a load for the common-source gain stage, as shown in Figure 4.53(a) for the case of an n-channel transistor. It is assumed that both transistors operate in the saturation region. For the transistor T_1, we can obtain

$$V_{DS_1} \geq V_{DS_1(sat)} = V_{GS_1} - V_{T_n} \tag{4.371}$$

where $V_{DS_1} = V_0 - V_{SS}$ and $V_{GS_1} = V_i - V_{SS}$. Therefore, the output voltage is such that $V_0 \geq V_i - V_{T_n}$. The transistor T_2 is saturated because its gate and drain are connected together. The currents flowing through the transistors T_1 and T_2 are respectively given by

$$I_1 = K_1(V_{GS_1} - V_{T_n})^2 \tag{4.372}$$

and

$$I_2 = K_2(V_{GS_2} - V_{T_n})^2 \tag{4.373}$$

where $V_{GS_2} = V_{DD} - V_0$. Applying Kirchhoff's current law at the output node gives

$$I_2 = I_1 + I_0 \tag{4.374}$$

Substituting Equations (4.372) and (4.373) into Equation (4.374), we obtain

$$V_0 = V_{DD} - V_{T_n} - \sqrt{[K_1(V_i - V_{SS} - V_{T_n})^2 + I_0]/K_2} \tag{4.375}$$

When the amplifier stage is connected to a high resistance load, the output current can be considered negligible. Hence, $I_0 \simeq 0$ and

$$V_0 = V_{DD} - V_{T_n} - \sqrt{K_1/K_2}(V_i - V_{SS} - V_{T_n}) \tag{4.376}$$

The dc input voltage is then amplified by a factor equal to the square root of transconductance parameters.

The small-signal equivalent model is depicted in Figure 4.53(b). Using Kirchhoff's current law, we can obtain

$$g_{m_1} v_i + g_{m_2} v_0 + g_1 v_0 + g_2 v_0 = 0 \tag{4.377}$$

Hence,

$$A = \frac{v_0}{v_i} = \frac{-g_{m_1}}{g_{m_2}} \left(\frac{1}{1 + \dfrac{g_1 + g_2}{g_{m_2}}} \right) \tag{4.378}$$

The gain is lower than the one of the aforementioned amplifier. Assuming that $(g_1 + g_2)/g_{m_2} \ll 1$, and taking into account the fact that T_1 and T_2 are biased by the same current I_D, $g_{m_1} = 2\sqrt{K_1 I_D}$ and $g_{m_2} = 2\sqrt{K_2 I_D}$, the voltage gain is reduced to

$$A = \frac{v_0}{v_i} \simeq \frac{-g_{m_1}}{g_{m_2}} = -\sqrt{\frac{W_1/L_1}{W_2/L_2}} \tag{4.379}$$

Due to the gain dependence on the width-to-length ratios of transistors, the

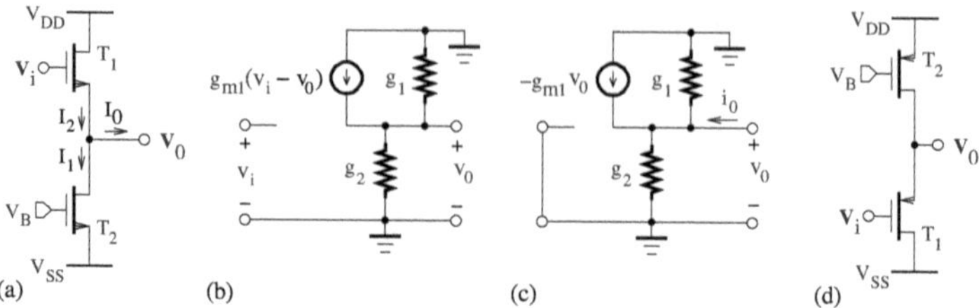

FIGURE 4.54
(a) nMOS source follower; (b) small-signal equivalent model; (c) small-signal equivalent model for the determination of the output resistance; (d) pMOS source follower.

requirement of a high gain is limited by the available amplifier area. Figure 4.53(c) shows the circuit diagram of a common-source gain stage with a *p*-channel transistor load.

A source follower, as shown in Figure 4.54(a) can be used as a buffer amplifier or dc level shifter. Both the input and load transistors are biased in the saturation region. Hence,

$$V_{DS_1} \geq V_{DS_1(sat)} = V_{GS_1} - V_{T_n} \tag{4.380}$$

and

$$V_{DS_2} \geq V_{DS_2(sat)} = V_{GS_2} - V_{T_n} \tag{4.381}$$

where $V_{GS_1} = V_i - V_O$ and $V_{GS_2} = V_B - V_{SS}$. Because $V_{DS_2} = V_O - V_{SS}$, the input range can be obtained as

$$V_{SS} + V_{DS_2(sat)} + V_{DS_1(sat)} + V_{T_n} \leq V_i \leq V_{DD} + V_{T_n} \tag{4.382}$$

The normal operation also requires that $V_O \geq V_B - V_{T_n}$. The currents that flow through the transistors T_1 and T_2 can respectively be written as

$$I_1 = K_1(V_{GS_1} - V_{T_n})^2 \tag{4.383}$$

and

$$I_2 = K_2(V_{GS_2} - V_{T_n})^2 \tag{4.384}$$

From Kirchhoff's current law at the output node, we obtain

$$I_2 = I_1 + I_O \tag{4.385}$$

To find the output voltage, Equation (4.385) can be rewritten using Equations (4.383) and (4.384) as

$$V_O = V_i - V_{T_n} - \sqrt{[K_2(V_B - V_{SS} - V_{T_n})^2 + I_O]/K_1} \tag{4.386}$$

The output voltage is simply equal to the input voltage minus a term determined by the amplifier and transistor characteristics. The source follower then provides a dc voltage gain of unity.

The small-signal equivalent model is depicted in Figure 4.54(b). Applying Kirchhoff's current law at the output node, we find

$$g_{m_1}(v_i - v_O) - g_1 v_O - g_2 v_O = 0 \tag{4.387}$$

The voltage gain is then obtained as

$$A = \frac{v_O}{v_i} = \frac{g_{m_1}/(g_1 + g_2)}{1 + g_{m_1}/(g_1 + g_2)} \tag{4.388}$$

For typical values of the transistor characteristics, $g_{m_1} \gg (g_1 + g_2)$ and $A \simeq 1$. To derive the output resistance, the input node is connected to the ground and the amplifier is driven by a voltage source applied to the output node. With reference to the small-signal equivalent model of Figure 4.54(c), it can be shown that

$$i_O = g_{m_1} v_O + g_1 v_O + g_2 v_O \tag{4.389}$$

Hence,

$$r_O = \frac{v_O}{i_O} = \frac{1}{g_{m_1} + g_1 + g_2} \simeq \frac{1}{g_{m_1}} \tag{4.390}$$

Because the small-signal output resistance, r_O, is generally less than 1 kΩ, it is considered small. Figure 4.54(d) shows the circuit diagram of a pMOS source follower.

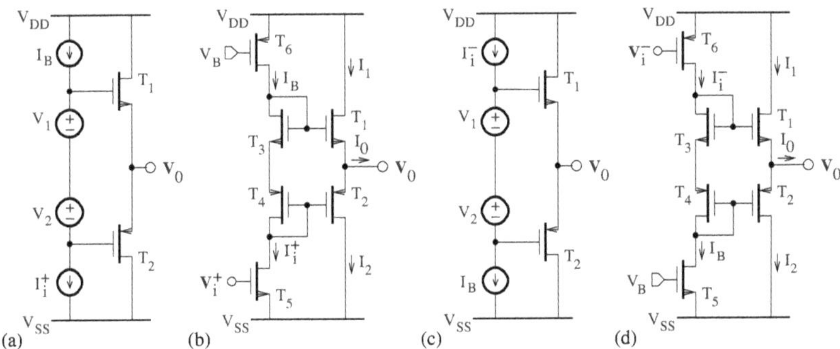

FIGURE 4.55
(a) Principle and (b) circuit diagram of the first version of the complementary source follower; (c) principle and (d) circuit diagram of the second version of the complementary source follower.

The aforementioned gain stages are suitable for class A operation, but output stages can also be designed using class AB circuit configuration. A class AB output stage should preferably provide a high maximum output current, while requiring only a low quiescent current. The principle and circuit diagram of complementary source followers are illustrated in Figures 4.55(a) and (b) in the case where the input voltage is applied to an n-channel transistor, and

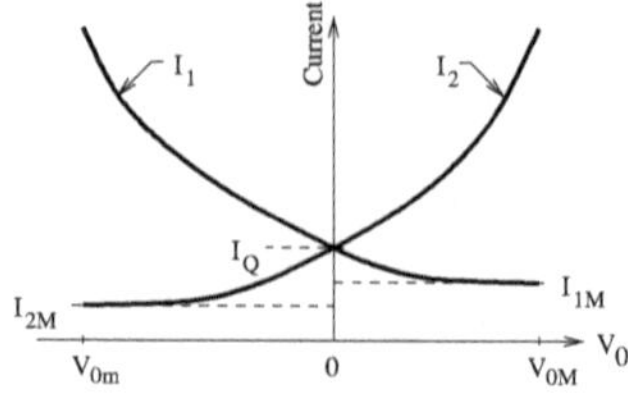

FIGURE 4.56
Plot of the currents I_1 and I_2.

in Figures 4.55(c) and (d) for designs requiring a p-channel input transistor. Two floating dc voltages are used to set the gate-source voltages of transistors T_1 and T_2, thereby defining the quiescent currents flowing through T_1 and T_2 so that crossover distortions are eliminated [36]. Applying Kirchhoff's voltage law to the translinear loop including $T_1 - T_4$, we obtain

$$V_{GS_3} + V_{SG_4} = V_{GS_1} + V_{SG_2} \tag{4.391}$$

where

$$V_{GS_1} = V_{T_n} + \sqrt{\frac{I_1}{K_{n_1}}} \tag{4.392}$$

$$V_{SG_2} = -V_{T_p} + \sqrt{\frac{I_2}{K_{p_2}}} \tag{4.393}$$

$$V_{GS_3} = V_{T_n} + \sqrt{\frac{I_B}{K_{n_3}}} \tag{4.394}$$

and

$$V_{SG_4} = -V_{T_p} + \sqrt{\frac{I_B}{K_{p_4}}} \tag{4.395}$$

Hence,

$$\sqrt{\frac{I_1}{K_{n_1}}} + \sqrt{\frac{I_2}{K_{p_2}}} = \left(\sqrt{\frac{1}{K_{n_3}}} + \sqrt{\frac{1}{K_{p_4}}}\right)\sqrt{I_B} \tag{4.396}$$

Note that the current I_B is set by the bias voltage V_B. The output current can be written as

$$I_0 = I_1 - I_2 \tag{4.397}$$

Considering that only dc voltages are applied to the output stage, so that $V_0 = 0$ and $I_0 = 0$, a current with the same value now flows through the transistors $T_1 - T_2$ and $T_3 - T_4$. That is, $I_1 = I_2 = I_Q$, where the quiescent current, I_Q, is given by

$$I_Q = I_B \frac{\left(\sqrt{\frac{1}{K_{n_3}}} + \sqrt{\frac{1}{K_{p_4}}}\right)^2}{\left(\sqrt{\frac{1}{K_{n_1}}} + \sqrt{\frac{1}{K_{p_2}}}\right)^2} \tag{4.398}$$

A class AB operation is achieved because both transistors T_1 and T_2 are biased to conduct when the output voltage is reduced to zero. Using the square-law characteristic of MOS transistors, the currents I_1 and I_2 can be expressed as

$$I_1 = K_{n_1}(V_{GS_1} - V_{T_n})^2 \tag{4.399}$$

and

$$I_2 = K_{p_2}(V_{SG_2} + V_{T_p})^2 \tag{4.400}$$

where

$$V_{GS_1} = V_{G_1} - V_0 = V_{DD} - V_{SD_6} - V_0 \tag{4.401}$$

and

$$V_{SG_2} = V_0 - V_{G_2} = V_0 - V_{DS_5} - V_{SS} \tag{4.402}$$

The swing of the output voltage, V_0, is limited due to the requirement of maintaining the transistors in the saturation region. Considering the path from the output node to the positive supply voltage, we can write

$$V_0 = V_{DD} - V_{SD_6} - V_{GS_1} \tag{4.403}$$

The maximum value of the output voltage is then given by

$$V_{0M} = V_{DD} - V_{SD_6(sat)} - V_{DS_1(sat)} - V_{T_n} \tag{4.404}$$

In the case of the path from the output node to the negative supply voltage, it can be shown that

$$V_0 = V_{SS} + V_{DS_5} + V_{SG_2} \tag{4.405}$$

Therefore, the maximum value of the output voltage is of the form

$$V_{0m} = V_{SS} + V_{DS_5(sat)} + V_{SD_2(sat)} + V_{T_p} \tag{4.406}$$

The graphical representations of currents I_1 and I_2 are shown in Figure 4.56, where the minimum values of I_1 and I_2 are respectively related to V_{0M} and V_{0m}.

An improved class AB operation can be achieved by an output stage based on the principle illustrated in Figure 4.57(a), where an independent loop is used to set each floating voltage required for the biasing of the output transistor. In the class AB output stage implementation shown in Figure 4.57(b) [37, 38], the common-source connected output transistors, T_1 and

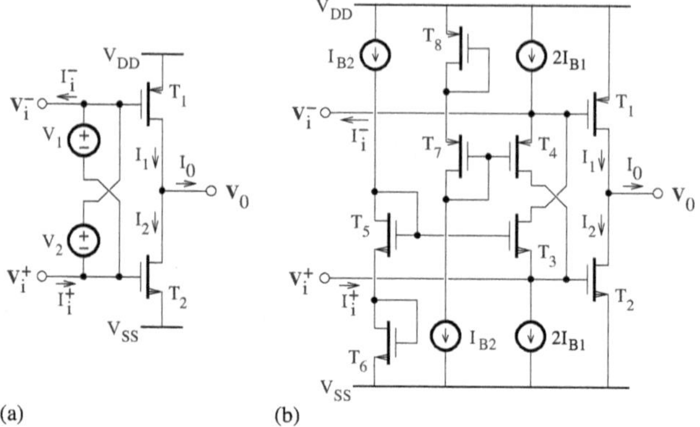

FIGURE 4.57
(a) Principle and (b) circuit diagram of a complementary common source gain stage.

T_2, are driven by in-phase input signals. The quiescent current flowing through the output transistors is determined by two independent translinear loops. Let the transistors of the same channel type be designed with identical threshold voltages and mobility parameters. Using Kirchhoff's voltage law for the loop including T_1,T_4, T_7, and T_8, we obtain

$$V_{SG_1} + V_{SG_4} = V_{SG_7} + V_{SG_8} \tag{4.407}$$

Based on the square law characteristic of transistors, the current I_1 can be written as

$$I_1 = K_{p_1}\left(\sqrt{\frac{I_{B_2}}{K_{p_7}}} + \sqrt{\frac{I_{B_2}}{K_{p_8}}} - \sqrt{\frac{I_{B_1}}{K_{p_4}}}\right)^2 \tag{4.408}$$

Considering the loop formed by T_2, T_3, T_5, and T_6, the voltage equation is of the form

$$V_{GS_2} + V_{GS_3} = V_{GS_5} + V_{GS_6} \tag{4.409}$$

It can then be shown that

$$I_2 = K_{n_2}\left(\sqrt{\frac{I_{B_2}}{K_{n_5}}} + \sqrt{\frac{I_{B_2}}{K_{n_6}}} - \sqrt{\frac{I_{B_1}}{K_{n_3}}}\right)^2 \tag{4.410}$$

To proceed further, we assume that

$$\frac{W_2/L_2}{W_1/L_1} = \frac{W_6/L_6}{W_8/L_8} = \frac{W_5/L_5}{W_7/L_7} = \frac{W_3/L_3}{W_4/L_4} = \frac{\mu_p}{\mu_n} \tag{4.411}$$

and

$$I_{B_1} = I_{B_2} \tag{4.412}$$

When the output current I_0 is set to zero, each of the currents I_1 and I_2 is reduced to the quiescent current, I_Q, flowing through the transistors T_1 and T_2. Hence,

$$I_Q = \frac{W_1/L_1}{W_8/L_8} I_{B_2} = \frac{W_2/L_2}{W_6/L_6} I_{B_2} \tag{4.413}$$

The maximum value of the current I_1 is obtained when T_4 is forced to operate in the cutoff region, that is, $V_{SG_4} \leq -V_{T_p}$, and the overall current $2I_{B1}$ flows through T_3. As a result,

$$I_1 = I_{max} = K_{p_1} \left(\sqrt{\frac{I_{B_2}}{K_{p_7}}} + \sqrt{\frac{I_{B_2}}{K_{p_8}}} \right)^2 = 4 \frac{W_1/L_1}{W_8/L_8} I_{B_2} = 4I_Q \tag{4.414}$$

The current I_2 is reduced to the minimum value given by

$$\begin{aligned} I_2 = I_{min} &= K_{n_2} \left(\sqrt{\frac{I_{B_2}}{K_{n_5}}} + \sqrt{\frac{I_{B_2}}{K_{n_6}}} - \sqrt{\frac{2I_{B_1}}{K_{n_3}}} \right)^2 \\ &= (2 - \sqrt{2})^2 \frac{W_2/L_2}{W_6/L_6} I_{B_2} = (2 - \sqrt{2})^2 I_Q \end{aligned} \tag{4.415}$$

Similarly, when the current I_1 becomes equal to its minimum value, due to the fact that T_3 is forced to operate in the cutoff region, that is, $V_{GS_3} \leq V_{T_n}$, the overall current $2I_{B1}$ flows through T_4. The current I_2 is then set to its maximum value. Thus, this output stage configuration has the advantage of maintaining a minimum current in the inactive output transistor.

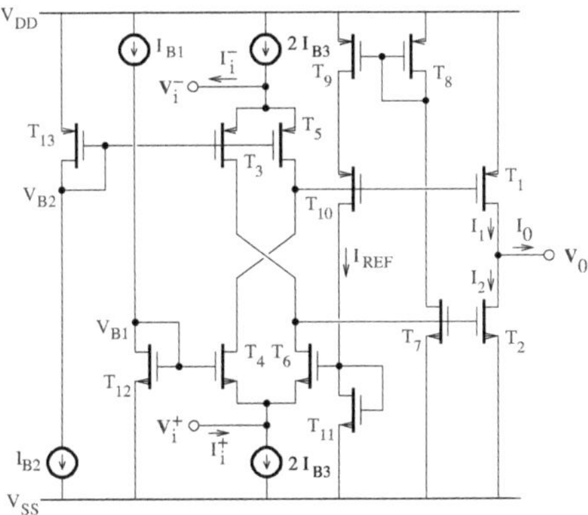

FIGURE 4.58
Circuit diagram of a low-voltage class AB output stage.

With reference to Figure 4.58 [39], a class AB output stage that can still operate with a low supply voltage is shown. In general, the minimum supply voltage is limited by the output stage, which uses transistors operating with sufficiently high gate-source voltages in order to drive high output currents. To reduce the minimum value of the supply voltage to the sum of one gate-source voltage and two saturation voltages, the gate voltages of the output transistors T_1 and T_2 are set by the folded mesh loop consisting of $T_3 - T_6$, which, together with the minimum current selector realized by $T_7 - T_{10}$, also regulates the minimum current flowing through the output transistors.

The transistors T_{12}, T_4, T_6, and T_{11} form a translinear loop, which defines the current I_{REF}. Applying Kirchhoff's voltage law around this loop gives

$$V_{GS_{12}} + V_{GS_4} = V_{GS_6} + V_{GS_{11}} \tag{4.416}$$

where

$$V_{GS_{12}} = V_{T_n} + \sqrt{I_{B1}/K_{n_{12}}} \tag{4.417}$$

$$V_{GS_4} = V_{T_n} + \sqrt{I_{B3}/K_{n_4}} \tag{4.418}$$

$$V_{GS_6} = V_{T_n} + \sqrt{I_{B3}/K_{n_6}} \tag{4.419}$$

and

$$V_{GS_{11}} = V_{T_n} + \sqrt{I_{REF}/K_{n_{11}}} \tag{4.420}$$

Hence,

$$I_{REF} = K_{n_{11}} \left(\sqrt{\frac{I_{B_1}}{K_{n_{12}}}} + \sqrt{\frac{I_{B_3}}{K_{n_4}}} - \sqrt{\frac{I_{B_3}}{K_{n_6}}} \right)^2 \tag{4.421}$$

When $K_{n_4} = K_{n_6}$, the expression of the current I_{REF} is reduced to

$$I_{REF} = \frac{K_{n_{11}}}{K_{n_{12}}} I_{B_1} \tag{4.422}$$

Let

$$I_1 = K_{p_1}(V_{SG_1} - V_{T_p})^2 \tag{4.423}$$

and

$$I_{REF} = K_{p_{10}}(V_{SG_{10}} - V_{T_p})^2 \tag{4.424}$$

$$= K_{p_9}[2(V_{SG_9} - V_{T_p})V_{SD_9} - V_{SD_9}^2] \tag{4.425}$$

where $0 < V_{SD_9} < V_{SG_9} - V_{T_p}$. For the loop including transistors T_1, T_{10}, and T_9, Kirchhoff's voltage law equation can be written as

$$V_{SG_1} = V_{SG_{10}} + V_{SD_9} \tag{4.426}$$

where

$$V_{SG_1} = V_{T_p} + \sqrt{I_1/K_{p_1}} \tag{4.427}$$

$$V_{SD_9} = V_{SG_9} - V_{T_p} - \sqrt{(V_{SG_9} - V_{T_p})^2 - I_{REF}/K_{p_9}} \tag{4.428}$$

and

$$V_{SG_{10}} = V_{T_p} + \sqrt{I_{REF}/K_{p_{10}}} \tag{4.429}$$

Because $V_{GS_2} = V_{GS_7}$, we have $I_{D_7} = I_2 = I_{D_8}$. Using the fact that

$$V_{SG_9} = V_{SG_8} = V_{T_p} + \sqrt{I_2/K_{p_8}} \tag{4.430}$$

we obtain

$$\sqrt{I_1/K_{p_1}} = \sqrt{I_{REF}/K_{p_{10}}} + \sqrt{I_2/K_{p_8}} - \sqrt{I_2/K_{p_8} - I_{REF}/K_{p_{10}}} \tag{4.431}$$

where $I_2 \geq (K_{p_8}/K_{p_{10}})I_{REF}$. Assuming that $I_1 = I_2 = I_Q$ and $K_{p_1} = K_{p_8}$, it can be deduced from Equation (4.431) that

$$I_Q = 2\frac{K_{p_1}}{K_{p_{10}}} I_{REF} \tag{4.432}$$

where I_Q is the quiescent current flowing through the output transistors.

During normal operation, the transistor T_9 operates in the linear region, where its drain current is a function of both the source-gate voltage set by the transistor T_8 and the source-drain voltage adjusted via the transistor T_{10}. The source-drain voltage of the transistor T_9, or the source voltage of the transistor T_{10}, can then be maintained sufficiently low such that the variations in the current I_1 can be tracked by the transistor T_{10}. The transistor T_7, which operates with the same gate-source voltage as the transistor T_2, is used to detect the current I_2. The minimum selector circuit $T_7 - T_{10}$ then evaluates the magnitudes of the currents I_1 and I_2 to help set a minimum current flowing through each of the output transistors as a function of the current I_{REF}.

However, as the drain current of the transistor T_1 increases such that its source-gate voltage becomes sufficiently high to provide enough headroom for the operation of T_9 in the saturation region, the transistors $T_8 - T_{10}$ realize a cascoded current mirror. When the current I_2 reaches its minimum value, $I_Q/2$, the maximum value of the current I_1 derived from Equation (4.431) is $2I_Q$. With $V_{SD_9} = V_{SG_9} - V_{T_p}$ and $V_{SG_9} = V_{SG_{10}}$, Equation (4.426) is reduced to $V_{SG_1} = 2V_{SG_9}$ and the drain current of T_9 is equal to $I_Q/2$. Because $V_{SG_8} = V_{SG_9}$, the bias current of the transistor T_7 is also set to $I_Q/2$. On the other hand, an increase in the current I_2 produces an augmentation of the current flowing through T_7 and T_8, and a decrease in the current I_1 leading to a reduction in the source-gate voltage of the transistor T_1. The source-gate voltage of the transistor T_1 can then be reduced until the source-drain voltage of the transistor T_9 becomes negligible. Hence, $V_{SG_1} \simeq V_{SG_{10}}$ and the current I_1 takes the minimum value $I_Q/2$, while the current I_2 is maximum.

It should be noted that the stability can be affected by poles associated with the folded mesh loop, the current mirror $T_8 - T_9$ and cascode transistor T_{10}. The frequency stabilization is achieved in practical implementations by using a pole-splitting compensation network. Furthermore, the control of the quiescent current can be limited by mismatches of transistors $T_8 - T_{10}$.

4.7.2 Two-stage amplifier

The two-stage amplifiers of Figures 4.59 and 4.60 consist of a differential input stage, an inverting output stage, and a biasing circuit. The single-ended signal is provided by the current mirror used as the load of the transistor differential pair and the amplifier output swing is $V_{sup} - 2V_{DS(sat)}$, where V_{sup} denotes the supply voltage. The Miller frequency compensation with a pole-zero cancellation is adopted to overcome the gain-bandwidth trade-off. A one-pole frequency response of the amplifier is obtained by including a zero in the left-half plane of the s-domain to cancel the first nondominant pole. The compensation section [36, 41, 44] can be implemented using a resistor in series with a capacitor, as shown in Figure 4.59, or a series connection of a MOS transistor operating in the triode region and a capacitor, as illustrated in Figure 4.60 [36].

The slew rate of a two-stage amplifier with RC frequency compensation can be approximatively given by,

$$SR \simeq \min\left(\frac{I_{B_1}}{C_C}, \frac{I_{B_2}}{C_C + C_L}\right) \tag{4.433}$$

where I_{B_1} is the bias current of the differential input transistor pair, I_{B_2} is the bias current of the output stage, C_C and C_L are the compensation and output load capacitances, respectively.

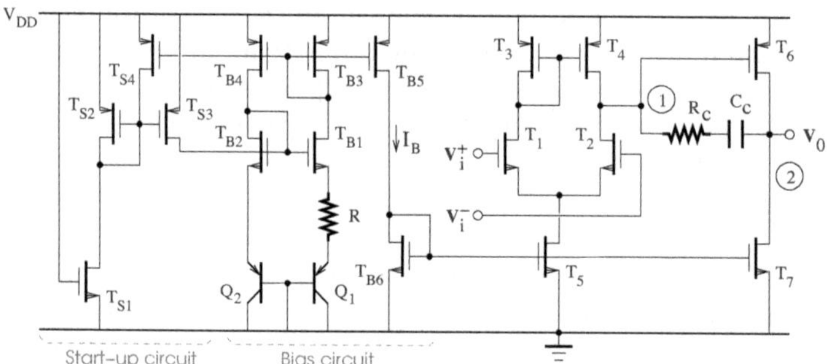

FIGURE 4.59
Circuit diagram of a two-stage amplifier with RC compensation.

An equivalent circuit of the RC-compensated two-stage amplifier is shown in Figure 4.61. Assuming that the parasitic coupling capacitances C_{p1} and C_{p2} can be neglected, the next nodal equations can be written

$$g_{m_1}V_i(s) + V_1(s)(g_1 + sC_1) - (V_O(s) - V_1(s))/(R_c + 1/sC_c) = 0 \tag{4.434}$$

and

$$g_{m_2}V_1(s) + V_O(s)(g_2 + sC_2) + (V_O(s) - V_1(s))/(R_c + 1/sC_c) = 0 \tag{4.435}$$

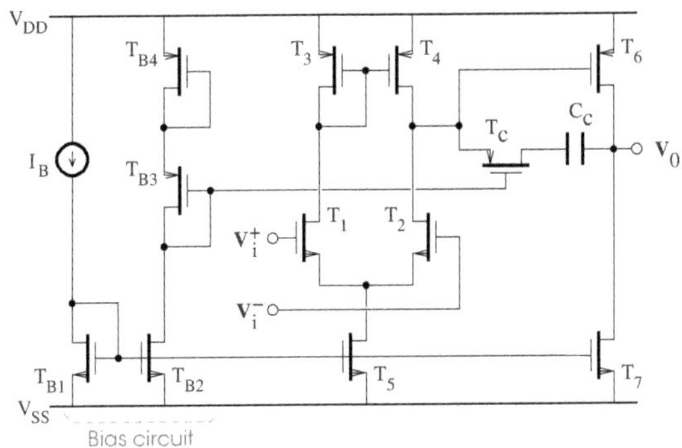

FIGURE 4.60
Circuit diagram of a two-stage amplifier with the compensation resistor implemented by a transistor.

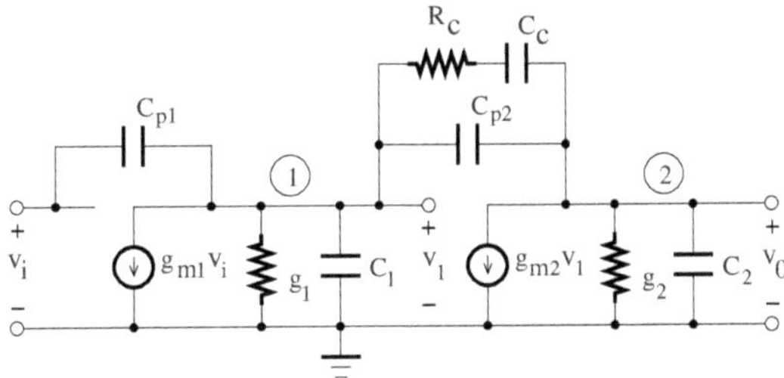

FIGURE 4.61
Small-signal equivalent of the RC-compensated two-stage amplifier.

for the node 1 and node 2, respectively, where g_{m_k}, g_k, and C_k $(k = 1, 2)$ denote the transconductance, output conductance, and output capacitor of the k stage, respectively, and R_c and C_c are the compensation resistor and capacitor, respectively. Solving the above system of equations gives

$$A(s) = \frac{V_0(s)}{V_i(s)} = A_0 \frac{1 + (R_c - 1/g_{m_2})\, sC_c}{1 + cs + bs^2 + as^3} \tag{4.436}$$

where

$$A_0 = \frac{g_{m_1} g_{m_2}}{g_1 g_2} \tag{4.437}$$

$$a = \frac{C_1 C_2 R_c C_c}{g_1 g_2} \tag{4.438}$$

$$b = \frac{C_1 C_c + C_1 C_2 + C_2 C_c}{g_1 g_2} + \left(\frac{C_1}{g_1} + \frac{C_2}{g_2}\right) R_c C_c \tag{4.439}$$

and

$$c = \frac{C_1 + C_c}{g_1} + \frac{C_2 + C_c}{g_2} + \frac{g_{m_2} C_c}{g_1 g_2} + R_c C_c \tag{4.440}$$

The capacitances C_1 and C_2 can be related to the transistor capacitances by $C_1 = C_{gd_4} + C_{db_4} + C_{gd_2} + C_{db_2} + C_{gs_2}$ and $C_2 = C_{db_6} + C_{db_7} + C_{gd_7} + C_L$, where C_L represents the load capacitance at the amplifier output and the contribution due to C_{gd_6} is assumed to be negligible because $C_{gd_6} \ll C_c$. The third-order transfer function, $A(s)$, can be put into the form

$$A(s) = \frac{V_0(s)}{V_i(s)} = A_0 \frac{1 - s/\omega_{z_1}}{(1 - s/\omega_{p_1})(1 - s/\omega_{p_2})(1 - s/\omega_{p_3})} \tag{4.441}$$

where ω_{z_1} is the frequency of the zero z_1, and ω_{p_1}, ω_{p_2}, and ω_{p_3} are the frequencies of the poles p_1, p_2, and p_3, respectively. In practice, $A_0 \gg 1$, $g_1 R_c \gg 1$, $g_2 R_c \gg 1$, $C_1/C_c \gg 1$, and $C_1/C_2 \gg 1$, and it is assumed that the poles are widely spaced. Thus, we can write

$$A(s) = \frac{V_0(s)}{V_i(s)} \simeq A_0 \frac{1 - s/\omega_{z_1}}{1 - s/\omega_{p_1} + s^2/\omega_{p_1}\omega_{p_2} - s^3/\omega_{p_1}\omega_{p_2}\omega_{p_3}} \tag{4.442}$$

where

$$\omega_{z_1} = \frac{1}{C_c(1/g_{m_2} - R_c)} \tag{4.443}$$

$$\omega_{p_1} = -\frac{g_1 g_2}{g_{m_2} C_c} = -\frac{g_{m_1}}{A_0 C_c} \tag{4.444}$$

$$\omega_{p_2} = -\frac{g_{m_2} C_c}{C_1 C_2 + C_c(C_1 + C_2)} \simeq -\frac{g_{m_2}}{C_2} \tag{4.445}$$

$$\omega_{p_3} = -\frac{1}{R_c C_1} \tag{4.446}$$

Note that p_1 represents the dominant pole. The compensation leads to a splitting of the initially close poles as depicted in Figure 4.62.

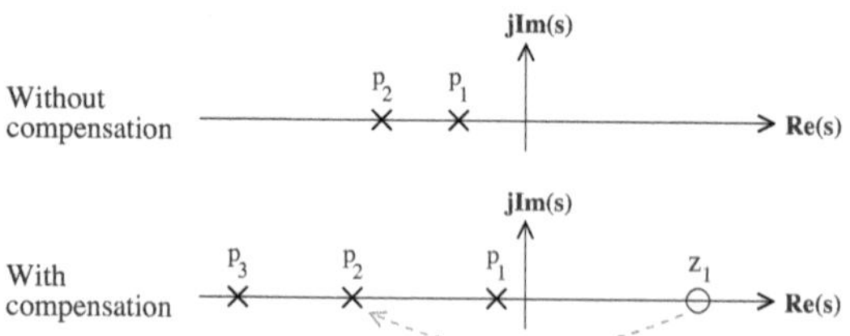

FIGURE 4.62
Pole-zero representations of a two-stage amplifier without and with frequency compensation.

By setting $R_c = 1/g_{m_2}$, the zero is rejected at infinity and the stability of the amplifier is guaranteed provided that p_2 and p_3 are located far beyond the unity-gain frequency, that is, $|\omega_{p_3}| > |\omega_{p_2}| > A_0|\omega_{p_1}|$. In this case, we have $C_c > g_{m_1}C_2/g_{m_2}$ and a large value of C_c or g_{m_2} is required to split the poles. As a result, the die area or power consumption must be increased.

It is also possible to cancel the pole p_2 using the zero z_1, which is moved from the right-half plane to the left-half plane (see Figure 4.62). Hence,

$$\omega_{p_2} = \omega_{z_1} \tag{4.447}$$

and the value of the compensation resistor is derived as

$$R_c = \frac{C_2 + C_c}{g_{m_2}C_c} \tag{4.448}$$

The compensation of the amplifier, whose frequency response is now determined by the two remaining poles, requires that $|\omega_{p_3}| > A_0|\omega_{p_1}|$. By using a small compensation capacitor to realize the pole-zero cancellation, the resulting amplifier can exhibit a wide bandwidth and a high slew rate.

In the case of the amplifier shown in Figure 4.59, the bias circuit is designed to be less sensitive to temperature and process variations. Initially, the start-up circuit [40] triggers the flow of a current used to activate the bandgap bias circuit. Due to the conduction of the transistor T_{S1}, a start-up current is mirrored into the loop $T_{B1} - T_{B4}$ of the bias circuit, via $T_{S2} - T_{S3}$. When the stationary state of the bias current is reached, the transistor T_{S4} starts conducting. The voltage level at the gates of transistors $T_{S2} - T_{S3}$ is approximately equal to the positive supply voltage, rendering T_{S2} and T_{S3} nonconducting. The generation of the start-up current is then interrupted.

The biasing circuit of the amplifier can be designed to ensure that the pole and zero track over process, voltage, and temperature variations. With reference to Figure 4.59, we can write

$$R_B I_B = V_{GS3} - V_{GS4} \tag{4.449}$$

and the bias current, I_B, is given by

$$I_B = \frac{1}{k_p R_B^2}\left(\sqrt{\frac{1}{W_3/L_3}} - \sqrt{\frac{1}{W_4/L_4}}\right)^2 \tag{4.450}$$

where $k_p = \mu_p C_{ox}/2$. If the transistors match accurately, the transconductance will be independent of K_p and determined by R_B.

The transistor T_c used in the circuit of Figure 4.60 operates in the triode region, where

$$I_D = K_p[2(V_{SG} - |V_{T_p}|)V_{SD} - V_{SD}^2] \tag{4.451}$$

and $K_p = \mu_p C_{ox} W/(2L)$. Its drain-source resistance is given by

$$R_{DS} = \left.\frac{\partial I_D}{\partial V_{SD}}\right|_{V_{SG}} = [2K_p(V_{SG} - |V_{T_p}| - V_{SD})]^{-1} \tag{4.452}$$

and is continuously adjusted to allow an adequate settling response.

The RC-based compensation network also introduces some limitations on the performance of the aforementioned two-stage amplifiers. The amplifier is stable in a feedback configuration only when the value of the load capacitor does not exceed the one of the compensation capacitor. Due to the extra path from the power supply voltage through the compensation network to the amplifier output, the positive and negative power supply rejections are degraded for frequencies greater than the pole frequency associated with the compensation capacitor in the case of n-channel and p-channel differential input transistor pairs, respectively.

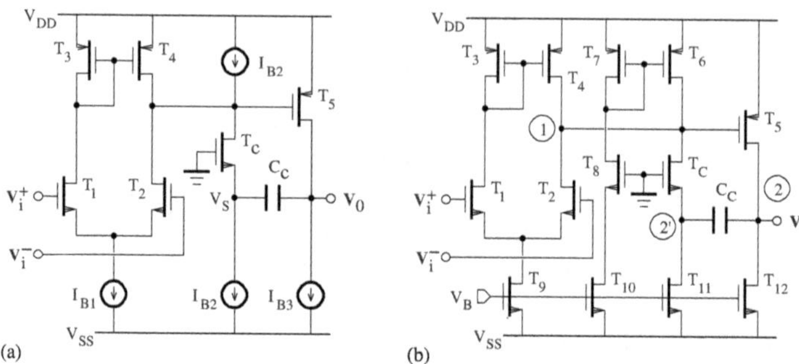

FIGURE 4.63
(a) Principle and (b) circuit diagram of a two-stage amplifier with a current buffer.

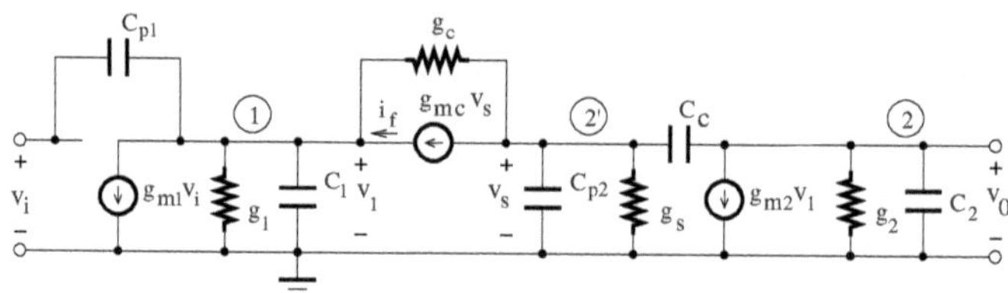

FIGURE 4.64
Small-signal equivalent model of the two-stage amplifier with a current buffer.

In the amplifier structures illustrated in Figure 4.63, the feedforward path is removed by inserting a current buffer between the input and output stages [41–43]. The small-signal equivalent model is depicted in Figure 4.64, where C_{p2} denotes the gate-source capacitor of T_C. To simplify the circuit analysis, the feedback current source can be replaced by its equivalent Y network consisting of the current source i_f connected between the node 1 and the ground, and a resistor with the admittance y_{22f} and connected between

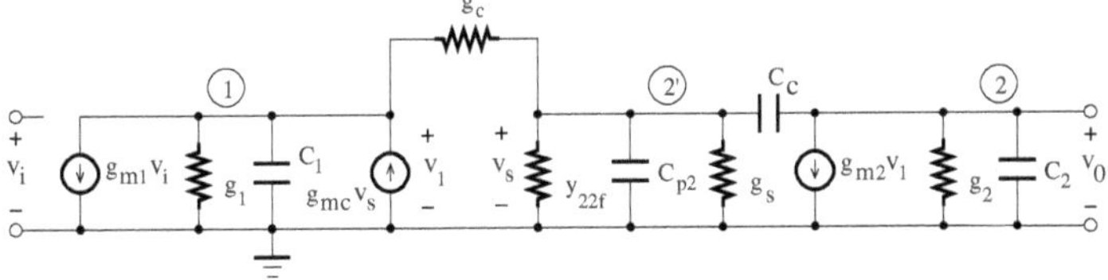

FIGURE 4.65
Shunt-shunt version of the amplifier small-signal equivalent model.

the node $2'$ and the ground. Because $i_f = g_{m_c} v_s$, we can obtain

$$y_{22f} = \left.\frac{i_f}{v_s}\right|_{v_i=0} = g_{m_c} \tag{4.453}$$

Assuming that the parasitic capacitor C_{p_1} is negligible, a shunt-shunt version of the small-signal equivalent model can be derived as shown in Figure 4.65. The use of Kirchhoff's current law at the circuit nodes 1, $2'$, and 2 gives

$$g_{m_1} V_i(s) + g_1 V_1(s) + sC_1 V_1(s) + g_c(V_1(s) - V_s(s)) = g_{m_c} V_s(s) \tag{4.454}$$

$$g_c(V_1(s) - V_s(s)) = g_{m_c} V_s(s) + sC_{p2} V_s(s) + g_s V_s(s) + sC_c(V_s(s) - V_0(s)) \tag{4.455}$$

and

$$sC_c(V_s(s) - V_0(s)) = g_{m_2} V_1(s) + g_2 V_0(s) + sC_2 V_0(s) \tag{4.456}$$

respectively, where

$$\begin{aligned} C_1 = C_{db_2} + C_{gd_2} + C_{db_4} + C_{gd_4} \\ + C_{gs_5} + C_{db_6} + C_{gd_6} + C_{db_c} + C_{gd_c} \end{aligned} \tag{4.457}$$

$$C_{p2} = C_{gs_c} + C_{sb_c} + C_{db_{11}} + C_{gd_{11}} \tag{4.458}$$

$$C_2 = C_{db_5} + C_{gd_5} + C_{db_{12}} + C_{gd_{12}} + C_L \tag{4.459}$$

$$g_1 = g_{ds_2} + g_{ds_4} \tag{4.460}$$

$$g_2 = g_{ds_5} + g_{ds_{12}} \tag{4.461}$$

and g_s can be reduced to the output conductance of the current source I_{B_2}. By solving the system of Equations (4.454), (4.455), and (4.456), the transfer function can obtained as

$$A(s) = \frac{V_0(s)}{V_i(s)} = A_0 \frac{1 + \dfrac{g_c + g_s}{g_{m_c}} + \left(\dfrac{C_c + C_{p2}}{g_{m_c}} - \dfrac{C_c g_c}{g_2 g_{m_c}}\right) s}{1 + \gamma s + \beta s^2 + \alpha s^3} \tag{4.462}$$

where

$$A_0 = \frac{g_{m_1} g_{m_2}}{g_1 g_2} \tag{4.463}$$

$$\alpha = C_1 \frac{C_2 C_c + C_2 C_{p2} + C_c C_{p2}}{g_{m_c} g_1 g_2} \tag{4.464}$$

$$\beta = C_1 \frac{C_2 + C_c}{g_1 g_2}\left(1 + \frac{g_c + g_s}{g_{m_c}}\right) + C_1 \frac{C_c + C_{p2}}{g_{m_c} g_1} + \frac{C_2 C_c + C_2 C_{p2} + C_c C_{p2}}{g_{m_c} g_2}\left(1 + \frac{g_c}{g_1}\right) \tag{4.465}$$

$$\gamma = \frac{g_{m_2} C_c}{g_1 g_2}\left(1 + \frac{g_c}{g_{m_c}}\right) + \frac{C_2 + C_c}{g_2}\left(1 + \frac{g_c}{g_{m_c}} + \frac{g_s}{g_{m_c}} + \frac{g_c g_s}{g_{m_c} g_1}\right) + \frac{C_1}{g_1}\left(1 + \frac{g_c + g_s}{g_{m_c}}\right) + \frac{C_c + C_{p2}}{g_{m_c}}\left(1 + \frac{g_c}{g_1}\right) \tag{4.466}$$

In general, it can be shown that

$$A(s) = \frac{V_0(s)}{V_i(s)} = A_0 \frac{1 - s/\omega_{z_1}}{(1 - s/\omega_{p_1})(1 - s/\omega_{p_2})(1 - s/\omega_{p_3})} \tag{4.467}$$

where the locations of the poles p_1, p_2, and p_3 can be determined using a computer program for symbolic analysis. The compensation capacitor can then be chosen to achieve a given settling time or phase margin specification. Assuming a dominant pole model, or $|\omega_{p_1}| \ll |\omega_{p_2}|, |\omega_{p_3}|$, and $|\omega_{p_2}| \ll |\omega_{p_3}|$, the transfer function can be approximated as

$$A(s) = \frac{V_0(s)}{V_i(s)} \simeq A_0 \frac{1 - s/\omega_{z_1}}{1 - s/\omega_{p_1} + s^2/\omega_{p_1}\omega_{p_2} - s^3/\omega_{p_1}\omega_{p_2}\omega_{p_3}} \tag{4.468}$$

Assuming that $g_{m_2}, g_{m_c} \gg g_1, g_2, g_c, g_s$; $C_c, C_2 \gg C_1, C_{p2}$; $g_{m_c}/g_c \gg g_s/g_1$, and $g_{m_2}/g_1 \gg C_2/C_c$, we obtain

$$\omega_{z_1} = -\frac{g_{m_c}}{C_c + C_{p2}} \tag{4.469}$$

$$\omega_{p_1} = -\frac{g_1 g_2}{C_c g_{m_2}} \tag{4.470}$$

$$\omega_{p_2} = -\frac{g_{m_2} C_c}{C_1(C_2 + C_c)} \tag{4.471}$$

$$\omega_{p_3} = -\frac{g_{m_c}(C_2 + C_c)}{C_2 C_c + C_2 C_{p2} + C_c C_{p2}} = -\frac{g_{m_c}}{\dfrac{C_2 C_c}{C_2 + C_c} + C_{p2}} \tag{4.472}$$

A cancellation between p_3 and z_1 is achieved only if $C_c/C_2 \ll 1$. It is incomplete in practice due to component mismatches, and p_3 and z_1 then form a pole-zero doublet. If now $|\omega_{p_2}| < |\omega_{z_1}| < |\omega_{p_3}|$, for instance, the stability requirement of the amplifier in feedback loop will be met.

The magnitude of the nondominant pole p_2 in this case is greater than the one of the conventional Miller compensated amplifier, which is given by $g_{m_2}C_c/[C_1C_2+C_c(C_1+C_2)]$. To achieve the same frequency response, the use of a current buffer then offers the advantage of reducing the required component values. Furthermore, the value of g_{m_c} can be determined by appropriately biasing and sizing the transistor T_c to set p_3 and z_1 well beyond p_2. However, the mismatch of the biasing currents of T_c may increase the amplifier offset voltage.

Note that if the current buffer is modeled as an ideal current source with a zero input resistance [41], or equivalently, g_{mc} is considered to be infinite, the right-half plane zero is cancelled and the transfer function only exhibits two poles.

4.7.3 Optimization of a two-pole amplifier for fast settling response

Amplifiers used in high-speed applications should preferably be designed to achieve the minimum settling time [46, 47]. In general, the settling time, t_s, is defined as the time required for the amplifier output voltage to settle to within an error tolerance, ϵ, around the final steady-state value.

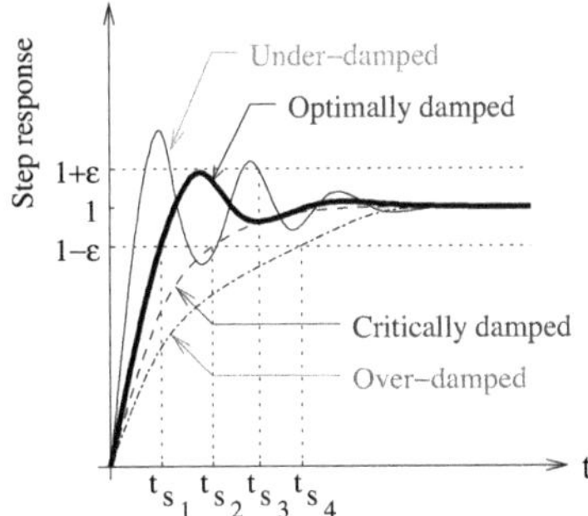

FIGURE 4.66
Step responses of a two-stage amplifier.

Consider a two-stage amplifier characterized by an open-loop transfer function of the form

$$A(s) = \frac{A_0}{(1 - s/\omega_{p_1})(1 - s/\omega_{p_2})} \tag{4.473}$$

where A_0 is the dc gain, and ω_{p_1} and ω_{p_2} are the frequency locations of the first and second poles (p_1 and p_2), respectively. In a unity-gain configuration, the closed-loop transfer function is

$$A_{CL}(s) = \frac{A(s)}{1 + A(s)} = \frac{A'_0}{(s/\omega_0)^2 + 2k(s/\omega_0) + 1} \tag{4.474}$$

where

$$A'_0 = \frac{A_0}{1 + A_0} \tag{4.475}$$

$$\omega_0 = [\omega_{p_1}\omega_{p_2}(1 + A_0)]^{1/2} \tag{4.476}$$

and

$$k = -\frac{\omega_{p_1} + \omega_{p_2}}{2\omega_0} \tag{4.477}$$

With $\beta = \omega_{p_2}/\omega_{p_1}$ being the pole separation factor, the damping factor k can be rewritten as

$$k = \frac{1 + \beta}{2[\beta(1 + A_0)]^{1/2}} \tag{4.478}$$

For a unit step input, $u(t)$, defined as

$$u(t) = \begin{cases} 1 & \text{for} \quad t \geq 0 \\ 0 & \text{for} \quad t < 0 \end{cases} \tag{4.479}$$

the closed-loop response in the time domain is given by

$$v_O(t) = \mathcal{L}^{-1}[H(s)] \tag{4.480}$$

where $H(s) = A_{CL}(s)/s$ and $\mathcal{L}^{-1}$ denotes the inverse Laplace transform. Depending on the value of k, there are three possible responses.

For $k > 1$, the closed-loop response is said to be over-damped. It shows no oscillation and is slow to settle. The output voltage is given by

$$v_O(t) = \left\{1 - \frac{1}{2(k^2 - 1)^{1/2}}\left[\frac{1}{k_1}\exp(-k_1\omega_0 t) - \frac{1}{k_2}\exp(-k_2\omega_0 t)\right]\right\}u(t) \tag{4.481}$$

where $k_1 = k - (k^2 - 1)^{1/2}$ and $k_2 = k + (k^2 - 1)^{1/2}$.

For $k = 1$, the closed-loop response is critically damped. It quickly converges to the steady-state value, and without oscillating. We then have

$$v_O(t) = [1 - (1 + \omega_0 t)\exp(-\omega_0 t)]u(t) \tag{4.482}$$

For $0 < k < 1$, the closed-loop response is under-damped and exhibits some oscillations before approaching the steady-state value. The output voltage can be written as

$$v_O(t) = \left\{1 - \left(\frac{k\sin[(1 - k^2)^{1/2}\omega_0 t]}{(1 - k^2)^{1/2}} + \cos[(1 - k^2)^{1/2}\omega_0 t]\right)\exp(-k\omega_0 t)\right\}u(t) \tag{4.483}$$

The over-damped, critically damped, and under-damped step responses are depicted in Figure 4.66. Also shown in Figure 4.66 is the optimally damped response. It is apparent that the amplifier exhibits a minimum settling time when the first peak of the step response just touches the upper settling error bound.

Let us consider the case of an under-damped response with the peak of the first overshoot occurring at the instant t_p. By setting the derivative of $v_0(t)$ equal to zero, we have

$$t = t_p = \frac{\pi}{\omega_0(1-k^2)^{1/2}} \tag{4.484}$$

The level of the first peak of the step response is of the form, $v_0(t_p) = 1 + \epsilon$, where

$$\epsilon = \exp[-k\pi/(1-k^2)^{1/2}] \tag{4.485}$$

With $\beta \gg 1$, we can combine the expression derived from Equation (4.478), that is, $k \simeq (1/2)\sqrt{\beta/(1+A_0)}$, and Equation (4.485) to get

$$\beta = \beta_{mst} \simeq \frac{4(1+A_0)}{1+(\pi/\ln\epsilon)^2} \tag{4.486}$$

where β_{mst} is the pole separation factor associated with the minimum settling time for a given error tolerance, ϵ. However, it is common to characterize an amplifier using the phase margin, ϕ_M, instead of the pole separation factor. When the amplifier poles are sufficiently separated, we have

$$\begin{aligned}\phi_M &= 180^\circ - \angle A(j\omega_u) \\ &= 180^\circ - \arctan(\omega_u/|\omega_{p_1}|) - \arctan(\omega_u/|\omega_{p_2}|)\end{aligned} \tag{4.487}$$

where ω_u is the unity-gain frequency. Assuming that $\omega_u \simeq A_0|\omega_{p_1}|$, the term $\arctan(\omega_u/|\omega_{p_1}|)$ is on the order of 90° as the dc gain A_0 is very high. Because $\omega_{p_2} = \beta\omega_{p_1}$, the substitution of Equation (4.486) into (4.487) then gives [47]

$$\phi_M = \phi_{M,mst} \simeq 90^\circ - \arctan\left[\frac{1+(\pi/\ln\epsilon)^2}{4}\right] \tag{4.488}$$

For very small values of ϵ, $(\pi/\ln\epsilon)^2$ becomes negligible and $\phi_{M,mst}$ is about 76°, which is in the same order as the phase margin of a critically damped system. Here, the minimum settling time is then achieved when the amplifier response is critically damped.

In the case of a two-stage amplifier with the pole-splitting frequency-compensation network, the transfer function is given by

$$A(s) = A_0\frac{1-s/\omega_{z_1}}{(1-s/\omega_{p_1})(1-s/\omega_{p_2})(1-s/\omega_{p_3})} \tag{4.489}$$

To obtain a two-pole frequency response and minimize the settling time, we need to have

$$\omega_{p_2} = \omega_{z_1} \tag{4.490}$$

and

$$\omega_{p_3} = \beta_{mst}\omega_{p_1} \tag{4.491}$$

The values of the components used in the compensation network should then be chosen such that the above requirements are met.

4.7.4 Three-stage amplifier

In low-voltage applications, amplifiers based on cascode or telescopic structure are no longer suitable for achieving a high *dc* gain and a wide bandwidth, due to their limited voltage swings. Multistage amplifiers capable of driving large capacitive loads are preferred under these circumstances. However, they suffer from stability problems caused by the presence of several poles and then require additional frequency compensation networks. Depending on the type of frequency compensation adopted, the non-dominant poles are either moved to high frequencies (resulting in pole splitting) or compensated using left half-plane zeros (leading to pole-zero compensation) in order to achieve a high gain-bandwidth product (GBW).

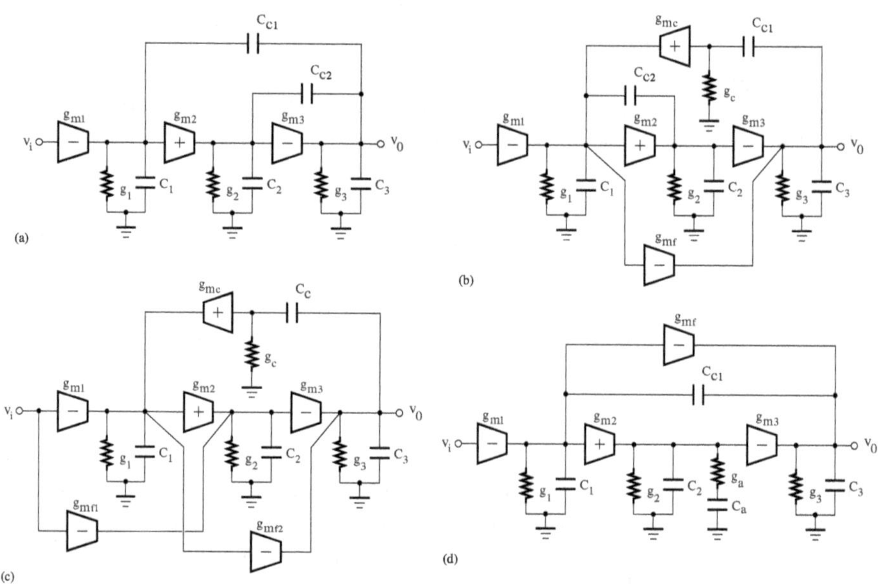

FIGURE 4.67
Block diagrams of three-stage amplifiers with (a) nested Miller compensation, (b) reversed active feedback frequency compensation, (c) cross feedforward cascode compensation, and (d) impedance adapting compensation.

The block diagrams of three-stage amplifiers with nested Miller compensation, reversed active feedback frequency compensation, cross feedforward cas-

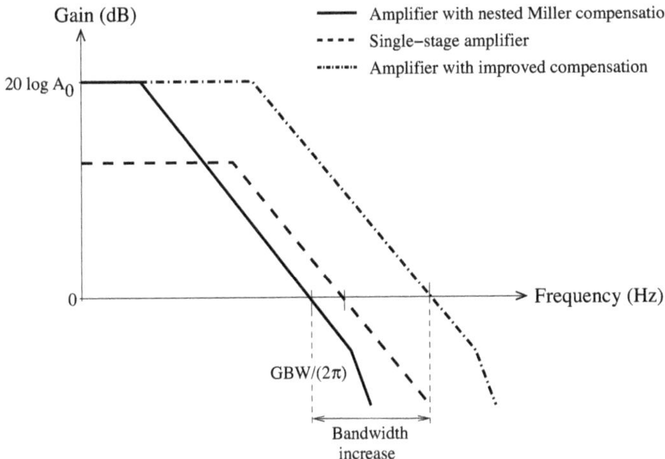

FIGURE 4.68
Bode plot illustrating the frequency responses of three-stage amplifiers.

code compensation, and impedance adapting compensation are represented in Figure 4.67, where $g_c \simeq g_{m_c}$.

A three-stage amplifier provides a sufficient *dc* gain, but the GBW of the nested Miller compensated amplifier can actually be less than that of a conventional single-stage amplifier, as shown in Figure 4.68. With reference to the nested Miller compensation, recent compensation techniques tend to improve the frequency response by increasing the GBW.

The performance of three-stage amplifiers can be compared using various figures of merit. The formulas that can be used for small-signal and large-signal assessments can be written as,

$$FOM_S = \frac{GBW \cdot C_3}{\text{Power}} \tag{4.492}$$

$$IFOM_S = \frac{GBW \cdot C_3}{I_{DD}} \tag{4.493}$$

$$FOM_L = \frac{SR \cdot C_3}{\text{Power}} \tag{4.494}$$

$$IFOM_L = \frac{SR \cdot C_3}{I_{DD}} \tag{4.495}$$

where I_{DD} is the supply current. The units of FOM_S and FOM_L are MHz · pF/mW and V/μs · pF/mW, respectively, while the units used to estimate the $IFOM_S$ and $IFOM_L$ are MHz · pF/mA and V/μs · pF/mA, respectively. The higher the figure of merit is, the better the frequency compensation topology will be.

The circuit diagram of a three-stage amplifier with the nested Miller frequency compensation [48, 49] is shown in Figure 4.69. The first stage is implemented by the differential input transistors, $T_1 - T_2$, biased by T_{15}, and loaded by $T_3 - T_8$, while the second and third ones consist of $T_9 - T_{12}$ and T_{13}, respectively. The compensation capacitors C_{c_1} and C_{c_2} are used to stabilize the amplifier.

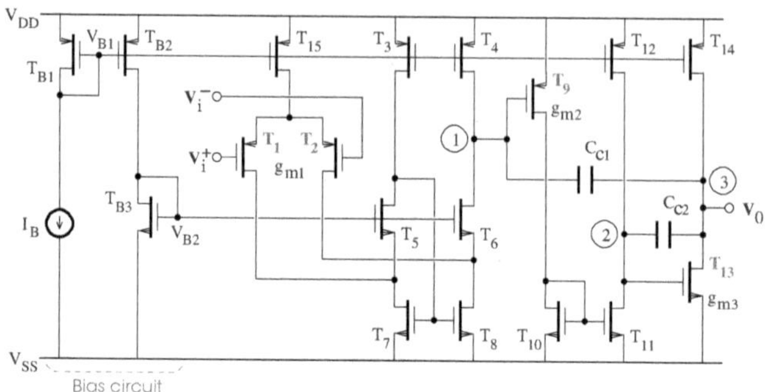

FIGURE 4.69
Circuit diagram of a three-stage amplifier with a nested Miller compensation.

The equivalent model of the three-stage amplifier with the nested Miller frequency compensation is shown in Figure 4.70. The first and third gain

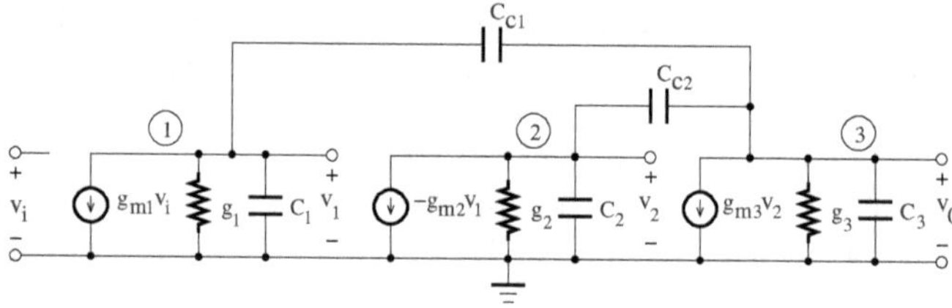

FIGURE 4.70
Small-signal equivalent circuit of the three-stage amplifier with a nested Miller compensation.

stages are of the inverting type. The noninverting second amplification section is required to ensure the negative feedback around the nested compensation paths. Applying Kirchhoff's current law at nodes 1, 2, and 3, we obtain

$$g_{m_1} V_i(s) + V_1(s)(g_1 + sC_1) - (V_O(s) - V_1(s))sC_{c_1} = 0 \tag{4.496}$$

$$-g_{m_2} V_1(s) + V_2(s)(g_2 + sC_2) - (V_O(s) - V_2(s))sC_{c_2} = 0 \tag{4.497}$$

and

$$g_{m_3} V_2(s) + V_O(s)(g_3 + sC_3) + (V_O(s) - V_1(s))sC_{c_1} + (V_O(s) - V_2(s))sC_{c_2} = 0 \tag{4.498}$$

where g_{m_k}, g_k, and C_k $(k = 1, 2, 3)$ denote the transconductance, output conductance, and output capacitor of the k stage, respectively. The node capacitances are given by

$$C_1 = C_{gd_4} + C_{gb_4} + C_{gd_6} + C_{gb_6} + C_{gs_9} \tag{4.499}$$
$$C_2 = C_{gd_{11}} + C_{gb_{11}} + C_{gd_{12}} + C_{gb_{12}} \tag{4.500}$$

and

$$C_3 = C_{gb_{13}} + C_{gd_{14}} + C_{gb_{14}} + C_L \tag{4.501}$$

where C_L is the external load capacitance that can be connected to the output node. Assuming that $g_{m_k} \gg g_k$ $(k = 1, 2, 3)$ and $C_{c_1}, C_{c_2}, C_3 \gg C_1, C_2$, the amplifier small-signal voltage gain can be computed as [50]

$$A(s) = \frac{V_0(s)}{V_i(s)} \simeq A_0 \frac{1 + ds + cs^2}{(1 + s/\omega_{p_1})(1 + bs + as^2)} \tag{4.502}$$

where

$$A_0 = \frac{g_{m_1} g_{m_2} g_{m_3}}{g_1 g_2 g_3} \tag{4.503}$$
$$\omega_{p_1} = \frac{g_1 g_2 g_3}{g_{m_2} g_{m_3} C_{c_1}} \tag{4.504}$$
$$a = \frac{C_{c_2} C_3}{g_{m_2} g_{m_3}} \tag{4.505}$$
$$b = \frac{C_{c_2}(g_{m_3} - g_{m_2})}{g_{m_2} g_{m_3}} \tag{4.506}$$
$$c = -\frac{C_{c_1} C_{c_2}}{g_{m_2} g_{m_3}} \tag{4.507}$$

and

$$d = -\frac{C_{c_2}}{g_{m_3}} \tag{4.508}$$

The zero frequencies are usually much greater than the unity-gain frequency of the amplifier. With the assumptions that $g_{m_3} \gg g_{m_1}, g_{m_2}$, and $b \simeq C_{c_2}/g_{m_2}$, the voltage transfer function can be further simplified to

$$A(s) = \frac{V_0(s)}{V_i(s)} \simeq A_0 \frac{1}{(1 + s/\omega_{p_1})(1 + bs + as^2)} \tag{4.509}$$

The dominant pole is related to the output node (node 3). The nested capacitor C_{c_1} splits the poles at nodes 2 and 3, while C_{c_1} splits the ones at nodes

1 and 3 (see Figure 4.71). With the poles being sufficiently separated, we can obtain

$$A(s) = \frac{V_0(s)}{V_i(s)} \simeq \frac{1}{(s/A_0\omega_{p_1})(1 + bs + as^2)} \tag{4.510}$$

The stabilization will be achieved if the amplifier exhibits a third-order Butterworth frequency response in the unity-gain feedback configuration [51]. That is,

$$A_{CL}(s) = \frac{A(s)}{1 + A(s)} = \frac{1}{1 + 2\frac{s}{\omega_c} + 2\frac{s^2}{\omega_c^2} + \frac{s^3}{\omega_c^3}} \tag{4.511}$$

where ω_c is the cutoff frequency. Consequently, the gain $A(s)$ should be of the form

$$A(s) = \frac{1}{2\frac{s}{\omega_c}\left(1 + \frac{s}{\omega_c} + \frac{1}{2}\frac{s^2}{\omega_c^2}\right)} \tag{4.512}$$

Comparing Equations (4.510) and (4.512) gives

$$\frac{2}{\omega_c} = \frac{1}{A_0\omega_{p_1}} \simeq \frac{C_{c_1}}{g_{m_1}} \tag{4.513}$$

$$\frac{1}{\omega_c} = b \simeq \frac{C_{c_2}}{g_{m_2}} \tag{4.514}$$

$$\frac{1}{2\omega_c^2} = a = \frac{C_{c_2}C_3}{g_{m_2}g_{m_3}} \tag{4.515}$$

This leads to the values of the compensation capacitors given by

$$C_{c_1} = 4\left(\frac{g_{m_1}}{g_{m_3}}\right)C_3 \tag{4.516}$$

$$C_{c_2} = 2\left(\frac{g_{m_2}}{g_{m_3}}\right)C_3 \tag{4.517}$$

Because the Butterworth response is said to be under-damped, it can also yield the minimum settling time. It is characterized by a unity-gain phase margin given by

$$\phi_M = 180^o - \angle A[j(A_0\omega_{p_1})] \tag{4.518}$$

$$= 180^o - \angle(2s/\omega_c) - \arctan\left(\frac{A_0\omega_{p_1}/\omega_c}{1 - (A_0\omega_{p_1}/\sqrt{2}\omega_c)^2}\right) \tag{4.519}$$

$$= 180^o - 90^o - \arctan(4/7) \simeq 60^o \tag{4.520}$$

With $A_0\omega_{p_1}$ and ω_c representing the amplifier unity-gain frequency and the cutoff frequency of the Butterworth response, respectively, it was assumed that $A_0\omega_{p_1}/\omega_c = 1/2$.

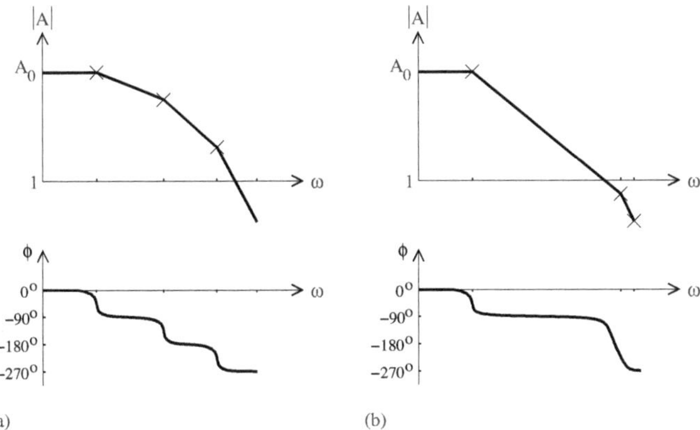

FIGURE 4.71
Magnitude and phase responses of the three-stage amplifier (a) before and (b) after the nested Miller compensation.

Note that the locations of the nondominant poles depend on C_3 and the amplifier stability can be affected by the output load capacitor. The gain-bandwidth product is given by

$$GBW \simeq A_0\omega_{p_1} = \frac{g_{m_1}}{C_{c_1}} = \frac{1}{4}\frac{g_{m_3}}{C_3} \tag{4.521}$$

The resulting GBW is four times smaller than the one of an uncompensated structure. The slew rate (SR) of the amplifier can be obtained as

$$SR = \min\left\{\frac{I_{B_1}}{C_{c_1}}, \frac{I_{B_2}}{C_{c_2}}, \frac{I_{B_3}}{C_3}\right\} \tag{4.522}$$

where I_{B_1}, I_{B_2}, and I_{B_3} are the bias current of the first, second, and third stage, respectively. The amplifier should exhibit a poor SR for large compensation capacitors.

Generally, an N-stage amplifier needs $N - 1$ compensation capacitors driven by the output stage. In this case, the requirement of a large bandwidth can be fulfilled only at the price of a high power consumption.

Due to the fact that both compensation capacitors affect the amplifier loading, the bandwidth achievable with the nested Miller compensation is limited. An improvement can be obtained using amplifiers based on the reversed nested Miller compensation, especially when driving large capacitive loads.

The circuit diagram of a three-stage reversed active feedback frequency compensation (RAFFC) amplifier [52] is depicted in Figure 4.72. It consists of an input stage with the folded-cascode structure, a common-source inverting stage, and a noninverting output stage. The feedforward transconductance is

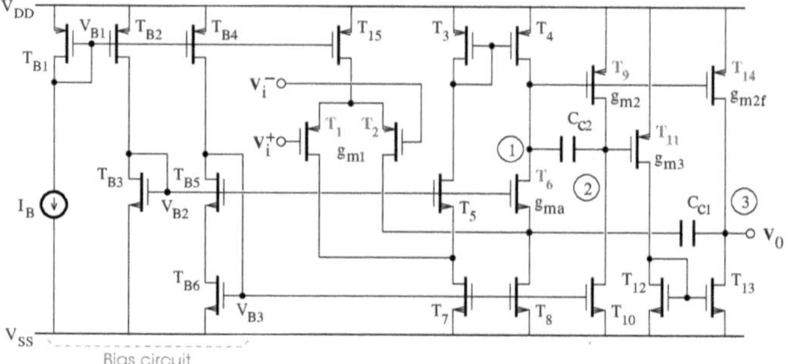

FIGURE 4.72
Circuit diagram of a three-stage RAFFC amplifier.

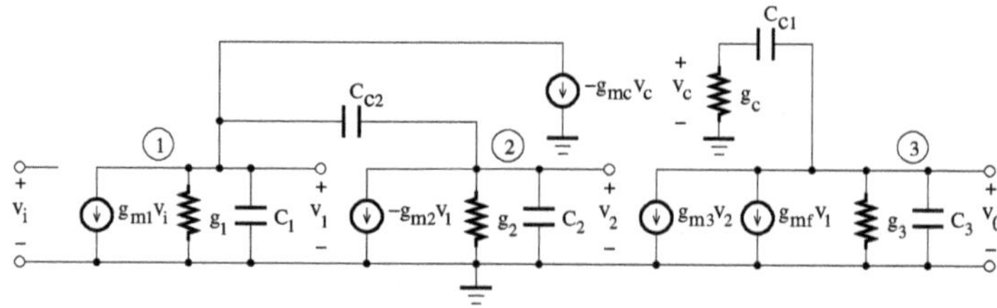

FIGURE 4.73
Small-signal equivalent circuit of the three-stage RAFFC amplifier.

realized by linking the gate of the load transistor T_{14} to the first stage output, while the feedback transconductance is obtained by connecting C_{c_1} to the source of T_6 instead of the first stage output. With reference to the small-signal equivalent circuit shown in Figure 4.73, where $g_c \simeq g_{m_c}$, the equations for nodes 1, 2, and 3 can be written as

$$g_{m_1} V_i(s) + V_1(s)(g_1 + sC_1) - g_{m_c} V_c(s) - (V_2(s) - V_1(s))sC_{c_2} = 0 \quad (4.523)$$

$$-g_{m_2} V_1(s) + V_2(s)(g_2 + sC_2) + (V_2(s) - V_1(s))sC_{c_2} = 0 \quad (4.524)$$

and

$$g_{m_3} V_2(s) + g_{m_f} V_1(s) + V_0(s)(g_3 + sC_3) + (V_0(s) - V_c(s))sC_{c_1} = 0 \quad (4.525)$$

respectively, where g_{m_f} is the feedforward transconductance, and g_{m_c} is the feedback compensation transconductance. The node capacitances can be expressed as

$$C_1 = C_{gd_4} + C_{gb_4} + C_{gd_6} + C_{gb_6} + C_{gs_9} \quad (4.526)$$

$$C_2 = C_{gd_9} + C_{gb_9} + C_{gd_{10}} + C_{gb_{10}} + C_{gs_{11}} \quad (4.527)$$

and

$$C_3 = C_{gd_{13}} + C_{gb_{13}} + C_{gd_{14}} + C_{gb_{14}} + C_L \quad (4.528)$$

where C_L is the external output load capacitance. Using the principle of voltage division, it can be shown that

$$V_c(s) = \frac{sC_{c_1}V_0(s)}{g_{m_c} + sC_{c_1}} \tag{4.529}$$

Assuming that $g_{m_k} \gg g_k$ ($k = 1, 2, 3$) and $C_{c_1}, C_{c_2}, C_3 \gg C_1, C_2$, the amplifier small-signal voltage gain can be computed as

$$A(s) = \frac{V_0(s)}{V_i(s)} \simeq A_0 \frac{1 + ds + cs^2}{(1 + s/\omega_{p_1})(1 + bs + as^2)} \tag{4.530}$$

where

$$A_0 = \frac{g_{m_1}g_{m_2}g_{m_3}}{g_1 g_2 g_3} \tag{4.531}$$

$$\omega_{p_1} = \frac{g_1 g_2 g_3}{g_{m_2}g_{m_3}C_{c_1}} \tag{4.532}$$

$$a = \frac{C_{c_2}C_3}{g_{m_c}g_{m_3}} \tag{4.533}$$

$$b = \frac{(g_{m_2} + g_{m_f} - g_{m_3})C_{c_1}C_{c_2} + g_{m_2}C_{c_2}C_3}{g_{m_2}g_{m_3}C_{c_1}} \tag{4.534}$$

$$c = \frac{(g_{m_f} - g_{m_3})C_{c_1}C_{c_2}}{g_{m_2}g_{m_3}g_{m_c}} \tag{4.535}$$

and

$$d = \frac{C_{c_1}}{g_{m_c}} + \frac{(g_{m_f} - g_{m_3})C_{c_2}}{g_{m_2}g_{m_3}} \tag{4.536}$$

To proceed further, it is assumed that $g_{m_f} = g_{m_3}$. As a result, the coefficient c of the transfer function is reduced to zero. The amplifier transfer function now exhibits one left-hand plane zero and three poles. By choosing g_{m_c} appropriately, the zero is at a much higher frequency than the dominant pole, and its effect can be neglected. The stability of the amplifier can then be ensured by considering that the denominator of the transfer function in the unity-gain closed-loop configuration is a third-order Butterworth polynomial with a cutoff frequency of ω_c. That is,

$$\frac{2}{\omega_c} = \frac{1}{A_0\omega_{p_1}} \simeq \frac{C_{c_1}}{g_{m_1}} \tag{4.537}$$

$$\frac{1}{\omega_c} = b \simeq \frac{(C_{c_1} + C_3)C_{c_2}}{g_{m_3}C_{c_1}} \tag{4.538}$$

and

$$\frac{1}{2\omega_c^2} = a = \frac{C_{c_2} C_3}{g_{m_c} g_{m_3}} \tag{4.539}$$

By solving the system of Equations (4.537), (4.538), and (4.539), the values of the compensation capacitors can be obtained as

$$C_{c_1} = \frac{1}{2}\left(\frac{4g_{m_1}}{g_{m_c}} - 1\right) C_3 \tag{4.540}$$

and

$$C_{c_2} = \frac{g_{m_c} g_{m_3}}{8}\left[\frac{1}{g_{m_1}}\left(\frac{4g_{m_1}}{g_{m_c}} - 1\right)\right]^2 C_3 \tag{4.541}$$

where $g_{m_1} > g_{m_c}/4$. The gain-bandwidth product of the amplifier is of the form

$$GBW \simeq \frac{g_{m_1}}{C_{c_1}} = N \frac{1}{4} \frac{g_{m_3}}{C_3} \tag{4.542}$$

where

$$N = \frac{2}{g_{m_3}\left(\frac{1}{g_{m_c}} - \frac{1}{4}\frac{1}{g_{m_1}}\right)} \tag{4.543}$$

Because $N > 1$, the GBW of the RAFFC amplifier is much greater than the one of the NMC amplifier. The phase margin of the RAFFC amplifier can be computed as

$$\phi_M = 180^o - \angle A[j(A_0 \omega_{p_1})] \tag{4.544}$$

$$\simeq 60^o + \arctan[A_0 \omega_{p_1} (C_{c_1}/g_{m_c})] \simeq 60^o + \arctan(g_{m_1}/g_{m_c}) > 74^o \tag{4.545}$$

Due to the contribution of the left-hand plane zero, the phase margin is greater than the value of 60° obtained in the case of the NMC amplifier [52,53]. The slew rate depends upon the value of compensation capacitors and the bias currents.

The circuit diagram of the three-stage amplifier with cross feedforward cascode compensation [54] is shown in Figure 4.74. In addition to the three transconductance stages, g_{m_1}, g_{m_2}, and g_{m_3}, it includes feedforward stages with transconductances $g_{m_{f_1}}$ and $g_{m_{f_2}}$ and a feedback network composed of the transconductance stage, g_{m_c}, and the compensation capacitor C_c. The amplifier stability should be improved by the left-half-plane zeros introduced, respectively, by the feedback network and the feedforward stage, $g_{m_{f_1}}$.

The small-signal equivalent model of the three-stage amplifier with cross

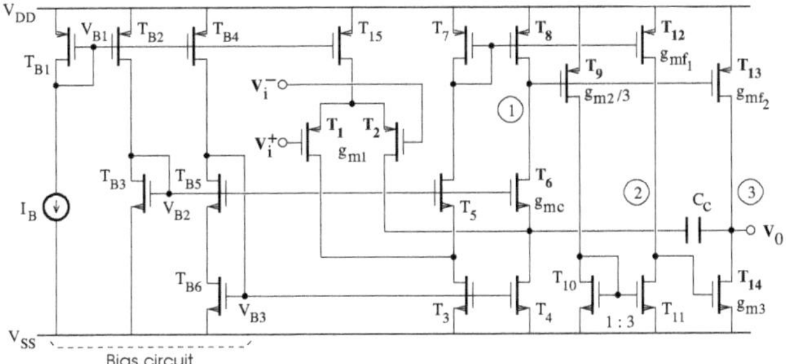

FIGURE 4.74
Circuit diagram of a three-stage amplifier with cross feedforward cascode compensation.

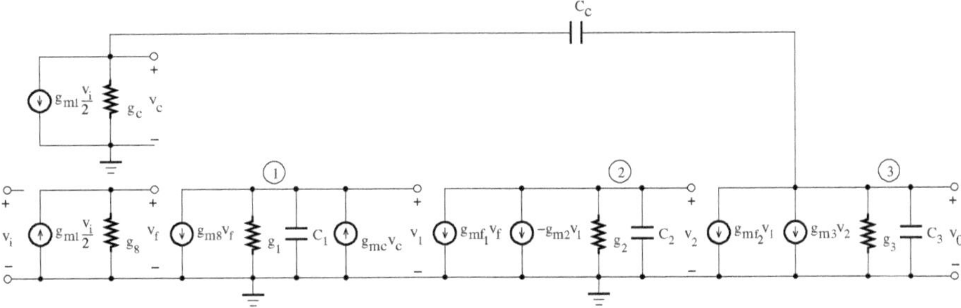

FIGURE 4.75
Equivalent circuit model of the three-stage amplifier with cross feedforward cascode compensation.

feedforward cascode compensation is depicted in Figure 4.75, where $g_c \simeq g_{m_c}$ and $g_8 \simeq g_{m_8}$. Using Kirchhoff's current law, it can be shown that

$$g_{m_1}\frac{V_i(s)}{2} + g_{m_c}V_c(s) + [V_c(s) - V_0(s)]sC_c = 0 \tag{4.546}$$

$$[V_c(s) - V_0(s)]sC_c - g_{m_3}V_2(s) - g_{m_{f_2}}V_1(s) - V_0(s)(g_3 + sC_3) = 0 \tag{4.547}$$

where

$$V_2(s) = -\frac{g_{m_{f_1}}V_f(s) - g_{m_2}V_1(s)}{g_2 + sC_2} \tag{4.548}$$

$$V_1(s) = \frac{g_{m_c}V_c(s) - g_{m_8}V_f(s)}{g_1 + sC_1} \tag{4.549}$$

and

$$V_f(s) = \frac{g_{m_1}}{g_{m_8}}\frac{V_i(s)}{2} \tag{4.550}$$

By solving the above equations and assuming that the gain g_{m_k}/g_k $(k = 1, 2, 3)$ of the kth stage is much greater than one, the output capacitance C_3 of the last stage and the compensation capacitance C_c are much greater than the parasitic capacitances C_1 and C_2, the amplifier gain transfer function can be obtained as

$$A(s) = \frac{V_0(s)}{V_i(s)} \simeq A_0 \frac{1 + \frac{s}{\omega_{z_1}} + \frac{s^2}{\omega_{z_1}\omega_{z_2}} - \frac{s^3}{\omega_{z_1}\omega_{z_2}\omega_{z_3}}}{\left(1 + \frac{s}{\omega_{p_1}} + \frac{s^2}{\omega_{p_1}\omega_{p_2}}\right)\left(1 + 2k\frac{s}{\omega_0} + \frac{s^2}{\omega_0^2}\right)} \tag{4.551}$$

where

$$A_0 = \frac{g_{m_1} g_{m_2} g_{m_3}}{g_1 g_2 g_3} \tag{4.552}$$

$$\omega_{z_1} = \frac{2g_{m_c}}{C_c} \tag{4.553}$$

$$\omega_{z_2} = \frac{g_{m_2} g_{m_8}}{g_{m_{f_1}} C_1} \tag{4.554}$$

$$\omega_{z_3} = \frac{g_{m_{f_1}} g_{m_3}}{g_{m_8} C_2} \tag{4.555}$$

$$\omega_{p_1} = \frac{g_1 g_2 g_3}{g_{m_2} g_{m_3} C_c} \tag{4.556}$$

$$\omega_{p_2} = \frac{g_2}{C_2} \tag{4.557}$$

$$\omega_0 = \sqrt{\frac{g_{m_2} g_{m_3} g_{m_c}}{g_2 C_1 C_3}} \tag{4.558}$$

$$k = \frac{1}{2}\sqrt{\frac{g_{m_c} g_2 C_1}{g_{m_2} g_{m_3} C_3}} \tag{4.559}$$

and

$$\frac{1}{g_{m_2}}\left(\frac{g_{m_{f_1}} g_1}{2g_{m_8}} + \frac{g_{m_{f_2}} g_2}{g_{m_3}}\right) \ll 1 \tag{4.560}$$

Assuming sufficiently spaced poles and zeros, the amplifier gain transfer function can be written as,

$$A(s) = \frac{V_0(s)}{V_i(s)} \simeq A_0 \frac{(1 + s/\omega_{z_1})(1 + s/\omega_{z_2})(1 - s/\omega_{z_3})}{(1 + s/\omega_{p_1})(1 + s/\omega_{p_2})[1 + 2k(s/\omega_0) + s^2/\omega_0^2]} \tag{4.561}$$

There are two left-half-plane zeros, z_1 and z_2, and one right-half-plane zero, z_3, whose effect can be neglected because it is located at a frequency higher than that of the fourth pole. The amplifier bandwidth can be increased by using the positive phase shift due to the left-half-plane zeros to compensate the negative phase shift caused by the non-dominant complex poles. Figure 4.76

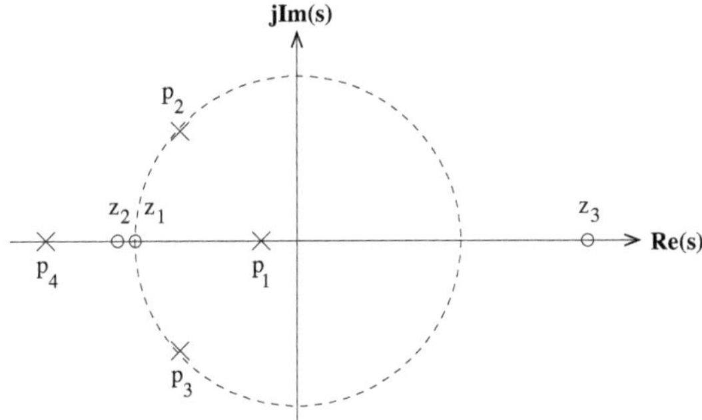

FIGURE 4.76
Pole-zero plot of the three-stage amplifier with cross feedforward cascode compensation.

shows the pole-zero plot of the three-stage amplifier with cross feedforward cascode compensation.

The placement of the amplifier poles and zeros should be performed so as to satisfy the stability requirements. Therefore, the amplifiers should be compensated to achieve an adequate phase margin, ϕ_M. It can be shown that

$$\phi_M = 180^o - \arctan\left(\frac{GBW}{\omega_{p_1}}\right) - \arctan\left(\frac{2k\dfrac{GBW}{\omega_0}}{1-\dfrac{GBW^2}{\omega_0^2}}\right) + \arctan\left(\frac{GBW}{\omega_{z_1}}\right) + \arctan\left(\frac{GBW}{\omega_{z_2}}\right) - \arctan\left(\frac{GBW}{\omega_{p_4}}\right) \quad (4.562)$$

Assuming that the poles of the amplifier in unity-feedback configuration have a third-order Butterworth (or maximally flat) frequency response, the damping factor k should be equal to $\sqrt{2}/2$ and the gain-bandwidth product GBW should be chosen as $\omega_0/2$ to meet the stability requirements. The amplifier bandwidth is increased by using both left-half-plane zeros to compensate for the non-dominant complex poles. Hence, $\omega_{z_1} = 2\times GBW$ and $\omega_{z_2} = 3\times GBW$. The phase margin can then be expressed as,

$$\phi_M \simeq 90^o - \arctan\left(\frac{GBW}{\omega_{p_4}}\right) \quad (4.563)$$

Furthermore, the following design equations can be derived:

$$g_{m_c} = g_{m_1} \quad (4.564)$$

$$C_c = 2\sqrt{\frac{g_{m_1} g_2 C_1 C_3}{g_{m_2} g_{m_3}}} \quad (4.565)$$

and

$$g_{m_{f_1}} = \frac{g_{m_2} g_{m_8}}{3 g_{m_1}} \frac{C_c}{C_1} \tag{4.566}$$

In practice, it may not be possible to achieve a complete cancellation of a complex pole-pair with left half-plane zeros. To reduce the effects of residual pole-zero doublets on the frequency response, the complex-pole compensation should be designed to maintain a low damping factor, k, while keeping ω_0 as high as possible.

The feedforward transconductance stage $g_{m_{f_2}}$ and the last transconductance stage g_{m_3} form a push-pull output stage, that is known to exhibit an improved transient response even with large output load capacitors. The slew rate is then primarily limited by the first stage that drives the compensation capacitor C_c. It is given by

$$SR = \frac{I_1}{C_c} \tag{4.567}$$

where I_1 denotes the current available to charge or discharge the capacitor C_c.

The amplifier with cross feedforward cascode compensation can be designed to achieve a GBW of 2.5 MHz, a phase margin of 60°, and a slew rate of 1.1 V/μS, when driving an output load capacitance of 300 pF.

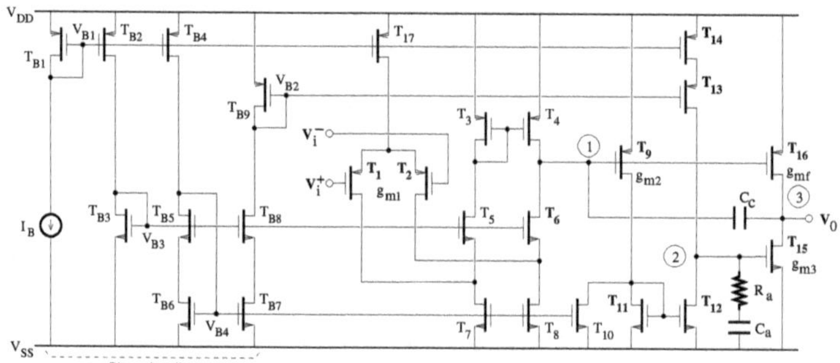

FIGURE 4.77
Circuit diagram of a three-stage amplifier with impedance adapting compensation.

The circuit diagram of the three-stage amplifier with impedance adapting compensation [55] is depicted in Figure 4.77. It is composed of two inverting gain stages, a non-inverting gain stage, a Miller compensation capacitor, C_c, a local impedance attenuation network (R_a and C_a), and a feedforward transconductor g_{m_f}. The use of the local impedance attenuation network helps reduce the high-frequency small-signal resistance at the second-stage output and allows the control of the non-dominant pole placement. The large-signal

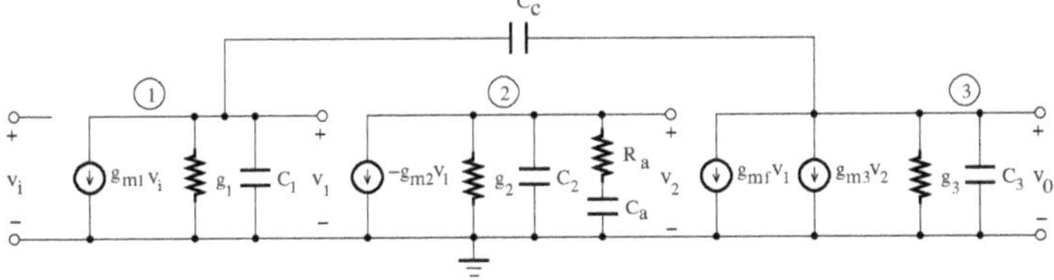

FIGURE 4.78
Small-signal equivalent model of the three-stage amplifier with impedance adapting compensation.

response is improved because the feedforward path set up by the transconductor g_{m_f} between the first-stage output and the last-stage input enables the push-pull operation of the last stage.

The small-signal equivalent model of the three-stage amplifier with impedance adapting compensation is depicted in Figure 4.78. Applying Kirchhoff's current law to each node, we obtain:

$$g_{m_1} V_i(s) + V_1(s)(g_1 + sC_1) + [V_1(s) - V_0(s)]sC_c = 0 \tag{4.568}$$

$$V_2(s) = g_{m_2} V_1(s) \left(g_2 + sC_2 + \frac{1}{\dfrac{1}{g_a} + \dfrac{1}{sC_a}} \right)^{-1} \tag{4.569}$$

$$[V_1(s) - V_0(s)]sC_c = g_{m_3} V_2(s) + g_{m_f} V_1(s) + V_0(s)(g_3 + sC_3) \tag{4.570}$$

To proceed further, it is assumed that the low-frequency gain of each stage is sufficiently larger than unity, or say $g_{m_k}/g_k \gg 1$ $(k = 1, 2, 3)$, the compensation capacitances are larger than the output parasitic capacitances, $C_c \gg C_1$, $C_a \gg C_2$, and smaller than the output load capacitance, $C_c, C_a \ll C_3$. The transfer function of the amplifier small-signal gain can then be derived as,

$$A(s) = \frac{V_0(s)}{V_i(s)} \simeq A_0 \frac{1 + \dfrac{s}{\omega_{z_1}} + \dfrac{s^2}{\omega_{z_1}\omega_{z_2}} + \dfrac{s^3}{\omega_{z_1}\omega_{z_2}\omega_{z_3}}}{\left(1 + \dfrac{s}{\omega_{p_1}}\right)\left(1 + \dfrac{s}{\omega_{p_2}} + \dfrac{s^2}{\omega_{p_2}\omega_{p_3}} + \dfrac{s^3}{\omega_{p_2}\omega_{p_3}\omega_{p_4}}\right)} \tag{4.571}$$

where

$$A_0 = \frac{g_{m_1} g_{m_2} g_{m_3}}{g_1 g_2 g_3} \tag{4.572}$$

$$\omega_{z_1} = \frac{1}{h}\frac{g_a}{C_a} \tag{4.573}$$

$$\omega_a = h\frac{g_{m_2} g_{m_3}}{g_a C_c} \tag{4.574}$$

$$\omega_b = \frac{g_a}{C_2} \tag{4.575}$$

$$\omega_{p_1} = \frac{g_1 g_2 g_3}{g_{m_2} g_{m_3} C_c} \tag{4.576}$$

$$\omega_{p_2} = \frac{1}{h} \frac{g_a}{C_a} \tag{4.577}$$

$$\omega_{p_3} = h \frac{g_{m_2} g_{m_3}}{g_a C_3} \tag{4.578}$$

$$\omega_{p_4} = \frac{g_a}{C_2} \tag{4.579}$$

and

$$h = 1 + \frac{g_{m_f} g_a}{g_{m_2} g_{m_3}} \simeq 1 \tag{4.580}$$

Assuming that the poles are well separated so that a dominant-pole roll-off frequency response can be achieved, the unity-gain frequency, GBW, is given by

$$GBW = A_0 \cdot \omega_{p_1} = g_{m_1}/C_c \tag{4.581}$$

where ω_{p_1} is the dominant pole. The non-dominant poles are placed by choosing the values of the conductance g_a and the capacitor C_a in order to satisfy the next condition:

$$\omega_{p_2} \ll \omega_{p_3} \ll \omega_{p_4} \tag{4.582}$$

The zeros of the gain transfer function are determined with the assumption that

$$\omega_{z_1} = \omega_{p_2} \ll |\omega_a| = \omega_{p_3} \frac{C_3}{C_c} \tag{4.583}$$

$$\omega_{z_1} = \omega_{p_2} \ll \omega_b = \omega_{p_4} \tag{4.584}$$

The first and second zeros are located at $-\omega_{z_1}$ and $-\omega_{z_2}$, respectively, in the left-half s-plane, while the third zero is placed at $-\omega_{z_3}$, in the right-half s-plane, where

$$\omega_{z_2} = \frac{1}{2}\left(\sqrt{4|\omega_a|\omega_b + \omega_b^2} + \omega_b\right) \tag{4.585}$$

$$\omega_{z_3} = -\frac{1}{2}\left(\sqrt{4|\omega_a|\omega_b + \omega_b^2} - \omega_b\right) \tag{4.586}$$

The zero, z_1, is at a low frequency and is compensated by the first non-dominant pole p_2, while the effects of other zeros, that are rejected at sufficiently higher frequencies than the second non-dominant pole p_3, can be neglected. Figure 4.79 shows the pole-zero plot of the three-stage amplifier with impedance adapting compensation. The zero z_1 is used to cancel the pole p_2.

Based on the amplifier transfer function obtained after the pole-zero cancellation, and disregarding the contributions of the pole p_4 and zeros (z_2 and

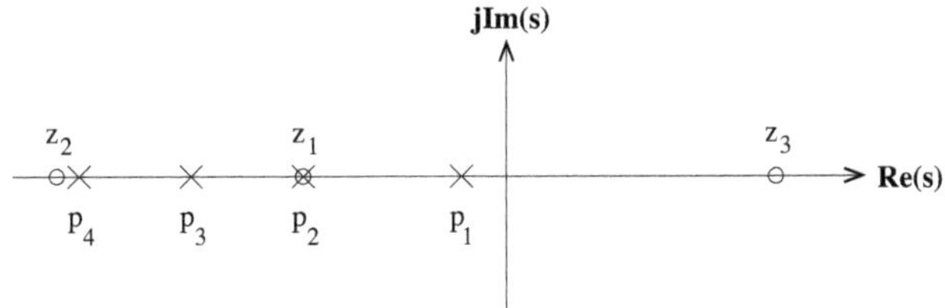

FIGURE 4.79
Pole-zero plot of the three-stage amplifier with impedance adapting compensation.

z_3) at very high frequencies, the phase margin can be derived as,

$$\phi_M \simeq 90^o - \arctan\left(\frac{GBW}{\omega_{p3}}\right) \tag{4.587}$$

To achieve a phase margin of at least 60°, the second non-dominant pole, p_3, should be on the order of $2 \cdot GBW$. Hence, the required value of the compensation capacitor is given by

$$C_c = \frac{2g_{m_1}}{\omega_{p3}} = \frac{2g_{m_1}g_a}{g_{m_2}g_{m_3}}C_3 \tag{4.588}$$

The feedforward transconductance stage, g_{m_f}, helps increase the slew rate of the third stage. The effect of the second stage on the slew rate can be considered negligible because the capacitor C_2 is much smaller than C_c and C_3, and g_{m_2}/g_a is greater than one. The amplifier slew rate is then primarily limited by the first stage and can be expressed as,

$$SR = \frac{I_1}{C_c} \tag{4.589}$$

where I_1 is the current provided by the first stage to charge the compensation capacitor.

Using $g_{m_1} = 16$ μS, $g_{m_2} = 157$ μS, $g_{m_3} = 59$ μS, $g_{m_f} = 60$ μS, $C_c = 0.5$ pF, $C_a = 1.1$ pF, and $R_a = 1/g_a = 750$ kΩ, the implementation of a three-stage amplifier with impedance adapting compensation can feature a GBW of 4.5 MHz, a phase margin of about 60°, and a slew rate of 1.8 V/μS, when driving an output load capacitance of 150 pF.

4.8 Rail-to-rail amplifiers

Generally, rail-to-rail amplifiers are useful in low-voltage applications, where it is necessary to efficiently use the limited span offered by the power supply.

While conventional amplifiers are capable of linear operation only for signals with a small excursion around the common-mode levels, rail-to-rail amplifiers are designed to allow signals to swing within millivolts of either power supply rail. The input and output voltage ranges are dependent on the amplifier topology, and the rail-to-rail operation can be achieved for either the input or the output, or both input and output.

4.8.1 Amplifier with a class AB input stage

The circuit diagram of the amplifier is depicted in Figure 4.80. It uses an input stage consisting of source cross-coupled transistors [56] as shown in Figure 4.81.

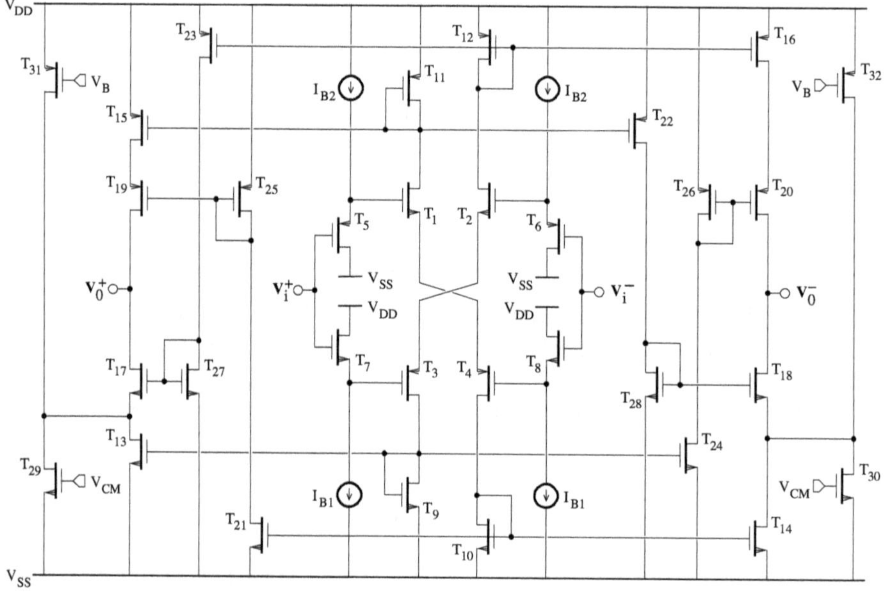

FIGURE 4.80
Cross-coupled transistor-based differential amplifier.

Using the square-law model for MOS devices in the saturation region, the drain currents are given by

$$i_D = \begin{cases} K_n(v_{GS} - V_{Tn})^2 & n\text{-channel transistor} \\ K_p(v_{GS} - V_{Tp})^2 & p\text{-channel transistor.} \end{cases} \tag{4.590}$$

Applying Kirchhoff's voltage law on the left and right sides of the transcon-

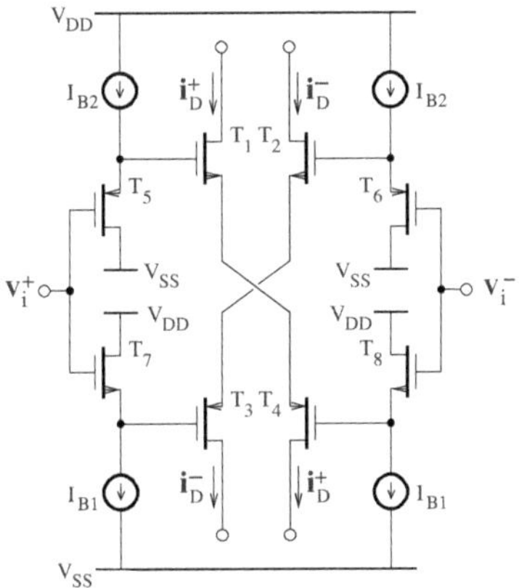

FIGURE 4.81
Source cross-coupled transistor pair.

ductor, we have

$$v_i = v_i^+ - v_i^- = v_{GS7} + v_{GS3} - (v_{GS2} + v_{GS6}) \tag{4.591}$$

$$-v_i = v_i^- - v_i^+ = v_{GS8} + v_{GS4} - (v_{GS1} + v_{GS5}) \tag{4.592}$$

With $i_{D5} = i_{D7} = i_{D6} = i_{D8} = I_{SS} = I_{DD} = I_B$, the drain currents are written as

$$i_D^+ = K_{eq}\left(v_i + \sqrt{\frac{I_B}{K_{eq}}}\right)^2 \tag{4.593}$$

$$i_D^- = K_{eq}\left(v_i - \sqrt{\frac{I_B}{K_{eq}}}\right)^2 \tag{4.594}$$

where

$$i_D^+ = i_{D1} = i_{D4} \tag{4.595}$$

$$i_D^- = i_{D2} = i_{D3} \tag{4.596}$$

and

$$K_{eq} = \frac{K_n K_p}{\left(\sqrt{K_n} + \sqrt{K_p}\right)^2} \tag{4.597}$$

The differential output current is given by

$$\Delta i = i_D^+ - i_D^- = 4\sqrt{K_{eq} I_B}\, v_i \tag{4.598}$$

The transfer characteristic of the differential stage is plotted in Figure 4.82.

The corresponding transconductance is $g_m = 4\sqrt{K_{eq} I_B}$. The transistors remain in the saturation region provided that

$$|v_i| \leq \sqrt{\frac{I_B}{K_{eq}}} \tag{4.599}$$

The class AB operation is achieved because the transconductor can generate an output current which is larger than the dc quiescent current flowing in the circuit.

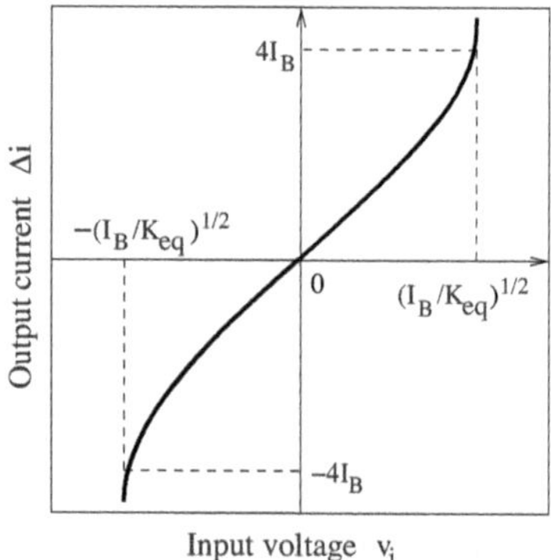

FIGURE 4.82
Transfer characteristic of the source cross-coupled transistor pair.

Due to the biasing condition realized by source followers $T_5 - T_8$, a current can flow through the input stage even for zero differential input. Given an increase in the voltage on the positive input and a corresponding decrease on the negative input, the drain currents of T_1 and T_4, and the ones of T_2 and T_3 increase and decrease from their initial values, as a result of a rise and reduction of their gate-source voltages, respectively. That is, the current in one side of the differential stage increases monotonically with the applied voltage and is limited by the power supply level, while the one in the other side decreases until a transistor turns off. The input currents are then directed to the output branches by current mirrors. The cascode transistors, $T_{17} - T_{20}$, are used to increase the amplifier gain. The conflicting requirements of high output current during the slewing period and large output swing during the settling are met by dynamically biasing the gates of cascode transistors so that the common-source transistors $T_{13} - T_{16}$ remain in the saturation region.

The common-mode feedback is realized by controlling the bias current of T_{29} and T_{30}. Transistors T_{31} and T_{32} are connected to the bias voltage, V_B, in order to deliver constant currents.

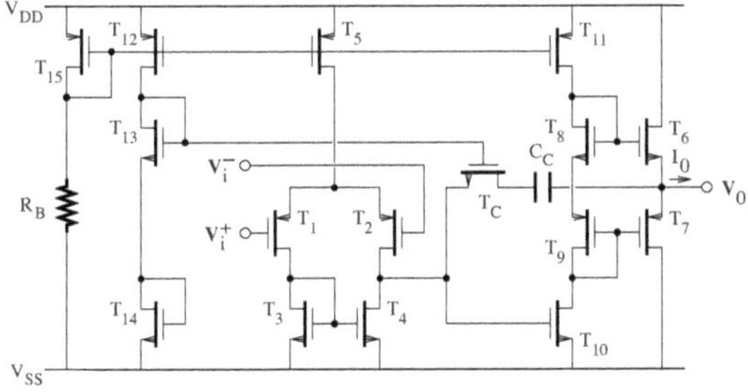

FIGURE 4.83
Circuit diagram of a two-stage amplifier with a class AB output stage.

4.8.2 Two-stage amplifier with class AB output stage

The circuit diagram of an amplifier based on the four-transistor class AB output stage is shown in Figure 4.83. The signal provided by the differential input stage is applied to the output stage consisting of common drain complementary transistors, whose quiescent point is set by a translinear loop. As a result, the conduction of the output transistors is maintained even in the absence of an incoming ac signal. Furthermore, the output stage has the capability to source or sink the output current. However, the available output dynamic range may be affected by fluctuations in the transistor threshold voltages. The amplifier stability is maintained by inserting a compensation network consisting of T_C and C_C between the input and output of the second stage. The transistor T_C is biased such that a proper compensation resistance can be maintained over IC process, temperature, and supply voltage variations.

4.8.3 Amplifier with rail-to-rail input and output stages

In low-voltage designs, a rail-to-rail input stage is required to improve the signal dynamic range for a given supply voltage [57, 58]. It is generally implemented, as shown in Figure 4.84, by a parallel connection of n-channel and p-channel transistor pairs. Due to the different dc behavior of each pair, the resulting transconductance shown in Figure 4.85 varies with the common-mode input voltage and is not constant.

The output current of the differential stage can be written as

$$i^+ = \frac{I_B}{2} + g_m \frac{v_i}{2} \tag{4.600}$$

$$i^- = \frac{I_B}{2} - g_m \frac{v_i}{2} \tag{4.601}$$

The transconductance is either

$$g_m = \frac{I_B}{\alpha U_T}, \qquad \text{for} \quad v_i < 2U_T \tag{4.602}$$

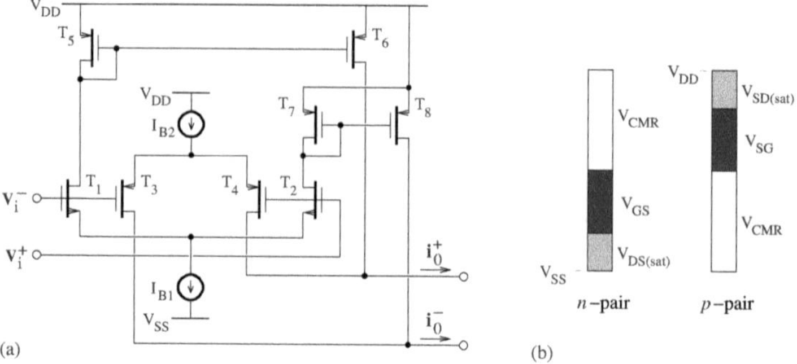

FIGURE 4.84
(a) Circuit diagram of a rail-to-rail input stage; (b) voltage swings of n- and p-channel transistor pairs.

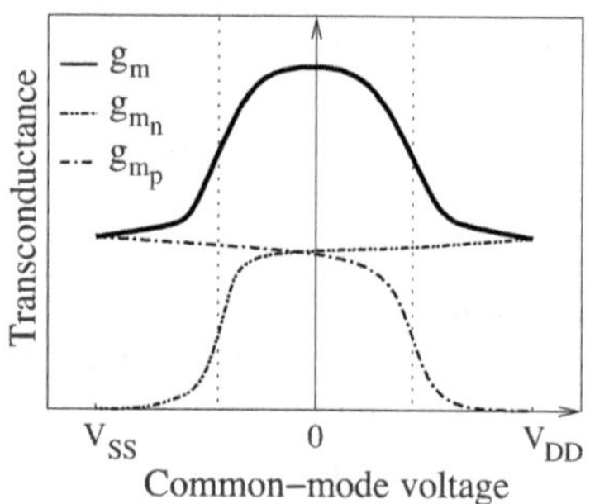

FIGURE 4.85
Transconductance variations versus the common-mode voltage.

in the weak inversion region, where α is the week inversion slope and U_T is the thermal potential, or

$$g_m = \sqrt{2KI_B}, \qquad \text{for} \quad |v_i| < (2I_B/K)^{1/2} \tag{4.603}$$

in the saturation region. The tail current I_B assumes the value of I_{B1} for the n-channel transistor pair while it is I_{B2} for the stage using p-channel transistor pair. The parameter K is given by

$$K = \begin{cases} K_n = (1/2)\mu_n C_{ox}(W/L) & n\text{-channel transistor} \\ K_p = (1/2)\mu_p C_{ox}(W/L) & p\text{-channel transistor.} \end{cases} \tag{4.604}$$

When the common-mode voltage, v_{CM}, is at mid-rail, both n- and p-channel transistor pairs operate normally. As a result, the total transconductance g_{mT}, is about two times greater than that obtained in the cases where v_{CM} is close to either of the supply voltages, V_{SS} or V_{DD}, and only one transistor pair type remains operational. This transconductance variation can be efficiently reduced by using differential pairs based on improved biasing circuits.

A constant transconductance can be obtained by inserting a constant voltage, V_Z, between the tails of the complementary input pairs, as shown in Figure 4.86. It is the result of stabilizing the gate-source voltages of the transistors.

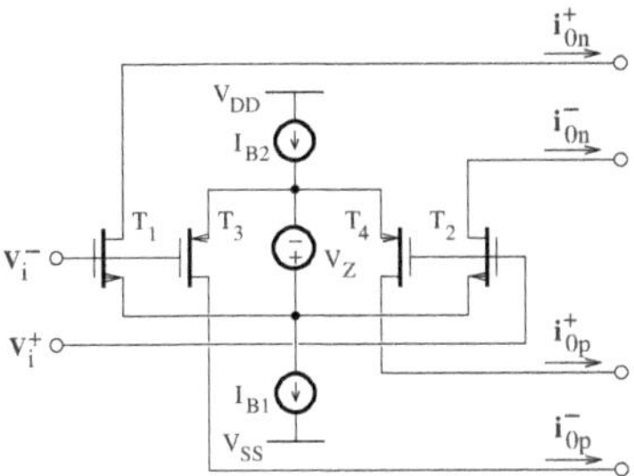

FIGURE 4.86
Circuit diagram of a rail-to-rail input stage with a constant voltage regulation.

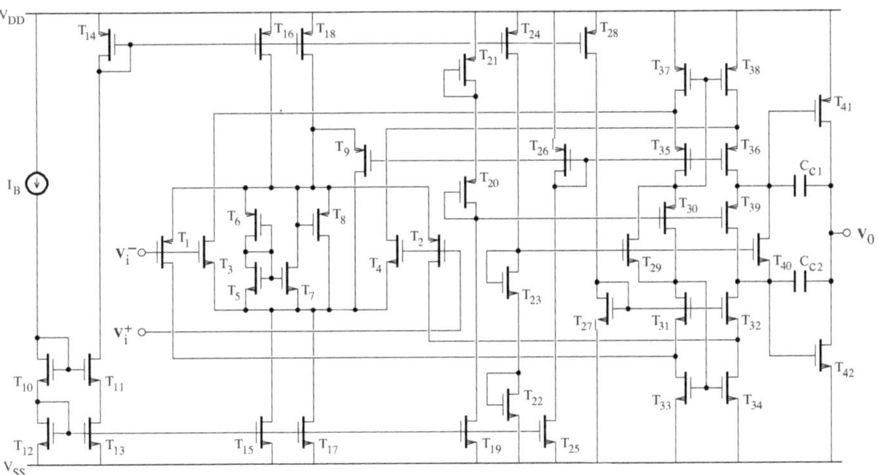

FIGURE 4.87
Circuit diagram of a rail-to-rail amplifier with the constant voltage regulation of the input stage.

The circuit diagram of the overall rail-to-rail amplifier is depicted in Figure 4.87. Transistors $T_5 - T_8$ and $T_{17} - T_{18}$ implement the voltage source V_Z. The parameters W/L of $T_5 - T_8$ can be sized to match the ones of the corresponding input transistors and $T_{15} - T_{16}$ should drive a current eight times greater than the one of $T_{17} - T_{18}$. The gate voltage of T_9 is used to limit the drain voltage of T_{18}, reducing in this way any additional variation of the current delivered by T_{15}.

The class AB output stage has the advantage of operating with a high speed and low supply voltages. The quiescent current flowing through the complementary output transistors, T_{41} and T_{42}, is minimized by the biasing structures established by $T_{19} - T_{21}$, T_{39} and $T_{22} - T_{24}$, T_{40}, respectively. For a negative output slew, the gate voltage of T_{42} increases. Because T_{40} is biased at a fixed voltage, it will turn off and the whole bias current now flows through T_{39}. As a result, the gate-source voltages of T_{39} and T_{41} decrease. A similar operation will be observed when the bias of T_{41} is reduced due to a positive change at the output. The floating current sources $T_{29} - T_{30}$ have the same structure as the ones used for the control of the output transistors, whose quiescent current is made less sensitive to supply voltage variations by using two current mirrors biased independently. The input signals of the current mirrors $T_{31} - T_{34}$ and $T_{35} - T_{38}$ are obtained from the p- and n-channel transistor pairs, $T_1 - T_2$ and $T_3 - T_4$, respectively. Assuming identical saturation and threshold voltages, the minimum supply voltage is about $3V_{DS(sat)} + 2V_T$.

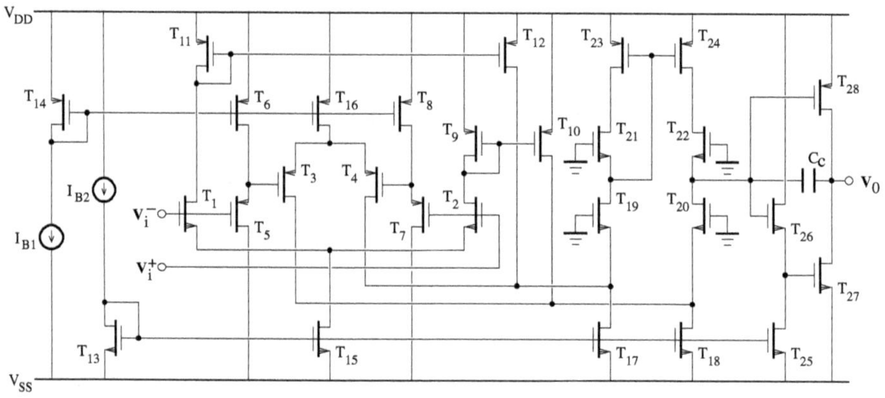

FIGURE 4.88
Circuit diagram of a rail-to-rail amplifier with dc level shifters.

The amplifier phase margin is determined by the two compensation capacitors, C_{c1} and C_{c2}, which split apart the amplifier poles to provide a first-order gain characteristic.

The amplifier architecture of Figure 4.88 is based on another simple technique, which can be adopted to achieve a constant transconductance. The dc level shifters, realized by the source followers $T_5 - T_6$ and $T_7 - T_8$, are required to overlap the transition region of the tail currents for the n- and p-channel transistor pairs $T_1 - T_2$ and $T_3 - T_4$. The input stage is loaded by the folded cascode structure, $T_{17} - T_{24}$, followed by a class AB output stage compensated by C_c. Note that the deviation of the transconductance due to layout mismatches and IC process variations can be reduced by tuning the bias currents I_{B1} and I_{B2}.

4.9 Amplifier characterization

The ideal model of an amplifier features an infinite gain and bandwidth. Furthermore, the input and output impedances are assumed to be infinite or zero. However, the performance characteristics of amplifiers are generally limited by various nonideal effects (finite gain and bandwidth, common-mode range, power supply rejection, input and output impedance, slew rate, offset) in practice.

4.9.1 Finite gain and bandwidth

The frequency response of an amplifier is shown in Figure 4.89. The first pole of the transfer function, which is characterized by ω_1, is generally made dominant using a suitable compensation technique to have a first-order frequency response. In this case, the 3-dB frequency of the amplifier is reduced to ω_1. The parasitic effects can then be modeled by a pole at ω_2. The parameter A_0 is the dc gain and ω_t represents the unity-gain frequency.

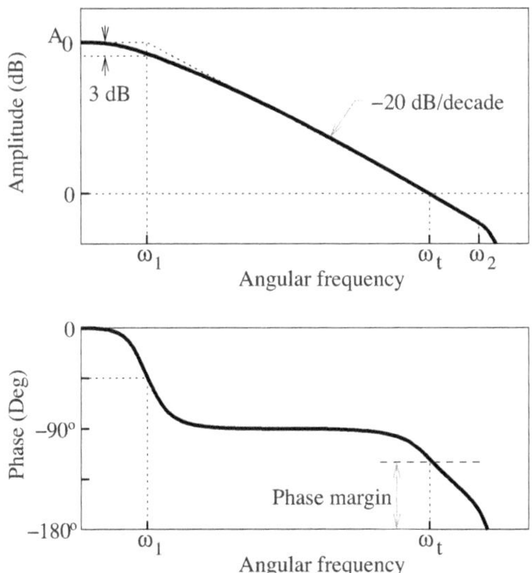

FIGURE 4.89
Frequency response of an amplifier.

The output voltage of an amplifier can be written as

$$v_O = A_{dm} v_d + A_{cm} v_c \tag{4.605}$$

where

$$v_d = v^+ - v^- \tag{4.606}$$

and

$$v_c = \frac{v^+ - v^-}{2} \tag{4.607}$$

are the differential and common-mode voltages, respectively. The common-mode rejection ratio (CMRR) is defined by

$$\mathrm{CMRR} = \frac{A_{dm}}{A_{cm}} \tag{4.608}$$

where A_{dm} and A_{cm} denote the small-signal differential mode and common mode gains, respectively. Ideally, the CMRR should be zero. It is often expressed in dB.

4.9.2 Phase margin

The phase margin, ϕ_M, is the amount by which the phase of the amplifier transfer function exceeds $-180°$ at the unity-gain frequency. It can be measured in degrees and indicates the relative stability of an amplifier in a closed-loop configuration. Typically, the minimum value of the phase margin is on the order of 45°.

4.9.3 Input and output impedances

Ideally, the input and output impedances of an amplifier can be either infinite or zero. However, they are determined in a practical amplifier by the resistors and capacitors associated with the input stage and output buffer, and can be finite and frequency dependent.

4.9.4 Power-supply rejection

The power-supply rejection ratio (PSSR) is used to characterize the amplifier sensitivity to variations in the supply voltage. It can be expressed as

$$\mathrm{PSSR} = \frac{V_O/V_i}{V_O/V_{sup}} \tag{4.609}$$

where V_{sup} can denote the positive or negative supply voltage. An ideal amplifier would feature an infinite PSSR. Due to the use of a compensation structure, the PSSR will be degraded as the number of amplification stages is increased.

4.9.5 Slew rate

The slew rate represents the maximum rate of change of the amplifier output in response to a step input signal. It is given by

$$SR = \max\left(\left|\frac{dv_O(t)}{dt}\right|\right) = \frac{I_B}{C_L} \tag{4.610}$$

where v_O denotes the amplifier output voltage, I_B is the bias current of the input differential transistor pair, and C_L is the output load capacitor. The SR worst-case value is obtained when the amplifier is in the noninverting unity gain configuration.

The slew rate can be improved by using an amplifier biased dynamically, as shown in Figure 4.90 [60]. This structure is based on the current subtractor depicted in Figure 4.91. By applying a signal at the amplifier input, the currents i_1 and i_2 become different. If i_2 is greater than i_1, a current $i = \alpha(i_2 - i_1)$, where $\alpha < 1$, will be generated and the overall bias current of the input stage will evolve into $I_B + i$. Note that in the cases where i_2 is less or equal to i_1, no current is mirrored in $T_3 - T_4$.

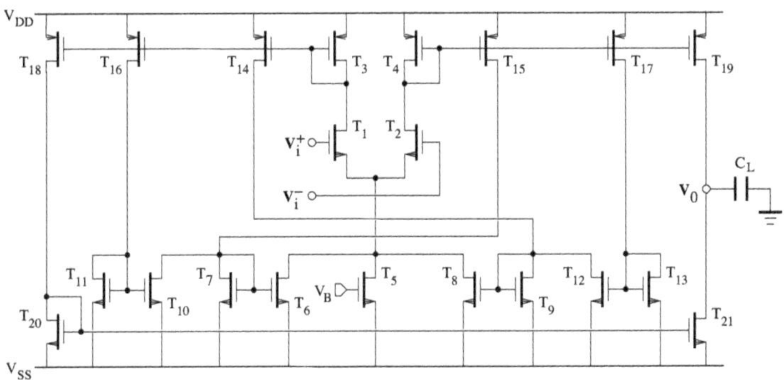

FIGURE 4.90
Single-stage amplifier with dynamic biasing.

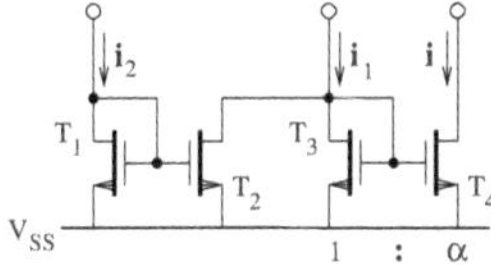

FIGURE 4.91
Current subtractor.

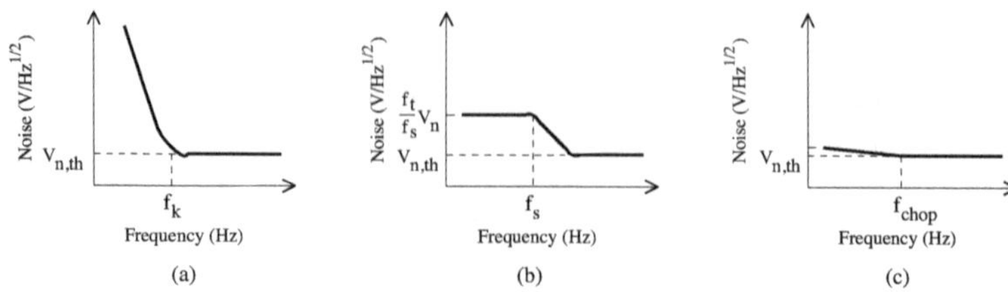

FIGURE 4.92
(a) CMOS amplifier, (b) auto-zeroed amplifier, and (c) chopper amplifier noise voltage spectra.

4.9.6 Low-frequency noise and dc offset voltage

The noise voltage spectrum of an amplifier is shown in Figure 4.92(a). It is dominated at low frequencies by the $1/f$ noise, which varies almost inversely with the frequency. The magnitude of the $1/f$ noise is dependent on the size of the input transistors and the IC process used. For high frequencies, the main contribution is due to the thermal noise, the spectrum of which is assumed to be flat. The frequency at which the thermal and $1/f$ noise components are equal corresponds to the knee frequency, f_k.

The straightforward approach to reduce the $1/f$ noise is to design the amplifier input differential pair using pMOS transistors, whose noise contribution is generally lower than the one of nMOS transistors. In a multistage amplifier, the gain of the input stage should be made as high as possible.

- **Offset voltage of a differential transistor pair**

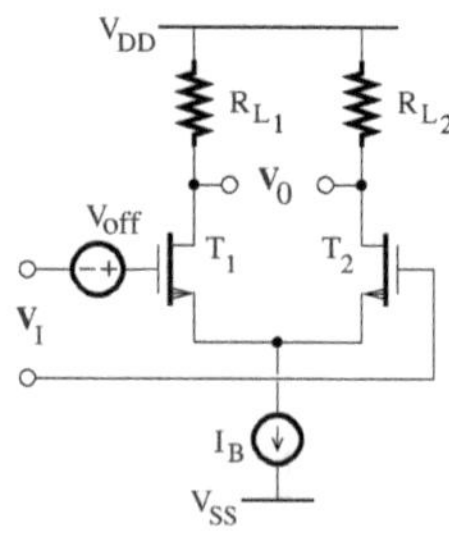

FIGURE 4.93
Circuit diagram of a differential transistor pair with the input offset voltage.

Amplifier stages may exhibit an input offset voltage, V_{off}, in the range of µV to mV due to mismatches of the transistor characteristics. Figure 4.93 shows a differential stage, where the offset voltage is modeled by a voltage source, V_{off}, in series with one of the input nodes. The differential input voltage can be expressed as

$$V_{ICM} = V_{GS_1} - V_{GS_2} \tag{4.611}$$

By exploiting the square-law I-V characteristic of MOS transistors in the saturation region, it can be shown that

$$V_{GS_1} = V_{T_1} + \sqrt{\frac{I_{D_1}}{K_1}} \tag{4.612}$$

and

$$V_{GS_2} = V_{T_2} + \sqrt{\frac{I_{D_2}}{K_2}} \tag{4.613}$$

To proceed further, the drain currents of transistors T_1 and T_2 can be expressed as

$$\triangle I_D = I_{D_1} - I_{D_2} \tag{4.614}$$

and

$$I_B = I_{D_1} + I_{D_2} \tag{4.615}$$

where I_B is the bias current. Solving this last system of equations gives

$$I_{D_1} = (I_B + \triangle I_D)/2 \tag{4.616}$$

and

$$I_{D_2} = (I_B - \triangle I_D)/2 \tag{4.617}$$

The output shift from zero experienced by the differential stage when the inputs are short-circuited to the ground can be described by the input offset voltage, which is the value of the input differential voltage, V_{ICM}, needed to reset the differential output voltage to zero. That is,

$$V_{off} = V_{T_1} - V_{T_2} + \sqrt{\frac{I_B}{2K_1}\left(1 + \frac{\triangle I_D}{I_B}\right)} - \sqrt{\frac{I_B}{2K_2}\left(1 - \frac{\triangle I_D}{I_B}\right)} \tag{4.618}$$

and

$$V_0 = R_{L_1} I_{D_1} - R_{L_2} I_{D_2} = 0 \tag{4.619}$$

With output load resistances of the form $R_{L_1} = (R_L + \triangle R_L)/2$ and $R_{L_2} = (R_L - \triangle R_L)/2$, it can be shown that

$$\frac{\triangle I_D}{I_B} = -\frac{\triangle R_L}{R_L} \tag{4.620}$$

Assuming that $(1 \pm x)^{1/2} \simeq 1 \pm x/2$ for $x \ll 1$, the input offset voltage of the differential stage is then given by

$$V_{off} = V_{T_1} - V_{T_2} + \sqrt{\frac{I_B}{2K_1}\left(1 - \frac{\Delta R_L}{R_L}\right)} - \sqrt{\frac{I_B}{2K_2}\left(1 + \frac{\Delta R_L}{R_L}\right)} \quad (4.621)$$

$$\simeq V_{T_1} - V_{T_2} + \sqrt{\frac{I_B}{2K_1}}\left(1 - \frac{\Delta R_L}{2R_L}\right) - \sqrt{\frac{I_B}{2K_2}}\left(1 + \frac{\Delta R_L}{2R_L}\right) \quad (4.622)$$

In practice, the mismatch errors are small so that the contribution of the square root terms is almost negligible, and V_{off} is primarily due to differences in the threshold voltage, V_T, caused by variations in the width, length, thickness, and doping levels of the transistor channels.

- **Noise in a differential transistor pair**

A differential stage including the input-referred noise, which is modeled by a voltage source, is depicted in Figure 4.94.

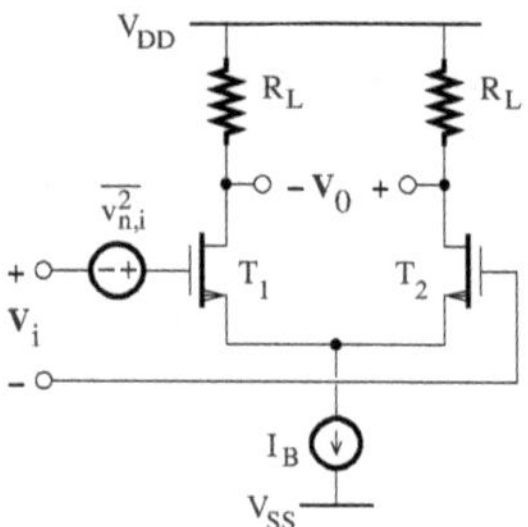

FIGURE 4.94
Circuit diagram of a differential MOS transistor stage with input-referred noises.

At low frequencies, the amplifier noise is dominated by the $1/f$ noise (or Flicker noise) contribution. By shorting the input nodes together, the total output noise voltage can be written as

$$\overline{v_{n,0}^2} = (\overline{i_{n_1}^2} + \overline{i_{n_2}^2})R_L^2 \quad (4.623)$$

where $\overline{i_{n_1}^2}$ and $\overline{i_{n_2}^2}$ are the power spectral densities of the noise currents of T_1 and T_2, respectively. Here, the noise current source of a given transistor T_j $(j = 1, 2)$ can be characterized by

$$\overline{i_{n_j}^2} \simeq \overline{i_{1/f,j}^2} = \frac{K_f}{C_{ox}L^2}\frac{I_{D_j}}{f} \quad (4.624)$$

where I_{D_j} is the drain current, f is the frequency, K_f is a process-dependent parameter, L is the transistor length, and C_{ox} represents the oxide capacitance per unit area.
Let g_m be the transconductance of the transistor T_1 or T_2. The power spectral density of the input-referred noise voltage is then given by

$$\overline{v_{n,i}^2} = \frac{\overline{v_{n,0}^2}}{g_m^2 R_L^2} \tag{4.625}$$

where $g_m R_L$ denotes the gain of the differential transistor pair. The unit of the voltage noise power spectral density is V^2/Hz. But, the input-referred noise voltage is usually in the $\mathrm{nV}/\sqrt{\mathrm{Hz}}$ range.
In high-precision applications, the level of the input-referred $1/f$ noise can be reduced using dynamic cancellation methods, such as auto-zeroing and chopping techniques.

Auto-zeroing and chopper techniques can be used to reduce the effect of the $1/f$ noise and dc offset in CMOS amplifiers [61, 63]. They operate with two clock phases. Although in the auto-zero method, the low-frequency noise is first estimated and then subtracted from the corrupted signal in the next phase, it is modulated to higher frequencies in the chopper approach. Figures 4.92(b) and (c) show the resulting amplifier noise voltage spectra provided by both techniques. The noise is reduced for the auto-zeroing sampling frequency, f_s, or chopping frequency, f_{chop}, greater than f_k. If the noise is stationary, it will be completely eliminated by the auto-zeroing; otherwise, the residual noise at low frequencies will be above the thermal noise floor. This is due to the noise components, which have a magnitude proportional to f_t/f_s, where f_t is the amplifier transition frequency, and are aliased by the sampling process into the signal baseband. In contrast to an amplifier using the auto-zero calibration, the one based on the chopper technique features a residual noise approximately equal to the thermal noise. Typically, the residual input noise level can be scaled from the millivolt range to the microvolt range, as a result of the use of either the auto-zeroing or chopper technique.

4.9.6.1 Auto-zero compensation scheme

The objective of the auto-zero scheme as shown in Figure 4.95 is to compensate the amplifier dc offset voltage induced by transistor mismatches.

The zeroing amplifier is implemented using a single-ended version of the main amplifier structure. The amplifier differential input stages ($T_1 - T_4$) have been modified to include auxiliary inputs for dc offset voltage correction. Its realization, as depicted in Figure 4.96, includes an additional transistor pair connected in parallel with the main differential input stage. If the potential

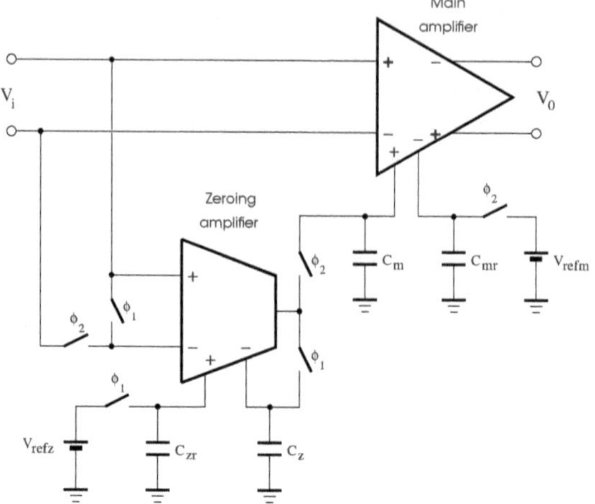

FIGURE 4.95
Circuit diagram of an auto-zero amplifier.

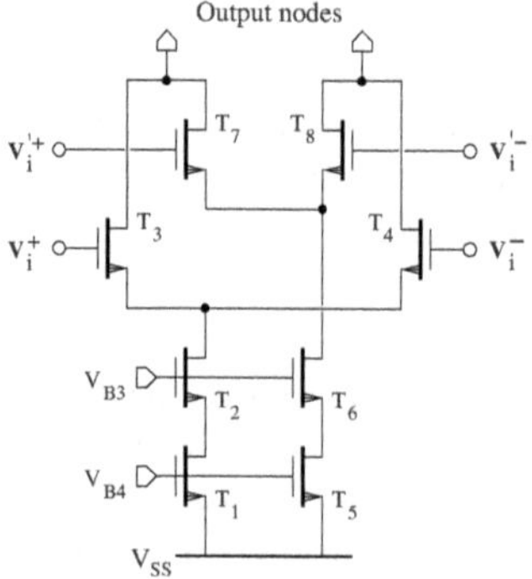

FIGURE 4.96
Differential input stage for the auto-zeroed amplifier.

differences V_i and V_i' exist between the primary and auxiliary inputs, respectively, the amplifier output voltage can be written as

$$V_0 = A_0(V_i + V_i'/\alpha + V_{off}) \tag{4.626}$$

where V_{off} is the amplifier offset voltage and the ratio of the open-loop dc gains between the primary and auxiliary inputs is defined as $\alpha = A_0/A_0'$. Basically, the role of the zeroing amplifier is to sense the dc offset voltage and to generate a correction voltage that is stored periodically on the capacitors and is used to drive the auxiliary input nodes of the amplifiers [65].

During the clock phase ϕ_1, the inputs of the zeroing amplifier are shorted and a feedback path is created between its output and inverting auxiliary input. The differential voltage applied to the auxiliary inputs of the zeroing amplifier can be expressed as

$$V_{C_{zr}} - V_{C_z} = V_{refz} - V_{0z} \tag{4.627}$$

According to Equation (4.626), we can obtain

$$V_{0z} = A_{0z}\left[\frac{V_{refz} - V_{0z})}{\alpha_z} + V_{offz}\right] \tag{4.628}$$

or equivalently,

$$V_{0z} = \frac{1}{1 + A_{0z}/\alpha_z}(A_{0z}V_{refz}/\alpha_z + A_{0z}V_{offz}) \tag{4.629}$$

where V_{offz} represents the zeroing amplifier dc offset voltage, and A_{0z} and $A'_{0z} = A_{0z}/\alpha_z$ are the open-loop dc gains of the zeroing amplifier with respect to the primary and auxiliary inputs, respectively. Combining Equations (4.627) and (4.629) gives

$$V_{C_{zr}} - V_{C_z} = \frac{1}{1 + A_{0z}/\alpha_z}(V_{refz} - A_{0z}V_{offz}) \tag{4.630}$$

In the clock phase ϕ_2, the main inputs of the amplifiers are connected together and the zeroing amplifier output is switched to the noninverting auxiliary input of the main amplifier. Due to the error voltage, $\triangle V_{C_{zr}} - \triangle V_{C_z}$, introduced by the opening of switches, the differential voltage maintained between the capacitors C_{zr} and C_z is

$$V_{C_{zr}} - V_{C_z} = \frac{1}{1 + A_{0z}/\alpha_z}(V_{refz} - A_{0z}V_{offz}) + (\triangle V_{C_{zr}} - \triangle V_{C_z}) \tag{4.631}$$

The output voltage of the main amplifier is given by

$$V_0 = A_{0m}(V^+ - V^- + V_{offm}) + A'_{0m}(V_{C_m} - V_{C_{mr}}) \tag{4.632}$$

where V_{offm} is the main amplifier offset voltage, and A_{0m} and $A'_{0m} = A_{0m}/\alpha_m$ are the open-loop dc gains of the main amplifier with respect to the primary and auxiliary inputs, respectively. The differential voltage driving the auxiliary inputs of the main amplifier can be written as

$$\begin{aligned} V_{C_m} - V_{C_{mr}} &= V_{0z} - V_{refm} && (4.633)\\ &= A_{0z}(V^+ - V^- + V_{offz}) + A'_{0z}(V_{C_{zr}} - V_{C_z}) - V_{refm} && (4.634) \end{aligned}$$

Note that the value of the above differential voltage stored on the capacitors during the next phase is affected by the switch error voltages, $\triangle V_{C_{mr}} - \triangle V_{C_m}$, according to the next equation:

$$\begin{aligned} V_{C_m} - V_{C_{mr}} = {} & A_{0z}(V^+ - V^- + V_{offz}) \\ & + A'_{0z}(V_{C_{zr}} - V_{C_z}) - V_{refm} + (\triangle V_{C_m} - \triangle V_{C_{mr}}) \end{aligned} \tag{4.635}$$

A first-order compensation of the switch error voltages is achieved by the differential structure and the substitution of Equation (4.631) into (4.635) can be reduced to

$$V_{C_m} - V_{C_{mr}} \simeq A_{0z}(V^+ - V^-) + \frac{A_{0z}V_{offz} + A'_{0z}V_{refz} - (1 + A'_{0z})V_{refm}}{1 + A'_{0z}} \tag{4.636}$$

Combining Equations (4.632) and (4.636), the output voltage, V_0, can then be expressed as

$$V_0 = (A_{0m} + A'_{0m}A_{0z})(V^+ - V^-) + A_{0m}V_{offm} + A'_{0m}\frac{A_{0z}V_{offz} + A'_{0z}V_{refz} - (1 + A'_{0z})V_{refm}}{1 + A'_{0z}} \tag{4.637}$$

By choosing the reference voltages such that $A'_{0z}V_{refz} = (1 + A'_{0z})V_{refm}$, and assuming that $A'_{0m}A_{0z} \gg A_{0m}$ and $A'_{0z} \gg 1$, Equation (4.637) can be put into the form

$$V_0 \simeq A'_{0m}A_{0z}(V^+ - V^- + V_{off,res}) \tag{4.638}$$

where the residual offset voltage, $V_{off,res}$, of the compensated amplifier is given by

$$V_{off,res} \simeq \frac{1}{A'_{0z}}\left[V_{offz} + \left(\frac{\alpha_m}{\alpha_z}\right)V_{offm}\right] \tag{4.639}$$

For a very low value of $V_{off,res}$, the gain A'_{0z} should be made very high. The effect of the offset voltage is similar to the one of the low-frequency or $1/f$ noise, which is also expected to be reduced in an auto-zeroed amplifier. Generally, the residual offset can be estimated to be in the range of 100 µV.

It is necessary that the zeroing amplifier has the structure of a transconductor. Then, it can implement, together with the load capacitor C_m, a lowpass filter, whose cutoff frequency can be very low for large values of C_m. The advantage of this structure is to reduce the effect of the parasitic signal caused by the sampling process.

4.9.6.2 Chopper technique

The principle of the chopper technique [59] is illustrated in Figure 4.97, where ϕ_1 and ϕ_2 are two square signals with nonoverlapping phases. Each array of cross-coupled switches, which is inserted at the input and output of the first amplifier stage, is then considered to be steered by a chopping square wave taking $+1$ or -1 values. Let us assume that the signal has a spectrum limited to half of the chopping frequency so that no aliasing occurs. After the first multiplication by V_{chop}, the input signal, V_i, is modulated and translated to odd harmonic frequencies. It is then amplified and translated back to the baseband, while the low-frequency noise voltage (offset voltage, $1/f$ noise), V_n, which is modulated only by the second multiplication stage, appears at odd harmonic frequencies of V_{chop} (see Figure 4.97(c)). The high-frequency

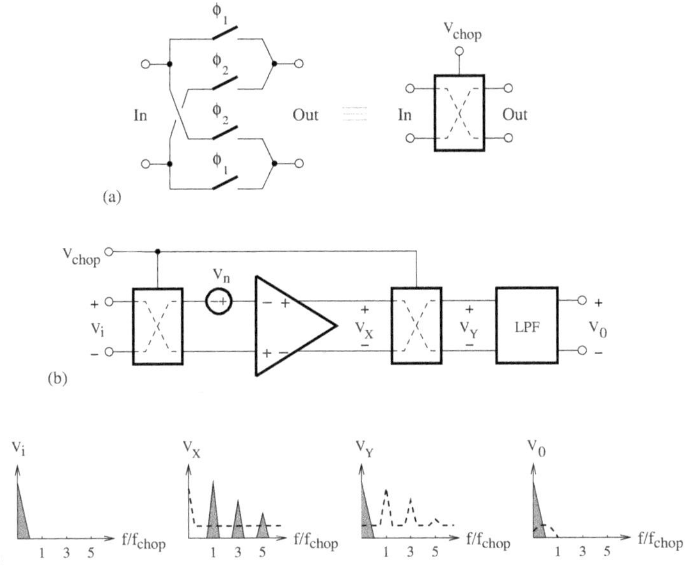

FIGURE 4.97
Principle of the chopper technique: (a) circuit symbol of the chopper section, (b) chopper amplifier, (c) signal spectra.

components are eliminated by the lowpass filter (LPF) included in the output stage of the chopper amplifier.

Let A be the amplifier gain. Taking into account the chopper effect, the amplifier output voltage can be written as

$$V_0(t) = [V_i(t) \cdot m(t) + V_n]A \cdot m(t) \tag{4.640}$$

where $m(t)$ represents the chopping signal alternating between 1 and -1 with a frequency f_{chop}, and V_n is the equivalent input noise of the amplifier. The signal multiplication by $m(t)$ induces a polarity change at the input nodes and a polarity restoration at the output nodes. That is,

$$V_0(t) = A \cdot V_i(t) + A \cdot V_n \cdot m(t) \tag{4.641}$$

The chopping signal $m(t)$ can be represented as a Fourier series given by

$$m(t) = \frac{4}{\pi}\sum_{k=1}^{\infty} \frac{1}{2k+1} \sin(2(2k+1)\pi f_{chop} t) \tag{4.642}$$

Using the Fourier transform, the spectrum of the signal $m(t)$ can be obtained as

$$M(f) = \frac{2}{\pi}\sum_{k=1}^{\infty} \frac{1}{j(2k+1)}[\delta(f-(2k+1)f_{chop}) - \delta(f+(2k+1)f_{chop})] \tag{4.643}$$

It can then be shown that the output power spectral density due to the noise source is of the form

$$S_{0,v_n}(f) = S_{v_n}(f) * |M(f)|^2 \tag{4.644}$$

where $*$ denotes the convolution operation, S_{v_n} is the power spectral density of the equivalent input noise, and

$$|M(f)|^2 = \left(\frac{2}{\pi}\right)^2 \sum_{k=1}^{\infty} \frac{1}{(2k+1)^2}[\delta(f + (2k+1)f_{chop}) - \delta(f - (2k+1)f_{chop})] \tag{4.645}$$

The low-frequency noise is then transposed to odd harmonic frequencies of the chopping signal $m(t)$, where it can be attenuated by a lowpass filter. Due to the $1/(2k+1)^2$ scaling factor, the contribution to the baseband of noise replicas is greatly attenuated.

Simulation results show that the output power spectral density increases with the ratio of the amplifier cutoff frequency to the chopper frequency, but always remains smaller than the power spectral density of the white noise inherent in the original amplifier. Due to the fact that each signal component is not sampled and held, but periodically inverted, the chopper technique has the advantage of not aliasing the broadband noise, in contrast to the auto-zero approach [62]. In practice, the frequency of the modulating signal should be greater than the sum of the signal bandwidth and the $1/f$ corner frequency to minimize the noise contribution in the baseband and relax the filtering specifications. As a result, the achievable frequency range may appear to be limited by the required high-frequency chopping signals.

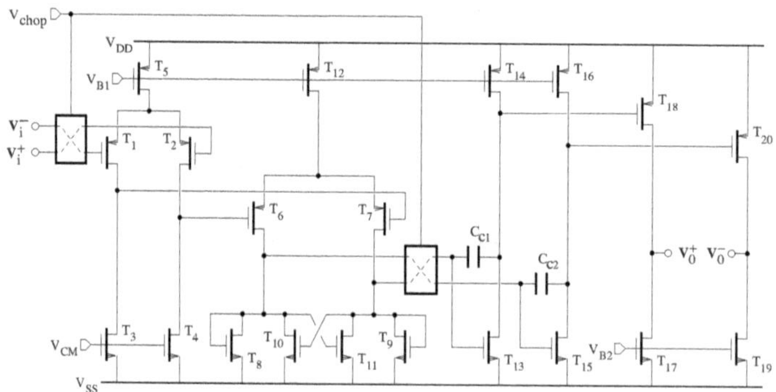

FIGURE 4.98
Circuit diagram of a low-noise amplifier using the chopper technique.

The circuit diagram of an amplifier using the chopper technique is shown in Figure 4.98. The first differential stage consisting of $T_1 - T_5$ should exhibit a low gain and noise. Transistors $T_6 - T_{12}$ form the second stage, which features a low unity-gain frequency and a high gain. These features can be useful for the reduction of the effect of the modulated offset and the offset due to the output stage comprising a gain inverting section and a source follower. The amplifier is compensated by the Miller capacitors, C_{c1} and C_{c2}. The voltage V_{CM} is generated by a CMF circuit and the cross-coupled connection of transistors $T_8 - T_{11}$ is used to define the CM level for the second differential stage. In this case, the residual offset voltage is about 10 µV [63]. It is generally due to spikes generated by mismatches between the charge injections of switches.

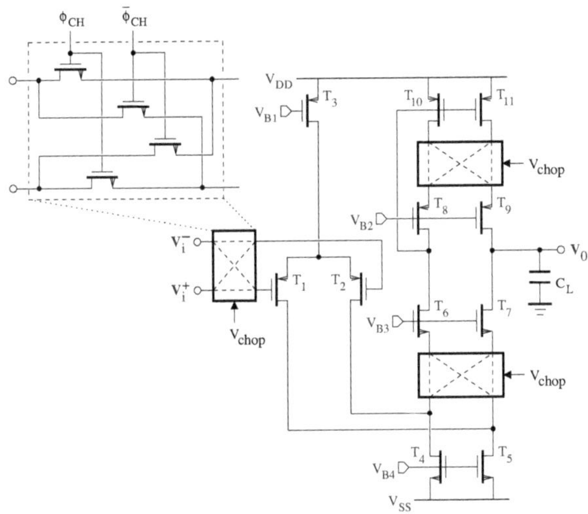

FIGURE 4.99
Circuit diagram of a folded-cascode amplifier with chopper technique-based offset voltage reduction.

A folded-cascode amplifier with chopper technique-based offset voltage reduction is shown in Figure 4.99, where V_{chop} represents a pulse sequence with two nonoverlapping phases, ϕ_{CH} and $\overline{\phi}_{CH}$. It uses three sets of chopper switches placed at the input and between the folded-cascode transistors [64]. Due to the fully differential signal path, the offset voltage contributions of transistors $T_1 - T_2$ and $T_4 - T_5$ are completely reduced. On the other hand, the mismatch errors of transistors $T_{10} - T_{11}$ cannot be completely cancelled because of the inherent asymmetry of the current mirror.

During the phase ϕ_{CH}, we can write the following transistor drain currents:

$$I_{D_{10}} = I_{D_8} = K(V_{GS_8} - V_{T_8})^2 \tag{4.646}$$

$$I_{D_{11}} = I_{D_9} = K(V_{GS_9} - V_{T_9})^2 \tag{4.647}$$

where K represents the transistor transconductance parameter, and V_{T_8} and V_{T_9} are the threshold voltages of transistors T_8 and T_9, respectively. To proceed further, the gate-source voltages of transistors T_8 and T_9 can be expressed as,

$$V_{GS_8} = \sqrt{\frac{I_{D_8}}{K}} + V_{T_8} \tag{4.648}$$

and

$$V_{GS_9} = \sqrt{\frac{I_{D_9}}{K}} + V_{T_9} \tag{4.649}$$

Using the fact that $V_{GS_8} = V_{GS_9}$, the drain current of the transistor T_{11} during the phase ϕ_{CH} is given by

$$I^{\phi}_{D_{11}} = I_{D_{10}} + K(V_{T_8} - V_{T_9})^2 + 2\sqrt{KI_{D_9}}(V_{T_8} - V_{T_9}) \tag{4.650}$$

During the phase $\overline{\phi}_{CH}$, the transistors T_8 and T_9 are swapped. Hence, we have $I_{D_{10}} = I_{D_9}$, $I_{D_{11}} = I_{D_8}$, $V_{GS_8} = V_{GS_9}$, and the current $I_{D_{11}}$ can be written as,

$$I^{\overline{\phi}}_{D_{11}} = I_{D_{10}} + K(V_{T_9} - V_{T_8})^2 + 2\sqrt{KI_{D_9}}(V_{T_9} - V_{T_8}) \tag{4.651}$$

The changes of the drain current of transistors T_8 and T_9 caused by mismatches are estimated as,

$$\triangle I_{D_8} = \frac{I^{\phi}_{D_{11}} + I^{\overline{\phi}}_{D_{11}}}{2} - I_{D_{10}} = K\triangle V^2_{T_{8,9}} = \frac{g^2_{m_8}\triangle V^2_{T_{8,9}}}{4I_{D_8}} \tag{4.652}$$

and

$$\triangle I_{D_9} = \frac{I^{\phi}_{D_{11}} + I^{\overline{\phi}}_{D_{11}}}{2} - I_{D_{10}} = K\triangle V^2_{T_{8,9}} = \frac{g^2_{m_9}\triangle V^2_{T_{8,9}}}{4I_{D_9}} \tag{4.653}$$

where $\triangle V_{T_{8,9}} = V_{T_8} - V_{T_9}$, I_{D_8} and g_{m_8} are the drain current and the transconductance of the transistor T_8, respectively, and I_{D_9} and g_{m_9} are the drain current and the transconductance of the transistor T_9, respectively. Assuming that transistors T_1 and T_2 are identical, the residual offset voltage of the amplifier is given by

$$V_{off,res} = \frac{\triangle I_{D_9}}{g_{m_1}} \simeq \frac{\triangle V^2_{T_{8,9}}}{g_{m_1}} = \frac{g_{m_8}\triangle V^2_{T_{8,9}}}{4g_{m_1}I_{D_8}} \tag{4.654}$$

where g_{m_1} is the transconductance of the transistor T_1.

In general, the use of input differential PMOS transistors with a high aspect ratio helps reduce not only the offset voltage to a few tens of μV (at a chopping frequency less than 200 kHz), but also the effect of flicker noise.

4.10 Summary

The available amplifier circuits can be divided into two main groups: single- and multistage architectures. Generally, the existing trade-off between speed and gain makes it difficult for CMOS amplifiers to exhibit a high bandwidth and dc gain simultaneously. Then, the need for a high gain leads to multistage designs with long-channel transistors biased at low current levels, whereas the requirement for a high unity-gain frequency is fulfilled by a single-stage topology with short-channel transistors biased at high current levels.

The choice of the amplifier architecture plays an important role in the design of low-voltage circuits. In addition to the amplifier characteristics such as the gain, bandwidth, and slew rate, the output swing also becomes critical. Generally, the power dissipation increases as the supply voltage is reduced. The two-stage amplifier appears to be suitable for low-voltage operation due to its larger output swing, while the telescopic structure achieves superior speed and power consumption.

4.11 Circuit design assessment

1. **Amplifier design challenges**
   Show that the transfer function of the RC circuit of Fig 4.100(a) can be written as

$$H(j\omega) = \frac{V_0(j\omega)}{V_i(j\omega)} = \frac{1}{1 + j\omega RC} \tag{4.655}$$

   Let k and T be the Boltzmann's constant and absolute temperature, respectively, and $\overline{v_{nf}^2}$ denote the output noise of the RC circuit over the frequency range from dc to infinity. The thermal noise can be modeled by adding a current source with the root mean squared value of $\sqrt{\overline{i_R^2}} = \sqrt{(4kT)/R}$ in parallel with the resistor R. Use the formula

$$\sqrt{\overline{v_{nf}^2}} = \left(\int_0^{+\infty} \overline{v_n^2} df \right)^{1/2} \tag{4.656}$$

   where $\overline{v_n^2} = R\,\overline{i_R^2}\,|H(j\omega)|^2$, to verify that

$$\sqrt{\overline{v_{nf}^2}} = \sqrt{\frac{kT}{C}} \tag{4.657}$$

   Note that

$$\int \frac{\mathrm{d}x}{x^2 + a^2} = \frac{1}{a} \arctan\left(\frac{x}{a}\right) + c \tag{4.658}$$

where a and c are two real constants.

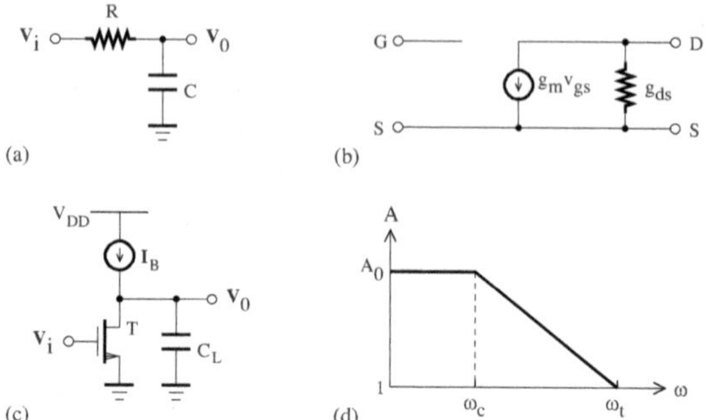

FIGURE 4.100
(a) First-order RC circuit; (b) MOS transistor model; (c) basic single-stage amplifier; (d) amplifier frequency response.

The transistor in the saturation region can be described by the square law given by

$$I_D = K(v_{GS} - V_T)^2(1 + \lambda v_{DS}) \tag{4.659}$$

where $K = \mu(C_{ox}/2)(W/L)$ and λ is the channel-length modulation parameter. The transconductance is defined by

$$g_m = \left.\frac{\partial I_D}{\partial v_{GS}}\right|_{v_{DS}} \simeq 2K(v_{GS} - V_T) = 2\sqrt{KI_D} \tag{4.660}$$

while the conductance can be written as

$$g_{ds} = \left.\frac{\partial I_D}{\partial v_{DS}}\right|_{v_{GS}} = K\lambda(v_{GS} - V_T)^2 = I_D\lambda \tag{4.661}$$

Show that the gain of the amplifier of Figure 4.100(c) can be written as

$$A(j\omega) = \frac{v_0(j\omega)}{v_i(j\omega)} = A_0 \frac{1}{1 + j\dfrac{\omega}{\omega_c}} \tag{4.662}$$

where $A_0 = g_m/g_{ds}$ and $\omega_c = g_{ds}/C_L$ denote the dc gain and bandwidth, respectively.

Add the asymptotic Bode representation of a gain A', which corresponds to $A'_0 > A_0$, to the graph of Figure 4.100(d). Can we meet simultaneously the high dc gain and bandwidth requirements?

Explain why the power consumption increases by α^2, as a result of the scaling by a factor α of the amplifier signal swing and thermal noise such that the dynamic range and bandwidth are maintained constant.

Hint: The dynamic range (DR), which is limited by the thermal noise, is given by

$$DR = \frac{S_S}{S_{kT/C}} = \frac{V_{max}^2}{kT/C} \tag{4.663}$$

where V_{max} is the maximum signal swing. By reducing the supply voltage by α, the DR and bandwidth become

$$DR' = \frac{V_{max}^2/\alpha^2}{kT/\alpha^2 C} \tag{4.664}$$

and

$$\omega_c' = \frac{\alpha^2 g_m}{\alpha^2 C_L} \tag{4.665}$$

respectively. The transconductance is given by

$$g_m = \sqrt{2\mu C_{ox}\frac{W}{L}I_D} \tag{4.666}$$

while C_{ox} is converted to αC_{ox} and I_D to $\alpha^3 I_D$. As a result, the power consumption contribution, $P = I_D V_{max}$, is magnified by α^2.

2. **Comparison of differential-stage and current-mirror amplifiers with single-ended outputs**
Consider the differential-stage amplifier and current-mirror amplifier shown in Figures 4.101(a) and (b). By sizing the transistors so that the frequency response of each single-stage amplifier is essentially determined by the pole due to the load capacitor C_L, a phase margin up to 90° can be achievable.

The pMOS input transistor pair of a differential stage is in conduction for low input common-mode voltages, or

$$V_{i,cm} < V_{DD} - V_{SD_3} - V_{SG_1} \tag{4.667}$$

Verify the results (slew rate, output swing, thermal noise, minimum supply) summarized in Table 4.1, where γ is the transistor noise coefficient, k is the Boltzmann constant, and T is the absolute temperature.

3. **Comparison of folded-cascode and two-stage amplifiers**
Consider the folded-cascode and two-stage amplifiers shown in Figures 4.102(a) and (b), respectively, where $V_{SS} = -V_{DD}$. With the assumption that symmetrical transistors are matched, and all transistors operate in the saturation region and exhibit the same drain-source saturation voltage, compare both amplifier structures and

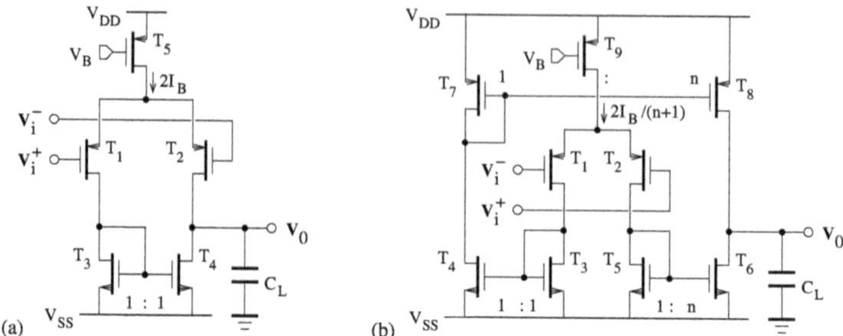

FIGURE 4.101
(a) Differential-stage amplifier; (b) current-mirror amplifier.

TABLE 4.1
Performance Characteristics of Differential-Stage and Current-Mirror Amplifiers

	Differential-Stage Amplifier	Current-Mirror Amplifier
Slew rate	$\frac{2I_B}{C_L}$	$\frac{n}{n+1}\frac{2I_B}{C_L}$
Unity-gain frequency	$\frac{g_{m_1}}{C_L}$	$\frac{n}{n+1}\frac{g_{m_1}}{C_L}$
Output swing	$V_{SS}+V_{DS(sat)} \le V_0$	$V_{SS}+V_{DS(sat)} \le V_0$
	$V_0 \le V_{DD} - 2V_{SD(sat)} + V_{T_p}$	$V_0 \le V_{DD} - V_{SD(sat)}$
Thermal noise	$\frac{8\gamma kT}{g_{m_1}}\left(1+\frac{g_{m_3}}{g_{m_1}}\right)$	$\frac{8\gamma kT}{g_{m_1}}\frac{(n+1)^2}{n}\left(1+\frac{g_{m_3}}{g_{m_1}}\right)$
Minimum supply	$V_{DS(sat)}+V_{SD(sat)}-V_{T_p}$	$V_{DS(sat)}+V_{SD(sat)}-V_{T_p}$

verify the results (slew rate, unity-gain frequency, lowest nondominant pole, output swing, input-referred thermal noise, minimum supply voltage) summarized in Table 4.2, where C_p represents a parasitic capacitance, γ is the transistor noise coefficient, k is the Boltzmann constant, and T is the absolute temperature.

Hint: Let $\overline{v_{n,i}^2}$ be the input-referred noise power spectral density of the amplifier, $\overline{i_{n,0}^2}$ be the output noise power spectral density of the amplifier, and $\overline{v_{nk}^2}$ be the input-referred noise power spectral density of the transistor T_k. It is convenient to assume that

$$\overline{v_{n,i}^2} = \overline{i_{n,0}^2}/g_{m_1}^2 \tag{4.668}$$

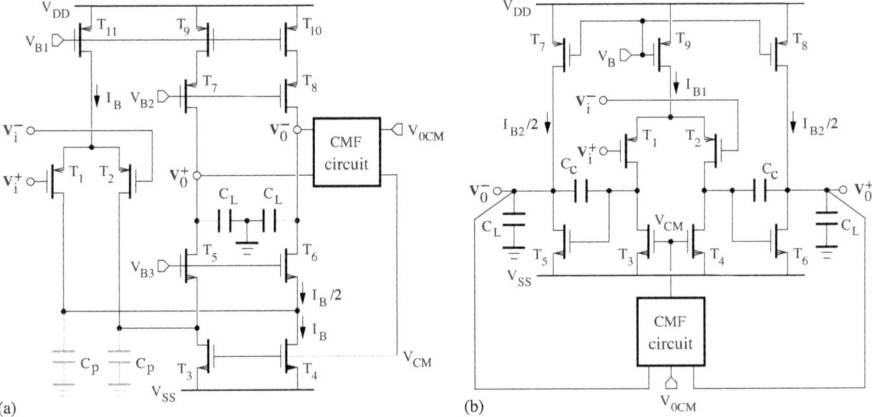

FIGURE 4.102
Circuit diagrams of (a) folded-cascode and (b) two-stage amplifiers.

TABLE 4.2
Performance Characteristics of Folded-Cascode and Two-Stage Amplifiers

	Folded-Cascode Amplifier	Two-Stage Amplifier
Slew rate	$\dfrac{I_B}{C_L}$	$\min\left(\dfrac{I_{B_1}}{C_C}, \dfrac{I_{B_2}}{C_C + C_L}\right)$
Unity-gain frequency	g_{m_1}/C_L	g_{m_1}/C_C
Nondominant pole	g_{m_5}/C_p	g_{m_5}/C_L
Output swing	$2V_{DD} - 8\lvert V_{DS(sat)}\rvert$	$2V_{DD} - 4\lvert V_{DS(sat)}\rvert$
Thermal noise	$\dfrac{8\gamma kT}{g_{m_1}}\left(1 + \dfrac{g_{m_3}}{g_{m_1}} + \dfrac{g_{m_5}}{g_{m_1}}\right)$	$\dfrac{8\gamma kT}{g_{m_1}}\left(1 + \dfrac{g_{m_3}}{g_{m_1}}\right)$
Minimum supply	$\lvert V_T\rvert + 2\lvert V_{DS(sat)}\rvert$	$\lvert V_T\rvert + 2\lvert V_{DS(sat)}\rvert$

and to model $\overline{v_{nk}^2}$ as the sum of the thermal noise and the $1/f$ noise, that is,

$$\overline{v_{nk}^2} = \left(\frac{2}{3}\right)\frac{4kT}{g_{m,k}} + \frac{KF}{C_{ox}W_kL_kf} \tag{4.669}$$

– Folded-cascode amplifier
Assuming that the noise due to the transistor T_{11} is cancelled by the circuit symmetry and input transistor matching, and the noise contribution of cascode transistors is negligible because they are source degenerated and exhibit a small effective transconductance, verify that the equivalent input-referred noise power spectral density, $\overline{v_{n,i}^2}$,

in V^2/Hz is of the form

$$\overline{v_{n,i}^2} = 2\left[\overline{v_{n1}^2} + \left(\frac{g_{m_3}}{g_{m_1}}\right)^2 \overline{v_{n3}^2} + \left(\frac{g_{m_9}}{g_{m_1}}\right)^2 \overline{v_{n9}^2}\right] \tag{4.670}$$

– Two-stage amplifier
With the assumption that the input-referred noise is mainly determined by the first stage, verify that

$$\overline{v_{n,i}^2} = 2\left[\overline{v_{n1}^2} + \left(\frac{g_{m_3}}{g_{m_1}}\right)^2 \overline{v_{n3}^2}\right] \tag{4.671}$$

4. **Design of a low-voltage amplifier**
For the amplifier shown in Figure 4.103, verify the following statements:
– The input common-mode is limited by:
V_{SS} and $V_{DD} - |V_T| - 2V_{DS(sat)}$
– The minimum supply voltage can be expressed as:
$\max\{V_{DS_3(sat)} + V_{DS_5(sat)} + V_{T_5}, V_{DS_3(sat)} + V_{DS_5(sat)} + V_{DS_7(sat)}\}$
Size of the amplifier components using a submicrometer IC process to meet the following specifications: 50 dB dc gain and 4 MHz unity-gain bandwidth.

Estimate the slew rate for a load capacitance of 10 pF and the power dissipation of the amplifier.

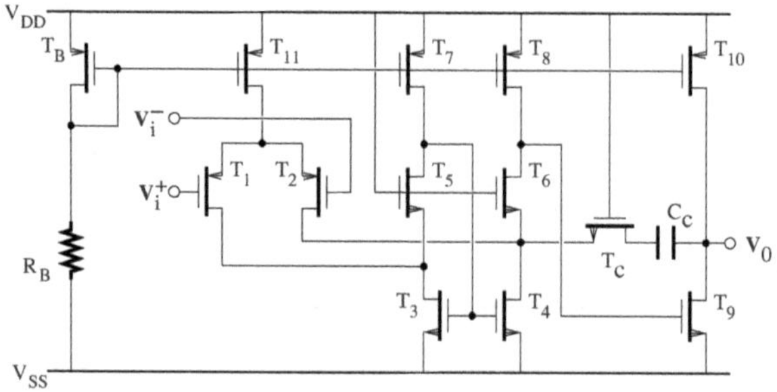

FIGURE 4.103
Circuit diagram of a low-voltage amplifier.

5. **Amplifier stage with a source degeneration resistance**
Use the small-signal equivalent model of the transistors to determine the voltage gain of the amplifier structure shown in Figure 4.104.

6. **Amplifier stage based on a series of differential pairs**
The linearity of a differential pair can be improved by using the amplifier stage shown in Figure 4.105 [9]. All transistors operate

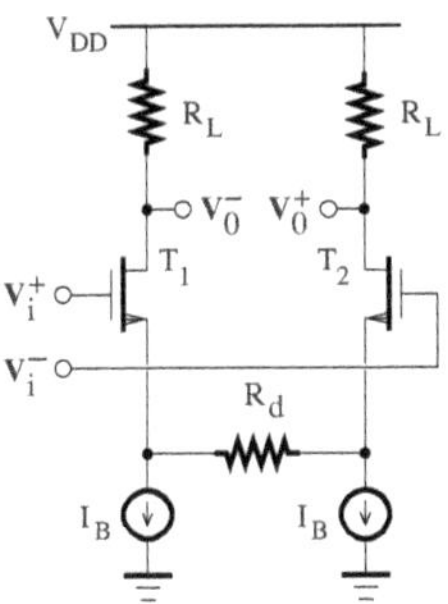

FIGURE 4.104
Differential amplifier with a source degeneration resistance.

in the saturation region. Find the relationship between the output current $\triangle i_0 = i_0^+ - i_0^-$ and the input voltage $V_i = V_i^+ - V_i^-$.

For a given CMOS technology, verify your calculations using SPICE simulations.

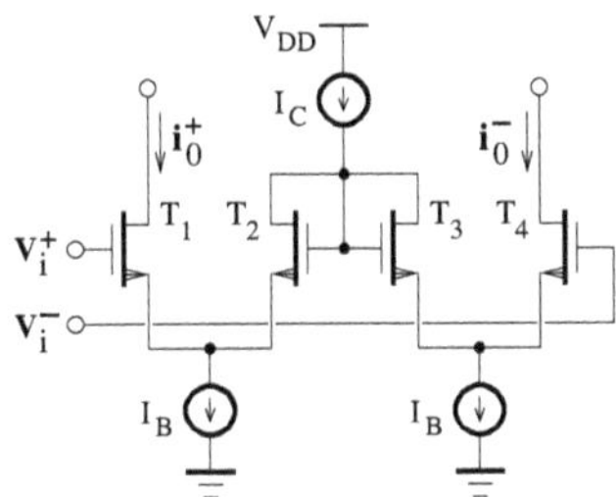

FIGURE 4.105
Differential amplifier based on a series of differential pairs.

7. **Amplifier with class AB input stage**
   Complete the design of the amplifier with class AB input stage of Figure 4.106 to meet the following specifications: a *dc* gain of at least 55 dB, a unity-gain frequency of 4 MHz, a phase margin of 70°, a slew rate of 5 V/μs at a 10 pF capacitive load, and a settling time of 10 ns. All transistors operate in the saturation region.

   Assuming the transistors T_{1a} and T_{1b}, T_{2a} and T_{2b}, and T_{3a} and T_{3b} are matched, the currents I_1 and I_2 can be expressed as,

$$I_1 = K_1\left(\sqrt{\frac{I_B}{K_1}} + V_i\right)^2 \quad \text{for} \quad I_2 < I_B, \quad V_i > 0 \tag{4.672}$$

$$I_2 = K_1\left(\sqrt{\frac{I_B}{K_1}} - V_i\right)^2 \quad \text{for} \quad I_1 < I_B, \quad V_i > 0 \tag{4.673}$$

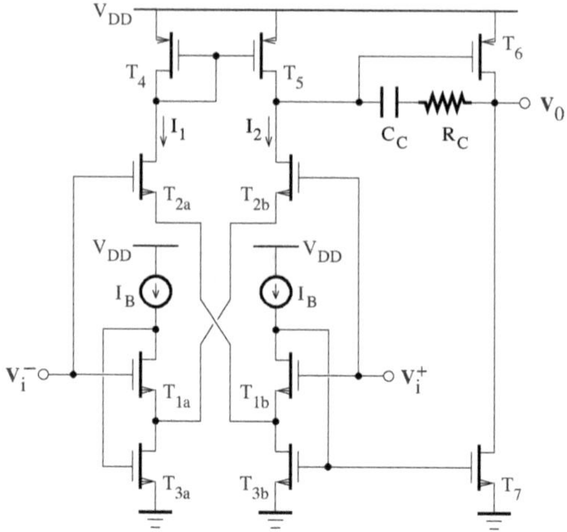

FIGURE 4.106
Circuit diagram of an amplifier with class AB input stage.

where $V_i = V_i^+ - V_i^-$. For small values of V_i, it can be shown that $I = I_1 - I_2 \simeq 4\sqrt{K_1 I_B} V_i$. The minimum supply voltage is equal to $2V_{DS(sat)} + V_T$, and the input common-mode voltage range is of the form $V_T - \sqrt{I_B/K_1}$, where $V_{DS(sat)}$ is the drain-source saturation voltage, V_T is the threshold voltage, and K_1 is the transconductance parameter of transistors T_{1a} (or T_{1b}), T_{2a} (or T_{2b}).

8. **Fully differential folded-cascode amplifier with a common-mode feedback (CMF) circuit**
   In a 0.13-μm CMOS technology, design the differential folded-cascode amplifier shown in Figure 4.107 to meet the specifications given in Table 4.3.

TABLE 4.3
Amplifier Specifications

Supply voltage	±1.25 V
Capacitive load	2 pF
Slew rate	100 V/µs
DC voltage gain	60 dB
Gain-bandwidth product	100 MHz
Phase margin	> 60°
Power consumption	< 15 mW)

Use SPICE simulations to adequately adjust the transistor aspect ratios.

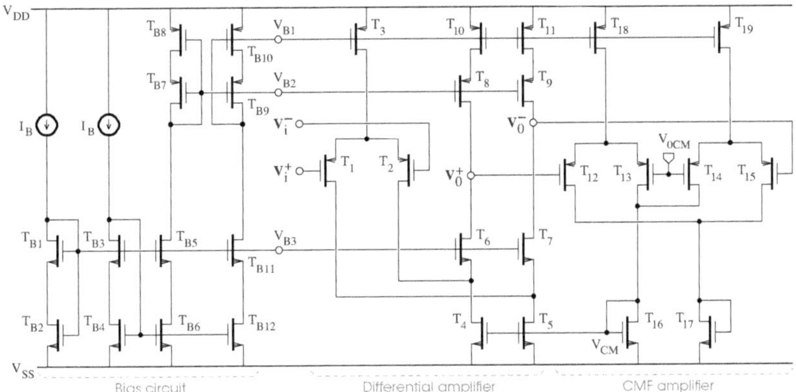

FIGURE 4.107
Differential folded-cascode amplifier with a common-mode feedback (CMF) circuit.

Provide a table including the simulated dc results.

Plot the transient step response and frequency response of the resulting amplifier in the unity feedback configuration.

Determine the settling time achieved by the resulting amplifier.

9. **Fully differential amplifier without a common-mode feedback circuit**
Due to the loading of the common-mode feedback circuit on the signal path, a fully differential amplifier can exhibit a limited speed and require more power. The amplifier structure of Figure 4.108 [66] can serve for the implementation of differential switched-capacitor (SC) circuits without a common-mode feedback stage. The SC gain stage shown in Figure 4.109, where V_{icm} and V_{0cm} are the input and output common-mode voltages, respectively, uses the clock phase 2 to define a common-mode voltage reference for the amplifier nodes. Use simulation results to analyze the effect of the common-mode signal at the amplifier input on the circuit behavior.

10. **Stability of a differential gain stage**
Consider the equivalent circuit of a differential gain stage shown in Figure 4.110, where $C_1 = 2$ pF, $C_2 = 1$ pF, and $C_L = 1$ pF.

Use the small-signal equivalent model of a fully differential amplifier depicted in Figure 4.111, where Z_i and Z_{cm} denote the impedance of a parallel combination of a resistor and a capacitor, to determine the transfer function V_0/V_i, where $V_0 = V_0^+ - V_0^-$ and $V_i = V_i^+ - V_i^-$.

Starting with the following initial component values,

$g_m = g_m' = 1$ mS, $C_i = C_0 = C_0' = C_{cm} = C_p = 1$ pF, and

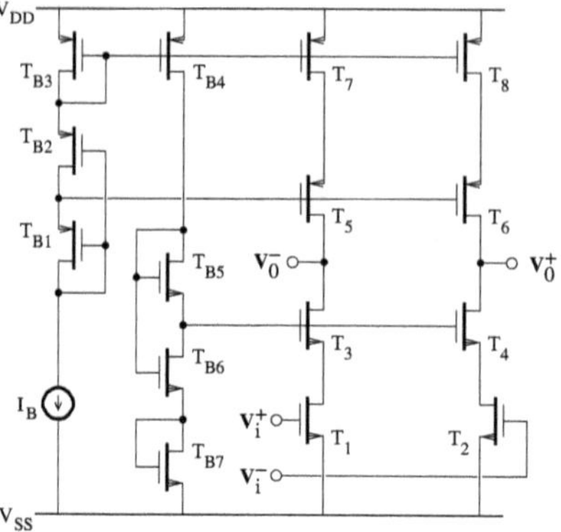

FIGURE 4.108
Circuit diagram of a pseudo fully differential amplifier.

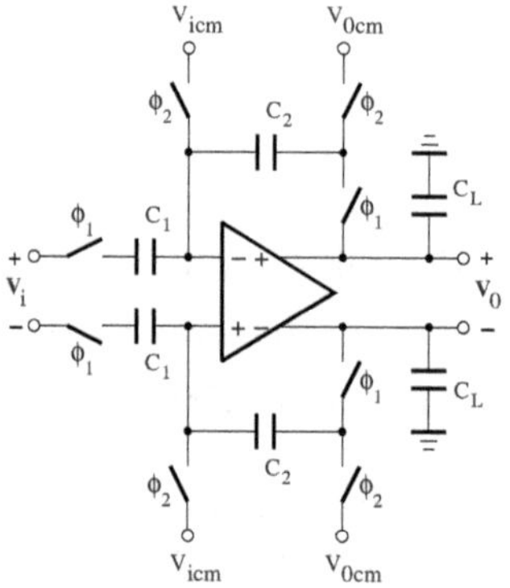

FIGURE 4.109
Circuit diagram of an SC gain stage.

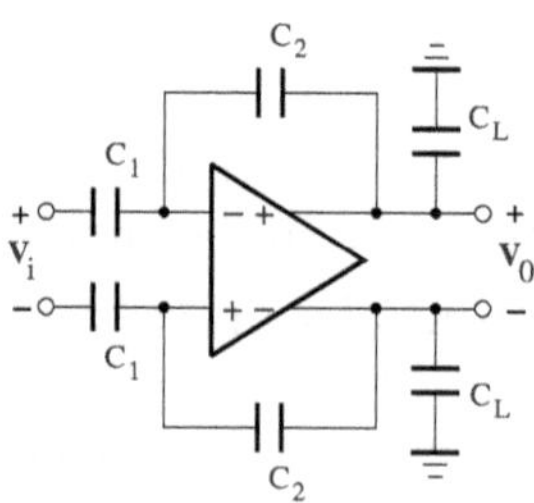

FIGURE 4.110
Equivalent circuit of a differential gain stage.

$R_i = R_0 = R_0' = R_{cm} = 1\ \text{M}\Omega$, analyze the stability of the gain stage using SPICE simulations.

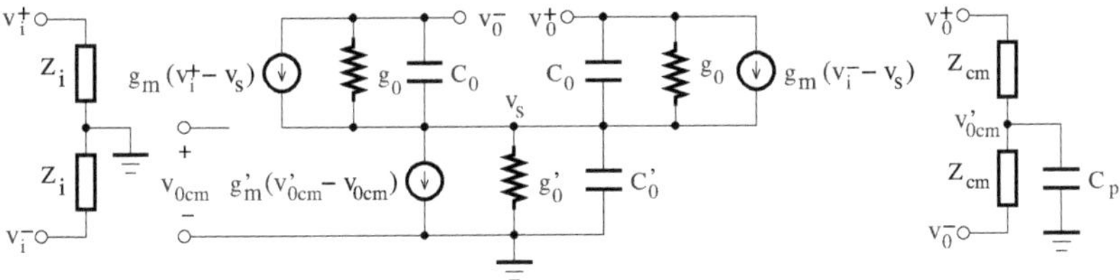

FIGURE 4.111
Small-signal equivalent model of a fully differential amplifier.

11. **Low-voltage amplifier with a class AB output**
Consider the low-voltage class AB amplifier shown in Figure 4.112. The transistor T_{11} operates in the triode region, while the remaining transistors are biased in the saturation region.

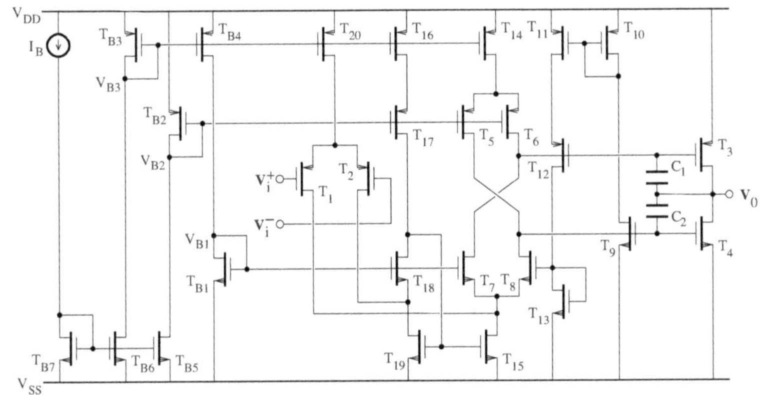

FIGURE 4.112
Circuit diagram of a low-voltage class AB amplifier.

Assuming that $V_{DD} = -V_{SS} = 1$ V and $I_B = 200$ µA, estimate the values of the width and length of each transistor to achieve a voltage gain of 10^4, a gain-bandwidth product of 3.5 MHz, and a phase margin of 70° in a given CMOS process. The capacitive and resistive loads of the amplifier are on the order of 2 pF and 10 kΩ, respectively.

12. **Two-stage amplifier with current-buffer compensation**
To improve the capacitive load range of a two-stage amplifier, compensation can be implemented using a current buffer as shown in Figure 4.113 [41]. A virtual ground is realized for the compensation capacitor as a result of the biasing of the transistor T_5 at a fixed dc potential by one of the two current sources, I_c.
Use the small-signal model of Figure 4.114 with the assumption

that the controlled current connected to the first stage is given by $I_c(s) = sC_cV_0(s)$ and show that the resulting transfer function of the compensated amplifier reads

$$A(s) = \frac{V_0(s)}{V_i(s)} = -A_0\frac{1}{1+bs+as^2} \tag{4.674}$$

where

$$A_0 = \frac{g_{m_1}g_{m_2}}{g_1g_2} \tag{4.675}$$

$$a = \frac{C_1(C_2+C_c)}{g_1g_2} \tag{4.676}$$

$$b = \frac{C_1}{g_1} + \frac{C_2+C_c}{g_2} + \frac{g_{m_2}C_c}{g_1g_2} \tag{4.677}$$

Verify that

$$A(s) = \frac{V_0(s)}{V_i(s)} = -A_0\frac{1}{(1+s/p_1)(1+s/p_2)} \tag{4.678}$$

where the poles p_1 and p_2 are computed as

$$\frac{1}{p_1} \quad \frac{1}{p_1} = \frac{1}{2}(-b \pm \sqrt{\Delta}) \tag{4.679}$$

and $\Delta = b^2 - 4a$ should be positive.
Let p_2 be much greater than p_1. The gain-bandwidth product, GBW, and phase margin, ϕ_M, are defined by

$$GBW \simeq g_{m_1}/C_c \tag{4.680}$$

and

$$\tan\phi_M = p_2/GBW \tag{4.681}$$

respectively. Relate C_c to ϕ_M and analyze the amplifier stability.

13. **Two-stage amplifier with a class AB output**
Consider the circuit diagram of a two-stage amplifier with a class AB output shown in Figure 4.115 [67]. To minimize cross-over distortion, the output transistors should be kept at the boundary of the conduction region in the absence of an incoming ac signal. This can be achieved through unbalanced current mirroring, thereby resulting in transistor sizes of the form

$W_1/L_1 = W_2/L_2$	$W_3/L_3 = W_4/L_4 = W_8/L_8$
$W_7/L_7 = \alpha(W_6/L_6)$	$W_{10}/L_{10} = \beta(W_9/L_9)$
$W_{12}/L_{12} = W_{13}/L_{13} = W_{15}/L_{15}$	$W_{11}/L_{11} = \gamma(W_{14}/L_{14})$

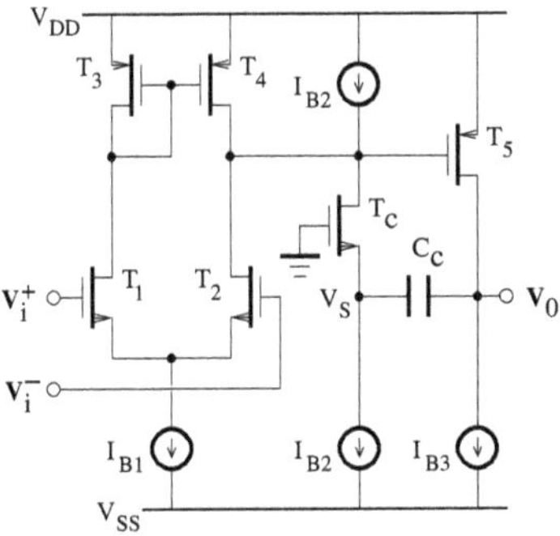

FIGURE 4.113
Circuit diagram of a two-stage amplifier with current buffer compensation.

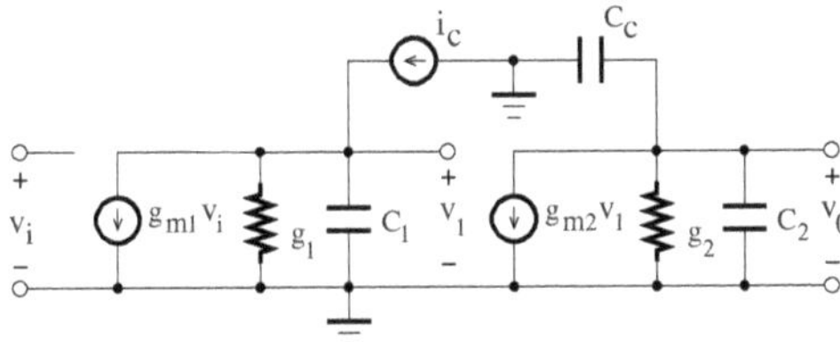

FIGURE 4.114
Small-signal equivalent of the two-stage amplifier.

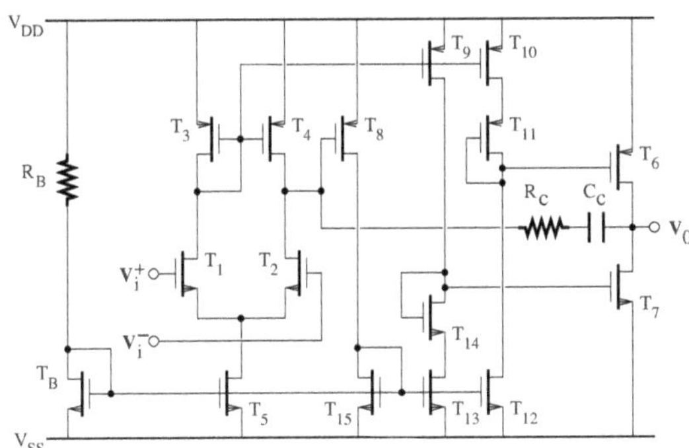

FIGURE 4.115
Circuit diagram of a two-stage amplifier with a class AB output stage.

where α, β, and γ are constant numbers. A frequency compensation network consisting of R_c and C_c is used to ensure closed-loop stability when the amplifier operates with a load.

Based on the equations

$$V_{SG_6} = V_{SG_{11}} + V_{SD_{10}} \tag{4.682}$$

and

$$V_{GS_7} = V_{GS_{14}} + V_{DS_{13}} \tag{4.683}$$

determine the quiescent current, I_Q, flowing through the output transistors.

Assuming that $V_{DD} = -V_{SS} = 2.5$ V and $R_B = 100$ kΩ, estimate the values of the width and length of each transistor to achieve a voltage gain of 10^3 and a gain-bandwidth product of 2 MHz in a given CMOS process.

14. Three-stage RNMC amplifier with a feedforward transconductance

Consider the circuit diagram of a three-stage reversed nested Miller compensation (RNMC) amplifier with a feedforward transconductance shown in Figure 4.116.

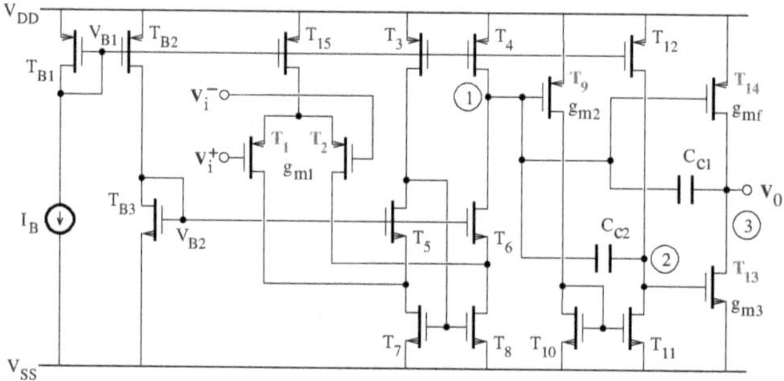

FIGURE 4.116
Circuit diagram of a three-stage RNMC amplifier with a feedforward transconductance.

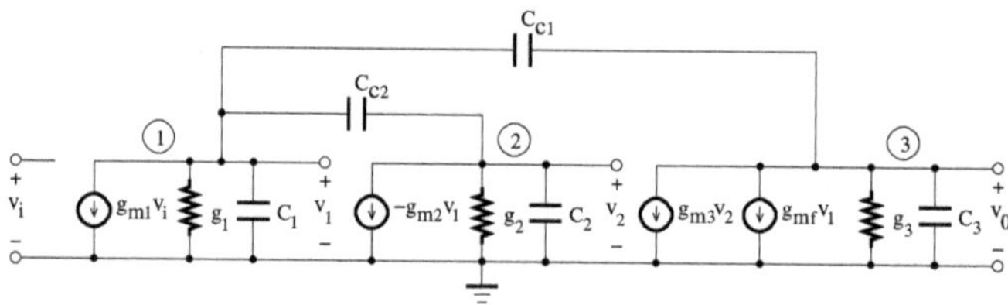

FIGURE 4.117
Small-signal equivalent circuit of the three-stage RNMC amplifier with a feedforward transconductance.

With reference to the small-signal equivalent circuit shown in Figure 4.117, verify that the equations for node 1, node 2, and node 3, can be written as

$$g_{m_1}V_i + V_1(g_1 + sC_1) - (V_0 - V_1)sC_{c_1} - (V_2 - V_1)sC_{c_2} = 0 \quad (4.684)$$

$$-g_{m_2}V_1 + V_2(g_2 + sC_2) + (V_2 - V_1)sC_{c_2} = 0 \quad (4.685)$$

and

$$g_{m_3}V_2 + g_{m_f}V_1 + V_0(g_3 + sC_3) + (V_0 - V_1)sC_{c_1} = 0 \quad (4.686)$$

respectively. Assuming that $g_{m_k} \gg g_k$ ($k = 1,2,3$) and $C_{c_1}, C_{c_2}, C_3 \gg C_1, C_2$, show that the amplifier small-signal voltage gain can be put into the form

$$A(s) = \frac{V_0(s)}{V_i(s)} \simeq A_0 \frac{1 + ds + cs^2}{(1 + s/\omega_{p_1})(1 + bs + as^2)} \quad (4.687)$$

where

$$A_0 = \frac{g_{m1}g_{m2}g_{m3}}{g_1g_2g_3} \quad (4.688)$$

$$\omega_{p_1} = \frac{g_1g_2g_3}{g_{m2}g_{m3}C_{c1}} \quad (4.689)$$

$$a = \frac{C_{c_2}C_3}{g_{m2}g_{m3}} \quad (4.690)$$

$$b = \frac{(g_{m_3} + g_{m_f} - g_{m_2})C_{c_1}C_{c_2} - g_{m_2}C_{c_2}C_3}{g_{m_2}g_{m_3}C_{c_1}} \quad (4.691)$$

$$c = -\frac{C_{c_1}C_{c_2}}{g_{m_2}g_{m_3}} \quad (4.692)$$

and

$$d = \frac{(g_{m_3} + g_{m_f})C_{c_2}}{g_{m_2}g_{m_3}} \quad (4.693)$$

Determine the gain-bandwidth product and slew rate of the amplifier.

15. **Three-stage amplifier with a feedforward compensation**
For the three-stage amplifier shown in Figure 4.118, verify that the small-signal equivalent model can be derived as depicted in Figure 4.119.

Show that

$$A(s) = \frac{V_0(s)}{V_i(s)} = A_0 \frac{1 + ds + cs^2}{(1 + s/\omega_{p_1})(1 + bs + as^2)} \quad (4.694)$$

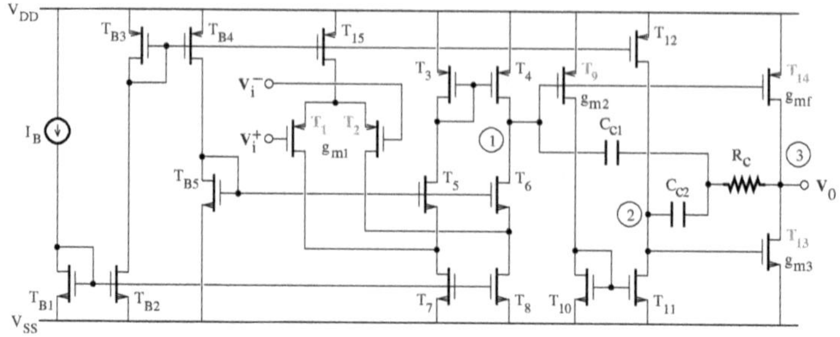

FIGURE 4.118
Circuit diagram of a three-stage amplifier.

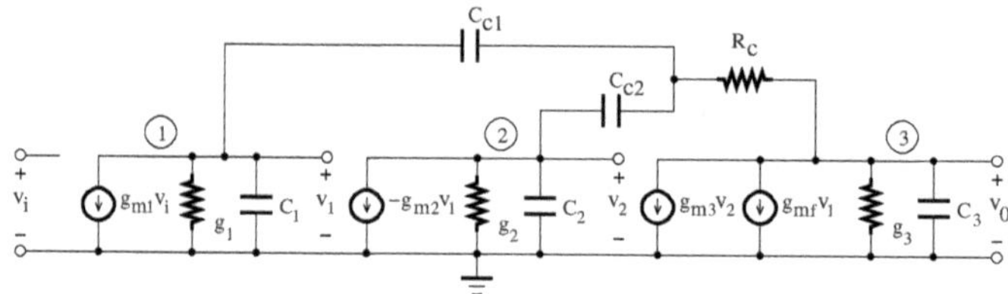

FIGURE 4.119
Small-signal equivalent model of the three-stage amplifier.

where

$$A_0 = \frac{g_{m_1} g_{m_2} g_{m_3}}{g_1 g_2 g_3} \tag{4.695}$$

$$\omega_{p_1} = \frac{g_1 g_2 g_3}{g_{m_2} g_{m_3} C_{c_1}} \tag{4.696}$$

$$a = \frac{C_3 C_{c_2}}{g_{m_2} g_{m_3}} \tag{4.697}$$

$$b = \frac{C_{c_2}(g_{m_3} + g_{m_f} - g_{m_2})}{g_{m_2} g_{m_3}} \tag{4.698}$$

$$c = \frac{C_{c_1} C_{c_2}[(g_{m_f} + g_{m_3})R_c - 1]}{g_{m_2} g_{m_3}} \tag{4.699}$$

and

$$d = (C_{c_1} + C_{c_2})R_c + \frac{C_{c_2}(g_{m_f} - g_{m_2})}{g_{m_2} g_{m_3}} \tag{4.700}$$

With the assumption that $R_c = 1/(g_{m_f} + g_{m_3})$, compute the gain-bandwidth product and slew rate of the amplifier.

16. Low-dropout linear regulator

Low-dropout (LDO) linear regulators are used to provide a regulated supply voltage with minimal ripple to supply-noise sensitive building blocks in battery-powered devices. By featuring a low-dropout voltage (typically 300 mV at 150 mA of load current), a low-power operation, and a low quiescent current (generally less than 140 μA), LDO regulators help prolong battery life. Another key parameter indicating the LDO regulator quality is the noise output, that is referred by the RMS noise measurement (a few tens of μV_{RMS}) or by the spectral noise density (about one $\mu V/\sqrt{\text{Hz}}$ at 100 kHz).

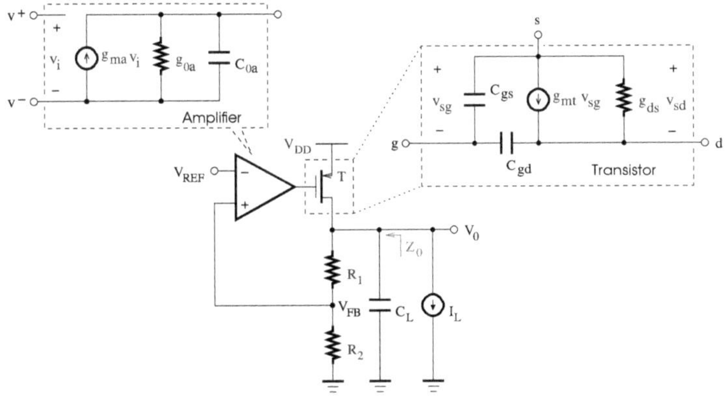

FIGURE 4.120
Circuit diagram of a low-dropout voltage regulator.

A pMOS LDO with on-chip load capacitor, implemented as shown in Figure 4.120, uses the feedback loop to generate a constant voltage that is independent of the load at the output and is given by,

$$V_0 = \left(1 + \frac{R_1}{R_2}\right) V_{REF} \tag{4.701}$$

where V_{REF} is the reference voltage. However, in practice, it can be affected by stability problems associated with the load current and capacitor. Hence, in addition to line/load regulation, line/load transient, and power-supply rejection, stability considerations at very low load currents should also be taken into account in the LDO regulator design.

(**i**) Assuming that the feedback path is broken between V^+ and V_{FB} and applying Miller's theorem to C_{gd}, show that the open

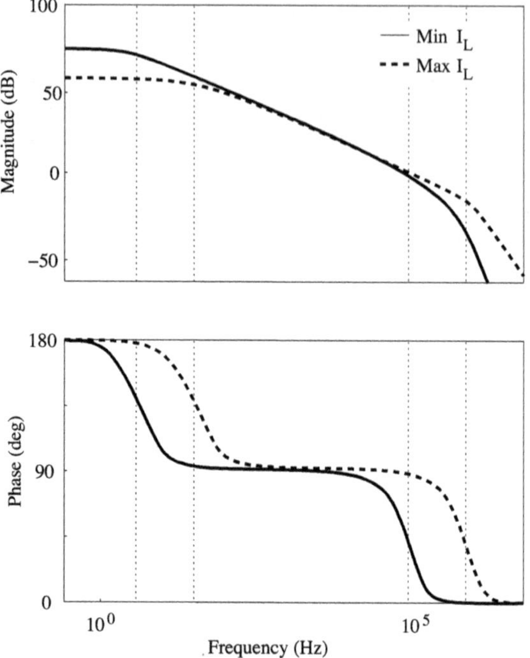

FIGURE 4.121
Bode plots of the transfer function $L(s)$.

loop transfer function can be obtained as,

$$L(s) = \frac{V_{FB}(s)}{V^+(s)} = \beta \frac{g_{ma} g_{mt} Z_L}{g_{0a}(1 + sC_p/g_{0a})} \tag{4.702}$$

where
$C_p = C_{0a} + C_{gs} + (1 + A_t)C_{gd}$, $A_t = g_{mt}/g_{ds}$, $\beta = R_2/(R_1 + R_2)$, and $Z_L \simeq (1/g_{ds}) \parallel (R_1 + R_2) \parallel (1/sC_L)$.

– Establish that the poles of the open loop transfer function $L(s)$ can be expressed as,

$$\omega_{p_1} = \frac{g_{0a}}{C_p} \tag{4.703}$$

$$\omega_{p_2} = \frac{g_{ds} + 1/(R_1 + R_2)}{C_L} \tag{4.704}$$

and the *dc* gain is of the form,

$$L_0 = \frac{\beta g_{ma} g_{mt}}{g_{0a}[g_{ds} + 1/(R_1 + R_2)]} \tag{4.705}$$

– Verify that Bode plots of the transfer function, $L(s)$, can be obtained as shown in Figure 4.121 because the transistor transconductance and drain-source conductance are dependent on the load current, I_L.

– The worst-case stability condition occurs for the minimum value of I_L. Assuming that $L_0 \gg 1$, the poles are sufficiently well separated and there is a dominant pole with the frequency ω_{p_1}, the unity frequency is given by $\omega_u \simeq L_0\omega_{p_1}$ and the phase margin can be derived as

$$\begin{aligned}\phi_M &= 180^o - \arctan\left(\frac{\omega_u}{\omega_{p_1}}\right) - \arctan\left(\frac{\omega_u}{\omega_{p_2}}\right) \\ &\simeq 90^o - \arctan\left(\frac{\omega_u}{\omega_{p_2}}\right) \end{aligned} \tag{4.706}$$

Determine the minimum value of C_L to achieve the phase margin of 70° required for loop stability.

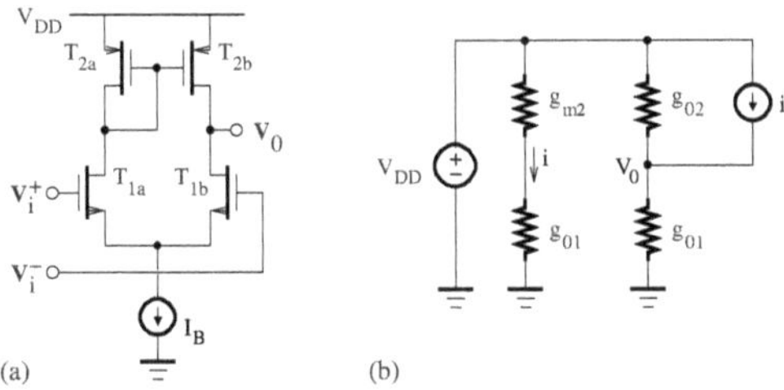

FIGURE 4.122
Amplifier (a) and its PSR equivalent model (b).

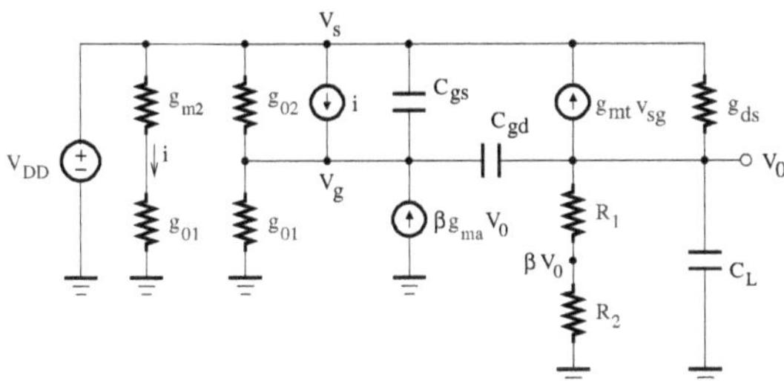

FIGURE 4.123
Equivalent circuit of the low-dropout voltage regulator for PSR determination.

(**ii**) In the case of single-stage structures, the amplifier with an nMOS differential input pair, as shown in Figure 4.122(a), provides better power-supply rejection (PSR) at low frequencies for the LDO regulator than the one with a pMOS differential input pair.

The PSR equivalent model of the amplifier is depicted in Figure 4.122(b), where the current mirror operation is modeled by the current-dependent current source, i, that reflects the current flowing through g_{01} into the output branch.

Using the superposition theorem, verify that

$$V_0 = \frac{g_{02}}{g_{01}+g_{02}} V_{DD} + \frac{i}{g_{01}+g_{02}} \tag{4.707}$$

$$= \frac{g_{02}}{g_{01}+g_{02}} V_{DD} + \frac{g_{01}}{g_{01}+g_{02}} V_{DD} = V_{DD} \tag{4.708}$$

Consider the equivalent circuit model of Figure 4.123. Assuming that $\beta g_{ma} \ll g_{mt}$, show that the PSR of the LDO can be written as,

$$PSR = \frac{V_0(s)}{V_{DD}(s)} \simeq \frac{1}{\beta(g_{ma}/g_{0a})A} \frac{(1+s/\omega_{z_1})(1+s/\omega_{z_2})}{(1+s/\omega_{p_1})(1+s/\omega_{p_2})} \tag{4.709}$$

where

$$\omega_{z_1} = g_{0a}g_{ds}/(g_{mt}C_{gd}), \quad \omega_{z_2} = g_{mt}/C_{gs}, \quad \omega_{p_1} = \beta g_{ma}/C_{gd},$$
$$\omega_{p_2} = g_{mt}/[C_L(1+C_{gd}/C_{gs}) + C_{gs}], \quad \text{and} \quad \beta = R_2/(R_1+R_2)$$

(iii) The load transient response is characterized by the magnitude of the overshoot or undershoot at the LDO regulator output under load current variations. Using small-signal perturbation analysis, it can be related to the output impedance according to the next relationship,

$$\triangle V_0 = Z_0(s) \cdot \triangle I_L(s) \tag{4.710}$$

where $\triangle I_L = I_{peak}/s$.

Assuming that $\beta g_{ma} \ll g_{mt}$, show that

$$Z_0(s) = \frac{R_0}{1+\dfrac{\beta g_{ma} g_{mt} R_0}{g_{0a}}} \cdot \frac{1+s(C_{gs}+C_{gd})/g_{0a}}{1+s\dfrac{C_{gd}}{\beta g_{ma}} + s^2 \dfrac{(C_{gs}+C_{gd})C_L + C_{gs}C_{gd}}{\beta g_{ma} g_{mt}}} \tag{4.711}$$

where $R_0 = (1/g_{ds}) \parallel (R_1+R_2)$.

Verify that the load regulation as measured in the steady state is given by

$$\text{Load regulation} \triangleq \left.\frac{\triangle V_0}{\triangle I_L}\right|_{t\to\infty} = Z_0(s)|_{s=0} = \frac{R_0}{1+\dfrac{\beta g_{ma} g_{mt} R_0}{g_{0a}}} \tag{4.712}$$

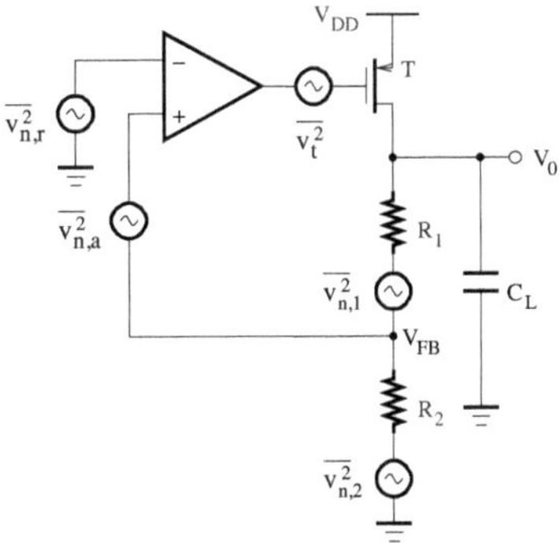

FIGURE 4.124
Circuit diagram of the LDO regulator illustrating the noise sources.

(**iv**) The line transient response is analyzed by measuring the output voltage variation in response to a voltage step at the supply voltage node of the LDO regulator. Using small-signal perturbation analysis, it can be related to the PSR according to the next relationship,

$$\triangle V_0 = PSR(s) \cdot \triangle V_{DD} \tag{4.713}$$

where $\triangle V_{DD}(s) = V_{peak}/s$.

Verify that the line regulation as measured in the steady state can be written as,

$$\text{Load regulation} \triangleq \left.\frac{\triangle V_0}{\triangle V_{DD}}\right|_{t\to\infty} = PSR(s)|_{s=0} = \frac{1}{\beta(g_{ma}/g_{0a})A} \tag{4.714}$$

(**v**) The circuit diagram of the LDO regulator with noise sources of the reference, $\overline{v_{n,r}^2}$, amplifier, $\overline{v_{n,a}^2}$, transistor, $\overline{v_{n,t}^2}$, and resistors, $\overline{v_{n,1}^2}$ and $\overline{v_{n,2}^2}$, is depicted in Figure 4.124. Show that the average mean-square value of the noise source voltage at the output is given by,

$$\overline{v_{n,0}^2} = \left|\frac{L(s)}{1+L(s)}\right|^2 \times \left[\left(\overline{v_{n,r}^2} + \overline{v_{n,a}^2} + \frac{\overline{v_{n,t}^2}}{(A_a(s))^2}\right)\left(1+\frac{R_1}{R_2}\right)^2 + \overline{v_{n,1}^2} + \overline{v_{n,2}^2}\left(\frac{R_1}{R_2}\right)^2\right] \tag{4.715}$$

where $A_a(s)$ is the voltage-gain transfer function of the amplifier.

Bibliography

[1] P. R. Gray and R. G. Meyer, "MOS operational amplifier design — A tutorial overview," *IEEE J. of Solid-State Circuits*, vol. 22, pp. 969–982, Dec. 1987.

[2] M. J. M. Pelgrom, A. C. J. Duinmaijer, and A. P. G. Welbers, "Matching properties of MOS transistors," *IEEE J. Solid-State Circuits*, vol. 24, no. 5, pp. 1433–1439, Oct. 1989.

[3] Y. Tsividis, Z. Czarnul, and S. C. Fang, "MOS transconductors and integrator with high linearity," *Electronics Letters*, vol 22, pp. 245–246, 1986.

[4] J. L. Pennock, P. Frith, and R. G. Barker, "CMOS triode transconductor continuous time filters," *Proc. of the IEEE CICC*, pp. 378–381, 1986.

[5] A. A. Fayed and M. Ismail, "A low-voltage, highly linear voltage-controlled transconductor," *IEEE Trans. on Circuits and Systems–II*, vol. 52, pp. 831–835, Dec. 2005.

[6] I. Mehr and D. R. Welland, "A CMOS continuous-time $G_m - C$ filter for PRML read channel applications at 150 Mb/s and beyond," *IEEE J. of Solid-State Circuits*, vol. 32, pp. 499–513, April 1997.

[7] J. N. Babanezhad, "A rail-to-rail CMOS op amp," *IEEE J. of Solid-State Circuits*, vol. 23, pp. 1414–1417, Dec. 1988.

[8] A. Nedungadi and T. R. Viswanathan, "Design of linear CMOS transconductance elements," *IEEE Trans. on Circuits and Systems*, vol. 31, pp. 891–894, Oct. 1984.

[9] R. R. Torrance, T. R. Viswanathan, and J. V. Hanson, "CMOS voltage to current transducers," *IEEE Trans. on Circuits and Systems*, vol. 32, pp. 1097–1104, Nov. 1985.

[10] K. Bult and H. Wallinga, "A CMOS four-quadrant analog multiplier," *IEEE J. of Solid-State Circuits*, vol. 21, pp. 430–435, June 1986.

[11] Z. Wang and W. Guggenbühl, "A voltage-controllable linear MOS transconductor using bias offset technique," *IEEE J. of Solid-State Circuits*, vol. 25, pp. 315–317, Feb. 1990.

[12] V. Peluso, P. Vancorenland, A. M. Marques, M. S. J. Steyaert, and W. Sansen, “A 900 mV low-power $\Delta - \Sigma$ A/D converter with 77-dB dynamic range,” *IEEE J. of Solid-State Circuits*, vol. 33, pp. 1887–1897, Dec. 1998.

[13] A. J. López-Martín, S. Baswa, J. Ramirez-Angulo, and R. G. Carvajal, “Low-voltage super class AB CMOS OTA cells with very high slew rate and power efficiency,” *IEEE J. of Solid-State Circuits*, vol. 40, pp. 1068–1077, May 2005.

[14] E. Seevinck and R. F. Wassenaar, “A versatile CMOS linear transconductor/square-law function circuit,” *IEEE J. of Solid-State Circuits*, vol. 22, pp. 366–377, June 1987.

[15] M. Milkovic, “Current gain high-frequency CMOS operational amplifiers,” *IEEE J. of Solid-State Circuits*, vol. 20, pp. 845–851, Aug. 1985.

[16] C.-C. Shih, and P. R. Gray, “Reference refreshing cyclic analog-to-digital and digital-to-analog converters,” *IEEE J. of Solid-State Circuits*, vol. 21, pp. 544–554, Aug. 1986.

[17] J. N. Babanezhad and R. Gregorian, “A programmable gain/loss circuit,” *IEEE J. of Solid-State Circuits*, vol. 22, pp. 1082–1090, Dec. 1987.

[18] S. M. Mallya and J. H. Nevin, “Design procedures for a fully differential folded-cascode CMOS operational amplifier,” *IEEE J. of Solid-State Circuits*, vol. 24, pp. 1737–1740, Dec. 1989.

[19] K. Bult and G. J. G. M. Geelen, “A fast-settling CMOS op-amp for SC circuits with 90 dB DC gain,” *IEEE J. of Solid-State Circuits*, vol. 25, pp. 1379–1384, Dec. 1990.

[20] J. Lloyd and H.-S. Lee, “A CMOS op-amp with fully differential gain-enhancement,” *IEEE Trans. on Circuits and Systems*, vol. 41, pp. 241–243, March 1994.

[21] D. Flandre, A. Viviani, J.-P. Eggermont, B. Gentine, and P. G. A. Jespers, “Improved synthesis of gain-boosted regulated-cascode CMOS stages using symbolic analysis and gm/ID methodology,” *IEEE J. of Solid-State Circuits*, vol. 32, pp. 1006–1012, July 1997.

[22] M. Das, “Improved design criteria of gain-boosted CMOS OTA with high-speed optimizations,” *IEEE Trans. on Circuits and Systems–II*, vol. 49, pp. 204–207, March 2002.

[23] M. M. Ahmadi, “A new modeling and optimization of gain-boosted cascode amplifier for high-speed and low-voltage applications,” *IEEE Trans. on Circuits and Systems–II*, vol. 53, pp. 169–173, March 2006.

[24] B. Y. Kamath, R. G. Meyer, and P. R. Gray, "Relationship between frequency response and settling time of operational amplifiers," *IEEE J. Solid-State Circuits*, vol. SC-9, pp. 347–352, Dec. 1974.

[25] M. Banu, J. M. Khoury, and Y. Tsividis, "Fully differential operational amplifiers with accurate output balancing," *IEEE J. Solid-State Circuits*, vol. 23, pp. 1410–1414, Dec. 1988.

[26] R. A. Whatley, "Fully differential operational amplifier with D.C. common-mode feedback," U.S. Patent 4,573,020, filed Dec. 18, 1984; issued Feb. 25, 1986.

[27] T. C. Choi, R. T. Kaneshiro, R. W. Brodersen, P. R. Gray, W. B. Jett, and M. Wilcox, "High-frequency CMOS switched-capacitor filters for communications application," *IEEE J. Solid-State Circuits*, vol. SC-18, pp. 652–664, Dec. 1983.

[28] Z. Czamul, S. Takagi, and N. Fuji, "Common-mode feedback circuit with differential-difference amplifier," *IEEE Trans. on Circuits and Systems–I*, vol. 41, pp. 243–246, March 1994.

[29] X. Zhang, and E. I. El-Masry, "A novel CMOS OTA based on body-driven MOSFETs and its applications in OTA-C filter," *IEEE Trans. on Circuits and Systems–I*, vol. 54, pp. 1204–1212, June 2007.

[30] D. Senderowicz, S. F. Dreyer, J. H. Huggins, C. F. Rahim and C. A. Laber, "A family of differential NMOS analog circuits for a PCM codec filter chip," *IEEE J. of Solid-State Circuits*, vol. 17, pp. 1014–1023, Dec. 1982.

[31] O. Choksi and L. R. Carley, "Analysis of switched-capacitor common-mode feedback circuit," *IEEE Trans. on Circuits and Systems–II*, vol. 50, pp. 906–917, Dec. 2003.

[32] D. A. Garrity and P. L. Rakers, "Common-Mode Output Sensing Circuit," U.S. Patent 5,894,284, filed Dec. 2, 1996; issued April 13, 1999.

[33] F. Krummenacher and N. Joehl, "A 4-MHz CMOS continuous-time filter with on-chip automatic tuning," *IEEE J. of Solid-State Circuits*, vol. 23, pp. 750–758, June 1988.

[34] H. Khorramabadi and P. R. Gray, "High-frequency CMOS continuous-time filters," *IEEE J. of Solid-State Circuits*, vol. SC-19, pp. 939–948, Dec. 1984.

[35] P. D. Walker and M. M. Green, "An approach to fully differential circuit design without common-mode feedback," *IEEE Trans. on Circuits and Systems–II*, vol. 43, pp. 752–762, Nov. 1996.

[36] W. C. Black, Jr., D. J. Allstot, and R. A. Reed, "A high performance low power filter," *IEEE J. of Solid-State Circuits,* vol. SC-15, pp. 929–938, Dec. 1980.

[37] D. M. Monticelli, "A quad CMOS single-supply opamp with rail-to-rail output swing," *IEEE J of Solid-State Circuits*, vol. SC-21, pp. 1026–1034, Dec. 1986.

[38] W.-C. S. Wu, W. J. Helms, J. A. Kuhn, and B. E. Byrkett, "Digital-compatible high-performance operational amplifier with rail-to-rail input and output ranges," *IEEE J. of Solid-State Circuits*, vol. SC-29, pp. 63–66, Jan. 1994.

[39] K. J. de Langen, and J. H. Huijsing, "Compact low-voltage power-efficient operational amplifier cells for VLSI," *IEEE J. of Solid-State Circuits*, vol. 33, pp. 1482–1496, Oct. 1998.

[40] P. Migliavacca, "Start-up aid circuit for a plurality of current sources," U.S. Patent 6,002,242, filed Aug. 4, 1998; issued Dec. 14, 1999.

[41] B. K. Ahuja, "An improved frequency compensation technique for CMOS operational amplifiers," *IEEE J. of Solid-State Circuits*, vol. 18, pp. 629–633, Dec. 1983.

[42] G. Palmisano and G. Palumbo, "A compensation strategy for two-stage CMOS opamps based on current buffer," *IEEE Trans. on Circuits and Systems–I*, vol. 44, pp. 257–262, March 1997.

[43] P. J. Hurst, S. H. Lewis, J. P. Keane, F. Aram, and K. C. Dyer, "Miller compensation using current buffers in fully differential CMOS two-stage operational amplifiers," *IEEE Trans. on Circuits and Systems–I*, vol. 51, pp. 275–285, Feb. 2004.

[44] H-T. Ng, R. M. Ziazadeh, and D. J. Allstot, "A multistage amplifier with embedded frequency compensation," *IEEE J. of Solid-State Circuits*, vol. 34, pp. 339–347, March 1999.

[45] Y.-F. Chou, "Low-power start-up circuit for a reference voltage generator," U.S. Patent 6,201,435, filed Aug. 26, 1999; issued March 13, 2001.

[46] C. T. Chuang, "Analysis of the settling behavior of an operational amplifier," *IEEE J. Solid-State Circuits*, vol. SC-17, pp. 74–80. Feb. 1982.

[47] H. C. Yang and D. J. Allstot, "Considerations for fast settling operational amplifiers," *IEEE Trans. on Circuits and Systems*, vol. 37, pp. 326–334, March 1990.

[48] S. Pernici, G. Nicollini, and R. Castello, "A CMOS low-distortion fully differential power amplifier with double nested Miller compensation," *IEEE J. of Solid-State Circuits*, vol. 28, pp. 758–763, July 1993.

[49] R. G. H. Eschauzier, R. Hogervorst, and J. H. Huijsing, "A programmable 1.5 V CMOS class-AB operational amplifier with hybrid nested Miller compensation for 120 dB gain and 6 MHz UGF," *IEEE J. of Solid-State Circuits*, vol. 29, pp. 1497–1504, Dec. 1994.

[50] K. N. Leung and P. K. T. Mok, "Analysis of multistage amplifier-frequency compensation," *IEEE Trans. on Circuits Systems–I*, vol. 48, pp. 1041–1056, Sept. 2001.

[51] R. G. H. Eschauzier, L. P. T. Kerklaan, and J. H. Huijsing, "A 100-MHz 100-dB operational amplifier with multipath nested Miller compensation structure," *IEEE J. Solid-State Circuits*, vol. 27, pp. 1709–1717, Dec. 1992.

[52] A. D. Grasso, G. Palumbo, and S. Pennisi, "Advances in reversed nested miller compensation," *IEEE Trans. on Circuits Systems–I*, vol. 54, pp. 1459–1470, July 2007.

[53] H. Lee, and P. K. T. Mok, "Active-feedback frequency-compensation technique for low-power multistage amplifiers," *IEEE J. of Solid-State Circuits*, vol. 38, pp. 511–520, March 2003.

[54] S. S. Chong and P. K. Chan, "Cross feedforward cascode compensation for low-power three-stage amplifier with large capacitive load," *IEEE J. of Solid-State Circuits*, vol. 47, no. 9, pp. 2227–2234, Sep. 2012.

[55] X. Peng, W. Sansen, L. Hou, J. Wang, and W. Wu, "Impedance adapting compensation for low-power multistage amplifiers," *IEEE J. of Solid-State Circuits*, vol. 46, no. 2, pp. 445–451, Feb. 2011.

[56] S. H. Lewis and P. R. Gray, "A pipelined 5-Msamples/s 9-bit analog-to-digital converter," *IEEE J. of Solid-State Circuits*, vol. 22, pp. 954–961, Dec. 1987.

[57] R. Hogervorst, J. P. Tero, and J. H. Huijsing, "Compact CMOS constant-g_m rail-to-rail input stage with g_m-control by an electronic Zener diode," *IEEE J. of Solid-State Circuits*, vol. 31, pp. 1035–1040, July 1996.

[58] M. Wang, T. L. Mayhugh, Jr., S. H. K. Embabi and E. Sànchez-Sinencio, "Constant-g_m rail-to-rail CMOS op-amp input stage with overlapped transition regions," *IEEE J. of Solid-State Circuits*, vol. 34, pp. 148–156, Feb. 1999.

[59] K.-C. Hsieh, P. R. Gray, D. Senderowicz, and D. G. Messerschmitt, "A low-noise chopper-stabilized differential switched-capacitor filtering technique," *IEEE J. of Solid-State Circuits*, vol. 16, pp. 708–715, Dec. 1981.

[60] M. G. Degrauwe, J. Rijmenants, E. A. Vittoz, and H. J. De Man, "Adaptive biasing CMOS amplifiers," *IEEE J. of Solid-State Circuits*, vol. 17, pp. 522–528, June 1982.

[61] M. C. W. Coln, "Chopper stabilization of MOS operational amplifiers using feed-forward techniques," *IEEE J. of Solid-State Circuits*, vol. 16, pp. 745–748, Dec. 1981.

[62] C. Enz and G. C. Temes, "Circuit techniques for reducing the effects of op-amp imperfections: Autozeroing, correlated double sampling, and chopper stabilization," *Proceedings of the IEEE*, vol. 84, pp. 1584–1614, Nov. 1996.

[63] C. C. Enz, E. A. Vittoz, and F. Krummenacher, "A CMOS chopper amplifier," *IEEE J. of Solid-State Circuits*, vol. 22, pp. 335–342, June 1987.

[64] W. Oh, B. Bakkaloglu, C. Wang, and S. K. Hoon, "A CMOS low noise, chopper stabilized low-dropout regulator with current-mode feedback error amplifier," *IEEE Trans. on Circuits and Systems-I*, vol. 55, no. 10, pp. 3006–3015, Nov. 2008.

[65] I. G. Finvers, J. W. Haslett, and F. N. Trofimenkoff, "A high temperature precision amplifier," *IEEE J. of Solid-State Circuits*, vol. 30, pp. 120–128, Feb. 1995.

[66] G. Nicollini, F. Moretti, and M. Conti, "High-frequency fully differential filter using operational amplifiers without common-mode feedback," *IEEE J. of Solid-State Circuits*, vol. 24, pp. 803–813, June 1989.

[67] D. K. Su, "Class AB CMOS output amplifier," U.S. Patent 5,039,953, filed May 18, 1990; issued Aug. 13, 1991.

5

Nonlinear Analog Components

CONTENTS

Nonlinear analog components in MOS technology essentially include comparators and multipliers. They can find applications in communication and instrumentation systems.

A comparator provides an output signal to indicate whether the input signal is higher or lower than a reference voltage, or equivalently, determines the sign of a voltage difference. Its output then assumes either the high or low level, depending on the relative magnitude of the input signals. In most applications, comparators are required to exhibit a high speed, a high gain, a high common-mode rejection, and a low offset voltage.

Multipliers or mixers are used in a variety of electronic systems to provide the product of two signals. The basic idea of the multiplier implementation relies on applying the input signals to a nonlinear device and cancelling the undesired output components. High linearity, or say, low intermodulation distortion, and low noise can be considered important performance characteristics in a multiplier, because they can affect the resulting dynamic range.

The next sections deal with the analysis and design of nonlinear active components (comparators and multipliers). Generally, the objective is to achieve a sufficient dynamic range and bandwidth using a submicrometer IC process with a low supply voltage.

5.1 Comparators

Comparators are used in applications such as data conversion and interfacing, where a decision regarding the relative value of the input signals is required. They can be realized using high-gain differential amplifiers, charge balancing techniques, and open-loop structures with a positive feedback [1, 2].

In general, the speed and accuracy of comparators are enhanced by reducing the probability of metastability. A metastable state occurs, for instance, when the input voltage difference is too small to be resolved unambiguously. Although the metastability cannot be completely avoided, its rate of occurrence is reduced by maximizing the available settling time, as by employing a cascade of output latches.

5.1.1 Amplifier-based comparator

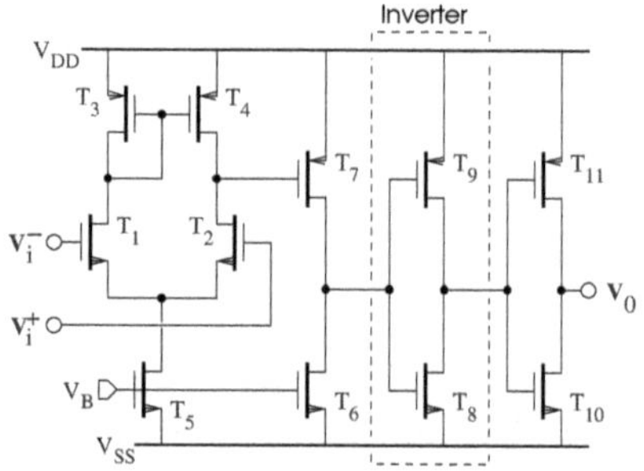

FIGURE 5.1
Circuit diagram of an amplifier-based comparator.

Amplifiers may appear suitable for comparing two signals. Figure 5.1 shows a comparator based on an uncompensated two-stage transconductance amplifier driving two inverters in series. The input and reference voltages are applied to either the positive node or negative node of the first stage of the comparator, which consists of a differential transistor pair loaded by a current mirror. The output of the second stage, which is a source follower, is used to drive the inverters. The compensation network of the amplifier is removed to somewhat increase the operation speed while the use of the inverter stage helps drive output capacitive loads with optimum speed.

In the noninverting configuration, the comparator output volt-

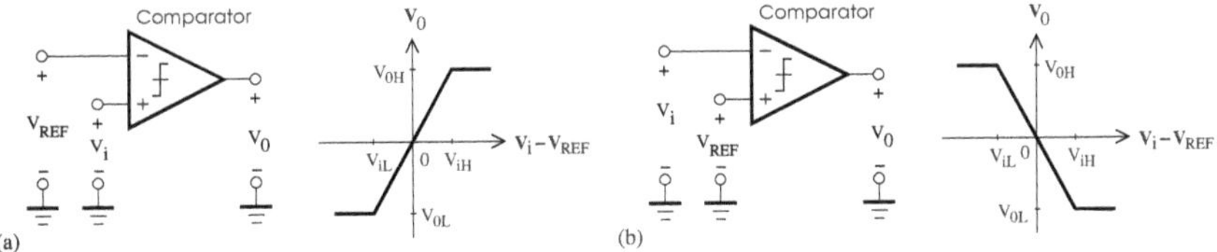

FIGURE 5.2
Practical transfer characteristic: (a) noninverting and (b) inverting comparators.

age can be written as,

$$V_0 = \begin{cases} V_{0H} & \text{if} \quad V_i - V_{REF} > V_{iH} \\ A_v(V_i - V_{REF}) & \text{if} \quad V_{iL} \leq V_i - V_{REF} \leq V_{iH} \\ V_{0L} & \text{if} \quad V_i - V_{REF} < V_{iL} \end{cases} \tag{5.1}$$

while in the inverting configuration it is given by

$$V_0 = \begin{cases} V_{0L} & \text{if} \quad V_i - V_{REF} > V_{iH} \\ -A_v(V_i - V_{REF}) & \text{if} \quad V_{iL} \leq V_i - V_{REF} \leq V_{iH} \\ V_{0H} & \text{if} \quad V_i - V_{REF} < V_{iL} \end{cases} \tag{5.2}$$

where A_v is the amplification gain; V_{iL} and V_{iH} denote the low and high level of the input thresholds, respectively; and V_{0L} and V_{0H} represent the lower and higher limits of the output thresholds, respectively.

The smallest voltage difference, which can be resolved by the comparator, is dependent on the gain, A_v, and

$$A_v = \frac{V_{0H} - V_{0L}}{\triangle V} \tag{5.3}$$

where $\triangle V$ is the resolution. In the ideal case, A_v is assumed to be infinite and $V_{iL} = V_{iH}$.

Due to transistor mismatches, the comparator exhibits an input-referred offset voltage that is equal to the value of the output signal when $V_i - V_{REF}$ is set to zero. Offsets, that are generally in the range of ± 10 mV, can be reduced to a few microvolts using trimming or auto-zero techniques.

In high-speed applications, the comparator performance is also dependent on the delay between the instant where the signals are applied at the inputs and the one where the output voltage reaches the steady state. In contrast to amplifiers, the speed requirement is achieved in multistage comparators by not using the frequency compensation.

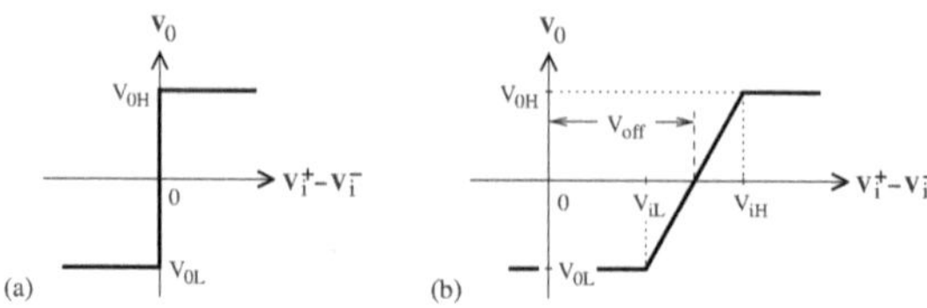

FIGURE 5.3
(a) Ideal transfer characteristic of the comparator; (b) comparator transfer characteristic showing the effect of the finite gain and offset voltage.

Ideally, the transition between logic levels at the output node will occur when both input signals have the same magnitude, as shown in Figure 5.3(a). This is not the case in practice due to the offset voltage caused by mismatches between nMOS and pMOS transistor characteristics. The comparator transfer characteristic showing the effect of the finite gain, A_v, and offset voltage, V_{off}, is illustrated in Figure 5.3(b). Furthermore, the transition between the output logic levels will occur with a time delay, which increases for large voltage swings on the inputs.

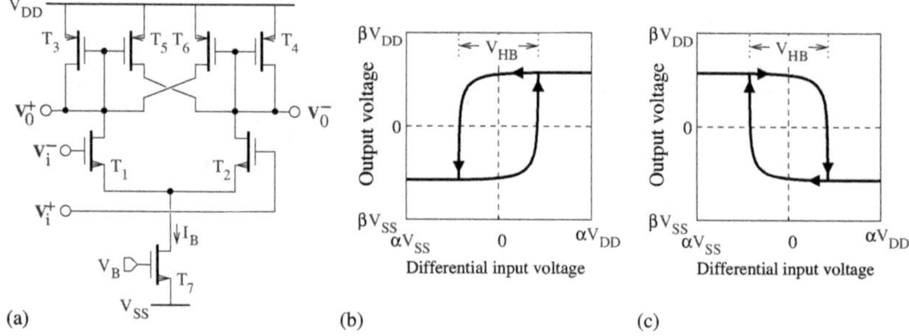

FIGURE 5.4
(a) Comparator circuit with hysteresis; comparator characteristics in the (b) noninverting and (c) inverting configurations.

The operation with a high gain can contribute to increasing the comparator sensitivity to the effect of the noise, which can corrupt the input signal. In the presence of noise or in the case where the comparator speed is much higher than the variation rate of the input signal around the reference signal level, the switching of the comparator output can become erratic. This problem is overcome by using a comparator with hysteresis. A common technique to generate the hysteresis is based on the use of positive feedback, which forces the threshold levels for the output switching to depend not only on the difference voltage at the input nodes, but also on the previous states of the input signals. The comparator will then exhibit a characteristic with hysteresis if the switching thresholds for low-to-high and high-to-low transitions of the input voltages are different from each other.

A hysteresis circuit is depicted in Figure 5.4(a) [4,5]. It consists of a differential transistor pair, $T_1 - T_2$, loaded by current mirrors, $T_3 - T_6$, with cross-coupled outputs. The current flowing through T_3 and T_4 are of the form $i_1 = i_3 + i_6$ and $i_2 = i_4 + i_5$. The extra current contributions of T_5 and T_6, which form the positive feedback, help increase not only the transconductance of the differential stage, but also the output slew rate.

Based on the small-signal assumption, the current flowing through each of the transistors T_1 and T_2 forming the differential input stage of the comparator shown in Figure 5.4(a) can be computed as

$$i_{d_1} \simeq \frac{I_B}{2} + g_{m_1}\frac{V_i}{2} \tag{5.4}$$

and

$$i_{d_2} \simeq \frac{I_B}{2} - g_{m_2}\frac{V_i}{2} \tag{5.5}$$

respectively, where $V_i = V_i^+ - V_i^-$ and I_B is the bias current. For the transistors T_5 and T_6, we have

$$i_{d_5} \simeq g_{m_5}\frac{V_0}{2} \tag{5.6}$$

and

$$i_{d_6} \simeq -g_{m_6}\frac{V_0}{2} \tag{5.7}$$

respectively, where $V_0 = V_0^+ - V_0^-$. By replacing the diode-connected transistor T_3 by its equivalent resistance given by $(1/g_{m3}) \parallel (1/g_3)$, the output voltages can be expressed as

$$V_0^+ = [(1/g_{m_3}) \parallel (1/g_3) \parallel (1/g_1)]i_{d_3} \tag{5.8}$$
$$V_0^- = [(1/g_{m_4}) \parallel (1/g_4) \parallel (1/g_2)]i_{d_4} \tag{5.9}$$

where $i_{d_3} = i_{d_1} - i_{d_6}$ and $i_{d_4} = i_{d_2} - i_{d_5}$. In practice, $g_{m_1} = g_{m_2}$, $g_{m_3} = g_{m_4}$, $g_{m_5} = g_{m_6}$, $g_1 = g_3$, and $g_2 = g_4$, while $1/g_1$ and $1/g_3$ are assumed to be negligible in comparison with $1/g_{m_3}$. The differential output voltage is given by

$$V_0 = \frac{1}{g_{m_3}}[(i_{d_1} - i_{d_2}) + (i_{d_5} - i_{d_6})] \tag{5.10}$$

where $i_{d_1} - i_{d_2} = g_{m_1} V_i$ and $i_{d_5} - i_{d_6} = g_{m_5} V_0$. The differential gain can then be written as

$$A_d = \frac{V_0}{V_i} = \frac{1}{1-\eta} \frac{g_{m_1}}{g_{m_3}} \tag{5.11}$$

where $\eta = g_{m_5}/g_{m_3}$. Note that the drain current for a transistor operating in the saturation region is of the form

$$i_{d_j} = \frac{1}{2} K' \left(\frac{W_j}{L_j} \right) (V_{gs_j} - V_T)^2 \tag{5.12}$$

where $K' = \mu_n C_{ox}$, and the transconductance around the bias point can be derived as

$$g_{m_j} = \frac{\partial i_{d_j}}{\partial V_{gs_j}} = \sqrt{\frac{2K'(W_j/L_j)}{i_{d_j}}} \tag{5.13}$$

where W_j and L_j denote the width and length of the transistor, respectively.

The operation mode of the comparator is determined by the value of the positive feedback factor, η. If $\eta < 1$, the comparator will operate as an improved gain stage [3]. If $\eta = 1$, the comparator will exhibit the behavior of a positive feedback latch. If $\eta > 1$, the operation of the comparator will be similar to the one of a Schmitt trigger circuit with the hysteresis width related to the value of the positive feedback factor.

In the case where $\eta > 1$, the comparator characteristics in the noninverting and inverting configurations are respectively shown in Figures 5.4(b) and (c), where $\alpha < 1$ and $\beta < 1$. To derive the positive trigger point, it is assumed that $V_i^+ = 0$. By connecting the inverting input node to a negative voltage, the transistor T_1 is turned on while T_2 is turned off. Because the current flowing through T_1 and T_3 is almost equal to the bias current, V_0^+ is set to the high logic level and V_0^- takes the low logic level. Hence, the current flowing through the transistors T_4, T_5, and T_6 is nearly zero.

By increasing the voltage at the inverting input node, T_2 can start to conduct and the load of the differential stage is reduced to the current mirror formed by T_3 and T_5. This electrical conduction is developed gradually until the drain currents of T_2 and T_5 are equal. It can be found that

$$i_{d_5} = \frac{W_5/L_5}{W_3/L_3} i_{d_3} \tag{5.14}$$

$$i_{d_2} = i_{d_5} \tag{5.15}$$

where $i_{d_3} = i_{d_1}$. Because the input differential pair consisting of transistors T_1 and T_2 is biased by the constant current I_B, we have

$$i_{d_1} + i_{d_2} = I_B \tag{5.16}$$

The drain current of T_1 and T_2 can then be given by

$$i_{d_1} = \frac{I_B}{1 + (W_5/L_5)/(W_3/L_3)} \tag{5.17}$$

and

$$i_{d_2} = \frac{(W_5/L_5)/(W_3/L_3)I_B}{1 + (W_5/L_5)/(W_3/L_3)} \tag{5.18}$$

The positive trigger level is defined as

$$V_{trig^+} = v_{gs_2} - v_{gs_1} \tag{5.19}$$

where

$$v_{gs_1} = \sqrt{\frac{i_{d_1}}{2K'(W_1/L_1)}} + V_T \tag{5.20}$$

and

$$v_{gs_2} = \sqrt{\frac{i_{d_2}}{2K'(W_2/L_2)}} + V_T \tag{5.21}$$

Assuming that the transistors T_1 and T_2 are matched, we obtain

$$V_{trig^+} = \sqrt{\frac{I_B}{2K'(W_1/L_1)}} \frac{\sqrt{(W_5/L_5)/(W_3/L_3)} - 1}{\sqrt{1 + (W_5/L_5)/(W_3/L_3)}} \tag{5.22}$$

As soon as the input voltage becomes greater than V_{trig^+}, the switching of comparator outputs will occur. As a result, V_0^+ changes to the low logic level and V_0^- now assumes the high logic level. With the transistor T_1 being off and T_2 on, the bias current primarily flows through T_2 and T_4 and there is no current flowing through transistors T_3, T_5, and T_6.

For the derivation of the negative trigger point, the voltage at the inverting input node is decreased to initiate the conduction of T_1. This is associated with a reduction in the current through T_2 and continues until the drain currents of T_1 and T_6 become equal. This is achieved at the trigger point. In a similar manner as previously, we find that

$$i_{d_6} = \frac{W_6/L_6}{W_4/L_4} i_{d_4} \tag{5.23}$$

$$i_{d_1} = i_{d_6} \tag{5.24}$$

$$i_{d_1} + i_{d_2} = I_B \tag{5.25}$$

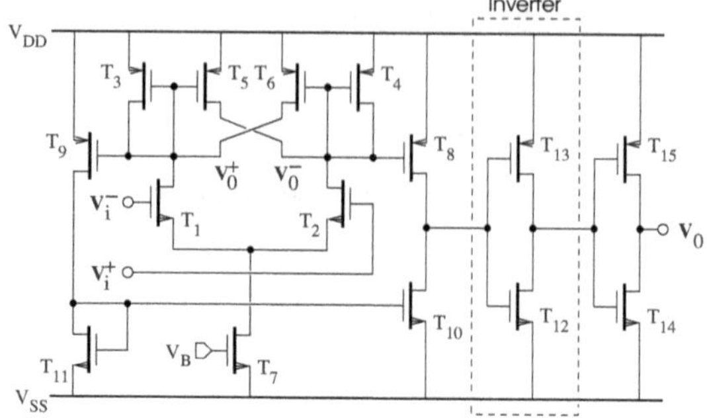

FIGURE 5.5
Circuit diagram of a comparator with hysteresis.

where $i_{d_4} = i_{d_2}$. This set of equations can then be solved for

$$i_{d_2} = \frac{I_B}{1 + (W_6/L_6)/(W_4/L_4)} \tag{5.26}$$

and

$$i_{d_1} = \frac{(W_6/L_6)/(W_4/L_4)I_B}{1 + (W_6/L_6)/(W_4/L_4)} \tag{5.27}$$

The negative trigger level is obtained as

$$V_{trig^-} = v_{gs_2} - v_{gs_1} \tag{5.28}$$

where

$$v_{gs_1} = \sqrt{\frac{i_{d_1}}{2K'(W_1/L_1)}} + V_T \tag{5.29}$$

and

$$v_{gs_2} = \sqrt{\frac{i_{d_2}}{2K'(W_2/L_2)}} + V_T \tag{5.30}$$

With $W_1/L_1 = W_2/L_2$, the expression for V_{trig^-} becomes

$$V_{trig^-} = \sqrt{\frac{I_B}{2K'(W_1/L_1)}} \frac{1 - \sqrt{(W_6/L_6)/(W_4/L_4)}}{\sqrt{1 + (W_6/L_6)/(W_4/L_4)}} \tag{5.31}$$

Due to the circuit symmetry, transistors T_3, T_4, T_5, and T_6 are matched. The difference between the positive and negative trigger points is the hysteresis band, which is given by

$$V_{HB} = V_{trig^+} - V_{trig^-} = 2\sqrt{\frac{I_B}{2K'(W_1/L_1)}} \frac{\sqrt{(W_5/L_5)/(W_3/L_3)} - 1}{\sqrt{1 + (W_5/L_5)/(W_3/L_3)}} \quad (5.32)$$

The complete circuit diagram of a comparator based on the hysteresis circuit is depicted in Figure 5.5 [4,5]. It also includes an output stage composed of $T_8 - T_{11}$ and two buffer inverters.

Unfortunately, the hysteresis introduced in the comparator characteristic has the inconvenience of limiting the speed of operation and the minimum level of the input voltage difference that can be accurately resolved. Due to the limitations of amplifier-based comparator structures, it may be better in practice to use comparators, which are designed to drive logic circuits, and operate in an open loop and with a high speed even when overdriven.

5.1.2 Comparator using charge balancing techniques

In switched capacitor circuits, the comparator can consist of an input stage based on charge balancing techniques [6] followed by a latch. Figure 5.6 shows the comparator input stages in the case of the single-ended and fully differential structures.

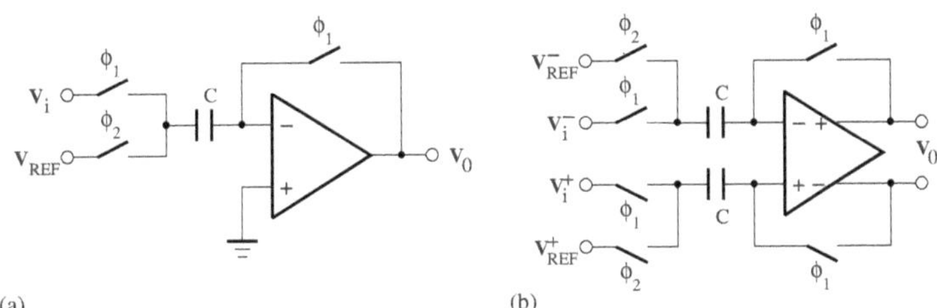

FIGURE 5.6
Circuit diagram of the comparator input stage: (a) single-ended and (b) differential configurations.

Let us consider the single-ended circuit of Figure 5.6(a), where V_i and V_{REF} are the input and reference voltages, respectively. The amplifier is in a unity-gain feedback loop during the clock phase ϕ_1, or during the time interval $(n-1)T < t \leq (n-1/2)T$, while it operates in an open loop during the clock phase ϕ_2, that is, $(n-1/2)T < t \leq nT$. With $V_0 = A_0(V^+ - V^-)$, $V^+ = V_{off}$ and $V^- = V_0$, the voltage at the inverting node of the amplifier can be written as

$$V^-((n-1/2)T) = \frac{A_0 V_{off}}{1 + A_0} \quad (5.33)$$

in the clock phase 1, and

$$V^-(nT) = \frac{C}{C + C_p}[V_{REF}(nT) - V_i((n-1/2)T)] + \frac{q_{inj}}{C + C_p} + \frac{A_0 V_{off}}{1 + A_0} \quad (5.34)$$

in the clock phase ϕ_2, where q_{inj} represents the error due to the charge injection, C_p is the total parasitic capacitance at the inverting node of the amplifier, V_{off} is the offset voltage, and A_0 is the amplifier dc gain. The output signal in the clock phase ϕ_2 is given by

$$V_0(nT) = -V_{sup} \cdot \text{sign}\{V^-(nT) - V^-((n-1/2)T)\} \quad (5.35)$$

where V_{sup} denotes the supply voltage of the amplifier. The resolution and settling speed of the comparator based on charge balancing techniques is limited by the bottom-plate parasitic of the capacitor and charge injections of switches.

5.1.3 Latched comparators

A latched comparator is configured with clock signals to sample the input signals and to generate the corresponding output signal. Its amplification gain and propagation delay time are improved by using a positive feedback loop for the signal regeneration into the full-scale logic level. The main advantage of a latched comparator is its high sensitivity to a small input difference.

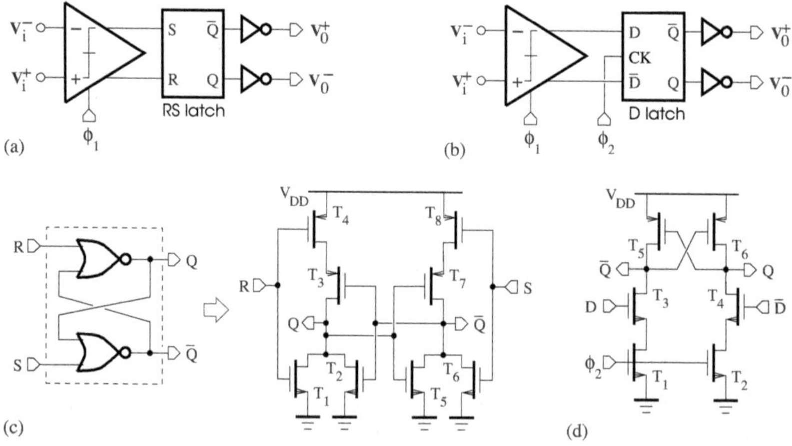

FIGURE 5.7
Circuit diagram of a latched comparator based on (a) an RS latch and (b) a D latch; circuit diagrams of (c) an RS latch and (d) a D latch.

The circuit diagrams of a latched comparator based on an RS latch and a D latch are depicted in Figure 5.7(a) and (b), respectively. These structures consist of a regenerative comparator and a latch loaded by two inverters, which provide the buffer function. The latch is used to store the result of the comparison. Figs. 5.7(c) and (d) show the circuit diagrams of a RS latch and D latch, respectively.

The overall power consumption of a comparator can be divided into static, dynamic, and short-circuit components. The static power is caused by the leakage current and dc current needed for the comparator operation. The dynamic power is associated with the charging and discharging of various switched capacitors. The short-circuit power is caused by the dc current flow from power rails as a result of the time delay to switch from one state to the other. Due to the relatively negligible value of the short-circuit power, circuit-level techniques are essentially exploited to reduce the static and dynamic power contributions. The regenerative comparator can then be implemented using either the dynamic or static logic circuits. The difference between both structures is basically related to the fact that the output state of a static logic circuit can change at any time in accordance with the input signal.

In a static comparator, there is always a low-impedance path between the output and both the supply voltage and ground, yielding an uninterrupted power consumption during the whole time that the circuit is powered up. Because the regeneration speed is typically proportional to the magnitude of this current, the power consumption can be considerable at high speeds.

A dynamic comparator uses a quiescent current to bias the output stage only during the time required for the generation of the output signals. As a result, it will quite likely dissipate less power than a static comparator. However, the speed of a dynamic comparator may be limited by the time delay required to charge capacitors before each comparison. Furthermore, the kick-back charge injection, which is caused by large voltage transitions associated with the regeneration mechanism, can be transmitted back to the input stage through capacitive coupling, thereby affecting the comparator accuracy.

5.1.3.1 Static comparator

The circuit diagram of a static comparator, which includes a preamplifier stage and a regenerative stage based on cross-coupled transistors, is depicted in Figure 5.8(a). The operation of the comparator is controlled by two nonoverlapping clock signals. During the clock phase ϕ_1, a bias current is supplied to the differential transistor pair of the preamplifier while the regenerative stage is disabled. Because the differential output voltage is an amplified version of the input voltage difference, the comparator is in the *tracking* mode. During the clock phase, ϕ_2, the preamplifier is disabled and the regenerative stage, which is now connected to a bias current, is enabled. The amplification of the tracked voltage difference to valid logic levels can then be achieved without being further affected by the actual signal levels at the input nodes. On the falling edge of the clock signal or at the end of the *latching* mode, the comparator outputs are reset, and the comparison process can start again on the next clock rising edge. Figure 5.8(b) shows the input, output, and clock waveforms explaining the comparator operation.

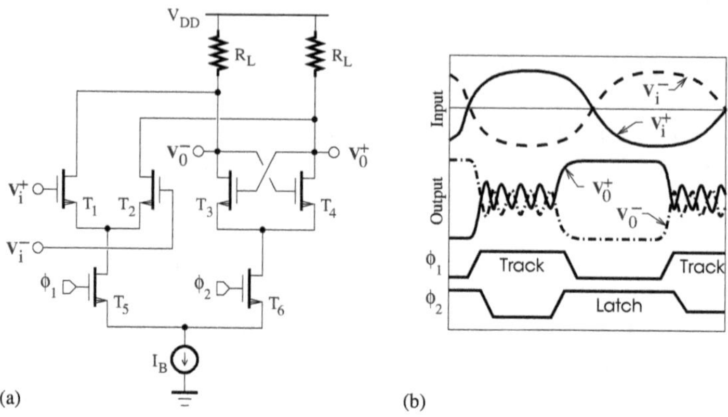

FIGURE 5.8
(a) Circuit diagram of a latched comparator; (b) input, output, and clock waveforms.

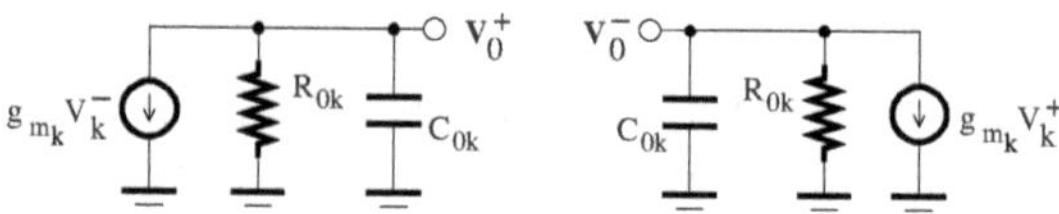

FIGURE 5.9
Small-signal equivalent model of the latched comparator.

For small-signal analysis, a latched comparator is often represented as two identical back-to-back inverting gain stages driving equal impedance loads [12, 13]. Figure 5.9 shows the small-signal equivalent model of the aforementioned latched comparator.

During the tracking mode, g_{m_k} is equal to g_{m_i}, which is the transconductance of the input differential transistor pair, R_{0k} and C_{0k} respectively denote the total resistance, R_{0i}, and capacitance, C_{0i}, at the output nodes. The output voltages, V_0^+ and V_0^-, satisfy the following differential equations

$$g_{m_i}V_i^- + \frac{V_0^+}{R_{0i}} + C_{0i}\frac{dV_0^+}{dt} = 0 \tag{5.36}$$

$$g_{m_i}V_i^+ + \frac{V_0^-}{R_{0i}} + C_{0i}\frac{dV_0^-}{dt} = 0 \tag{5.37}$$

Subtracting Equation (5.37) from (5.36), we obtain

$$\frac{dV_0}{dt} + \frac{V_0}{R_{0i}C_{0i}} = \frac{g_{m_i}V_i}{C_{0i}} \tag{5.38}$$

where $V_i = V_i^+ - V_i^-$ and $V_0 = V_0^+ - V_0^-$. Assuming that the initial condition is determined by the comparator state at the end of the previous latching mode, the differential output voltage can be computed as

$$V_0 = \overline{V}_{0r}\,\mathrm{e}^{-t/\tau_i} + g_{m_i} R_{0i} V_i (1 - \mathrm{e}^{-t/\tau_i}) \tag{5.39}$$

where $\overline{V}_{0r}$ represents the steady-state output voltage in the latching mode, and $\tau_i = R_{0i} C_{0i}$ is the preamplification time constant.

During the latching mode, the preamplifier is disabled and the operation of the comparator is determined by the regenerative stage. Hence, g_{m_k} is identical to the transconductance, g_{m_r}, of the cross-coupled transistors, and R_{0k} and C_{0k} respectively represent the total resistance, R_{0r}, and capacitance, C_{0r}, at the output nodes. The dynamic behavior of the output voltages, V_0^+ and V_0^-, can be modeled by the next differential equations,

$$g_{m_r} V_0^- + \frac{V_0^+}{R_{0r}} + C_{0r} \frac{\mathrm{d}V_0^+}{\mathrm{d}t} = 0 \tag{5.40}$$

$$g_{m_r} V_0^+ + \frac{V_0^-}{R_{0r}} + C_{0r} \frac{\mathrm{d}V_0^-}{\mathrm{d}t} = 0 \tag{5.41}$$

The difference between Equations (5.40) and (5.41) can be put into the form

$$\frac{\mathrm{d}V_0}{\mathrm{d}t} - \frac{(g_{m_r} - 1/R_{0r})V_0}{C_{0r}} = 0, \tag{5.42}$$

where $V_0 = V_0^+ - V_0^-$. Solving for the differential output voltage gives

$$V_0 = \overline{V}_{0i}\,\mathrm{e}^{t/\tau_r} \tag{5.43}$$

where the regenerative time constant can be written as

$$\tau_r = \frac{C_{0r}}{g_{m_r} - 1/R_{0r}} \tag{5.44}$$

and $\overline{V}_{0i}$ is the steady-state output voltage in the tracking mode. When g_{m_r} is greater than $1/R_{0r}$, the time constant is positive, and the output voltage can increase exponentially until it reaches the valid logic level.

For high-frequency applications, the comparator should be designed to exhibit a short signal propagation delay. Hence, the operation speed of the comparator can be increased by minimizing the preamplification and regeneration time constants.

• The circuit diagram of the comparator with coupling capacitors is shown in Figure 5.10 [14]. During the resetting phase, that is, when the clock phase 2 is high, the difference of the input signal is amplified by $T_1 - T_2$ and applied to the parasitic capacitors connected at the drains of these transistors, and the comparator output nodes are shorted by the switch T_6. When the clock

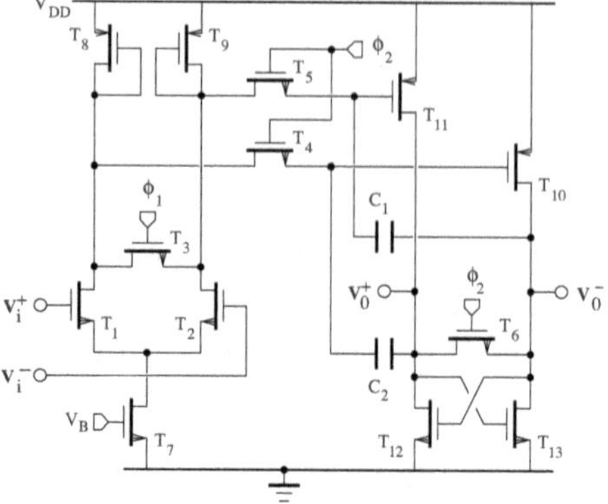

FIGURE 5.10
Circuit diagram of a static comparator with coupling capacitors.

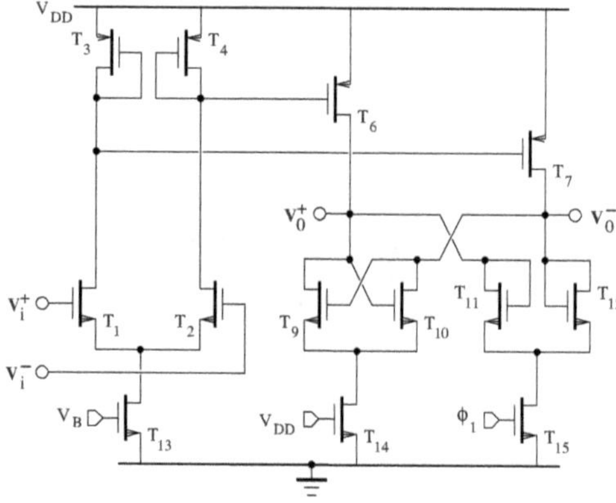

FIGURE 5.11
Circuit diagram of a static comparator with diode-connected transistor-based output initialization.

phase 1 is high, the preamplifier is disabled and the regeneration is achieved in the output stage based on the signal acquired during the previous phase. By establishing a cross-coupling between $T_{10}-T_{11}$ and $T_{12}-T_{13}$, the positive feedback realized by the capacitors C_1 and C_2 increases the regeneration speed. However, because the kickback noise can also be transmitted through this connection, switches T_3-T_4 are required to uncouple the input and output stages.

- The circuit diagram of a static comparator with diode-connected transistors is depicted in Figure 5.11 [15]. This structure is based on a preamplifier and latch output stage. The clock phase 1 is high during the resetting period and the transistors $T_{11}-T_{12}$, which are now activated, affect the amplification gain. The latch output resets at a speed, which can be optimized by sizing the biasing transistors. When the clock signal goes low, transistors $T_{11} - T_{12}$ are disconnected and the comparator operation is determined by the positive feedback due to the cross-coupled transistors $T_9 - T_{10}$. The delay required by the output to reach the low and high level is shortened because the connection between T_9 and T_{10} is maintained during both phases. However, this also results in more power dissipation. The kickback noise is reduced by the isolation of the input stage provided by the current mirrors.

5.1.3.2 Dynamic comparator

In general, the operation of a dynamic comparator can be divided into successive phases determined by clock signals. After each comparison of voltage levels periodically supplied to the regenerative stage by the preamplifier connected to the input signals, the comparator outputs should be reset.

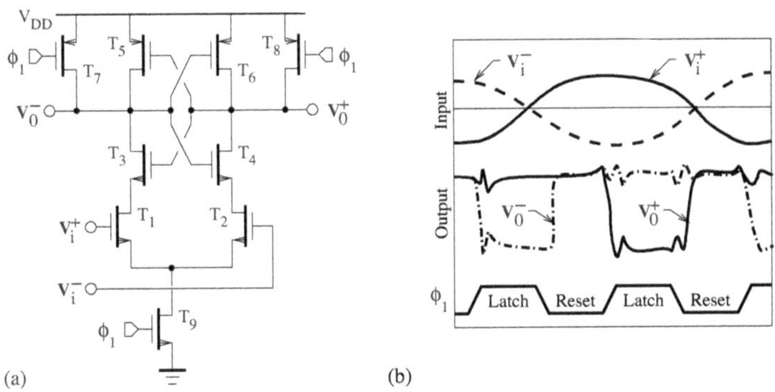

FIGURE 5.12
(a) Circuit diagram of a dynamic comparator based on the current-controlled sense amplifier; (b) input, output, and clock waveforms.

• The circuit diagram of a dynamic comparator based on the current-controlled sense amplifier is shown in Figure 5.12(a) [11]. When the input differential transistor pair is disabled, the comparator outputs are pulled up to the positive supply voltage, V_{DD}, by switch transistors, $T_7 - T_8$. During the clock phase, ϕ, the sense amplifier is enabled, and the current generated by the input differential transistor pair, $T_1 - T_2$, is used to drive the serially connected regenerative stage, $T_3 - T_6$. Due to the positive feedback provided by cross-coupled transistor pairs, a small input difference can be translated to full logic output voltages. If V_i^+ is greater than V_i^-, V_0^+ will remain at V_{DD} and V_0^- will be set to the ground. Conversely, if V_i^+ is smaller than V_i^-, V_0^+ will be set to the ground and V_0^- will remain at V_{DD}. Because the dc current flow is restricted to the relatively short period that is allocated to the transistor switching, the static power consumption of the comparator is al-

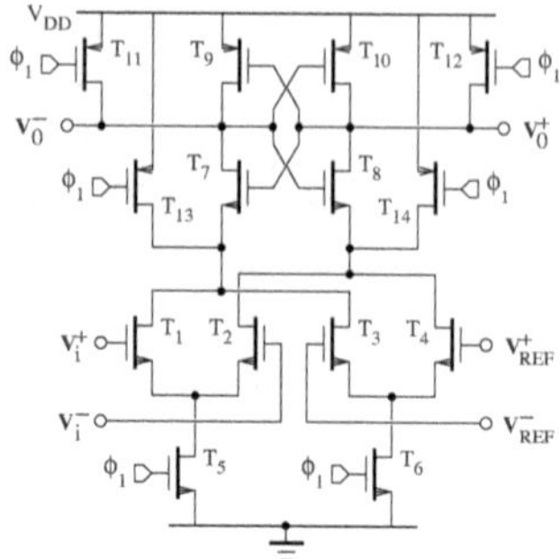

FIGURE 5.13
Circuit diagram of a fully differential dynamic comparator based on the current-controlled sense amplifier.

most negligible. Figure 5.12(b) shows the input, output, and clock waveforms during normal operation.

- **Comparator time delay**

For the aforementioned dynamic comparator, Figure 5.14 shows the waveforms illustrating the response delay. When the voltage V_i^+ is greater than V_i^-, the drain current of the transistor T_1 increases in comparison to that of transistor T_2, and the capacitive load at the node V_0^- is discharged more quickly than the one at the node V_0^+. The comparator delay is composed of two components, t_0 and t_{latch} [16]. The term t_0 is due to the capacitive discharge of the output load capacitance C_L until one of the p-channel transistors, T_5 and T_6, turns on, while t_{latch} represents the latching delay of the two cross-coupled inverters.

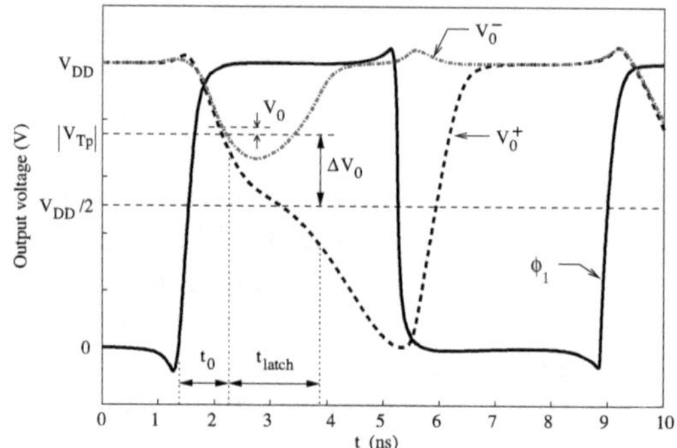

FIGURE 5.14
Comparator waveforms illustrating the response delay.

Assuming that each of the transistors, T_1 and T_2, exhibits the transconductance, $g_m = g_{m_1} = g_{m_2}$, and is driven by the bias current $I_B/2$ provided by the transistor T_9 ($I_{D_9} = I_B$), the charge conservation equation can be written as,

$$I_B \cdot t_0/2 = C_L|V_{T_p}| \tag{5.45}$$

where V_{T_p} is the threshold voltage of the p-channel transistor. Hence,

$$t_0 = 2\frac{C_L|V_{T_p}|}{I_B} \tag{5.46}$$

The sensing behavior of the comparator is determined by the cross-coupled inverters, whose response is governed by an exponential equation. Hence,

$$\triangle V_0 = V_0 \exp\left[\left(\frac{g_m - 1/R_0}{C_L}\right)t_{latch}\right] \tag{5.47}$$

or equivalently,

$$t_{latch} = \frac{C_L}{g_m - 1/R_0} \cdot \ln\left(\frac{\triangle V_0}{V_0}\right) \tag{5.48}$$

where $\triangle V_0$ represents the output swing and the initial voltage V_0 can be obtained as,

$$V_0 = |V_0^+(t = t_0) - V_0^-(t = t_0)| \tag{5.49}$$

$$= |V_{T_p}| - \frac{I_{D_2}t_0}{C_L} = |V_{T_p}|\left(1 - \frac{I_{D_2}}{I_{D_1}}\right) = |V_{T_p}|\frac{\triangle I_i}{I_{D_1}} \tag{5.50}$$

Considering the differential transistor pair, T_1 and T_2, biased by the drain current of the transistor T_9, it can be shown that

$$\triangle I_i = |I_{D_1} - I_{D_2}| = g_m \triangle V_i \tag{5.51}$$

The latch delay then becomes

$$t_{latch} = \frac{C_L}{g_m - 1/R_0} \cdot \ln\left(\frac{1}{|V_{T_p}|(g_m/I_{D_1})}\frac{\triangle V_0}{\triangle V_i}\right) \tag{5.52}$$

Taking into account the fact that each of the currents I_{D_1} and I_{D_2} can be assumed to be almost equal to $I_B/2$ for small changes of the differential input voltage $\triangle V_i$, we obtain

$$t_{latch} = \frac{C_L}{g_m - 1/R_0} \cdot \ln\left(\frac{1}{2|V_{T_p}|(g_m/I_B)}\frac{\triangle V_0}{\triangle V_i}\right) \tag{5.53}$$

Finally, the comparator time delay is given by

$$t_{delay} = t_0 + t_{latch} \tag{5.54}$$

$$\simeq 2\frac{C_L|V_{T_p}|}{I_B} + \frac{C_L}{g_m - 1/R_0} \cdot \ln\left(\frac{1}{2|V_{T_p}|(g_m/I_B)}\frac{\triangle V_0}{\triangle V_i}\right) \tag{5.55}$$

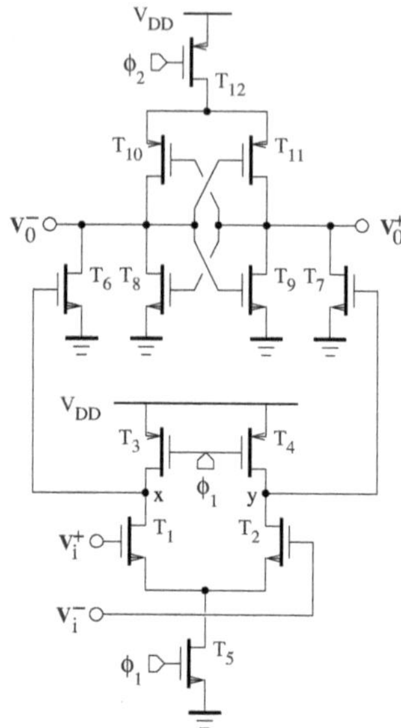

FIGURE 5.15
Double-tail dynamic comparator.

Note that the value of the output swing can, for instance, be estimated as $\triangle V_0 = V_{DD}/2 - |V_{T_p}|$. The comparator delay is proportional to the capacitive load at the output nodes.

The dynamic comparator shown in Figure 5.12(a) uses stacked transistors that can operate properly only if the supply voltage is high enough to satisfy the requirement for a large voltage headroom. Furthermore, the latch speed and the input common-mode voltage are determined by the same bias current and cannot be adjusted independently if necessary.

The double-tail dynamic comparator [17] of Figure 5.15 is suitable for low-voltage operation and allows flexible control of the latch speed and the input common-mode voltage by using one tail (or bias) current for the input stage and another for the latching stage. It operates in two phases: reset and comparison. During the reset phase, $\phi_1 = 0$, T_5 and T_{12} are turned off, and the nodes x and y are pre-charged through transistors T_3 and T_4 to V_{DD}, causing the connection of the output nodes V_0^+ and V_0^- to the ground because transistors T_6 and T_6 are switched on. Here, dedicated reset transistors are not required at the output nodes.

During the comparison phase, $\phi_1 = V_{DD}$, T_5 and T_{12} are turned on, and the differential input voltage is applied to the cross-coupled inverters, which then regenerate the voltage difference at the output nodes.

- **Comparator offset voltage**

A fully differential dynamic comparator based on the current-controlled sense amplifier is depicted in Figure 5.13 [7, 8]. The input stage consists of two differential pairs connected, respectively, to the input and reference voltages with the same polarity. The regenerative output stage is driven by the sum of currents caused by the voltages at the input nodes. The transistors T_{11} – T_{14} operate as switches and are used to pre-charge the outputs and internal nodes of the comparator, so that each comparison is made independent of the previous one. They should be designed with minimum sizes to keep the discharging time of parasitic capacitors as low as possible.

A dynamic comparator often exhibits a higher offset voltage than its static counterpart. This is due to the fact that in addition to the static offset voltage associated with mismatches of transistor characteristics such as the threshold voltage and charge mobility, it is also affected by the dynamic offset that can be caused by the imbalance of parasitic and load capacitances [18, 19].

To estimate the offset voltage of the comparator shown in Figure 5.13, it is assumed that all transistors operate in the saturation region. The drain currents of transistors $T_1 - T_4$ can be written as,

$$I_{D_1} = \frac{\mu_{n_1} C_{ox}}{2} \left(\frac{W_1}{L_1}\right) (V_{GS_1} - V_{T_{n_1}})^2 \tag{5.56}$$

$$I_{D_2} = \frac{\mu_{n_2} C_{ox}}{2} \left(\frac{W_2}{L_2}\right) (V_{GS_2} - V_{T_{n_2}})^2 \tag{5.57}$$

$$I_{D_3} = \frac{\mu_{n_3} C_{ox}}{2} \left(\frac{W_3}{L_3}\right) (V_{GS_3} - V_{T_{n_3}})^2 \tag{5.58}$$

$$I_{D_4} = \frac{\mu_{n_4} C_{ox}}{2} \left(\frac{W_4}{L_4}\right) (V_{GS_4} - V_{T_{n_4}})^2 \tag{5.59}$$

Under a balanced condition, the drain current of the transistor T_7 should be equal to that of T_8. That is,

$$I_{D_1} + I_{D_3} = I_{D_2} + I_{D_4}\,. \tag{5.60}$$

To proceed further, all parameter variations are considered as random and uncorrelated variables with a normal distribution and zero mean. Using the error propagation method and taking into account only first-order variation terms, it can be shown that

$$\triangle I_{D_1} + \triangle I_{D_3} = \triangle I_{D_2} + \triangle I_{D_4} \tag{5.61}$$

where

$$\triangle I_{D_1} = C_{ox}\left(\frac{W_1}{L_1}\right)\left[\frac{\triangle\mu_{n1}}{2}(V_{GS_1} - V_{T_n})^2 - \mu_n(V_{GS_1} - V_{T_n})(\triangle V_{T_{n_1}} - \triangle V_i)\right] \tag{5.62}$$

$$\triangle I_{D_2} = C_{ox}\left(\frac{W_2}{L_2}\right)\left[\frac{\triangle\mu_{n2}}{2}(V_{GS_2} - V_{T_n})^2 - \mu_n(V_{GS_2} - V_{T_n})\triangle V_{T_{n2}}\right] \tag{5.63}$$

$$\triangle I_{D_3} = C_{ox}\left(\frac{W_3}{L_3}\right)\left[\frac{\triangle\mu_{n3}}{2}(V_{GS_3} - V_{T_n})^2 - \mu_n(V_{GS_3} - V_{T_n})\triangle V_{T_{n3}}\right] \tag{5.64}$$

$$\triangle I_{D_4} = C_{ox}\left(\frac{W_4}{L_4}\right)\left[\frac{\triangle\mu_{n4}}{2}(V_{GS_4} - V_{T_n})^2 - \mu_n(V_{GS_4} - V_{T_n})\triangle V_{T_{n4}}\right] \tag{5.65}$$

and $\triangle V_i$ denotes the variation voltage to be added to the current gate-to-source voltage, V_{GS_1}, of the transistor T_1 in order to preserve the current balance. The input-referred offset voltage contribution, $V_{off,1-4}$, due to mismatches between transistors $T_1 - T_4$ is given by

$$V_{off,1-4} = \triangle V_i \tag{5.66}$$

$$\begin{aligned}
&= \frac{(W_2/L_2)(V_{GS_2} - V_{T_n})^2}{2(W_1/L_1)(V_{GS_1} - V_{T_n})}\frac{\triangle\mu_{n2}}{\mu_n} - \frac{(W_2/L_2)(V_{GS_2} - V_{T_n})}{(W_1/L_1)(V_{GS_1} - V_{T_n})}\triangle V_{T_{n2}} \\
&+ \frac{(W_4/L_4)(V_{GS_4} - V_{T_n})^2}{2(W_1/L_1)(V_{GS_1} - V_{T_n})}\frac{\triangle\mu_{n4}}{\mu_n} - \frac{(W_4/L_4)(V_{GS_4} - V_{T_n})}{(W_1/L_1)(V_{GS_1} - V_{T_n})}\triangle V_{T_{n4}} \\
&- \frac{(W_3/L_3)(V_{GS_3} - V_{T_n})^2}{2(W_1/L_1)(V_{GS_1} - V_{T_n})}\frac{\triangle\mu_{n3}}{\mu_n} + \frac{(W_3/L_3)(V_{GS_3} - V_{T_n})}{(W_1/L_1)(V_{GS_1} - V_{T_n})}\triangle V_{T_{n3}} \\
&- \frac{(V_{GS_1} - V_{T_n})}{2}\frac{\triangle\mu_{n1}}{\mu_n} + \triangle V_{T_{n1}}
\end{aligned} \tag{5.67}$$

where $\mu_{n_k} = \mu_n + \triangle\mu_{n_k}$ and $V_{T_{n_k}} = V_{T_n} + \triangle V_{T_{n_k}}$ $(k = 1, 2, 3, 4)$. Its variance is found to be

$$\sigma^2_{V_{off,1-4}} = \sigma^2_{\triangle V_i} = \sum_{k=1}^{4}\left(\frac{V_{GS_k} - V_{T_n}}{V_{GS_1} - V_{T_n}}\right)^2\left[\sigma^2_{V_{T_{n_k}}} + \left(\frac{V_{GS_k} - V_{T_n}}{2}\right)^2\sigma^2_{\mu_{n_k}}\right] \tag{5.68}$$

To determine the offset voltage contribution caused by mismatches of transistors T_5 and T_6, Equation (5.60) can be rewritten as,

$$I_{D_1} - I_{D_2} = I_{D_4} - I_{D_3} \tag{5.69}$$

Because $I_{D_1} + I_{D_2} = I_{D_5}$ and $V_{GS_1} - V_{GS_2} = V_i$, $I_{D_4} + I_{D_3} = I_{D_6}$ and $V_{GS_4} - V_{GS_3} = V_{REF}$, we can obtain

$$I_{D_1} - I_{D_2} = KV_i\sqrt{\frac{2I_{D_5}}{K} - V_i^2} \tag{5.70}$$

$$I_{D_4} - I_{D_3} = KV_{REF}\sqrt{\frac{2I_{D_6}}{K} - V_{REF}^2} \tag{5.71}$$

where $K_1 = K_2 = K_3 = K_4 = K = (\mu_n C_{ox}/2)(W_1/L_1)$. By writing the drain currents of transistors T_5 and T_6 as,

$$I_{D_5} = \frac{\mu_{n_5} C_{ox}}{2}\left(\frac{W_5}{L_5}\right)(V_{GS_5} - V_{T_{n_5}})^2 \tag{5.72}$$

$$I_{D_6} = \frac{\mu_{n_6} C_{ox}}{2}\left(\frac{W_6}{L_6}\right)(V_{GS_6} - V_{T_{n_6}})^2 \tag{5.73}$$

Equation (5.69) becomes

$$\begin{aligned}\frac{\mu_{n_5} C_{ox}}{K}\left(\frac{W_5}{L_5}\right)V_i^2(V_{GS_5} - V_{T_{n_5}})^2 - V_i^4 \\ = \frac{\mu_{n_6} C_{ox}}{K}\left(\frac{W_6}{L_6}\right)V_{REF}^2(V_{GS_6} - V_{T_{n_6}})^2 - V_{REF}^4\end{aligned} \tag{5.74}$$

The application of the error propagation method to (5.69) yields

$$\triangle(I_{D_1} - I_{D_2}) = \triangle(I_{D_4} - I_{D_3}) \tag{5.75}$$

where

$$\begin{aligned}\triangle(I_{D_1} - I_{D_2}) = {} & 2\frac{\triangle\mu_{n_5}}{\mu_n}\frac{(W_5/L_5)}{W_1/L_1}V_i^2(V_{GS_5} - V_{T_n})^2 \\ & - 4\frac{(W_5/L_5)}{W_1/L_1}V_i^2(V_{GS_5} - V_{T_n})\triangle V_{T_{n_5}} \\ & - 4\left[\frac{(W_5/L_5)}{W_1/L_1}V_i(V_{GS_5} - V_{T_n})^2 - V_i^3\right]\triangle V_i\end{aligned} \tag{5.76}$$

and

$$\begin{aligned}\triangle(I_{D_4} - I_{D_3}) = {} & 2\frac{\triangle\mu_{n_6}}{\mu_n}\frac{(W_6/L_6)}{W_1/L_1}V_{REF}^2(V_{GS_6} - V_{T_n})^2 \\ & - 4\frac{(W_6/L_6)}{W_1/L_1}V_{REF}^2(V_{GS_6} - V_{T_n})\triangle V_{T_{n_6}}\end{aligned} \tag{5.77}$$

The input-referred offset voltage, $V_{off,5-6}$, contributed by mismatches of T_5 and T_6 is given by

$$\begin{aligned}V_{off,5-6} = \triangle V_i = G\Bigg[& \frac{W_5/L_5}{W_1/L_1}\triangle V_{T_{n_5}} - \frac{W_6/L_6}{W_1/L_1}\triangle V_{T_{n_6}} \\ & + \frac{V_{GS_5} - V_{T_n}}{2}\left(\frac{W_6/L_6}{W_1/L_1}\frac{\triangle\mu_{n_6}}{\mu_n} - \frac{W_5/L_5}{W_1/L_1}\frac{\triangle\mu_{n_5}}{\mu_n}\right)\Bigg]\end{aligned} \tag{5.78}$$

where

$$G = \left(\frac{V_{GS_5} - V_{T_n}}{V_{REF}} - \frac{(W_1/L_1)}{(W_5/L_5)}\frac{V_{REF}}{V_{GS_5} - V_{T_n}}\right)^{-1} \tag{5.79}$$

Its variance can be written as,

$$\sigma^2_{V_{off,5-6}} = G^2\left\{\left(\frac{W_5/L_5}{W_1/L_1}\right)^2 \sigma^2_{V_{T_{n_5}}} + \left(\frac{W_6/L_6}{W_1/L_1}\right)^2 \sigma^2_{V_{T_{n_6}}} + \left(\frac{V_{GS_5} - V_{T_n}}{2\mu_n}\right)^2 \left[\left(\frac{W_5/L_5}{W_1/L_1}\right)^2 \sigma^2_{\mu_{n_5}} + \left(\frac{W_6/L_6}{W_1/L_1}\right)^2 \sigma^2_{\mu_{n_6}}\right]\right\} \tag{5.80}$$

Mismatches of the transistors composing the cross-coupled inverters primarily contribute to dynamic offset voltage. The remaining transistors operate as switches and do not really affect the offset voltage.

In the worst case, the variance of the overall static offset voltage is then given by

$$\sigma^2_{V_{soff}} = \sigma^2_{V_{off,1-4}} + \sigma^2_{V_{off,5-6}} \tag{5.81}$$

The standard deviation of the threshold voltage and charge mobility between two identically drawn transistors can be predicted using the next formula [20],

$$\sigma^2_{V_{T_{n_k}}} = \frac{A^2_{V_{T_{n_k}}}}{W \cdot L} + S^2_{V_{T_{n_k}}} D^2 \tag{5.82}$$

$$\sigma^2_{\mu_{n_k}} = \frac{A^2_{\mu_{n_k}}}{W \cdot L} + S^2_{\mu_{n_k}} D^2 \tag{5.83}$$

where W and L are the width and length of the transistor pair, D is the distance on chip between the matching transistor, $A_{V_{T_{n_k}}}$ and $A_{\mu_{n_k}}$ are the area proportionality constants for the threshold voltage and charge mobility, respectively, and $S_{V_{T_{n_k}}}$ and $S_{\mu_{n_k}}$ $(k = 1, 2, 3, 4, 5, 6)$ represent the variations of the threshold voltage and charge mobility with the spacing, respectively.

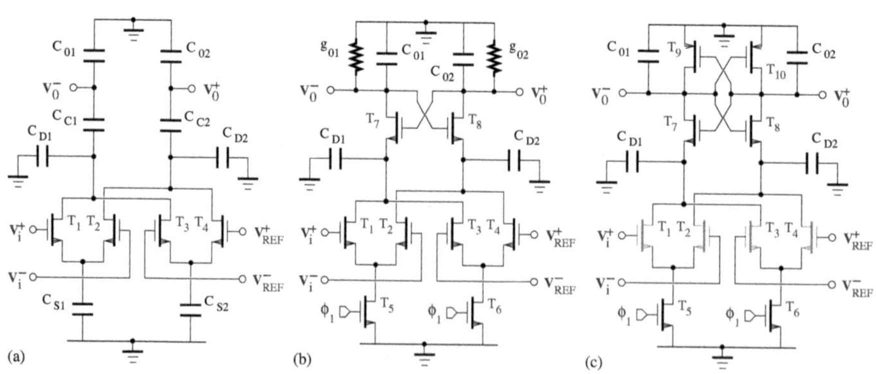

FIGURE 5.16
Dynamic comparator equivalent circuits: (a) phase 1, (b) phase 2, (c) phase 3.

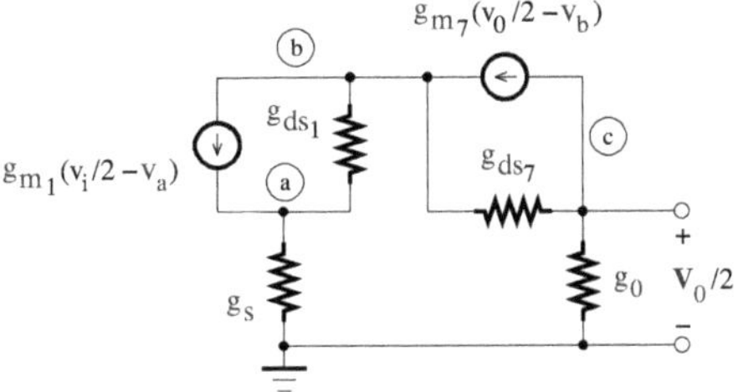

FIGURE 5.17
Dynamic comparator equivalent circuit for the *dc* gain determination during phase 2.

The dynamic comparator of Figure 5.13 goes through a set of distinct operating phases in each cycle, namely: reset, sampling, and regeneration or decision phases (or simply phase 1, phase 2, and phase 3). The comparator equivalent circuits for each phase are shown in Figure 5.16, where C_{S_k}, C_{D_k}, and C_{C_k} represent the equivalent parasitic capacitances between the drain and source of the transistors in the cut-off state, and C_{0_k} $(k = 1, 2)$ is the sum of the load capacitances and transistor parasitic capacitances.

Initially, the comparator is in the reset phase (or phase 1) that is defined as the time interval during which the reset switches are turned on and the output nodes are pulled up to the supply voltage, V_{DD}. With the clock signal ϕ_1 taking the low level, no *dc* current can flow through the input transistors because the bias transistor is turned off.

In phase 2, the input transistors start discharging the capacitors C_{P_1} and C_{P_2} depending on the input voltage difference. The voltage at the output nodes then begins to decrease due to the start of the conduction of transistors T_7 and T_8. The input voltages are sampled and stored onto the capacitors associated with the internal and output nodes until the output voltages drop below $V_{DD} - |V_{T_p}|$, and the operation of the pMOS transistors of the cross-coupled inverters goes from the saturation region to the triode region.

The equivalent circuit model for the determination of the *dc* voltage gain is shown in Figure 5.17. Here, the capacitors are considered open. Applying Kirchhoff's current law at each of the nodes yields

$$g_{m_1}(v_i/2 - v_a) + g_{ds_1}(v_b - v_a) = g_s v_a \tag{5.84}$$

$$g_{m_1}(v_i/2 - v_a) + g_{ds_1}(v_b - v_a) + g_{ds_7}(v_b - v_0/2) = g_{m_7}(v_0/2 - v_b) \tag{5.85}$$

$$g_{m_7}(v_0/2 - v_b) + g_0 v_0/2 = g_{ds_7}(v_b - v_0/2) \tag{5.86}$$

This set of three equations can be solved for the voltage gain, A_0, given by

$$A_0 = \frac{v_0}{v_i} = \frac{-(1-\alpha)g_{m_1}(g_{m_7} + g_{ds_7})}{(1-\alpha)g_{ds_1}(g_{m_7} + g_{ds_7} + g_0) + (g_{m_7} + g_{ds_7})g_0} \tag{5.87}$$

where $\alpha = (g_{m_1} + g_{ds_1})/(g_{m_1} + g_{ds_1} + g_s)$.

In phase 3, the output voltage is sufficiently reduced to now allow the operation of the pMOS transistors of the cross-coupled inverters in the saturation region. With the assumption that the input transistors are in the triode region with very large conductances compared to that of other transistors, the internal nodes can be considered almost short-circuited to ground. The resulting positive feedback then helps achieve the regeneration of the voltage difference stored on the output nodes.

The *dc* analysis of the equivalent cross-coupled inverters for phase 3 helps obtain the switching threshold voltage, V_{S_k}, $(k = 1, 2)$, that is defined as the voltage at which the drain currents of the pMOS and nMOS transistors are equal. At the transition point, $V_0^- = V_{S_1}$ and $V_0^+ = V_{S_2}$. By equating the currents through the transistors, assumed to operate in the saturation region, we have

$$I_{D_7} = I_{D_9} \tag{5.88}$$

$$\frac{1}{2}\mu_{n_7}\left(\frac{W_7}{L_7}\right)(V_{S_1} - V_{T_{n_7}})^2 = \frac{1}{2}\mu_{p_9}\left(\frac{W_9}{L_9}\right)(V_{DD} - V_{S_1} - |V_{T_{p_9}}|)^2 \tag{5.89}$$

and

$$I_{D_8} = I_{D_{10}} \tag{5.90}$$

$$\frac{1}{2}\mu_{n_8}\left(\frac{W_8}{L_8}\right)(V_{S_2} - V_{T_{n_8}})^2 = \frac{1}{2}\mu_{p_{10}}\left(\frac{W_{10}}{L_{10}}\right)(V_{DD} - V_{S_2} - |V_{T_{p_{10}}}|)^2 \tag{5.91}$$

Solving for V_{S_1} and V_{S_2} yields

$$V_{S_1} = \frac{V_{DD} - |V_{T_{p_9}}| + V_{T_{n_7}}\sqrt{\dfrac{\mu_{n_7}(W_7/L_7)}{\mu_{p_9}(W_9/L_9)}}}{1 + \sqrt{\dfrac{\mu_{n_7}(W_7/L_7)}{\mu_{p_9}(W_9/L_9)}}} \tag{5.92}$$

and

$$V_{S_2} = \frac{V_{DD} - |V_{T_{p_{10}}}| + V_{T_{n_8}}\sqrt{\dfrac{\mu_{n_8}(W_8/L_8)}{\mu_{p_{10}}(W_{10}/L_{10})}}}{1 + \sqrt{\dfrac{\mu_{n_8}(W_8/L_8)}{\mu_{p_{10}}(W_{10}/L_{10})}}} \tag{5.93}$$

The voltages V_{S_1} and V_{S_2} then depend on the supply voltage, the transistor threshold voltage, aspect ratio, and charge mobility. In practice, it is generally desirable for V_{S_1} and V_{S_2} to be set around the middle of the available voltage swing (or at $V_{DD}/2$ when a single supply power is used).

Applying Kirchhoff's current law at the output nodes without taking into account the output resistances, the cross-coupled inverters can be modeled by the following differential equations:

$$C_{01}\frac{dv_0^-(t)}{dt} = (g_{m_{p2}} - g_{m_{n2}})[v_0^+(t) - V_{S_2}] \tag{5.94}$$

$$C_{02}\frac{dv_0^+(t)}{dt} = (g_{m_{p1}} - g_{m_{n1}})[v_0^-(t) - V_{S_1}] \tag{5.95}$$

where g_{m_p} and g_{m_n} are the transconductances of the pMOS and nMOS transistors, respectively, V_{S_1} and V_{S_1} are the switching threshold voltages, and C_{01} and C_{02} denote the total output capacitances that are equal to the sum of the load capacitance and transistor parasitic capacitances.

The comparator decision is made once either of the voltages v_0^- and v_0^+ reaches the switching threshold voltage. Assuming that this occurs for v_0^- after a time interval, t_s, the voltage V_{S_1} can be expressed through a first-order Taylor series expansion as,

$$V_{S_1} = v_0^-(t_s) = v_0^-(0) + (v_0^+(0) - V_{S_2})\frac{g_{m_{p1}} - g_{m_{n1}}}{C_{01}}t_s \tag{5.96}$$

and

$$t_s = \frac{V_{S_1} - v_0^-(0)}{v_0^+(0) - V_{S_2}}\frac{C_{01}}{g_{m_{p1}} - g_{m_{n1}}} \tag{5.97}$$

where $v_0^-(0)$ and $v_0^+(0)$ denote the initial output voltages. Depending on the input voltage difference, a correct sensing is achieved in the case that the voltage v_0^- reaches V_{S_1} first. That is,

$$V_{S_2} < v_0^+(t_s) = v_0^+(0) + [v_0^-(0) - V_{S_1}]\frac{g_{m_{p2}} - g_{m_{n2}}}{C_{02}}t_s \tag{5.98}$$

Combining (5.98) and (5.97), we obtain

$$V_{S_2} < v_0^+(0) + [v_0^-(0) - V_{S_1}]\frac{(g_{m_{p2}} - g_{m_{n2}})[V_{S_1} - v_0^-(0)]C_{01}}{(g_{m_{p1}} - g_{m_{n1}})[v_0^+(0) - V_{S_2}]C_{02}} \tag{5.99}$$

or equivalently,

$$[V_{S_2} - v_0^+(0)]^2 > [V_{S_1} - v_0^-(0)]^2\frac{(g_{m_{p2}} - g_{m_{n2}})C_{01}}{(g_{m_{p1}} - g_{m_{n1}})C_{02}} \tag{5.100}$$

and

$$v_0^+(0) - V_{S_2} > [v_0^-(0) - V_{S_1}]\sqrt{\frac{(g_{m_{p2}} - g_{m_{n2}})C_{01}}{(g_{m_{p1}} - g_{m_{n1}})C_{02}}} \tag{5.101}$$

For small variations of transconductances and capacitances, a linear approximation of the right-handed term can be obtained using a first-degree Taylor polynomial. Hence,

$$v_0^+(0) - V_{S_2} > [v_0^-(0) - V_{S_1}]\left(1 + \frac{1}{2}\frac{G_{m_2}C_{01} - G_{m_1}C_{02}}{G_{m_1}C_{02}}\right) \tag{5.102}$$

and

$$v_0^+(0) - v_0^-(0) > \triangle V_0 = \triangle V_S + \frac{1}{2}[v_0^-(0) - V_{S_1}]\frac{G_{m_2}C_{01} - G_{m_1}C_{02}}{G_{m_1}C_{02}} \tag{5.103}$$

where $G_{m_1} = g_{m_{p1}} - g_{m_{n1}}$, $G_{m_2} = g_{m_{p2}} - g_{m_{n2}}$, and $\triangle V_S = V_{S_2} - V_{S_1}$. In practice, the main contributor to the difference between V_{S_1} and V_{S_1} comes from the mismatch in the transistor threshold voltages. When the comparator speed requirement can be easily met, the effect of the dynamic offset voltage can be attenuated by adding capacitors with very good matching properties at the output nodes to increase the output capacitive loads.

The input-referred dynamic offset voltage can be obtained as

$$V_{doff} = \triangle V_0/|A_0| \tag{5.104}$$

and its variance is given by

$$\sigma^2_{V_{doff}} = \triangle V_0^2/|A_0|^2 \tag{5.105}$$

For an ideal comparator, V_{doff} is equal to zero. Note that the dynamic offset voltage is only present during the regeneration phase. In practice, a dynamic comparator with mid-level sensing provides not only a speed advantage but can also help reduce the effect of capacitive load mismatches. Typical offset voltage values range from a few mV up to several tens of mV.

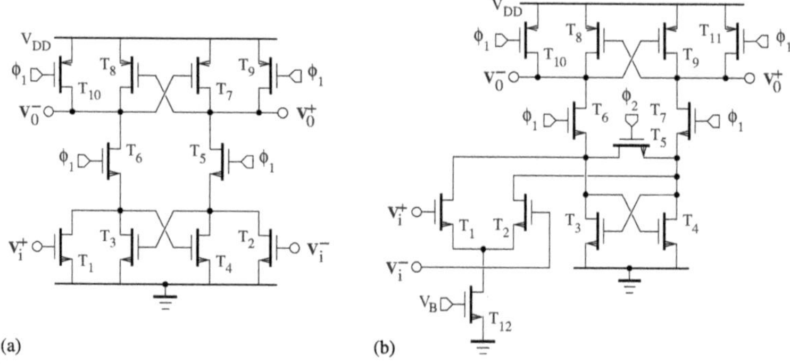

FIGURE 5.18
Circuit diagram of a dynamic comparator with two cross-coupled transistor pairs and an input transistor pair (a) without and (b) with a bias current.

• The dynamic comparator with two cross-coupled transistor pairs, as shown in Figure 5.18(a) [9, 10], can detect the difference between the input signals within a few millivolts and regeneratively switch to the appropriate output levels. The outputs rise simultaneously to the same logic state related to the supply-voltage level when the clock signal goes low. The cross-coupled transistors $T_7 - T_8$ are used to speed the pull-up of the outputs on the rising

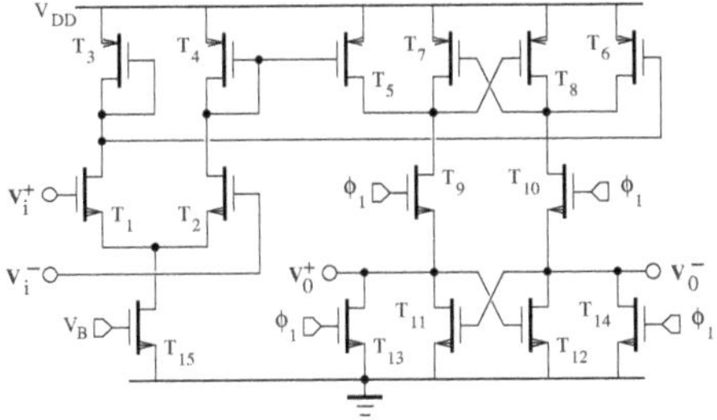

FIGURE 5.19
Circuit diagram of a dynamic comparator including a preamplifier.

edge of the clock signal. When one of the output nodes drops below $V_{DD} - |V_T|$, the transistor whose gate is connected to this node turns on, linking the other node to V_{DD}. The offset voltage can be minimized by designing the cross-coupled transistors with non-minimum gate lengths. The power dissipation is reduced because the static current flows only during the transition phase of the outputs. This is due to the fact that the nMOS and pMOS transistors are turned off at the end of the pull-up and pull-down operations, respectively.

In the dynamic comparator implementation of Figure 5.18(b), the differential transistor pair is biased by a current to ensure a high transconductance and, consequently, a high gain required to detect a small voltage difference between the inputs. When the transistor switch T_5 is closed by the high logic level of the clock signal ϕ_2, the charges previously stored on parasitic capacitors at the input of the regenerative stage are equalized and made independent of the input voltages because the low logic level of the clock signal ϕ_1 further turns off the pass transistors, T_6 and T_7, and turns on the pre-charge transistor switches, T_{10} and T_{11}.

A modified version of the above comparator is depicted in Figure 5.19. It includes an input preamplifier, which can isolate the input signal from the latch kickback noise. The output current provided by the differential pair $T_1 - T_2$ is mirrored from T_3 and T_4 to T_5 and T_6, respectively, and fed to the output latch.

5.2 Multipliers

Analog multipliers generally produce an output signal that is proportional in magnitude to the product of the two input signals. They can be designed to perform either a four-quadrant multiplication, when both input signals can be bipolar (i.e., there is no restriction on the sign of the input signals), or a two-quadrant multiplication in the case where one of the input signals is unipolar and the other can be bipolar.

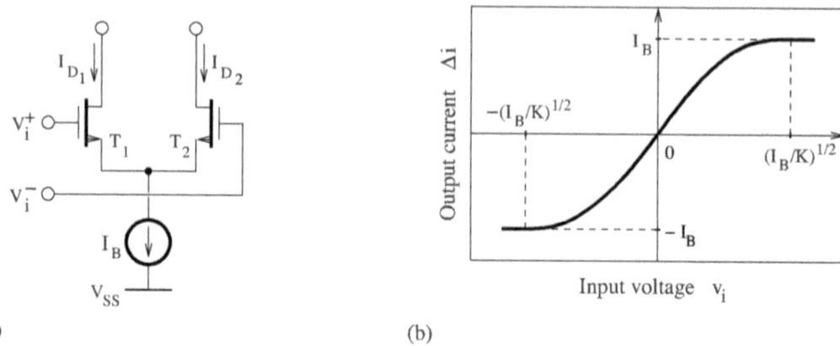

FIGURE 5.20
(a) Differential source-coupled transistor pair; (b) transfer characteristic.

Various techniques can be used for the implementation of multipliers. Let the drain current, i_D, of an nMOS transistor be expressed as,

$$I_D = \begin{cases} 0 & \text{if} \quad V_{GS} \leq V_T \\ K[2(V_{GS} - V_T)V_{DS} - V_{DS}^2] & \text{if} \quad V_{GS} > V_T;\ 0 < V_{DS} \leq V_{DS(sat)} \\ K(V_{GS} - V_T)^2 & \text{if} \quad V_{GS} > V_T;\ V_{DS} \geq V_{DS(sat)} \end{cases} \tag{5.106}$$

in the cutoff, triode, and saturation regions, respectively, where the drain-source saturation voltage is of the form, $V_{DS(sat)} = V_{GS} - V_T$, V_{GS} is the gate-source voltage, V_T is the threshold voltage, $K = \mu_n(C_{ox}/2)(W/L)$ is related to the transconductance parameter, μ_n is the effective surface carrier mobility, C_{ox} is the gate oxide capacitance per unit area, and W and L are the channel width and length, respectively. Figure 5.20(a) shows a differential source-coupled transistor pair, which can be used in the multiplier design. Assuming that the transistors are matched and operate in the saturation region, we have

$$V_{GS_1} = V_T + \sqrt{\frac{I_{D_1}}{K}} \tag{5.107}$$

and

$$V_{GS_2} = V_T + \sqrt{\frac{I_{D_2}}{K}} \tag{5.108}$$

Kirchhoff's voltage law equation for the input loop can be written as

$$V_i^+ - V_{GS_1} + V_{GS_2} - V_i^- = 0 \tag{5.109}$$

Substituting Equations (5.107) and (5.108) into Equation (5.109), we can obtain

$$\sqrt{I_{D_1}} - \sqrt{I_{D_2}} = \sqrt{K}V_i \tag{5.110}$$

where $V_i = V_i^+ - V_i^-$. Using Kirchhoff's current law at the source node, we find

$$I_{D_1} + I_{D_2} = I_B \tag{5.111}$$

where I_B is the bias current. The system involving Equations (5.110) and (5.111) can be solved for I_{D_1}. That is,

$$I_{D_1} = \frac{I_B}{2} \pm K\frac{v_i}{2}\sqrt{\frac{2I_B}{K} - V_i^2} \tag{5.112}$$

where $|V_i| \leq \sqrt{I_B/K}$. For positive values of V_i, the current I_{D_1} should be greater than $I_B/2$. Hence,

$$I_{D_1} = \frac{I_B}{2} + K\frac{V_i}{2}\sqrt{\frac{2I_B}{K} - V_i^2} \tag{5.113}$$

The substitution of Equation (5.113) into (5.111) yields

$$I_{D_2} = \frac{I_B}{2} - K\frac{V_i}{2}\sqrt{\frac{2I_B}{K} - V_i^2} \tag{5.114}$$

The difference between I_{D_1} and I_{D_2} is then given by

$$\triangle i = I_{D_1} - I_{D_2} = KV_i\sqrt{\frac{2I_B}{K} - V_i^2} \tag{5.115}$$

A plot of the transfer characteristic is shown in Figure 5.20(b). As illustrated by the characteristic curvature, the linear range is limited because it can only be extended by increasing the magnitude of the bias current, or equivalently, the power consumption.

In general, the multiplication of two voltages can be achieved using the transistor characteristic in the saturation and triode regions. Various circuit techniques have been proposed to implement multipliers in CMOS technology, which is known to feature a high integration density. For instance, they can exploit the variation in the transconductance of differential transistor stages or be based on the subtraction of the sum-squared and difference-squared of two input signals. The most critical design specification is the linear dynamic range, which is limited in some multiplier structures even for high supply voltages.

5.2.1 Multiplier cores

Analog multiplication can be performed by a multiplier core loaded by an output stage. Multiplier cores are designed by exploiting various mathematical and circuit principles.

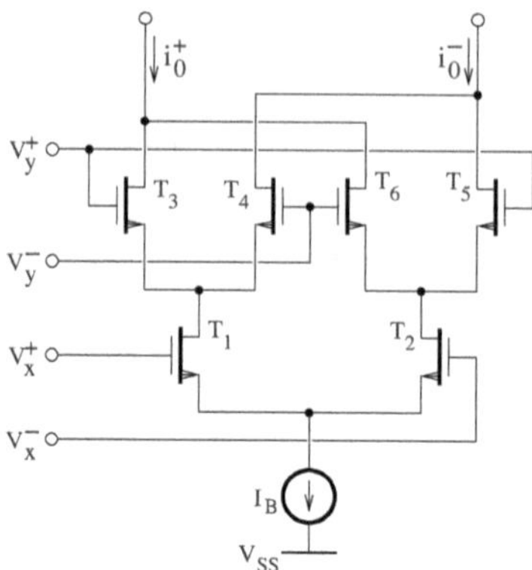

FIGURE 5.21
Multiplier core based on cross-coupled transconductance stages.

5.2.1.1 Multiplier core based on externally controlled transconductances

Various analog multiplier architectures based on externally controlled transconductances are available. In general, they consist of differential voltage-to-current converters and provide an output proportional to the product of two input signals.

- The multiplier core shown in Figure 5.21 [21,22] is based on cross-coupled transconductance stages. Assuming that all transistors operate in the saturation region, the difference between the output currents is given by

$$\begin{aligned} \triangle i_0 &= i_0^+ - i_0^- \qquad (5.116) \\ &= (I_{D_3} + I_{D_6}) - (I_{D_4} + I_{D_5}) \\ &= (I_{D_3} - I_{D_4}) - (I_{D_5} - I_{D_6}) \qquad (5.117) \end{aligned}$$

where

$$I_{D_3} - I_{D_4} = KV_y\sqrt{\frac{2I_{D_1}}{K} - V_y^2} \qquad (5.118)$$

and

$$I_{D_5} - I_{D_6} = KV_y\sqrt{\frac{2I_{D_2}}{K} - V_y^2} \qquad (5.119)$$

To proceed further, it can be shown that

$$I_{D_1} = \frac{I_{SS}}{2} + K\frac{V_x}{2}\sqrt{\frac{2I_B}{K} - V_x^2} = \frac{K}{4}\left(\sqrt{\frac{2I_B}{K} - V_x^2} + V_x\right)^2 \qquad (5.120)$$

and

$$I_{D_2} = \frac{I_B}{2} - K\frac{V_x}{2}\sqrt{\frac{2I_B}{K} - V_x^2} = \frac{K}{4}\left(\sqrt{\frac{2I_B}{K} - V_x^2} - V_x\right)^2 \tag{5.121}$$

Substituting Equations (5.121), (5.120), (5.119), and (5.118) into Equation (5.117), we obtain

$$\begin{aligned}\triangle i_0 = KV_y & \left[\sqrt{\frac{1}{2}\left(\sqrt{\frac{2I_B}{K} - V_x^2} + V_x\right)^2 - V_y^2}\right. \\ & \left. - \sqrt{\frac{1}{2}\left(\sqrt{\frac{2I_B}{K} - V_x^2} - V_x\right)^2 - V_y^2}\right]\end{aligned} \tag{5.122}$$

Note that $\triangle i_0$ has a nonlinear relationship with V_x and V_y. Nevertheless, when V_x and V_y are small, and

$$|V_y| \leq \sqrt{\frac{I_B}{K} - \frac{1}{2}V_x^2} - \frac{1}{\sqrt{2}}V_x \tag{5.123}$$

it can be shown that

$$\triangle i_0 \simeq KV_y\left[\sqrt{\frac{1}{2}\left(\sqrt{\frac{2I_B}{K} - V_x^2} + V_x\right)^2} - \sqrt{\frac{1}{2}\left(\sqrt{\frac{2I_B}{K} - V_x^2} - V_x\right)^2}\right] \tag{5.124}$$

and finally,

$$\triangle i_0 \simeq \sqrt{2}KV_xV_y \tag{5.125}$$

Because the multiplier core is composed of differential transistor pairs, it will operate in the linear region only if we have $|V_x| \leq (I_B/K)^{1/2}$ and $|V_y| \ll (2\min(I_{D_1}, I_{D_2})/K)^{1/2}$, where I_{D_1} and I_{D_2} are the drain currents of the transistors T_1 and T_2, respectively.

The circuit diagram of a high-linearity multiplier core is depicted in Figure 5.22 [23]. It is composed of nMOS transistors, T_1, T_2, T_3, T_4, T_5, and T_6, and pMOS transistors, T_7 and T_8, which can be partitioned into two sections. A constant current sink, I_{B1}, is connected to the sources of T_1, T_2, and T_5, and the drain of T_7, while the drain of T_5, which is connected to the gate of T_7, is driven by a constant current source, I_{B2}. Similarly, a constant current sink, I_{B1}, is connected to the sources of T_3, T_4, and T_6 and the drain of T_8, while the drain of T_6, which is connected to the gate of T_8, is driven by a constant current source, I_{B2}.

All transistors operate in the saturation region. Let V_I be the dc component

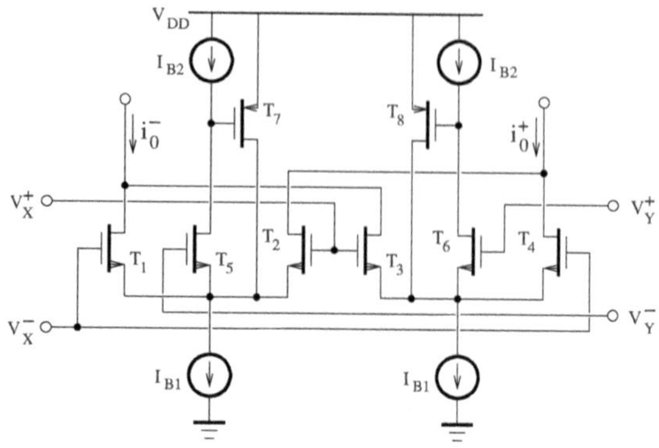

FIGURE 5.22
High-linearity multiplier core.

associated with the input voltages. Using $V_x^+ = V_I + V_x/2$, $V_x^- = V_I - V_x/2$, $V_y^+ = V_I + V_y/2$, and $V_y^- = V_I - V_y/2$, we can obtain

$$V_{GS_5} = V_{G_5} - V_{S_5} = V_I - V_y/2 - V_{S_5} \tag{5.126}$$

$$V_{GS_6} = V_{G_6} - V_{S_6} = V_I + V_y/2 - V_{S_6} \tag{5.127}$$

and

$$I_{D_5} = K(V_{GS_5} - V_T)^2 = K(V_I - V_y/2 - V_{S_5} - V_T)^2 = I_{B2} \tag{5.128}$$

$$I_{D_6} = K(V_{GS_6} - V_T)^2 = K(V_I + V_y/2 - V_{S_6} - V_T)^2 = I_{B2} \tag{5.129}$$

Hence,

$$V_{S_5} = V_I - \frac{V_y}{2} - V_T - \sqrt{\frac{I_{B2}}{K}} \tag{5.130}$$

$$V_{S_6} = V_I + \frac{V_y}{2} - V_T - \sqrt{\frac{I_{B2}}{K}} \tag{5.131}$$

The gate-source voltages and the drain currents of transistors $T_1 - T_4$ can also be expressed as

$$V_{GS_1} = V_{G_1} - V_{S_1} = V_I - V_x/2 - V_{S_5} \tag{5.132}$$

$$V_{GS_2} = V_{G_2} - V_{S_2} = V_I + V_x/2 - V_{S_5} \tag{5.133}$$

$$V_{GS_3} = V_{G_3} - V_{S_3} = V_I + V_x/2 - V_{S_6} \tag{5.134}$$

$$V_{GS_4} = V_{G_4} - V_{S_4} = V_I - V_x/2 - V_{S_6} \tag{5.135}$$

and

$$I_{D_1} = K(V_{GS_1} - V_T)^2 = K(V_I - V_x/2 - V_{S_5} - V_T)^2 \quad (5.136)$$
$$I_{D_2} = K(V_{GS_2} - V_T)^2 = K(V_I + V_x/2 - V_{S_5} - V_T)^2 \quad (5.137)$$
$$I_{D_3} = K(V_{GS_3} - V_T)^2 = K(V_I + V_x/2 - V_{S_6} - V_T)^2 \quad (5.138)$$
$$I_{D_4} = K(V_{GS_4} - V_T)^2 = K(V_I - V_x/2 - V_{S_6} - V_T)^2 \quad (5.139)$$

Note that $I_{B1} = I_{D_1} + I_{D_2} + I_{D_5} + I_{D_7}$ and $I_{B1} = I_{D_3} + I_{D_4} + I_{D_6} + I_{D_8}$. Combining Equations (5.130), (5.131), and (5.136) through (5.139) and simplifying, we find

$$I_{D_1} = K\left(-\frac{V_x}{2} + \frac{V_y}{2} + \sqrt{\frac{I_{B2}}{K}}\right)^2 \quad (5.140)$$

$$I_{D_2} = K\left(\frac{V_x}{2} + \frac{V_y}{2} + \sqrt{\frac{I_{B2}}{K}}\right)^2 \quad (5.141)$$

$$I_{D_3} = K\left(\frac{V_x}{2} - \frac{V_y}{2} + \sqrt{\frac{I_{B2}}{K}}\right)^2 \quad (5.142)$$

$$I_{D_4} = K\left(-\frac{V_x}{2} - \frac{V_y}{2} + \sqrt{\frac{I_{B2}}{K}}\right)^2 \quad (5.143)$$

The difference of the output currents is given by

$$\begin{aligned}
\triangle i_O &= i_0^+ - i_0^- && (5.144)\\
&= (I_{D_2} + I_{D_4}) - (I_{D_1} + I_{D_3})\\
&= -(I_{D_1} - I_{D_2}) - (I_{D_3} - I_{D_4}) && (5.145)\\
&= 2KV_xV_y && (5.146)
\end{aligned}$$

where

$$I_{D_1} - I_{D_2} = -2KV_x\left(\frac{V_y}{2} + \sqrt{\frac{I_{B2}}{K}}\right) \quad (5.147)$$

$$I_{D_3} - I_{D_4} = 2KV_x\left(-\frac{V_y}{2} + \sqrt{\frac{I_{B2}}{K}}\right) \quad (5.148)$$

The output dc characteristic, $\triangle i_O$, which is proportional to the product of the differential input voltages, is linear over a wide input range.

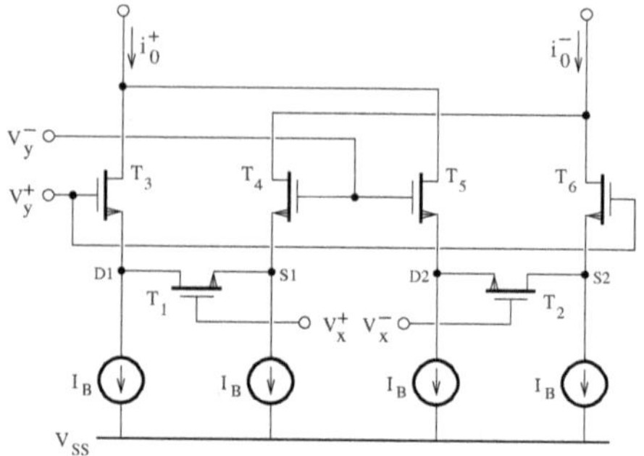

FIGURE 5.23
Multiplier core based on transconductance stages with source degeneration.

• The circuit diagram of a multiplier core based on transconductance stages with source degeneration is depicted in Figure 5.23 [24]. The currents I_1 and I_2 flowing respectively through the transistors T_1 and T_2, which are assumed to be identical and operate in the triode region, are given by

$$I_1 = K[2(V_x^+ - V_{S1} - V_T)V_y - V_y^2] \tag{5.149}$$
$$I_2 = K[2(V_x^- - V_{S2} - V_T)V_y - V_y^2] \tag{5.150}$$

where $V_{DS_1} = V_{DS_2} = V_y$. The transistors $T_3 - T_6$ have the same geometry and operate in the saturation region. The next relation can be written

$$V_{GS_4} = \sqrt{\frac{I_B}{K}} + V_T = V_y^- - V_{S1} \tag{5.151}$$
$$V_{GS_6} = \sqrt{\frac{I_B}{K}} + V_T = V_y^+ - V_{S2} \tag{5.152}$$

where I_B is the bias current of the transistor. The currents i_0^+ and i_0^- can be obtained as

$$i_0^+ = 2I_B - (I_1 - I_2) = 2I_B + 2KV_xV_y \tag{5.153}$$
$$i_0^- = 2I_B + (I_1 - I_2) = 2I_B - 2KV_xV_y \tag{5.154}$$

and the resulting differential current is

$$\Delta i_0 = 4KV_xV_y \tag{5.155}$$

The dynamic rage is limited by the fact that T_1 and T_2 should be in the triode region, that is, $V_{DS} \leq V_{DS(sat)} = V_{GS} - V_T$.

• The circuit diagram of a multiplier core based on transconductance stages with active cascode structure is depicted in Figure 5.24. Transistors $T_1 - T_4$

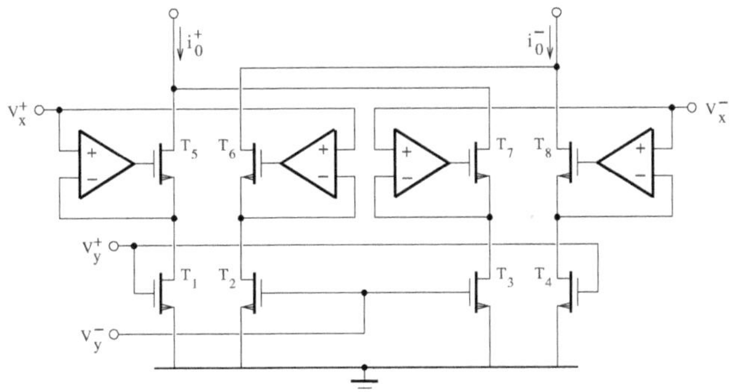

FIGURE 5.24
Multiplier core based on transconductance stages with active cascode structure.

operate in the triode region and the currents i_0^+ and i_0^- can be written as

$$i_0^+ = K[2(V_x^+ - V_T)V_{D1} - V_{D1}^2] + K[2(V_x^- - V_T)V_{D3} - V_{D3}^2] \tag{5.156}$$

$$i_0^- = K[2(V_x^+ - V_T)V_{D4} - V_{D4}^2] + K[2(V_x^- - V_T)V_{D2} - V_{D2}^2] \tag{5.157}$$

Using the gate voltages given by

$$V_{Gi} = A(V_x^+ - V_{D(i-4)}), \quad \text{if} \quad i = 5, 6 \tag{5.158}$$

$$V_{Gj} = A(V_x^- - V_{D(j-4)}), \quad \text{if} \quad j = 7, 8 \tag{5.159}$$

where A represents the gain of the amplifier, and the fact that $T_5 - T_8$ operate in the saturation region, we obtain

$$V_{Dk} = \frac{A}{1+A}V_x^+ - \frac{1}{1+A}\left(\sqrt{\frac{I_{Dk}}{K}} + V_T\right) \simeq V_x^+, \quad \text{if} \quad k = 1, 2 \tag{5.160}$$

$$V_{Dl} = \frac{A}{1+A}V_x^- - \frac{1}{1+A}\left(\sqrt{\frac{I_{Dl}}{K}} + V_T\right) \simeq V_x^-, \quad \text{if} \quad l = 3, 4 \tag{5.161}$$

The difference of the output currents can be expressed as

$$\triangle i_0 = i_0^+ - i_0^- = 2KV_xV_y \tag{5.162}$$

It should be noted that the effective transconductance of $T_5 - T_8$ is increased due to the presence of amplifiers. As a result, the linearity of the multiplier is improved.

5.2.1.2 Multiplier core based on the quarter-square technique

Multipliers based on stacking transconductance stages have poor linearity when they operate with low supply voltages. A way to circumvent this problem is to use the quarter-square technique.

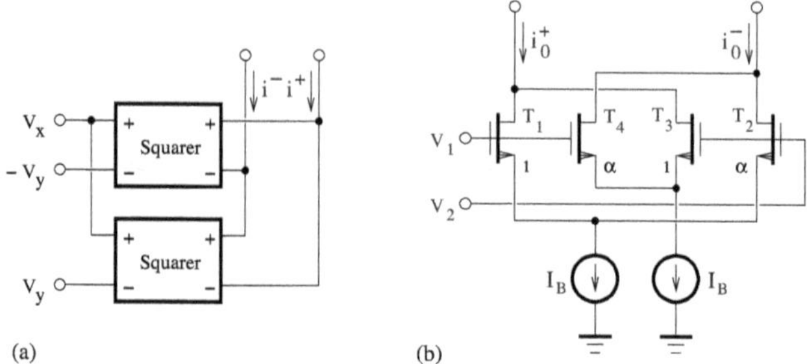

FIGURE 5.25
(a) Block diagram of a multiplier based on the quarter-square technique; (b) circuit diagram of a squarer.

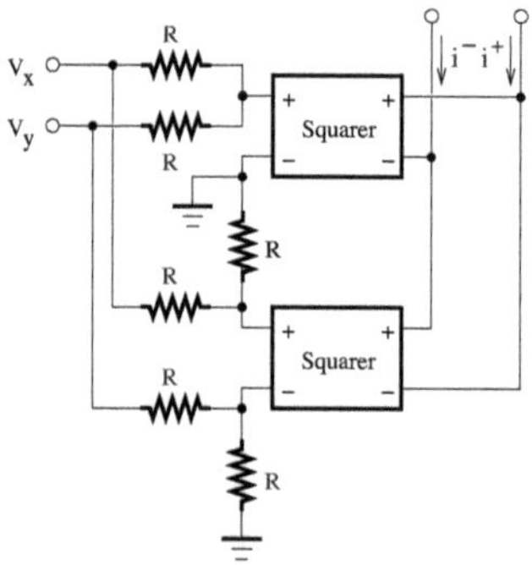

FIGURE 5.26
Quarter-square multiplier core with an input-resistive network.

• Using two squarer stages whose output nodes are cross-coupled, a multiplier based on the quarter-square technique can be implemented as shown in Figure 5.25(a). The multiplication of the input voltages, V_x and V_y, is achieved by taking the difference between the output currents due to $(V_x + V_y)^2$ and $(V_x - V_y)^2$. The circuit diagram of a squarer is depicted in Figure 5.25(b) [29,30]. It consists of two unbalanced source-coupled transistor pairs with the W/L ratio of α. The drain currents of transistors $T_1 - T_4$ are, respectively, given by

$$I_{D_1} = K'\frac{W}{L}(V_{GS_1} - V_T)^2 \tag{5.163}$$

$$I_{D_2} = K'\alpha\frac{W}{L}(V_{GS_2} - V_T)^2 \tag{5.164}$$

$$I_{D_3} = K'\frac{W}{L}(V_{GS_3} - V_T)^2 \tag{5.165}$$

and

$$I_{D_4} = K'\alpha \frac{W}{L}(V_{GS_4} - V_T)^2 \tag{5.166}$$

where $K' = \mu_n C_{ox}/2$. Using Kirchhoff's voltage law, we can obtain

$$V_{GS_1} - V_{GS_2} = V_{GS_4} - V_{GS_3} = V_1 - V_2 \tag{5.167}$$

Combining Equations (5.163) through (5.166) and (5.167), it can be shown that

$$\sqrt{I_{D_1}} - \sqrt{I_{D_2}/\alpha} = \sqrt{K}(V_1 - V_2) \tag{5.168}$$

and

$$\sqrt{I_{D_3}} - \sqrt{I_{D_4}/\alpha} = -\sqrt{K}(V_1 - V_2) \tag{5.169}$$

By applying Kirchhoff's current law, we find

$$I_{D_1} + I_{D_2} = I_{D_3} + I_{D_4} = I_B \tag{5.170}$$

Solving the system of Equations (5.168), (5.169), and (5.170) gives

$$I_{D_1}, I_{D_3} = \frac{1}{\left(1 + \frac{1}{\alpha}\right)^2} \left[\left(1 + \frac{1}{\alpha}\right) \frac{I_B}{\alpha} + \left(1 - \frac{1}{\alpha}\right) K(V_1 - V_2)^2 \right. \\ \left. \pm 2K(V_1 - V_2) \sqrt{\left(1 + \frac{1}{\alpha}\right) \frac{I_B}{K\alpha} - \frac{(V_1 - V_2)^2}{\alpha}} \right] \tag{5.171}$$

where the upper plus sign is for I_{D_1}, while the lower minus sign is for I_{D_3}, and

$$I_{D_2}, I_{D_4} = \frac{1}{\left(1 + \frac{1}{\alpha}\right)^2} \left[\left(1 + \frac{1}{\alpha}\right) I_B - \left(1 - \frac{1}{\alpha}\right) K(V_1 - V_2)^2 \right. \\ \left. \mp 2K(V_1 - V_2) \sqrt{\left(1 + \frac{1}{\alpha}\right) \frac{I_B}{K\alpha} - \frac{(V_1 - V_2)^2}{\alpha}} \right] \tag{5.172}$$

where the upper minus sign is for I_{D_2} while the lower plus sign is for I_{D_4}. The difference in the output currents can then be expressed as

$$\triangle I_0 = I_0^+ - I_0^- \tag{5.173}$$

$$= (I_{D_1} + I_{D_3}) - (I_{D_2} + I_{D_4}) \tag{5.174}$$

$$= (I_{D_1} - I_{D_2}) + (I_{D_3} - I_{D_4}) \tag{5.175}$$

$$= \frac{-2\left(1+\dfrac{1}{\alpha}\right)\left[\left(1-\dfrac{1}{\alpha}\right)I_B - 2K(V_1 - V_2)^2\right]}{\left(1+\dfrac{1}{\alpha}\right)^2} \tag{5.176}$$

For the normal operation of the squarer circuit, the input voltage, $V_1 - V_2$, should be in the range $|V_1 - V_2| \leq \sqrt{I_B/(K\alpha)}$.

The aforementioned quarter square can also be implemented as shown in Figure 5.26. Here, the inversion of one of the inputs is not required. The input resistive network operates as a voltage divider with a factor of 2 to produce the voltages $(V_x + V_y)/2$ and $(V_x - V_y)/2$, which are applied at the input nodes of the multiplier core.

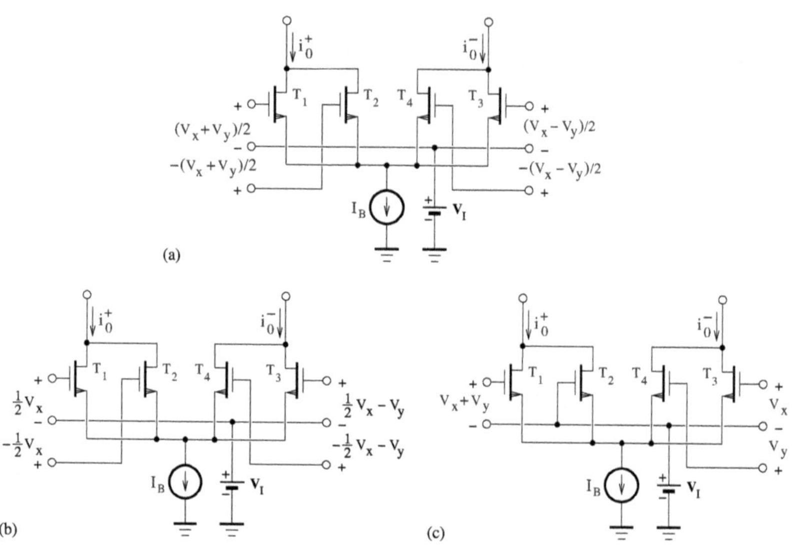

FIGURE 5.27
Multiplier core based on the quarter-square technique.

- Other structures of the multiplier core based on the quarter-square technique are depicted in Figure 5.27(a) [32], (b) [33,34], and (c) [35]. They consist of four source-coupled transistors biased by a constant current source, but differ in the required combination of the input voltages. The voltage V_I represents the *dc* component of the input voltages.

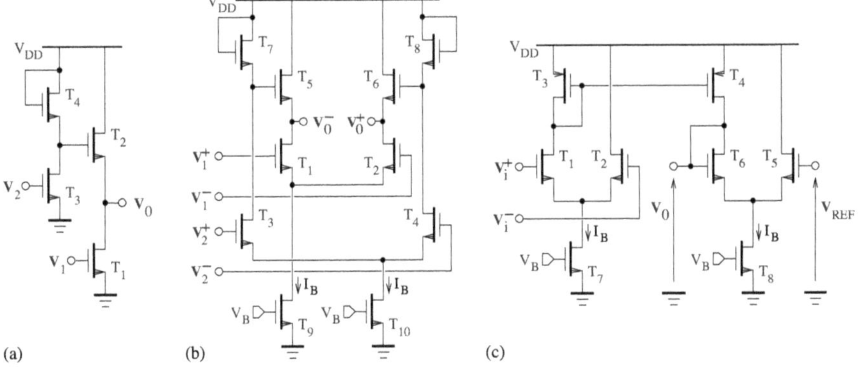

FIGURE 5.28
Summer circuit diagrams.

Let us consider the structure of Figure 5.27(a). Assuming the current-voltage characteristic of MOS transistors operating in the saturation region, the drain currents are given by

$$I_{D_1} = K\left(V_I + \frac{V_x + V_y}{2} - V_S - V_T\right)^2 \tag{5.177}$$

$$I_{D_2} = K\left(V_I - \frac{V_x + V_y}{2} - V_S - V_T\right)^2 \tag{5.178}$$

$$I_{D_3} = K\left(V_I + \frac{V_x - V_y}{2} - V_S - V_T\right)^2 \tag{5.179}$$

and

$$I_{D_4} = K\left(V_I - \frac{V_x - V_y}{2} - V_S - V_T\right)^2 \tag{5.180}$$

The difference in the output currents can be written as

$$\triangle i_0 = i_0^+ - i_0^- = (I_{D_1} + I_{D_2}) - (I_{D_3} + I_{D_4}) = 2KV_xV_y \tag{5.181}$$

Applying Kirchhoff's current law at the source node, we have

$$I_B = (I_{D_1} + I_{D_2}) + (I_{D_3} + I_{D_4}) = i_0^+ + i_0^- \tag{5.182}$$

Combining Equations (5.181) and (5.182), it can be shown that

$$i_0^+ = I_{D_1} + I_{D_2} = \frac{I_B}{2} + KV_xV_y \tag{5.183}$$

and

$$i_0^- = I_{D_3} + I_{D_4} = \frac{I_B}{2} - KV_xV_y \tag{5.184}$$

The range of input voltages over which the transistors still operate in the saturation region can be determined by the next worst-case requirement,

$$V_{GS_2} = V_I - \frac{V_x + V_y}{2} - V_S \geq V_T \tag{5.185}$$

Substituting Equations (5.177) and (5.178) into Equation (5.183), we obtain

$$V_I - V_S - V_T = \sqrt{\frac{I_B}{4K} - \frac{V_x^2 + V_y^2}{4}} \tag{5.186}$$

Combining Equations (5.186) and (5.185) gives

$$V_x^2 + V_yV_x + V_y^2 - \frac{I_B}{2K} \leq 0 \tag{5.187}$$

or equivalently,

$$|V_x| \leq -\frac{V_y}{2} + \sqrt{\frac{I_B}{2K} - \frac{3V_y^2}{4}}, \quad |V_y| \leq \sqrt{\frac{2I_B}{3K}} \tag{5.188}$$

A similar analysis of the multiplier cores shown in Figures 5.27(b) and (c) also results in the output current characteristic and dynamic range given by Equations (5.181) and (5.187), respectively.

Depending on the implementation of the input stage, the resulting multiplier can have the advantage of exhibiting the same transfer characteristic with respect to any of the differential input voltages.

A single-ended summer structure is depicted in Figure 5.28(a) [25]. It is based on two nMOS inverting stages. Using Kirchhoff's voltage law for the loop including the transistors T_2 and T_2, the output voltage can be expressed as

$$V_0 = V_{DD} - V_{GS_2} - V_{GS_4} \tag{5.189}$$

The transistors are assumed to have the same threshold voltage, V_T, and to operate in the saturation region. Because $I_{D_1} = I_{D_2}$ and $I_{D_3} = I_{D_4}$, we obtain

$$V_{GS_2} = \sqrt{\frac{(W_1/L_1)}{(W_2/L_2)}}(V_{GS_1} - V_T) + V_T \tag{5.190}$$

and

$$V_{GS_4} = \sqrt{\frac{(W_3/L_3)}{(W_4/L_4)}}(V_{GS_3} - V_T) + V_T \tag{5.191}$$

where $V_{GS_1} = V_1$ and $V_{GS_3} = V_2$. Assuming that $(W_1/L_1) = (W_3/L_3)$ and $(W_2/L_2) = (W_4/L_4)$ and combining Equations (5.189), (5.190), and (5.191), we find

$$V_0 = V_0' - \sqrt{\frac{(W_1/L_1)}{(W_2/L_2)}}(V_1 + V_2) \tag{5.192}$$

where

$$V_0' = V_{DD} + 2V_T\left(\sqrt{\frac{(W_1/L_1)}{(W_2/L_2)}} - 1\right) \tag{5.193}$$

The dc component, V_0', of the output voltage can be cancelled by adopting the differential summer structure of Figure 5.28(b). Performing a similar analysis as previously, it can be shown that

$$V_0^+ = V_0' - \sqrt{\frac{(W_1/L_1)}{(W_2/L_2)}}(V_1^- + V_2^-) \tag{5.194}$$

and

$$V_0^- = V_0' - \sqrt{\frac{(W_1/L_1)}{(W_2/L_2)}}(V_1^+ + V_2^+) \tag{5.195}$$

The differential output voltage is then given by

$$V_0 = V_0^+ - V_0^- = \sqrt{\frac{(W_1/L_1)}{(W_2/L_2)}}(V_1 + V_2) \tag{5.196}$$

where $V_1 = V_1^+ - V_1^-$ and $V_2 = V_2^+ - V^-$.

An alternative summer circuit is shown in Figure 5.28(c) [26]. It is composed of two differential stages, which are coupled by a current mirror in such a way that $I_{D_1} = I_{D_6}$. Hence, the gate-source voltages of transistors T_1 and T_6 have the same value. Because both differential stages are assumed to be biased by a constant current, I_B, we can write

$$I_{D_1} + I_{D_2} = I_B = I_{D_5} + I_{D_6} \tag{5.197}$$

The drain currents of transistors T_2 and T_5 should also be identical, thereby forcing the gate-source voltages of transistors T_2 and T_5 to be matched. It can then be shown that

$$V_0 - V_{REF} = V_{GS_6} - V_{GS_5} = V_{GS_1} - V_{GS_2} = V_i \tag{5.198}$$

where $V_i = V_i^+ - V_i^-$.

5.2.1.3 Design issues

In practice, the accuracy of the output voltage can be affected by the input offset voltage due to mismatches between transistor characteristics, and can become degraded at high frequencies. Without resorting to special matching methods, it may be difficult to preserve the advantage of the quarter-square technique, such as the symmetry of the output characteristic.

For low-voltage applications, the aforementioned multiplier core can be designed without the bias current I_B. In this case, the transistor sources are directly connected to either the ground or a supply voltage terminal. A disadvantage of the resulting circuit is that the dc components of the input voltages are simultaneously the bias voltages. As a result, it may be impossible to independently control the common-mode levels at the inputs.

5.2.2 Design examples

The circuit diagram of a multiplier with single-ended output is shown in Figure 5.29. The transistors operate in the saturation region and the output

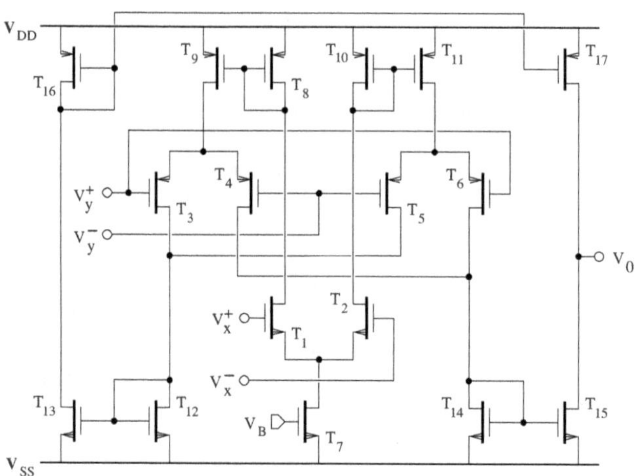

FIGURE 5.29
Four-quadrant multiplier circuit with a single-ended output.

current is given by

$$i_0 = \sqrt{2mK_nK_p}v_x v_y \tag{5.199}$$

where m is the scaling ratio of the current mirrors $T_8 - T_9$ and $T_{10} - T_{11}$, and K_n and K_p are the transconductance of the n-channel and p-channel transistors, respectively.

The structure of a multiplier with differential outputs is depicted in Figure 5.31. It is based on the multiplier core of Figure 5.21. The common-mode feedback is implemented by the transistors $T_{F1} - T_{F3}$ and resistors R_1 and

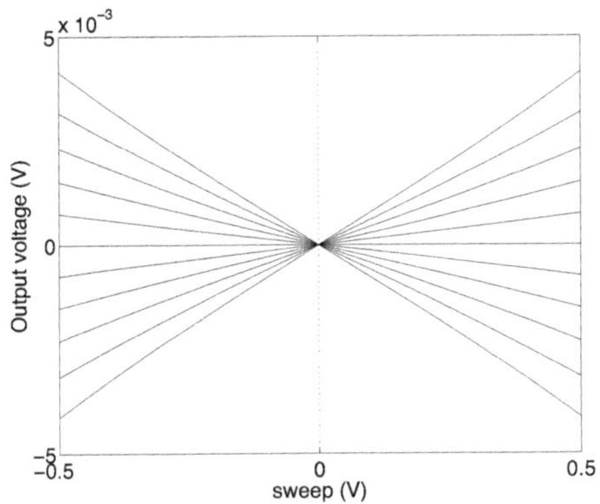

FIGURE 5.30
Plot of dc transfer characteristics of the four-quadrant multiplier.

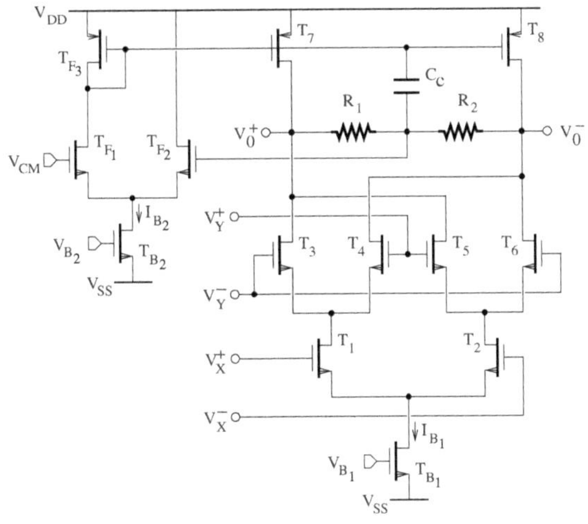

FIGURE 5.31
Four-quadrant multiplier circuit with differential outputs.

R_2. The reference level is defined by the voltage V_{CM} and the frequency compensation is provided by C_c.

The dc characteristics of the multiplier are computed based on 0.5-µm CMOS transistor parameters, by performing SPICE [31] simulations, and are shown in Figure 5.30.

Another multiplier implementation is shown in Figure 5.32 [32]. It is based on the quarter-square technique. A current I_B generated by a transistor current source is used to bias each set of transistors, $T_1 - T_4$ and $T_5 - T_8$, whose sources are connected together. Assuming that transistors $T_1 - T_{12}$ are iden-

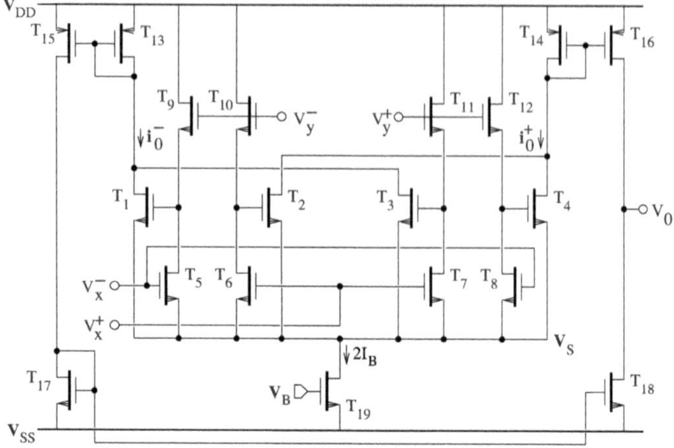

FIGURE 5.32
Four-quadrant multiplier based on the quarter-square technique.

tical and operate in the saturation region, we can write

$$V_{G_1} = V_{DD} - V_{DS_9} = V_{DD} - (V_{GS_9} - V_T) \tag{5.200}$$
$$V_{G_2} = V_{DD} - V_{DS_{10}} = V_{DD} - (V_{GS_{10}} - V_T) \tag{5.201}$$
$$V_{G_3} = V_{DD} - V_{DS_{11}} = V_{DD} - (V_{GS_{11}} - V_T) \tag{5.202}$$

and

$$V_{G_4} = V_{DD} - V_{DS_{12}} = V_{DD} - (V_{GS_{12}} - V_T) \tag{5.203}$$

where

$$V_{GS_9} = V_y^- - (V_{DS_5} + V_S) = V_y^- - (V_{GS_5} - V_T) - V_S \tag{5.204}$$
$$V_{GS_{10}} = V_y^- - (V_{DS_6} + V_S) = V_y^- - (V_{GS_6} - V_T) - V_S \tag{5.205}$$
$$V_{GS_{11}} = V_y^+ - (V_{DS_7} + V_S) = V_y^+ - (V_{GS_7} - V_T) - V_S \tag{5.206}$$

and

$$V_{GS_{12}} = V_y^+ - (V_{DS_8} + V_S) = V_y^+ - (V_{GS_8} - V_T) - V_S \tag{5.207}$$

where V_S is the common source voltage. Also, it can be shown that

$$V_{GS_5} = V_x^- - V_S \tag{5.208}$$
$$V_{GS_6} = V_x^+ - V_S \tag{5.209}$$
$$V_{GS_7} = V_x^+ - V_S \tag{5.210}$$

and

$$V_{GS_8} = V_x^- - V_S \tag{5.211}$$

Combining Equations (5.208) through (5.211), (5.204) through (5.207), and (5.200) through (5.203) gives

$$V_{G_1} = V_{DD} + (V_x^- - V_y^-) \tag{5.212}$$

$$V_{G_2} = V_{DD} + (V_x^+ - V_y^-) \tag{5.213}$$

$$V_{G_3} = V_{DD} + (V_x^+ - V_y^+) \tag{5.214}$$

and

$$V_{G_4} = V_{DD} + (V_x^- - V_y^+) \tag{5.215}$$

Let $V_x^+ = V_I + v_x/2$, $V_x^- = V_I - v_x/2$, $V_y^+ = V_I + v_y/2$, and $V_y^- = V_I - v_y/2$, where V_I is the common dc component of the input voltage. The drain currents of transistors $T_1 - T_4$ are, respectively, given by

$$I_{D_1} = K[V_{DD} - (v_x - v_y)/2 - V_S - V_T]^2 \tag{5.216}$$

$$I_{D_2} = K[V_{DD} + (v_x + v_y)/2 - V_S - V_T]^2 \tag{5.217}$$

$$I_{D_3} = K[V_{DD} + (v_x - v_y)/2 - V_S - V_T]^2 \tag{5.218}$$

and

$$I_{D_4} = K[V_{DD} - (v_x + v_y)/2 - V_S - V_T]^2 \tag{5.219}$$

The difference of output currents can be expressed as

$$\triangle i_0 = i_0^+ - i_0^- \tag{5.220}$$

$$= (I_{D_2} + I_{D_4}) - (I_{D_1} + I_{D_3}) \tag{5.221}$$

$$= 2K v_x v_y \tag{5.222}$$

Transistors $T_{13} - T_{18}$ are required for the generation of a single-ended version of the output. The linear operation of the multiplier is achieved only for the range of input voltages over which the transistors remain in the saturation region.

5.3 Summary

Several circuit structures are available for the implementation of nonlinear functions, such as comparison and multiplication. They can be used in a given application depending on the design requirement to be met, that is, low supply voltage, low power consumption, high bandwidth, and low sensitivity to component nonidealities. In most cases, a trade-off exists between the performance characteristics.

5.4 Circuit design assessment

1. **Analysis of a comparator circuit**
   The comparator block diagram, as shown in Figure 5.33(a), consists of an input amplifier followed by the dynamic and storage latches. Let g_{mn} and g_{mp} be the transconductance of the n-channel and p-channel transistors, respectively. Verify that the dynamic latch of Figure 5.33(b) can be described by equations of the form

$$C_{pi}\frac{dV_O(t)}{dt} = g_m V_O(t) \tag{5.223}$$

$$V_O(t) = (V_i \pm V_{off})\exp\left[\frac{g_m}{C_{pi}}(t - T_S)\right] \tag{5.224}$$

   where $g_m = g_{m_p} - g_{m_n}$, g_m is the equivalent transconductance of a single inverter, C_{pi} is the parasitic capacitance at the input node, T_S is the sampling period, and V_{off} denotes the offset voltage.
   Estimate the response time of the latch.
   Determine the equivalent bit error rate of the comparator.

   Hint: Assuming that LSB is the voltage level corresponding to the least-significant bit of a converter, the bit error rate (BER) can be defined as

$$\text{BER} = \frac{\text{SLMR}}{\text{Input amplifier gain} \times (1\ \text{LSB}/2)} \tag{5.225}$$

   with SLMR being the input referred metastable region of the storage latch given by

$$\text{SLMR} = \Delta V_O \exp\left(-\frac{T_S/2 - t_d}{\tau_{dlatch}}\right) \tag{5.226}$$

   where ΔV_O is the initial output metastable region, τ_{dlatch} is the time constant of the dynamic latch, and t_d is the delay between the output signals of the dynamic and storage latches.

2. **Comparator design**
   A double tail dynamic comparator is shown in Figure 5.34, where ϕ_s is the strobe signal used to activate the output during each comparison. It consists of an input stage followed by a regenerative latch stage and an output buffer stage based on current starved inverters.

   In a given CMOS process, determine the transistor sizes to meet the following specifications: rising time and falling time of 2 ns and 3.5 ns of the output pulse, propagation delay of 10 ns for an input slope of 32 V/ms.

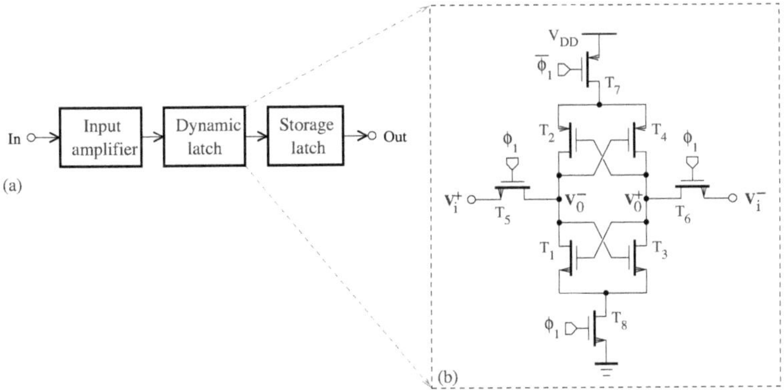

FIGURE 5.33
(a) Comparator block diagram; (b) circuit diagram of a dynamic latch.

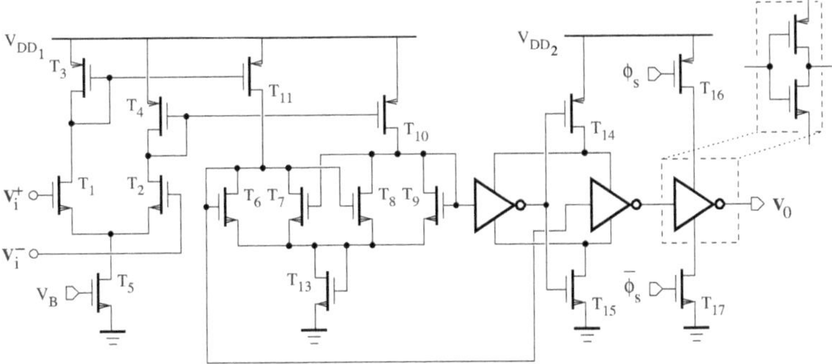

FIGURE 5.34
Comparator circuit diagram.

3. **Two-stage comparator circuit design**

   Consider the two-stage dynamic comparator shown in Figure 5.35. To reduce the input-referred offset voltage, the regenerative latch stage is preceded by a preamplifier.

   During the reset phase (or clock phase ϕ_2), the tail current of the preamplifier is disabled and the output nodes of the preamplifier are connected to the ground, while the comparator output nodes are pre-charged to the supply voltage.

   During the active phase (or clock phase ϕ_1), the tail transistor of the preamplifier is turned on to allow the sensing of the input difference voltage, while the preamplifier load transistors and pre-charge switch transistors are turned off. At the preamplifier output nodes, the common-mode voltage then increases while the currents caused by the input difference voltage are integrated. The regeneration will

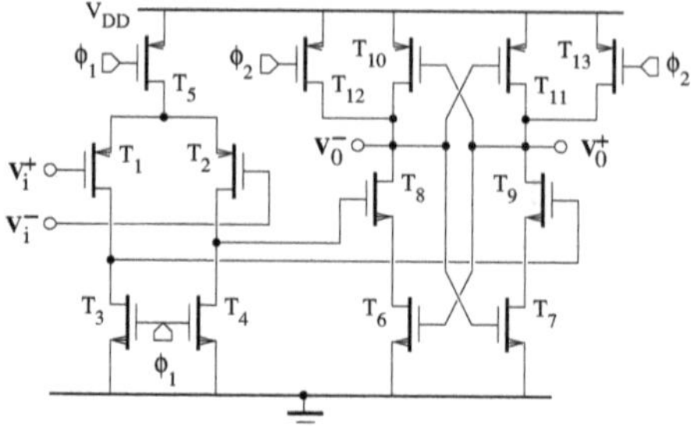

FIGURE 5.35
Circuit diagram of a two-stage dynamic comparator.

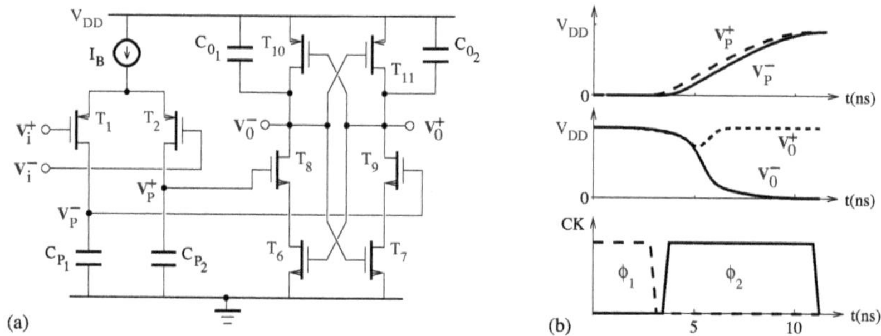

FIGURE 5.36
(a) Comparator equivalent circuit during the active phase; (b) waveforms when $V_i^+ - V_i^- > 0$.

begin as soon as the common-mode voltage exceeds the switching voltage of the regenerative latch. Figure 5.36 shows the comparator equivalent circuit and its waveforms.

Determine the transistor sizes in a given CMOS process to achieve a propagation delay t_p less than 50% of the active clock pulse or ϕ_1 (e.g., $t_p \leq 12.5$ ns for a 20-MHz clock signal with 50% duty cycle). Use SPICE to simulate the resulting circuit.

4. **Fully differential comparator circuit design**
A comparator can be characterized by the propagation delay, that is defined as the time required for the output to reach the 50% point of a transition, after the differential input signal crosses the offset voltage, when driven by a square wave (typically 100 mV) to a prescribed value of input overdrive (around 20 mV), as shown in Figure 5.38.

A. Complete the design of the comparator shown in Figure 5.37

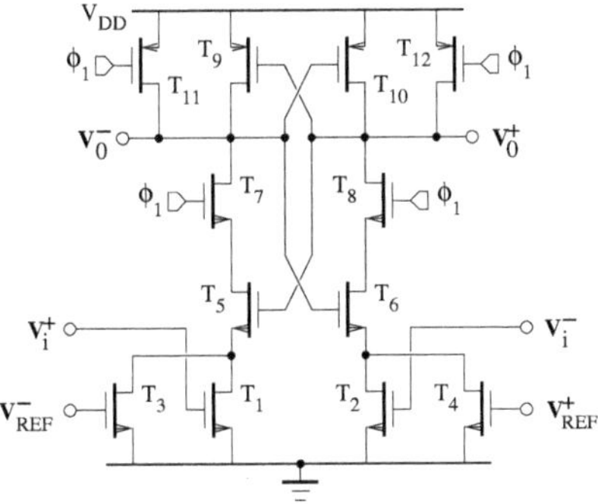

FIGURE 5.37
Circuit diagram of a fully differential comparator.

so that the propagation delay can be on the order of 200 ps for a capacitive load of 10 fF and a clock signal frequency of 250 MHz. Transistors $T_1 - T_4$ operate in the triode region, while the others are biased in the saturation region.
Use SPICE simulations to obtain the transient response of the comparator. The signals are connected to the input nodes during the clock phase ϕ_2 ($\phi_2 = \overline{\phi_1}$).

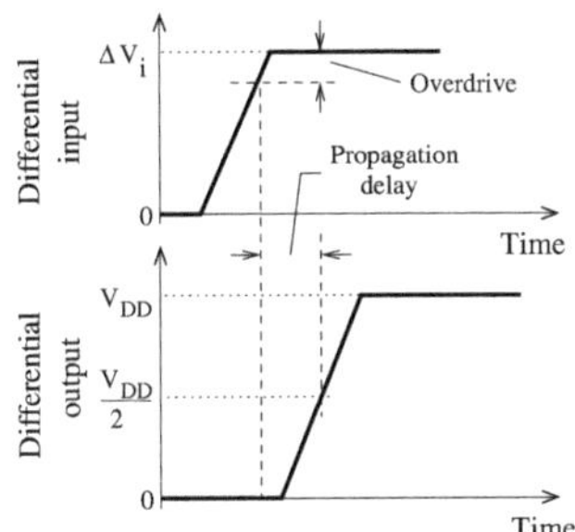

FIGURE 5.38
Comparator specifications.

B. A dynamic comparator can be implemented by connecting a latch at the outputs of a sense amplifier. During a normal operation, a low-swing differential input is detected by the sense amplifier, while full-swing outputs are generated by the latch.

Figure 5.39 shows a dynamic comparator with nMOS input. The latch is based on NAND gates and the input is sampled when the clock signal ϕ_1 goes to the high level.

Figure 5.40 shows a dynamic comparator with pMOS input. The

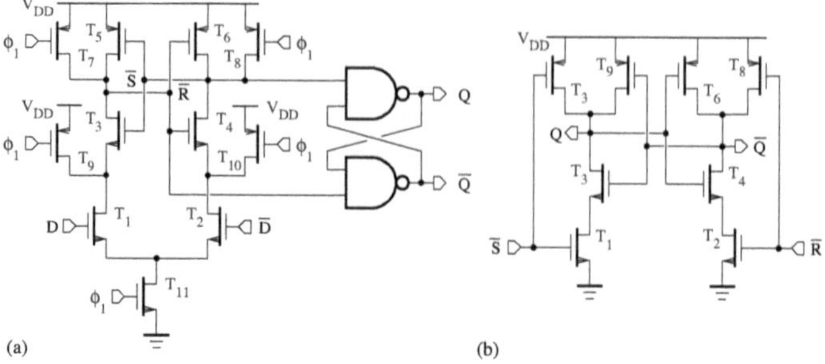

FIGURE 5.39
(a) Dynamic comparator with nMOS input; (b) latch based on NAND gates.

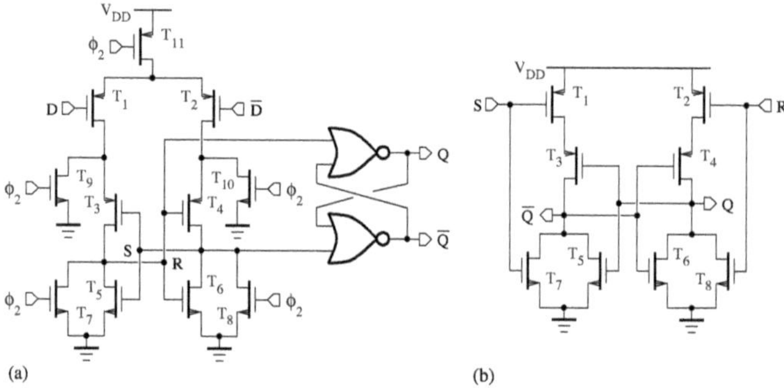

FIGURE 5.40
(a) Dynamic comparator with pMOS input; (b) latch based on NOR gates.

latch is based on NOR gates and the input is sampled when the clock signal ϕ_1 goes to the low level.

Size the comparator transistors to achieve a propagation delay of 5 ns for a capacitive load of 10 pF and a clock signal frequency of at least 150 MHz.

5. **Two-stage comparator design**
Consider the comparator of Figure 5.41, which can operate with a high common-mode voltage. The differential input stages are loaded by transistors, whose gates are driven by the clock signal. The output stage is a self-timed latch, that provides a rail-to-rail output swing.

Use SPICE simulations to size the comparator transistors in a given CMOS technology.

6. **Squaring circuit**
Consider the circuit shown in Figure 5.42(a) [27], where the transistor T_1 operates in the triode region, while T_2 is biased in the

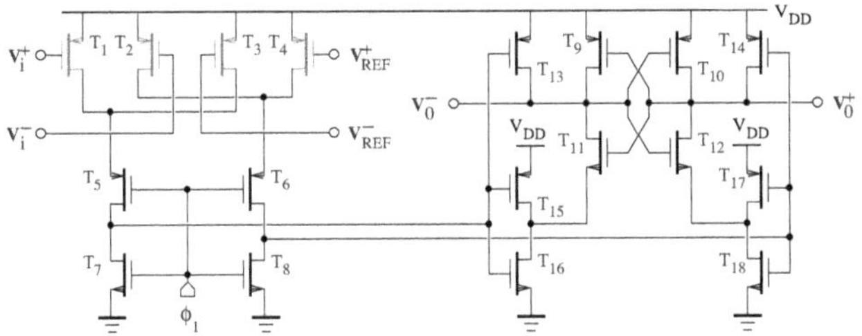

FIGURE 5.41
Circuit diagram of a two-stage comparator.

saturation region.
Assuming that $V_{GS1} = V + V_{GS2}$, show that

$$I - 2K_1V\sqrt{\frac{I}{K_2}} - K_1V^2 = 0 \tag{5.227}$$

and

$$I = \left(1 + 2\frac{K_1}{K_2} \pm 2\sqrt{\frac{K_1}{K_2}}\sqrt{1 + \frac{K_1}{K_2}}\right) K_1V^2 \tag{5.228}$$

where $K_1 = \mu_n C_{ox} W_1/(2L_1)$ and $K_2 = \mu_n C_{ox} W_2/(2L_2)$.
The circuit of Figure 5.42(b) can provide the square of a positive input signal. Let $0 < V \leq V_m$ and show that

$$V_m = \frac{\sqrt{(1/K_1) + (1/K_2)} - \sqrt{(1/K_2)}}{\sqrt{(1/K_1) + (1/K_2)} + \sqrt{(1/K_3)}}(V_{DD} - V_{Tn} - |V_{Tp}|) \tag{5.229}$$

where $K_3 = \mu_p C_{ox} W_3/(2L_3)$, V_{Tn}, and V_{Tp} are the threshold voltages of the n-channel and p-channel transistors, respectively.
The operating range can be defined by noting that the transistor T_3 will be shut down if

$$V_{GS1} = V_{SG3} = V_{DD} \tag{5.230}$$

where

$$V_{GS1} = V_{Tn} + \sqrt{(I/K_1) + (I/K_2)} \tag{5.231}$$

and

$$V_{SG3} = |V_{Tp}| + \sqrt{\frac{I}{K_3}} \tag{5.232}$$

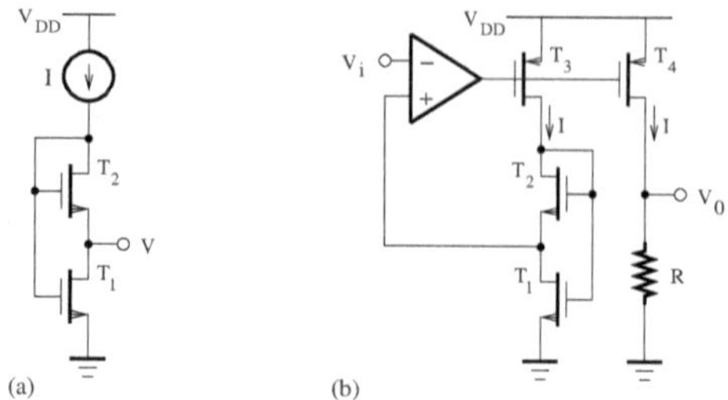

FIGURE 5.42
(a) Two transistor circuit; (b) squaring circuit.

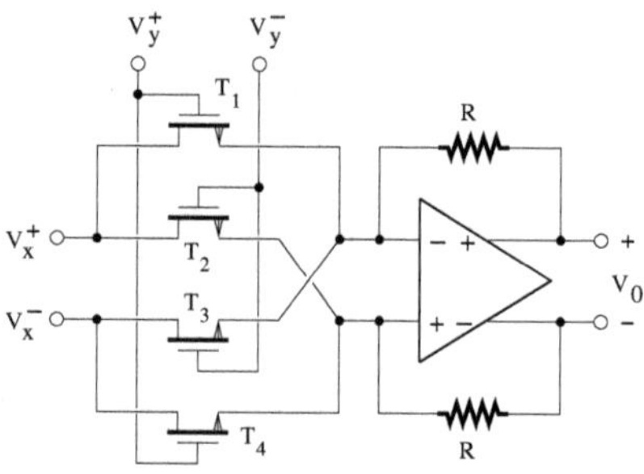

FIGURE 5.43
Multiplier circuit based on four transistors.

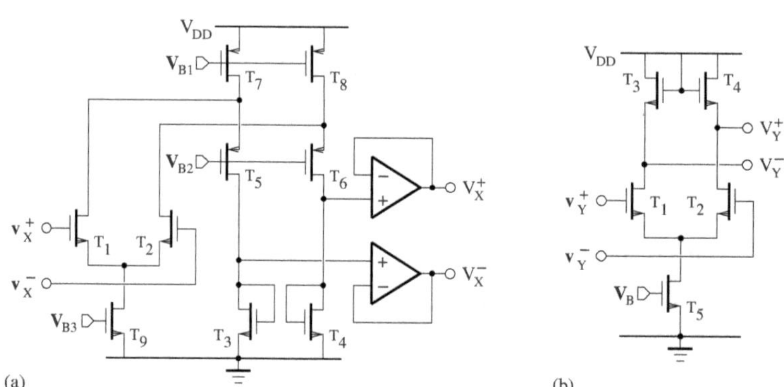

FIGURE 5.44
(a) Level shifter with a buffer for the input nodes X; (b) level shifter for the input nodes Y.

7. **Multiplier design**
   In the circuit of Figure 5.43 [32], the transistors operate in the triode region. Let I_{D_i} $(i = 1, 2, 3, 4)$ be the current flowing through the

transistor T_i. Write the current $\Delta i = (I_{D_1} + I_{D_3}) - (I_{D_2} + I_{D_4})$ in the form

$$\Delta i = 2K[(V_x^+ - V_x^-)(V_y^+ - V_y^-) - (V_P - V_N)(V_y^+ + V_y^-) + 2V_T(V_P - V_N) + V_P^2 - V_N^2] \tag{5.233}$$

where V^+ and V^- are the voltages at the noninverting and inverting nodes of the amplifier, respectively.
What condition must be met to realize a multiplier?
Show that

$$V_0 = \frac{R\Delta i}{1 + 1/A} \tag{5.234}$$

where A is the amplifier dc gain.
The common-mode characteristic of the structure shown in Figure 5.43 can be improved by connecting two level shifters (see Figure 5.44) at its inputs.
Explain why buffers are required to drive the X inputs.
Verify that the gain of level shifters will be unity if the design requirement, $V_{GS_1} - V_{GS_2} = \pm(V_{GS_4} - V_{GS_3})$, is met.

8. **Multiplier circuit**
The circuit shown in Figure 5.45 [35] is the generalized structure for the implementation of multiplier based on the quarter-square technique. The bias current is represented by I_B and V_I is the dc component of the input voltage. Transistors $T_1 - T_4$ operate in the saturation region and are designed with the same transconductance parameter, $K = \mu_n(C_{ox}/2)(W/L)$, and threshold voltage, V_T.
Show that $\Delta i = i_0^+ - i_0^- = 2KV_xV_y$, for any numbers p, q, and r.

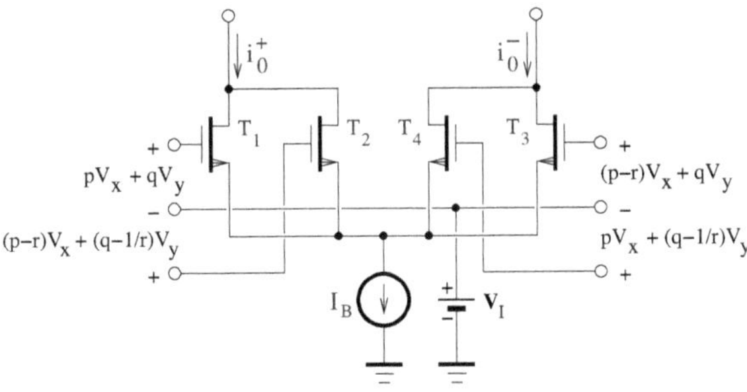

FIGURE 5.45
Generalized multiplier core based on the quarter-square technique.

9. **Multiplier based on two squaring circuits**
The squaring circuit of Figure 5.46 is based on unbalanced differential transistor pairs. The transistors operate in the saturation region and are sized so that $W_2/L_2 = W_4/L_4 = \alpha(W_3/L_3) = \alpha(W_3/L_3)$.

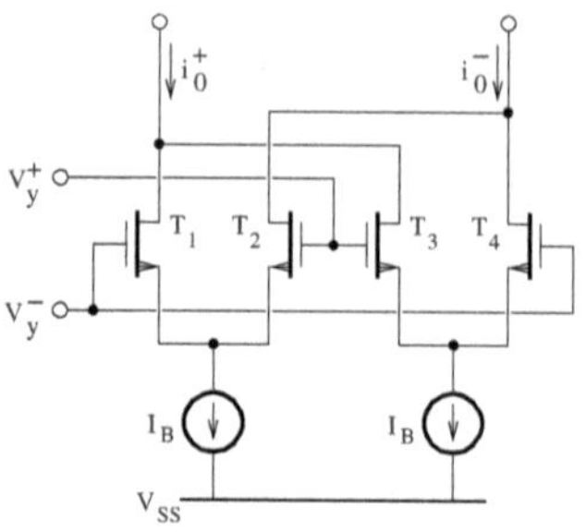

FIGURE 5.46
Squaring circuit based on four transistors.

Estimate the differential output current $\triangle i_0 = i_0^+ - i_0^-$ as a function of the input voltage $V_i = V_i^+ - V_i^-$.
Based on the relation $(V_x + V_y)^2 - (V_x - V_y)^2 = 4V_xV_y$, design a multiplier circuit.

Hint: The output current, $\triangle i$, of a single unbalanced differential pair connected to the input voltage, V_i, is given by

$$\triangle i = \begin{cases} i - \dfrac{1-\alpha}{1+\alpha} I_B & \text{for } -\sqrt{\dfrac{I_B}{K}} < V_i < \sqrt{\dfrac{I_B}{\alpha K}} \\ I_B \operatorname{sign}(V_i) & \text{for } V_i \leq -\sqrt{\dfrac{I_B}{K}} \text{ and } \sqrt{\dfrac{I_B}{\alpha K}} \leq V_i \end{cases} \tag{5.235}$$

where

$$i = \frac{2\alpha(1-\alpha)KV_i^2 + 4\alpha\sqrt{KI_B}V_i\sqrt{1+\alpha-\alpha KV_i^2/I_B}}{(1+\alpha)^2} \tag{5.236}$$

and K is the transconductance parameter of the transistors.

Bibliography

[1] R. Gregorian, *Introduction to CMOS Op-Amps and Comparators*, New York, NY: John Wiley & Sons, 1999.

[2] P. E. Allen and D. R. Holberg, *CMOS Analog Circuit Design*, 2nd ed., New York, NY: Oxford University Press, 2002.

[3] R. Wang and R. Harjani, "Partial positive feedback for gain enhancement of low-power CMOS OTAs," *Analog Integrated Circuits and Signal Processing*, vol. 8, no 1, pp. 21–35, July 1995.

[4] K. B. Ohri and M. J. Callahan, Jr., "Integrated PCM codec," *IEEE J. of Solid-State Circuits*, vol. 14, pp. 38–46, Feb. 1979.

[5] D. J. Allstot, "A precision variable-supply CMOS comparator," *IEEE J. of Solid-State Circuits*, vol. 17, pp. 1080–1087, Dec. 1982.

[6] B. J. Hosticka, W. Brockherde, U. Kleine, and R. Schweer, "Design of nonlinear analog switched-capacitor circuits using building blocks," *IEEE Trans. on Circuits and Systems*, vol. 31, pp. 354–368, April 1984.

[7] M. Matsui, H. Hara, Y. Uetani, L.-S. Kim, T. Nagamatsu, Y. Watanabe, A. Chiba, K. Matsuda, and T. Sakurai, "A 200 MHz 13 mm^2 2-D DCT macrocell using sense-amplifying pipeline flip-flop scheme," *IEEE J. of Solid-State Circuits*, vol. 29, pp. 1482–1490, Dec. 1994.

[8] B. Nikolić, V. G. Oklobdžija, V. Stojanović, W. Jia, J. K.-S. Chiu, and M. M.-T. Leung, "Improved sense-amplifier-based flip-flop: Design and measurements," *IEEE J. of Solid-State Circuits*, vol. 35, pp. 876–884, June 2000.

[9] A. Yukawa, "A CMOS 8-bit high-speed A/D converter IC," *IEEE J. of Solid-State Circuits*, vol. 22, pp. 775–779, June 1985.

[10] B. P. Brandt, D. E. Wingard, and B. A. Wooley, "Second-order sigma-delta modulation for digital-audio signal acquisition," *IEEE J. of Solid-State Circuits*, vol. 26, pp. 618–627, April 1991.

[11] T. Kobayashi, K. Nogami, T. Shirotori, and Y. Fujimoto, "A current-controlled latch sense amplifier and a static power-saving input buffer for low-power architecture," *IEEE J. of Solid-State Circuits*, vol. 28, pp. 523–527, April 1993.

[12] G. M. Yin, F. Op't Eynde, and W. Sansen, "A high-speed CMOS comparator with 8-b resolution," *IEEE J. of Solid-State Circuits*, vol. 27, pp. 208–211, Feb. 1992.

[13] S. Park and M. P. Flynn, "A regenerative comparator structure with integrated inductors," *IEEE Trans. on Circuits and Systems*, vol. 53, pp. 1704–1711, Aug. 2006.

[14] J. Robert, G. C. Temes, V. Valencic, R. Dessoulavy, and P. Deval, "A 16-bit low-voltage CMOS A/D converter," *IEEE J. of Solid-State Circuits*, vol. 22, pp. 157–163, April 1987.

[15] B.-S. Song, S.-H. Lee, and M. F. Tompsett, "A 10-b 15-MHz CMOS recycling two-step A/D converter," *IEEE J. of Solid-State Circuits*, vol. 25, pp. 1328–1338, Dec. 1990.

[16] B. Wicht, T. Nirschl, and D. Schmitt-Landsiedel, "Yield and speed optimization of a latch-type voltage sense amplifier," *IEEE J. of Solid-State Circuits*, vol. 39, no. 7, pp. 1148–1158, Jul. 2004.

[17] D. Shinkel, E. Mensink, E. Klumperink, E. van Tuijl, and B. Nauta, "A double-tail latch-type voltage sense amplifier with 18ps Setup+Hold time," in *Proc. IEEE Int. Solid-State Circuits Conf., Dig. Tech. Papers*, Feb. 2007, pp. 314–315.

[18] J. He, S. Zhan, D. Chen, and R. L. Geiger, "Analyses of static and dynamic random offset voltages in dynamic comparators," *IEEE Trans. Circuits Systems-I*, vol. 56, no. 5, pp. 911–919, May 2009.

[19] A.-T. Do, Z.-H. Kong, and K.-S. Yeo, "Criterion to evaluate input-offset voltage of a latch-type sense amplifier," *IEEE Trans. Circuits Systems-I*, vol. 57, no. 1, pp. 83–92, Jan. 2010.

[20] M. J. M. Pelgrom, A. C. J. Duinmaijer, and A. P. G. Welbers, "Matching properties of MOS transistors," *IEEE J. Solid-State Circuits*, vol. 24, no. 5, pp. 1433–1439, Oct. 1989.

[21] B. Gilbert, "A precise four-quadrant multiplier with subnanosecond response," *IEEE J. of Solid-State Circuits*, vol. 3, pp. 365–373, Dec. 1968.

[22] J. N. Babanezhad and G. C. Temes, "A 20-V four-quadrant CMOS analog multiplier," *IEEE J. of Solid-State Circuits*, vol. 20, pp. 1158–1168, Dec. 1985.

[23] Y. H. Kim and S. B. Park, "Four-quadrant CMOS analogue multiplier," *Electronics Letters*, vol. 28, no. 7, pp. 649–650, March 1992.

[24] C. W. Kim and S. B. Park, "New four-quadrant CMOS analog multiplier," *Electronics Letters*, vol. 23, no. 24, pp. 1268–1270, Nov. 1987.

[25] J. S. Peña-Finol and J. A. Connelly, "A MOS four-quadrant analog multiplier using the quarter-square technique," *IEEE J. of Solid-State Circuits*, vol. 22, pp. 1064–1073, Dec. 1987.

[26] R. R. Torrance, T. R. Viswanathan, and J. V. Hanson, "CMOS voltage to current transducers," *IEEE Trans. Circuits and Systems*, vol. CAS-32, pp. 1097–1104, Nov. 1985.

[27] I. M. Filanovsky and H. P. Baltes, "Simple CMOS square-rooting and squaring circuits," *IEEE Trans. on Circuits and Systems*, vol. 39, pp. 312–315, April 1992.

[28] B.-S. Song, "CMOS RF circuits for data communications applications," *IEEE J. of Solid-State Circuits*, vol. 21, pp. 310–317, April 1986.

[29] A. Nedungadi and T. R. Viswanathan, "Design of linear CMOS transconductance elements," *IEEE Trans. Circuits and Systems*, vol. CAS-31, pp. 891–894, Sept. 1984.

[30] K. Kimura, "Some circuit design techniques for low-voltage analog functional elements using squaring circuits," *IEEE Trans. on Circuits and Systems*, vol. 43, pp. 559–576, July 1996.

[31] *HSPICE Users' Guide*, Campbell, CA: Meta-Software, Inc., Feb. 1996.

[32] K. Bult and H. Wallinga, "A CMOS four-quadrant analog multiplier," *IEEE J. of Solid-State Circuits*, vol. 21, pp. 430–435, June 1986.

[33] K. Bult, "Analog CMOS square-law circuits," Ph.D. dissertation, University of Twente, The Netherlands, 1988.

[34] Z. Wang, "A CMOS four-quadrant analog multiplier with single-ended voltage output and improved temperature performance," *IEEE J. of Solid-State Circuits*, vol. 26, pp. 1293–1301, Sept. 1991.

[35] K. Kimura, "An MOS four-quadrant analog multiplier based on the multitail technique using a quadritail cell as a multiplier core," *IEEE Trans. on Circuits and Systems*, vol. 42, pp. 448–454, Aug. 1995.

6

Continuous-Time Circuits

CONTENTS

Real-world data are generally converted by sensors into analog electrical signals corrupted by noise. In this case, the extraction of useful information, which can be displayed by actuators, requires the use of complex algorithms, whose efficient implementation exploits the programmability and flexibility of digital circuits. The suitable level and representation of sensor output information are provided by signal conditioning and interface structures. In communication systems, the tasks of signal processing generally encountered in the front-end and back-end sections of a digital signal processor, namely, amplification of the desired channel to the full-scale of the data converters, automatic gain control, and filtering to remove the interference of the adjacent channels, are preferably implemented using continuous-time (CT) circuits, which offer the advantages of high-speed operation and low-power consumption. Other applications include anti-aliasing and smoothing filters, channel equalization in magnetic disk drives, and high-speed data links.

Due to factors such as the fabrication tolerance, temperature variation, and aging, the values of components can drift to about 10% to 50% from their nominal specifications. The automatic tuning scheme included in CT filters provides a means to solve this problem. When the requirement of a high dynamic range results in a large power consumption and the noise level can be reduced only at the price of a large chip, it is common to adopt a solution based on multiple integrated circuits (ICs) and discrete components instead of a monolithic IC. With the down-scaling of MOS transistors into the submicrometer regime, analog and digital circuits may share the same die on a single chip in mixed-signal design. The result is the onset of several problems related to the signal integrity, substrate noise, crosstalk, interconnect parasitic impedances, and electromagnetic interference. By increasing the dynamic range, differential architectures can reduce the sensitivity to some of these effects and supply voltage variations.

In addition to a survey of various CT building blocks (integrator, summer, gain stage), conventional design techniques of CT circuits will be addressed.

6.1 Wireless communication system

A transceiver consists of a receiver and transmitter. It includes the antenna, and radio frequency (RF), intermediate frequency (IF), and baseband and bit-stream processing functions. This partition is justified by the large change in bandwidth due to the decimation or interpolation within each section and the flexibility and cost of the hardware implementation. The RF section, which is generally implemented with gallium arsenide (GaAs), silicon germanium (SiGe), indium phosphide (InP), and bi-complementary metal-oxide semiconductor (BiCMOS), performs the frequency conversion.

In the receiver, the RF signal delivered by a low-noise amplifier (LNA) is down-converted into an IF waveform, which is then sent to the demodulator. It should be noted that the incoming signal at the antenna can be lower than a half microvolt root-mean square (μV_{rms}) and is modulated in the gigahertz (GHz) frequency range.

The transmitter follows the reverse path. The IF of the modulator output signal is up-converted into RF and a power amplifier (PA) is required to adjust the level of the signal to be emitted by the antenna.

Basically, a communication system can be reduced to a channel decoder, source decoder, source encoder, and channel encoder.

The trend in software-defined radios [1] is to move the analog-to-digital converter (ADC) and digital-to-analog converter (DAC) closer to the antenna so that more functions can be realized digitally. Such an approach is necessary to meet the requirements (high degree of adaptation, easy reconfiguration) of a multi-standard communication environment based on a multi-band antenna and RF conversion. It relies on the use of high-speed and high-accuracy data converters. Figure 6.1 shows the distribution of converter resolutions versus the signal bandwidth for some common applications.

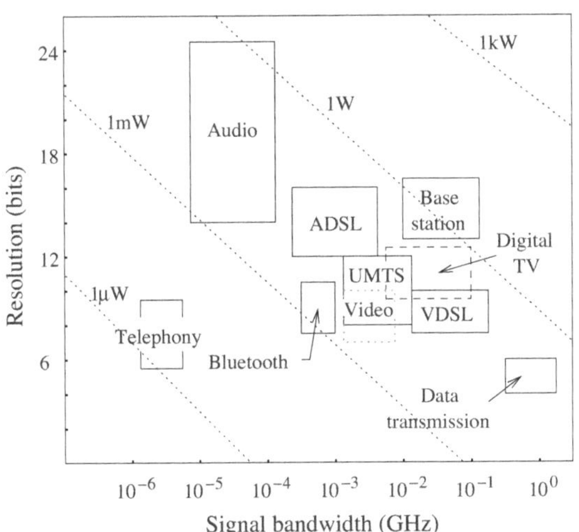

FIGURE 6.1
Specifications of CMOS data converters.

The high-speed attribute of the submicrometer CMOS process can offer the opportunity for a single-chip implementation of transceivers [2, 3], which actually require costly RF discrete components. To achieve the complete integration of such a system, suitable architectures and circuit techniques are required to overcome the limitations inherent in the low-Q components available in CMOS IC technology. Specifically, RF devices involve complex noise, intermodulation, and electromagnetic interference effects.

6.1.1 Receiver and transmitter architectures

RF transceivers find applications in bidirectional communication systems. They include a receiver section, which down-converts the detected RF signal to a baseband frequency (or a low frequency around 0 Hz), and a transmitter section, which up-converts the baseband signal to a radio (or very high) frequency. Transceivers should be designed to achieve a low power consumption and to feature a high IC density in order to be suitable for portable devices. Other criteria such as the number of off-chip components and cost may play an important role in the transceiver architecture choice.

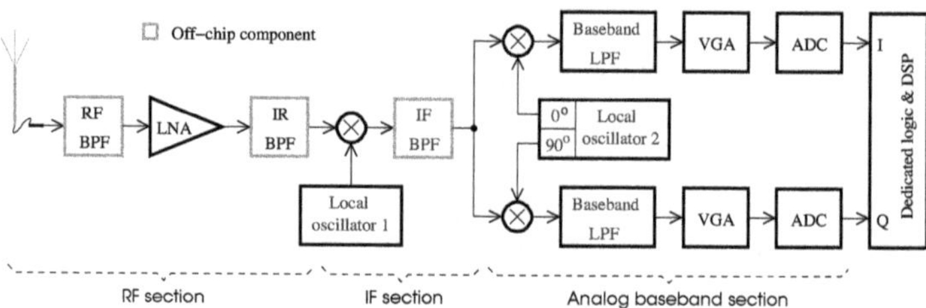

FIGURE 6.2
Block diagram of a superheterodyne receiver.

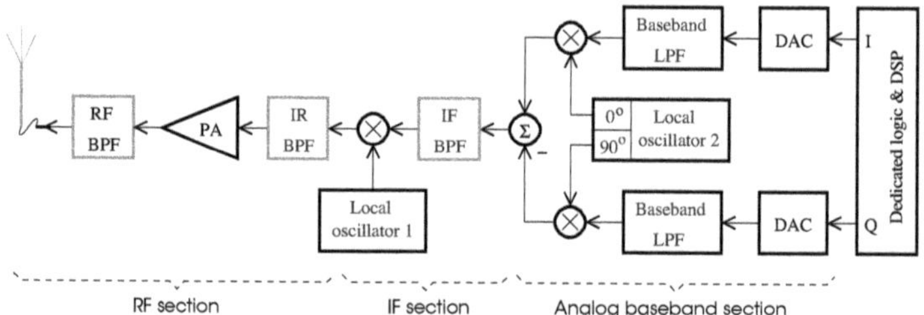

FIGURE 6.3
Block diagram of a superheterodyne transmitter.

The superheterodyne architecture [4,5] is commonly used due to its capacity to detect a low-level RF signal in the presence of strong interferers. Note that the term "hetero" means "different," while "dyne" stands for "power." Hence, "heterodyne" is related to the fact that the conversion to an intermediate frequency is achieved by mixing a signal frequency with a locally generated frequency; "super" comes from the fact that these frequencies may be above the audible spectrum.

The block diagram of a superheterodyne receiver is shown in Figure 6.2. The RF bandpass filter immediately after the antenna is used to select the desired RF band. The low-noise amplifier (LNA) is designed to boost the signal level while introducing as little of its own noise as possible. The resulting signal is passed through an image-reject bandpass filter, which also attenuates the LNA noise contribution present in the image band. The down-conversion to an intermediate frequency, which is equal to the difference between the frequency of the incoming RF signal and the frequency of the local oscillator, is then achieved by the first mixer and IF bandpass filter. The small transition band of each of the IF bandpass filters needed for channel selection results in the requirement of a high Q-factor usually satisfied by using an off-chip surface acoustic wave or ceramic circuits. The choice of the IF is determined by the trade-off to be achieved between image rejection and channel selection, or say, between the sensitivity and selectivity of the receiver. The down-conversion from the IF to the baseband frequency is realized by quadrature mixers followed by lowpass filters. The frequency of the local oscillator that feeds the second mixer section should be equal to the IF. The last section of the receiver front end includes variable gain amplifiers (VGAs) and ADCs. It is placed just before the digital signal processor (DSP), which performs the digital demodulation and decoding.

At the system level, the superheterodyne transmitter is configured to realize the reverse operation of the corresponding receiver. The block diagram of a superheterodyne transmitter is depicted in Figure 6.3. A quadrature mixer followed by an IF filter is used to translate the baseband signal, which is provided by the DAC, to an intermediate frequency. The IF-to-RF conversion, which then follows, is achieved with the help of another mixer and filter stage. Finally, the power amplifier (PA) increases the signal level and the RF filter transmits only the signal components satisfying the spectral mask requirements.

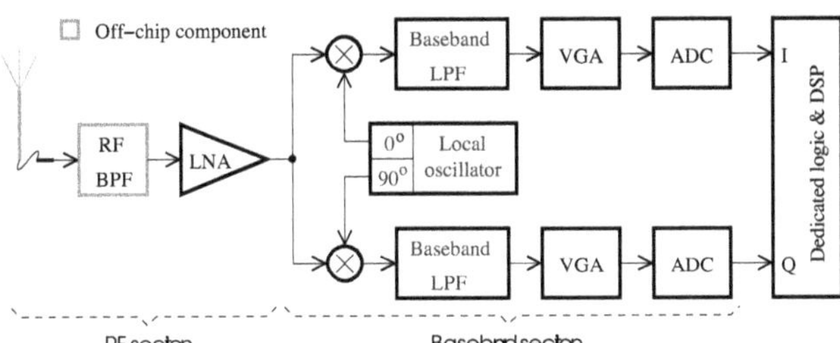

FIGURE 6.4
Block diagram of a homodyne receiver.

Homodyne architectures [6] for receivers and transmitters, as illustrated in Figures 6.4 and 6.5, respectively, have the advantage of eliminating many off-chip components in the signal paths. They are sometimes referred to as direct-conversion or zero-IF receivers and transmitters. Here, the conversion

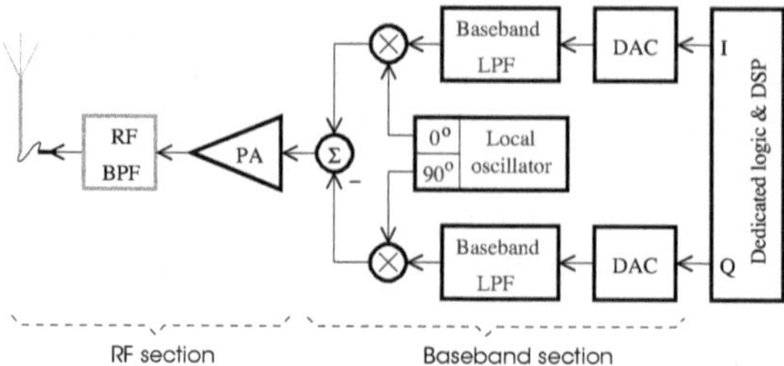

FIGURE 6.5
Block diagram of a homodyne transmitter.

from the radio frequency to the baseband frequency, and vice versa, is achieved using only one mixer stage. The word "homo" implies "same"; this is equivalent to having an identical frequency for the signal of interest and local oscillator signal. Even though the IF frequency is zero, the mirrored image of the desired signal will be superimposed on the down-converted signal, which is generated by the mixer. The image problem is solved by performing the frequency conversion in quadrature. The amplitude and phase of the desired signal can readily be determined from the in-phase and quadrature components. In homodyne architectures, the channel select filtering at the baseband is simply performed by lowpass filters, suitable to monolithic IC integration. Hence, the requirements associated with different communication standards can be met using programmable filters and high-dynamic range data converters. Overall, this approach is excellent at saving cost, die area, and power consumption. However, the effects of dc offset, device flicker noise, even-order harmonics, and local-oscillator leakage can critically limit the accuracy of the signal detection.

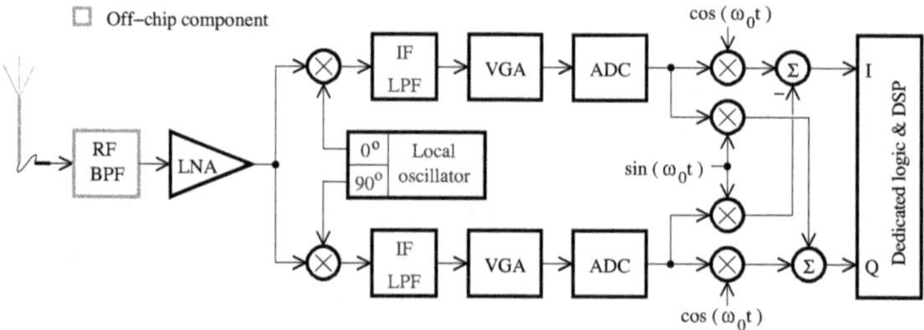

FIGURE 6.6
Block diagram of a low-IF receiver.

Low-IF architectures exploit the advantages of both heterodyne and homodyne structures. They include quadrature frequency conversions that can be both done in the analog domain or be split into analog and digital stages. However, the realization of data conversions at the IF instead of the baseband

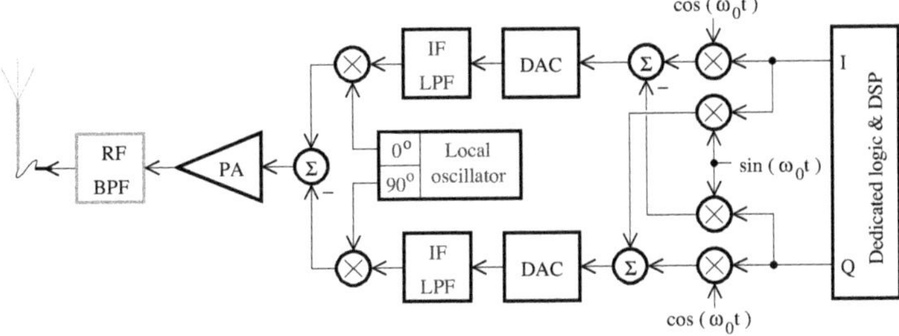

FIGURE 6.7
Block diagram of a low-IF transmitter.

may result in a power consumption increase and could make the overall performance sensitive to clock jitter, distortion, and noise errors. Typically, the IF should be a half to a few channel spacings from dc.

In the low-IF receiver of Figure 6.6, the RF signal is first translated to a low IF, before being down-converted to baseband frequency in the digital domain. Conversely, the low-IF transmitter, as depicted in Figure 6.7, uses both digital and analog steps to perform the up-conversion from the baseband frequency to RF. The image rejection is achieved by summing the output signals provided by a pair of quadrature mixers such that image-band signals ideally cancel out while the desired signals add together coherently. The use of harmonic rejection mixers excludes the need for discrete IF filters, thus making low-IF architectures better suited for single-chip integration than superheterodyne structures. In contrast to direct-conversion architectures, low-IF structures use local oscillators operating at frequencies that are lower than that of the incoming RF signal, to reduce the LO re-transmission, thereby attenuating the dc offset level. It should be noted that the static errors associated with the baseband section can generally be cancelled by suitable calibration techniques.

6.1.2 Frequency translation and quadrature multiplexing

Frequency translations are generally used in transmitters to reduce the cost associated with the processing of radio-frequency signals and to meet the transmission specifications. Basically, the frequency translation along with sideband modulation and demodulation relies on the multiplication of two signals. Given a carrier $\cos(\omega_c t)$, with the angular frequency ω_c, the modulated signal can be derived from the waveform $m(t)$ containing the message to be transmitted, as

$$x(t) = m(t)\cos(\omega_c t) \tag{6.1}$$

The Fourier transform of $x(t)$ is

$$X(\omega) = \frac{1}{2}[M(\omega - \omega_c) + M(\omega + \omega_c)] \tag{6.2}$$

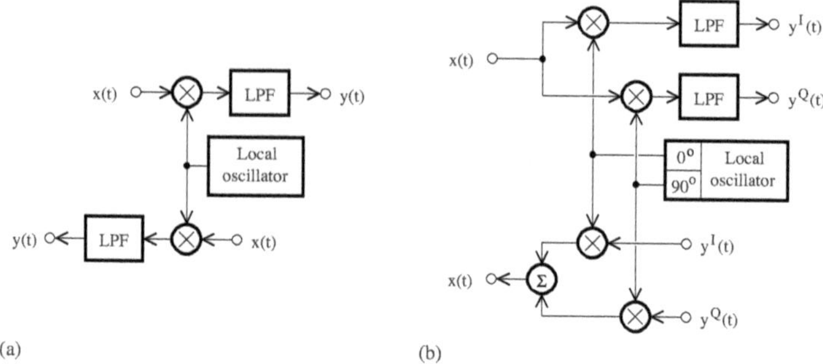

FIGURE 6.8
Functional block diagram of (a) basic and (b) quadrature mixers.

where $M(\omega)$ is the spectrum of the message signal $m(t)$. Assuming that $\omega = 0$, the spectrum of $m(t)$ is shifted to the left and right of the origin by ω_c. This process, known as double-sideband modulation (see Figure 6.8), exhibits a transmission bandwidth that is two times the bandwidth ω_B of $m(t)$. To recover the message signal by the demodulation, the spectra located at $\pm\omega_c$ should not overlap as is the case for $\omega_c \geq \omega_B$. The multiplication of $x(t)$ by the carrier provides a signal of the form

$$\begin{aligned} y(t) &= x(t)\cos(\omega_c t) \\ &= \frac{1}{2}m(t)[1 + \cos(2\omega_c t)] \end{aligned} \tag{6.3}$$

The spectrum of $y(t)$ is given by

$$Y(\omega) = \frac{1}{2}M(\omega) + \frac{1}{4}[M(\omega - 2\omega_c) + M(\omega + 2\omega_c)] \tag{6.4}$$

A version of the desired signal, that is, $m(t)/2$, is obtained by eliminating the spectral components with the frequency $2\omega_c$ using a lowpass filter.

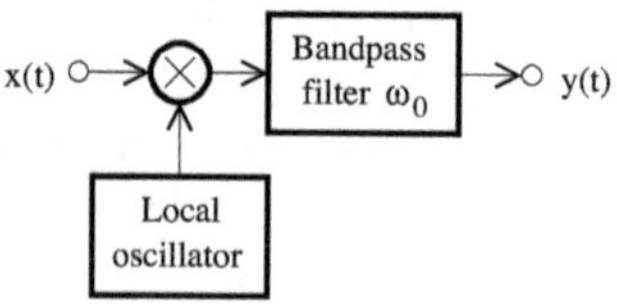

FIGURE 6.9
Block diagram of a frequency translation mixer.

The quadrature multiplexing scheme (see Figure 6.8(b)) makes use of the orthogonality of sine and cosine waves to transmit and receive two different signals simultaneously on the same carrier frequency. The transmission

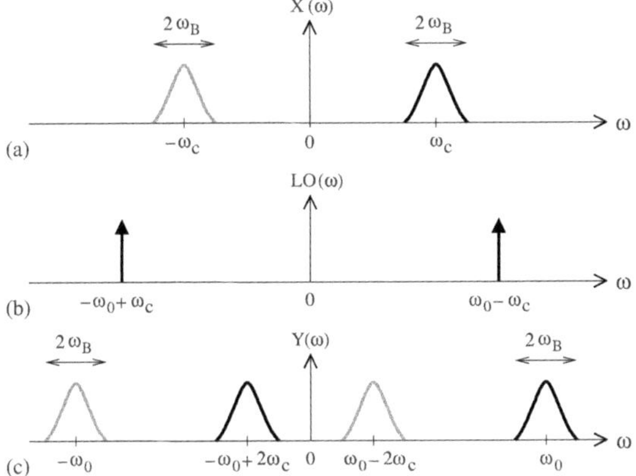

FIGURE 6.10
Plot of spectra of (a) the message, (b) LO signal, and (c) frequency-translated signals for the down-conversion.

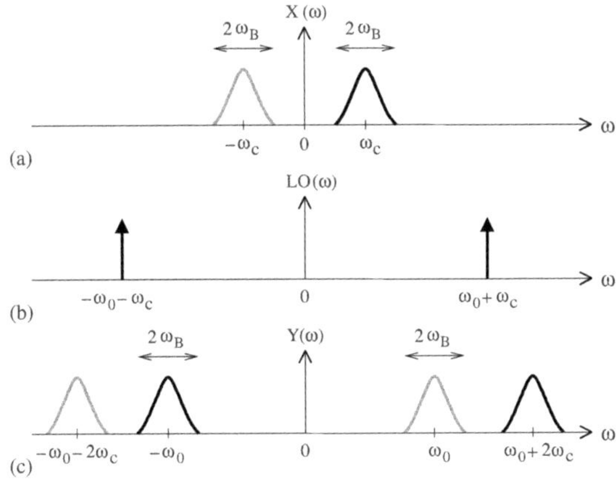

FIGURE 6.11
Plot of spectra of (a) the message, (b) LO signal, and (c) frequency-translated signals for the up-conversion.

bandwidth remains $2\omega_B$, and the modulated and demodulated signals can be written as

$$x(t) = Q(t)\sin(\omega_c t) + I(t)\cos(\omega_c t) \tag{6.5}$$

and

$$\begin{aligned} y^I(t) &= x(t)\cos(\omega_c t) \\ &= \frac{1}{2}I(t) + \frac{1}{2}[I(t)\cos(2\omega_c t) + Q(t)\sin(2\omega_c t)] \end{aligned} \tag{6.6}$$

$$y^Q(t) = x(t)\sin(\omega_c t)$$
$$= \frac{1}{2}Q(t) + \frac{1}{2}[I(t)\sin(2\omega_c t) - Q(t)\cos(2\omega_c t)] \quad (6.7)$$

respectively. The terms at $2\omega_c$ are suppressed by a lowpass filter, yielding $y^I(t) = I(t)/2$ and $y^Q(t) = Q(t)/2$ for the in-phase and quadrature channels, respectively. In general, the local carriers at the demodulator and modulator should be synchronized in frequency and phase for efficient demodulation.

The frequency translation, which is the process of shifting a signal from one frequency to another, can be realized using a mixer. It is then also referred to as frequency mixing. Figure 6.9 shows the block diagram of a frequency translation mixer. Let

$$x(t) = m(t)\cos(\omega_c t) \quad (6.8)$$

First, the multiplication of the signal $x(t)$ by a locally generated sine wave, $2\cos(\omega_{LO} t)$, is realized. Thus,

$$y(t) = 2x(t)\cos(\omega_{LO} t)$$
$$= m(t)\cos[(\omega_c - \omega_{LO})t] + m(t)\cos[(\omega_c + \omega_{LO})t] \quad (6.9)$$

With the assumption that $\omega_{LO} = \omega_c \pm \omega_0$, we can obtain

$$y(t) = m(t)\cos(\omega_0 t) + m(t)\cos[(2\omega_c \pm \omega_0)t] \quad (6.10)$$

A bandpass filter, whose center frequency is ω_0 and bandwidth is equal to or greater than $2\omega_B$, is used to detect the signal component at the angular frequency ω_0. The carrier frequency is then translated from ω_c to ω_0. The plus and minus signs in the ω_{LO} expression correspond to an up-conversion and a down-conversion, respectively. The spectra of signals are shown in Figures 6.10 and 6.11 for a message signal featuring a bandwidth of $2\omega_B$.

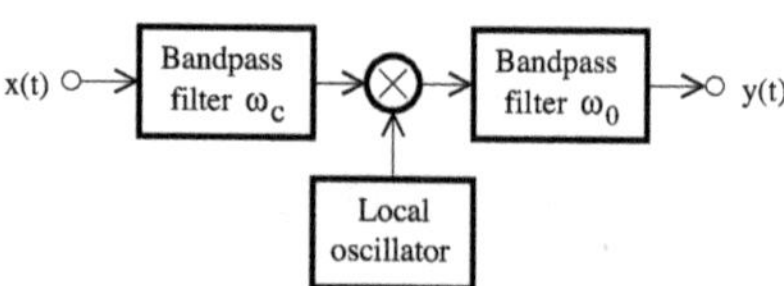

FIGURE 6.12
Block diagram of a frequency translation mixer with the bandpass pre-filter.

By multiplying the input signal and a reference oscillator signal, spectral images can be produced at either the sum or difference frequencies. In particular, interference signals or harmonic components at $\omega_c \pm 2\omega_0$ are also translated to the frequency ω_0, due to the fact that

$$2\cos[(\omega_c \pm 2\omega_0)t]\cos[(\omega_c \pm \omega_0)t] = \cos(\omega_0 t) + \cos[(2\omega_c \pm 3\omega_0)t] \quad (6.11)$$

The problem of spectral images can be addressed by placing a high-quality pre-filter at the mixer input, as shown in Figure 6.12. In order to pass only the signal of interest, this filter should have a bandwidth equal to or greater than the bandwidth of the message signal and a center frequency ω_c. However, for wideband applications, the pre-filter may become complex due to the required tunable range.

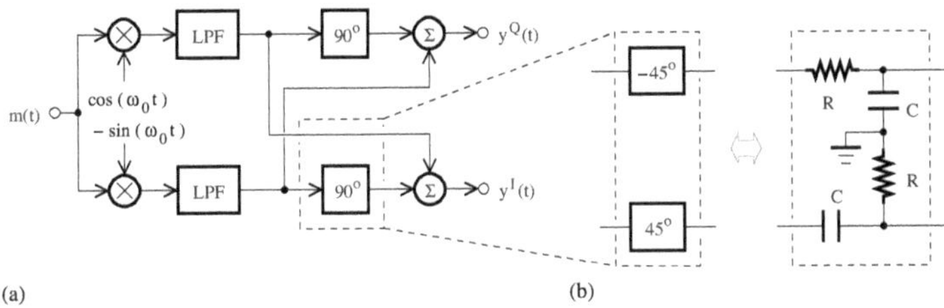

FIGURE 6.13
(a) Block diagram of Hartley image-reject down-converter; (b) phase shifter implementation.

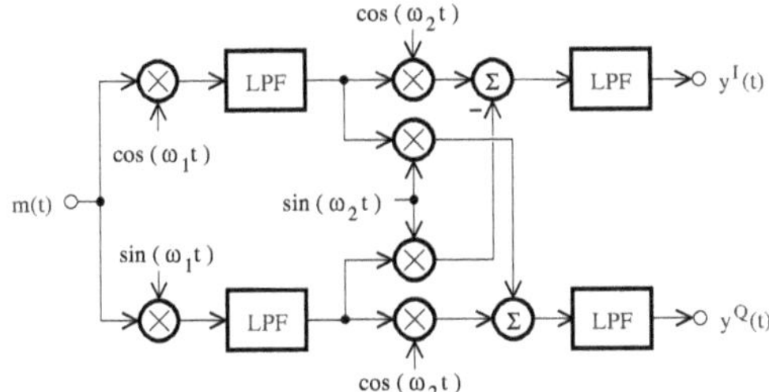

FIGURE 6.14
Block diagram of Weaver image-reject down-converter.

Other mechanisms of dealing with the image rejection problem involve the use of special image-rejection architectures, such as the Hartley architecture [7] shown in Figure 6.13, and the Weaver architecture [8] depicted in Figure 6.14. The exploitation of quadrature conversions in the separation of the signal of interest from image components makes these architectures more suitable for IC implementations than conventional structures requiring a high-quality filter with sharp cutoff characteristics. In Hartley and Weaver architectures, an opposite phase difference is introduced between the image band signals, which are then cancelled out by a summing operation, while the desired signals, whose phase difference remains equal to zero, adds up coherently. The main difference between both architectures is the implementation of the 90° phase shift required on the signal path. Hartley architectures typically use a passive phase shifter, while a second mixing stage is used in the Weaver architecture.

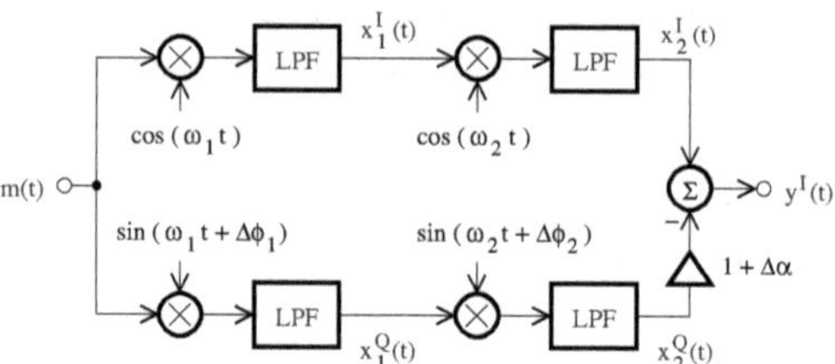

FIGURE 6.15
Equivalent model of a Weaver converter with mismatches.

The image rejection ratio (IRR) of the Weaver converter can be determined in accordance with the equivalent model shown in Figure 6.15. The message signal is assumed to be of the form

$$m(t) = \cos \omega t \tag{6.12}$$

After each mixer stage, the signal component at the frequency $\omega + \omega_1$ is eliminated by the lowpass filter. The in-phase and quadrature signals, $x_1^I(t)$ and $x_1^Q(t)$, can be respectively expressed as

$$x_1^I(t) = \frac{1}{2}\cos(\omega - \omega_1)t \tag{6.13}$$

and

$$x_1^Q(t) = -\frac{1}{2}\sin[(\omega - \omega_1)t - \triangle\phi_1] \tag{6.14}$$

where $\triangle\phi_1$ denotes the phase error of the first local oscillator. The converter output signal is given by

$$y^I(t) = x_2^I(t) - (1 + \triangle\alpha)x_2^Q(t) \tag{6.15}$$

where

$$x_2^I(t) = \frac{1}{4}\cos(\omega - \omega_1 - \omega_2)t \tag{6.16}$$

and

$$\begin{aligned} x_2^Q(t) = -\frac{1}{4}\{&\cos[(\omega - \omega_1 - \omega_2)t]\cos(\triangle\phi_1 - \triangle\phi_2) \\ &- \sin[(\omega - \omega_1 - \omega_2)t]\sin(\triangle\phi_1 - \triangle\phi_2)\} \end{aligned} \tag{6.17}$$

where $\triangle\phi_2$ represents the phase error of the second local oscillator and $\triangle\alpha$ is the gain mismatch. Assuming that $\omega = \omega_{RF}$ for the message of interest, and $\omega = \omega_{IM}$, where $\omega_{IM} = 2\omega_1 - \omega_{RF}$, for the image signal, the IRR (in dB) can be computed as

$$\text{IRR} = 10\log_{10}\left[\frac{|y^I(t)|^2_{\omega=\omega_{RF}}}{|y^I(t)|^2_{\omega=\omega_{IM}}}\right] \tag{6.18}$$

$$= 10\log_{10}\left[\frac{1 + (1+\Delta\alpha)^2 + 2(1+\Delta\alpha)\cos(\Delta\phi_1 + \Delta\phi_2)}{1 + (1+\Delta\alpha)^2 - 2(1+\Delta\alpha)\cos(\Delta\phi_1 - \Delta\phi_2)}\right] \tag{6.19}$$

In this case, the image rejection performance can then be improved by making both phase errors equal. Note that an expression identical to Equation (6.19) can also be derived for the IRR of the Hartley converter.

Various nonideal effects are present in a practical transceiver. In the following, we review the most important of them.

- The dc offset is mainly due to transistor mismatches, self-mixing of a strong interference signal coupled to LO nodes by parasitic capacitors, and the in-band signal associated with the leakage of the LO signal into the antenna. It can be mitigated by ac coupling or self-calibration.
- The mismatch between the signal path and the phase error of the LO signals results in undesired spectral components in the signal spectrum. Differential circuit structures and adaptive algorithms can be required to improve the image rejection capability of the transceiver.

6.1.3 Architecture of a harmonic-rejection transceiver

Basic frequency translation techniques (see Figure 6.8) can be combined in different ways to optimize the transceiver performance. Figure 6.16 shows the block diagram of a harmonic-rejection transceiver [9–11]. The transmitter and receiver can operate simultaneously using separate channels, as in a duplex system. Their isolation from the undesirable noise, which can arise at the antenna, is achieved by RF bandpass filters.

A low-noise amplifier (LNA) adjusts the level of the incoming RF signal. It is followed by a quadrature down-conversion stage, which includes two mixers and the lowpass filters necessary to deliver a wideband IF signal and remove all up-converted frequency components. The conversion from IF to baseband is achieved by an image-reject structure. After the signal is processed by the mixers, the useful frequency components are added, while the undesired interference is cancelled by subtraction. The first local oscillator (LO) can operate at a fixed frequency, while a tunable frequency characteristic is required for the second LO so that all channel information can be accurately transferred

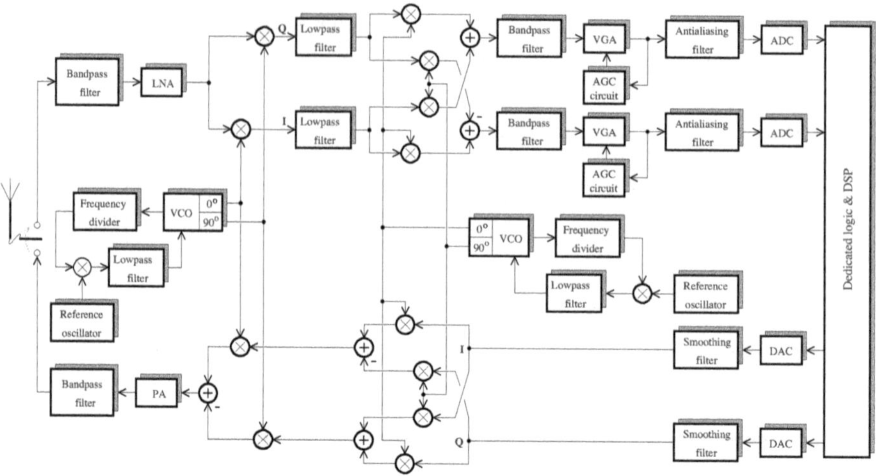

FIGURE 6.16
Block diagram of a harmonic-rejection transceiver.

to the baseband. The use of digitally programmable filters for channel selection provides the flexibility, which is essential in multi-standard receivers. The suitable signal level is fixed by the following variable gain amplifier (VGA), which is steered by an automatic gain control (AGC) circuit. The resulting I and Q signals are then digitized by the ADC and demodulated in the digital domain using dedicated logic circuits and digital signal processor (DSP).

In the transmit path, the baseband signal provided by the DAC is up-converted in two steps. The signal is first translated from the baseband to IF, and the channel tuning is performed using four mixers and a quadrature LO based on a voltage-controlled oscillator (VCO) structure. The next stage achieves the IF-to-RF conversion using an LO operating at a fixed frequency. The specifications of the RF filter, which follows the power amplifier (PA), are relaxed due to the image rejection feature of the frequency translation.

In general, any mismatches in the I and Q signal paths and fluctuations of the LO signals limit the image reject capability and therefore the sensitivity of the transceiver. However, a total interference rejection of at least 60 dB can still be achieved provided the IF is chosen high enough and well-suited RF filters are employed.

6.1.4 Amplifiers

Amplifiers are used to boost the amplitude level of a signal. A broad range of circuit architectures is available to meet various amplification requirements. Nevertheless, there are common performance characteristics for almost all amplifiers, such as the linearity, bandwidth, power efficiency, noise figure, and impedance matching. The appropriate amplifier structure for a given application is determined by the requirement of optimizing a specific characteristic and performance trade-offs (e.g., power amplifier, low-noise amplifier).

6.1.4.1 Power amplifier

Amplifiers [12,13] are required in the transmitter to scale the signal amplitude or power to the desired level.

- Principle and architectures

For wireless systems, the power is on the order of tens to hundreds of milliwatts (mW) and can be expressed in dBm. That is,

$$P(\text{in dBm}) = 10\log_{10}\left(\frac{P(\text{in W})}{10^{-3}}\right) \tag{6.20}$$

The efficiency, η, which characterizes the useful power consumption during the amplifier operation, is defined as

$$\eta = \frac{\text{Power delivered to load}}{\text{Power drawn from supply}} \tag{6.21}$$

The power-added efficiency (PAE) is defined as the difference between the output power, $P_{RF,out}$, and the input power, $P_{RF,in}$, divided by the supply power, P_{DC}, that is,

$$\text{PAE} = \frac{P_{RF,out} - P_{RF,in}}{P_{DC}} \tag{6.22}$$

An amplifier is then suitable for a given application due to its high PAE.

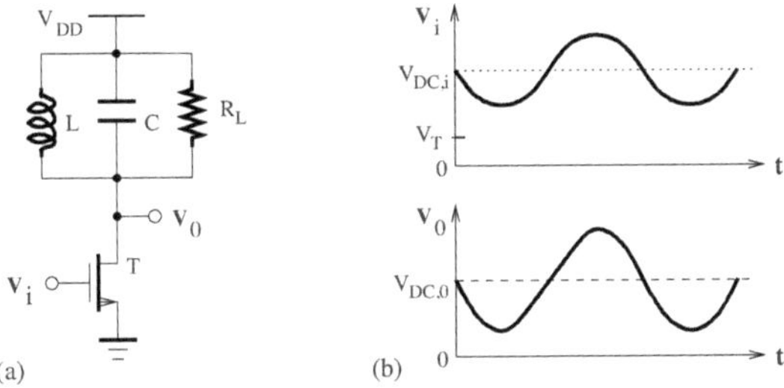

FIGURE 6.17
(a) Circuit diagram of a class A amplifier; (b) input and output voltages.

The operating point of the transistor determines whether an amplifier belongs to class A, B, AB, or C. Given a sinusoidal input voltage, the transistor conduction in a class A amplifier (see Figure 6.17(a)) should be guaranteed for the overall signal period. The value of the input bias, $V_{DC,i}$, as shown in Figure 6.17(b), is chosen so that the maximum swing of the input signal is kept above the threshold voltage necessary to maintain the transistor on. The average value of the output voltage, $V_{DC,0}$, is generally near V_{DD} and the output peak-to-peak swing is on the order of V_{DD}. Because the maximum power delivered to the load is expressed as $V_{DD}I_D/2$, and the power drawn from the supply voltage is of the form, $V_{DD}I_D$, where I_D is the transistor drain current, the maximum efficiency is 50%. This value is reduced in practical implementations due to the nonzero drain-source saturation voltage and the extra power dissipated by parasitic resistors. By consuming dc power regardless of the input signal level, class A amplifiers may appear to be inefficient for some applications.

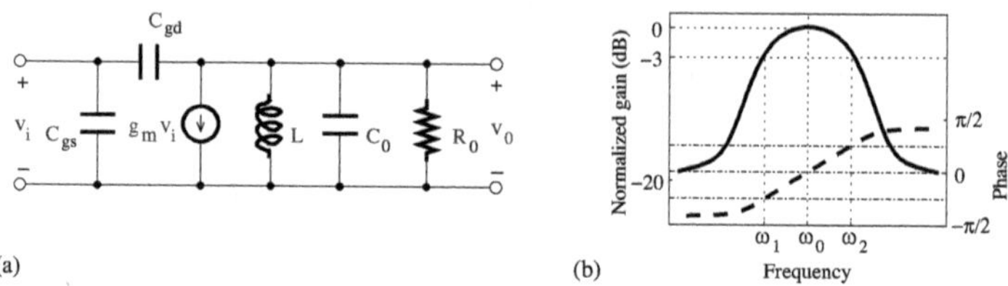

FIGURE 6.18
(a) Small-signal equivalent model of the class A amplifier; (b) frequency response characteristics.

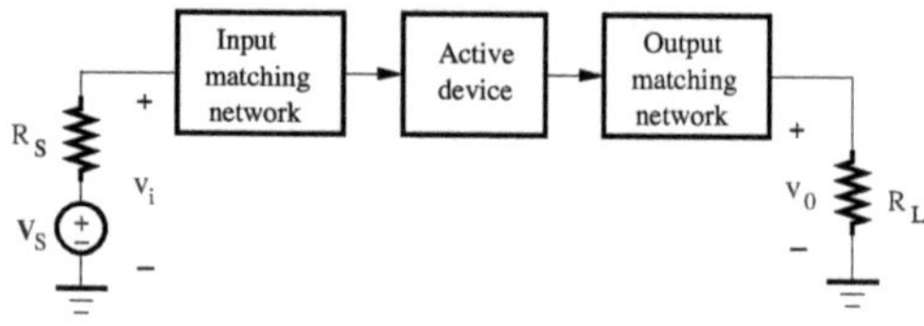

FIGURE 6.19
Generic block diagram of a power amplifier.

A class A tuned amplifier, as shown in Figure 6.17(a), is often used in applications where a narrow band of frequencies around a center frequency is of interest. The small-signal equivalent model is depicted in Figure 6.18(a). It is assumed that R_0 and C_0 denote the total resistance and capacitance at the output node, respectively. Applying Kirchhoff's current law at the output node gives

$$(V_0 - V_i)sC_{gd} + g_m V_i + V_0(G_0 + sC_0 + 1/sL) = 0 \tag{6.23}$$

where $G_0 = 1/R_0 = G_L + g_{ds}$, $C_0 = C_L + C_{db}$, and G_L and C_L are the load conductance and capacitance, respectively. This last equation can be solved for the voltage gain, which is expressed as

$$A(s) = \frac{V_0(s)}{V_i(s)} = \frac{sC_{gd} - g_m}{G_0 + s(C_0 + C_{gd}) + 1/sL} \tag{6.24}$$

Assuming that $g_m \gg \omega C_{gd}$ and $C_0 \gg C_{gd}$, it can be shown that

$$A(s) = \frac{V_0(s)}{V_i(s)} \simeq -g_m Z_0(s) \tag{6.25}$$

where the output impedance, Z_0, is given by

$$Z_0(s) = \frac{1}{G_0 + sC_0 + 1/sL} = \frac{1}{G_0 + C_0(s + \omega_0^2/s)} \tag{6.26}$$

At the center (or resonant) frequency, $\omega_0 = 1/\sqrt{LC_0}$, the gain is reduced to

$$A_0 = -g_m R_0 \tag{6.27}$$

The frequencies at which $|A(\omega)| = |A_0|/\sqrt{2}$, or the half-power frequencies, can be found as

$$\omega_1 = -\frac{1}{2R_0C_0} + \sqrt{\omega_0^2 + \frac{1}{4R_0^2C_0^2}} \tag{6.28}$$

and

$$\omega_2 = \frac{1}{2R_0C_0} + \sqrt{\omega_0^2 + \frac{1}{4R_0^2C_0^2}} \tag{6.29}$$

Note that $\omega_0 = \sqrt{\omega_1\omega_2}$. The -3 dB bandwidth of the amplifier is given by

$$BW = \omega_2 - \omega_1 = \frac{1}{R_0C_0} \tag{6.30}$$

The quality factor of the amplifier is defined as

$$Q = \frac{\omega_0}{BW} = \omega_0 R_0 C_0 = \frac{R_0}{L\omega_0} \tag{6.31}$$

The frequency response characteristics are illustrated in Figure 6.18(b). It can be observed that the phase is zero for $\omega = \omega_0$, and equal to $-\pi/4$ and $\pi/4$ for $\omega = \omega_1$ and $\omega = \omega_2$, respectively.

Considering a situation in which a class A amplifier with 1.5-V supply voltage should transmit a power of 500 mW, the output resistance will be given by

$$R_0 = \frac{V_{DD}^2}{2P_{max}} = \frac{1.5^2}{2 \times 0.5} = 2.25\,\Omega \tag{6.32}$$

Because the antenna resistance is generally chosen to be 50 Ω, an impedance matching network should be inserted between the amplifier and antenna to transform the load resistance to the required value. Hence, a high output power can be delivered without having to increase the value of the supply voltage. Figure 6.19 shows the generic block diagram of a power amplifier. A matching network is ideally lossless, and it is composed of inductors and capacitors.

For the untuned and tuned amplifiers shown in Figures 6.20(a) and (b), respectively, the transistor can be biased for operation in class A, B, or C mode. In a class B amplifier, the transistor is biased such that it operates as a voltage-controlled current source for a half signal cycle and is in the cutoff region during the next half cycle. The output signal of a class AB amplifier is available for a duration between the half and full period. The class C operation is achieved by biasing the transistor to have a conduction for less than a half signal period. Due to the distortion of the output waveform, this mode of operation is mainly used in tuned amplifiers (see Figure 6.20), where a pulsed current can be filtered to extract the fundamental frequency component. In both cases, the amplification is sustained by the current drawn from the supply voltage using a radio-frequency choke (RFC), which is an inductor designed to block high-frequency (or RF) signals while passing signal components at the low frequency (or dc). Generally, the output power can be increased by reducing the conduction time of the transistor for a given input power level. However, the device matching constraint required to get a linear reproduction of the input waveform can become difficult to meet.

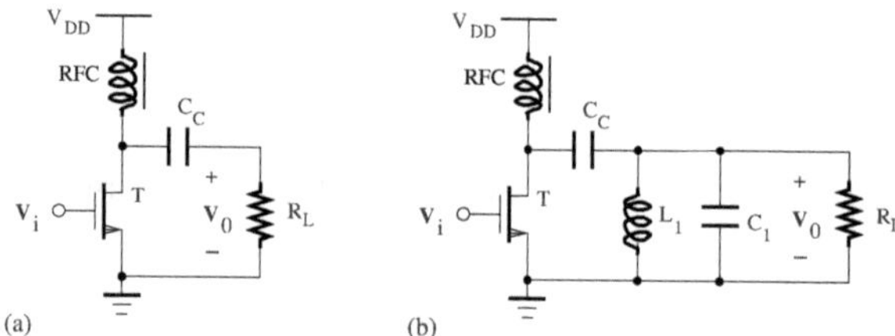

FIGURE 6.20
Circuit diagrams of (a) untuned and (b) tuned amplifiers.

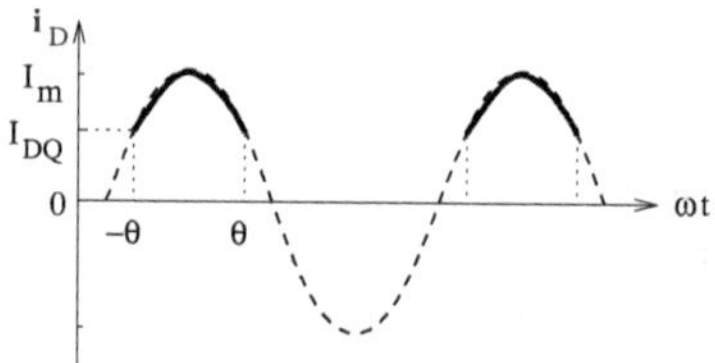

FIGURE 6.21
Representation of the drain current waveform.

Assuming that the conduction angle of the transistor, as shown in Figure 6.21, is such that $-\theta < \omega t < \theta$, the *dc* component of the drain current can be obtained as

$$I_{DC} = \frac{1}{2\pi}\int_{-\theta}^{\theta} i_D \mathrm{d}\omega t \tag{6.33}$$

where

$$i_D = I_m \cos\omega t - I_{CQ} = I_m \cos\omega t - I_m \cos\theta \tag{6.34}$$

Because the cosine function is even, it can be shown that

$$\begin{aligned} I_{DC} &= \frac{I_m}{2\pi}\int_{-\theta}^{\theta}(\cos\omega t - \cos\theta)\mathrm{d}\omega t \\ &= \frac{I_m}{\pi}\int_{0}^{\theta}(\cos\omega t - \cos\theta)\mathrm{d}\omega t \\ &= \frac{I_m}{\pi}(\sin\theta - \theta\cos\theta) \end{aligned} \tag{6.35}$$

The dc power consumption can be written as

$$P_{DC} = V_{DD}I_{DC} = \frac{V_{DD}I_m}{\pi}(\sin\theta - \theta\cos\theta) \tag{6.36}$$

The fundamental component of the drain current is given by

$$\begin{aligned} I_{D0} &= \frac{1}{2\pi}\int_{-\theta}^{\theta} i_D \cos\omega t \,\mathrm{d}\omega t \\ &= \frac{I_m}{\pi}\int_{0}^{\theta}(\cos\omega t - \cos\theta)\cos\omega t\,\mathrm{d}\omega t = \frac{I_m}{2\pi}(2\theta - \sin 2\theta) \end{aligned} \tag{6.37}$$

The maximum power delivered to the output load can then be expressed as

$$P_{0,max} = \frac{1}{2}V_{DD}I_{D0} = \frac{V_{DD}I_m}{4\pi}(2\theta - \sin 2\theta) \tag{6.38}$$

Thus, the maximum efficiency of the amplifier is of the form

$$\eta_{max} = \frac{P_{0,max}}{P_{DC}} = \frac{2\theta - \sin 2\theta}{4(\sin\theta - \theta\cos\theta)} \tag{6.39}$$

Equation (6.39) can be used for amplifiers of the class A, B, AB, or C. Note that $\theta = \pi$ for a class A amplifier, $\theta = \pi/2$ for a class B amplifier, $\pi/2 < \theta < \pi$ for a class AB amplifier, and $\theta < \pi/2$ for a class C amplifier. The efficiency increases as the conduction angle decreases.

In the case where the input signal is constant-envelope modulated such that the amplitude remains fixed and the frequency or phase is time varying (e.g., pulse-width modulated signals, sigma-delta modulated signals, etc.), a better approach to achieve an optimal drain voltage waveform is to rely on a transistor operating as a switch. The input signal then synchronizes the switching times of the transistor so that the frequency and phase information is transferred to the output. Switching-mode power amplifiers of class D, E, and F can feature a higher efficiency than class A, B, and C amplifiers that are based on transistors operating as voltage-controlled current sources. They

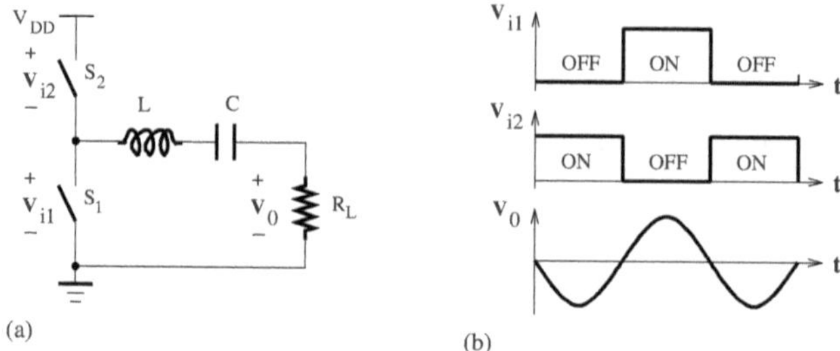

FIGURE 6.22
(a) Circuit diagram of an ideal class D amplifier; (b) voltage waveforms.

can exhibit an efficiency almost equal to 100% provided the switching devices are ideal.

The circuit diagram of an ideal class D amplifier is shown in Figure 6.22(a). The voltages V_{i1} and V_{i2} (see Figure 6.22(b)) are assumed to be square waveforms. The transistors used for the switch implementation operate either in the cutoff region or in the saturation region. They conduct on alternate half periods of the input signal. A series LC circuit tuned at a desired frequency is necessary to recover the output signal, which will be a sinusoid. Due to the nonzero on-resistance and the parasitic capacitances of the transistor, the maximum attainable efficiency can be less than 100%. A power loss occurs at the switching transitions due to the discharge of the capacitors connected at the transistor nodes. In practice, this power dissipation is mitigated by introducing a dead time of a few nanoseconds between the turn-on and turn-off of the transistors. The gate voltages are now two sinusoids, which are out of phase by 180°.

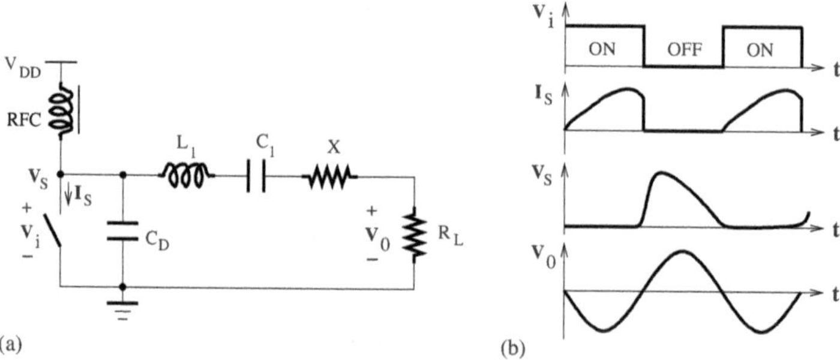

FIGURE 6.23
(a) Circuit diagram of an ideal class E amplifier; (b) voltage and current waveforms.

The efficiency can be improved by reducing the power dissipation in the transistor, as is the case in a class E amplifier. The circuit diagram of an ideal class E amplifier is shown in Figure 6.23(a). The switch, whose implementation is based on transistors, is controlled by the input signal and operates with a duty cycle of 50%. The L_1C_1 circuit is supposed to resonate at the frequency of the first harmonic of the input signal. The suitable phase shift between the voltage across the switch and the output voltage is introduced by the reactive element, X. Ideally, the switch power dissipation is zero, because the voltage and current waveforms of the switch, as shown in Figure 6.23(b), do not overlap. In order to avoid the power dissipation due to the discharging of the capacitor C_D when it is connected to the ground, the circuit must be designed such that the voltage across the switch returns to zero with zero slope (that is, $V_s = 0$ and $dV_S/dt = 0$) right before the switch is turned on. All the power from the dc supply voltage can then be transmitted to the output load.

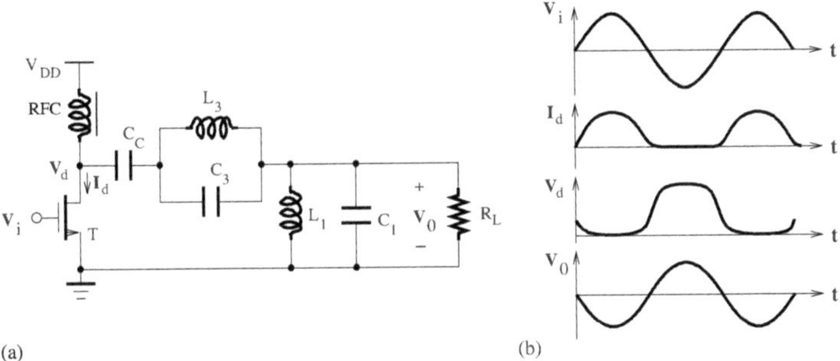

FIGURE 6.24
(a) Circuit diagram of a class F1 amplifier; (b) voltage and current waveforms.

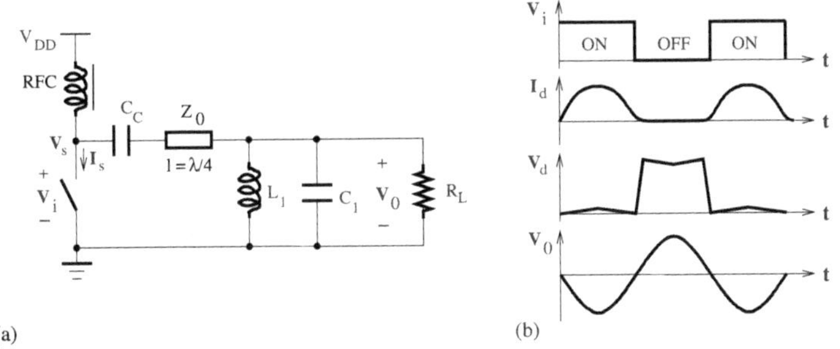

FIGURE 6.25
(a) Circuit diagram of a class F2 amplifier; (b) voltage and current waveforms.

A class F PA enhances the efficiency by using harmonic resonators in the output network to shape the output waveforms in such a way as to minimize the device power dissipation. There are two types of class F amplifiers.

With reference to the class F1 amplifier shown in Figure 6.24, the drain current and voltage of the transistor are processed by two LC circuits, which are connected in series and are tuned to the first and third harmonics of the output signal, respectively. The transistor should be biased in the same way as in a class B PA. However, the drain voltage swing of the class F1 PA is compressed with respect to the fundamental component because the ac component of the drain voltage is equal to the difference between the fundamental and third components. As a result, the class F1 PA exhibits a better power efficiency than a PA of the class B type due to the fact that the compressed drain voltage waveform has less overlap with the drain current and thus less power is dissipated in the device. By increasing the number of harmonics used in the PA, the efficiency can be further improved. For instance, the addition of a fifth harmonic resonator can increase the theoretical power conversion efficiency from about 88.4% to 92.0%.

In the case of the class F2 PA depicted in Figure 6.25, the input device is assumed to operate as a switch rather than a transconductor. Here, a quarter wavelength ($\lambda/4$) transmission line acting as an impedance transformer is used instead of a third-order harmonic resonator. At the carrier frequency or the resonance frequency of the LC network, the transmission line loaded by the impedance Z_L exhibits an input impedance of the form, Z_0^2/Z_L, where Z_0 is the characteristic impedance, because its length is equal to $\lambda/4$. The ac contributions superposed to the dc and fundamental components of the current I_s and voltage V_s are associated with even and odd harmonics, respectively, as the resistance at the switch output node is finite for even harmonics and infinite for odd harmonics. Ideally, the current I_s and voltage V_s should be insensitive to the input signal amplitude. This requires that the input signal must be large enough to turn on the transistor. By sizing the switch such that its on-resistance is much smaller than the input impedance of the matching network, the waveforms of the current I_s and voltage V_s are essentially determined by the matching network termination, which can then be designed to minimize the overlapping time between the current I_s and voltage V_s for improved power efficiency.

- Design examples

The circuit diagram of a class AB PA is depicted in Figure 6.26. The PA isolation from the mixer is ensured by the cascode transistors ($T_3 - T_4$) included in the pre-amplification stage, whose outputs are ac coupled to a class AB differential pair. The inductively degenerated output stage, shunt inductors, and coupling capacitors contribute to the fulfillment of the 50-Ω impedance matching requirement. A balun using a pair of two short coupled microstrip lines and tuning capacitors may be required for the conversion of the differential signal into the single-ended one, which can be processed by the following filter and antenna.

The circuit diagram of a class E PA is shown in Figure 6.27. It is based on the differential configuration, which has the advantage of increasing the signal swing and reducing the substrate coupling effects. The overall amplifier

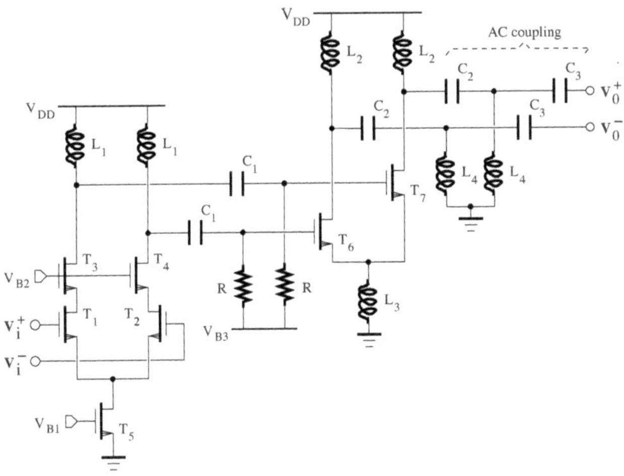

FIGURE 6.26
Circuit diagram of a class AB power amplifier.

consists of a driver and class E amplification stages with cross-coupled switching transistors. The current flowing through an inductive load is also used to control the switching of the other circuit half. Ideally, the class E amplifier should operate with sharp input pulses having a 50% duty cycle, and the peak voltage on the switch can be higher than three times V_{DD}, putting a limit on the maximum possible supply voltage. To overcome this limitation, the input capacitors of the class E amplifier are tuned out by the dc-feed inductors. The driver is designed to deliver sinusoid switching signals with a peak-to-peak voltage on the order of two times V_{DD}. The 50-Ω output impedance is provided by an output matching network, which involves off-chip capacitors and inductors realized with the bond wires. Note that the self-oscillation of the amplifier in the absence of the input signal can be prevented by switching the common source to the supply voltage. An integrated PA can deliver 1 W with up to 48% PAE.

For the design of efficient integrated PAs, the inductor and its associated parasitic components are dependent upon each other and should be included in an iterative optimization process. A CAD program is then necessary to reduce the effect of loss in the matching network.

Note that class D, E, and F PAs, which are switching-mode amplifiers, generally rely on transistors operating in the triode region to achieve the optimum efficiency and output power, in contrast to class A, B, and C PAs, whose transistors should not operate in the triode region.

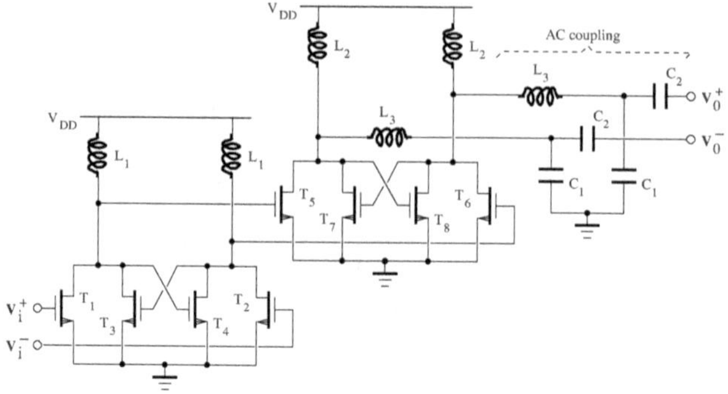

FIGURE 6.27
Circuit diagram of a class E power amplifier.

6.1.4.2 Low-noise amplifier

A low-noise amplifier (LNA) helps increase the amplitude of a very low-power signal without adding significant noise or distortion components.

- Fundamentals

In the receiver, the LNA should increase by 10 to 100 times the level of the signal coming from the antenna. It is generally designed to feature a low additive noise and to maintain linearity under large-signal conditions. An estimation of the amplifier contribution to the overall noise is provided by the noise figure, which is given by

$$\mathrm{NF} = 10\log_{10}(\mathrm{F}) \tag{6.40}$$

where F is the noise factor defined as

$$\mathrm{F} = \frac{\text{Overall output noise}}{\text{Output noise due to the source resistor}} \tag{6.41}$$

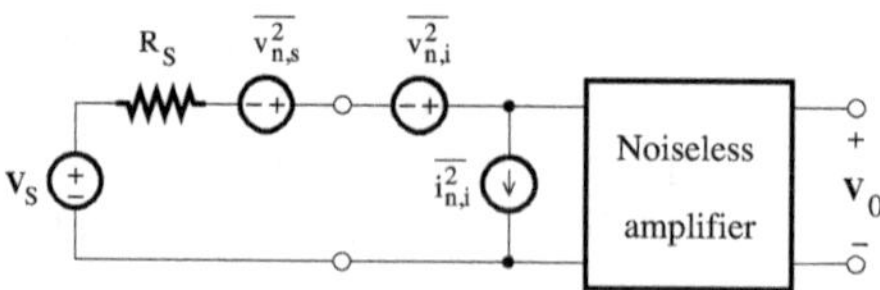

FIGURE 6.28
Amplifier equivalent model including noise sources.

The amplification stage, as shown in Figure 6.28, is assumed to be driven by a voltage source with the resistance R_s, whose noise at the output nodes of a noiseless amplifier model should be proportional to the gain factor. The source resistor gives rise to a thermal noise per unit bandwidth with the mean-square voltage of $\overline{v_{n,s}^2} = 4kTR_s$, where k is the Boltzmann constant and T is the absolute temperature. Let A_v be the amplifier gain. The noise factor, F, can be computed as

$$\mathrm{F} = \frac{\overline{v_{n,0}^2}}{A_v \overline{v_{n,s}^2}} \tag{6.42}$$

where $\overline{v_{n,0}^2}$ is the mean-square output noise voltage estimated with $v_s = 0$. To proceed further,

$$\mathrm{F} = 1 + \frac{\overline{v_{n,i}^2}}{4kTR_s} + \frac{\overline{i_{n,i}^2} R_s}{4kT} \tag{6.43}$$

where $\overline{v_{n,i}^2}$ and $\overline{i_{n,i}^2}$ are the input-referred voltage and current noises per unit bandwidth, respectively.

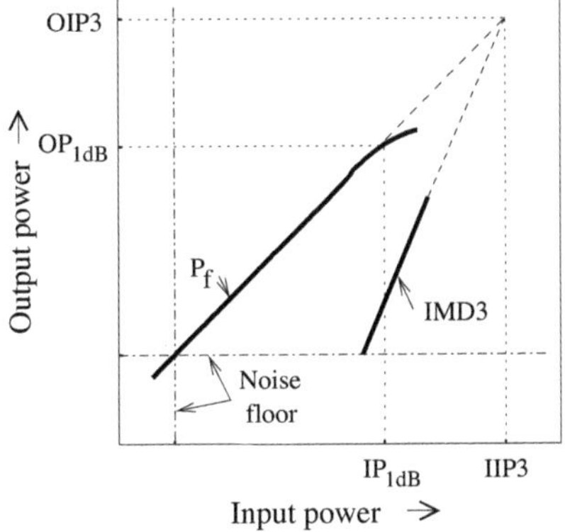

FIGURE 6.29
Plot of P_f and $IMD3$ versus the input power.

The amplifier will exhibit a nonlinear transfer function when it processes input voltages with a large magnitude. That is, two close frequencies, f_1 and $f_2 = f_1 + \triangle f$, contained in a signal can interact and produce third-order intermodulation distortion (IMD3) at $2f_1 - f_2$ and $2f_2 - f_1$. These interference products cannot be easily eliminated by a filter. The output power, P_f, of the signal component at the fundamental frequency and IMD3 versus the input power are represented in Figure 6.29. The IMD3 characteristic increases with a higher slope than P_f. The input-referred intercept point (IIP3) and output-referred intercept point (OIP3) correspond to the location, where the extrapolated IMD3 and the tangent to the P_f curve meet each other. These figures of merit provide a relative comparison of P_f and IMD3 up to the start of the power compression. The output compression point (OP_{1dB}) is defined as the output power level, which is 1 dB lower than it should be by applying the equivalent input level to an amplifier assumed to be linear. The input compression point (IP_{1dB}) corresponds to the input level that produces an output power 1 dB lower than it should be in an amplifier operating linearly. The spurious free dynamic range (SFDR) specifies the range over which the input level must be increased over the minimum detectable signal (MDS) level so that the third-order intermodulation products and the MDS take the same magnitude. The blocking dynamic range (BDR) lies between the MDS and the power compression point. It is determined by the noise and large-signal limitations.

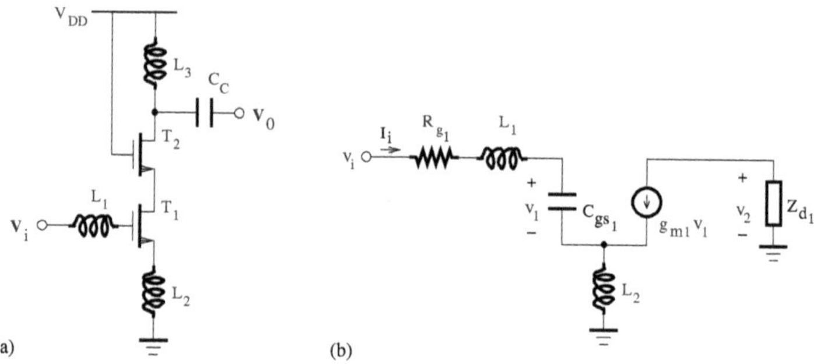

FIGURE 6.30
(a) Circuit diagram of a cascode LNA with source degeneration inductor; (b) small-signal equivalent circuit for the calculation of the input impedance.

In the design of LNAs, the cascode structure is generally adopted because it exhibits better isolation, improved bandwidth, and higher gain even at millimeter-wave frequencies when compared to architectures based on a single transistor. LNA circuits that are based on the inductively degenerated common source topology [19–21] have demonstrated a relatively low noise figure, and adequate gain, power consumption, and impedance matching for narrowband and wideband applications. The circuit diagram of an LNA based on such a topology is shown in Figure 6.30(a). The coupling capacitor C_C should be sized such that only the ac signal is transferred to the output load while the dc voltage component is blocked. Because the Miller effect of the gate-drain capacitance of T_1 is reduced by the cascode transistor T_2, the small-signal equivalent circuit for the calculation of the input impedance can be obtained as shown in Figure 6.30(b). Using Kirchhoff's voltage law, the input voltage can be derived as

$$V_i = \left(R_{g_1} + sL_1 + \frac{1}{sC_{gs_1}} \right) I_i + sL_2(I_i + +g_{m1}V_1) \tag{6.44}$$

where $V_1 = I_i/sC_{gs_1}$ and R_{g_1} is the equivalent series resistance available at the gate of T_1. The input impedance is then given by

$$Z_i = \frac{V_i}{I_i} \simeq R_{g_1} + g_{m1}\frac{L_2}{C_{gs_1}} + s(L_1 + L_2) + \frac{1}{sC_{gs_1}} \tag{6.45}$$

At the series resonance of the input circuit, $\omega^2 = \omega_0^2 = 1/(L_1 + L_2)C_{gs_1}$, the impedance becomes purely real and equal to $R_{g_1} + g_{m1}L_2/C_{gs_1}$. By carefully dimensioning the inductors L_1 and L_2, and the transistor T_1, the real term of the input impedance can be made equal to 50 Ω.

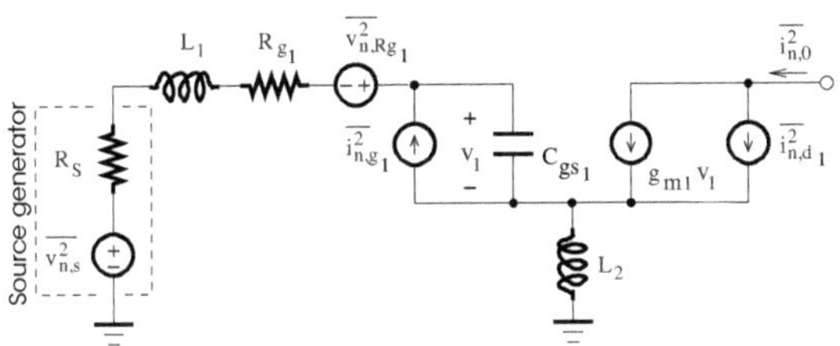

FIGURE 6.31
Small-signal equivalent circuit for the noise analysis.

The small-signal equivalent model of the source degenerated LNA shown in Figure 6.31 is considered for the noise analysis. The noise factor, F, is defined as the ratio of the total output noise power to the output noise power due to the signal source. In the case where the noise contribution due to the load of transistor T_1 can be ignored, we have

$$F = \frac{\overline{i^2_{n,0s}} + \overline{i^2_{n,0Rg_1}} + \overline{i^2_{n,0g_1}} + \overline{i^2_{n,0d_1}} + \overline{i^2_{n,0T_2}}}{\overline{i^2_{n,0s}}} \tag{6.46}$$

where $\overline{i^2_{n,0s}}$ is the output noise power contribution of the signal source, $\overline{i^2_{n,0Rg_1}}$ denotes the output noise power due to the overall resistance at the gate of T_1, and $\overline{i^2_{n,0g_1} + i^2_{n,0d_1}}$ is the output noise power contribution due to the gate and drain noise currents of T_1. To proceed further,

$$F = 1 + F_1 \tag{6.47}$$

where

$$F_1 = \frac{\overline{i^2_{n,0Rg_1}} + \overline{i^2_{n,0g_1}} + \overline{i^2_{n,0d_1}}}{\overline{i^2_{n,0s}}} \tag{6.48}$$

Let G, F, and E be the transfer functions relating the noise sources, $\overline{v^2_{n,Rg_1}}$, $\overline{i^2_{n,g_1}}$, and $\overline{i^2_{n,d_1}}$, respectively, to the output noise current. It can be shown that

$$\begin{aligned} F_1 = \frac{1}{\overline{v^2_{n,s}}|G(\omega)|^2} \Big[& \overline{v^2_{n,Rg_1}}|G(j\omega)|^2 + \overline{i^2_{n,g_1}}|F(j\omega)|^2 + \overline{i^2_{n,d_1}}|E(j\omega)|^2 \\ & + \overline{i_{n,g_1} i^*_{n,d_1}} F(j\omega)E^*(j\omega) + \overline{i^*_{n,g_1} i_{n,d_1}} F^*(j\omega)E(j\omega) \Big] \end{aligned} \tag{6.49}$$

For the determination of G, F, and E, it is assumed that the output noise contribution is solely caused by the noise source of interest. Applying the principle of voltage division to determine the ratio of V_1 to either V_s or $V_{R_{g1}}$, we get

$$\frac{V_1}{V_{R_{g1}}} = \frac{V_1}{V_s} = \frac{1/sC_{gs_1}}{R_s + R_{g1} + g_{m1}\dfrac{L_2}{C_{gs_1}} + s(L_1 + L_2) + \dfrac{1}{sC_{gs_1}}} \tag{6.50}$$

The output noise current is given by

$$I_{n,0} = g_{m_1} V_1 \tag{6.51}$$

Assuming that the input impedance matching condition is fulfilled, Equations (6.50) and (6.51) can be solved for

$$G(j\omega_0) = \frac{I_{n,0}(j\omega_0)}{V_s(j\omega_0)} = \frac{I_{n,0}(j\omega_0)}{V_{R_{g1}}(j\omega_0)} = \frac{g_{m1}}{\omega_0 C_{gs_1}(R_s + R_{g1} + g_{m1}L_2/C_{gs_1})} \tag{6.52}$$

Similarly, for the noise source i_{n,d_1}, it can be shown that

$$(R_s + R_{g1} + sL_1 + 1/sC_{gs_1})sC_{gs_1}V_1 = -sL_2(sC_{gs_1}V_1 + g_{m1}V_1 + I_{n,d_1}) \tag{6.53}$$

and

$$I_{n,0} = g_{m_1}V_1 + I_{n,d_1} \tag{6.54}$$

Combining Equations (6.53) and (6.54) gives

$$E(j\omega_0) = \frac{I_{n,0}(j\omega_0)}{I_{n,d_1}(j\omega_0)} = \frac{R_s + R_{g1}}{R_s + R_{g1} + g_{m1}L_2/C_{gs_1}} \tag{6.55}$$

The equations considered in the case of the noise source i_{n,g_1} are

$$(R_s + R_{g1} + sL_1)(I_{n,g_1} - sC_{gs_1}V_1) = V_1 + sL_2(sC_{gs_1}V_1 + g_{m1}V_1 - I_{n,g_1}) \tag{6.56}$$

and

$$I_{n,0} = g_{m_1}V_1 \tag{6.57}$$

Using Equations (6.56) and (6.57), the next transfer function can be derived:

$$F(j\omega_0) = \frac{I_{n,0}(j\omega_0)}{I_{n,g_1}(j\omega_0)} = \frac{g_{m1}[R_s + R_{g1} + j\omega_0(L_1 + L_2)]}{\omega_0 C_{gs_1}(R_s + R_{g1} + g_{m1}L_2/C_{gs_1})} \tag{6.58}$$

The input source has a power spectral density

$$\overline{v_{n,s}^2} = 4kTR_s\triangle f \tag{6.59}$$

and the power spectral density of the noise due to the parasitic resistance at the gate of the transistor T_1 is of the form

$$\overline{v_{n,Rg1}^2} = 4kTR_{g1}\triangle f \tag{6.60}$$

where $\triangle f$ is the noise bandwidth (in Hz), k is Boltzmann's constant, and T denotes the absolute temperature. For MOS transistors, the power spectral density of the gate-induced noise can be written as

$$\overline{i_{n,g_1}^2} = 4kT\delta g_g \triangle f \tag{6.61}$$

and the power spectral density of the channel noise is

$$\overline{i_{n,d_1}^2} = 4kT(\gamma/\alpha)g_{m1}\triangle f = 4kT\gamma g_{d10}\triangle f \tag{6.62}$$

where $g_g = \omega_0^2 C_{gs_1}^2/5g_{d10}$, g_{d10} is the conductance of the transistor T_1 when the drain-source voltage is equal to zero, and δ and γ are the coefficients of channel noise and gate-induced noise, respectively. Furthermore,

$$\overline{i_{n,g_1} i_{n,d_1}^*} = c\sqrt{\overline{i_{n,g_1}^2}}\sqrt{\overline{i_{n,d_1}^2}} \tag{6.63}$$

The parameter F_1 can then be expressed as

$$F_1 = \frac{R_{g_1}}{R_s} + \frac{\gamma}{\alpha}\frac{\chi}{Q_L}\frac{\omega_0}{\omega_T} \tag{6.64}$$

where

$$\chi = 1 - 2|c|\sqrt{\frac{\delta\alpha^2}{5\gamma}} + \frac{\delta\alpha^2}{5\gamma}(1 + Q_L^2) \tag{6.65}$$

$$Q_L = \frac{\omega_0(L_1 + L_2)}{R_s + R_{g_1}} \simeq \frac{\omega_0(L_1 + L_2)}{R_s} \simeq \frac{1}{\omega_0 C_{gs_1} R_s} \tag{6.66}$$

and

$$\omega_T = \frac{g_{m1}}{C_{gs_1}} \tag{6.67}$$

For long channel transistors, $\gamma = 2/3$, $\delta = 4/3$, $\alpha = 1$, and $c = 0.395\jmath$. It can be observed that the gate noise contribution is reduced by decreasing Q_L, while the channel noise contribution is attenuated by increasing Q_L. Therefore, there is an optimum value of Q_L that minimizes the noise figure. That is, $\partial F/\partial Q_L|_{Q_L=Q_{L,opt}} = 0$ at fixed ω_T, and

$$Q_{L,opt} = \sqrt{1 + 2|c|\sqrt{\frac{5\gamma}{\delta\alpha^2}} + \frac{5\gamma}{\delta\alpha^2}} \tag{6.68}$$

A typical value of $Q_{L,opt}$ is about 1.5 to 3. Generally, a low-Q input-matching network is preferred to make the design less sensitive to variations in the inductance and parasitic capacitances. Furthermore, the noise figure can also be minimized with respect to the width of T_1, which is related to g_{m1} and g_{d10}.

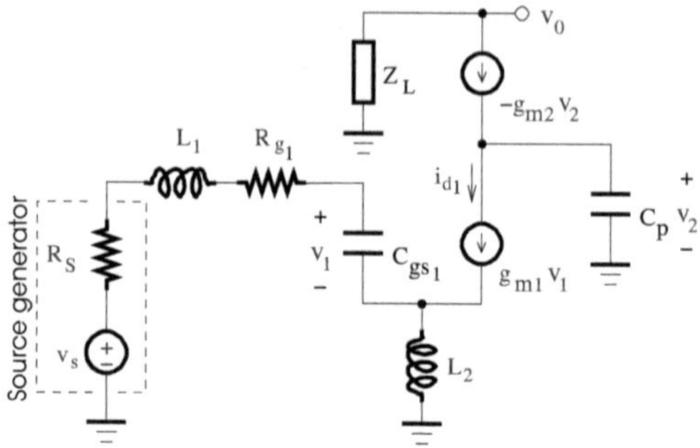

FIGURE 6.32
Small-signal equivalent circuit of the LNA.

With reference to the small-signal equivalent model of the LNA depicted in Figure 6.32, it can be shown that

$$V_2 = sC_p(-g_{m2}V_2 - I_{d_1}) \tag{6.69}$$

and

$$g_{m2}V_2 = V_0/Z_L \tag{6.70}$$

where $C_p \simeq C_{gs_2} + C_{sb_2} + C_{db_1}$. The output voltage is then given by

$$V_0 = \frac{-\,sC_p}{1 + sg_{m2}C_p} g_{m2} Z_L I_{d_1} \tag{6.71}$$

where $I_{d_1} = G(j\omega_0)V_s$. In practice, the parasitic capacitances connected to the common lower-drain-to-upper-source node can reduce the gain of the input transistor T_1 in such a way that the noise contribution of T_2 becomes relatively significant. They may be lowered by utilizing the dual-gate layout technique [22, 23] based on a transistor with two gates sharing one channel disposed between a source and a drain. Specifically, when T_1 and T_2 are of an identical size, the cascode structure is realized using the primary gate as the input node and biasing the second gate appropriately. Because the gate nearest the drain is ac grounded, it acts as a shield between the input gate and the output drain, thereby minimizing the parasitic coupling capacitance.

Considering a low-noise amplifier (LNA) as a two-port network whose input and output are connected to a generator with the output resistance R_0 and a load impedance Z_L, respectively, the input and output reflection coefficients can be written as

$$\Gamma_i = \frac{Z_i - R_0}{Z_i + R_0} \quad \text{and} \quad \Gamma_0 = \frac{Z_0 - R_0}{Z_0 + R_0} \tag{6.72}$$

where Z_i and Z_0 are the input and output impedances of the LNA, respectively. Let the generator and load reflection coefficients be defined as follows:

$$\Gamma_G = \frac{Z_G - R_0}{Z_G + R_0} \quad \text{and} \quad \Gamma_L = \frac{Z_L - R_0}{Z_L + R_0} \tag{6.73}$$

where Z_G is the impedance of the equivalent Thevenin circuit of the generator. It can be shown that

$$\Gamma_i = S_{11} + \frac{S_{12}S_{21}\Gamma_L}{1 - S_{22}\Gamma_L} \tag{6.74}$$

and

$$\Gamma_0 = S_{22} + \frac{S_{12}S_{21}\Gamma_G}{1 - S_{11}\Gamma_G} \tag{6.75}$$

where S_{11}, S_{12}, S_{22} and S_{21} are referred to as the scattering parameters (or S-parameters).
The impedance matching quality can generally be evaluated using the magnitude of the reflection coefficient. A good matching required a value of the reflection coefficient less than −10 dB which corresponds to the case where more than 90% of the signal power is transferred from the driving generator to the LNA or from the LNA to the loading stage.

Note that a more accurate expression of the noise figure can be derived by taking into account the parasitic capacitances and the noise contribution of T_2. In general, an LNA should be designed with a 50-Ω input impedance, minimum noise contribution, and maximum gain and IIP3. The inductances used in the LNA implementation are generally lower than a few hundred picohenries, making the amplification characteristics sensitive to package parasitic components as the frequency is increased.

To provide low-cost devices, RF circuits are realized using packages that are generally designed for low-frequency analog and digital ICs. At high frequencies, the parasitic components of these packages may be the cause of signal attenuation, and poor isolation of signal and ground. The effect of the pad capacitance (a few hundred femtofarads) and the lead and bonding wire inductance (a few nanohenries) should then be taken into account in the matching requirement for the input impedance. Figure 6.33(a) shows an implementation of a cascode LNA with a source degeneration inductor. The bias circuit for the transistor T_1 includes the resistors, R_{B_1} and R_{B_2}, the decoupling capacitor C_D, and the transistor T_B. The sum of the off-chip inductance, L_G, and the bond wire inductance, L_{BOND}, is equal to the inductance L_1. To overcome the lack of precise on-chip passive elements, the noise figure of the LNA is improved by using a high Q off-chip inductor in the input matching network.

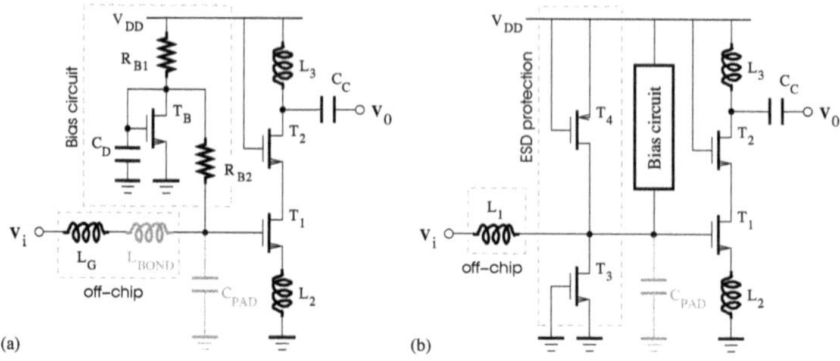

FIGURE 6.33
(a) Implementation of a cascode LNA with source degeneration inductor; (b) LNA implementation having ESD protection.

Transistors become more sensitive to electrostatic discharge (ESD) as the oxide thickness is lowered as the result of the CMOS process scaling toward short channel lengths. Figure 6.33(b) shows an LNA implementation having ESD protection. A low resistance path is associated with the input pin such that the ESD current mainly flows through the protection device (e.g., a power clamp). The transistors T_3 and T_4 should have thicker gate oxide than the transistors of the LNA core to exhibit higher threshold and breakdown voltages [14]. However, an increase in the LNA noise level can be caused by the parasitic components introduced by the ESD protection transistors. It is then important to keep the size of these parasitic components very small.

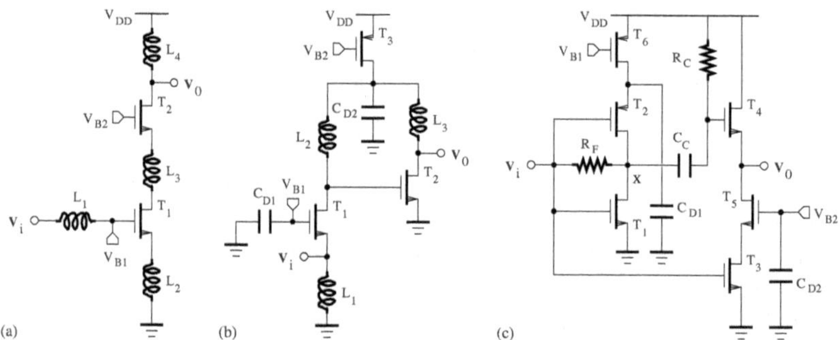

FIGURE 6.34
Circuit diagrams of low-noise amplifiers with (a) middle inductor and inductively degenerated, (b) common gate, and (c) resistive feedback input stages.

The impact of the parasitic capacitances connected to the common lower-drain-to-upper-source node on the cascode LNA performance at high frequencies can also be attenuated using an extra series inductor [15]. With reference to Figure 6.34(a), the inductor L_3 is inserted to create the appropriate series resonance for the compensation of the parasitic capacitance pole. In contrast to the approach where an inductor should be connected in parallel with the parasitic capacitance, this compensation method has the advantage of preventing the use of a large and accurate bypass capacitor. However, the inductor L_3 may provoke a reduction of the real part of the input impedance [16]. Assuming that $g_{m1} = g_{m2} = g_m$, $L_2 = L_3 = L$, $C_{gs_1} \gg C_{gd_1}$, and $g_m^2 L/C_{gs_1}$ is approximately equal to unity, the real part of the input impedance remains positive provided $L < 4C_{gs_1}/g_m^2$.

A common-source LNA exhibits relatively higher gain and a lower noise figure than the common-gate counterpart, which is limited by the absence of any degree of freedom in the choice of the input transistor transconductance that is essentially defined by the impedance-matching condition. However, the common-gate architecture has the highest potential to achieve a wideband input matching, especially at frequencies greater than 5 GHz. The circuit diagram of an LNA with a common gate input stage [3, 17] is shown in Figure 6.34(b). Because the resistance looking into the source terminal of the input transistor is on the order of $1/g_{m_1}$, the input matching condition, which is determined at the resonance of L_1 with $C_{gs_1} + C_{sb_1}$ and the pad capacitance, is reduced to $R_s \simeq 1/g_{m_1}$. The first stage of the LNA is inductively loaded by L_2. Transistor T_2, along with the inductor L_3, constitutes the second amplification stage, which also improves the output drive capability. The first stage is biased such that ac coupling can be avoided between the two stages of the LNA, thereby improving the amplifier response at higher frequencies. Note that C_{D1} and C_{D2} are decoupling capacitors.

To reduce the chip area, a wideband LNA can replace several LC-tuned LNAs typically used in multi-band and multi-mode narrowband receivers. This design solution can be adopted for analog cable (50 MHz to 850 MHz), satellite system (950 MHz to 2150 MHz), terrestrial digital video broadcasting (450 MHz to 850 MHz), and any application requiring reconfigurability for agile service switching. By using a feedforward noise-canceling technique, which can attenuate the noise and distortion contributions of the input (or matching) transistors, the noise and impedance-matching requirements can be met simultaneously. This allows the circuit components to be sized such that the resulting wideband impedance-matching LNA can exhibit a sufficiently large gain, adequate linearity, and a noise figure well below 3 dB. Figure 6.34(c) shows the circuit diagram of an LNA with a resistive feedback input stage [18]. With the assumption that $g_{m1} + g_{m2} \gg 1/R_{L_1}$ and $1 \gg R_F/R_{L_1}$, the shunt feedback around the CMOS inverter makes the input impedance equal to about $1/(g_{m1} + g_{m2})$. Here, R_{L_1} is the load resistance seen at the inverter output node. By adopting a current source to bias the inverter and connecting the source of T_2 to the ground via the decoupling capacitor C_{D_1}, the effects of supply voltage variations on the gain and input impedance is attenuated. To improve the overall noise figure, a highpass filter, $C_C - R_C$, is inserted between the inverter output and the gate of the transistor T_4. In the output stage, the transistor T_4 operates as a source follower while T_3, which is configured as a common-source stage, provides the gain. The transistor T_5, whose gate is ac coupled by the decoupling capacitor C_{D_2} to the ground, attenuates the Miller effect due to the gate-drain capacitance of T_3 and thus improves input-output isolation. Hence, two feedforward paths leading to the output node are implemented such that the signal contributions are combined in phase while the noise contribution from each input matching transistor equally counterbalances another. The common-source output stage including the transistors $T_3 - T_4$ provides a gain of about g_{m3}/g_{m4}. Considering the LNA with a source voltage v_s and a source resistance R_S, the superposition principle is used to show that

$$v_0 = v_x - (g_{m3}/g_{m4})v_i \tag{6.76}$$

where $v_i = v_s - R_S i_d$ and $v_x = v_s - (R_S + R_F)i_d$. Hence,

$$v_0 = \left(1 - \frac{g_{m3}}{g_{m4}}\right)v_s + \left[R_S\frac{g_{m3}}{g_{m4}} - (R_S + R_F)\right]i_d \tag{6.77}$$

Ideally, the fluctuations of i_d due to noise is cancelled provided $g_{m3}/g_{m4} = 1 + R_F/R_S$. Note that the aforementioned noise canceling approach can also be applied to other LNA architectures.

- Differential LNA architectures

The circuit diagram of an LNA with inductively degenerated input stage [19] is shown in Figure 6.35(a). It exhibits a pseudo-differential configuration. The low noise requirement is met using large input transistors with bias currents on the order of a few milliamperes. The LNA input impedance depends on L_2 and the gate-source capacitance of the transistor, and its value at the resonance is to be matched to 50 Ω. The cascode transistors T_3 and T_4 reduce the Miller effect of the gate-drain capacitors, and improve the amplifier reverse isolation. The output load inductors should resonate with the capacitors connected at the transistor drains. Furthermore, the network consisting of L_3 and C_C is sized to ensure an optimal power transfer to the following stages.

For a better rejection of on-chip interferences and a lower sensitivity to the substrate and supply voltage noises, a differential LNA architecture can be selected. Figure 6.35(b) shows the circuit diagram of a fully differential LNA with inductively degenerated input stage [24]. In order to stabilize the dc bias voltage at the output nodes, the actual common-mode voltage is detected and compared to a given reference voltage by the common-mode feedback circuit. In comparison with a single-ended structure, a differential LNA offers a twofold increase in dynamic range. However, its power consumption is somewhat increased.

A common-source LNA is known to feature better noise performance than a common-gate LNA at operating frequencies well below the transistor transition frequency, but at the cost of a higher power consumption and use of

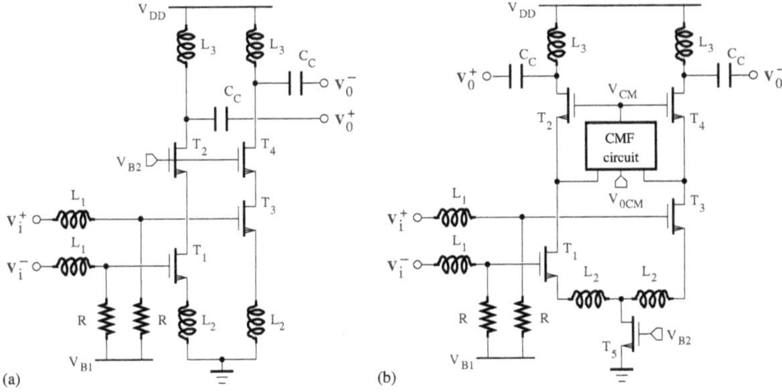

FIGURE 6.35
(a) Circuit diagram of a pseudo-differential LNA with inductively degenerated input stage; (b) circuit diagram of a fully differential LNA with inductively degenerated input stage.

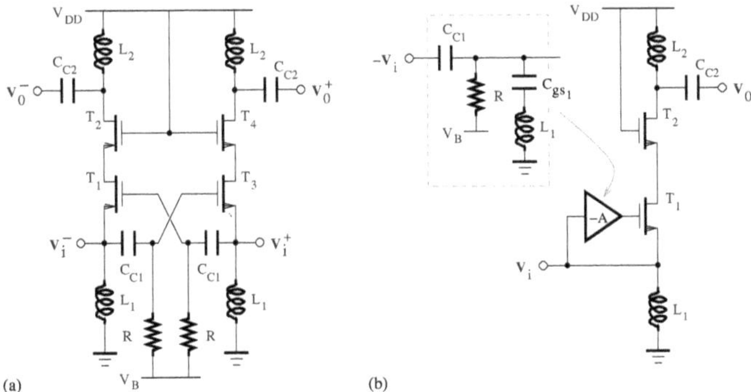

FIGURE 6.36
(a) Circuit diagram of a pseudo-differential, low-noise amplifier with capacitor cross-coupled input stage; (b) single-ended equivalent model.

off-chip matching components. The noise figure of the common-gate LNA can be improved without affecting the power consumption by using a pseudo-differential cross-coupled LNA, as shown in Figure 6.36(a). The achieved improvement is due to the transistor transconductance boosting, which is realized by inserting an inverting amplification stage between the source and gate nodes of the input transistors [25], as illustrated in Figure 6.36(b) using a single-ended equivalent model. The effective transconductance looking into the source node is now $(1 + A)g_m$, where A is the gain of the amplification stage inserted between the source and gate and g_m denotes the transconductance of each of the matched transistors T_1 and T_2. Hence, the input matching condition is of the form $R_s \simeq 1/g_m(1 + A)$. The contribution of the drain-induced noise to the amplifier noise factor is then reduced by the factor $1+A$ when compared to the case of a conventional common-gate LNA.

6.1.5 Mixer

Mixers are important components for transceivers. They are necessary for the RF-to-IF down-conversion and IF-to-RF up-conversion of the signals. Figure 6.37(a) shows the circuit diagram of a single-balanced mixer. Inductive

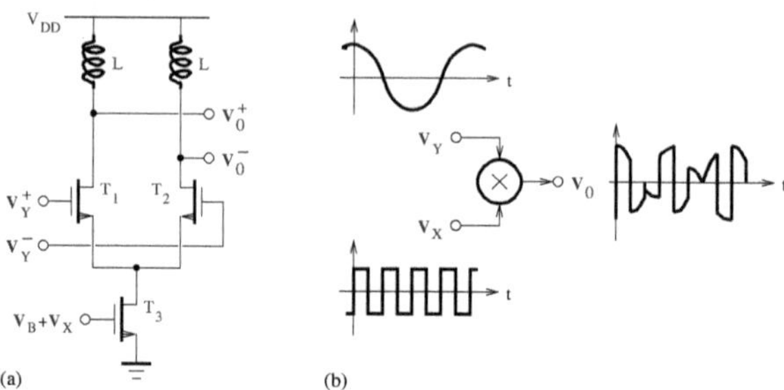

FIGURE 6.37
(a) Circuit diagram of a single-balanced mixer; (b) operation principle of a switching mixer.

loads are used to minimize the noise level. Transistors T_1 and T_2 should be sized based on the trade-off to be achieved between the switching time and the conversion gain, which can be degraded by parasitic capacitors. The third intercept point is determined by the overdrive voltage of the transistor T_1.

In communication applications, the single-balanced mixer [26] operates as a switching device to perform the signal multiplication, as shown in Figure 6.37(b). Let V_X be the input RF signal and V_Y be a square wave with amplitude of ± 1 and 50% duty cycle. The RF signal is a sinusoid of the form

$$V_X = A\cos(\omega_{RF} t) \tag{6.78}$$

where A is the signal amplitude. The operating point is set by a bias voltage superposed to V_X. The Fourier series representation of the square wave generated by the local oscillator (LO) is given by

$$V_Y = \sum_{n=1}^{+\infty} B_n \cos(n\omega_{LO} t) \tag{6.79}$$

where $\omega_{LO} = 2\pi/T$ denotes the LO fundamental frequency, T is the waveform period, and

$$B_n = \frac{4}{T}\int_0^{T/2} \cos(n\omega_{LO} t)\mathrm{d}t = \frac{\sin(n\pi/2)}{n\pi/4} \quad \text{for} \quad n \neq 0 \tag{6.80}$$

The output signal, which is obtained by periodically switching the drain current of T_3 to either the noninverting or inverting output node, can be expressed as

$$V_0 = KV_X V_Y = \frac{KA}{2} \sum_{n=1}^{+\infty} B_n[\cos(n\omega_{LO} + \omega_{RF})t + \cos(n\omega_{LO} - \omega_{RF})t] \quad (6.81)$$

where K is the mixer gain. Due to the spectral contents of the square waveform, the spectrum of the mixer output exhibits components around the LO fundamental frequency and its odd harmonics. Ideally, these components should represent only the sum and difference frequencies of the two input signals.

To determine the mixer gain, it is assumed that the transistors T_1 and T_2 operate as ideal switches. The transistor T_3, whose transconductance is equal to g_m, is biased such that the input voltage is converted into a current, $g_m V_X$, which, together with the bias current, is transferred to either the noninverting node for one half of the switching signal period or the inverting node for the other half. Hence, the mixer gain is of the form $K = g_m R_L$, where R_L is the equivalent output load resistance. Because the important contribution to the mixer output is associated with the fundamental Fourier coefficient, B_1, of the switching signal, the conversion gain can be reduced to

$$K_{CG} = \frac{g_m R_L}{2} B_1 = \frac{2 g_m R_L}{\pi} \quad (6.82)$$

However, the nonlinearity of the transconductor and mismatches of transistor switches have the effect of producing intermodulation products and dc offset at the mixer output. Double-balanced mixers feature the advantage of suppressing output spurs caused by the local oscillator as well as some high-order products.

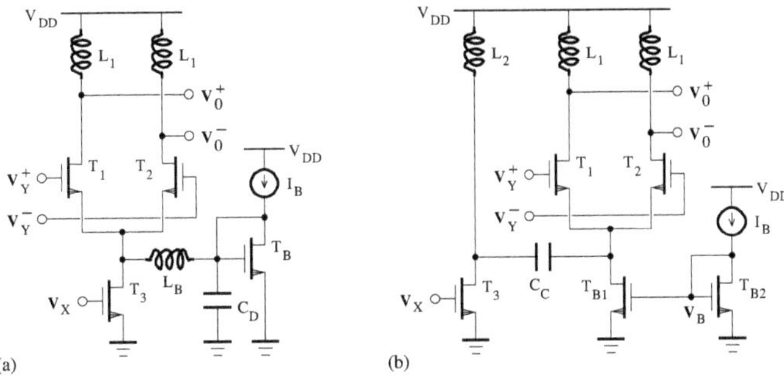

FIGURE 6.38
Circuit diagram of a single-balanced mixer (a) using auxiliary bias current path and (b) with capacitive coupling.

For the conventional single-balanced mixer, the frequency response is limited by the pole due to the parasitic capacitance at the drain of the transistor

T_3, and the switching speed remains low because all the bias current has to flow through either T_1 or T_2, and the noise contribution of the switching transistors may be folded into the frequency domain of the desired signal. Furthermore, to achieve an acceptable level of linearity, a drain-source voltage well above the saturation voltage is necessary for T_3. The choice of operating T_1 and T_2 in the saturation region requires a significant voltage headroom, thereby limiting the dynamic range available for the load and hence, the conversion gain [26].

The performance characteristics can be improved by adopting the mixer structure shown in Figure 6.38(a). This is achieved with the help of the inductor L_B which can resonate with the parasitic capacitance at the drain of the transistor T_3 and can drive almost half of the bias current. Although the aforementioned technique can help to improve the switching speed, the accuracy of the bias current can be degraded due to the impedance variations of the network consisting of C_D, L_B, and T_3.

Another design approach consists of using the mixer topology depicted in Figure 6.38(b). The coupling capacitor, C_C, provides isolation between the bias current of the switching transistors, T_1 and T_2, and that of the input transistor, T_3. As a result, an increase of the mixer gain and a reduction of the noise factor can be observed.

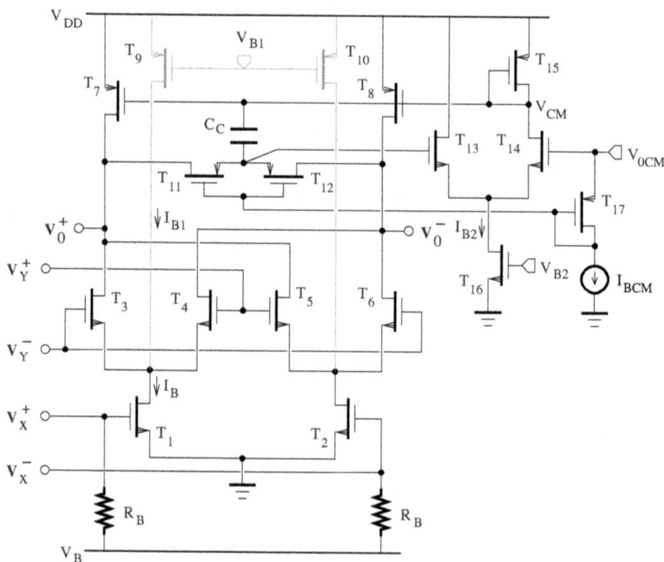

FIGURE 6.39
Circuit diagram of a double-balanced mixer with the common-mode feedback circuit.

Double-balanced mixers are widely used because they can reject the noise from the LO circuit and can minimize the LO signal feed-through, which can affect the performance of single-balanced structures. The circuit diagram [27] of a double-balanced mixer is shown in Figure 6.39. Transistors $T_1 - T_2$ realize a transconductor, while $T_3 - T_6$ operate as a switch controlled by the signal provided by the LO. The actual output common-mode (CM) voltage is detected using two matched transistors, T_{11} and T_{12}, which are biased by the diode-connected transistor T_{17} to operate in the triode region, and compared to the desired common-mode reference by the amplifier consisting of transistors $T_{13} - T_{16}$. The common-mode voltage is then set by varying the bias voltage of transistors $T_7 - T_8$ in order to reduce the comparison difference. The compensation capacitor C_C is used to provide an appropriate bandwidth to the CM feedback loop. The contribution of the switching transistors to the mixer noise is dominant during the time period in which both transistors in the source-coupled pair are conducting. It can be reduced using an LO waveform with sharpened transitions and minimizing the parasitic capacitors of transistors. Transistors driven by a sine wave, whose amplitude is raised to sharpen its transitions, can be forced to operate in the triode region. As a result, an increase in the mixer distortion can be observed due to the nonlinear output resistance of the transistors. The reduction in the overdrive of the LO signals necessary for the switching of $T_3 - T_6$ and the increase in the conversion gain are simultaneously achieved using the current sources I_{B_1}. This latter is set to about $3I/4$ to preserve the linearity of the mixer, which can be worsened for a value of I_{B_1} approaching the one of the bias current I_B of T_1 or T_2.

Mixers with a grounded-source differential transistor pair feature a better linearity than the ones based on a differential stage biased by a constant tail current. Furthermore, the current consumption, which is on the order of a few milliamperes, should be reduced as the transistor is scaled down.

6.1.6 Voltage-controlled oscillator

A voltage-controlled oscillator (VCO) generates an output signal whose oscillation frequency is set by a control voltage. Generally, the phase noise is a key specification for the VCO design.

- **Noise**

The signal generated by an ideal oscillator can be written as

$$v_0(t) = A\cos(\omega_0 t + \phi) \tag{6.83}$$

where A denotes the amplitude, ω_0 is the angular oscillation frequency, and ϕ represents the phase. Due to the different noise sources, the amplitude and phase can become a time-varying function. As a result, the spectrum of a practical oscillator exhibits sidebands close to ω_0. The noise caused by the amplitude and phase fluctuations can be characterized by

$$\mathcal{L}\{\triangle\omega\} = 10\log_{10}\left(\frac{P_{sideband}(\omega_0 + \triangle\omega)}{P_{carrier}}\right) \tag{6.84}$$

where $P_{sideband}(\omega_0 + \triangle\omega)$ is the power per unit bandwidth of a single sideband at a frequency offset of $\triangle\omega$ from the carrier, and $P_{carrier}$ is the carrier power. The spectral density $\mathcal{L}$ is expressed in units of decibels below the carrier per hertz (dBc/Hz). Note that the amplitude noise can be minimized by an appropriate limiter and the overall oscillator noise is then dominated by the phase noise. An oscillator based on a lossless LC network should have no phase noise and exhibit a noise factor equal to one.

- **Differential *LC* oscillator**

Although various oscillator configurations can be used in wireless communication systems, a differential architecture is more suitable for integrated-circuit implementations, especially when the effects of power supply noise and substrate noise coupling are to be minimized. A voltage-controlled oscillator (VCO) consisting of a cross-coupled differential pair of transistors loaded by inductors is shown in Figure 6.40(a). To sustain the oscillation, the resistive losses in the passive elements are compensated by the negative resistance provided by the cross-coupled transistors (T_1 and T_2). The tuning of the oscillation frequency is achieved by varying the control voltage, which determines the capacitances, C, realized by MOS transistors (T_3 and T_4) in the inversion or accumulation mode. In addition to the reduction in parasitic resistances in the inductors and MOS variable capacitors, an acceptable level of phase noise can be achieved by sizing the transistors appropriately. Note that the minimum supply voltage is $V_{DS(sat),T_B} + V_{GS,T_1}$. A small value of $V_{GS} - V_T$ for T_1 and T_2 can then provide a large transconductance and a small power consumption. However, in this case, the transistor sizes can become very large, resulting in large parasitic capacitances.

Assuming that the tail current is equal to a constant bias current, I_B, voltage and current waveforms of the VCO can be represented as shown in Figure 6.40(b), where V_{CM} is the output common-mode voltage. Ideally, crossed-coupled transistors operate as switches. The drain currents, I_{D_1} and I_{D_2}, can then be considered as periodic square signals, with a *dc* component, $I_B/2$, and a magnitude of the first harmonic, $2I_B/\pi$, at the oscillation frequency, ω_0. The resistance of the tank circuit is small only at ω_0 and the bias current flowing through each inductor is $I_B/2$. Output signals oscillate around V_{CM} with an amplitude $A_0/2$, where $A_0 = 2I_B R_{tank}/\pi$ and R_{tank} is the equivalent resistance of the tank circuit. Note that the start-up condition for oscillations can be reduced to $g_m/2 > 1/R_{tank}$, where g_m is the transistor transconductance.

With reference to the small-signal equivalent model of the VCO depicted in Figure 6.40(c), the nodal equations at both outputs can be written as

$$i^+ = \frac{v_0^+}{Z_{ds1}} + \frac{v_0^+}{Z_{gs2}} + g_{m1}v_0^- \tag{6.85}$$

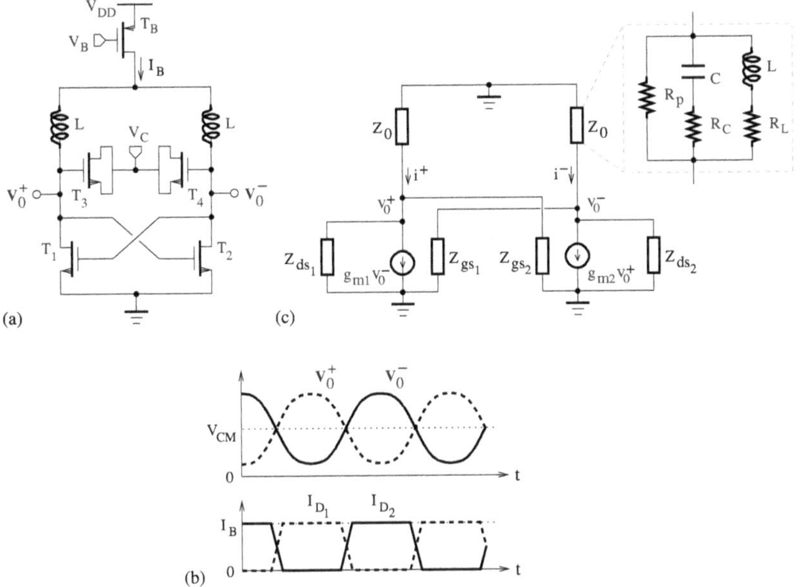

FIGURE 6.40
(a) Circuit diagram, (b) waveforms, and (c) small-signal equivalent model of a voltage-controlled oscillator.

and

$$i^- = \frac{v_0^-}{Z_{ds2}} + \frac{v_0^-}{Z_{gs1}} + g_{m2}v_0^+ \tag{6.86}$$

In practice, the transistors T_1 and T_2 are matched, that is, $g_{m1} = g_{m2} = g_m$, $Z_{ds2} = Z_{ds1} = Z_{ds}$, and $Z_{gs2} = Z_{gs1} = Z_{gs}$. Assuming that $v_0^+ = v_0/2$ and $v_0^- = -v_0/2$, the impedance provided by the cross-coupled section is given by

$$Z_i = \frac{v_0}{i^+} = -\frac{v_0}{i^-} = \frac{-2}{g_m - \dfrac{Z_{ds} + Z_{gs}}{Z_{ds}Z_{gs}}} \tag{6.87}$$

Because the magnitude of the impedance Z_{ds} is very high, it can be shown that $(Z_{ds} + Z_{gs})/Z_{ds}Z_{gs} \simeq 1/Z_{gs}$. For $Z_{gs} = 1/(sC_{gs})$, we obtain

$$Z_i = \frac{v_0}{i^+} = -\frac{v_0}{i^-} \simeq \frac{-2}{g_m(1 - s/\omega_t)} \tag{6.88}$$

where $\omega_t = g_m/C_{gs}$. When $\omega \ll \omega_t$, $Z_i = R_i = -2/g_m$ and the cross-coupled section realizes a negative resistance that can sustain oscillations by compensating loss in the LC tank. The use of Kirchhoff's current law leads to

$$(Z_i + Z_0)i^+ = 0 \tag{6.89}$$

and

$$(Z_i + Z_0)i^- = 0 \tag{6.90}$$

For the VCO to oscillate, the currents i^+ and i^- must be nonzero. Hence,

$$Z_i + Z_0 = 0 \tag{6.91}$$

To proceed further, it can be assumed that the impedance Z_0 is realized by the parallel connection of the parasitic resistance, R_p, in parallel with the LC-tank, the capacitor C with its parasitic resistance R_C, and the inductor L with its parasitic resistance R_L. The resonant frequency at which oscillations will occur is derived from $\mathrm{Im}(Z_0) = -\mathrm{Im}(Z_i)$ to be

$$\omega_0 = \frac{1}{\sqrt{LC}}\sqrt{\frac{L/C - R_L^2}{L/C - R_C^2}} \tag{6.92}$$

At the oscillation frequency, the magnitude of the output voltage is either equal to the product of the tail current and the LC tank equivalent resistance (current-limited operation regime) or the minimum value between the supply voltage and the voltage at which there is a change in the operating regions of transistors (voltage-limited operation regime). On the other hand, by setting $\mathrm{Re}(Z_i) = -\mathrm{Re}(Z_0)$, the value of transconductance, g_m, for each of the transistors T_1 and T_2 is obtained as

$$g_m = \frac{2}{R_p} + \frac{2R_C}{R_C^2 + (1/\omega_0 C)^2} + \frac{2R_L}{R_L^2 + \omega_0^2 L^2} \tag{6.93}$$

$$= \frac{2}{R_p} + \frac{2}{R_C(1+Q_C^2)} + \frac{2}{R_L(1+Q_L^2)} \tag{6.94}$$

where $Q_C = 1/(\omega_0 C R_C)$ and $Q_L = \omega_0 L/R_L$ denote the quality factor of the capacitor and inductance, respectively. To guarantee the start-up of the oscillator, the transconductance g_m should be sufficiently high to overcome all resistive losses in the oscillator circuit. Ideally, $L/C \gg R_C, R_L$, and the oscillation frequency is reduced to $\omega_0 = 1/\sqrt{LC}$.

Due to the signal phase fluctuations caused by the noise contribution of oscillator components, the oscillation criterion may not hold perpetually. The phase noise then appears to be a performance characteristic of the oscillator. It provides a measure of the stability of the output frequency over a given duration. In the case where R_C and R_L are negligible and the overall resistance, R, at the output node is compensated by the negative resistance of the cross-coupled transistors, the current noise flows through a lossless LC

network with the impedance given by

$$Z(\omega_0 + \triangle\omega) = \left[j(\omega_0 + \triangle\omega)C + \frac{1}{j(\omega_0 + \triangle\omega)L} \right]^{-1} \tag{6.95}$$

$$= \frac{j(\omega_0 + \triangle\omega)L}{1 - (\omega_0^2 + 2\omega_0\triangle\omega + \triangle\omega^2)LC} \tag{6.96}$$

$$= \frac{j\omega_0(1 + \triangle\omega/\omega_0)L}{2(\triangle\omega/\omega_0) + (\triangle\omega/\omega_0)^2} \tag{6.97}$$

where $\omega_0^2 = 1/LC$. Because $\omega_0 \gg \triangle\omega$, it can be shown that

$$Z(\omega_0 + \triangle\omega) \simeq -j\frac{\omega_0 L}{2\triangle\omega/\omega_0} = -j\frac{R}{2Q\triangle\omega/\omega_0} \tag{6.98}$$

where $Q = R/(\omega_0 L)$. In the absence of amplitude saturation, the power contribution due to the phase noise can be written as

$$P_{sideband} = \frac{1}{2}\frac{\overline{v_n^2}}{B \cdot R} = \frac{1}{2}\frac{\overline{i_n^2} \cdot |Z|^2}{B \cdot R} = 2kT\left(\frac{\omega_0}{2Q\triangle\omega}\right)^2 \tag{6.99}$$

where $\overline{i_n^2}$ and $\overline{v_n^2}$ are the mean-square noise current and voltage, respectively; B denotes the bandwidth; ω_0 is the oscillation frequency; Q is the effective quality factor of the LC tank; k is Boltzmann's constant; T is the temperature in degrees Kelvin; and $\triangle\omega$ represents the offset from the carrier. The use of the $1/2$ factor is justified by the equipartition theorem of thermodynamics, which predicts an even distribution of the noise power among all of the quadratic degrees of freedom (here, the amplitude and phase) in thermal equilibrium.

A first-order approximation of the single-sided noise spectral density, or phase noise, for the oscillator is of the form

$$\mathcal{L}\{\triangle\omega\} = 10\log_{10}\left[\frac{2kT}{P_{carrier}}\left(\frac{\omega_0}{2Q\triangle\omega}\right)^2\right] \tag{6.100}$$

where $P_{carrier} = \overline{v_0^2}/R$. The phase noise is determined by considering only the resistor thermal noise and can be reduced by increasing the quality factor, Q, of the LC tank section. While a high Q is desirable, it should also be noted that Q is a function of L, and thereby the available circuit area. The values of the phase noise computed from Equation (6.100) exhibit a different behavior and are smaller than the ones measured between the output nodes of a practical circuit due to the effect of other noise sources.

By considering the VCO as a linear, time-invariant system, a semi-empirical model [28, 29] can be adopted for the derivation of the phase noise as follows

$$\mathcal{L}\{\triangle\omega\} = 10\log_{10}\left\{\frac{2FkT}{P_{carrier}}\left[1 + \left(\frac{\omega_0}{2Q\triangle\omega}\right)^2\right]\left(1 + \frac{\omega_{1/f^3}}{\triangle\omega}\right)\right\} \tag{6.101}$$

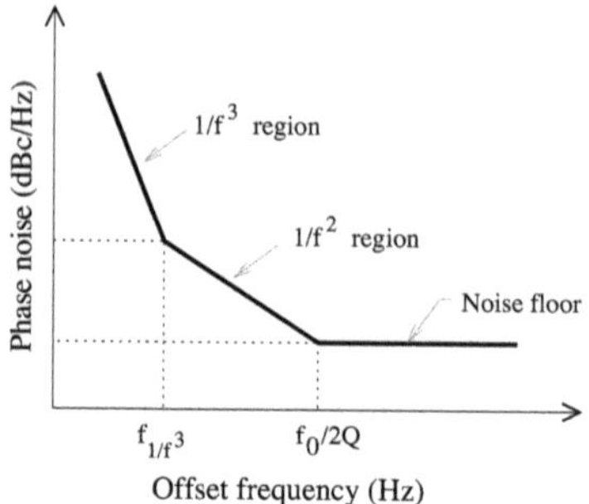

FIGURE 6.41
Plot of the phase noise according to the empirical model.

where F is an empirical parameter, which is also known as the device excess noise number, and ω_{1/f^3} is the frequency of the corner between the $1/f^2$ and $1/f^3$ regions. Figure 6.41 shows the phase noise versus the offset frequency. It appears that the phase noise spectrum exhibits a $1/f^3$ region in addition to the $1/f^2$ region and a noise floor at large frequency offsets. The use of Equation (6.101) for the computation of the phase noise is limited by the fact that fitting parameters (F, ω_{1/f^3}), which should be determined from measurements, are required.

In practice, the circuit parameters required for the phase noise computation change as the bias point of the transistors fluctuates during each oscillation cycle. For each noise source, the VCO can then be considered a linear, time-variant system.

Let $h_\phi(t,\tau)$ be the impulse response at time τ for the excess phase ϕ, and $\Gamma(\omega_0\tau)$ be the 2π-periodic impulse sensitivity function (ISF). The excess phase can be obtained as the superposition of impulse responses for all τ [30]. That is,

$$\phi(t) = \int_{-\infty}^{\infty} h_\phi(t,\tau)\mathrm{d}\tau = \frac{1}{q_{max}} \int_{-\infty}^{t} \Gamma(\omega_0\tau)i(\tau)\mathrm{d}\tau \tag{6.102}$$

where

$$h_\phi(t,\tau) = \frac{\Gamma(\omega_0\tau)}{q_{max}} u(t-\tau) \tag{6.103}$$

Here, $u(t)$ is the unit step function, and q_{max} is the maximum charge swing across the capacitor on the output node. The ISF is normalized to q_{max} in order to make $h_\phi(t,\tau)$ independent of the amplitude. For each noise source, the impulse response $h_\phi(t,\tau)$ can be determined by SPICE simulations based on the extracted VCO model. The ISF can be expressed as a Fourier series of the form

$$\Gamma(\omega_0\tau) = \frac{c_0}{2} + \sum_{n=1}^{\infty} c_n \cos(n\omega_0\tau + \theta_n) \tag{6.104}$$

Substituting Equation (6.104) into (6.102) gives

$$\phi(t) = \frac{1}{q_{max}} \left[\frac{c_0}{2} \int_{-\infty}^{t} i(\tau)\,\mathrm{d}\tau + \sum_{n=1}^{\infty} c_n \int_{-\infty}^{t} i(\tau)\cos(n\omega_0\tau)\mathrm{d}\tau \right] \tag{6.105}$$

We observe that the excess phase $\phi(t)$ can be computed for an arbitrary input current injected into any circuit node using the corresponding Fourier coefficients of the ISF. Let the current to be injected in a given node be a sinusoidal function, for instance. Thus,

$$i(t) = I_M \cos[(m\omega_0 + \triangle\omega)t] \tag{6.106}$$

where I_M represents the maximum amplitude. Combining Equations (6.105) and (6.106), we obtain

$$\begin{aligned} \phi(t) = {} & \frac{c_0 I_M}{2q_{max}} \int_{-\infty}^{t} \cos[(m\omega_0 + \triangle\omega)t]\mathrm{d}\tau \\ & + \frac{I_M}{q_{max}} \sum_{n=1}^{\infty} c_n \int_{-\infty}^{t} \cos[(m\omega_0 + \triangle\omega)\tau]\cos(n\omega_0\tau)\,\mathrm{d}\tau \end{aligned} \tag{6.107}$$

For causal signals, recall that

$$\begin{aligned} & \int_{-\infty}^{t} \cos[(m\omega_0 + \triangle\omega)\tau]\cos(n\omega_0\tau)\,\mathrm{d}\tau \\ & \quad = \frac{1}{2}\left[\frac{\sin(m\omega_0 + n\omega_0 + \triangle\omega)t}{m\omega_0 + n\omega_0 + \triangle\omega} + \frac{\sin(m\omega_0 - n\omega_0 + \triangle\omega)t}{m\omega_0 - n\omega_0 + \triangle\omega} \right] \end{aligned} \tag{6.108}$$

is nonnegligible only when $n = m$. Because $\triangle\omega$ is close to any integer multiple of the oscillation frequency, and $\sin(x)/x$ decays to zero as x tends to infinity, has a maximum value of unity at $x = 0$, and is zero at $x = k\pi$ ($k = \pm1, \pm2, \ldots$), the excess phase can be approximated as

$$\phi(t) \simeq \frac{c_m I_M}{2q_{max}\triangle\omega}\sin(\triangle\omega\, t) \tag{6.109}$$

Combining Equation (6.109) with the oscillator output signal considered to be of the form

$$v_0(t) = A\cos[\omega_0 t + \phi(t)] \tag{6.110}$$

we have

$$v_0(t) = A\cos\left[\omega_0 t + \frac{c_m I_M}{2q_{max}\triangle\omega}\sin(\triangle\omega\, t)\right] \tag{6.111}$$

If $c_m I_M / 2q_{max}\triangle\omega \ll 1$, it can be found that

$$v_0(t) \simeq A\left[\cos(\omega_0 t) - \frac{c_m I_M}{2q_{max}\triangle\omega}\sin(\omega_0 t)\sin(\triangle\omega\, t)\right] \tag{6.112}$$

or equivalently,

$$v_0(t) \simeq A\cos(\omega_0 t) + \frac{c_m I_M A}{4q_{max}\triangle\omega}[\cos(\omega_0 + \triangle\omega)t - \cos(\omega_0 - \triangle\omega)t] \quad (6.113)$$

Hence, the injection of a current at $n\omega_0 + \triangle\omega$ into an oscillator node produces a pair of equal sidebands at $\omega_0 \pm \triangle\omega$ with a sideband power relative to the carrier given by

$$\frac{P_{sideband}(\triangle\omega)}{P_{carrier}} = \left(\frac{c_m I_M}{4q_{max}\triangle\omega}\right)^2 \quad (6.114)$$

where $I_M^2/2 = \overline{i_n^2}/\triangle f$. To proceed further, Parseval's theorem can be used to show that

$$\sum_{m=0}^{\infty} c_m^2 = \frac{1}{\pi}\int_0^{\pi} |\Gamma(x)|dx = 2\Gamma_{rms}^2 \quad (6.115)$$

and the phase noise expression for the $1/f^2$ region can then be written as

$$\mathcal{L}(\triangle\omega) = 10\log\left[\frac{(\overline{i_n^2}/\triangle f)\sum_{m=0}^{\infty} c_m^2}{8q_{max}^2\triangle\omega^2}\right] = 10\log\left(\frac{\Gamma_{rms}^2}{q_{max}^2}\frac{\overline{i_n^2}/\triangle f}{4\triangle\omega^2}\right) \quad (6.116)$$

where $\overline{i_n^2}/\triangle f$ is the power spectral density of the current noise source, and Γ_{rms} is the root-mean-square value of the ISF. In the simple case where a noise-free sinusoid waveform is used for the ISF determination, we have $\Gamma_{rms}^2 = 1/2$, and the noise contribution due to the parasitic resistance R_p alone can be derived from Equation (6.116), assuming that $\overline{i_n^2}/\triangle f = 4kT/R_p$ and $q_{max} = CV_{max}$, where V_{max} is the maximum voltage swing across the LC tank. The time-variant noise analysis approach is limited by the fact that the determination of the ISF can require complex simulations. In practice, simulation programs such as SpectreRF can be used to compute the phase noise directly.

For a 5-GHz oscillator, the computed phase noise, $\mathcal{L}\{\triangle f\}$, at $\triangle f = 100$ kHz is on the order of 103 dBc/Hz. The performance of various oscillator architectures can be compared using a figure of merit (FOM) defined as

$$FOM = \mathcal{L}(\triangle f) + 10\log_{10}\left[\left(\frac{\triangle f}{f_0}\right)^2 P\right] \quad (6.117)$$

where $\mathcal{L}(\triangle f)$ is the single sideband phase noise at offset frequency $\triangle f$ from the oscillation frequency, f_0; and P represents the total power consumption of the oscillator in milliwatts (mW). In CMOS designs whose power consumption is to be minimized, the oscillator performance is improved as the absolute value of FOM is increased.

Due to the IC process and temperature variations, the accuracy requirement of the oscillation frequency may no longer be satisfied. In practice, the

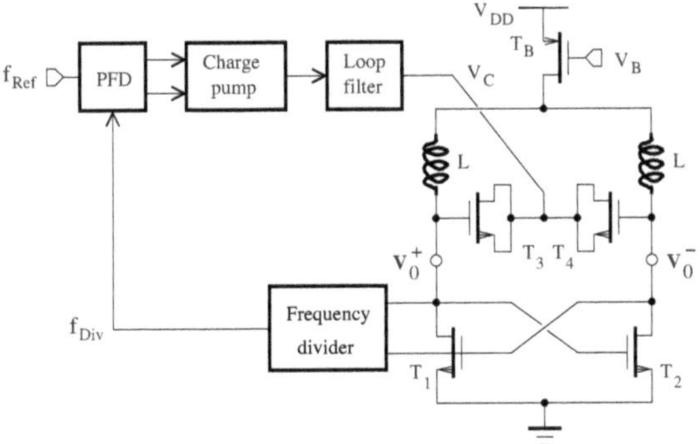

FIGURE 6.42
Circuit diagram of a voltage-controlled oscillator with the frequency control loop.

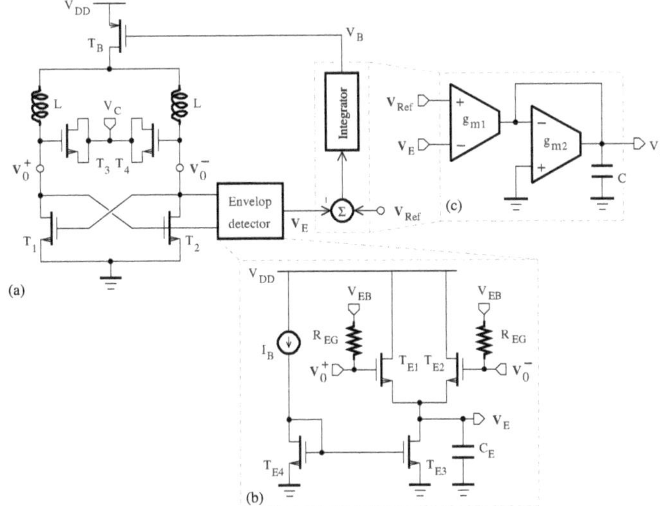

FIGURE 6.43
Circuit diagram of a voltage-controlled oscillator with the amplitude control loop.

oscillation frequency is tuned by making the capacitor C variable and controllable through a voltage. The circuit diagram of a VCO with the frequency control loop [31] is shown in Figure 6.42. The VCO output is connected to a frequency divider, whose output signal is compared to a reference signal, using a phase and frequency detector (PFD) followed by a charge pump circuit. If the frequency of the reference signal is higher than that of the feedback signal, the control voltage of the VCO available at the loop filter output will be increased; otherwise, it will be reduced.

Furthermore, because the oscillation amplitude changes with the biasing condition of transistors and the additive amplitude noise may be detected by varactors, resulting in the reduction of the tuning range and sensitivity, practical implementations must often include a loop for the automatic control of the output level [32]. With reference to Figure 6.43(a), the oscillation amplitude is sensed by the peak detector circuit (rectifier and filter), whose output drives the error amplifier stage. The resulting signal is then processed by a lowpass filter (here, an integrator), before being used to control the magnitude of the VCO bias current. Note that, in [33], by combining discrete and continuous tuning, it was also possible to lower the phase noise sensitivity while maintaining a wide tuning range. In this case, the capacitor is realized using varactors, whose capacitance depends continuously on the control voltage, and a binary-weighted switched-capacitor array with a given number of control bits. First, the oscillator center frequency is digitally set to one of the possible discrete frequencies, and varactors are then interpolated continuously around this frequency.

Consider now the envelope detector circuit depicted in Figure 6.43(b). The input signal is sensed by a pair of transistors, T_{E1} and T_{E2}, which is biased by the current I_B flowing through the current mirror composed of T_{E3} and T_{E4}. If the input peak voltage is greater than $V_E + V_T$, where V_T is the transistor threshold voltage, the transistors T_{E1} and T_{E2} will operate in the strong inversion region, thereby supplying a charge current to the capacitor C_E so that the output voltage, V_E, is increased. In contrast, if the input peak voltage becomes less than $V_E + V_T$, transistors T_{E1} and T_{E2} will be biased in the subthreshold region, and the discharge of C_E induced by leakage currents yields a reduction in the output signal level. The time required to track the input signal amplitude depends on the value of the capacitor C_E and the bias current source I_B. In the subthreshold region, the I-V characteristic of the n-channel transistor with the source connected to the substrate is approximated by

$$I_d = I_t \frac{W}{L} \exp\left(\frac{qV_{gs}}{nkT}\right) \tag{6.118}$$

where I_t is a technology-dependent positive constant; n is the subthreshold swing parameter, typically ranging from 1 to 2; and kT/q represents the thermal voltage. Assuming that the input signal whose amplitude is to be detected is sinusoidal, the drain current for each of the transistors T_{E1} and T_{E2} satisfies the next relation [34]

$$I_d = I_t \frac{W}{L} \exp\left(\frac{qV_{GS}}{nkT}\right) \exp\left(\frac{q}{nkT} v_i \cos \omega t\right) \tag{6.119}$$

where V_{GS} is the gate-source dc voltage. It is then possible to expand the drain current as follows,

$$I_d = I_t \frac{W}{L} \exp\left(\frac{qV_{GS}}{nkT}\right) \left[I_0\left(\frac{qv_i}{nkT}\right) + 2\sum_{k=1}^{\infty} I_j\left(\frac{qv_i}{nkT}\right) \cos(j\omega t) \right] \quad (6.120)$$

where I_j is the j-th order modified Bessel functions of the first kind. The average of the drain current is of the form

$$\overline{I_d} = I_t \frac{W}{L} \exp\left(\frac{qV_{GS}}{nkT}\right) I_0\left(\frac{qv_i}{nkT}\right) \quad (6.121)$$

where

$$I_0\left(\frac{qv_i}{nkT}\right) \simeq \frac{\exp\left(\frac{qv_i}{nkT}\right)}{\sqrt{2\pi \frac{qv_i}{nkT}}} \quad (6.122)$$

This last approximation results in less than 1% error for $qv_i/nkT > 15$. Because $V_{GS} = V_G - V_S = V_G - V_0$ and $\overline{I_d} = I_B/2$, the output voltage can be written as

$$V_0 = v_i + V_G - \frac{nkT}{q} \log\left[\frac{I_B}{2I_t(W/L)}\sqrt{2\pi \frac{qv_i}{nkT}}\right] \quad (6.123)$$

Note that the logarithmic term is generally assumed negligible, especially for large input signals.

One approach to realize the integrator section relies on the use of transconductors and capacitors. The circuit diagram of the integrator is illustrated in Figure 6.43(c). The output node equation can be written as

$$g_{m1}(V_E - V_{Ref}) + g_{m2}V_B + sCV_B = 0 \quad (6.124)$$

or equivalently,

$$V_B = -\frac{g_{m1}(V_E - V_{Ref})}{sC + g_{m2}} \quad (6.125)$$

Equation (6.125) corresponds to that of a lossy integrator with the dc gain, g_{m1}/g_{m2}, and the 3-dB cutoff frequency, g_{m2}/C. However, the resulting frequency response may be limited by the parasitic capacitors and resistances associated with real transconductors.

The VCO performance is primarily determined by the phase noise, tuning range, and power dissipation. In addition to the aforementioned LC VCO, various other architectures can be used for the oscillator design. Figure 6.44 shows the circuit diagram of complementary VCOs using both nMOS and pMOS cross-coupled transistors to compensate for the losses in the LC tank. A reduction in the phase noise is achieved for these last structures because the symmetry of the oscillation waveform generally facilitates the cancellation of

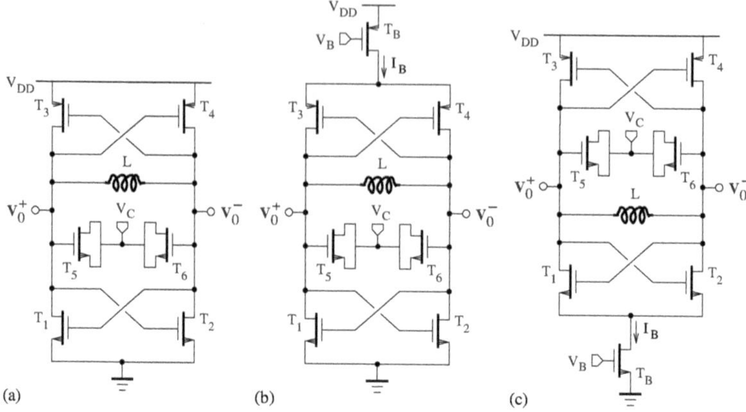

FIGURE 6.44
Circuit diagrams of complementary VCOs (a) without current source, (b) with pMOS current source, and (c) with nMOS current source.

the phase noise dc component. The tail current source, which can significantly increase the phase noise, was omitted in the VCO structure of Figure 6.44(a) to maximize the signal swing. However, by using nMOS or pMOS current sources, as shown in Figures 6.44(b) and (c), either the ground or the power supply is isolated from the output nodes so that the contribution of the substrate or supply voltage to the phase noise is further reduced.

Here, we have a start-up condition for oscillations that is of the form, $(g_{mn} + g_{mp})/2 > 1/R_{tank}$, where R_{tank} is the equivalent resistance of the tank circuit, and g_{mn} and g_{mp} are the transconductances of the nMOS and pMOS transistors, respectively. With the tail current being equal to a constant bias current, I_B (see Figures 6.44(b) and (c)), the oscillation amplitude is $A_0/2$, where $A_0 = 4I_B R_{tank}/\pi$, because the bias current flowing through the inductor can go as high as the value I_B.

- **Class C VCO**

The design of LC VCOs with high spectral purity and low power consumption may be challenging, especially for frequency synthesizers used in wireless applications.

The conventional (or class B) LC VCO, as shown in Figure 6.45(a), is extremely attractive for CMOS implementation due to its simplicity and robustness. However, it can be affected by the noise from the biasing current source and the deterioration of the phase noise due to a low power efficiency.

One attractive solution to improve the phase noise of LC VCOs is to use the class C architecture shown in Figure 6.45(b). The biasing of the cross-coupled transistors for class C operation results in more efficient generation of

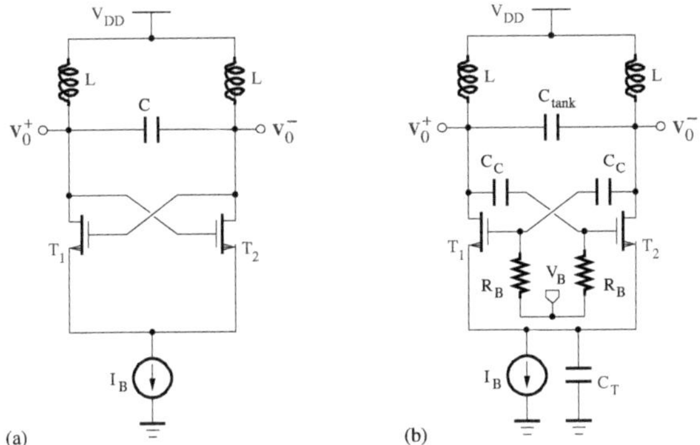

FIGURE 6.45
Circuit diagrams of conventional and class C VCOs.

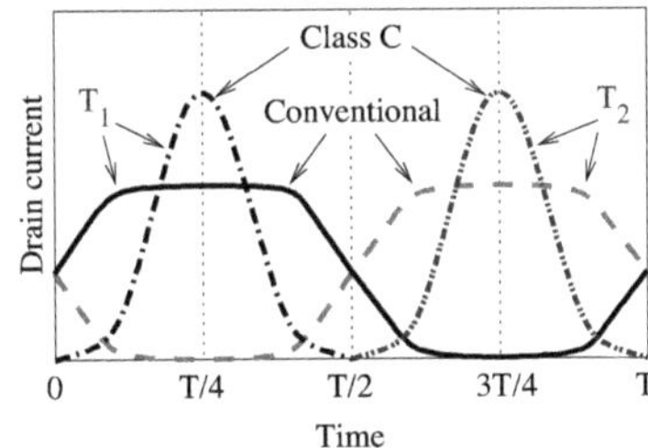

FIGURE 6.46
Drain currents for conventional and class C VCOs.

oscillation currents, making it possible to save as much as 36% of the power to achieve the same phase noise level as in conventional LC VCOs.

Figure 6.46 shows the drain currents for conventional and class C VCOs. At the resonance, the amplitude of the output oscillation voltage is $R_{tank}I_{\omega 0}$, where R_{tank} is the equivalent tank resistance seen at the transistor output and $I_{\omega 0}$ is the amplitude of the fundamental harmonic of the transistor current. When there is no tail capacitor, C_T, each of the cross-coupled transistors is in conduction for half of the oscillation period, and the tank current looks like a square wave with 50% duty cycle, so that $I_{\omega 0}$ is almost equal to $2I_B/\pi$ in a conventional LC VCO. With a large tail capacitor, C_T, however, the conduction duration is much lower than half of the oscillation period, the drain current waveforms are made of tall and narrow pulses, and the corresponding $I_{\omega 0}$ is, in a class C LC VCO, quite equal to I_B, which can be 3.9 dB higher than in the former case. Hence, the oscillation amplitude is likewise increased, leading to an improved phase noise characteristic. In order to not worsen the phase noise, the tail capacitor, C_T, should be sized such that the transistors are always operating in the active region.

However, implementations of class C LC VCOs can be limited by the required trade-off between maximum oscillation amplitude and start-up robustness. A constraint is put on the oscillation amplitude by the level of the gate bias voltage, V_B, due to the requirement of maintaining the cross-coupled transistors in the active region. A low-voltage V_B is necessary to increase the maximum oscillation amplitude, while at the same time, lowering V_B hinders the start-up of the oscillator. It is then required to design class C LC VCOs with a negative feedback circuit that can adaptively adjust the bias voltage or a dynamic biasing circuit.

- **Generation of I and Q signals**

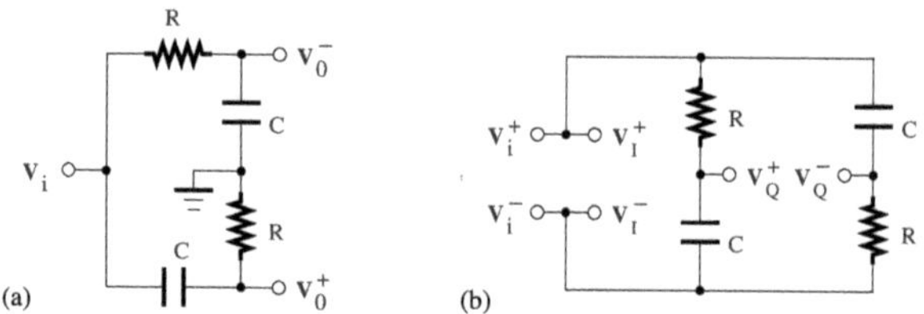

FIGURE 6.47
(a) Circuit diagram of an RC-CR phase shift network; (b) circuit diagram of an allpass-based phase shifter.

One simple approach for generating quadrature signals is to use a set of RC and CR networks. Figure 6.47(a) shows the circuit diagram of an RC-CR phase shift network. Using the voltage division principle gives the next transfer function

$$T^-(s) = \frac{V_0^-(s)}{V_i(s)} = \frac{1/sC}{R + 1/sC} \tag{6.126}$$

The magnitude of the transfer function is of the form

$$|T^-(\omega)| = \frac{1}{\sqrt{1 + (\omega RC)^2}} \tag{6.127}$$

and the phase can be expressed as

$$\angle T^-(\omega) = \arctan(-\omega RC) = -\arctan\left(\frac{1}{\omega RC}\right) \tag{6.128}$$

Similarly, it can be shown that

$$T^+(s) = \frac{V_0^+(s)}{V_i(s)} = \frac{R}{R + 1/sC} \tag{6.129}$$

The magnitude of the transfer function is obtained as

$$|T^+(\omega)| = \frac{\omega RC}{\sqrt{1 + (\omega RC)^2}} \tag{6.130}$$

and the phase is given by

$$\angle T^+(\omega) = \arctan\left(\frac{1}{\omega RC}\right) \tag{6.131}$$

At the 3-dB frequency, $\omega_c = 1/RC$, the magnitudes, $|T^-|$ and $|T^+|$, are attenuated by $1/\sqrt{2}$, while the phases, $\angle T^-$ and $\angle T^+$, are equal to -45° and 45°, respectively. Hence, a 90° phase shift is realized between the outputs V_0^+ and V_0^-.

The magnitude of signals can be maintained constant using the allpass-based phase shifter depicted in Figure 6.47(b). The I signal is derived directly from the input signal, while the Q signal is obtained at the output of the allpass filter. Using the superposition principle, we can obtain

$$V_Q^+ = \frac{V_i^+ + sRCV_i^-}{1 + sRC} \tag{6.132}$$

and

$$V_Q^- = \frac{sRCV_i^+ + V_i^-}{1 + sRC} \tag{6.133}$$

Hence,

$$T(s) = \frac{V_Q^+ - V_Q^-}{V_i^+ - V_i^-} = \frac{1 - sRC}{1 + sRC} \tag{6.134}$$

The transfer function, $T(s)$, is of the allpass type. Its magnitude is unity and its phase is of the form

$$\angle T(\omega) = \arctan(-\omega RC) - \arctan(\omega RC) = -2\arctan(\omega RC) \tag{6.135}$$

At the 3-dB cutoff frequency, $\omega_c = 1/RC$, the phase difference between the I and Q signals is 90°.

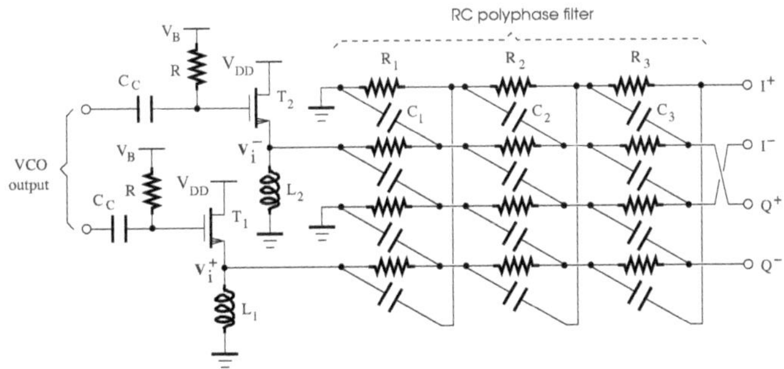

FIGURE 6.48
Circuit diagram of a VCO output buffer followed by a polyphase RC filter-based phase shifter for the generation of I and Q signals.

The allpass-based phase shifter is designed to feature a gain of unity at any frequency, while the RC-CR phase shift network should ideally provide a constant 90° phase shift at all frequencies. However, the performance of the aforementioned phase shifters is limited by component mismatches due to process variations. To overcome this problem, the RC-CR network is preferably configured in the form of a polyphase filter. Fully differential phase shifters are adopted to further attenuate the effect of component mismatches on the achievable accuracy. Figure 6.48 shows the circuit diagram of an amplifier buffer and a phase shifter based on a polyphase RC filter for the I and Q signal generation [10]. The VCO outputs are connected to source follower-based buffers using large dc-blocking capacitors, C_C. The inductive loads of the buffer are sized to resonate with the equivalent input capacitors of the polyphase filter at the frequency of interest, providing high ac impedances. In this way, transistors with small aspect ratios can be used to provide output currents, as high as required, while still loading the VCO core appropriately. The buffer output signals are applied to the polyphase RC filter, which is less sensitive to the absolute variations of the R and C values due to the adopted multistage filter structure.

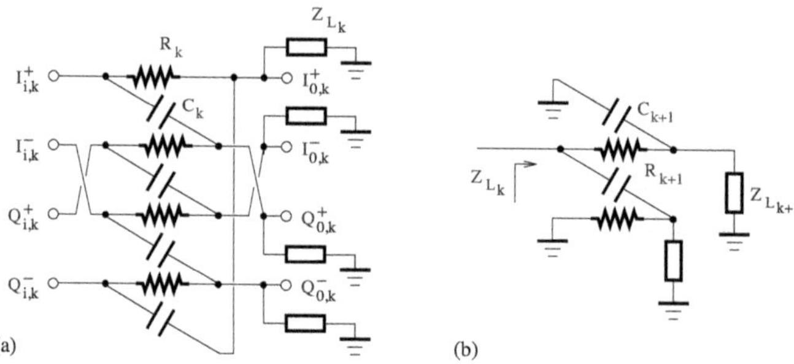

FIGURE 6.49
(a) Circuit diagram of the k-th stage of a phase shifter based on a polyphase RC filter; (b) circuit diagram of the Z_{L_k} loading network.

For the analysis [35, 36] of the phase shifter based on a polyphase RC filter, the k-th stage illustrated in Figure 6.49(a) can be considered. The load impedance, Z_{L_k}, of each stage except the last one, can be derived from the network depicted in Figure 6.49(b). Let $\mathbf{V}_{i,k} = [V_{I^+_{i,k}} \; V_{Q^+_{i,k}} \; V_{I^-_{i,k}} \; V_{Q^-_{i,k}}]^T$ and $\mathbf{V}_{0,k} = [V_{I^+_{0,k}} \; V_{Q^+_{0,k}} \; V_{I^-_{0,k}} \; V_{Q^-_{0,k}}]^T$. Using the division and superposition principles with the assumption that the input nodes, which are not driven by a signal source, are virtual grounds, we arrive at

$$\mathbf{V}_{0,k} = \mathbf{T}_k \mathbf{V}_{i,k} \tag{6.136}$$

where

$$\mathbf{T}_k = \begin{bmatrix} Z_k & 0 & 0 & Z'_k \\ Z'_k & Z_k & 0 & 0 \\ 0 & Z'_k & Z_k & 0 \\ 0 & 0 & Z'_k & Z_k \end{bmatrix} \tag{6.137}$$

$$Z_k = \frac{(1/sC_k) \parallel Z_{L_k}}{R_k + (1/sC_k) \parallel Z_{L_k}} \tag{6.138}$$

$$Z'_k = \frac{R_k \parallel Z_{L_k}}{R_k \parallel Z_{L_k} + (1/sC_k)} \tag{6.139}$$

and

$$Z_{L_k} = [(1/sC_{k+1}) + R_{k+1} \parallel Z_{L_{k+1}}] \parallel [R_{k+1} + (1/sC_{k+1}) \parallel Z_{L_{k+1}}] \tag{6.140}$$

$$= \frac{R_{k+1} + Z_{L_{k+1}} + C_{k+1}R_{k+1}Z_{L_{k+1}}s}{1 + C_{k+1}(R_{k+1} + 2Z_{L_{k+1}})s} \tag{6.141}$$

The differential in-phase and quadrature voltages can then be obtained by putting Equation (6.136) into the form

$$\begin{bmatrix} V_{I_{0,k}} \\ V_{Q_{0,k}} \end{bmatrix} = \frac{Z_{L_k}}{R_k + Z_{L_k} + sR_kZ_{L_k}C_k} \begin{bmatrix} 1 & -sR_kC_k \\ sR_kC_k & 1 \end{bmatrix} \begin{bmatrix} V_{I_{i,k}} \\ V_{Q_{i,k}} \end{bmatrix} \tag{6.142}$$

where $V_{Q_{0,k}} = V_{Q^+_{0,k}} - V_{Q^-_{0,k}}$, $V_{I_{0,k}} = V_{I^+_{0,k}} - V_{I^-_{0,k}}$, $V_{Q_{i,k}} = V_{Q^+_{i,k}} - V_{Q^-_{i,k}}$, and $V_{I_{i,k}} = V_{I^+_{i,k}} - V_{I^-_{i,k}}$. Applying Equation (6.142) successively to the first, second, and third stage of the RC polyphase filter shown in Figure 6.48, the ratio of the quadrature to the in-phase outputs is derived as

$$\frac{V_{Q_{0,3}}}{V_{I_{0,3}}} = \frac{(R_1C_1 + R_2C_2 + R_3C_3)s - R_1R_2R_3C_1C_2C_3s^3}{1 - (R_1R_2C_1C_2 + R_1R_3C_1C_3 + R_2R_3C_2C_3)s^2} \tag{6.143}$$

The phase difference between the outputs is 90° regardless of the frequency, because $V_{Q_{0,3}}/V_{I_{0,3}}$ is purely imaginary. However, the magnitudes of the outputs are equal to unity at the resonant frequencies of each stage, $\omega_{c_k} = 1/R_kC_k$ $(k = 1, 2, 3)$. A minimum image-reject ratio can be achieved provided the resonant frequency of the second stage is almost equal to the geometric mean of the other two resonant frequencies, that is, $\omega_{c_2} \simeq \sqrt{\omega_{c_1}\omega_{c_3}}$. For any input signal with a frequency within the band delimited by lowest and highest resonant frequencies, the quadrature outputs exhibit almost identical gains and phases, making the multistage RC polyphase filter suitable for wideband applications. The error resulting from all of the stages prior to the last stage is averaged out by the subsequent stages. Each additional stage then improves the matching accuracy of the signals to be generated. On the other hand, a single-stage *RC-CR* phase shifter, whose ratio of the quadrature to the in-phase outputs is reduced to sRC and the amplitudes of the quadrature outputs are equal only at the frequency $\omega_c = 1/RC$, is generally adopted for narrowband applications.

- **Quadrature voltage-controlled oscillator**

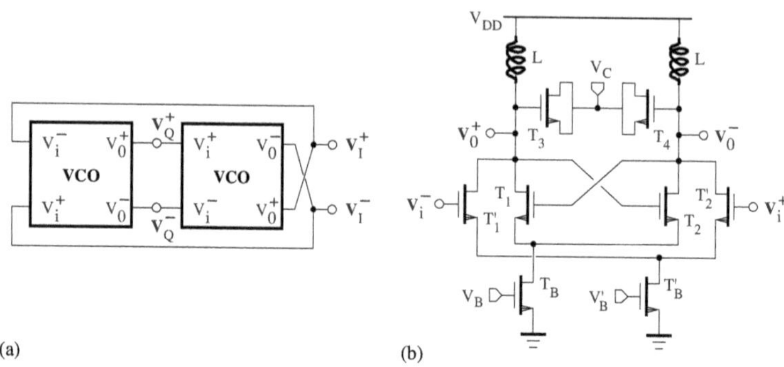

FIGURE 6.50
(a) Quadrature coupled LC VCO; (b) circuit diagram of the LC VCO with coupling transistors.

The use of a VCO driving an RC-CR polyphase filter to generate quadrature signals can be limited by mismatches of passive elements and the high power consumption of the required output buffers. An alternative approach, which consists of coupling two identical oscillators in such a way that their outputs are forced to oscillate either 90° or −90° out of phase, can be adopted to achieve some performance improvements. This can be implemented as shown in Figure 6.50. Each VCO includes an additional transistor pair, which forms the coupling network [37]. In the steady state, the coupled oscillators are synchronized to the same frequency, which is proportional to the resonance frequency of the unloaded tank network, while their output phases are in quadrature. Quadrature coupled oscillators can exhibit a higher phase noise than the corresponding stand-alone VCO. In practice, coupling transistors can then be designed to be a few times smaller than other transistors to limit their noise contribution, while still providing the required coupling factor. Furthermore, a 90° phase shifter can be inserted in each coupling path [38–40] to reduce the effects of mismatches on the oscillator output phases.

6.1.7 Automatic gain control

In the presence of signals with variable amplitude, a variable gain amplifier (VGA) and automatic gain control (AGC) is required to keep the dynamic range constant. The AGC loop shown in Figure 6.51 includes a VGA, a peak (or envelope) detector, and an integrator (or lowpass filter). The gain of the VGA is controlled by the integrator output signal based on the difference between the actual output amplitude and the reference voltage V_{REF}. The gain control feedback then forces the estimated peak amplitude to track the dc reference voltage V_{REF}.

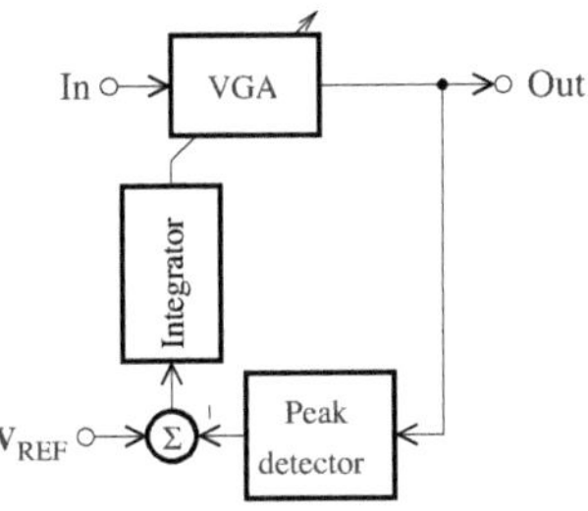

FIGURE 6.51
Block diagram of a variable-gain amplifier with automatic gain control.

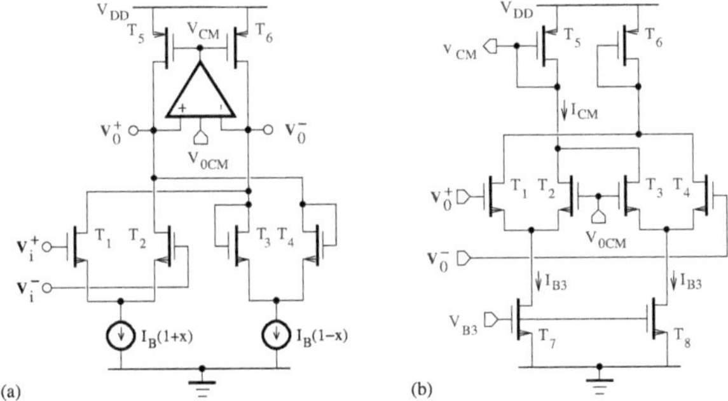

FIGURE 6.52
(a) Circuit diagram of a variable-gain fully differential amplifier; (b) circuit diagram of the common-mode feedback amplifier.

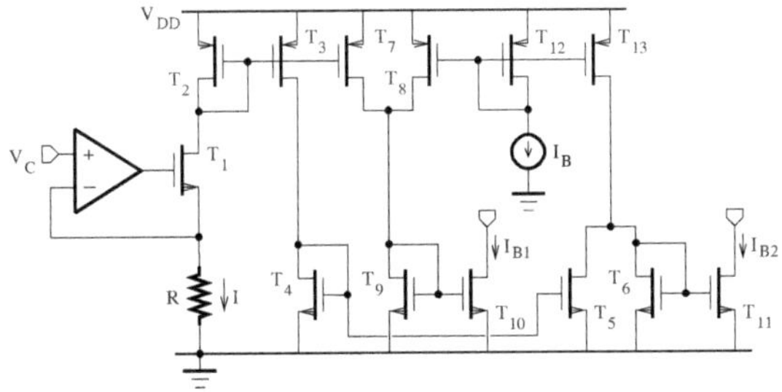

FIGURE 6.53
Circuit diagram of the control and bias generator.

A variable-gain amplifier [41] and its common-mode feedback amplifier can be implemented as shown in Figures 6.52(a) and (b), respectively. The amplification stage consists of a source-coupled input transistor pair and diode-connected transistors biased by the currents $I_B(1+x)$ and $I_B(1-x)$, respectively. It is assumed that the transistors $T_1 - T_4$ are matched. The drain currents of transistors T_1 and T_2 can be written as

$$I_{D_1} = K(V_{GS_1} - V_T)^2 \tag{6.144}$$

and

$$I_{D_2} = K(V_{GS_2} - V_T)^2 \tag{6.145}$$

Applying Kirchhoff's current and voltage laws at the source node of transistors T_1 and T_2, we have

$$I_{D_1} + I_{D_2} = I_B(1+x) \tag{6.146}$$

and

$$V_{GS_1} - V_{GS_2} = V_i \tag{6.147}$$

Solving Equations (6.144) through (6.147), we arrive at

$$I_{D_1} = \frac{1}{2}\left[I_B(1+x) + KV_i\sqrt{\frac{2I_B(1+x)}{K} - V_i^2}\right] \tag{6.148}$$

$$I_{D_2} = \frac{1}{2}\left[I_B(1+x) - KV_i\sqrt{\frac{2I_B(1+x)}{K} - V_i^2}\right] \tag{6.149}$$

Similarly, for transistors T_3 and T_4, the drain currents are given by

$$I_{D_3} = K(V_{GS_3} - V_T)^2 \tag{6.150}$$

and

$$I_{D_4} = K(V_{GS_4} - V_T)^2 \tag{6.151}$$

To proceed further, it can be shown that

$$I_{D_3} + I_{D_4} = I_B(1-x) \tag{6.152}$$

and

$$V_{GS_3} - V_{GS_4} = V_0 \tag{6.153}$$

Combining Equations (6.144) through (6.147) gives

$$I_{D_3} = \frac{1}{2}\left[I_B(1-x) + KV_0\sqrt{\frac{2I_B(1-x)}{K} - V_0^2}\right] \tag{6.154}$$

$$I_{D_4} = \frac{1}{2}\left[I_B(1-x) - KV_0\sqrt{\frac{2I_B(1-x)}{K} - V_0^2}\right] \tag{6.155}$$

Because $I_{D_1} + I_{D_3} = I_B$ and $I_{D_2} + I_{D_4} = I_B$, it can be found that

$$V_i\sqrt{\frac{2I_B(1+x)}{K} - V_i^2} = -V_0\sqrt{\frac{2I_B(1-x)}{K} - V_0^2} \tag{6.156}$$

Assuming that $V_i \ll \sqrt{2I_B(1+x)/K}$ and $V_0 \ll \sqrt{2I_B(1-x)/K}$, the amplifier gain is derived as

$$A = \frac{V_0}{V_i} \simeq \frac{\sqrt{\frac{2I_B(1+x)}{K}}}{\sqrt{\frac{2I_B(1-x)}{K}}} = \sqrt{\frac{1+x}{1-x}} \tag{6.157}$$

The circuit diagram of the control and bias generator, which can provide the currents, $I_B(1+x)$ and $I_B(1-x)$, is shown in Figure 6.53. With the assumption that the amplifier is ideal, that is, $V_C = V^+ = V^-$ and $I = V_C/R$, and that each of the current mirrors (T_2, T_3, T_7), (T_4, T_5), (T_6, T_{11}), (T_9, T_{10}), and (T_8, T_{12}, T_{13}) is realized using transistors of identical size, we can obtain

$$I_{B_1} = I_B + V_C/R = I_B(1+x) \tag{6.158}$$

and

$$I_{B_2} = I_B - V_C/R = I_B(1-x) \tag{6.159}$$

where $x = V_C/RI_B$. Because the approximation exploited here to obtain a linear variation of the amplifier gain in dB is of the form

$$e^x \simeq \sqrt{\frac{1+x}{1-x}} \tag{6.160}$$

the tuning range is limited to about 15 dB with a linearity error of less than 0.5 dB. A multistage amplifier configuration may then be needed in applications requiring a wide tuning range.

The block diagram of an AGC amplifier [42] with offset compensation is presented in Figure 6.54. It is composed of amplifier 1, which operates as a VGA and is combined with the exponential voltage generator to realize a linear-in-dB characteristic between the gain and the control voltage; amplifier 2 with offset voltage tuning inputs connected to the feedback lowpass filter (LPF); amplifier 3, which helps extend the bandwidth; and the gain tuning feedback including a peak detector and an integrator.

The circuit diagram of amplifier 1 and the exponential voltage generator is shown in Figure 6.55. For small variations of the input voltages, it can be shown that

$$V_0^+ - V_0^- \simeq K_v(V_C^+ - V_C^-)(V_i^+ - V_i^-) \tag{6.161}$$

where the voltage gain K_v is a function of the transistor characteristics. The

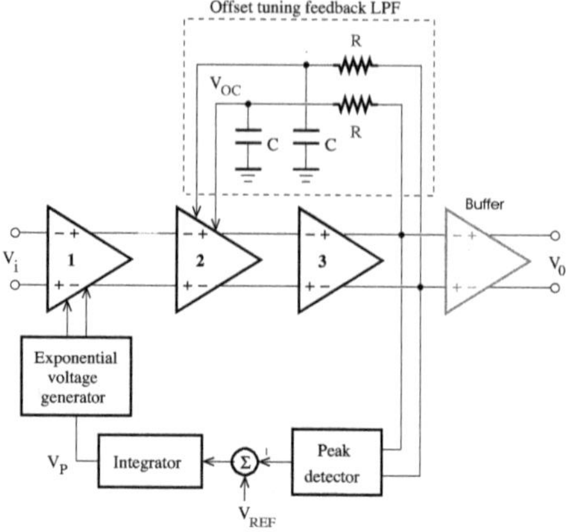

FIGURE 6.54
Block diagram of an AGC amplifier with offset compensation.

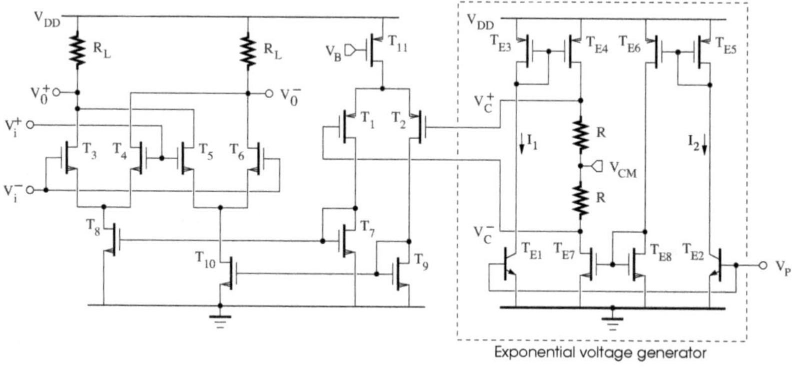

FIGURE 6.55
Circuit diagram of amplifier 1 and the exponential voltage generator.

bipolar transistors, T_{E1} and T_{E2}, operate in the active region and their collector currents are of the form, $I_1 = I_2 = I_S \exp(V_P/U_T)$, where I_S is the saturation current and U_T is the thermal voltage. Assuming that the sizes of the current mirror transistors $T_{E3} - T_{E4}$, $T_{E5} - T_{E6}$, and $T_{E7} - T_{E8}$ are chosen to be identical, we can obtain

$$V_C^+ - V_C^- = R(I_1 + I_2) = 2RI_S \exp(V_P/U_T) \tag{6.162}$$

Hence, the resulting gain in decibels (dB) of amplifier 1 can be written as,

$$G(\text{in dB}) = K_1 + K_2 V_P \tag{6.163}$$

where $K_1 = 20\log_{10}(2K_v RI_S)$ and $K_2 = 20\log_{10}(\mathrm{e})/U_T$. Using dB units, the relationship between the gain and the control voltage is linear.

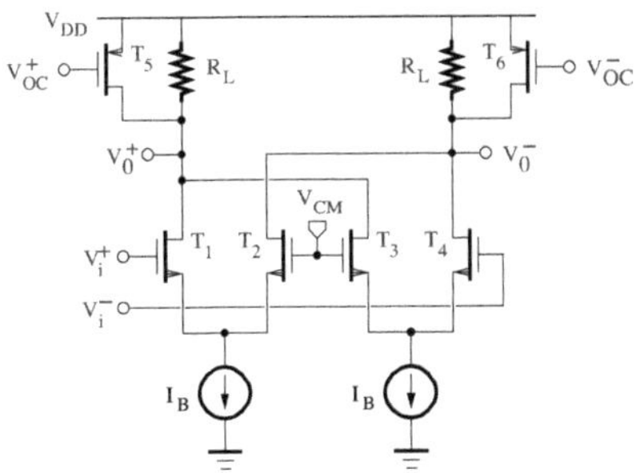

FIGURE 6.56
Circuit diagram of amplifier 2.

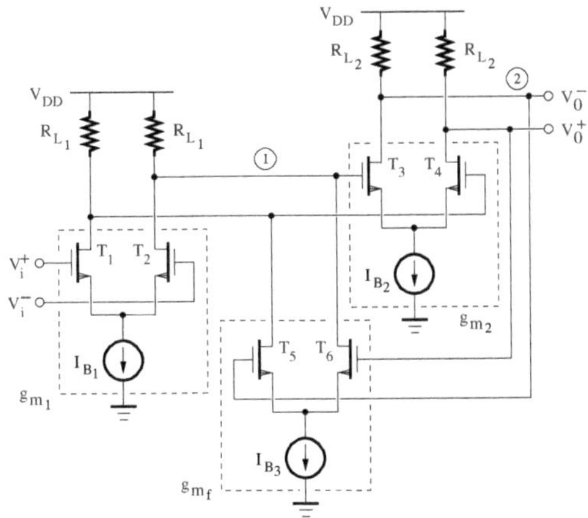

FIGURE 6.57
Circuit diagram of amplifier 3.

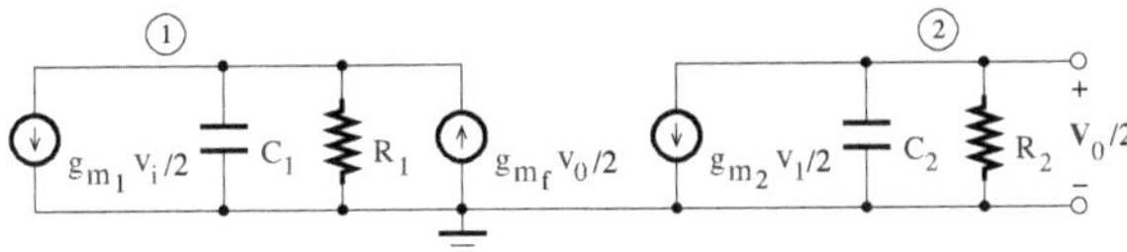

FIGURE 6.58
Equivalent circuit of amplifier 3.

The circuit diagram of amplifier 2 is shown in Figure 6.56, where V_{CM} is the common-mode voltage that is commonly set at the midpoint of the supply voltage level. The *dc* offset voltage effect is reduced by using the feedback LPF to estimate the correction voltage V_{OC}.

If we let g_{m_i} and g_{m_f} be the transconductances associated with the direct and feedback paths, respectively, $V_{i,off}$ and $V_{f,off}$ be the input-referred offset voltages associated with the direct and feedback paths, respectively, the output offset voltage can be derived as,

$$V_{0,off} = g_{m_i} R_0 V_{i,off} + g_{m_f} R_0 V_{f,off} - g_{m_f} R_0 V_{OC} \tag{6.164}$$

where R_0 represents the output resistance. At the steady state, $V_{0,off} = V_{OC}$ and the output offset voltage becomes

$$V_{0,off} = \frac{g_{m_i} R_0 V_{i,off} + g_{m_f} R_0 V_{f,off}}{1 + g_{m_f} R_0} \tag{6.165}$$

Assuming that $g_{m_f} R_0 \gg 1$, we obtain

$$V_{0,off} \simeq \frac{g_{m_i} V_{i,off}}{g_{m_f}} + V_{f,off} \tag{6.166}$$

Usually, $g_{m_f} < g_{m_i}$ and the offset voltage $V_{f,off}$ is small so that $V_{i,off}$ represents the main contribution to the output offset.

Amplifier 3, as shown in Figure 6.57, relies on active feedback for bandwidth improvement. The node equations derived from the equivalent circuit shown in Figure 6.58 can be written as,

$$g_{m_1} V_i - g_{m_f} V_0 + V_1 s C_1 + V_1/R_1 = 0 \tag{6.167}$$

$$g_{m_2} V_1 + V_0 s C_2 + V_0/R_2 = 0 \tag{6.168}$$

By solving the system of equations (6.167) and (6.168), the voltage gain can be derived as follows:

$$A(s) = \frac{V_0(s)}{V_i(s)} = A_0 \frac{1}{\alpha s^2 + \beta s + 1} \tag{6.169}$$

where

$$A_0 = \frac{g_{m_1} g_{m_2} R_1 R_2}{1 + g_{m_f} g_{m_2} R_1 R_2} \tag{6.170}$$

$$\alpha = \frac{R_1 R_2 C_1 C_2}{1 + g_{m_f} g_{m_2} R_1 R_2} \tag{6.171}$$

$$\beta = \frac{R_1 C_1 + R_2 C_2}{1 + g_{m_f} g_{m_2} R_1 R_2} \tag{6.172}$$

The amplifier bandwidth is proportional to the pole frequency, that is given by

$$\omega_0 = \sqrt{\frac{1 + g_{m_f} g_{m_2} R_1 R_2}{R_1 R_2 C_1 C_2}} \tag{6.173}$$

In the case where the amplifier is designed without the active feedback, the pole frequency is found to be $1/(R_1 R_2 C_1 C_2)^{1/2}$. Hence, the use of the active feedback leads to a pole frequency improvement by a factor of $(1 + g_{m_f} g_{m_2} R_1 R_2)^{1/2}$.

The amplifier stages following amplifier 2 essentially provide a fixed gain for the AGC. But, the gain peaking due to the feedback transconductance, g_{m_f}, can be exploited to further improve the gain-bandwidth response in applications with less stringent requirements on the signal distortion due to group delay variations.

6.2 Continuous-time filters

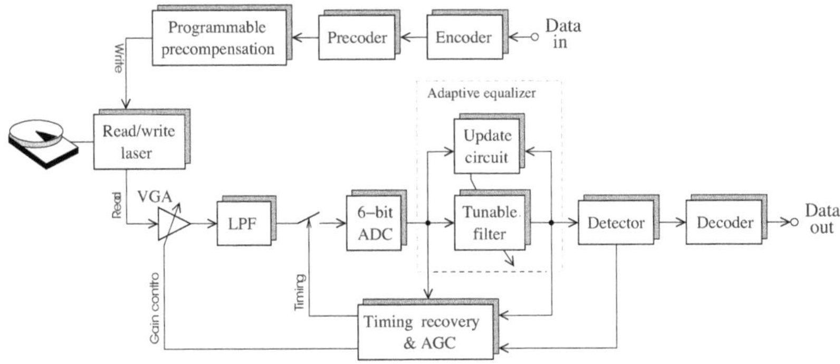

FIGURE 6.59
Block diagram of a read/write channel integrated circuit.

Continuous-time (CT) circuits find applications in the direct processing of analog signals and interfacing of digital signal processors. The block diagram of a read/write channel integrated circuit is shown in Figure 6.59. It is a mixed-signal system, which includes a variable gain amplifier (VGA), a lowpass filter (LPF), an analog-to-digital converter (ADC), and an automatic gain control (AGC) in addition to various digital building blocks. Due to the hysteresis effects in the magnetic disk, the number of signal levels injected through the channels can be limited to two (e.g., ± 1).

Error-correcting encodings are used to correct burst errors (or error patterns) associated with media defects or generated in the write or read process. During the write process, magnetic fields are converted into binary waveforms using modulation or encoding methods, known to reduce the timing uncertainty that can affect the decoding of the stored data. Due to noise errors still remaining in the stream of digital samples after the read sensing, amplification, and conversion and equalization operations, there is a need for a detector and decoder to efficiently recover the data bits.

The increase in the storage capacity is primarily related to the technological improvements in the design of magnetic media and heads. However, as the storage density increases, the performance and speed of the different signal processing blocks (encoder, modulator, equalizer, and decoder) play an important role in the reduction of inter-symbol interference and noise emerging in magnetic disk systems.

Various methods can be used for the synthesis of CT filters. In the cascade design, the filter transfer function is realized as a connection of first- and second-order sections. The sensitivity in the passband can be reduced using LC ladder-based design approaches. An active filter is derived either by simulating the equations of a passive LC prototype, or by replacing the inductors of a passive LC prototype with impedance converters.

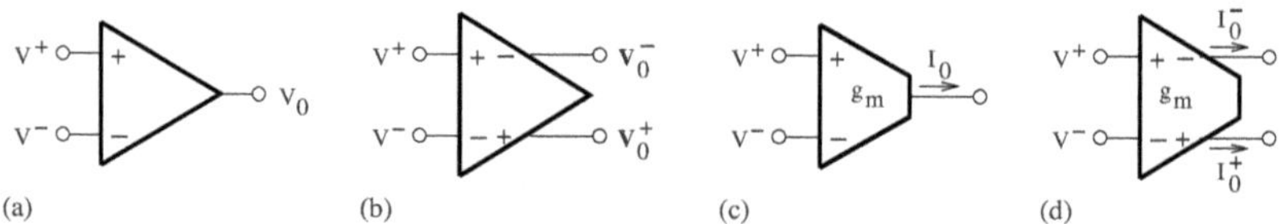

FIGURE 6.60
(a) Single-ended and (b) fully differential voltage amplifier structures; (c) single-ended and (d) fully differential transconductance amplifier structures.

In general, integrated filters are implemented using resistors, capacitors, and amplifiers. The filter design is commonly based on two types of amplifiers. The first ones (see Figures 6.60(a) and (b)) operate as voltage-controlled voltage sources and are described by

$$V_0 = A(V^+ - V^-) \tag{6.174}$$

and

$$V_0^+ - V_0^- = A(V^+ - V^-) \tag{6.175}$$

where A is the amplifier gain, while the second structures (see Figures 6.60(c) and (d)), known as operational transconductance amplifiers, are equivalent to a voltage-controlled current source of the form,

$$I_0 = g_m(V^+ - V^-) \tag{6.176}$$

and

$$I_0^+ - I_0^- = g_m(V^+ - V^-) \tag{6.177}$$

respectively, where g_m is the amplifier transconductance. For differential structures, it was assumed that $V_0^+ = A(V^+ - V^-)/2$, $V_0^- = -A(V^+ - V^-)/2$, $I_0^+ = g_m(V^+ - V^-)/2$, and $I_0^- = -g_m(V^+ - V^-)/2$. Note that V^+ and V^- denote the voltages at the noninverting and inverting input nodes of the amplifier, respectively.

The integrator and gyrator are the fundamental building blocks used for a transfer function description and their different implementations will be reviewed in the next section.

6.2.1 RC circuits

RC circuits consist of operational amplifiers (OAs), resistors, and capacitors. Applying Kirchhoff's current law at the inverting node of the single-ended

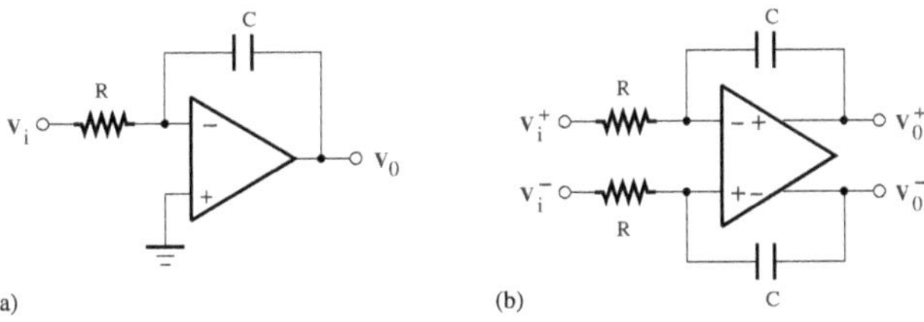

FIGURE 6.61
(a) Single-ended and (b) fully differential RC integrator structures.

integrator shown in Figure 6.61(a) yields

$$\frac{V_i(s) - V^-(s)}{R} = sC[V^-(s) - V_0(s)] \tag{6.178}$$

where $V_0(s) = -A(s)V^-(s)$. The transfer function is then derived as

$$\frac{V_0(s)}{V_i(s)} = -\frac{1}{sRC}\frac{1}{1 + \dfrac{1}{A(s)}\left(1 + \dfrac{1}{sRC}\right)} \tag{6.179}$$

Ideally, $|A(j\omega)| \gg 1$ and we arrive at $V_0(s)/V_i(s) \simeq -1/(sRC)$. Hence, the gain is inversely proportional to the frequency, while the phase remains equal to 90°. Equation (6.179) can also be applied to the fully differential structure depicted in Figure 6.61(b), if $V_0 = (V_0^+ - V_0^-)/2$ and $V_i = (V_i^+ - V_i^-)/2$. The noninverting version of the differential integrator is obtained by permuting the polarities of the input signal.

Generally, it is assumed that the integrator operates with ac signals. But, in the case where the input also includes dc components, the feedback capacitors, which act like an open circuit at dc, should be periodically discharged to prevent saturation of the amplifier outputs. Alternatively, resistors with large values can be added in parallel with the feedback capacitor, as shown in Figure 6.62(a), to provide the necessary dc feedback. Figure 6.62(b) shows the plot of the magnitude of the integrator transfer functions. The dc gain of the integrator is now reduced to $A_0 = 20\log_{10}(R_0/R)$, where $R_0 > R$. Assuming that the amplifier is ideal, the cutoff and transition frequencies are of the form $\omega_c = 1/R_0C$ and $\omega_u = 1/RC$, respectively. At high frequencies, the integrator gain rolls off at a rate of -20 dB/decade.

The straightforward integration of these structures, however, suffers from the imprecise RC product realizable in MOS technology. Since both the resistors and capacitors are designed with random processing variations on the

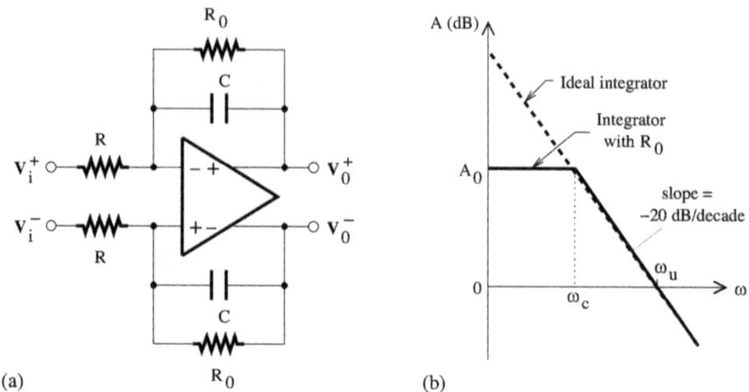

FIGURE 6.62
(a) Integrator with dc feedback; (b) plot of the magnitude of the integrator transfer functions.

order of 10 to 20%, the overall accuracy of the RC time constant can be as high as 20%. This imprecision can also be observed in the frequency response of the integrator and is unacceptable in most applications.

In some applications, it may be necessary to reset the integrator.

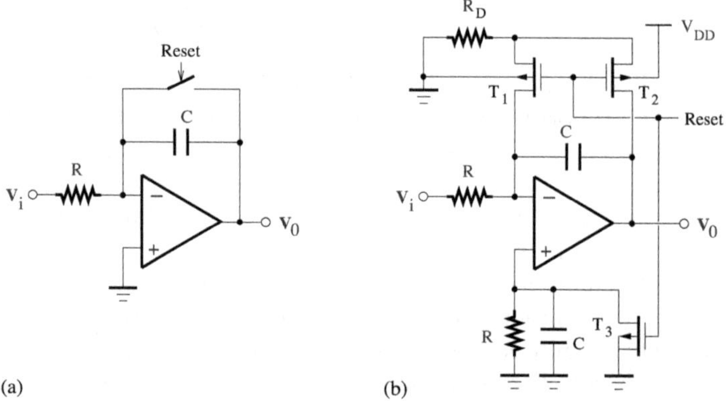

FIGURE 6.63
(a) RC integrator with a reset switch, (b) low drift RC integrator with a reset mechanism.

Figure 6.63(a) shows an RC integrator with a reset switch, that can be implemented using a MOS transistor whose source is connected at the inverting node in order to prevent the gate-source voltage bias from being affected by the input signal. Furthermore, the transistor switch should be sized and biased to reduce the effects of charge injection through stray capacitors and charge leakage that can cause integration errors. Note that the integrator precision can also be affected by the amplifier imperfections, such as the offset voltage and input bias currents. A practical implementation of a low drift RC integrator with a reset mechanism is shown in Figure 6.63(b). With the resistor $R_D = 100$ kΩ, a worst-case drift less than 500 μV/s over the temperature range from -55^oC to 125^oC can be achieved.

6.2.2 MOSFET-C circuits

In this approach, the metal oxide semiconductor field-effect transistors (MOSFET) operating in the triode region are used as variable resistors that are automatically adjusted to provide accurate RC products by an on-chip control circuit. According to the n-channel transistor shown in Figure 6.64, if V_{GS}, V_{DS}, V_T, and V_B are the gate-source, drain-source, threshold, and substrate voltages, respectively, the drain current, I, in the triode region will be given by

$$I = K[2(V_{GS} - V_T)V_{DS} - V_{DS}^2] \tag{6.180}$$

where $K = \mu_n C_{ox}(W/2L)$, $V_{GS} > V_T$, $V_{DS} < V_{DS(sat)} = V_{GS} - V_T$, and $V_B \leq V_S$. Assuming that $V_{GS} = V_C$, the current, I, can be written as [43]

$$I = \frac{V_{DS}}{R} + f(V_{DS}) \tag{6.181}$$

where $f(V_{DS}) = -KV_{DS}^2$ and the tunable resistor, R, is of the form

$$R = \frac{1}{2K(V_C - V_T)} = \frac{1}{\mu_n C_{ox}(W/L)(V_C - V_T)} \tag{6.182}$$

where W and L are the channel width and length, respectively; C_{ox} is the gate capacitance per unit area; and μ_n is the effective mobility of the transistor. The MOSFET behaves as a voltage-controlled resistor only for small signals,

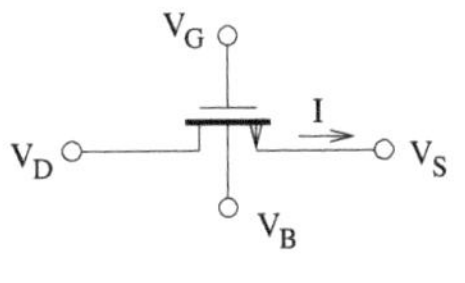

FIGURE 6.64
Symbol of an n-channel transistor.

and the higher-order terms introduced by the nonlinear function $f(V_{DS})$ must be considered in the context of large signals.

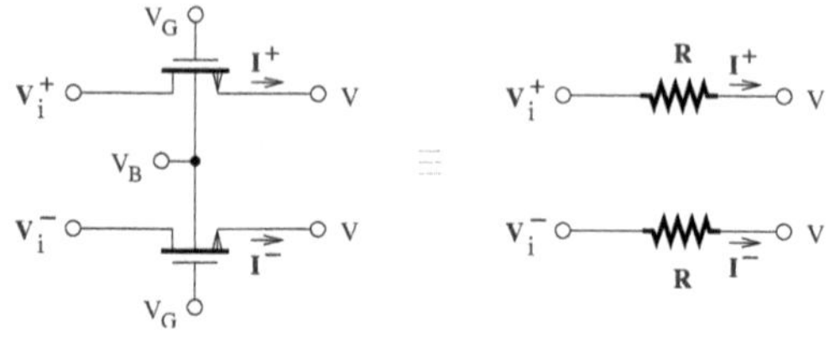

FIGURE 6.65
Resistor based on a balanced transistor configuration.

By driving two transistor devices in balanced form by $V_i^+ = V_i/2$ and $V_i^- = -V_i/2$ at the input terminals and the same voltage V at the other terminals [44] (see Figure 6.65), the effects of transistor nonlinearities are considerably reduced especially when the voltage V is almost equal to zero, and the difference of the output currents is reduced to

$$I^+ - I^- = \frac{V_1^+ - V_1^-}{R} \tag{6.183}$$

where $R = 1/[2K(V_C - V_T)]$. The magnitude of the control voltage, V_C, can be changed to adjust the resistor value. Thus, the differential configuration has the advantage of canceling even nonlinearities in terms of the input signal.

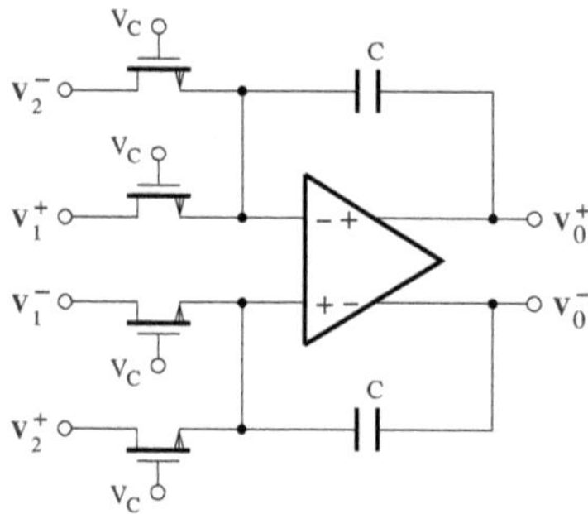

FIGURE 6.66
Differential MOSFET-C integrator.

Resistors based on a balanced transistor configuration find applications in the design of gain stages and integrators. The analysis of the differential integrator shown in Figure 6.66 results in the following expressions in the time and frequency domain:

$$v_0(t) = -\frac{1}{RC}\int_{-\infty}^{t} [v_1(\tau) - v_2(\tau)]\,\mathrm{d}\tau \Rightarrow V_0(s) = -\frac{1}{sRC}[V_1(s) - V_2(s)] \tag{6.184}$$

where $V_1 = V_1^+ - V_1^-$, $V_2 = V_2^+ - V_2^-$, and $V_0 = V_0^+ - V_0^-$. The fact that the allowable signal swing is limited by the triode region of transistors can be a drawback for the low-voltage operation.

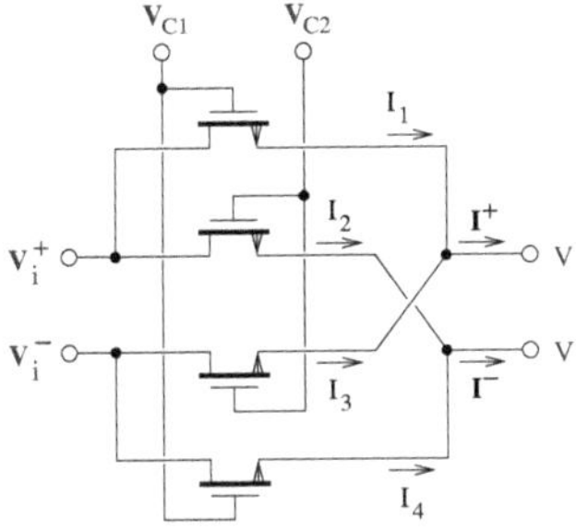

FIGURE 6.67
Four-transistor implementation of the resistor.

Typically, the remaining nonlinearities arising from device mismatches can be estimated to be about 0.1% for 1 V peak-to-peak signals. If a high linearity is needed, the MOSFET pair will be replaced by the four-MOSFET cross-coupled structure of Figure 6.67 [45–47]. In this case, the difference of the output currents can be written as

$$I^+ - I^- = (I_1 + I_3) - (I_2 + I_4) \tag{6.185}$$

$$= (I_1 - I_2) + (I_3 - I_4)$$

$$= 2K(V_{C1} - V_{C2})(V_i^+ - V_i^-) \tag{6.186}$$

The resulting resistance is then given by

$$R = \frac{V_i^+ - V_i^-}{I^+ - I^-} = \frac{1}{2K(V_{C1} - V_{C2})} \tag{6.187}$$

All transistors remain in the triode region, if $V_{DS} < V_{DS(sat)} = V_{GS} - V_T$, or equivalently, $V_i^+, V_i^- < \min[V_{C1} - V_T, V_{C2} - V_T]$. Ideally, the nonlinear contributions to the output current are cancelled, and the equivalent resistor is controlled by the differential voltage, $V_{C1} - V_{C2}$, such that its value remains independent of the threshold voltage, V_T.

6.2.3 g_m-C circuits

The active component of a g_m-C structure is a transconductor. This latter is a voltage-to-current converter, which gives out a current proportional to a frequency-dependent transconductance, $g_m(s)$. Generally, its value can be regarded as constant, that is, $g_m(s) = g_m$, as long as the transconductor bandwidth is assumed to be sufficiently large. For the single-ended integrator shown in Figure 6.68(a), a transconductor and a capacitor are used to provide the amplification and the integration, respectively [48, 49].

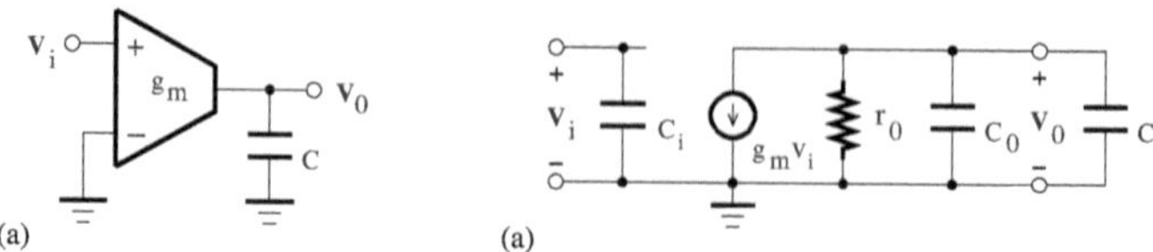

FIGURE 6.68
(a) Single-ended g_m-C integrator; (b) small-signal equivalent model.

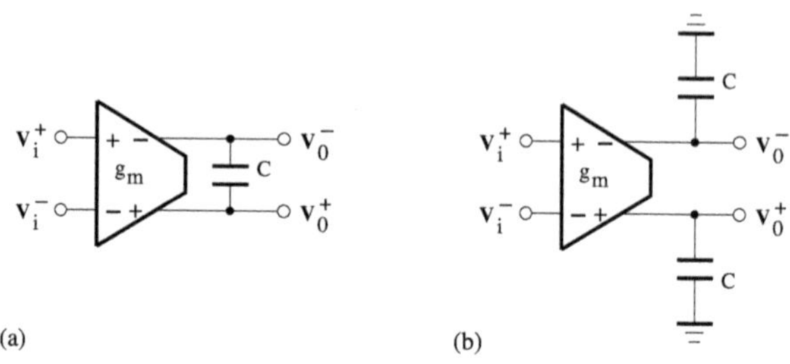

FIGURE 6.69
Fully differential g_m-C integrators.

With reference to the small-signal equivalent model of Figure 6.68(b), $V_0(s) = Z\,I(s)$, where $I(s) = g_m V_i(s)$ and $Z = r_0 \parallel C_0 \parallel C$. The transfer function of the integrator can then be written as

$$\frac{V_0(s)}{V_i(s)} = \frac{\omega_u}{s\left(1 + \dfrac{C_0}{C}\right) + \dfrac{1}{r_0 C}} \tag{6.188}$$

Ideally, the output capacitance, C_0, should be small, and the output resistance, r_0, should be large, that is, $C_0/C \ll 1$ and $r_0 C \gg 1$. Therefore, we obtain

$$\frac{V_0(s)}{V_i(s)} = \frac{\omega_u}{s} \tag{6.189}$$

where $\omega_u = g_m/C$ is the unity-gain frequency. Although the transconductance amplifier can be designed to exhibit a sufficiently high output resistance, r_0, the output parasitic capacitance, C_0, can generally not be assumed to be negligible. Theoretically, parasitic capacitances should be absorbed into a total integrating capacitance value, but their effect becomes dominant in circuits for high-frequency applications. In this case, the integrator unity-gain frequency is found to be $\tilde{\omega}_u = \omega_u/(1 + C_0/C)$. Because the value of C_0 is not accurately known, the value of $\tilde{\omega}_u$ can be scaled only by adjusting the parameters g_m and C. However, this solution implies an augmentation or a reduction of device dimensions, and thus an increase in C_0 or in mismatches and distortions.

The total integrating capacitance of the fully differential integrator shown in Figure 6.69(a) is two times smaller than the one needed in the integrator configuration of Figure 6.69(b), which, however, has the advantage of exhibiting a low sensitivity to parasitic capacitances in the case where C_0 becomes large relative to C.

Practically, due to parasitic capacitances, g_m-C circuits are only suitable for low or medium linearity applications and their high-frequency performance tends to be limited.

6.2.4 g_m-C operational amplifier (OA) circuits

In g_m-C OA circuits, the problem of parasitic capacitances at the transconductor output is minimized by using an OA. These capacitors are connected between the OA inputs (virtual ground) and the ground and thus do not carry any charge. In addition, the OA plays the role of a buffer at the outputs of the transconductor and the structure of this latter can be very simple. But the use of two active components can still mean significant power consumption and large chip area.

Let C_p be the total equivalent capacitance connected at each of the transconductance amplifier outputs. Figure 6.70 shows the circuit diagram of g_m-C OA inverting and noninverting integrators. Applying Kirchhoff's current law to the integrator of Figure 6.70(a) [31] gives

$$I^-(s) = sC_pV^+(s) + sC(V^+(s) - V_0^-(s)) \tag{6.190}$$

and

$$I^+(s) = sC_pV^-(s) + sC(V^-(s) - V_0^+(s)) \tag{6.191}$$

where

$$I^+(s) - I^-(s) = g_m[V_i^+(s) - V_i^-(s)] \tag{6.192}$$

$$V_0^+(s) - V_0^-(s) = A(s)[V^+(s) - V^-(s)] \tag{6.193}$$

The transfer function can then be computed as

$$\frac{V_0(s)}{V_i(s)} = -\frac{\omega_u}{s}\frac{1}{1+(1+C_p/C)/A(s)} \tag{6.194}$$

where $\omega_u = g_m/C$, $V_0(s) = [V_0^+(s) - V_0^-(s)]/2$, and $V_i(s) = [V_i^+(s) - V_i^-(s)]/2$.

When the OA gain is modeled as $A(s) = \omega_t/s$, where ω_t is the OA unity-gain frequency, it can be shown that a pole at $s = -\omega_t/(1 + C_P/C)$ is introduced in the integrator transfer function. The price to be paid in order to reduce the error caused by this parasitic pole is an increase in the integrator's dc gain. However, for high-frequency applications, a special compensation technique may be needed. This can be achieved by inserting tunable resistors in series with the capacitors, as shown in Figure 6.71(a). The resistor r can

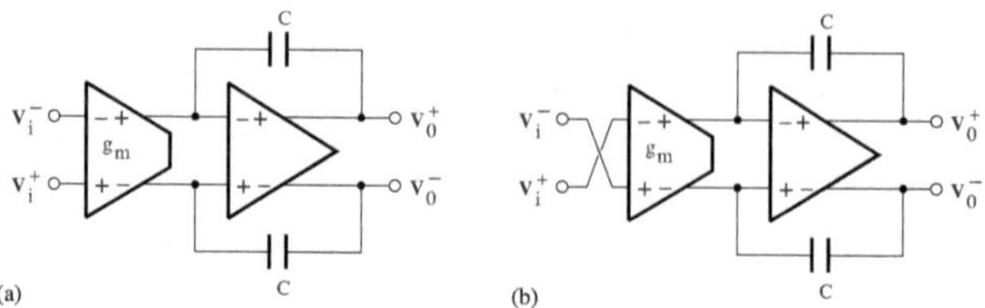

FIGURE 6.70
(a) g_m-C OA inverting integrator; (b) g_m-C OA noninverting integrator.

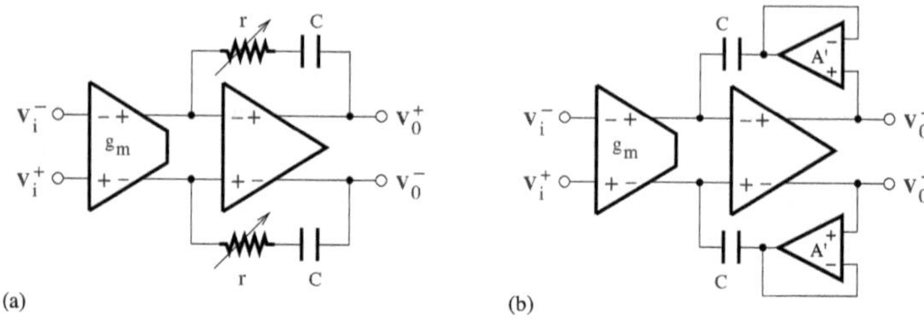

FIGURE 6.71
(a) g_m-C OA integrator with passive compensation; (b) g_m-C OA integrator with active compensation.

be realized, for instance, by a MOS transistor operating in the triode region. The application of Kirchhoff's current law to the input nodes of the OA yields

$$I^-(s) = sC_pV^+(s) + \frac{V^+(s) - V_0^-(s)}{r + \dfrac{1}{sC}} \tag{6.195}$$

and

$$I^+(s) = sC_pV^-(s) + \frac{V^-(s) - V_0^+(s)}{r + \dfrac{1}{sC}} \tag{6.196}$$

The voltage transfer function can readily be written as

$$\frac{V_0(s)}{V_i(s)} = -\frac{\omega_u}{s}\frac{1 + srC}{1 - \dfrac{\omega^2 rC_p}{\omega_t} + \dfrac{s}{\omega_t}\left(1 + \dfrac{C_p}{C}\right)} \simeq -\frac{\omega_u}{s}\frac{1 + srC}{1 + \dfrac{s}{\omega_t}\left(1 + \dfrac{C_p}{C}\right)} \tag{6.197}$$

The pole-zero cancellation can be achieved provided

$$r = \frac{1}{\omega_t C}\left(1 + \frac{C_p}{C}\right) \tag{6.198}$$

Taking into consideration the value of r, the assumption of neglecting the term $\omega^2 r C_p/\omega_t$ may be justified by the fact that, in practice, $C_p/C \ll 1$ and $\omega/\omega_t \ll 1$. The passive compensation appears to be limited by matching errors and process variations affecting the value of the resistor r.

An improved tracking performance can be achieved by adopting the integrator structure depicted in Figure 6.71(b). In this approach, the frequency responses of two unity-gain buffers used to isolate the integrating capacitors from the OA outputs are exploited to minimize the integrator losses [50]. The equations for the OA input nodes are given by

$$I^-(s) = sC_pV^+(s) + sC[V^+(s) - V_0'^-(s)] \tag{6.199}$$

and

$$I^+(s) = sC_pV^-(s) + sC[V^-(s) - V_0'^+(s)] \tag{6.200}$$

where $V_0'^-(s) = V_0^-(s)/[1+1/A'(s)]$ and $V_0'^+(s) = V_0^+(s)/[1+1/A'(s)]$. It can then be shown that

$$\frac{V_0(s)}{V_i(s)} = -\frac{\omega_u}{s}\frac{1}{\dfrac{1}{1+1/A'(s)} + \dfrac{s}{\omega_t}\left(1+\dfrac{C_p}{C}\right)} \tag{6.201}$$

Because $1/|A'(\omega)| \ll 1$, we have

$$\frac{1}{1+1/A'(\omega)} \simeq 1 - \frac{1}{A'(\omega)} + \frac{1}{[A'(\omega)]^2} - \cdots \tag{6.202}$$

This suggests that the integrator quality factor, which is computed as the ratio of the imaginary part to the real part of the transfer function denominator, can be maximized by making the gain, A', of the buffer amplifier identical to A. That is,

$$\frac{V_0(j\omega)}{V_i(j\omega)} \simeq -\frac{\omega_u}{j\omega\left(1-\dfrac{\omega^2}{\omega_t^2}\right)}\frac{1}{1+j\dfrac{1}{Q_I}} \tag{6.203}$$

where

$$Q_I = \frac{C}{C_p}\frac{\omega_t}{\omega}\left(1-\frac{\omega^2}{\omega_t^2}\right) = \frac{C}{C_p}|A(\omega)|\left(1-\frac{1}{|A(\omega)|^2}\right) \tag{6.204}$$

Here, the matching can be efficiently realized since the involved components are both of the same type, that is, active, and the integrator quality factor, Q_I, may be made high. Note that Q_I is infinite for an integrator with no losses.

Structures based on fully differential amplifiers are easily derived from single-ended circuits consisting of amplifiers with one input node connected to the ground. When this requirement is not met (see Figure 6.72(a)), the differential input and output are realized by coupling two single-ended circuits. With $V_0(s) = I(s)/sC$ and $I(s) = g_m[V_i(s) - V_0(s)]$, the circuit of Figure 6.72(a) exhibits the next transfer function,

$$H(s) = \frac{V_0(s)}{V_i(s)} = \frac{g_m}{sC + g_m} \tag{6.205}$$

and its differential version (see Figure 6.72(b)) should have a common-mode rejection ratio (CMRR) given by

$$\text{CMRR} = \frac{1}{2}\frac{H_1(s) + H_2(s)}{H_1(s) - H_2(s)} \tag{6.206}$$

where $H_j(s) = g_{mj}/(sC_j + g_{mj})$, $j = 1, 2$, is the transfer function for each signal path. Ideally, $H_1(s) = H_2(s)$ and the CMRR is infinite. However, due to fabrication tolerances, $H_1(s)$ and $H_2(s)$ are different. As a result, the CMRR is finite.

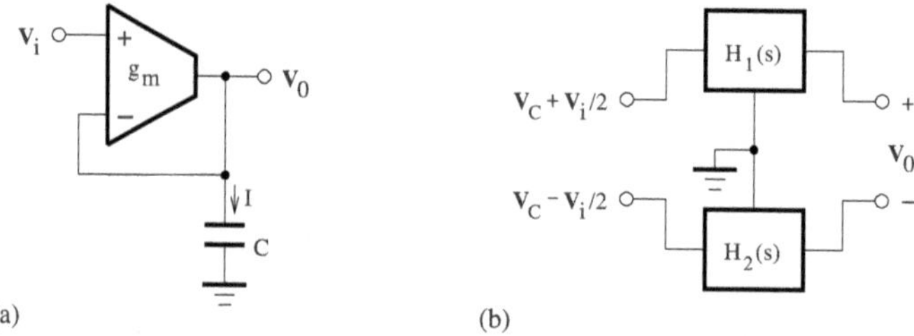

FIGURE 6.72
First-order lowpass filter: (a) Single-ended and (b) differential structures.

6.2.5 Summer circuits

A summer circuit is used to add, or subtract, analog voltage signals. It can be based on a noninverting or inverting topology.

The circuit diagram of an inverting summer based on a voltage amplifier is depicted in Figure 6.73(a). The currents due to the voltages V_1 and V_2 and flowing through the resistors R_1 and R_2, respectively, are added at the virtual ground terminal of the amplifier, and the sum is converted to a voltage with the reversed polarity by the feedback resistor, R. Ideally, $V^- = 0$ and the application of Kirchhoff's current law at the inverting node of the amplifier gives

$$\frac{V_1}{R_1} + \frac{V_2}{R_2} = -\frac{V_0}{R} \tag{6.207}$$

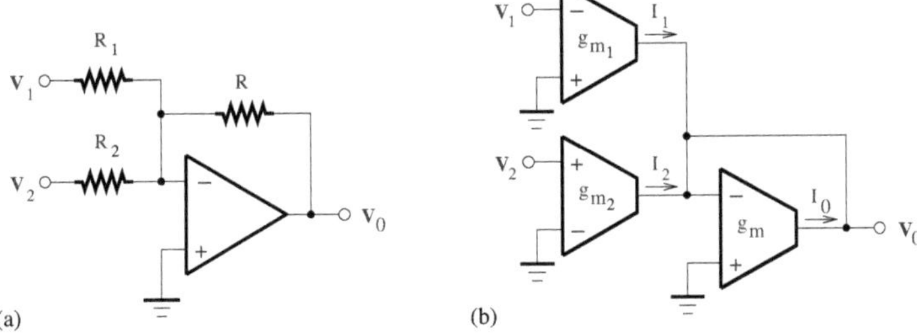

FIGURE 6.73
Summer circuits based on (a) voltage amplifiers and (b) transconductance amplifiers.

Hence,

$$V_0 = -\frac{R}{R_1}V_1 - \frac{R}{R_2}V_2 \tag{6.208}$$

This circuit can be extended to more than two inputs because the input signals are isolated from each other due to the virtual ground.
A summer can also be implemented using transconductance amplifiers, as illustrated in Figure 6.73(b). The input transconductors operate as a voltage-to-current converter, while the output transconductor is configured as a current-to-voltage converter. It can be shown that

$$I_1 + I_2 = -I_0 \tag{6.209}$$

where $I_1 = -g_{m_1}V_1$, $I_2 = g_{m_2}V_2$, and $I_0 = -g_mV_0$. The output voltage can then be written as

$$V_0 = \frac{g_{m2}}{g_m}V_2 - \frac{g_{m1}}{g_m}V_1 \tag{6.210}$$

Note that the polarity of an input signal connected to the inverting node of the input transconductor remains unchanged, and the polarity inversion is achieved by connecting the input signal to the noninverting node of the input transconductor.

6.2.6 Gyrator

In the design of integrated filters, gyrators can be used to eliminate the inductors of the LC prototype [51]. The symbol and small-signal equivalent model of a gyrator are shown in Figure 6.74. For an ideal gyrator, $V_i = z_{12}I_0$ and $V_0 = z_{21}I_i$. Provided that $z_{12} = -z_{21} = r$, we get

$$\begin{bmatrix} V_0 \\ V_i \end{bmatrix} = \mathbf{Z} \begin{bmatrix} I_0 \\ I_i \end{bmatrix} \tag{6.211}$$

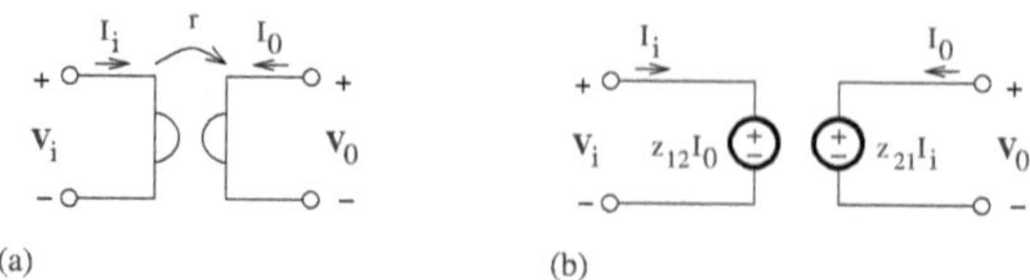

FIGURE 6.74
(a) Symbol and (b) small-signal equivalent model of a gyrator.

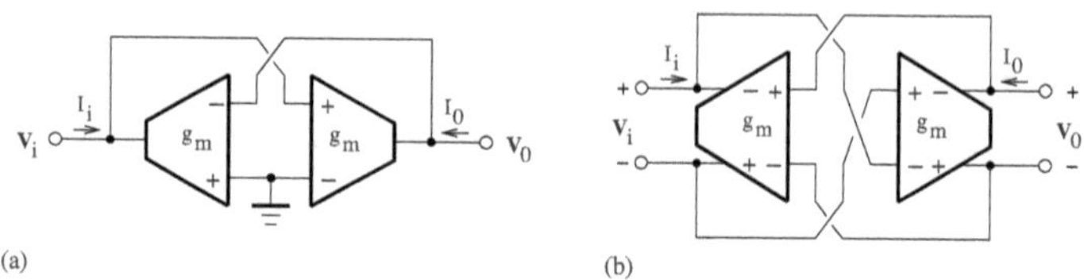

FIGURE 6.75
(a) Single-ended and (b) differential implementations of a gyrator.

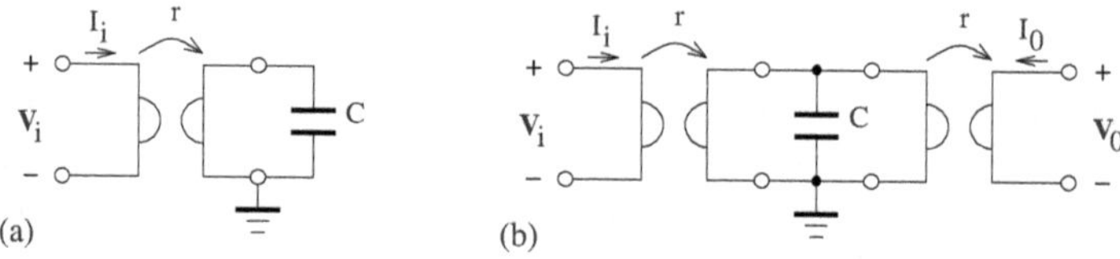

FIGURE 6.76
Equivalent circuit models of gyrator-based implementations of (a) grounded and (b) floating inductors.

where the open-circuit impedance of a gyrator is of the form

$$\mathbf{Z} = \begin{bmatrix} 0 & -r \\ r & 0 \end{bmatrix} \tag{6.212}$$

and r is the gyrator resistance. A gyrator is then a linear two-port network, the input impedance of which is given by

$$Z_i(s) = \frac{r^2}{Z_L(s)} \tag{6.213}$$

where Z_L is the load impedance. Let $Z_L(s) = 1/sC$, that is, the gyrator is loaded by a capacitor C. We can obtain

$$Z_i(s) = r^2 Cs = Ls \tag{6.214}$$

where $L = r^2 C$ is the value of the simulated inductor. Single-ended and differential implementations of a gyrator are shown in Figures 6.75(a) and (b), respectively. They are based on inverting and noninverting transconductors designed to exhibit the same value $g_m = 1/r$. Here, the accuracy of the gyrator can be limited by the parasitic input and output capacitances, and finite output conductance of transconductors.

The equivalent circuit models of grounded and floating inductors are shown in Figures 6.76 (a) and (b), respectively. In the first configuration, an inductor with the impedance Z is realized, while the second network can be characterized by the following short-circuit admittance matrix:

$$\mathbf{Y} = \frac{1}{Z}\begin{bmatrix} 1 & -1 \\ -1 & 1 \end{bmatrix} \tag{6.215}$$

6.3 Filter characterization

A filter can be characterized by its transfer function, which is the ratio of two s-domain polynomials with real coefficients and is given by

$$H(s) = \frac{Y(s)}{U(s)} = \frac{\sum_{j=0}^{M} b_j s^j}{1 + \sum_{i=1}^{N} a_i s^i} \tag{6.216}$$

where $U(s)$ and $Y(s)$ are the Laplace transforms of the input and the output variables. The zeros of the denominator are called poles of the filter, while the zeros of the numerator are referred to as transmission zeros. The filter will be realizable if $N \geq M$ and will not oscillate provided $a_i > 0$. For filter stability, the real part of all poles should be lower than zero.

On the $j\omega$-axis, the filter transfer function can expressed as

$$H(j\omega) = |H(j\omega)|e^{j\phi(\omega)} \tag{6.217}$$

The magnitude of the transfer function can be written either in the form of a gain

$$G(\omega) = 20 \log |H(j\omega)| \tag{6.218}$$

or an attenuation

$$A(\omega) = -20 \log |H(j\omega)| \tag{6.219}$$

using decibel (dB) units. The phase response, $\phi(\omega)$, is expressed in radians. The group delay of the filter is defined as

$$\tau(\omega) = -\frac{d\phi(\omega)}{d\omega} \tag{6.220}$$

It is often used as a specification in applications where the behavior in the time domain is of importance, for example, the allpass equalizer. Note that the transfer function of an allpass filter is of the form $H(s) = k \cdot D(-s)/D(s)$, where k is the dc gain and $D(s)$ is the transfer function denominator.

6.4 Filter design methods

Continuous-time filters process high-speed signals in applications where the linearity and accuracy specifications are relaxed, such as disk drives, communication devices, and video systems. In CMOS technology, the integration of active RC filters, which are realized using voltage amplifiers, resistors, and capacitors, may be impractical due the high resistance values required. This problem is solved in implementation approaches, such as g_m-C filters, which are only based on transconductance amplifiers and capacitors.

A filter can be categorized according to the type of function used to approximate its gain or attenuation specifications. Butterworth, Chebyshev, or Cauer approximation functions are often used in the synthesis of a stable and realizable filter prototype. To simplify the computation of filter coefficients, the transfer function is scaled to have a maximum magnitude of unity, and the frequency, ω, is normalized with respect to the passband edge frequency, ω_p, such that $\Omega = \omega/\omega_p$. The order of the transfer function is determined from the prescribed magnitude specifications in the passband,

$$1/\sqrt{1+\delta_p^2} \leq |H(\Omega)| \leq 1, \quad \text{if} \quad 0 \leq |\Omega| \leq 1 \tag{6.221}$$

and in the stopband,

$$|H(\Omega)| \leq 1/\sqrt{1+\delta_s^2}, \quad \text{if} \quad |\Omega| \geq \Omega_s \tag{6.222}$$

where Ω_s is the normalized stopband edge frequency. The passband ripple, δ_p, and stopband ripple, δ_s, can be computed as

$$\delta_p^2 = 10^{A_p/10} - 1 \tag{6.223}$$

and

$$\delta_s^2 = 10^{A_s/10} - 1 \tag{6.224}$$

where the passband and stopband attenuations in dB are respectively defined as $A_p = -\min(20\log_{10}|H(\Omega)|)$ for Ω in the filter passband, and $A_s = -\max(20\log_{10}|H(\Omega)|)$ for Ω in the filter stopband.

The determination of the filter transfer function focuses primarily on lowpass prototype filters, as highpass, bandpass, or bandstop prototypes can be derived from lowpass prototypes through appropriate spectral transformations. Note that, in the case of Bessel filters, the design is often performed based on the delay or phase specifications, that is, the maximally acceptable delay errors in the frequency band of interest. The synthesis of the transfer function, which is also referred to as the approximation problem, is performed using tables of classical filter functions or software tools such as MATLAB®.

There are several types of analog filters with various important characteristics. The most often used filter types are based on Butterworth, Chebyshev,

Cauer (or elliptic), and Bessel (or Thomson) approximation functions. The choice of a given filter is generally made depending on the application.

Butterworth filters: They are characterized by a magnitude response that is maximally flat in the passband and is monotonically decreasing in both the passband and stopband. However, the transition band of Butterworth filters is somewhat wide.

Chebyshev filters: They are known to have a steeper roll-off than Butterworth filters. Their magnitude responses either have equal ripples in the passband and decrease monotonically in the stopband (type I) or decrease monotonically in the passband and exhibit equal ripples in the stopband (type II or inverse Chebyshev).

Cauer filters: Their magnitude responses have equal ripples in both the passband and stopband. In comparison with other filter types, Cauer filters provide the sharpest transition band and are then the most efficient from the viewpoint of requiring the smallest order to realize a given magnitude specification. However, the sharp transition band is obtained at the price of a more nonlinear phase response in the passband, especially near the passband edge.

Bessel filters: They exhibit a maximally linear phase response or maximally flat group delay in the passband. However, Bessel filters can be plagued by their largest transition band.

Butterworth, Chebyshev, Cauer, and Bessel filters are synthesized with the focus on one primordial characteristic. In practice, they are often used directly, but their zero and pole placement can also be further optimized to simultaneously satisfy several characteristic specifications.

High-order filters that simulate the behavior of doubly resistively terminated LC networks are known to exhibit minimum sensitivity to component variations, in contrast to cascade realizations based on first-order and second-order (or biquadratic) filter structures. The frequency normalization, which corresponds to a frequency scale, should be taken into account in the determination of the nominal component values for the filter implementation. Because it is performed by dividing the variable, s, by the passband edge frequency, ω_p, the normalized variable is then of the form, $s' = s/\omega_p$, and the normalized element values can be related to actual values (R, C, and L) of the filter components as follows

$$R' = \frac{R}{\tilde{R}}, \qquad C' = \tilde{R} C \omega_p \,, \qquad L' = L \frac{\omega_p}{\tilde{R}} \tag{6.225}$$

where the normalizing resistance, $\tilde{R}$, is generally chosen to obtain well-suited and practical element values. The filter synthesis results in a circuit that should be analyzed to take into account the effect of real component imperfections and dynamic range on the overall performance.

6.4.1 First-order filter design

In general, the transfer function of a first-order filter section can be written as

$$H(s) = \frac{k_1 s + k_0}{s + \omega_c} \tag{6.226}$$

Table 6.1 presents the different types of first-order filters, which can be implemented using g_m-C circuits. The transfer function coefficients are then determined by the ratios of component values.

TABLE 6.1
First-Order Filter Section Classification

First-Order Filter Type	Transfer Function Coefficients
Lowpass filter	$k_1 = 0$
Highpass filter	$k_0 = 0$
Allpass filter	$k_1 = k,\ k_0 = -k\,\omega_c$

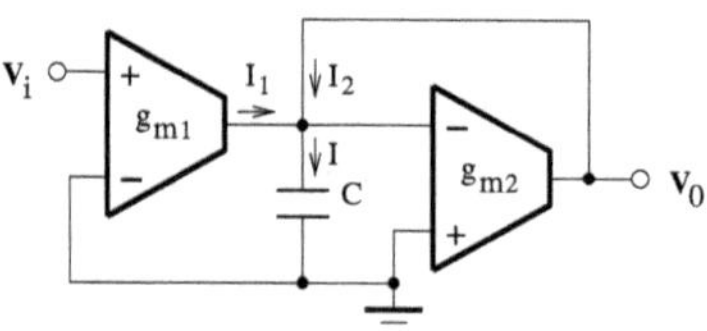

FIGURE 6.77
g_m-C implementation of a first-order lowpass filter.

Applying Kirchhoff's current law to the first-order filter depicted in Figure 6.77, it can be shown that

$$I_1 + I_2 = I' \tag{6.227}$$

Since $I_1 = g_{m1} V_i$ and $I_2 = -g_{m2} V_0$, we arrive at

$$g_{m1} V_i - g_{m2} V_0 = sCV_0 \tag{6.228}$$

The transfer function can then be computed as

$$H(s) = \frac{V_0(s)}{V_i(s)} = \frac{g_{m1}}{sC + g_{m2}} \tag{6.229}$$

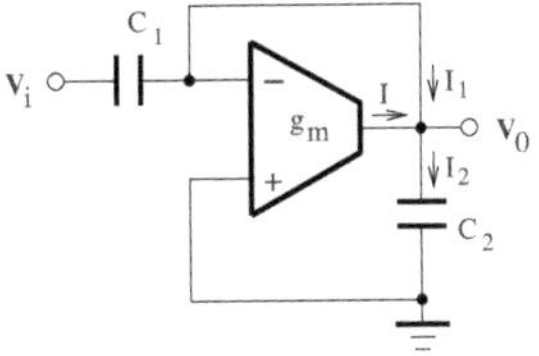

FIGURE 6.78
g_m-C implementation of a first-order highpass filter.

We see that $H(s)$ has a dc gain with the value, g_{m1}/g_{m2}, and a single pole at $-g_{m2}/C$. Equation (6.229) therefore characterizes a first-order lowpass filter.

Proceeding with the circuit of Figure 6.78, we find an output node equation of the form

$$I_1 + I = I_2 \tag{6.230}$$

In terms of the device parameters, this equation can be rewritten as

$$sC_1(V_i - V_0) - g_m V_0 = sC_2 V_0 \tag{6.231}$$

The transfer function is then derived as

$$H(s) = \frac{V_0(s)}{V_i(s)} = \frac{sC_1}{s(C_1 + C_2) + g_m} \tag{6.232}$$

In this case, a zero and a pole are located at 0 and $-g_m/(C_1+C_2)$, respectively. Because the dc gain is zero, and the gain at high frequencies is of the form, $C_1/(C_1+C_2)$, the circuit realizing $H(s)$ is referred to as a first-order highpass filter.

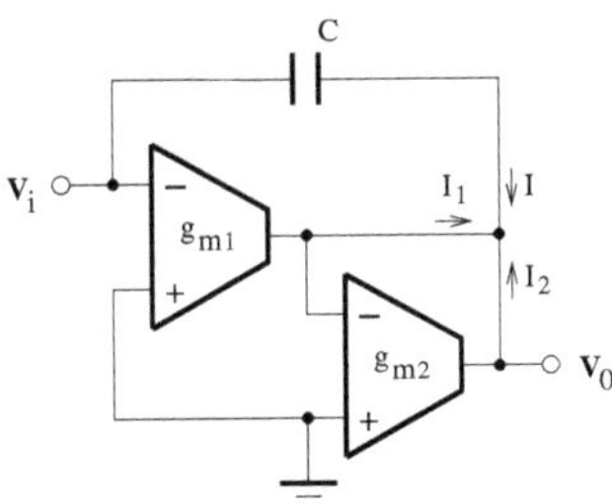

FIGURE 6.79
g_m-C implementation of a first-order allpass filter.

The circuit shown in Figure 6.79 includes two transconductance amplifiers and a capacitor. The output node equation can be written as

$$I + I_1 + I_2 = 0 \tag{6.233}$$

or equivalently,

$$sC(V_i - V_0) + g_{m1}V_i - g_{m2}V_0 = 0 \tag{6.234}$$

The transfer function is then given by

$$H(s) = \frac{V_0(s)}{V_i(s)} = \frac{sC - g_{m1}}{sC + g_{m2}} \tag{6.235}$$

Assuming that $g_{m1} = g_{m2} = g_m$, the magnitude of $H(s)$ is equal to unity for all frequencies, and a first-order allpass filter is realized. The function $H(s)$ exhibits a pole and a zero located symmetrically on either side of the $j\omega$ axis. Hence, the phase response is readily found to be

$$\theta(\omega) = \arctan\left(-\frac{\omega}{\omega_c}\right) - \arctan\left(\frac{\omega}{\omega_c}\right) = \pi - 2\arctan\left(\frac{\omega}{\omega_c}\right) \tag{6.236}$$

where $\omega_c = g_m/C$. Note that the phase is π at $\omega = 0$, $\pi/2$ at $\omega = \omega_c$, and zero at high frequencies.

6.4.2 Biquadratic filter design methods

The design of a biquadratic or second-order filter can be based on signal-flow graphs (SFGs) or gyrators. In some cases, the resulting filter can retain the low sensitivity properties of a doubly terminated lossless network.

6.4.2.1 Signal-flow graph-based design

The SFG block diagram realizing a general biquadratic transfer function is shown in Figure 6.80. It consists of integrators and summing stages. Its transfer function is given by

$$H(s) = \frac{V_0(s)}{V_i(s)} = \frac{k_2 s^2 + k_1 s + k_0}{s^2 + \left(\frac{\omega_p}{Q}\right)s + \omega_p^2} \tag{6.237}$$

where ω_p and Q denote the pole frequency and quality factor, respectively. For the filter stability, the parameters ω_p and Q should assume only positive values.

The transfer function of the g_m-C biquad circuit shown in Figure 6.81 [52] can be computed as

$$H(s) = \frac{V_0(s)}{V_i(s)} = \frac{k_2 s^2 + k_1 s + k_0}{s^2 + \left(\frac{\omega_p}{Q}\right)s + \omega_p^2} \tag{6.238}$$

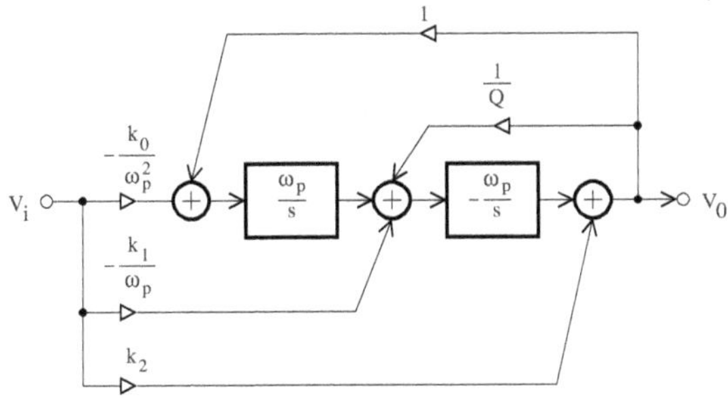

FIGURE 6.80
Signal-flow graph representation of a general biquadratic filter.

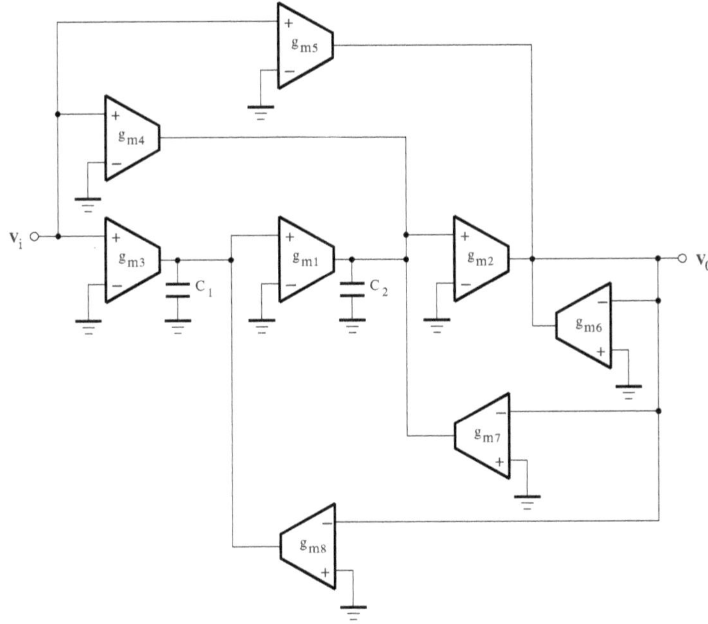

FIGURE 6.81
g_m-C implementation of a general biquadratic filter.

where

$$k_0 = \frac{g_{m1} g_{m2} g_{m3}}{g_{m6} C_1 C_2} \tag{6.239}$$

$$k_1 = \frac{g_{m2} g_{m4}}{g_{m6} C_2} \tag{6.240}$$

$$k_2 = \frac{g_{m5}}{g_{m6}} \tag{6.241}$$

$$\omega_p = \sqrt{\frac{g_{m1} g_{m8}}{g_{m7} C_1}} \tag{6.242}$$

and

$$Q = \frac{g_{m6}C_2}{g_{m2}g_{m7}}\sqrt{\frac{g_{m1}g_{m8}}{g_{m7}C_1}} \tag{6.243}$$

Different biquad types can be realized depending on the choice of the transfer

TABLE 6.2
Biquad Classification

Biquad Type	Transfer Function Coefficients
Lowpass filter	$k_0 = k\omega_p^2$, $k_1 = k_2 = 0$
Highpass filter	$k_0 = k_1 = 0$, $k_2 = k$
Bandpass filter	$k_0 = k_2 = 0$, $k_1 = k(\omega_p/Q_p)$
Lowpass notch filter	$k_0 = k\omega_z^2$, $k_1 = 0$, $k_2 = k$, and $\omega_z > \omega_p$
Highpass notch filter	$k_0 = k\omega_z^2$, $k_1 = 0$, $k_2 = k$, and $\omega_z < \omega_p$
Symmetrical notch filter	$k_0 = k\omega_z^2$, $k_1 = 0$, $k_2 = k$ and $\omega_z = \omega_p$
Allpass filter	$k_0 = k\omega_p^2$, $k_1 = -k(\omega_p/Q)$, $k_2 = k$

function coefficients (see Table 6.2).

• In the case of the second-order lowpass filter, the magnitude response is maximum at the frequency where $d|H(\omega)|/d\omega = 0$. This frequency is given by

$$\omega_0 = \omega_p\sqrt{1 - 1/(2Q^2)} \tag{6.244}$$

and the maximum gain is of the form

$$H_{max} = |H(\omega_0)| = \frac{kQ}{\sqrt{1 - 1/(4Q^2)}} \tag{6.245}$$

The dc gain is k and the magnitude response decreases as $1/\omega^2$ at high frequencies.

• Considering the highpass filter, the maximum gain is still given by Equation (6.245), but occurs at the frequency

$$\omega_0 = \frac{\omega_p}{\sqrt{1 - 1/(2Q^2)}} \tag{6.246}$$

The magnitude response increases as ω^2 for low frequencies and the high-frequency gain is k.

• For the bandpass filter, the −3 dB bandwidth is computed as the difference between the passband edge frequencies, that is,

$$\text{BW} = \omega_2 - \omega_1 = \frac{\omega_p}{Q} \tag{6.247}$$

where $\omega_p^2 = \omega_2\,\omega_1$. The maximum value of the magnitude response, which is equal to k, occurs at the frequency ω_p. The increase and decrease in the magnitude response both follow a $1/\omega$ function.

• The notch filter exhibits a magnitude response with transmission zero at $\omega = \omega_z$. For the lowpass and highpass notch filters, the maximum gain can be computed as $H_{max} = |H(\omega_0)|$, where

$$\omega_0 = \omega_p \sqrt{\frac{\dfrac{\omega_z^2}{\omega_p^2}\left(1 - \dfrac{1}{2Q^2}\right) - 1}{\dfrac{\omega_z^2}{\omega_p^2} + \dfrac{1}{2Q^2} - 1}} \tag{6.248}$$

The dc gain and high-frequency gain are equal to $k(\omega_z^2/\omega_p^2)$ and k, respectively. In the case of the symmetrical notch filter, the −3 dB bandwidth is ω_p/Q_p, and the dc gain and high-frequency gain are equal to the same value, k.

• The second-order allpass filter is characterized by a constant gain equal to k, but its phase response changes from 0° to −360° and takes the value −180° at ω_p.

6.4.2.2 Gyrator-based design

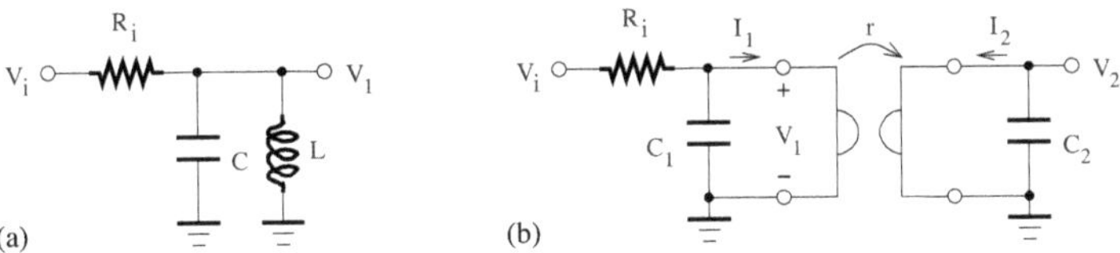

FIGURE 6.82
(a) Second-order bandpass RLC prototype; (b) circuit diagram of a second-order bandpass filter based on a gyrator.

To achieve a low sensitivity to the variations of component values, g_m-C filters can be designed by simulating passive *RLC* prototypes. A second-order bandpass *RLC* prototype filter is shown in Figure 6.82(a). By replacing the inductor with the corresponding gyrator network, the circuit diagram of Fig-

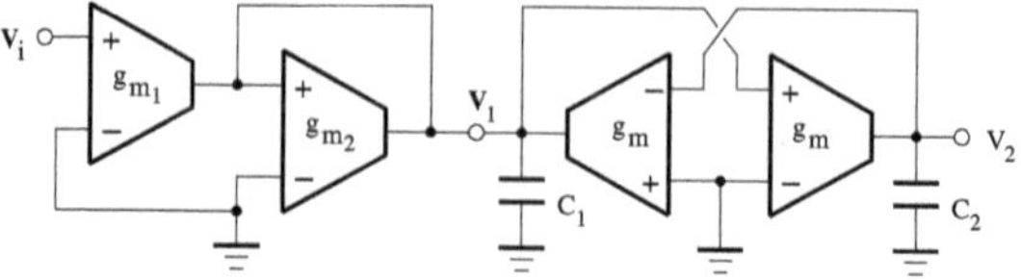

FIGURE 6.83
Single-ended g_m-C implementation of the second-order bandpass filter.

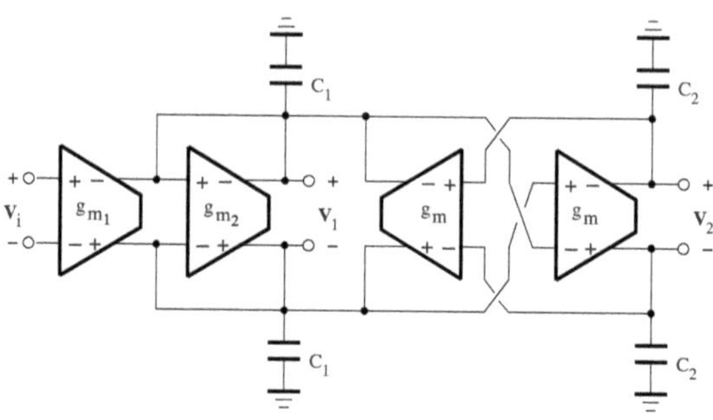

FIGURE 6.84
Differential g_m-C implementation of the second-order bandpass filter.

ure 6.82(b) is derived. Using Kirchhoff's current law, we can obtain

$$\frac{V_i - V_1}{R_i} = V_1 s C_1 + I_1 \tag{6.249}$$

$$I_2 + V_2 s C_2 = 0 \tag{6.250}$$

where $V_1 = rI_2$ and $V_2 = -rI_1$. The system of Equations (6.249) and (6.250) can be solved either for a bandpass transfer function of the form

$$H_1(s) = \frac{V_1(s)}{V_i(s)} = \frac{s(r^2/R_i)C_2}{s^2 r^2 C_1 C_2 + s(r^2/R_i)C_2 + 1} \tag{6.251}$$

or a lowpass transfer function given by

$$H_2(s) = \frac{V_2(s)}{V_i(s)} = -\frac{r/R_i}{s^2 r^2 C_1 C_2 + s(r^2/R_i)C_2 + 1} \tag{6.252}$$

The single-ended and differential g_m-C implementations of the filter are shown in Figures 6.83 and 6.84, respectively. They realize the transfer functions H_1 and H_2 with $R_i = 1/g_{m_1} = 1/g_{m_2}$ and $r = 1/g_m$. In practice, it may be preferable to use only one filter output at a time because a resistive load at the node V_1 can affect the transfer function associated with the node V_2, and vice versa.

The circuit diagram of a second-order allpass network based on gyrators is

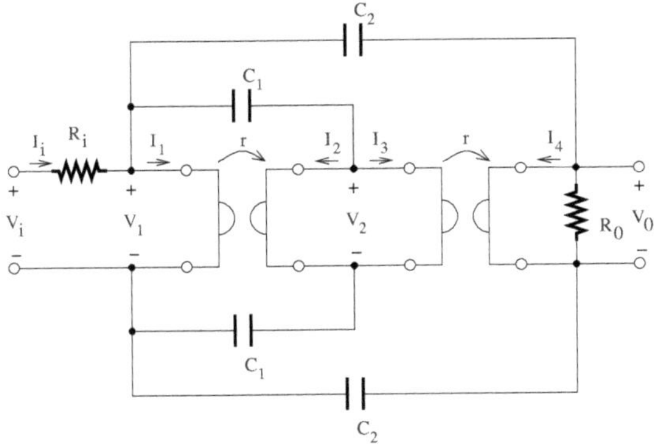

FIGURE 6.85
Circuit diagram of a second-order allpass network based on gyrators.

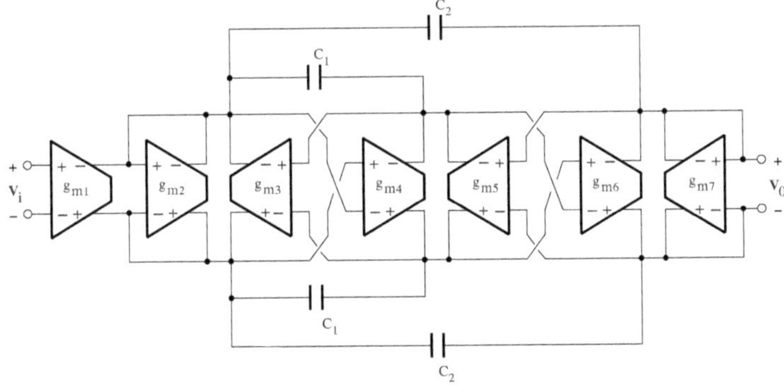

FIGURE 6.86
Differential implementation of the allpass filter based on gyrators.

shown in Figure 6.85. It is based on a nonreciprocal[1] network, which cannot be directly derived from LC filter prototypes. The nodal equations can be written as

$$\frac{V_i - V_1}{R_i} = I_1 + (V_1 - V_2)sC_1 + (V_1 - V_0)sC_2 \tag{6.253}$$

$$(V_1 - V_2)sC_1 = I_2 + I_3 \tag{6.254}$$

$$(V_1 - V_0)sC_2 = I_4 + \frac{V_0}{R_0} \tag{6.255}$$

where $I_1 = -V_2/r = -I_4$, $I_2 = V_1/r$, and $I_3 = -V_0/r$. Solving the system of

[1]A two-pair network described by the chain matrix $\mathbf{M}$ is nonreciprocal if $\det(\mathbf{M}) \neq \pm 1$. In the special case of LC networks, $\det(\mathbf{M}) = 1$.

Equations (6.253) through (6.255) gives

$$H(s) = \frac{V_0(s)}{V_i(s)} = \frac{s^2r^2C_1C_2 - srC_1 + 1}{s^2r^2C_1C_2\left(1 + \frac{R_i}{R_0}\right) + sC_1\left(R_i + \frac{r^2}{R_0}\right) + 1 + \frac{R_i}{R_0}} \tag{6.256}$$

To realize a second-order allpass transfer function, it is required to set either $r = R_i$ or $r = R_0$. Hence,

$$H(s) = \frac{V_0(s)}{V_i(s)} = k\frac{\frac{s^2}{\omega_0^2} - \frac{s}{\omega_0 Q} + 1}{\frac{s^2}{\omega_0^2} + \frac{s}{\omega_0 Q} + 1} \tag{6.257}$$

where

$$\frac{1}{k} = 1 + \frac{R_i}{R_0}, \quad \frac{1}{\omega_0^2} = r^2C_1C_2, \quad \text{and} \quad \frac{1}{\omega_0 Q} = rC_1 \tag{6.258}$$

The differential implementation of the allpass filter is depicted in Figure 6.86. The gyrators and resistors are replaced by the corresponding g_m-C structures.

The design of high-order filters is generally achieved after the pole-zero pairing and gain assignment related to each low-order section obtained from the decomposition of the target transfer function. Once a suitable topology is found and the filter coefficient mapping is carried out, the scaling of the component values can be performed to obtain the desired dynamic range [54] and chip area. Note that the distortion level at the filter output can be set under a prescribed value (e.g., 1% of the input signal magnitude) by biasing adequately the amplifiers.

For cascade designs, there are many connection possibilities for low-order filter sections. The ordering sequence (for example, sections ordered according to the increasing values of Q, lowpass and bandpass filters as first stages, highpass and bandpass filters as last stages) should be chosen to maximize the resulting dynamic range.

6.4.3 Ladder filter design methods

High-order active filters can be derived either from LC ladder networks or using signal-flow graph techniques. The first approach consists of using various component-simulation methods, which can be based on gyrator elements or generalized immittance converters, to replace inductors (and sometimes resistors), while the second approach relies on simulating equations characterizing the operation of a lossless network prototype.

6.4.3.1 LC ladder network-based design

Active filters derived from LC ladder networks should feature a lower sensitivity to component variations in the passband in comparison to the cascade of low-order structures. The LC ladder shown in Figure 6.87 realizes a third-order lowpass elliptic filter. It can be described by

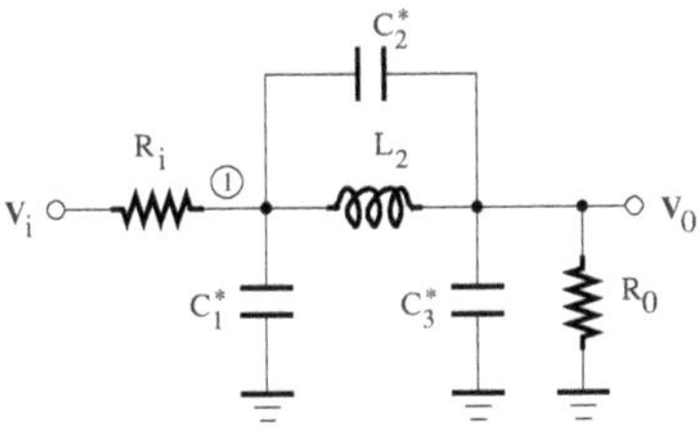

FIGURE 6.87
Third-order elliptic lowpass filter.

$$V_1 = \frac{1}{s(C_1^* + C_2^*)}\left(sC_2^* V_0 - I_{L_2} + \frac{V_i - V_1}{R_i}\right) \tag{6.259}$$

$$I_{L_2} = \frac{V_1 - V_0}{sL_2} \tag{6.260}$$

$$V_0 = \frac{1}{s(C_2^* + C_3^*)}\left(sC_2^* V_1 + I_{L_2} - \frac{V_0}{R_0}\right) \tag{6.261}$$

Let $R_i = R_0 = R$, $C_1^* = C_3^* = C^*$, and $\hat{s} = s/\omega_c$, where ω_c is the cutoff frequency. The normalized transfer function of the LC filter of Figure 6.87 can be written as

$$H(\hat{s}) = \frac{V_0(\hat{s})}{V_i(\hat{s})} = k\frac{\hat{s}^2 + \alpha}{(\hat{s} + \beta)(\hat{s}^2 + \beta_1\hat{s} + \beta_2)} \tag{6.262}$$

where

$$k = \frac{C_2^*}{RC^*(C^* + 2C_2^*)} \tag{6.263}$$

$$\alpha = \frac{1}{L_2 C_2^*} \tag{6.264}$$

$$\beta = \frac{1}{RC^*} \tag{6.265}$$

$$\beta_1 = \frac{1}{R(C^* + 2C_2^*)} \tag{6.266}$$

$$\beta_2 = \frac{2}{L_2(C^* + 2C_2^*)} \tag{6.267}$$

and the resistor R can be chosen equal to one due to the impedance-level normalization.

Starting with the normalized LC network prototype derived from a given

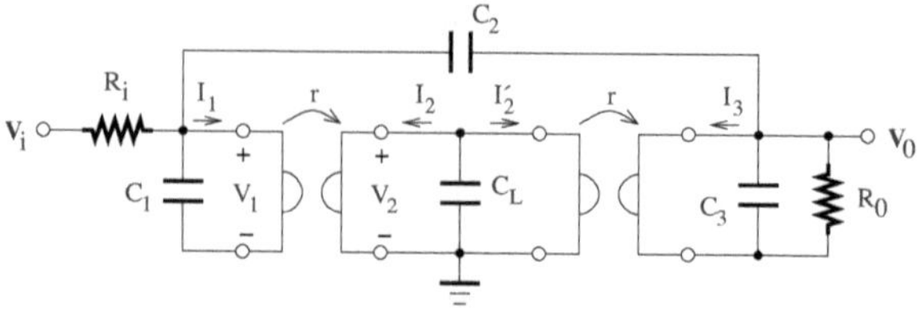

FIGURE 6.88
Single-ended circuit of the third-order elliptic lowpass filter based on gyrators.

filter specification, the inductors can be eliminated with the help of gyrators. Figures 6.88 and 6.89 show the single-ended and differential circuit diagrams of the third-order elliptic lowpass filter of Figure 6.87 after the inductor substitution.

Using Kirchhoff's current laws, the analysis of the circuit depicted in Figure 6.88 results in

$$\frac{V_i - V_1}{R_i} = V_1 s C_1 + (V_1 - V_0) s C_2 + I_1 \tag{6.268}$$

$$(V_1 - V - 0) s C_2 = V_0 s C_3 + \frac{V_0}{R_0} + I_3 \tag{6.269}$$

$$V_2 s C_L + I_2 + I_2' = 0 \tag{6.270}$$

where $V_1 = rI_2$, $V_2 = -rI_1 = rI_3$, and $V_2 = -rI_2'$. To proceed further, it can be assumed that $C_1 = C_3 = C$ and $R_i = R_0 = R$. Solving the system of Equations (6.268) through (6.270) for the output-input voltage transfer function gives

$$\begin{aligned} H(s) &= \frac{V_0(s)}{V_i(s)} \\ &= \frac{C_2}{RC(C+2C_L)} \frac{s^2 + \dfrac{1}{r^2 C_2 C_L}}{\left(s + \dfrac{1}{RC}\right)\left[s^2 + \dfrac{s}{R(C+2C_L)} + \dfrac{2}{r^2 C_L (C+2C_L)}\right]} \end{aligned} \tag{6.271}$$

Comparing Equations (6.262) and (6.271), the component values for a single-ended implementation are $C = C^*/R\omega_c$, $C_2 = C_2^*/r\omega_c$, and $C_L = L_2/r\omega_c$.

The fully differential g_m-C realization of the filter is shown in Figure 6.90, where $g_{m1} = g_{m3} = g_{m6} = 2g_m$, $g_{m2} = g_{m4} = g_{m5} = g_{m7} = g_m$, and $C_1 = C_3 = C$. The resistor R of the LC network and the resistor r are implemented by transconductors. Their values are of the form, $1/g_m$. Due to the inherent two-times increase of the dynamic range in the differential configuration, the sizes of the circuit components can be scaled by a factor of

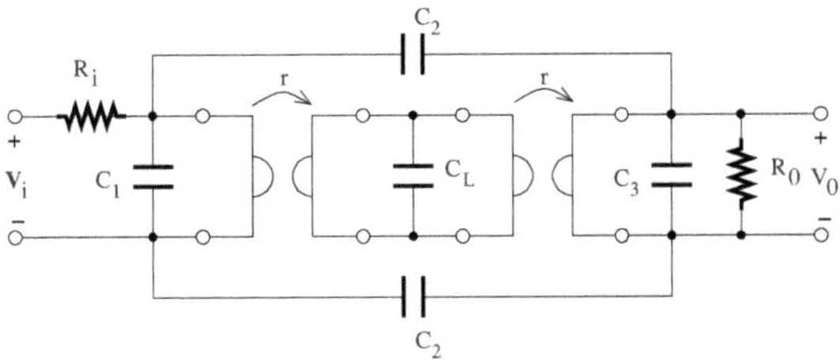

FIGURE 6.89
Differential circuit of the third-order elliptic lowpass filter based on gyrators.

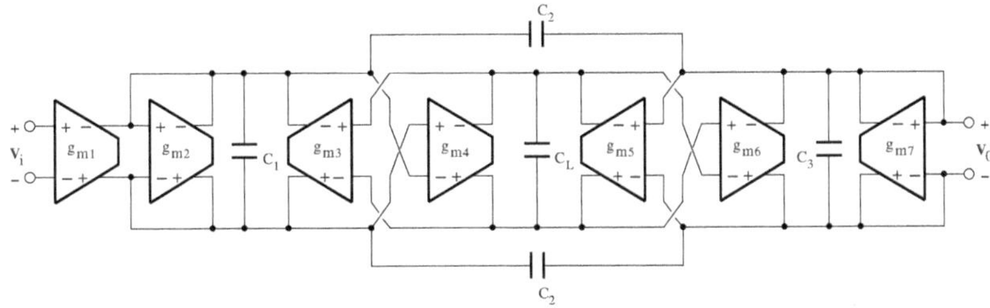

FIGURE 6.90
Differential implementation of the third-order elliptic lowpass filter based on gyrators.

1/2 and are then given by $C_j = g_m C_j^*/2\omega_c$, $j = 1, 3$, $C_2 = g_m C_2^*/2\omega_c$, and $C_L = g_m L_2^*/2\omega_c$. The resulting transfer function of the g_m-C filter reads

$$H(\hat{s}) = \frac{V_0(\hat{s})}{V_i(\hat{s})} = K \frac{\hat{s}^2 + a}{(\hat{s} + b)(\hat{s}^2 + b_1 \hat{s} + b_2)} \tag{6.272}$$

where

$$K = \frac{2 g_m C_2}{C(C + 2C_2)} \tag{6.273}$$

$$a = \frac{g_m^2}{C_L C_2} \tag{6.274}$$

$$b = \frac{g_m}{C} \tag{6.275}$$

$$b_1 = \frac{g_m}{C + 2C_2} \tag{6.276}$$

and

$$b_2 = \frac{2 g_m^2}{C_L (C + 2C_2)} \tag{6.277}$$

The filter gain and cutoff frequency are determined by the values of the transconductors and capacitors. However, an automatic tuning is generally required to control the transconductance fluctuations due to fabrication tolerance and temperature variations, and the filter parameters are preferably made programmable using capacitor arrays.

6.4.3.2 Signal-flow graph-based design

Another approach to design continuous-time ladder filters is based on the signal-flow graph (SFG) representation of nodal equations of an LC network. The passive filter structure is decomposed into integrator and gain stages, which can then be realized using circuits of a given type (active RC, MOSFET-C, g_m-C, etc.). The number of amplifiers is expected to be at least equal to the order of the filter to be designed.

The SFG, as shown in Figure 6.91, is derived from the LC network of Figure 6.87. The circuit diagram shown in Figure 6.92, where $C_X = C_1 + C_2$ and $C_Z = C_2 + C_3$, is then obtained using RC circuits. It consists essentially of inverting and noninverting integrators. The filter gain of an active RC filter can be actually set to k times the one of the LC network provided $R_X = R_i/k$, where k is a real number. The values of the elements connected to the noninverting integrator are multiplied by a factor R, and the value of the feedback capacitor is actually $C_Y = L_2/R^2$. The circuit diagram of the

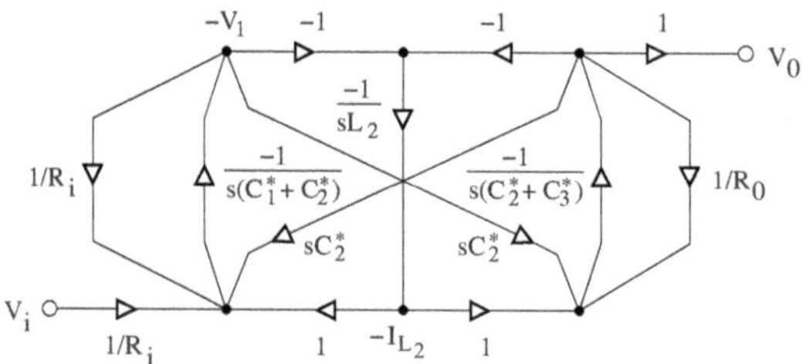

FIGURE 6.91
Signal-flow graph of a third-order elliptic lowpass filter.

g_m-C filter is depicted in Figure 6.93. Due to the absence of a summing node in g_m-C integrators, the number of active components is reduced using amplifiers with two differential inputs. Here, the resistors simulated by transconductors assume the value $1/g_m$, where g_m is the amplifier transconductance, and the value of the capacitor related to the inductor, L_2, of the passive network is $C_3 = L_2 g_m^2$. The compensation of the 6-dB loss, which is proper to LC networks, is achieved by sizing the input transconductor to exhibit the value $2g_m$, while the transconductance of the remaining amplifiers is g_m.

Filter structures can also be derived from SFGs, which are not directly related to LC prototypes. This approach is suitable for the design of high-order allpass filters [49, 55]. Let us consider the transfer function of an allpass

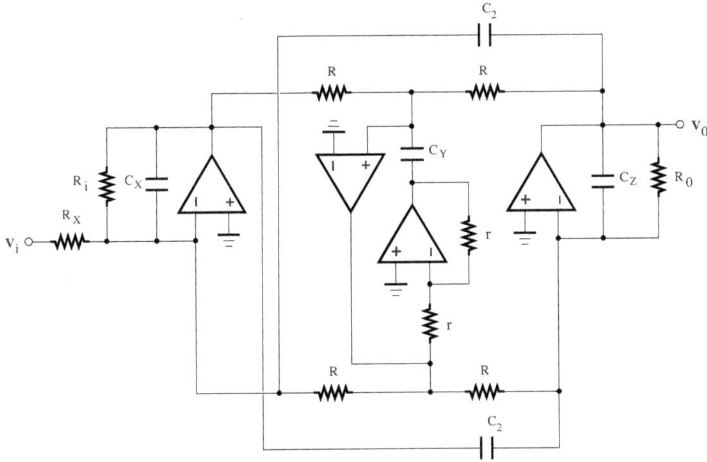

FIGURE 6.92
Active RC realization of a third-order elliptic lowpass filter.

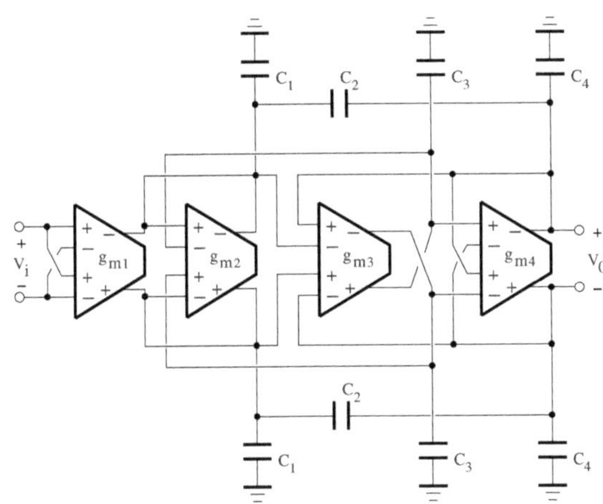

FIGURE 6.93
g_m-C implementation of the third-order elliptic lowpass filter based on the SFG technique.

filter given by

$$T(s) = \pm k \frac{D(-s)}{D(s)} \tag{6.278}$$

where k is the desired constant gain and $D(s)$ is the transfer function denom-

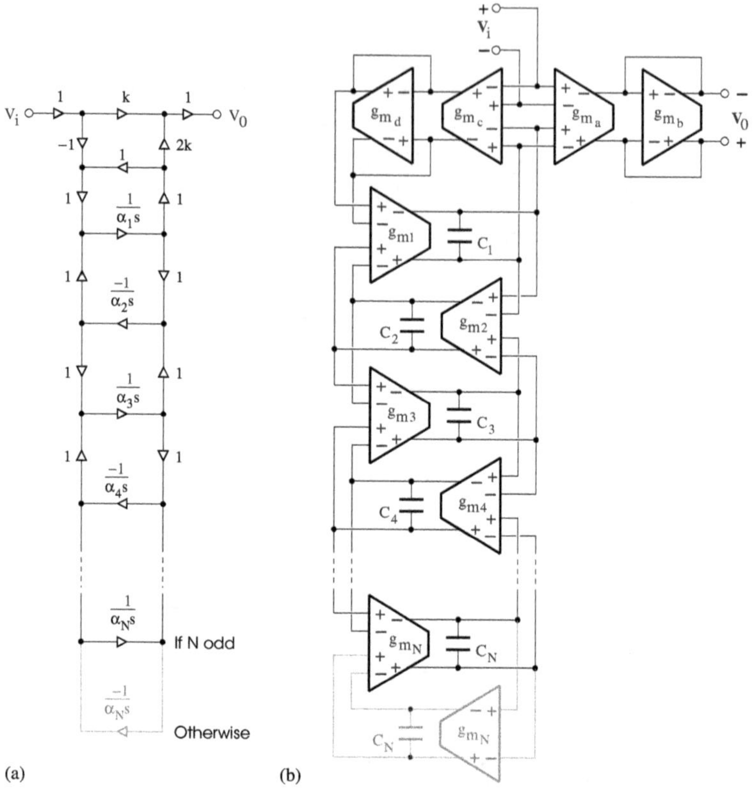

FIGURE 6.94
(a) Signal-flow graph-based allpass filter; (b) g_m-C implementation of the allpass filter.

inator. Assuming that $D(s) = 1 + Y(s)$ and $D(-s) = 1 - Y(s)$, we obtain

$$T(s) = \mp\left[k - \frac{2k}{1 + Y(s)}\right] \tag{6.279}$$

where $Y(s)$ is the admittance function. It can then be found that

$$Y(s) = \begin{cases} \dfrac{D(s) - D(-s)}{D(s) + D(-s)} & \text{if N is odd} \\ \dfrac{D(s) + D(-s)}{D(s) - D(-s)} & \text{if N is even,} \end{cases} \tag{6.280}$$

where $D(s)$ is a Hurwitz polynomial. To proceed further, the function $Y(s)$ is expanded as continued fractions. That is,

$$Y(s) = \alpha_1 s + \cfrac{1}{\alpha_2 s + \cfrac{1}{\alpha_3 s + \cdots + \cfrac{1}{\alpha_N s}}} \tag{6.281}$$

The SFG that realizes Equations (6.279) and (6.281) is depicted in Figure 6.94(a). Its g_m-C implementation is shown Figure 6.94(b), where the capacitor values are of the form

$$C_k = \frac{1}{\alpha_k} \qquad k = 1, 2, 3, \cdots, N \tag{6.282}$$

Table 6.3 gives the capacitor values provided by the synthesis of allpass filters

TABLE 6.3
Allpass Filter Design Equations for $N = 2$ and 3 ($a > c/b$)

Transfer Function	Capacitor Values
$T(s) = -\frac{1 - as + bs^2}{1 + as + bs^2}$	$C_1 = a/b \quad C_2 = 1/a$
$T(s) = -\frac{1 - as + bs^2 - cs^3}{1 + as + bs^2 + cs^3}$	$C_1 = b/c \quad C_2 = \frac{a}{b} - \frac{c}{b^2} \quad C_3 = \frac{1}{a - c/b}$

with $N = 2$ and 3. Note that another circuit topology can be derived by decomposing $Y(s)$ into partial fractions. In any case, the resulting allpass filter can be realized such that its amplitude response remains insensitive to variations of some component values. This low-sensitivity property is especially useful in high-Q applications.

By scaling the component values of continuous-time filters conceived by one of the above design methods, the dynamic performance can be improved and the component values are set to practical IC sizes. The signal dynamic range is limited by the noise and saturation level of the amplifiers. The scaling relies on the fact that the transfer function from the filter input to the output of an amplifier will remain unchanged if the components in its forward and feedback paths are appropriately multiplied or divided by a given factor. The scaling generally ensures that the output signals of the amplifiers saturate at the same level of the input signal. The scaling factor can be made equal to either a constant or $V_{0j,max}/V_{0,max}$, where $V_{0j,max}$ and $V_{0,max}$ denote the maximum output voltage of the j-th amplifier and the filter, respectively.

6.5 Design considerations for continuous-time filters

The responses of continuous-time (CT) filters are determined by the values of the capacitors, C, resistors, R, and transconductors, g_m. Due to factors such as fabrication tolerances, temperature variation, and aging, time constants (RC products or g_m/C ratios) can drift to about 10% to 50% from their nominal values. As a consequence, the electrical characteristics of the filters do not meet the design specifications.

6.5.1 Automatic on-chip tuning of continuous-time filters

The tasks of signal processing generally encountered in the front-end section of a signal processor, namely, amplification of the desired channel to the full scale of the analog-to-digital converter, automatic gain control, and filtering to remove the interference of the adjacent channels, are preferably implemented using CT circuits. An analog equalizer for a read channel in magnetic recording, for instance, operates at the given data rate using a lower area and power than that required by an equivalent digital structure.

In a cellular phone system, the channels can be spaced at 25-kHz intervals, and each channel needs a bandwidth of 21 kHz. The desired channel can be selected by a bandpass filter whose passband is one channel wide, only if the center frequency accuracy is strictly less than 4 kHz. For a center frequency fixed at 450 kHz, this corresponds to a precision requirement better than 0.8% (that is, 4 kHz/450 kHz). The specifications of a bandpass filter required for channel selection in communication systems are illustrated in Figure 6.95, where the parameters can be specified as follows: $f_{02} = 450$ kHz, $2f_p = 21$ kHz, and $\triangle f = 25$ kHz.

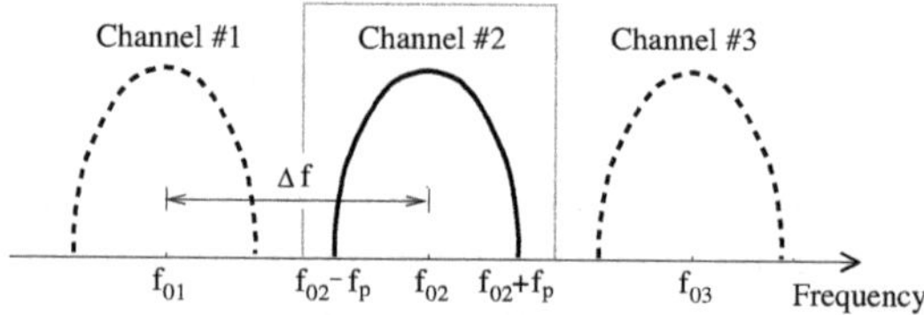

FIGURE 6.95
Bandpass filter specifications required for channel selection in communication systems.

Sigma-delta (ΣΔ) modulators using a CT filter have the advantage of sampling the signals at higher frequencies than the ones based on discrete-time circuits. They are then suitable for the design of wireless receivers operating at the intermediate frequency. However, the noise transfer function of CT modulators can effectively suppress the quantization noise at the desired frequency only if the pole locations of the loop filter are accurately defined. Digitizing a 200-kHz band at 70 MHz, for instance, would require a center frequency precision better than 200 kHz/70 MHz, or equivalently about 0.3%.

The solution that is generally adopted is to design CT filters with an associated automatic tuning scheme, as shown in Figure 6.96 [56, 57]. Note that the reference signal (REF) provided by a crystal oscillator is connected to the frequency-tuning master only when this latter is a voltage-controlled filter. Here, the tuning scheme is based on the master-slave technique. It consists of four major blocks: the main filter or slave, master, frequency control, and Q control circuits. The input signal is processed and filtered by the slave. The frequency-tuning master can be either a voltage-controlled oscillator (VCO) [31, 58] or a voltage-controlled filter (VCF) [59–61] and it models the slave's behavior that is essential for the tuning. The control circuits sense the master output signal and compare it to the one supplied by the reference source. The master is tuned until these two quantities are similar. The correction signals generated by the control circuits and used for this operation are simultaneously applied to the master and to the slave, which is then indirectly tuned.

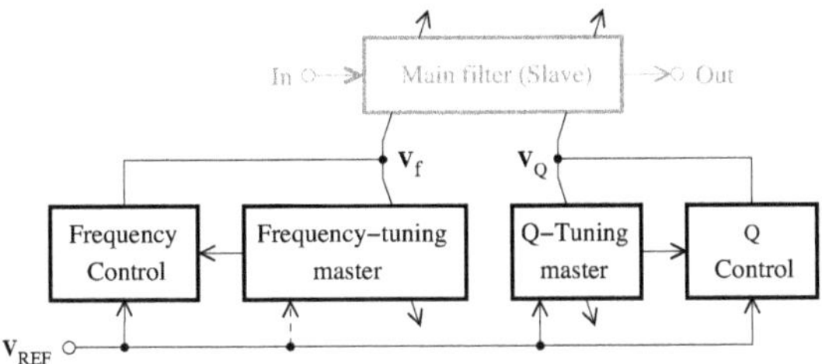

FIGURE 6.96
Block diagram of the master-slave tuning scheme.

The tuning circuit can principally be divided into two parts: the frequency and Q-tuning system. Generally, the automatic tuning of CT filters consists of locking the filter's response to an external and accurate reference. Next, various implementations of on-chip tuning approaches will be reviewed and the errors and performance limitations due to nonideal effects will also be discussed.

6.5.2 Nonideal integrator

Integrators are the principal building blocks of fully integrated filters. It can then be expected that the integrator nonidealities will affect the overall filter performance. Figure 6.97 shows the ideal and real frequency responses of an integrator. Ideally, an integrator has a pole at the origin and a phase shift of $-\pi/2$ at all frequencies. It is described by a transfer function of the form, $T_i(s) = -\omega_u/s$, where ω_u is the unity-gain frequency. However, due to the

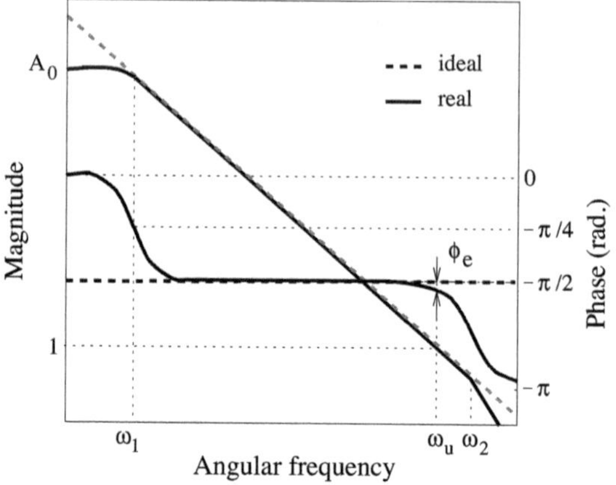

FIGURE 6.97
Frequency response of an ideal and real integrator.

parasitic effects in real amplifiers, the frequency response exhibits further deviations for frequencies greater than ω_u. Therefore, an integrator model, which takes into account the second-order effects of additional poles, is characterized by the following transfer function,

$$T_r(s) = \frac{A_0}{\left(1 + \dfrac{s}{\omega_1}\right)\left(1 + \dfrac{s}{\omega_2}\right)} \tag{6.283}$$

where A_0 is the finite dc gain, and $\omega_1 = \omega_u/A_0$ and ω_2 denote the dominant and nondominant poles, respectively. It is assumed that $\omega_1 \ll \omega_u \ll \omega_2$. As depicted in Figure 6.97, the phase response decreases to values below $-\pi/2$ for higher frequencies, and ϕ_e denotes the phase error at the unity-gain frequency. This can also be confirmed by analyzing the phase variations of the function T_r given by[2]

$$\phi(\omega) = -\frac{\pi}{2} + \Delta\phi(\omega) \tag{6.284}$$

where $\Delta\phi(\omega) = \arctan(\omega_1/\omega) - \arctan(\omega/\omega_2)$. Low dc gains and parasitic poles result, respectively, in leading and lagging phase errors. The phase tolerance in worst cases depends on the specifications (quality factor, pole frequency, etc.) of the filter to be designed.

[2]It can be verified that $\arctan(1/x) = -\arctan(x) + [\pi \text{sgn}(x)]/2$ provided $x \neq 0$ and $\arctan(-x) = -\arctan(x)$.

6.6 Frequency-control systems

Frequency-control systems can be implemented using the phase-locked-loop-based technique or charge comparison-based technique.

6.6.1 Phase-locked-loop-based technique

The filter tuning is performed by locking a loop to an external reference signal.

6.6.1.1 Operation principle

The frequency-control system, as depicted in Figure 6.98, where V_f is the frequency tuning signal, includes a master and a phase detector. Here, the variable of interest is the phase of the filter transfer function.

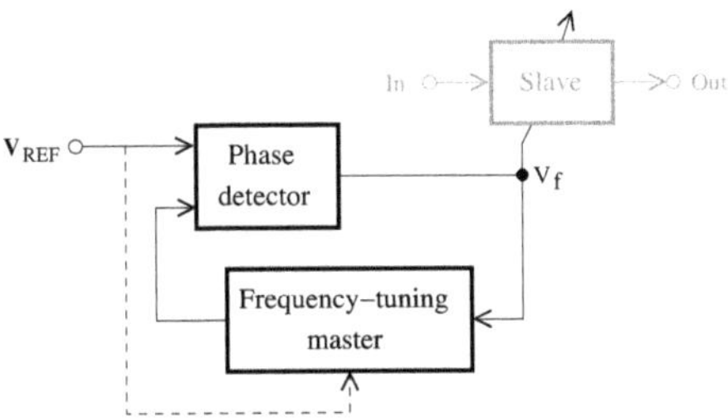

FIGURE 6.98
Frequency tuning scheme based on the phase-locked loop principle.

The entire system will operate like a phase-locked loop when the master is a VCO. This latter attempts to produce a signal that tracks the phase of the reference signal. The phase detector can consist of an analog multiplier or an exclusive OR (XOR) gate driven by a pair of comparators, followed by a loop filter (or an integrator). It measures the phase difference between the above two waveforms and produces the control signal that drives the VCO and changes its oscillation frequency in order to reduce the phase error. Because the integrators in the slave are identical to those of the VCO and are tuned by the same voltage, the frequency response of the slave is also made accurate and stable.

Another alternative consists of using a master with a VCF structure. In this case, the reference signal is also applied to the master (dashed line). Because the filter phase characteristic is monotonic, it is appropriate for the determination of the resonance frequency location with respect to the reference frequency. The phase difference between the VCF output and an accurate external clock signal is used to generate a tuning signal that alters the resonance frequency of the VCF until it equals the reference frequency.

6.6.1.2 Architecture of the master: VCO or VCF

For efficient tuning, the master should be able to model the pertinent filter characteristic. It can be realized by either a VCO or a VCF including the basic building blocks of the filter.

- Voltage-controlled oscillator (VCO)

A tuning scheme based on a VCO operates as a phase-locked loop. The roll-off frequency of the filter is then controlled by locking the frequency of the VCO output to the one of a reference signal. Specifically, in the case where the slave is obtained by an interconnection of biquads, the VCO can have the structure of a second-order section, the poles of which are always on the imaginary axis. The Q factor of the VCO is then infinite and harmonic oscillations can occur. In order to control the amplitude of the oscillations, nonlinear circuit components must be included in the VCO structure. This introduces extra parasitic components at the nodes where these components are connected and can significantly worsen the required matching, especially at high frequencies.

- Voltage-controlled filter (VCF)

A biquadratic filter with a center frequency that can be controlled by a voltage is commonly used as the VCF. Its topology must be similar to the one of the slave. The tuning operation essentially exploits the output phase characteristics of the VCF. Because the control voltage supplied by the phase detector changes the filter's pole frequency rather than the phase angle, as is in the case of a VCO master, any offset in the detection stage will result in a frequency tuning error. In order to give a relevant relative phase estimation between the input and output signals, the VCF requires a reference signal with low harmonic content, for example, a sinusoid. Therefore, a square wave cannot be used for this purpose. It should be noted that the Q-factor of the VCF must be high enough to provide a sufficient tuning sensitivity. For a master center frequency of a few megahertz, a Q-value on the order of 10 can meet the requirement. Furthermore, accurate tuning will be achieved only if the dc offsets in the loop and the nonidealities of the phase detector are reduced.

6.6.1.3 Phase detector

A phase detector can be constructed around an analog multiplier or an XOR gate.

- In the first case, let $v_{REF}(t) = V_r \sin(\omega_i t)$ and $v_m(t) = V_m \cos(\omega_i t - \phi)$ be the reference and the master output signals, respectively. The voltage at the output of the analog multiplier shown in Figure 6.99 can be written as

$$v_x(t) = \frac{1}{2} k_x V_r V_m [\sin(\phi) + \sin(2\omega_i t - \phi)] \tag{6.285}$$

where k_x is the multiplier gain. Assuming that the high-frequency part of this

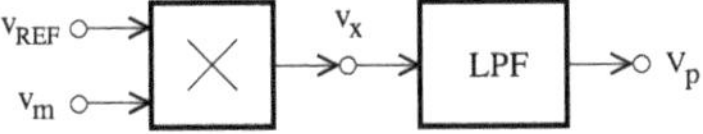

FIGURE 6.99
Block diagram of an analog multiplier-based phase detector.

signal is suppressed by the lowpass filter (LPF), the phase detector output signal is given by

$$V_p = k_p \sin(\phi) \tag{6.286}$$

where $k_p = k_x H(0) V_r V_m / 2$ is a constant and $H(0)$ is the LPF dc gain.

• In the second case, two comparators are needed to transform the reference and master output signals (v_{REF} and v_m) into square waveforms with a 50% duty cycle. The output voltage, $\overline{v}_x$, of the XOR gate, as shown in Figure 6.100, is high whenever both input signals are different. It is processed by the LPF,

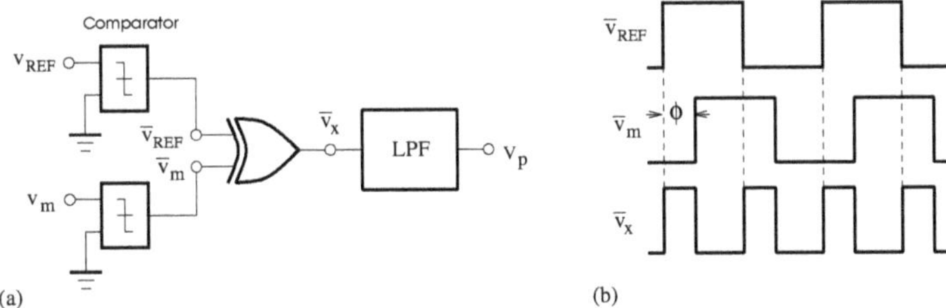

FIGURE 6.100
(a) Block diagram of a phase detector based on XOR gate; (b) input and output waveforms of the XOR gate.

and the average voltage at the output of the phase detector can be expressed as

$$V_p = k_\phi \phi \tag{6.287}$$

where $0 < \phi < \pi$, $k_\phi = H(0)V_{DD}/\pi$ is the gain of the phase detector, and V_{DD} is the supply voltage.

Note that the value of the phase detector gain, $k_\phi = \partial V_p / \partial \phi$, must be large enough to make the capture and lock ranges insensitive to the amplitudes of the reference and master output signals.

6.6.1.4 Implementation issues

The initial lock of the PLL can be obtained over a frequency range, $\triangle f_c$, which is the capture range of the loop. Once the PLL is locked to the reference signal, the frequency range in which the frequency variations of the reference signal due to change in the operating conditions can still be tracked is called the lock range, $\triangle f_l$. Generally, $\triangle f_c$ is smaller than $\triangle f_l$ and is determined by the cutoff frequency of the LPF. Because the PLL can lock to any harmonic of the reference frequency that can pass through the loop, the cutoff frequency of the LPF must be set just below the second harmonic of the lowest frequency to be tracked. The lock range is related to the PLL dynamic behavior.

The capture and lock ranges of the PLL can be investigated using circuit simulators. This method is advantageous particularly when the macromodels used for the simulation include the different parasitic components and bias-dependent parameters [62], which can affect the tuning and the high-frequency behavior of the filter. Figure 6.101 shows the plots of the tuning voltage versus the frequency. As evidenced by the arrows in these illustrations, the simulation

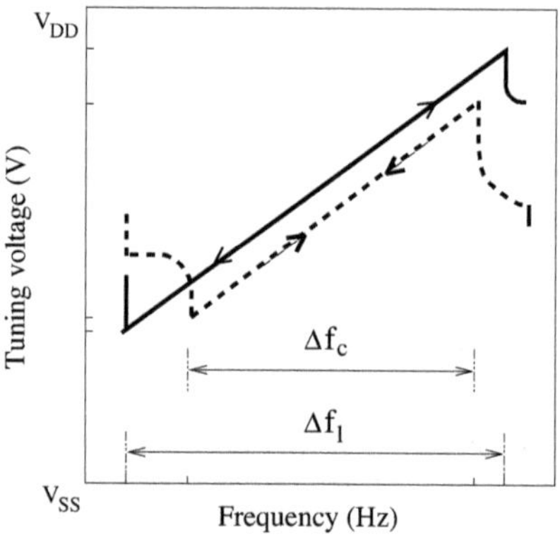

FIGURE 6.101
Lock range, $\triangle f_l$, and capture range, $\triangle f_c$, of the PLL tuning scheme.

of the capture range is done with the assumption that the PLL is initially in its locked state, in contrast to the one of the lock range.

6.6.2 Charge comparison-based technique

The effect of process variations on the resonant frequency of a g_m-C filter can be compensated for using the tuning scheme shown in Figure 6.102. It is based on the charge balancing principle as proposed in [32, 63] and consists of a charge comparator (CC) and an LPF. The dc value of the CC output voltage is extracted by the LPF and used to control the g_m value of transconductors. The tuning performance is primarily determined by the characteristics of the CC, which can be implemented by the following two circuit architectures.

Consider the circuit diagram shown in Figure 6.103. During the first clock phase, the capacitor C is charged to the voltage V_I, which is dependent on the input dc current I_{REF} ($V_I = I_{REF}/g_m$). The difference between the charges produced by the emptying of C and the dc current NI_{REF} is transferred onto

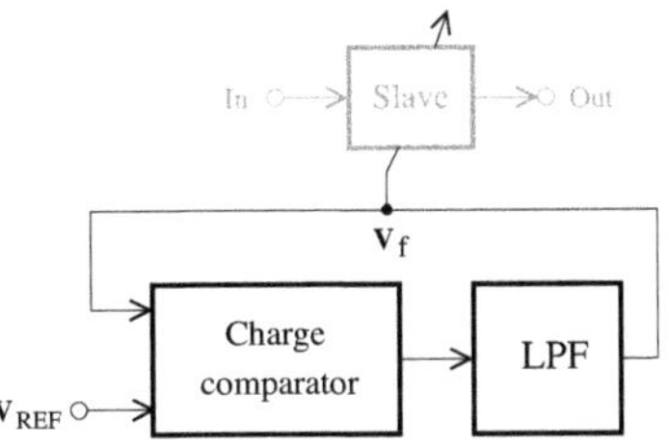

FIGURE 6.102
Frequency tuning loop based on charge comparison.

the feedback capacitor C_F during the second clock phase. Consequently,

$$\triangle q_{C_F} = TNI_{REF} - CV_I \tag{6.288}$$

where T is the period of the clock signal and N is the ratio of *dc* current sources. At the steady state, the average output voltage of the CC is constant. This means that there is no charge variation from the actual clock phase to the next one, and

$$\frac{g_m}{C} = \frac{f_c}{N} \tag{6.289}$$

where $f_c = 1/T$ is the clock frequency.

An alternative circuit diagram for the comparator is shown in Figure 6.104 [65]. Here, a reference voltage is connected to a transconductor and a

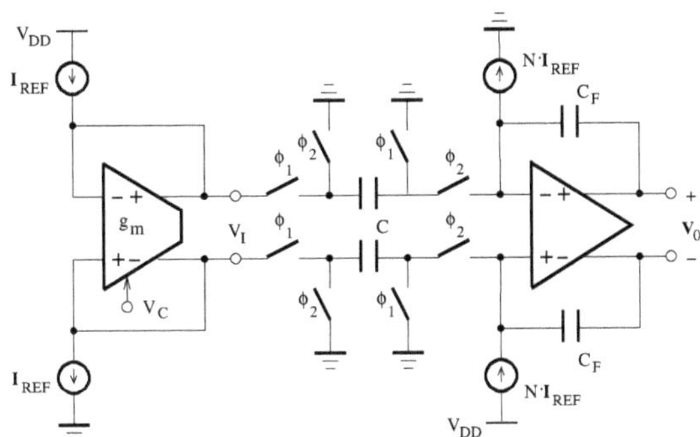

FIGURE 6.103
Circuit diagram of the charge comparator I.

switched-capacitor branch with an indirect path. The total charge transferred onto C_F can be expressed as

$$\triangle q_{C_F} = Tg_m V_{REF} - CV_{REF} \tag{6.290}$$

With the same assumption, as previously at the steady state, the next tuning

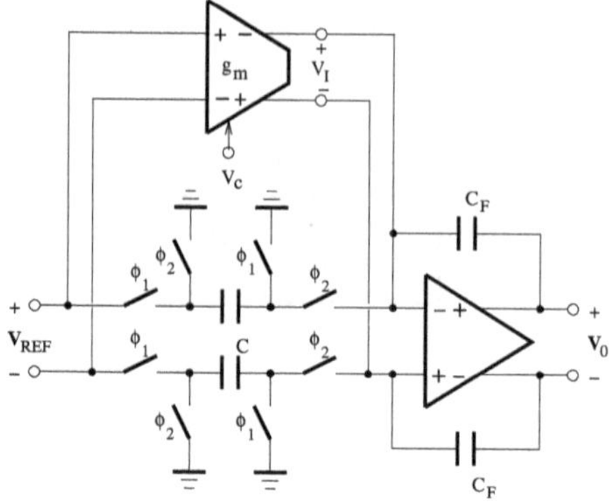

FIGURE 6.104
Circuit diagram of the charge comparator II.

condition is met, that is,

$$\frac{g_m}{C} = f_c \tag{6.291}$$

The tuning accuracy of the frequency is influenced by the level of matching of the g_m/C ratios achievable between the slave and the tuning circuits and the precision of the tuning circuit itself. It may be necessary to model the parasitic capacitors of the main filter and include their effects in the charge comparator capacitor C. In order to reduce the error due to the tuning loop, the offset voltages of the transconductor and the amplifier of the charge comparator must be low, and a reference signal with a sufficiently high magnitude is advisable.

6.7 Quality-factor and bandwidth control systems

Quality-factor and bandwidth control systems can be realized using the magnitude-locked-loop-based technique or envelope detection-based technique.

6.7.1 Magnitude-locked-loop-based technique

The errors of the pole quality factor, Q, which is sensitive to parasitic components at high frequencies, can be reduced by the tuning scheme shown in Figure 6.105. This tuning scheme is based on a magnitude-locked loop and is a suitable method to automatically control the shape of the transfer function. Generally, the master has a biquadratic transfer function, whose magnitude is proportional to Q at the oscillation (or center) frequency. The amplitude

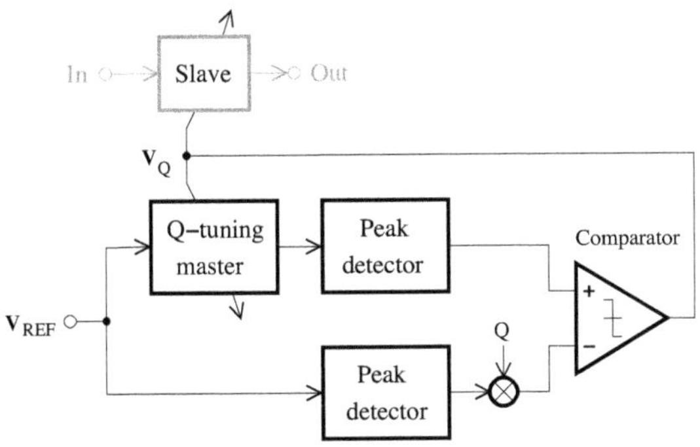

FIGURE 6.105
Magnitude-locked loop for Q tuning.

of the reference signal is estimated by the peak detector. It is amplified by a gain factor Q and then compared to the detected amplitude of the master output voltage. This results in a signal that is used to tune the master and slave until these two quantities are equal.

An implementation of the peak detector is shown in Figure 6.106. The input voltage is first fully rectified and then lowpass filtered to generate a signal, which is related to the amplitude of the input signal. Note that the switches can be implemented using CMOS analog gates.

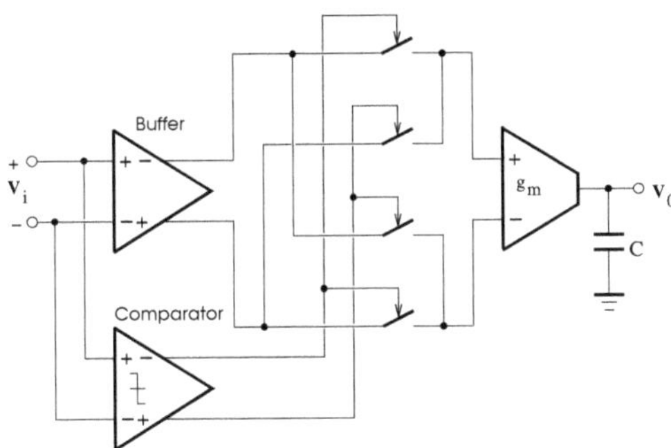

FIGURE 6.106
Circuit diagram of a peak detector.

The VCF, the output signal of which exhibits a reduced harmonic distortion in comparison to the one of the VCO, is preferred for the master implementation. Furthermore, the Q tuning seems to work appropriately for biquadratic sections and filters with cascade topology, but it may not be useful for high-order ladder filters.

In order to reduce the errors that can be introduced by the poor high-frequency behavior of the peak detector and the offset voltage of the comparator, the alternative scheme [66] shown in Figure 6.107 can be adopted. Its

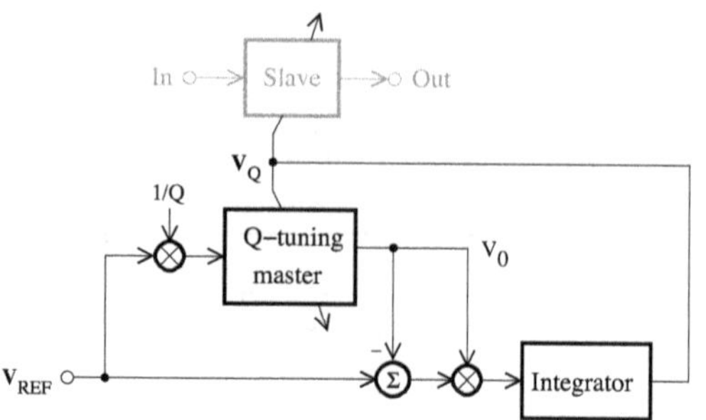

FIGURE 6.107
Magnitude-locked loop for Q tuning based on an adaptive technique.

operation is based on an adaptive algorithm of the least-mean square type. The Q-control signal, V_Q, is updated according to the following equation

$$\frac{dV_Q(t)}{dt} = \mu[V_{REF}(t) - V_0(t)]V_0(t) \tag{6.292}$$

where V_{REF} is the reference signal that has a frequency equal to the center frequency of the filter, and V_0 is the output voltage of the master (bandpass filter). The least-mean square (LMS) algorithm will try to minimize the error signal, $V_{REF} - V_0$. Ideally, V_{REF} is equal to V_0 after the tuning because the phase shift of V_0 is zero at the center frequency.

6.7.2 Envelope detection-based technique

The bandwidth of a CT filter can be controlled using the tuning architecture shown in Figure 6.109. The tuning circuit, which is based on the envelope detection technique, is composed of a first-order LPF, a tunable biquadratic filter, two envelope detectors, and a one-phase integrator followed by a sample-and-hold (S/H) circuit [63]. The circuit diagram of the envelope detector is shown in Figure 6.110(a). The input amplifiers followed by current mirrors $T_1 - T_4$ can be modeled by a transconductance g_m, as shown in Figure 6.110(b). During the signal detection, the capacitor C_d is charged according to the following equation,

$$C_d \frac{dV_0}{dt} = I_d - g_m V \tag{6.293}$$

where $V = V_0 - V_i$. The range of detection is limited to the negative transitions of the signal due to the unidirectional characteristic of the current mirror. For positive transitions, g_m is reduced to zero and the output voltage increases linearly with a constant slope of value I_d/C_d.

An envelope detector can also be implemented using a full-wave rectifier followed by a lowpass RC filter. Figure 6.108(a) shows the circuit diagram

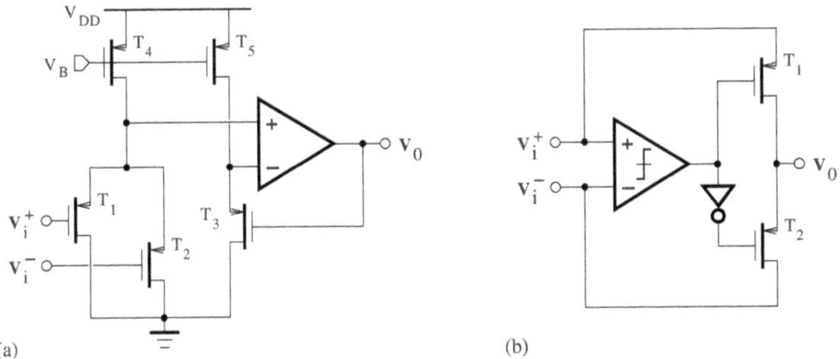

FIGURE 6.108
Circuit diagram of rectifiers ($V_i^- = -V_i^+$).

of a rectifier [61]. Due to the high gain of the amplifier, the voltage at the inverting and noninverting nodes should be made equal. This is achieved when a rectified version of the input signal is reproduced at the gate of T_3 or output. The rectifier shown in Figure 6.108(b) [64] is based on a comparator. Each of the transistors T_1 and T_2, which operate as a switch, is closed or open according to the input signal polarity.

An implementation of the integrator section is shown in Figure 6.111. The

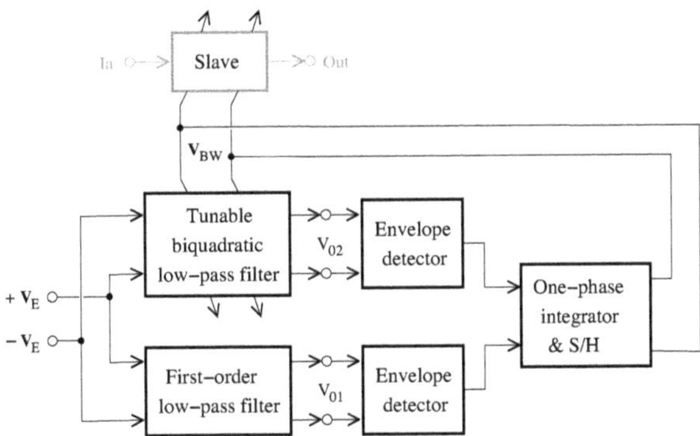

FIGURE 6.109
Bandwidth tuning loop based on envelope detection.

transconductor is connected either to the ground or to the amplifier inputs during the first and second clock phases, respectively. The charge stored on the amplifier feedback capacitor is transferred onto the hold capacitor, C_h, when there is no signal injected into the amplifier input nodes.

The detection of signal envelopes is carried out in the time domain. Assuming that the output voltage of the filter is initially set to zero, the step

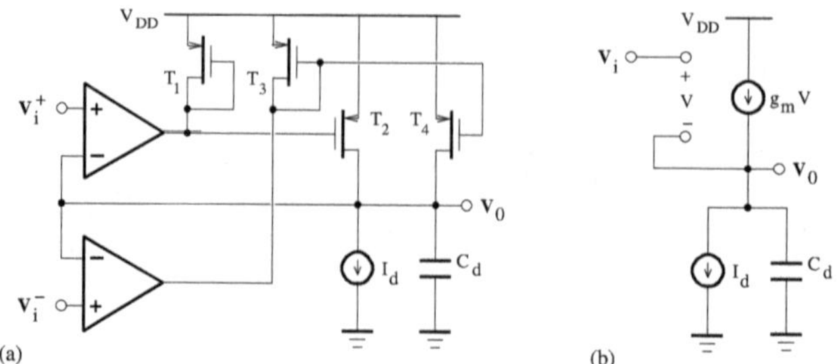

FIGURE 6.110
(a) Circuit diagram and (b) equivalent model of an envelope detector.

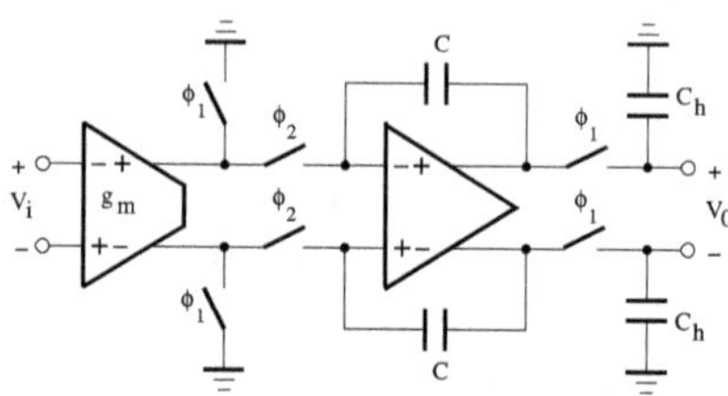

FIGURE 6.111
Circuit diagram of a one-phase integrator including the sample-and-hold function.

response can be computed as

$$v_O(t) = \mathcal{L}^{-1}\left\{\frac{H(s)}{s}\right\} \tag{6.294}$$

$$= \int_0^t h(\tau)\,\mathrm{d}\tau \tag{6.295}$$

where $\mathcal{L}^{-1}$ denotes the inverse Laplace transform, and $H(s)$ and $h(t)$ are the transfer function and the impulse response of the filter, respectively. For a first-order LPF with a unity dc gain, this results in

$$v_{O1}(t) = E[1 - \exp(-\omega_c t)] \tag{6.296}$$

where ω_c and E are the -3 dB cutoff frequency and the amplitude of the input step voltage, respectively. In the case of the second-order LPF, the next output voltage can be written as

$$v_{O2}(t) = E[1 - h(t)] \tag{6.297}$$

Note that the filter dc gain and the initial conditions on the voltage are one and zero, respectively. The filter impulse response, h, is given by

$$h(t) = \frac{1}{\sqrt{1 - \frac{1}{4Q^2}}} \exp\left(-\frac{BW}{2}t\right) \cos\left(\sqrt{1 - \frac{1}{4Q^2}}\omega_0 t - \phi\right) \tag{6.298}$$

and

$$\phi = \arctan \frac{1/2Q}{\sqrt{1 - \frac{1}{4Q^2}}} \tag{6.299}$$

where $BW = \omega_0/Q$ is an approximation of the LPF bandwidth, which is related to the center frequency, ω_0, and the Q-pole factor. This latter must be larger than 1/2. For high Q, the only difference between the above step responses is the harmonic term that appears in the expression of h. Therefore, the envelopes measured by both detectors will have similar shapes if ω_c takes the value $BW/2$. The input signal V_E is a train of pulse with the amplitude E. In its positive transition, the detected signal at the output of the first-order and second-order LPFs correspond to an exponential charging response, whose final value is E and an exponential decay from $2E$ to E, respectively. For this reason, and because V_E is a pulse train, the second clock phase of the integrator used in the tuning loop is synchronized with the negative transition of V_E, which is the beginning of the time period where the signal envelopes can be successfully compared (see Figure 6.112). In this approach, some practical

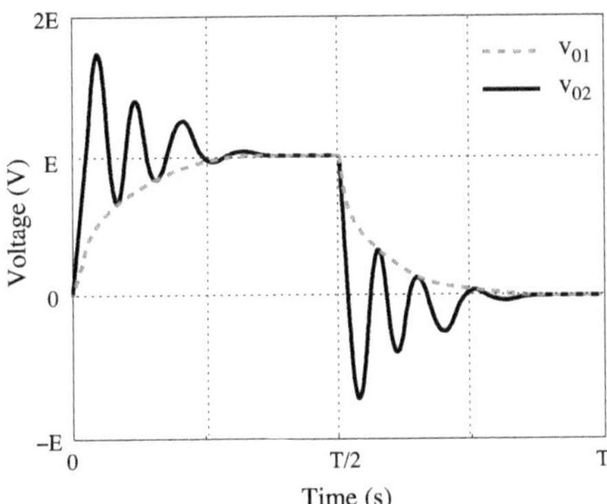

FIGURE 6.112
Plot of the transient responses (outputs $v_{01}(t)$ and $v_{02}(t)$) for $\omega_c = \mathrm{BW}/2$.

problems can limit the tuning accuracy: the level of the residual offset voltage in the transconductor required for the comparison of the filter envelopes and the precision of the envelope detectors.

6.8 Practical design considerations

Let us assume that the main filter in the master-slave tuning scheme (see Figure 6.96) is based on a biquadratic filter section with the following bandpass transfer function,

$$T(s) = k\frac{\frac{\omega_o}{Q}s}{s^2 + \left(\frac{\omega_o}{Q}\right)s + \omega_o^2} \tag{6.300}$$

where k is a gain factor, ω_0 denotes the center frequency, and Q is the quality factor. The phase difference between the reference and the filter output signal, ϕ, is required in the frequency-tuning loop while the magnitude of the filter output signal, M, is needed in the Q-tuning circuit. Assuming that the input signal of the filter is at the same frequency as the reference signal, that is, $s = j\omega_r$, we have

$$\phi(\omega_o, Q) = \arg[T(s)] = \frac{\pi}{2} - \arctan\left(\frac{\frac{1}{Q}\frac{\omega_r}{\omega_o}}{1 - \frac{\omega_r^2}{\omega_o^2}}\right) \tag{6.301}$$

$$M(\omega_o, Q) = |T(s)| = k\frac{Q\frac{\omega_r}{\omega_o}}{\sqrt{\left(1 - \frac{\omega_r^2}{\omega_o^2}\right)^2 + \left(\frac{1}{Q}\frac{\omega_r}{\omega_o}\right)^2}} \tag{6.302}$$

where ω_r is the frequency of the reference signal. The coupled nature of the above parameters appears in Figures 6.113 and 6.114, where the phase and magnitude surfaces are drawn. For a high Q factor, a small error in the pole frequency can result in a low output voltage, which directly translates to an inappropriate increase in the filter quality factor by the Q-tuning loop.

Ideally, the loops can be considered independent if

$$\frac{\partial\phi(\omega_0, Q)}{\partial Q} = 0 \tag{6.303}$$

$$\frac{\partial M(\omega_0, Q)}{\partial \omega_0} = 0 \tag{6.304}$$

The next approaches that can be used to this end consist basically of making quasi-independent the two tuning loops. In the first solution, the Q-loop is designed to be much slower than the frequency loop. As a result, the loop interaction, which may lead to instability, is reduced. The second one consists of using an oscillator [67] or filter structures [68] with a reduced coupling of magnitude and frequency parameters.

For the choice of the reference signal frequency, a trade-off must be made

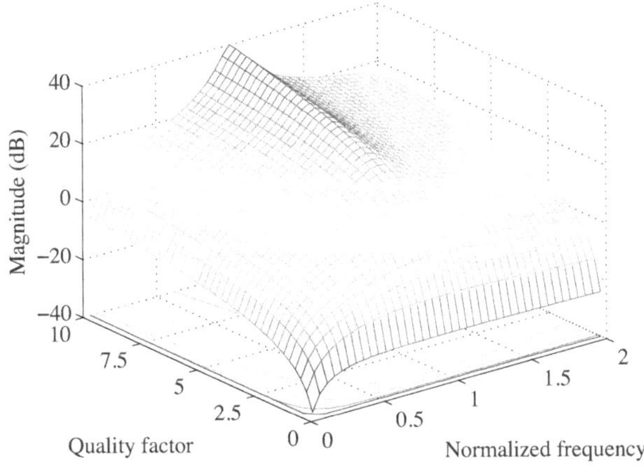

FIGURE 6.113
Magnitude surfaces.

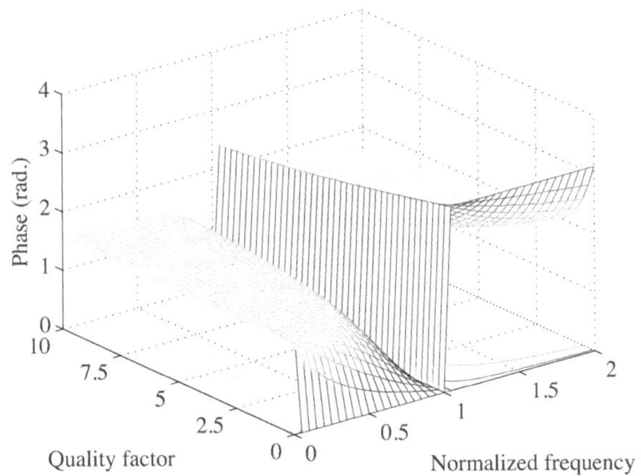

FIGURE 6.114
Phase surfaces.

between the achievable level of the master-slave matching and the amount of reference signal feedthrough that can still be coupled to the output signal of the slave. When the reference signal is at the passband edge, that is, very close to the unity-gain frequency of the integrators in the slave, a best matching will be observed. But, this selection also results in a worse immunity to the reference signal feedthrough. On the contrary, the feedthrough will be minimum for a reference signal frequency located in the stopband due to the high attenuation of the filter, and the matching will be very poor.

6.9 Other tuning strategies

In applications where the limitations of a master-slave architecture become critical, alternative tuning strategies (tuning scheme using an external resistor, self-tuned filters) can be adopted. They exhibit the advantage of providing a tuning performance independent of the filter topology.

6.9.1 Tuning scheme using an external resistor

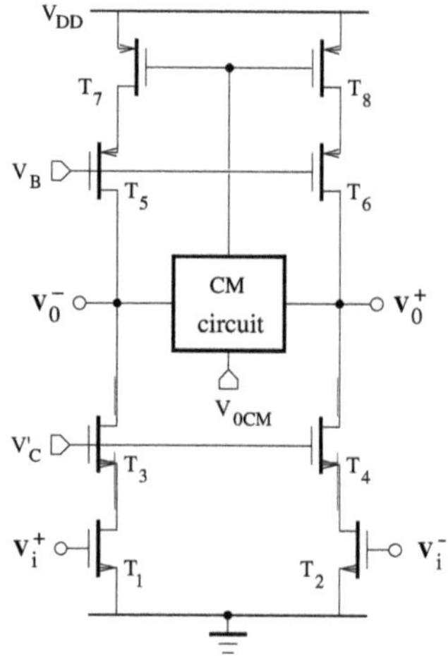

FIGURE 6.115
Circuit diagram of a differential transconductor with the common-mode circuit.

The transconductance can be set in g_m-C filters by locking the filter amplifiers to an external reference resistor [71, 72]. An example CMOS transconductor is shown in Figure 6.115; V_B is the bias voltage and V'_C is the internal control voltage used for the automatic tuning. This latter is supplied by the circuit shown in Figure 6.116. Here, a known voltage, V_{REF}, and a known current, I, are applied to a replica input stage of the transconductor $(T_1 - T_4)$ in a feedback loop and the transconductance is then defined as the ratio of the above current and voltage. The output current, I, of the transconductor is related to the external control voltage, V_C, and the value of the transconductance is

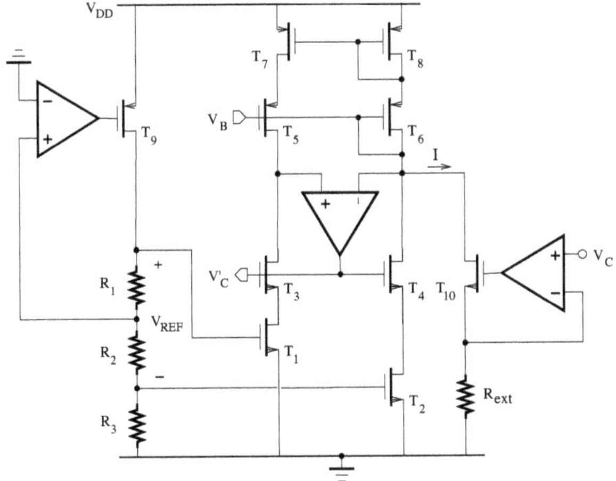

FIGURE 6.116
Tuning circuit for the transconductor using an external resistor.

given by

$$g_m = \frac{I}{V_{REF}} = \frac{1}{R_{ext}} \times \frac{V_C}{V_{REF}} \tag{6.305}$$

This tuning scheme can then be used to improve the accuracy of the transconductance provided that the temperature coefficients of the resistor, R_{ext}, and capacitors are sufficiently small.

6.9.2 Self-tuned filter

Generally, the matching between the master and slave is limited by the variations of the device characteristics. As a result, some percentage of uncertainty (1 to 2%) with respect to the filter parameters remains after the tuning. An improvement can be observed by increasing the device size. But this leads to higher power consumption.

A tuning scheme [69, 70], the performance of which is not related to the matching accuracy, is shown in Figure 6.117. The filter is first tuned and then connected to the signal path. During the tuning process, the filter is coupled to the step-signal generator and is unavailable for the processing of the signal of interest. By comparing the square wave version of the filter output provided by the inverter-based buffer and a reference signal, the phase and frequency detector (PFD) can generate either an Up signal or a Down signal used to drive the Up/Down counter. The control circuit receives the adjustment signal and initiates the filter tuning, which is achieved via the N-bit digital-to-analog converter (DAC).

Here, an input step response is applied to the filter; its ringing frequency is measured and tuned to the design value. For a second-order bandpass filter

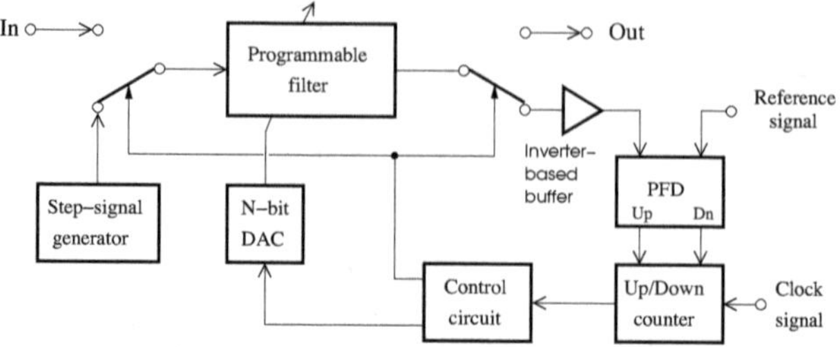

FIGURE 6.117
Block diagram of a self-tuned filter.

with the transfer function $T(s)$ (see Equation (6.300)), the step response is given by

$$s(t) = \mathcal{L}^{-1}\left\{\frac{H(s)}{s}\right\}$$
$$= \frac{k/Q}{\sqrt{1-\frac{1}{4Q^2}}}\exp\left(-\frac{\omega_0 t}{2Q}\right)\sin\left(\omega_0\sqrt{1-\frac{1}{4Q^2}}t\right)u(t) \tag{6.306}$$

where k is the filter gain, $u(t)$ is the unit step response, and the initial conditions were assumed to be zero. This signal crosses zero when

$$\sin\left(\omega_0\sqrt{1-\frac{1}{4Q^2}}t\right) = 0 \tag{6.307}$$

and we can write

$$\frac{f_r}{f_0} = 2\sqrt{1-\frac{1}{4Q^2}} \tag{6.308}$$

Note that Q is the quality factor of the filter, f_r is the ringing frequency, and $\omega_0 = 2\pi f_0$, where f_0 is the center frequency. The circuit section, consisting of the PFD and Up/Down counter, estimates the number of cycles N_i of the input waveform in a period, which corresponds to a number of counts, M, of the clock signal with the frequency f_c. In this way, the actual ringing frequency, f_{ra}, can be computed as

$$f_{ra} = \frac{Mf_c}{N_i} \tag{6.309}$$

The target ringing frequency, f_{rt}, is related to the count value N_t and can be written as

$$f_{rt} = \frac{Mf_c}{N_t} \tag{6.310}$$

The objective of the tuning is to adjust the ringing frequency of the filter so that the next condition is fulfilled, that is,

$$|N_i - N_t| \leq \epsilon \tag{6.311}$$

where ϵ denotes the residual tuning error. In this way, the overall accuracy of the resulting filter characteristics depends on the level of ϵ and frequency measurement errors.

A filter with the above tuning scheme, or self-tuned filter, can find applications in personal digital cellular systems with a spaced channel. However, in situations where a CT operation of the circuit is required, two filters must be associated in a parallel configuration as shown in Figure 6.118. One filter

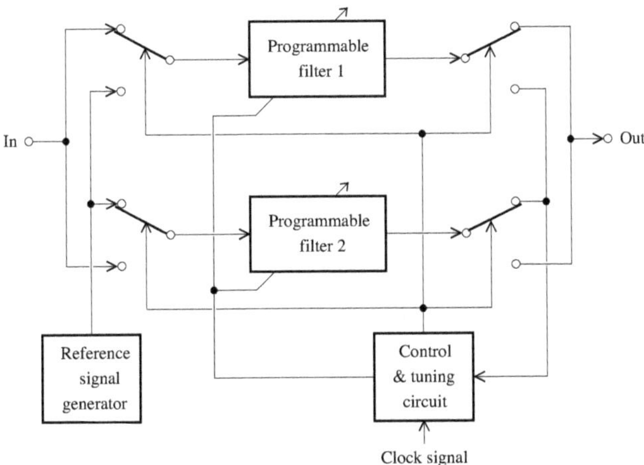

FIGURE 6.118
Self-tuned filter for the continuous-time operation.

will be tuned when the other processes the input signal, and vice versa.

6.9.3 Tuning scheme based on adaptive filter technique

Here, the tuning objective [73] must be formulated in terms of a function to be minimized. This goal can be accomplished using an adaptive algorithm to update the filter coefficients (see Figure 6.119).

Although a white noise source is commonly used as the input signal in the system identification application, a sum of sinusoids, whose frequencies are chosen within the passband of the desired transfer function, H, seems to be more appropriate. That is,

$$x(t) = \sum_{l=1}^{L} \sin(2\pi f_l t) \tag{6.312}$$

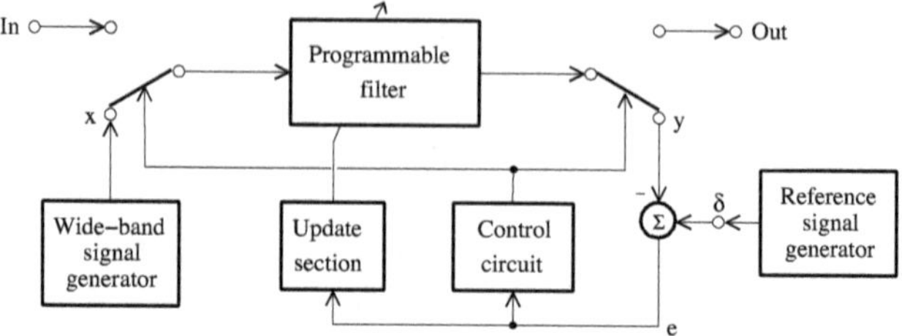

FIGURE 6.119
Tuning scheme based on adaptive filter technique.

The reference signal, δ, can then be obtained as the ideal filter response to the excitation, x. Hence,

$$\delta(t) = \sum_{l=1}^{L} a_l \sin(2\pi f_l t + \phi_l) \tag{6.313}$$

where $a_l = |H(2\pi j f_l)|$ and $\phi_l = \arg\{H(2\pi j f_l)\}$. The tuning of the filter to meet the passband specifications can be achieved by updating the variable filter coefficients according to the least-mean-square (LMS) algorithm. During the adaptation process, the error signal, e, is given by

$$e(t) = k[\delta(t) - y(t)] \tag{6.314}$$

where k is the error amplifier gain, and δ and y are the reference signal and filter output, respectively. Each variable coefficient of the programmable filter, labeled w_i, are changed to minimize the mean-squared error signal denoted as $E[e^2]$. Thus, for a filter with N variable coefficients,

$$\frac{\mathrm{d}}{\mathrm{d}t} w_i(t) = -\mu \frac{\partial e^2(t)}{\partial w_i} \tag{6.315}$$

$$= 2\mu e(t)\Gamma_i(t) \qquad i = 1, 2, \cdots, N \tag{6.316}$$

where μ is a small positive step size that determines the trade-off between the speed of the algorithm and the residual convergence error, and Γ_i denotes the gradient signals defined as

$$\Gamma_i(t) = k \left. \frac{\partial y(t)}{\partial w_i} \right|_{w_i = w_i(t)} \tag{6.317}$$

The generation of Γ_i can require additional structures, which are driven by signals associated with filter states or nodes. It is worth noting that filter architectures with orthogonal states provide the advantage of improving the LMS algorithm performance.

In practice, a $\Delta\Sigma$ modulator-based oscillator may be utilized to generate the sinusoids used as input and reference signals [74, 75]. It combines a digital resonator structure having poles on the unit circle and a $\Delta\Sigma$ modulator. The oscillation frequency is set by the loop gain and the external clock frequency. The amplitude and phase of the signal are determined by the initial conditions. This technique results in high-quality sinusoids with a spurious-free dynamic range larger than 90 dB.

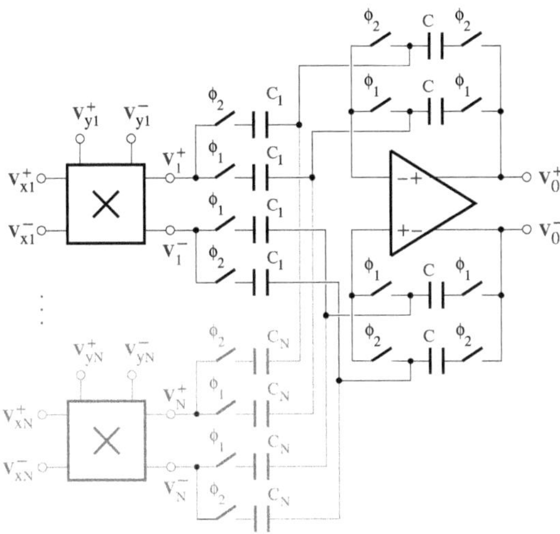

FIGURE 6.120
Multiplier with offset voltage cancellation.

The level of the tuning error can be limited by the different nonidealities of components, such as the offset voltages of the amplifier and multiplier, which can be on the order of ± 10 mV. A solution for the cancellation of the multiplier offset voltage is provided by cascading multiplier circuits and a gain stage, as shown in Figure 6.120 [76], where ϕ_1 and ϕ_2 are two complementary clock phases. The offset voltages stored on the capacitors C_k $(k = 1, 2, \cdots, N)$ during the previous clock phase ϕ_1 (or ϕ_2) are used during the actual clock phase ϕ_1 (or ϕ_2) to cancel the current offset voltage contributions. The output voltage of the multi-input offset-free tunable gain stage is proportional to the ratio C_k/C and the multiplier gain. Auto-zero or chopper schemes can also be adopted to reduce the amplifier and multiplier sensitivity to offset voltages [77, 78]. With the use of these compensation techniques, the residual misadjustment can be reduced to about 60 dB, guaranteeing a reasonable accuracy for the filter characteristics.

By analyzing the filter structures, the tunability range of the characteristics can be determined. Note that the evaluation of the closed-form relations between the parameters of high-order filters is more tedious than the one of first- and second-order structures and may require the use of circuit analysis and computation programs like SPICE, MATLAB, and Hardware Description Language (HDL).

6.10 Summary

An overview of high-performance CT circuits was provided. In advanced applications, CT circuits should be designed to exhibit a high dynamic range, a programmable bandwidth, precise tuning, high speed, low power, and a small chip area. The choice of the architecture and design techniques is generally determined by the cost and performance.

Because CT filters are prone to fluctuations in their electrical parameters, an insight into the principles of operation of on-chip tuning loops is provided. A comparison of the performance is carried out in order to analyze the suitability of each tuning scheme to the high-frequency and high-Q filter applications and to choose the more convenient topology for a given design purpose. Generally, the tuning maintains a low drift of the filter parameters over power supply, temperature, and IC process variations. As a result, CT filters with less than $\pm 1\%$ pole frequency and quality factor accuracies become realizable.

6.11 Circuit design assessment

1. **Single-stage phase shift network**
   Consider each of the phase shift networks shown in Figure 6.121. Use the voltage divider principle to find the transfer functions

   $$T(s) = [V_0^+(s) - V_0^-(s)]/V_i(s)$$

   Determine and plot the magnitude and phase of T versus the frequency.

   Deduce the phase of T at the 3-dB cutoff frequency.

2. **Two-stage RC-CR phase shifters**
   For the two-stage RC-CR phase shifters of Figure 6.122, assume that the coupling and decoupling capacitors, C_C and C_D, act as a short-circuit for the entire range of operating frequencies.

   Show that, for the circuit of Figure 6.122(a),

   $$\frac{V_Q(s)}{V_I(s)} = \frac{s(R_1C_1 + R_2C_2)}{1 - R_1R_2C_1C_2s^2} \tag{6.318}$$

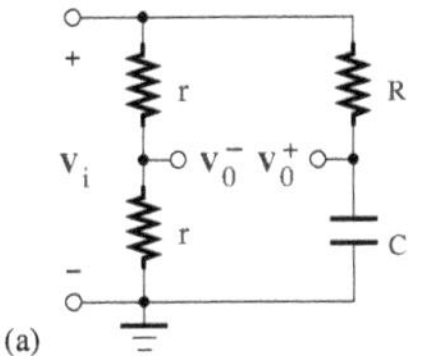

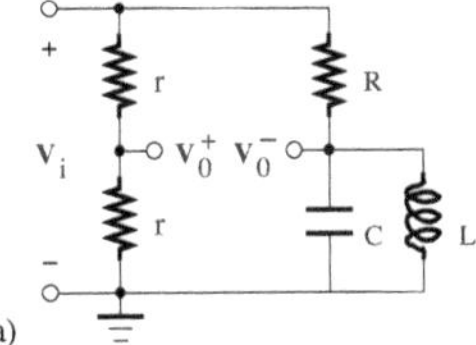

FIGURE 6.121
Circuit diagram of phase shift networks.

and for the circuit of Figure 6.122(b),

$$\frac{V_Q(s)}{V_I(s)} = \frac{1 + (R_1C_1 + R_2C_2)s - R_1R_2C_1C_2s^2}{1 - (R_1C_1 + R_2C_2)s - R_1R_2C_1C_2s^2} \tag{6.319}$$

where $V_Q = V_Q^+ - V_Q^-$ and $V_I = V_I^+ - V_I^-$.

Plot the magnitude and phase of the function $V_Q(s)/V_I(s)$ as a function of the frequency.

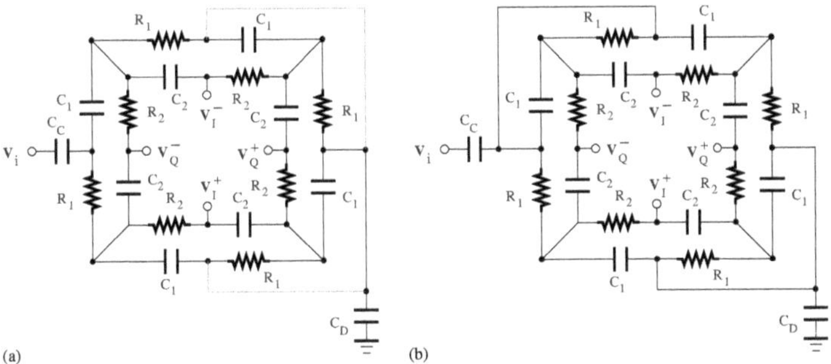

FIGURE 6.122
Circuit diagrams of two-stage RC-CR phase shifters (a) without phase imbalance and (b) without gain imbalance.

3. **Common-source amplifier stages**

   Determine the transfer function of the circuit shown in Figure 6.123 using the small-signal equivalent transistor model of Figure 6.124.

   Provided that the transistor operates as an ideal transconductor with the value g_m, show that the transfer function of the first and second amplifiers can, respectively, be reduced to

$$\frac{V_0(s)}{V_i(s)} = \frac{A_0}{1 + \tau s} \tag{6.320}$$

where $A_0 = -g_m R$ and $\tau = RC$, and

$$\frac{V_0(s)}{V_i(s)} = A_0 \omega_0 Q \frac{s + \dfrac{\omega_0}{Q}}{s^2 + \left(\dfrac{\omega_0}{Q}\right) s + \omega_0^2} \tag{6.321}$$

where $A_0 = -g_m R$, $Q = \omega_0 L/R$ and $\omega_0^2 = 1/LC$.

Verify that the zero and poles introduced by the shunt inductor result in an increase in the amplifier bandwidth.

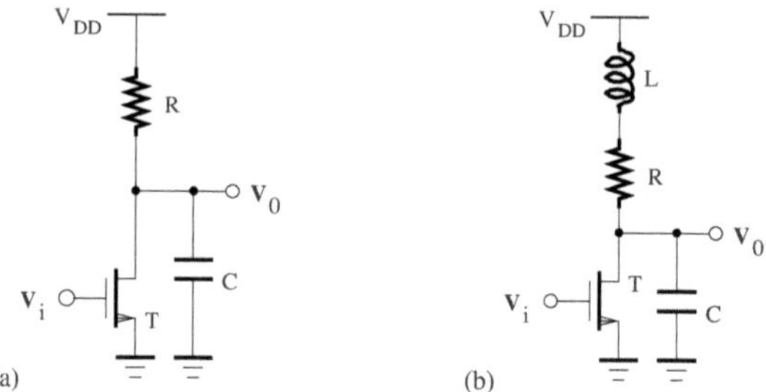

FIGURE 6.123
Circuit diagrams of (a) a simple common-source amplifier and (b) a common-source amplifier with shunt peaking.

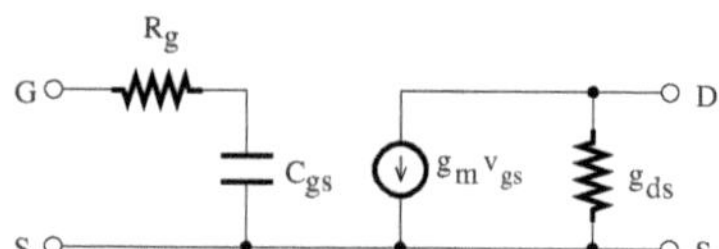

FIGURE 6.124
Small-signal equivalent transistor model.

4. **Single-stage tuned amplifier**
   The circuit diagram of a tuned amplifier is depicted in Figure 6.125. Assuming that the transistor is equivalent to the transconductor with the value g_m, show that

$$A(s) = \frac{V_0(s)}{V_i(s)} = -\frac{g_m}{C} \frac{s}{s^2 + \left(\dfrac{1}{RC}\right) s + \dfrac{1}{LC}} \tag{6.322}$$

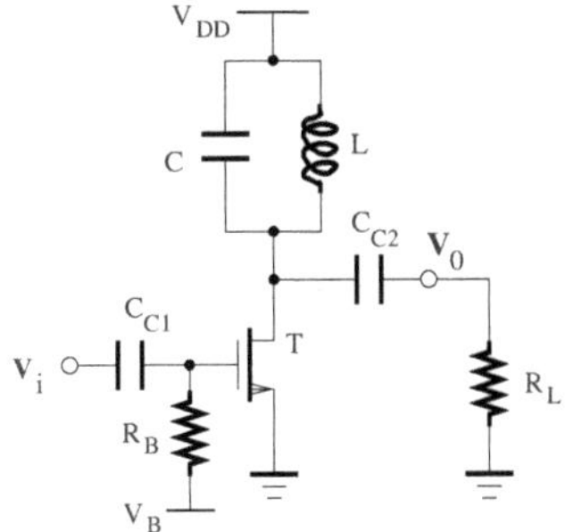

FIGURE 6.125
Circuit diagram of a tuned amplifier.

Verify that

$$A(s) = A_0 \frac{\left(\frac{\omega_0}{Q}\right) s}{s^2 + \left(\frac{\omega_0}{Q}\right) s + \omega_0^2} \tag{6.323}$$

and

$$A(\omega) = A_0 \frac{1}{1 + jQ\left(\frac{\omega}{\omega_0} - \frac{\omega_0}{\omega}\right)} \tag{6.324}$$

where $A_0 = -g_m R$ and $Q = RC\omega_0 = R/(L\omega_0)$.

5. **SPICE analysis of low-noise amplifiers**
   Analyze the input impedance and noise of the amplifiers shown in Figure 6.126 using the SPICE program.

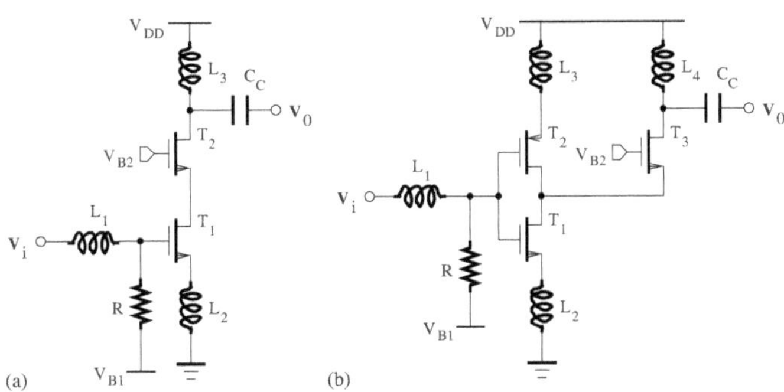

FIGURE 6.126
Circuit diagrams of single-ended low-noise amplifiers with (a) nMOS, (b) pMOS-nMOS, and (c) inductively coupled input stages.

6. **Low-noise amplifiers with improved noise factor**

(a) In the low-noise amplifier shown in Figure 6.127(a) [14], the RLC input network is merged with the resistive feedback to increase the gain and reduce the noise factor.

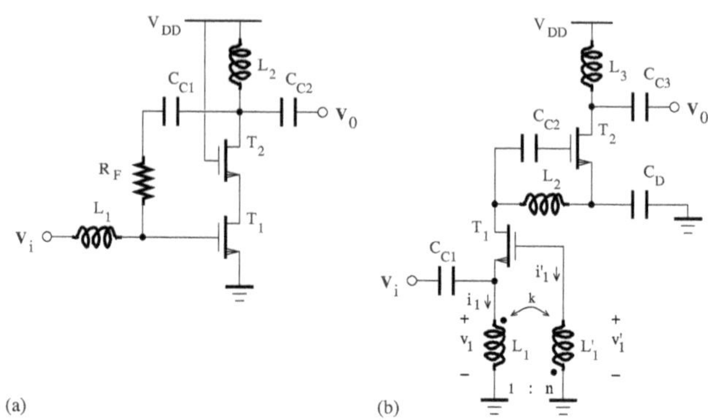

FIGURE 6.127
(a) Circuit diagram of a low-noise amplifier with RC feedback; (b) circuit diagram of a g_m-boosted low-noise amplifier using current reuse.

Assuming that the coupling capacitors, C_{C_1} and C_{C_2}, are equivalent to a short-circuit, determine the input impedance and noise factor.

(b) Consider the low-noise amplifier circuit of Figure 6.127(a) [21]. This architecture, which is realized by cascading the g_m-boosted input stage with a common-source output stage such that there is a current reuse, has the advantage of providing a high gain even with a low power consumption. The coupling capacitors C_{C_1}, C_{C_2}, and C_{C_3}, and the decoupling capacitor, C_D, can be considered as a short-circuit. The i-v laws for the coupled coils, $L_1 - L_1'$, with zero initial conditions are as follows:

$$i_1 = \frac{Mi_1'}{L_1} + \frac{v_1}{sL_1} \tag{6.325}$$

and

$$i_1' = \frac{v_1'}{sL_1'} + \frac{Mi_1}{L_1'} \tag{6.326}$$

where the coupling coefficient is given by

$$k = \frac{M}{\sqrt{L_1 L_1'}} \tag{6.327}$$

and M is the mutual inductance between the primary and secondary windings.

Assuming that $v_{gs_1} = -(1 + A)v_i$, where $A = kn$, show that the frequency response of the low-noise amplifier is the product of two second-order bandpass transfer functions due to the parallel resonant circuits at the drains of both transistors.

Determine the total noise figure of the amplifier using Friis's formula of the form

$$F = F_1 + \frac{F_2 - 1}{A_1} \tag{6.328}$$

where F_1 and F_2 denote the noise figures of the first and second stages, respectively, and A_1 is the voltage gain of the first stage.

7. **Single-stage amplifier with parasitic coupling capacitors**
Let us consider the single-stage amplifier of Figure 6.128(a). Determine the voltage transfer function and input impedance using the transistor equivalent model shown in Figure 6.128(b).

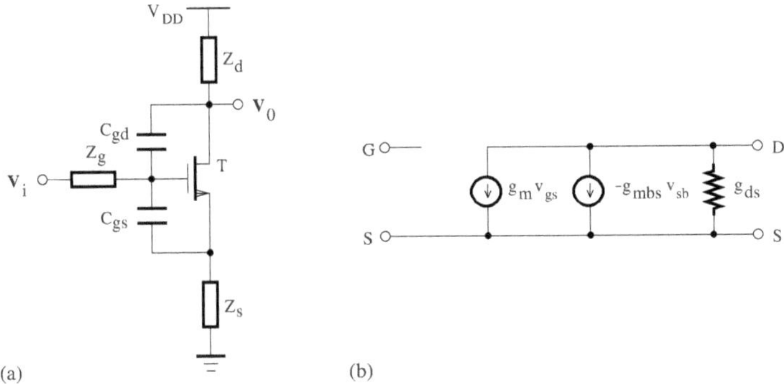

FIGURE 6.128
(a) Circuit diagram of a single-stage amplifier; (b) MOS transistor small-signal equivalent model.

8. **Pseudo-differential low-noise amplifier**
Consider the circuit diagram of a pseudo-differential low-noise amplifier with the capacitor cross-coupled input stage shown in Figure 6.129(a), where C_d represents a decoupling capacitor. The input transistors should be biased to operate in the saturation region.

Use the small-signal equivalent model depicted in Figure 6.129(b), where the input source generator consists of the voltage source v_s in series with the resistor R_s, and Z_L is the impedance of the output load, to determine the voltage gains, $G = V_0/V$ and $G_s = V_0/V_s$, and the input impedance, $Z_i = V/I$.

Compare the magnitude and phase of Z_i obtained by hand calculation to SPICE simulation results.

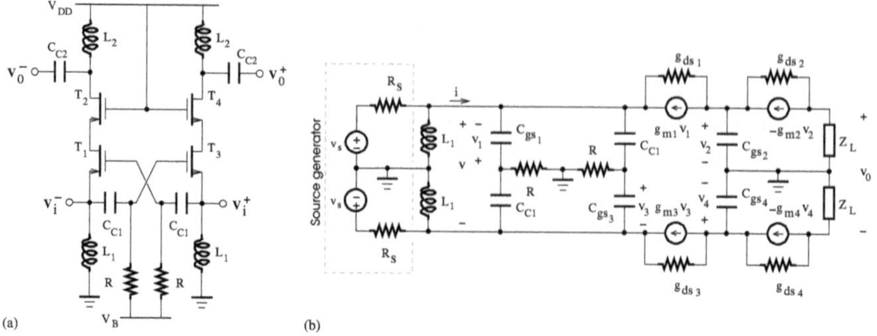

FIGURE 6.129
(a) Circuit diagram of a pseudo-differential low-noise amplifier with a capacitor cross-coupled input stage; (b) small-signal equivalent model.

9. **Cascaded VGA**
A cascaded VGA, as shown in Figure 6.130, is generally required to provide a sufficient gain and achieve a wide gain tuning range. Assuming a cascade of n identical amplifier stages, each with the bandwidth BW, the overall bandwidth is given by [79],

$$BW_T = BW\,\sqrt[m]{2^{1/n} - 1} \tag{6.329}$$

where m is equal to 2 for a first-order amplifier stages and 4 for a second-order amplifier stages.

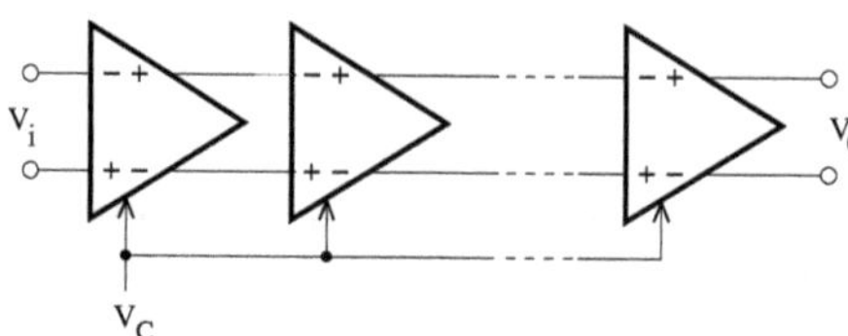

FIGURE 6.130
Circuit diagram of a cascaded VGA.

In order to obtain a total gain A_T, the required gain-bandwidth product of each amplifier stage can be computed as

$$GBW = \frac{GBW_T}{A_T^{(n-1)/n} \cdot \sqrt[m]{2^{1/n} - 1}} \tag{6.330}$$

where $GBW = A_T^{1/n} \cdot BW$ and $GBW_T = A_T \cdot BW_T$.

The level of the overall input-referred noise increases with n, so that the value of n is generally limited to about 5 for a reasonable noise figure.

Assuming $A_T = 50$ dB and $BW_T = 2.5$ GHz, plot GBW as a function of n for GBW comprised between 5 GHz and 30 GHz and $n \leq 15$, in the case where $m = 2, 4$.

For a cascaded VGA with $n = 5$, determine the value to be exceeded by the bandwidth of each (first and second order) amplifier stage.

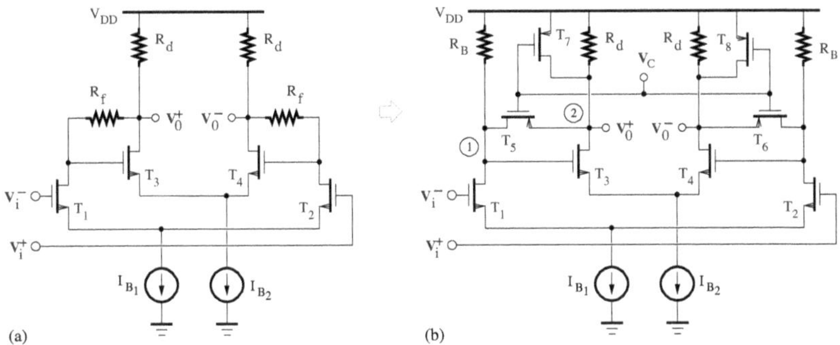

FIGURE 6.131
VGA: (a) Principle, (b) implementation.

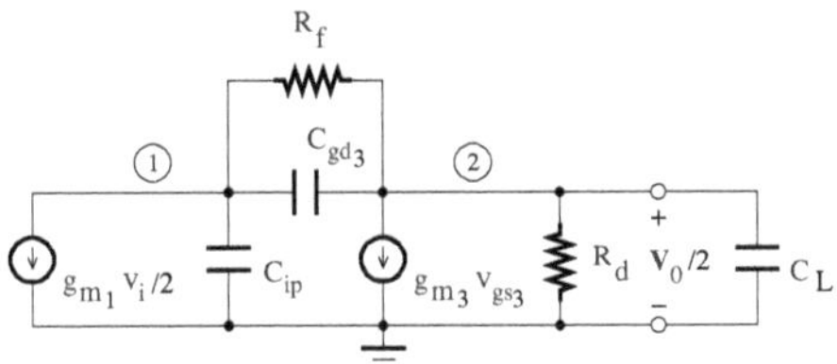

FIGURE 6.132
Small-signal equivalent circuit of the VGA.

A variable gain amplifier can be designed as shown in Figure 6.131 [80]. This structure exhibits a good linearity and a high gain, and can operate without a common-mode feedback circuit.

Based on the small-signal equivalent circuit of figure 6.131, where C_{ip} is the total parasitic capacitance at the node 1, and C_L is the total capacitance at the node 2, show that

$$A(s) = \frac{V_0(s)}{V_i(s)} = A_0 \frac{1 - s(C_{gd3}/g_{m3})}{s^2 \alpha (R_f/g_{m3}) + s\beta + 1} \tag{6.331}$$

where

$$\alpha = C_{ip}(C_{gd_3} + C_L) + C_{gd_3} C_L \tag{6.332}$$

$$\beta = C_{gd_3} R_f + C_{ip} \frac{R_f + R_d}{g_{m_3} R_d} + \frac{C_L}{g_{m_3}} \tag{6.333}$$

and $A_0 = g_{m_1} R_f$ provided $g_{m_3} R_f \gg 1$, $g_{m_3} R_d \gg 1$, and $R_d \gg R_f$.

10. **Class C VCO**

The phase noise characteristic of LC VCOs can be improved by increasing the oscillation amplitude. Figure 6.133 shows the circuit diagram of a class C VCO with an amplitude feedback loop that operates as follows:

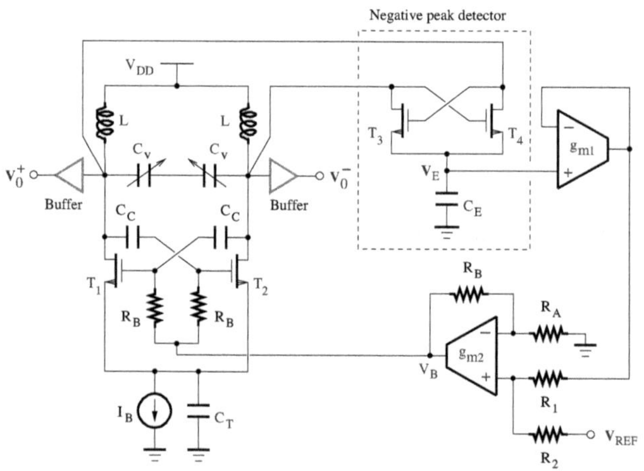

FIGURE 6.133
Circuit diagram of a class C VCO with amplitude feedback loop.

When the VCO does not oscillate, the voltage V_{DD} is applied to the drain and gate of transistors T_3 and T_4. The output voltage V_E begins to increase because these two transistors are turned on. This leads to a growth of the bias voltage V_B, followed by that of the transconductance of the cross-coupled transistors, T_1 and T_2, until the start-up condition is satisfied.

When the VCO begins to oscillate, and the difference between the voltages V_0^+ and V_0^- exceeds the transistor threshold voltage, one of the transistors T_3 and T_4 is turned on while the other is turned off, so that the capacitor C_E is charged by the negative output signal. There is also a phase, in which the difference between the voltages V_0^+ and V_0^- is inferior to the transistor threshold voltage, and both transistors, T_3 and T_4, are turned on so that the capacitor C_E is charged by the negative and positive output signals. But it is fairly short, and the resulting change of the voltage V_E remains minor. Hence, the circuit consisting of transistors T_3 and T_4 and the capacitor C_E can be considered as a negative peak detector.

Show that the bias voltage V_B can be expressed as,

$$V_B = \left(1 + \frac{R_B}{R_A}\right)\left(V_{REF}\frac{R_1}{R_1 + R_2} + V_E\frac{R_2}{R_1 + R_2}\right) \qquad (6.334)$$

Determine the relation between the resistances such that $V_B = V_{REF} + V_E$.

11. Peak detector

Peak detectors [81], as shown in Figure 6.134, find applications in automatic gain control systems. Figure 6.135 shows the input and output waveforms illustrating the behavior difference between peak detectors with a constant decay rate and with a time-constant decay.

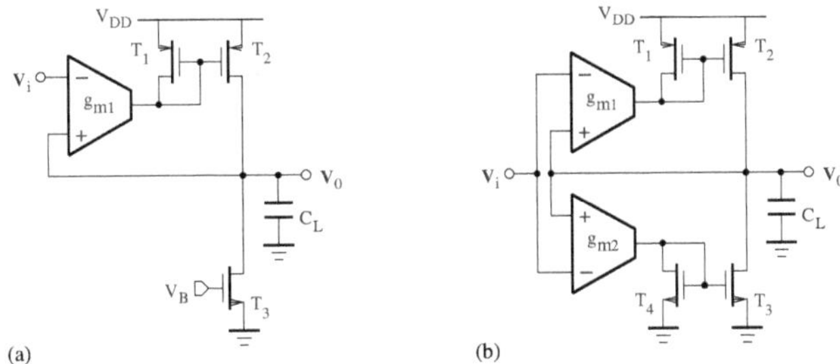

FIGURE 6.134
Circuit diagrams of peak detectors with a constant decay rate (a) and with a time-constant decay (b).

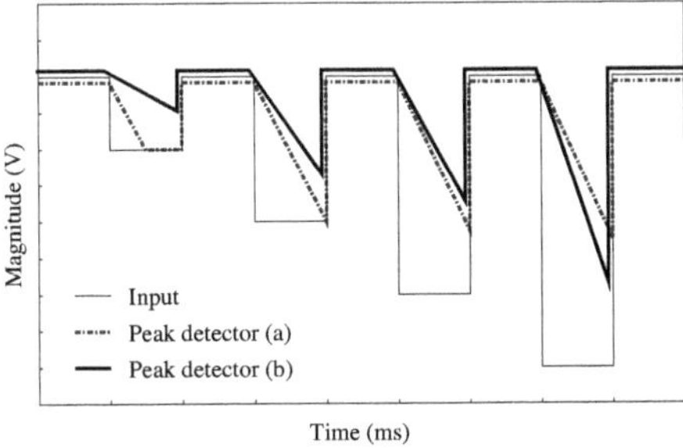

FIGURE 6.135
Peak detector input and output waveforms.

In the circuit of Figure 6.134(a), when $V_i > V_0$, the input peak voltage is tracked with a time constant of C_L/g_{m_1}, and when $V_i < V_0$, the capacitor C_L is discharged through the transistor T_3 at a constant rate regardless of the level of the input voltage decrease because no current then flows through T_2.

To improve the precision, the peak detector can be implemented as shown in Figure 6.134(b). The transconductance amplifiers are designed such that $g_{m_1} > g_{m_2}$ to speed the tracking process of the peak level. The downward-going segment of the input is followed with a time constant of C_L/g_{m_2}. The differential equation of the peak detector can be written as,

$$C_L \frac{dV_0}{dt} = \begin{cases} g_{m_1}(V_i - V_0), & V_i > V_0 \\ g_{m_2}(V_i - V_0), & V_i \leq V_0 \end{cases} \tag{6.335}$$

Let $R_0 = V_{0,p}/V_{i,p}$ be the ripple ratio defined as the amplitude of the output ripple normalized to the amplitude of the input, and $A_t = V_{0,dc}/V_{i,p}$ be the tracking level defined as the ratio of the *dc* output level to the input amplitude ($A_t = 1/\sqrt{2}$ is selected for rms tracking). With the output voltage assumed to be of the form,

$$V_0 = V_{i,p}(R_0 \sin(\omega t + \phi) + A_t) \tag{6.336}$$

where ϕ is the phase shift, show that, for $\omega = 0$ (or at *dc*), we have

$$\frac{1}{2} \ln\left(\frac{g_{m_1}}{g_{m_2}}\right) = \coth^{-1}\left[\frac{2}{\pi}\left(\frac{R_e}{A_t}\sqrt{1 - \frac{A_t^2}{R_e^2}} + \sin^{-1}\left(\frac{A_t}{R_e}\right)\right)\right] \tag{6.337}$$

where

$$R_e = \sqrt{1 + R_0^2 - R_0 \cos(\phi)} \tag{6.338}$$

and for the fundamental frequency, ω, we can obtain

$$\frac{\omega C_L R_0}{g_{m_2} R_e} = \frac{R_g + 1}{2} + \frac{1 - R_g}{\pi} \sin^{-1}\left(\frac{A_t}{R_e}\right) + \frac{1 - R_g}{\pi} \frac{A_t}{R_e} \sqrt{1 - \frac{A_t^2}{R_e^2}} \tag{6.339}$$

where $R_g = g_{m_1}/g_{m_2}$.

12. Noise in lossy integrators

The noise performance of a MOS operational amplifier can be specified by giving the input-referred voltage, $\overline{v_{n_a}^2}$. A resistor R_j generates a noise voltage characterized by the spectral density $\overline{v_{n_{r_j}}^2} = 4kTR_j$.

Show that the output noise spectral density of the integrator depicted in Figure 6.136(a) can be written as

$$\overline{v_{n,0}^2}(f) \simeq \overline{v_{n_{r_2}}^2} + \overline{v_{n_{r_1}}^2} \, |H_1(s)|^2_{s=j2\pi f} + \overline{v_{n_a}^2} \, |H_a(s)|^2_{s=j2\pi f} \tag{6.340}$$

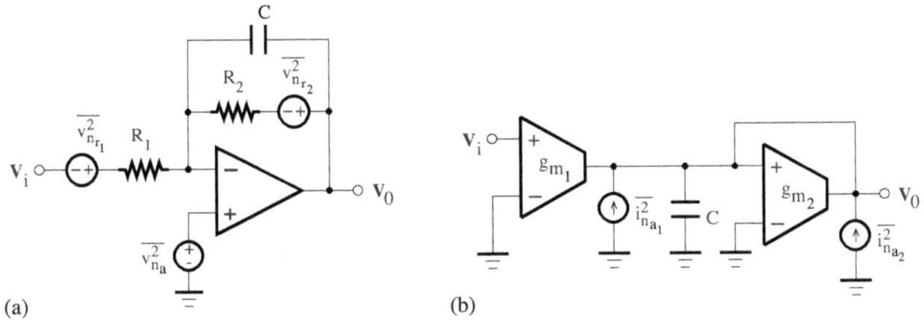

FIGURE 6.136
RC (a) and $g_m - C$ (b) lossy integrators.

where

$$H_1(s) = -\frac{R_2}{R_1}\frac{1}{1+sR_2C} \tag{6.341}$$

$$H_a(s) = 1 + \frac{R_2}{R_1}\frac{1}{1+sR_2C} \tag{6.342}$$

and $H_1(s)$ ($H_a(s)$) is the transfer function from $\overline{v^2_{n_{r_1}}}$ ($\overline{v^2_{n_a}}$) to the output of the integrator.

The noise due to the transconductance amplifier can be modeled by a current source connected at the output node with (one-sided) spectral density, $\overline{i^2_{n_{a_j}}} = 4kT\gamma g_{m_j}$, where k is Boltzmann's constant, T denotes the absolute temperature, and γ is the noise factor whose typical value is larger than 1, and g_{m_j} represents the transconductance.

Verify that the output noise spectral density of the integrator shown in Figure 6.136(b) can be expressed as

$$\overline{v^2_{n,0}}(f) = \overline{i^2_{n_{a_1}}}\,|H_1(s)|^2_{s=j2\pi f} + \overline{i^2_{n_{a_2}}}\,|H_2(s)|^2_{s=j2\pi f} \tag{6.343}$$

and $H_s(s)$ ($j = 1, 2$) is the transfer function from $\overline{i^2_{n_{a_j}}}$ to the output of the integrator, $H_1(s) = H_2(s) = -1/(g_{m_2} + sC)$.

Let the output noise mean-squared value be given by

$$N_0 = \int_0^{+\infty} \overline{v^2_{n,0}}(f)df \tag{6.344}$$

For each integrator, determine the maximum signal-to-noise ratio (SNR_{max}) defined as

$$\text{SNR}_{max} = \frac{|V_{i,max}|^2 \max\limits_f |H_i(s)|^2_{s=j2\pi f}}{2N_0} \tag{6.345}$$

where

$$\max_f |H_i(s)|^2_{s=j2\pi f} = \begin{cases} (R_2/R_1)^2 & \text{for RC integrator} \\ (g_{m_1}/g_{m_2})^2 & \text{for } g_m\text{-C integrator,} \end{cases} \quad (6.346)$$

$V_{i,max}$ is the maximum input voltage that guarantees the linearity at a given electrical power consumption, and $H_i(s)$ is the transfer function from the input V_i to the output V_0.

Hint: The following integral derivation can be useful:

$$\int_0^{+\infty} \left| \frac{1}{1 + s/\omega_c} \right|^2_{s=j2\pi f} df = \frac{\omega_c}{4} \quad (6.347)$$

13. **Dynamic range of integrator**
The dynamic range can be defined as

$$\text{DR} = 20 \log_{10} \frac{V_{max}^2}{\overline{v_{in}^2}}, \quad (6.348)$$

where V_{max} is the maximum undistorted rms value of the input voltage, for which the total harmonic distortion (THD) equals 1%, and $\overline{v_{in}^2}$ is the rms value of the input referred noise integrated over the desired signal bandwidth, Δf.

Compare the dynamic range of the different continuous-time integrators using SPICE simulations

14. **Low-frequency lowpass filter**
Consider the low-frequency lowpass filter of Figure 6.137. Determine the transfer function $V_0(s)/V_i(s)$.

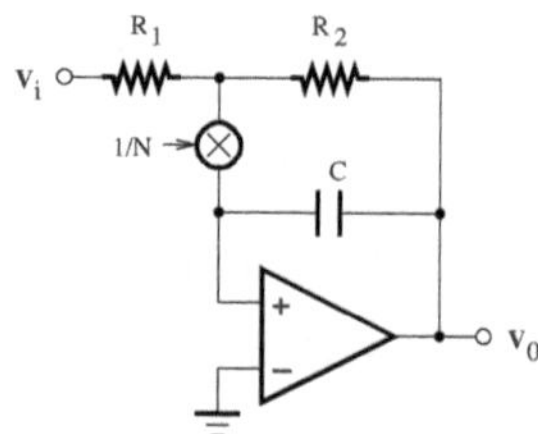

FIGURE 6.137
Circuit diagram of a low-frequency lowpass filter.

Show that the input referred noise can be written as

$$\overline{v_{in}^2} = 4kT(R_1 + R_2) + \overline{v_{in,1/N}^2} \quad (6.349)$$

where $\overline{v_{in,1/N}^2}$ denotes the noise contribution due to the $1/N$ scaling

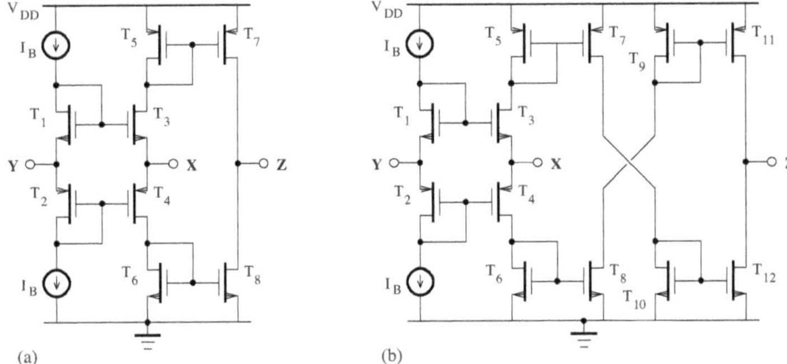

FIGURE 6.138
Circuit diagrams of (a) positive and (b) negative current conveyors.

block, and compare $\overline{v_{in}^2}$ to the input referred noise, which can be generally observed in a similar filter structure without the current down-scaling.

Verify that the current conveyors (CCs) shown in Figure 6.138 can be ideally characterized by the transfer matrix; the plus and minus signs apply to the positive and negative CCs, respectively.

$$\begin{bmatrix} i_Y \\ v_X \\ i_Z \end{bmatrix} = \begin{bmatrix} 0 & 0 & 0 \\ 1 & 0 & 0 \\ 0 & \pm 1 & 0 \end{bmatrix} \begin{bmatrix} v_Y \\ i_X \\ v_Z \end{bmatrix} \tag{6.350}$$

Verify that the 1/N scaling can be realized by the negative cur-

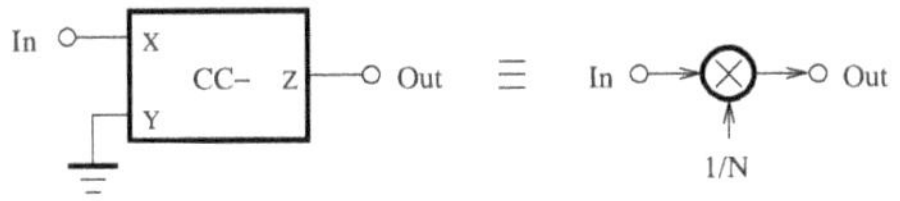

In X CC− Z Out Y ≡ In Out 1/N

FIGURE 6.139
Current conveyor-based implementation of the scaling by 1/N.

rent conveyor with two N:1 output current mirrors (i.e., $T_9 - T_{10}$: N(W/L), $T_{11} - T_{12}$: W/L) (see Figure 6.139).

15. **Grounded resistors and inductor based on transconductors**
In integrated filter design, active networks based on transconductance amplifiers can be used to simulate the behavior of various passive elements.

Let g_m be the amplifier transconductance. Verify that the circuits of Figures 6.140(a) and (b) can be used to realize positive and negative resistors with the value $1/g_m$, respectively.

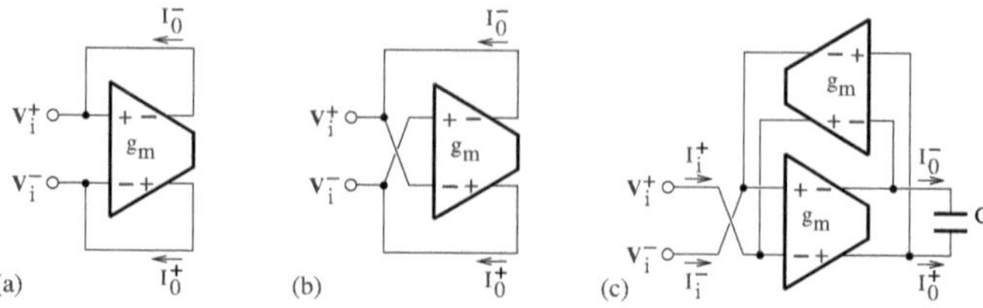

FIGURE 6.140
Circuit diagrams of transconductor-based (a) (b) resistors and (c) inductor.

Show that the g_m-C circuit of Figure 6.140(c) is equivalent to a grounded inductor with the value $L = C/g_m^2$.

Analyze the effect of transconductor parasitic elements (input capacitor, output capacitor and resistor) on the simulated resistors and inductor.

16. **Lowpass and highpass filter transformations**
Let

$$H(s) = \frac{1}{s^2 + \sqrt{2}s + 1} \tag{6.351}$$

be the transfer function of a second-order Butterworth lowpass filter with the normalized passband edge frequency $\Omega_p = 1$.

Use the following spectral transformation,

$$s \to \frac{\Omega_p}{\Omega'_p} s \tag{6.352}$$

to derive the transfer function of a lowpass filter with the normalized passband edge frequency $\Omega'_p = 3\Omega_p/2$.

Determine the transfer function of the highpass filter obtained using the transformation given by

$$s \to \frac{\Omega_p \Omega'_p}{s} \tag{6.353}$$

where $\Omega'_p = 3\Omega_p/2$ is the normalized passband edge frequency of the highpass filter.

17. **Bandpass and bandstop filter transformations**
Consider a first-order lowpass filter with a transfer function of the form

$$H(s) = k \frac{\omega_c}{s + \omega_c} \tag{6.354}$$

where k is the gain factor and ω_c is the cutoff frequency. It is assumed that the passband edge frequency is $\Omega_p = \omega_c$.

Show that the transfer function of the bandpass filter derived using the following spectral transformation,

$$s \to \Omega_p \frac{s^2 + \Omega_l \Omega_u}{s(\Omega_u - \Omega_l)} \tag{6.355}$$

where Ω_l and Ω_u denote the lower and upper passband edge frequencies, respectively, takes the form

$$H_{BP}(s) = k \frac{\frac{\omega_0}{Q} s}{s^2 + \frac{\omega_0}{Q} s + \omega_0^2} \tag{6.356}$$

where ω_0 and Q are parameters to be determined.

Let $H_{AP}(s)$ be the transfer function of an allpass filter. Verify that

$$H_{AP}(s) = 1 - 2H_{BP}(s) \tag{6.357}$$

The transfer function of a second-order bandstop filter can be written as

$$H_{BS}(s) = k \frac{s^2 + \omega_0^2}{s^2 + \frac{\omega_0}{Q} s + \omega_0^2} \tag{6.358}$$

Find the relation between the bandstop filter parameters (ω_0 and Q) and ω_c using the next transformation

$$s \to \Omega_p \frac{s(\Omega_u - \Omega_l)}{s^2 + \Omega_l \Omega_u} \tag{6.359}$$

where Ω_l and Ω_u denote the lower and upper passband edge frequencies, respectively.

Verify that

$$-H_{AP}(s) = 1 - 2H_{BS}(s) \tag{6.360}$$

where $H_{AP}(s)$ is the transfer function of an allpass filter.

18. Analysis of second-order bandpass filter

Find the transfer function, $H = V_0/V_i$, of the second-order bandpass filter shown in Fig 6.141 and put it into the form

$$H(s) = \frac{V_0(s)}{V_i(s)} = k \frac{\frac{\omega_0}{Q} s}{s^2 + \frac{\omega_0}{Q} s + \omega_0^2} \tag{6.361}$$

where

$$k = \frac{g_{m1}}{g_{m4}} \tag{6.362}$$

$$\omega_0 = \sqrt{\frac{1}{g_{m2}g_{m3}C_1C_2}} \tag{6.363}$$

$$Q = \frac{1}{g_{m4}}\sqrt{g_{m2}g_{m3}\frac{C_1}{C_2}} \tag{6.364}$$

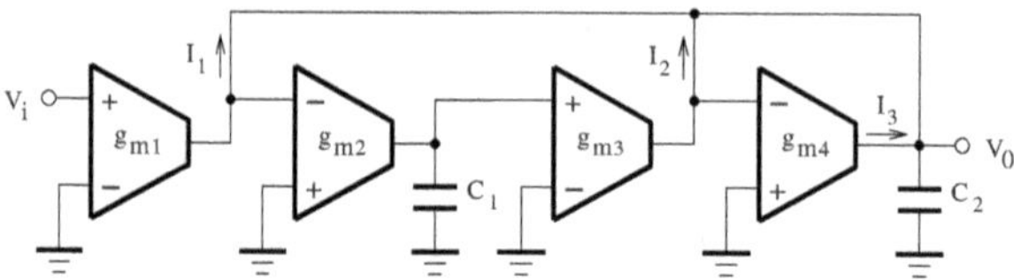

FIGURE 6.141
Circuit diagram of a second-order bandpass filter.

Hint: At the output node, we have $V_0 = sC_2(I_1 + I_2 + I_3)$.

19. **Analysis of a general biquad**
For the biquad structure of Figure 6.142, verify that the transfer function is given by

$$H(s) = \frac{V_0(s)}{V_i(s)} = \frac{s^2 + \dfrac{g_{m5}}{C_2}s - \dfrac{g_{m2}g_{m4}}{C_1C_2}}{s^2 + \dfrac{g_{m3}}{C_2}s + \dfrac{g_{m2}g_{m4}}{C_1C_2}} \tag{6.365}$$

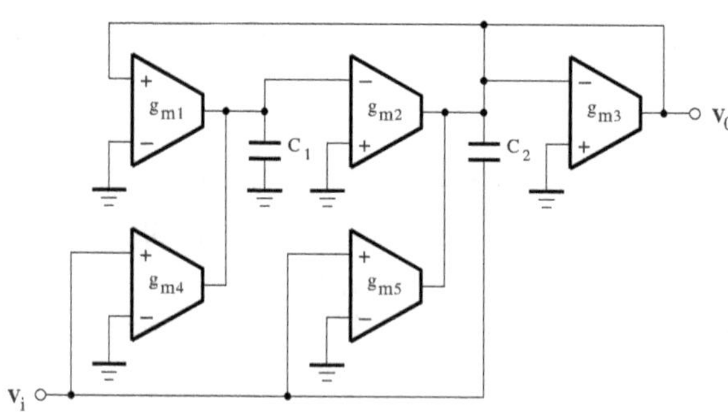

FIGURE 6.142
Circuit diagram of a biquadratic filter section.

20. Transfer function synthesis

Consider the network shown in Figure 6.143. Verify that

$$V_1 = H_1(s)V_i(s) \tag{6.366}$$

$$V_2 = H_2(s)V_0(s) \tag{6.367}$$

and

$$g_{m1}V_1(s) - g_{m2}V_2(s) = 0 \tag{6.368}$$

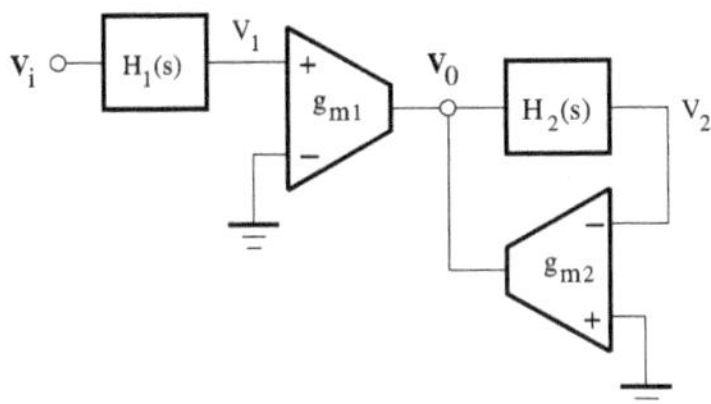

FIGURE 6.143
Circuit diagram of a network for transfer function synthesis.

Deduce that

$$H(s) = \frac{V_0(s)}{V_i(s)} = \frac{H_1(s)}{H_2(s)} \tag{6.369}$$

provided that $g_{m1} = g_{m2}$.

Let

$$H_2(s) = \frac{1}{s^2 + \dfrac{\omega_p}{Q_p}s + \omega_p^2} \tag{6.370}$$

and

$$H_2(s) = \frac{1}{k(s^2 + \omega_z^2)} \tag{6.371}$$

Propose two circuits realizing the above transfer functions, and use them to build a band-reject filter.

21. Bump equalizer

Show that the bump equalizer depicted in Figure 6.144 realizes a transfer function of the form

$$T(s) = \frac{V_0(s)}{V_i(s)} = \frac{1 \mp kH(s)}{1 \pm kH(s)} \tag{6.372}$$

where k is a variable gain.

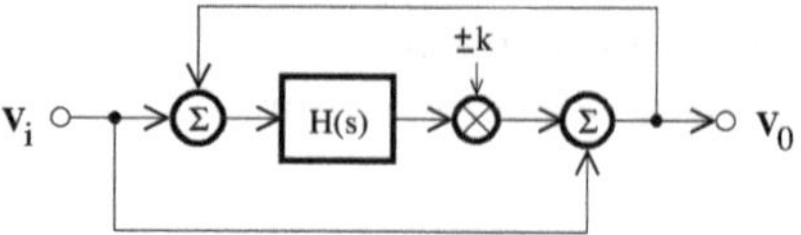

FIGURE 6.144
Block diagram of a bump equalizer.

In general, $|H(s)| \leq 1$ and $|k| \leq 1$. Assuming that

$$H(s) = \frac{s^2 - \frac{\omega_0}{Q}s + \omega_0^2}{s^2 + \frac{\omega_0}{Q}s + \omega_0^2} \tag{6.373}$$

where $Q = 1/\sqrt{3}$ and $\omega_0 = 1$ rad/s, plot the magnitude of $T(s)$ for various values of k.

22. Active synthesis of all-pole filter

The circuit diagram of a fifth-order all-pole LC filter is shown in Figure 6.145. Verify that the node equations can be written as

$$V_1 = \frac{1}{sC_1}\left(\frac{V_i - V_1}{R_i} - I_{L2}\right) \tag{6.374}$$

$$I_{L2} = \frac{1}{sL_2}(V_1 - V_2) \tag{6.375}$$

$$V_2 = \frac{1}{sC_3}(I_{L2} - I_{L4}) \tag{6.376}$$

$$I_{L4} = \frac{1}{sL_4}(V_2 - V_0) \tag{6.377}$$

$$V_0 = \frac{1}{sC_5}\left(I_{L4} - \frac{V_0}{R_0}\right) \tag{6.378}$$

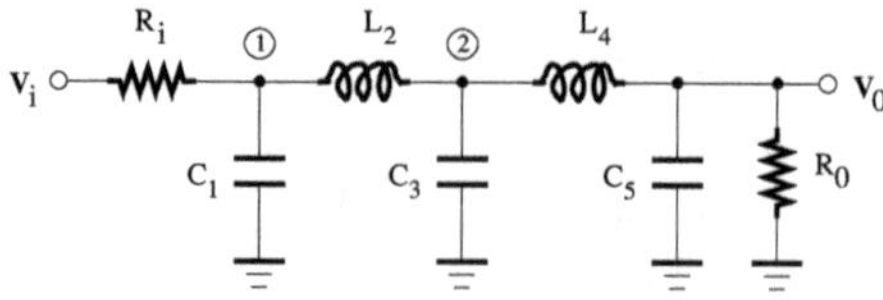

FIGURE 6.145
Circuit diagram of an all-pole LC filter.

Draw a signal-flow graph representation of the filter.

Design RC and g_m-C realizations of the all-pole filter prototype.

23. Analysis of a fourth-order Chebyshev highpass filter
The circuit shown in Figure 6.146 [54] is used to design a Chebyshev highpass filter with the transfer function,

$$H(s) = k\frac{s^4}{s^4 + p_3\omega_p s^3 + p_2\omega_p^2 s^2 + p_1\omega_p^3 s + p_0\omega_p^4} \tag{6.379}$$

where ω_p is the passband edge frequency, and k, p_0, p_1, p_2, and p_3 are real coefficients. Based on its signal-flow graph depicted in Figure 6.147, determine the transfer function $V_0(s)/V_i(s)$.

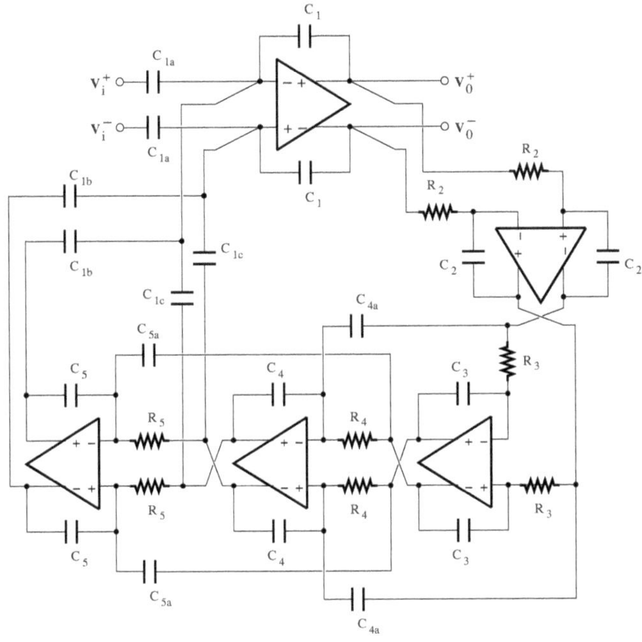

FIGURE 6.146
Circuit diagram of a fourth-order highpass RC filter.

Assuming that

$$R_2 = \frac{1}{q_1 C_2 \omega_p} \qquad R_3 = \frac{q_1}{q_2 C_3 \omega_p} \qquad R_4 = \frac{q_2}{q_3 C_4 \omega_p}\frac{p_1 - p_3}{p_1}$$
$$\text{and} \quad R_5 = \frac{q_3}{q_4 C_4 \omega_p}\frac{p_1(p_2 - p_0)}{p_0(p_1 - p_3)} \tag{6.380}$$

where q_j ($j = 1, 2, 3, 4$) are scaling coefficients, find the transfer functions, $V_j(s)/V_i(s)$, from the filter input to the output of each amplifier.

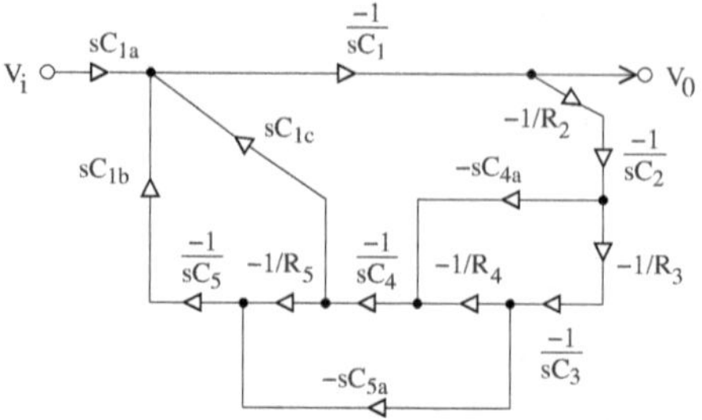

FIGURE 6.147
Signal-flow diagram of a fourth-order highpass filter.

24. **Third-order elliptic filter design**
Realize the third-order elliptic lowpass filter with the transfer function

$$H(s) = \frac{1.53210(s^2 + 1.69962)}{(s + 1.84049)(s^2 + 0.308389s + 1.41484)} \tag{6.381}$$

as a cascade connection of first-order and second-order g_m-C circuit sections, and as a g_m-C ladder circuit.

Analyze the effect of transconductor imperfections on the frequency response of each realized filter circuit using SPICE simulations.

25. **Element simulation-based filter design**
The design of a lowpass filter with a 3-dB frequency equal to 1 rad/s results in the LC circuit prototype shown in Figure 6.148.

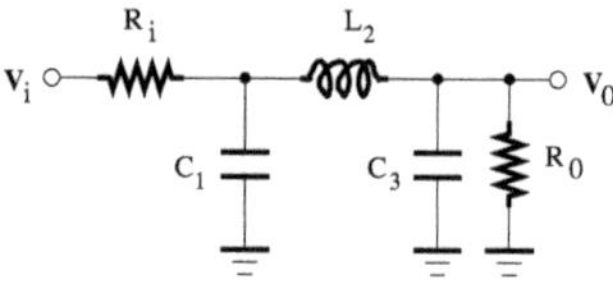

FIGURE 6.148
Lowpass LC filter prototype.

Determine the transfer function $H(s) = V_0(s)/V_i(s)$.

For $R_i = R_0 = 1\ \Omega$, $C_1 = C_3 = 1$ F, and $L_2 = 2$ H, verify that

$$H(s) = \frac{V_0(s)}{V_i(s)} = \frac{1}{2}\frac{1}{s^3 + 2s^2 + 2s + 1} \tag{6.382}$$

Let $\phi(\omega)$ be the phase of the transfer function, $H(s)$. Show that the group delay can be written as

$$\tau(\omega) = -\frac{d\phi(\omega)}{d\omega} = \frac{2 + \omega^2 + 2\omega^4}{1 + \omega^6} \tag{6.383}$$

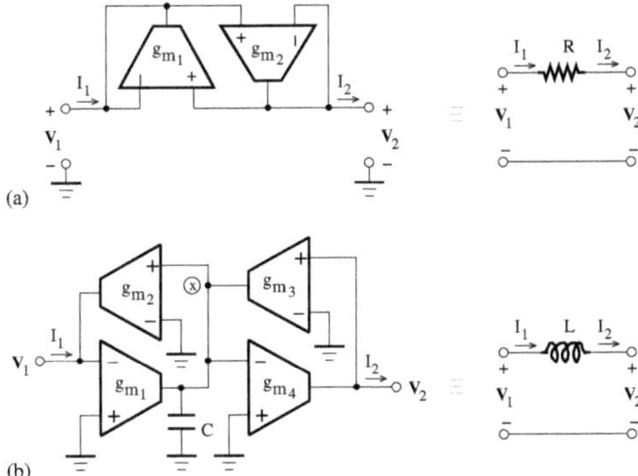

FIGURE 6.149
Floating (a) resistor and (b) inductor realized by a g_m-C circuit.

For the design of integrated filters for low-frequency applications, it is often necessary to substitute resistors and inductors, which tend to be large and sensitive to process variations, by an equivalent active network.

For ideal transconductors, verify that the circuit shown in Figure 6.149(a) can be described by

$$\begin{bmatrix} I_1 \\ I_2 \end{bmatrix} = \begin{bmatrix} g_{m_1} & -g_{m_1} \\ g_{m_2} & -g_{m_2} \end{bmatrix} \begin{bmatrix} V_1 \\ V_2 \end{bmatrix} \tag{6.384}$$

and realizes a floating resistor of the form $R = 1/g_m$, where $g_m = g_{m_1} = g_{m_2}$.

Using the current equation at node x, show that the circuit depicted in Figure 6.149(b) is characterized by

$$\begin{bmatrix} I_1 \\ I_2 \end{bmatrix} = \frac{1}{sC} \begin{bmatrix} g_{m_1} g_{m_2} & -g_{m_2} g_{m_3} \\ g_{m_1} g_{m_4} & -g_{m_3} g_{m_4} \end{bmatrix} \begin{bmatrix} V_1 \\ V_2 \end{bmatrix} \tag{6.385}$$

and is then equivalent to a floating inductor with the value $L = C/(g_{m_1} g_{m_2})$ provided that $g_{m_1} = g_{m_3}$ and $g_{m_2} = g_{m_4}$.

Use SPICE to plot the frequency responses of the filter circuit obtained by substituting the resistors and inductor with their respective equivalent active network. The largest realizable capacitor is 50 pF. Each transconductor can be characterized by a single-pole model with the output capacitance $C_0 = 0.2$ pF, the output resistance $R_0 = 12$ MΩ, the input capacitance $C_i = 0.6$ pF, and a very high input resistance ($R_i = \infty$).

Bibliography

[1] J. Mitola, "The software radio architecture," pp. 26–38, *IEEE Communications Magazine*, May 1995.

[2] A. Rofougaran, G. Chang, J. J. Rael, J. Y.-C. Chang, M. Rofougaran, P. J. Chang, M. Djafari, M.-K. Ku, E. W. Roth, A. A. Abidi, and H. Samueli, "A single-chip 900-MHz spread-spectrum wireless transceiver in 1-μm CMOS–Part I: Architecture and transmitter design," *IEEE J. of Solid-State Circuits*, vol. 33, pp. 513–534, April 1998.

[3] A. Rofougaran, G. Chang, J. J. Rael, J. Y.-C. Chang, M. Rofougaran, P. J. Chang, M. Djafari, M.-K. Ku, E. W. Roth, A. A. Abidi, and H. Samueli, "A single-chip 900-MHz spread-spectrum wireless transceiver in 1-μm CMOS–Part II: Receiver design," *IEEE J. of Solid-State Circuits*, vol. 33, pp. 535–547, April 1998.

[4] L. Levi, "Electrical transmission of energy," U.S. Patent 1,734,038, filed Aug. 12, 1918; issued Nov. 5, 1929.

[5] E. H. Armstrong, "The superheterodyne — Its origin, development, and some recent improvements," *Proc. IRE.*, vol. 12, pp. 539–552, Oct. 1924.

[6] F. M. Colebrook, "Homodyne," *Wireless World and Radio Review*, vol. 13, pp. 645–648, 1924.

[7] R. Hartley, "Single-sideband modulator," U.S. Patent 1,666,206, filed Jan. 15, 1925; issued April 17, 1928.

[8] D. K. Weaver, "A third method of generation and detection of single-sideband signals," *Proc. IRE*, vol. 44, pp. 1703–1705, 1956.

[9] J. Crols and M. Steyaert, "A single-chip 900 MHz CMOS receiver front-end with a high performance low-IF topology," *IEEE J. of Solid-State Circuits*, vol. 30, pp. 1483–1492, Dec. 1995.

[10] J. C. Rudell, J.-J. Ou, T. B. Cho, G. Chien, F. Brianti, J. A. Weldon, and P. Gray, "A 1.9 GHz wide-band IF double conversion CMOS receiver for cordless telephone applications," *IEEE J. of Solid-State Circuits*, vol. 32, pp. 2071–2088, Dec. 1997.

[11] J. A. Weldon, R. S. Narayanaswani, J. C. Rudell, L. Li, M. Otsuka, S. Dedieu, L. Tee, K.-C. Tsai, C.-W. Lee, and P. R. Gray, "A 1.75-GHz highly integrated narrow-band CMOS transmitter with harmonic-rejection mixers," *IEEE J. of Solid-State Circuits*, vol. 36, pp. 2003–2015, Dec. 2001.

[12] T. Sowlati, C. A. T. Salama, J. Sitch, G. Rabjohn, and D. Smith, "Low voltage, high efficiency GaAs class E power amplifiers for wireless transmitters," *IEEE J. of Solid-State Circuits*, vol. 30, pp. 1074–1080, Oct. 1995.

[13] K.-C. Tsai and P. R. Gray, "A 1.9-GHz, 1-W CMOS class-E power amplifier for wireless communications," *IEEE J. of Solid-State Circuits*, vol. 34, pp. 962–970, July 1999.

[14] S. Joo, T.-Y. Choi, and B. Jung, "A 2.4-GHz resistive feedback LNA in 0.13 μm CMOS," *IEEE J. of Solid-State Circuits*, vol. 44, pp. 3019–3029, Nov. 2009.

[15] W. S. Kim, X. Li, and M. Ismail, "A 2.4 GHz CMOS low noise amplifier using an inter-stage matching inductor," in *Proc. 42nd Midwest Symp. Circuits and Systems*, Aug. 1999, vol. 2, pp. 1040–1043.

[16] A. Parsa and B. Razavi, "A new transceiver architecture for the 60-GHz band," *IEEE J. of Solid-State Circuits*, vol. 44, pp. 751–762, March 2009.

[17] B. Razavi, "A 60-GHz CMOS receiver front-end," *IEEE J. of Solid-State Circuits*, vol. 41, pp. 17–22, Jan. 2006.

[18] F. Bruccoleri, E. A. M. Klumperink, and B. Nauta, "Wide-band CMOS low-noise amplifier exploiting thermal noise canceling," *IEEE J. of Solid-State Circuits*, vol. 39, pp. 275–282, Feb. 2004.

[19] D. K. Shaeffer and T. H. Lee, "A 1.5-V, 1.5-GHz CMOS low noise amplifier," *IEEE J. of Solid-State Circuits*, vol. 32, pp. 745–759, May 1997.

[20] D. K. Shaeffer and T. H. Lee, "Corrections to "A 1.5-V, 1.5-GHz CMOS low noise amplifier"," *IEEE J. of Solid-State Circuits*, vol. 40, pp. 1397–1398, June 2005.

[21] S. Shekhar, J. S. Walling, S. Aniruddhan, and D. J. Allstot, "CMOS VCO and LNA using tuned-input tuned-output circuits," *IEEE J. of Solid-State Circuits*, vol. 43, pp. 1177–1186, May 2008.

[22] F. Stubbe, S. V. Kishore, C. Hull, and V. Dellatorre, "A CMOS RF-receiver front-end for 1-GHz applications," *1998 Symposium on VLSl Circuits Digest of Technical Papers*, Honolulu, HI, pp. 80–83, June 1998.

[23] F. Behbahani, J. C. Leete, Y. Kishigami, A. Roithmeier, K. Hoshino, and A. A. Abidi, "A 2.4 GHz low-IF receiver for wideband WLAN in 0.6 μm CMOS – Architecture and front-end," *IEEE J. Solid-State Circuits*, vol. 35, pp. 1908–1916, Dec 2000.

[24] D. K. Shaeffer, A. R. Shahani, S. S. Mohan, H. Samavati, H. R. Rategh, M. d. M. Hershenson, M. Xu, C. P. Yue, D. J. Eddleman, and T. H. Lee, "A 115-mW, 0.5-µm CMOS GPS receiver with wide dynamic-range active filters," *IEEE J. of Solid-State Circuits*, vol. 33, pp. 2219–2231, Dec. 1998.

[25] W. Zhuo, X. Li, S. Shekhar, S. H. K. Embabi, J. Pineda de Gyvez, D. J. Allstot, and E. Sanchez-Sinencio, "A capacitor cross-coupled common-gate low-noise amplifier," *IEEE Trans. on Circuits and Systems–II*, vol. 52, pp. 875–879, Dec. 2005.

[26] B. Razavi, "A millimeter-wave CMOS heterodyne receiver with on-chip LO and divider," *IEEE J. of Solid-State Circuits*, vol. 43, pp. 477–485, Feb. 2008.

[27] T.-P. Liu and E. Westerwick, "5-GHz CMOS radio transceiver front-end chipset," *IEEE J. of Solid-State Circuits*, vol. 35, pp. 1927–1933, Dec. 2000.

[28] L. S. Cutler and C. L. Searle, "Some aspects of the theory and measurement of frequency fluctuations in frequency standards," *Proc. of the IEEE*, vol. 54, pp. 136–154, Feb. 1966.

[29] D. B. Leeson, "A simple model of feedback oscillator noises spectrum," *Proc. of the IEEE*, vol. 54, pp. 329–330, Feb. 1966.

[30] A. Hajimiri and T. H. Lee, "A general theory of phase noise in electrical oscillators," *IEEE J. of Solid-State Circuits*, vol. 33, pp. 179–194, Feb. 1998.

[31] K.-S. Tan and P. R. Gray, "Fully integrated analog filters using bipolar-JFET technology," *IEEE J. of Solid-State Circuits*, vol. 13, pp. 814–821, Dec. 1978.

[32] J. Silva-Martinez, M. S. J. Steyaert, and W. Sansen, "A 10.7-MHz 68-dB SNR CMOS continuous-time filter with on-chip automatic tuning," *IEEE J. of Solid-State Circuits*, vol. 27, pp. 1843–1853, Dec. 1992.

[33] E. Hegazi, H. Sjöland, and A. A. Abidi, "A filtering technique to lower LC oscillator phase noise," *IEEE J. of Solid-State Circuits*, vol. 36, pp. 1921–1930, Dec. 2001.

[34] R. G. Meyer, "Low-power monolithic RF peak detector analysis," *IEEE J. of Solid-State Circuits*, vol. 30, pp. 65–67, Jan. 1995.

[35] S. H. Galal, H. F. Ragaie, and M. S. Tawfik, "RC sequence asymmetric polyphase networks for RF integrated transceivers," *IEEE Trans. on Circuits and Systems–II*, vol. 47, pp. 18–27, Jan. 2000.

[36] J. Kaukovuori, K. Stadius, J. Ryynänen, and K. A. I. Halonen, "Analysis and design of passive polyphase filters," *IEEE Trans. on Circuits and Systems–I*, vol. 55, pp. 3023–3037, Nov. 2008.

[37] A. Rofougaran, J. Rael, M. Rofougaran, and A. Abidi, "A 900 MHz CMOS LC-oscillator with quadrature outputs," *1996 IEEE ISSCC Digest of Technical Papers*, pp. 392–393, Feb. 1996.

[38] P. Vancorenland and M. Steyaert, "A 1.57-GHz fully integrated very low-phase noise quadrature VCO," *IEEE J. Solid-State Circuits*, vol. 37, pp. 653–656, May 2002.

[39] J. van der Tang, P. van de Ven, D. Kasperkovitz, and A. van Roermund, "Analysis and design of an optimally coupled 5-GHz quadrature LC oscillator," *IEEE J. Solid-State Circuits*, vol. 37, pp. 657–661, May 2002.

[40] A. Mirzaei, M. E. Heidari, R. Bagheri, S. Chehrazi, and A. A. Abidi, "The quadrature LC oscillator: A complete portrait based on injection locking," *IEEE J. of Solid-State Circuits*, vol. 42, pp. 1916–1932, Sept. 2007.

[41] K. Kimura, "Variable gain amplifier circuit," U.S. Patent 6,867,650, filed Dec. 9, 2002; issued March 15, 2005.

[42] C. Liu, Y.-P. Yan, W.-L. Goh, Y.-Z. Xiong, L.-J. Zhang, and M. Madihian, "A 5-Gb/s automatic gain control amplifier with temperature compensation," *IEEE Journal of Solid-State Circuits*, vol. 47, no. 6, pp. 1323–1333, Jun. 2012.

[43] R. Unbehauen and A. Cichocki, *MOS Switched-Capacitor and Continuous-Time Integrated Circuits and Systems*, Communications and Control Engineering Series, Berlin, Germany: Springer-Verlag, 1989.

[44] Y. Tsividis, "Continuous-time filters," In Y. Tsividis and P. Antognetti, Editors, *Design of MOS VLSI Circuits for Telecommunications*, Chap. 11, pp. 334–371, Upper Saddle River, New Jersey: Prentice Hall, 1985.

[45] B.-S. Song, "CMOS RF circuits for data communication applications," *IEEE J. of Solid-State Circuits*, vol. 21, pp. 310–317, April 1987.

[46] Z. Czarnul, "Modification of Banu-Tsividis continuous-time integrator structure," *IEEE Trans. on Circuits and Systems*, vol. 33, pp. 714–716, July 1986.

[47] T. Ndjountche, "Linear voltage-controlled impedance architecture," *Electronics Letters*, vol. 32, pp. 1528–1529, Aug. 1996.

[48] R. L. Geiger and E. Sánchez-Sinencio, "Active filters using operational transconductance amplifiers: A tutorial," *IEEE Circuits and Dev. Mag.*, vol. 1, pp. 20–32, March 1985.

[49] T. Ndjountche and A. Zibi, "On the design of OTA-C structurally allpass filters," *Int. J. of Circuit Theory and Applications*, vol. 23, pp. 525–529, Sept.-Oct. 1995.

[50] P. O. Brackett and A. S. Sedra, "Active compensation for high-frequency effects in op-amp circuits with applications to active RC filters," *IEEE Trans. on Circuits and Systems*, vol. 23, no. 2, pp. 68–72, Feb. 1976.

[51] T. Ndjountche, R. Unbehauen, and F.-L. Luo, "Electronically tunable generalized impedance converter structures," *Int. J. of Circuit Theory and Applications*, vol. 27, pp. 517–522, 1999.

[52] R. Nawrocki and U. Klein, "New OTA-capacitor realisation of a universal biquad," *Electronics Letters*, vol. 22, pp. 50–51, Jan. 1986.

[53] A. Wyszyński, R. Schaumann, S. Szczepański, and P. V. Halen, "Design of 2.7-GHz linear OTA and a 250-MHz elliptic filter in bipolar transistor-array technology," *IEEE Trans. on Circuits and Systems*, vol. 40, pp. 19–31, Jan. 1993.

[54] F. Lin, X. Yu, S. Ranganathan, and T. Kwan, "A 70 dB MPTR integrated programmable gain/bandwidth fourth-order Chebyshev highpass filter for ADSL/VDSL receivers in 65 nm CMOS," *IEEE J. of Solid-State Circuits*, vol. 44, pp. 1290–1297, April 2009.

[55] L. Ping and J. I. Sewell, "Active and digital ladder-based allpass filters," *IEE Proceedings*, vol. 137, Pt. G., no. 6, pp. 439–445, Dec. 1990.

[56] K. R. Rao, V. Sethuraman, and P. K. Neelakantan, "A novel "follow the master" filter," *Proceedings of the IEEE*, vol. 65, pp. 1725–1726, Dec. 1977.

[57] R. Schaumann and M. A. Tan, "The problem of on-chip automatic tuning in continuous-time integrated filters," in *IEEE Proc. of the ISCAS*, 1989, pp. 106–109

[58] M. Banu and Y. Tsividis, "An elliptic continuous-time CMOS filter with on-chip automatic tuning," *IEEE J. of Solid-State Circuits*, vol. 20, pp. 1114–1121, Dec. 1985.

[59] H. Khorramabadi and P. R. Gray, "High-frequency CMOS continuous-time filters," *IEEE J. of Solid-State Circuits*, vol. 19, pp. 939–948, Dec. 1984.

[60] V. Gopinathan, Y. P. Tsividis, K.-S. Tan, and R. K. Hester, "Design considerations for high-frequency continuous-time filters and implementation of an antialiasing filter for digital video," *IEEE J. of Solid-State Circuits*, vol. 25, pp. 1368–1378, Dec. 1990.

[61] C. Yoo, S.-W. Lee, and W. Kim, "A ±1.5-V, 4 MHz CMOS continuous-time filter with a single-integrator based tuning," *IEEE J. of Solid-State Circuits*, vol. 33, pp. 18–27, Jan. 1998.

[62] C. Plett and M. A. Copeland, "A study of tuning for continuous-time filters using macromodels," *IEEE Trans. on Circuits and Systems–II*, vol. 39, pp. 524–531, Aug. 1992.

[63] T. R. Viswanathan, S. Murtuza, V. H. Syed, J. Berry, and M. Staszel, "Switched-capacitor frequency control loop," *IEEE J. of Solid-State Circuits*, vol. 17, pp. 775–778, Aug. 1982.

[64] Y. Kuraishi, K. Nakayama, K. Miyadera, and T. Okamura, "A single-chip 20-channel speech spectrum analyzer using a multiplexed switched-capacitor filter bank," *IEEE J. of Solid-State Circuits*, vol. 19, pp. 964–970, Dec. 1984.

[65] Z. Y. Chang, D. Haspeslagh, and J. Verfaillie, "A highly linear CMOS g_m-C bandpass filter with on-chip frequency tuning," *IEEE J. of Solid-State Circuits*, vol. 27, pp. 388–397, March 1997.

[66] J.-M. Stevenson and E. Sànchez-Sinencio, "An accurate quality factor tuning scheme for IF and high-Q continuous-time filters," *IEEE J. of Solid-State Circuits*, vol. 33, pp. 1970–1978, Dec. 1998.

[67] S. Pavan and Y. P. Tsvidis, "An analytical solution for a class of oscillators, and its application to filter tuning," *IEEE Trans. on Circuits and Systems–I*, vol. 45, pp. 547–556, May 1998.

[68] O. Shana'a and R. Schaumann, "Low-voltage high-speed current-mode continuous-time IC filters with orthogonal ω_0-Q tuning," *IEEE Trans. on Circuits and Systems–II*, vol. 46, pp. 390–400, April 1999.

[69] Y. Tsividis, "Self-tuned filters," *Electronics Letters*, vol. 11, pp. 406–407, June 1981.

[70] H. Yamazaki, K. Oishi, and K. Gotoh, "An accurate center frequency tuning scheme for 450-kHz CMOS g_m-C bandpass filters," *IEEE J. of Solid-State Circuits*, vol. 34, pp. 1691–1697, Dec. 1999.

[71] C. A. Laber and P. R. Gray, "A 20-MHz sixth-order BiCMOS parasitic-insensitive continuous-time filter and second-order equalizer optimized for disk-drive read channels," *IEEE J. of Solid-State Circuits*, vol. 28, pp. 462–470, April 1993.

[72] C.-S. Kim, G.-O. Cho, Y.-H. Kim, and B.-S. Song, "A CMOS 4× speed DVD read channel IC," *IEEE J. of Solid-State Circuits*, vol. 33, pp. 1168–1178, Aug. 1998.

[73] P. M. Vanpeteghem and R. Song, "Tuning strategies in high-frequency integrated continuous-time filters," *IEEE Trans. on Circuits and Systems*, vol. 36, pp. 136–139, Jan. 1989.

[74] K. A. Kozma, D. A. Johns, and A. S. Sedra, "An approach for tuning high-Q continuous-time bandpass filter," in the *IEEE Int. Symp. for Circuits and Systems*, pp. 1037–1040, 1995.

[75] M. F. Toner and G. W. Roberts, "A BIST scheme for a SNR, gain tracking and frequency response test of a sigma-delta ADC," *IEEE Trans. on Circuits and Systems*, vol. 42, pp. 1–15, Jan. 1995.

[76] T. Ndjountche and R. Unbehauen, "Improved structure for programmable filters: Application in a switched-capacitor adaptive filter design," *IEEE Trans. on Circuits and Systems*, vol. 46, pp. 1137–1147, Sept. 1999.

[77] M. C. Coln, "Chopper stabilization of MOS operational amplifiers using feed-forward techniques," *IEEE J. of Solid-State Circuits*, vol. 16, pp. 745–748, Dec. 1981.

[78] T. Ndjountche, *Dynamic Analog Circuit Techniques for Real-Time Adaptive Networks*, Aachen, Germany: Shaker Verlag, 2000.

[79] S. Galal and B. Razavi, "10-Gb/s Limiting amplifier and laser/modulator driver in 0.18-μm CMOS technology," *IEEE J. of Solid-State Circuits*, vol. 38, no. 12, pp. 2138–2146, Dec. 2003.

[80] Y. Wang, B. Afshar, L. Ye, V. C. Gaudet, A. M. Niknejad, "Design of a low power, inductorless wideband variable-gain amplifier for high-speed receiver systems," *IEEE Trans. on Circuits and Systems–I*, vol. 59, no. 4, pp. 696–707, Apr. 2012.

[81] B. Rumberg and D. W. Graham, "A low-power magnitude detector for analysis of transient-rich signals," *IEEE J. of Solid-State Circuits*, vol. 47, no. 3, pp. 676–685, Mar. 2012.

7

Switched-Capacitor Circuits

CONTENTS

The hardware implementation of switched-capacitor (SC) circuits [1] must have the following characteristics:

– Low power consumption

– Low chip area

However, real components are subject to several nonidealities that affect the circuit performance. Because SC structures can be configured to reduce these limitations, they appear to be suitable for interfacing and implementation of signal processing operations.

Basically, capacitors, switches, and amplifiers are necessary for the realization of SC circuits. The design is often modular, consisting of a combination of small-sized blocks (sample-and-hold, integrator, gain stage, etc.). Building blocks must first be optimized with respect to their transfer function sensitivities to nonideal effects before they can be connected together into the overall system.

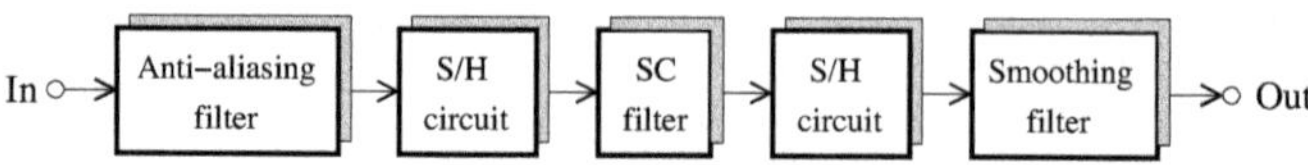

FIGURE 7.1
Block diagram of a signal processor based on a switched-capacitor filter.

The block diagram of the SC filter-based processor, which can be used with analog signals, is shown in Figure 7.1. It includes additional building blocks such as sample-and-hold (S/H) or track-and-hold (T/H) circuits, anti-aliasing and smoothing filters, whose purpose is not to deliver the specified frequency shaping but rather to overcome the limitations related to the sampled-data processing. The overall filter response is primarily determined by the SC filter, provided the ratio of the sampling frequency to the cutoff frequency is much greater than one.

The anti-aliasing filter is first described in the context of sampled-data systems. In addition to describing the various properties of the basic elements, we also analyze the different compensation techniques (dummy switch, bootstrapped switch, correlated double sampling) that result in high-performance circuits. The trend toward lower supply voltages while maintaining a high dynamic range and the need for enhanced power-supply noise rejection make the use of fully differential structures mandatory.

7.1 Anti-aliasing filter

According to the Nyquist sampling theorem, a continuous-time signal can only be recovered from its samples provided the maximum frequency component of the input signal of interest is less than or equal to half of the sampling frequency. In practice, the frequency content of the signal to be sampled is limited by an anti-aliasing filter, which is a suitable continuous-time lowpass or bandpass filter [2, 3] placed before the sample-and-hold circuit as shown in Figure 7.2. In this way, aliasing is avoided by filtering out unwanted high-frequency signal components, which can be folded back into the baseband. Ideally, the anti-aliasing filter should exhibit unity gain in the passband from dc to $f_s/2$, where f_s denotes the sampling frequency, and zero gain in the stopband. But, this type of filter is difficult to implement, and it is necessary to sample the signal at a rate higher than twice the highest frequency component to relax the anti-aliasing filter specifications in practical cases.

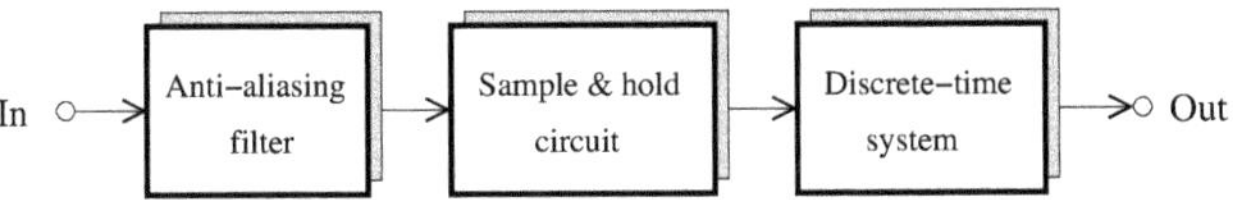

FIGURE 7.2
Building blocks of a discrete-time system.

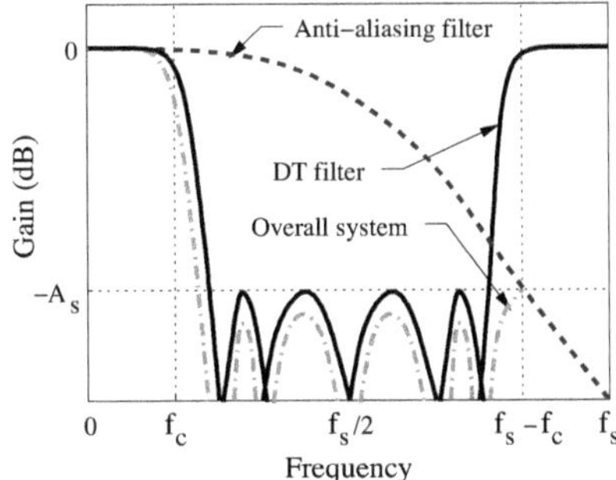

FIGURE 7.3
Discrete-time (DT) filter, anti-aliasing filter, and overall system frequency responses.

The frequency responses of a discrete-time filter, an anti-aliasing filter, and the overall system are shown in Figure 7.3. The discrete-time filter has a low-pass characteristic with a cutoff frequency f_c and a stopband attenuation A_s. Note that the spectrum of every discrete-time system is replicated at multiples of the sampling frequency. The transition from the passband to the stopband of the anti-aliasing filter consists of the frequency region located between $f_s/2$ and $f_s - f_c$. A lowpass filter prototype is characterized by a cutoff (or pass-band) frequency, a stopband frequency, a maximum attenuation (or ripple) in the passband, and a minimum attenuation in the stopband. Depending on the type of application, the filter transfer function can be approximated by functions known as the Butterworth, Bessel, Chebyshev, or elliptic response, each of which has its own advantages or disadvantages. The Butterworth filter exhibits the flattest passband and lowest attenuation in the stopband. The Bessel filter has a more gradual roll-off and features a linear phase response, resulting in a constant time delay over a wide range of frequencies through the passband. The Chebyshev filter has a steeper roll-off near the cutoff frequency and ripples in the passband. The elliptic filter has the steepest roll-off and equal ripples in both passband and stopband.

As the transition band becomes smaller, the complexity of the filter architecture is increased due to the requirement for a quality factor with a high value. In this case, a multi-stage filter can then be required to meet the anti-aliasing specifications.

7.2 Capacitors

MOS capacitors are usually formed between two layers of polycrystalline silicon or metal, or between polycrystalline silicon and a heavily doped crystalline silicon [4]. Silicon dioxide (SiO_2), which is one of the most stable dielectrics, is usually used as insulator.

Ideally, the value of a MOS capacitor is given by

$$C = \frac{\varepsilon_{ox} A_p}{t_{ox}} = \frac{\varepsilon_o \varepsilon_{rox} WL}{t_{ox}} \tag{7.1}$$

where $A_p = WL$ is the area of each capacitor plate, t_{ox} is the thickness of the SiO_2 layer, and $\varepsilon_{ox} = \varepsilon_o \varepsilon_{rox} \simeq 35$ pF/m is the permittivity of the SiO_2. Typical values of MOS capacitances range from 0.25 to 0.5 fF/µm^2 (1 fF$= 10^{-15}$ F). The size of such capacitors is dependent on the accuracy requirements and the signal frequencies used. Capacitors are rarely made smaller than 0.1 pF. The stability of MOS capacitors with respect to the temperature and voltage difference between the input and output nodes is characterized by the temperature and voltage coefficients. Both coefficients are on the order of 100 to 110 ppm (remember that ppm means parts per million, that is 10^{-4}%) and will therefore have a negligible effect on the overall distortion of the circuits.

The smallest capacitor is generally realized in the form of a unit capacitor with a square shape ($W = L$). Larger capacitor values are formed by parallel connection of unit capacitors and possibly one fractional-valued capacitor with the same area-to-perimeter ratio as the unit capacitor. Such a design style eliminates the effect of undercutting the capacitor plates during etching on capacitance ratios and improves the matching accuracy. It also allows for regular and area efficient layout for all network capacitors.

There are unavoidable parasitic capacitances associated with MOS capacitors. The parasitic capacitance of the upper (top) plate and the lower (bottom) plate are typically 0.1 to 2% and 5 to 30% of the nominal capacitance, respectively. The effect of these parasitic capacitances can be eliminated if clever design techniques are used. To this end, in order to minimize the injection of the substrate noise into circuit nodes, the bottom plates of the capacitors should be connected only to low-impedance nodes (i.e., to ground voltage sources or to the output of the amplifier) and not to virtual inputs of the amplifiers.

7.3 Switches

SC circuits exploit the principles of signal sampling and charge transfer that are generally implemented using voltage-controlled switches.

7.3.1 Switch description

Switches can consist of MOS transistors as shown in Figure 7.4. CMOS switches can be realized with complementary (i.e., *p*- and *n*-channel) transistors connected in parallel. The *p*- and *n*-channel transistors are controlled by appropriate positive and negative supply voltages, respectively.

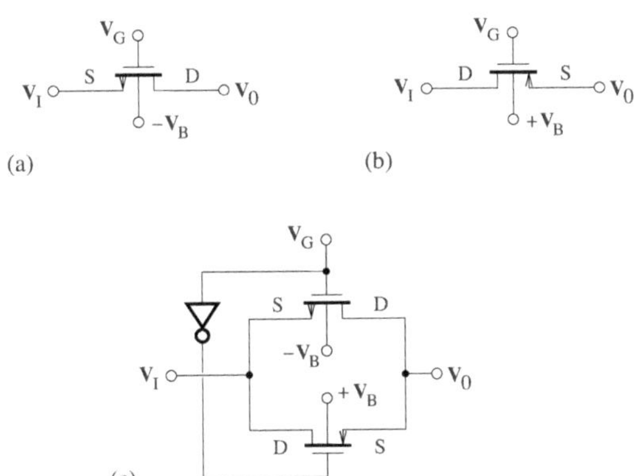

FIGURE 7.4
MOS switches: (a) nMOS transistor switch, $V_I = V_S$ and $V_0 = V_D$, (b) pMOS transistor switch, $V_I = V_D$ and $V_0 = V_S$, (c) CMOS transistor switch.

The terminal at the lower potential is considered the source for the nMOS transistor and the drain for the pMOS transistor. Therefore, the location of the drain and the source depends on the direction of the current through the channel. To avoid current dissipation through the substrate, the bulk potential must be kept lower for the nMOS device and higher for the pMOS device than the one of the source and drain. The gate voltage is used to determine the switch state. The transistor of nMOS type is in the on-state when $V_{GS} > V_T$ and the pMOS type when $V_{GS} < V_T$, where V_{GS} and V_T are the gate-source and threshold voltages, respectively. If V_{DS} is the drain-source voltage, the drain current of a transistor operating in the triode region, will be given by [5]

$$I_D = \pm K' \frac{W}{L} \left[(V_{GS} - V_T) V_{DS} - \frac{1}{2} V_{DS}^2 \right] \tag{7.2}$$

where $K' = \mu C_{ox}$ is the gain for the square-sized transistor, and W and L are the width and length of the channel, respectively. It should be noted that $V_{DS} = -V_{SD}$, $V_{GS} = -V_{SG}$, $V_T > 0$ for an nMOS transistor and $V_T < 0$ for a pMOS transistor. The operating conditions of nMOS and pMOS transistors are expressed as $0 < V_{DS} < V_{GS} - V_T$ and $V_{GS} - V_T < V_{DS} < 0$, respectively. The threshold voltage, V_T, is determined according to the following equation

$$V_T = V_{T0} \pm \gamma \left(\sqrt{\pm V_{SB} + |\phi_B|} - \sqrt{|\phi_B|} \right) \tag{7.3}$$

where V_{SB} denotes the substrate-bulk voltage, V_{T0} is the threshold voltage for $V_{SB} = 0$, γ is the body effect factor, and ϕ_B is the approximate surface potential in strong inversion for zero back-gate bias.

For a fixed gate voltage, V_G, and for a fixed bulk voltage, V_B, the common-mode voltage, V_{CM}, and the differential-mode voltage, V_{DM}, of the switch are defined as

$$V_{CM} \hat{=} \frac{V_I + V_0}{2}, \qquad V_{DM} \hat{=} V_I - V_0 \tag{7.4}$$

where V_I and V_0 represent the voltages at the input and output terminals, respectively. The voltage V_{GS} can take the form

$$V_{GS} = \pm V_G \mp V_{CM} \pm \frac{1}{2} V_{DM}$$

and the current I_D can be rewritten as

$$I_D = \pm K' \frac{W}{L} \left(\pm V_G \mp V_{CM} - V_T \right) V_{DM} \tag{7.5}$$

Note that, in each case, the upper and lower signs have to be applied for nMOS and pMOS devices, respectively. The above current is linearly dependent on the differential-mode voltage, V_{DM}. The current flowing through the CMOS switch can be obtained by summing the current I_D of the nMOS transistor and the one of the pMOS transistor, that is,

$$I_{CMOS} = I_{D,nMOS} + I_{D,pMOS} \tag{7.6}$$

The on-conductances can then be defined as

$$G_{on,nMOS} = \frac{I_{D,nMOS}}{V_{DM}} \tag{7.7}$$

$$G_{on,pMOS} = \frac{I_{D,pMOS}}{V_{DM}} \tag{7.8}$$

and

$$G_{on,CMOS} = \frac{I_{CMOS}}{V_{DM}} \tag{7.9}$$

for the nMOS, pMOS, and CMOS switches, respectively. They are represented in Figure 7.5 with respect to V_{CM} based on 0.5-μm process parameters. It

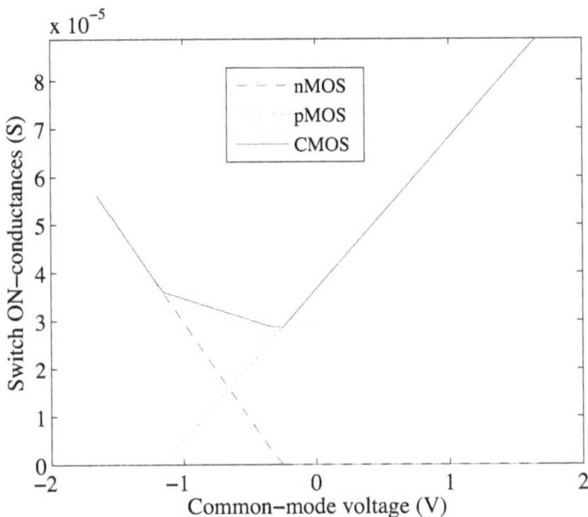

FIGURE 7.5
Switch on-conductances.

can be observed that nMOS and pMOS switches conduct, respectively, for lower and higher voltages and the conductivity of CMOS switches is ensured over the whole signal dynamic range. However, in low-voltage applications where transistors can be critically biased, there is no improvement in the characteristics of CMOS switches in comparison to that of nMOS and pMOS switches.

7.3.2 Switch error sources

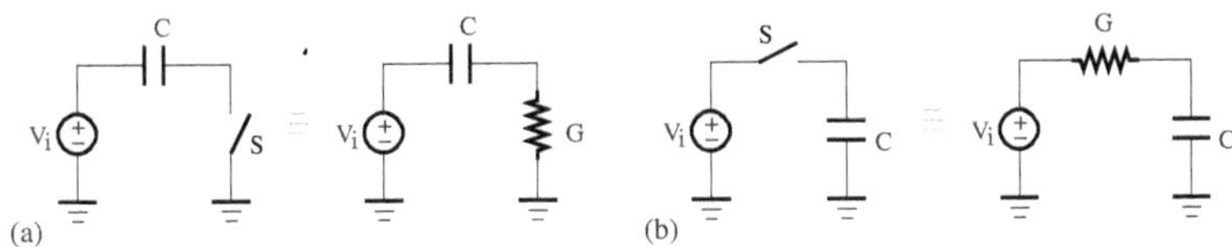

FIGURE 7.6
Simple equivalent circuit models of a (a) grounded and (b) floating switch in the on-state.

Simple equivalent circuit models of a grounded and floating switch are shown in Figures 7.6(a) and (b), respectively. The input signal is assumed to be a unit step, that is, $V_i(t) = V$ for $t \geq 0$ and zero, otherwise. The loop equation is

$$C\frac{dV_C(t)}{dt} - G(V_{CM})[V_i - V_c(t)] = 0 \tag{7.10}$$

where $G(V_{CM})$ denotes the switch on-conductance given by [6]

$$G(V_{CM}) = G_{on} + k_G|V_{CM}| \tag{7.11}$$

where k_G is an empirical constant and V_{CM} denotes the switch common-mode voltage. Taking into account that $V_{CM} = [V - V_C(t)]/2$ in the grounded switch configuration, while $V_{CM} = [V + V_C(t)]/2$ in the floating one, we obtain the next Riccati differential equation

$$\frac{dV_C(t)}{dt} + \{\alpha + \beta[V_c(t) - V]\}\,[V_c(t) - V] = 0 \tag{7.12}$$

The coefficients α and β are expressed for the grounded switch by

$$\alpha = \frac{G_{on}}{C} \tag{7.13}$$

and

$$\beta = \mp\frac{k_G}{2C} \tag{7.14}$$

where the minus sign is for $V_C(0) \leq V$ and the plus sign corresponds to $V_C(0) > V$. For a floating switch, we have

$$\alpha = \frac{G_{on}}{C}\left(1 \pm \frac{K_G V}{G_{on}}\right) \tag{7.15}$$

and

$$\beta = \pm\frac{k_G}{2C} \tag{7.16}$$

where the minus sign is valid for $V < 0$ and $V_C(0) < -V$, and the plus sign is to be used for $V \geq 0$ and $V_C(0) \geq -V$. Given the particular integral, $V_C(t) = V$, the transformation $V_C(t) \mapsto V + 1/V_C(t)$ yields a linear form of the above differential equation with the general solution

$$V_C(t) = \left[1 + \frac{(\alpha\gamma/V)\exp(-\alpha t)}{1 - \beta\gamma\exp(-\alpha t)}\right] V \tag{7.17}$$

where γ is the integrating constant. With the assumption that $V_C(0) = 0$, the capacitor voltage may be written in the form

$$V_C(t) = [1 - \epsilon(t)]V \tag{7.18}$$

where the relative charge transfer error, $\epsilon(t)$, with respect to the ideal case (i.e., $G(V_{CM}) = 0$) is given by

$$\epsilon = \frac{(1-\rho)\exp(-\alpha t)}{1-\rho\exp(-\alpha t)} \tag{7.19}$$

and

$$\rho = \frac{\beta V}{\beta V - \alpha} \tag{7.20}$$

Note that $0 < |\rho| < 1$. The initial value of V_{CM} is $V/2$. For a grounded switch, the common-mode voltage can change toward 0 and

$$\exp\left[-\frac{G(V/2)}{C}t\right] \leq \epsilon(t) \leq \exp\left[-\frac{G(0)}{C}t\right] \tag{7.21}$$

while for the floating one, it can move toward V and

$$\exp\left[-\frac{G(V)}{C}t\right] \leq \epsilon(t) \leq \exp\left[-\frac{G(V/2)}{C}t\right] \tag{7.22}$$

Thus, the charge transfer in the grounded switch structure is affected by a lower error than in the floating switch configuration. However, the floating switch may be more sensitive to the variation of the input signal.

For the circuit design, the minimum worst-case on-conductance over the signal swings should be used in order to determine the sizes of the switches. To this end, the output signal is required to approximate its final value to within an error of 0.1%.

In addition to the on-conductance, the behavior of a switch is affected by the following effects:

- Clock feedthrough
- Charge injection
- Leakage current
- Noise

Let us consider the lumped model of a single switch shown in Figure 7.7. It includes the gate-source and gate-drain overlap capacitors C_{ov}. This model can be used for the single-transistor switch, which is loaded by the capacitors C_i and C_0 and controlled by a clock phase ϕ_k ($k = 1$ or 2), as shown in the circuit of Figure 7.8.

A clock signal with a voltage swing $V_{DD} - V_{SS}$ is applied to the gate of the transistor. Because there is a voltage divider between the overlap and load capacitors during the off-state, a voltage error proportional to $C_{ov}/(C_{ov}+C_0)$

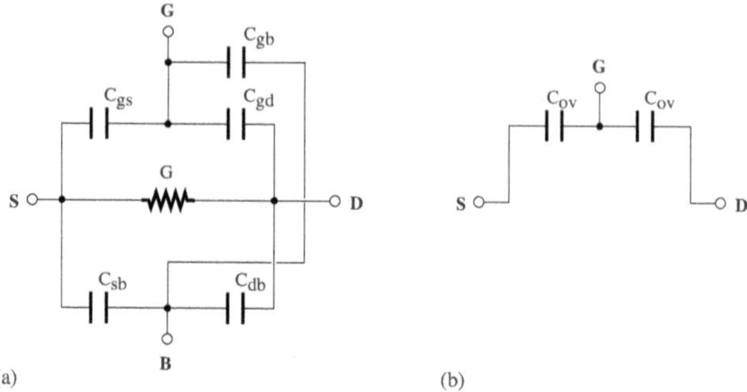

FIGURE 7.7
Equivalent circuit model for the switches: (a) on-state, (b) off-state.

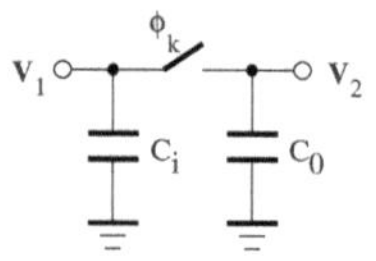

FIGURE 7.8
Single switch environment in a switched-capacitor circuit.

can be observed between the circuit output nodes. This error due to clock feedthrough is given by

$$\triangle V_{CF} = (V_{DD} - V_{SS})\frac{C_{ov}}{C_{ov} + C_0} \tag{7.23}$$

Ideally, a switch is used to connect and disconnect two circuit nodes. However, during the turn-on phase of the MOS transistor switch, a finite amount of mobile charges is trapped in the channel. When the switch is turned off, these charges exit through the transistor terminals. A voltage change, $\triangle V_{CI}$, can then be observed due to the fraction of channel charge, q_{inj}, injected onto C_0 [7,8],

$$\triangle V_{CI} = \frac{q_{inj}}{C_0} \tag{7.24}$$

where $q_{inj} = Q_a + Q_b + Q_c$ and is dependent on the input signal and the clock signal falling rate. The component Q_a is due to the charges in the strong inversion region, Q_b represents the channel charges in the weak inversion region, and Q_c represents the charges coupled through the gate-to-diffusion overlap capacitance of the transistor.

At high temperatures, the switch operation can be affected by the leakage current, I_{leak}, associated with the drain-bulk junction of the MOS transistors. For a hold time T_h, the voltage stored in the capacitor C_0 is perturbed by an amount

$$\triangle V_{leak} = \frac{I_{leak} T_h}{C_0} \tag{7.25}$$

This voltage can be reduced if the value of C_0 is chosen in the picofarad range. Note that the leakage current has no effect on circuit operation at room temperature.

The noise due to the switch is also stored on the output capacitor. It is reduced to a voltage of $\sqrt{kT/C_0}$ in the simplified case, where the switch is equivalent in the on-state to a resistor. A sufficiently large capacitor is then required to meet the low-noise design requirement.

7.3.3 Switch compensation techniques

Many circuit techniques have been proposed for the reduction of errors in the switch operation.

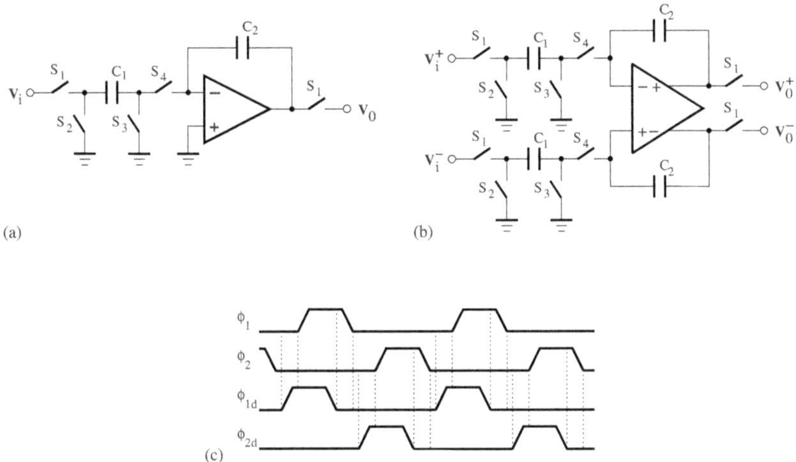

FIGURE 7.9
(a) Circuit diagram of a single-ended SC integrator; (b) circuit diagram of a differential SC integrator; (c) clocking scheme with delayed clock phases.

Consider the circuit diagram of single-ended and fully differential SC integrators shown in Figures 7.9(a) and (b), where switches are used to connect and disconnect capacitors to circuit nodes.

Basically, the operation of SC circuits requires a clock signal with two nonoverlapping phases, ϕ_1 and ϕ_2. This is realized for the aforementioned SC integrators using the clock phase ϕ_1 for the control of switches S_1 and S_3, and the clock phase ϕ_2 for the control of switches S_2 and S_4. In practice, due to the nonideal behavior, or say charge errors, of switches, the use of minimum-sized switches controlled by a combination of two-phase clock signals with a trapezoidal shape and their delayed versions, as shown in Figure 7.9(c), may be necessary for a proper circuit operation [9]. The switch timing for the implementation of inverting and noninverting SC integrators is now as follows:

- Inverting configuration: $S_1(\phi_1)$, $S_2(\phi_2)$, $S_3(\phi_{2d})$, and $S_4(\phi_{1d})$;
- Noninverting configuration: $S_1(\phi_1)$, $S_2(\phi_2)$, $S_3(\phi_{1d})$, and $S_4(\phi_{2d})$.

Here, the switch compensation objective is to maintain almost constant the signal-independent total charge that is trapped and released by each switch during its operation. Note that the use of high-gain amplifiers with current-source output stages prevents the switch charge errors in an SC circuit to be signal dependent. In this case, because the value of the amplifier output impedance is high, the impedance load seen by switches is dominated by one of the surrounding capacitors, which are generally linear.

Switch errors can also be compensated by associating the following strategies.

One is to adopt clock signals with a very short transition time for switches in such a way that the transistor channel charges are divided equally between the source and drain and the charge injections are cancelled by half-sized dummy switches [10, 11] as shown in Figure 7.10. Note that the input dummy switch can be omitted for moderate value of the ratio C_i/C_0 and $\phi_2 = \overline{\phi}_1$.

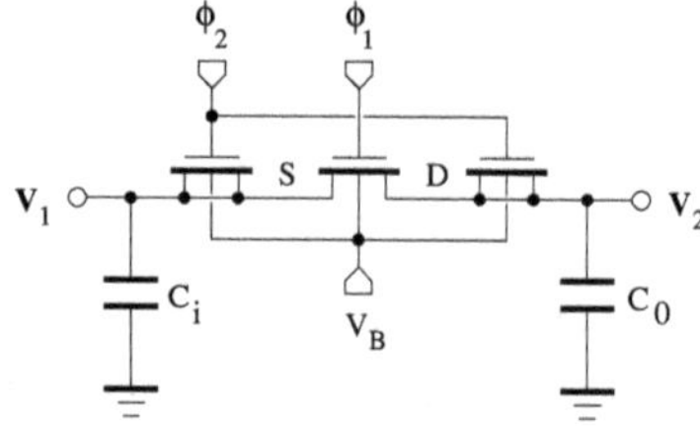

FIGURE 7.10
Charge injection cancellation using dummy switches.

The other relies on the use of differential circuit structures. For a good matching of capacitors, the voltage errors due to nonidealities are injected into the common-mode signal path and the differential signal remains unaffected in the first order.

It is interesting to note that the charge error cancellation that can be achieved in CMOS switches is not complete because the matching between the channel charges of the nMOS and pMOS transistors is poor and signal dependent.

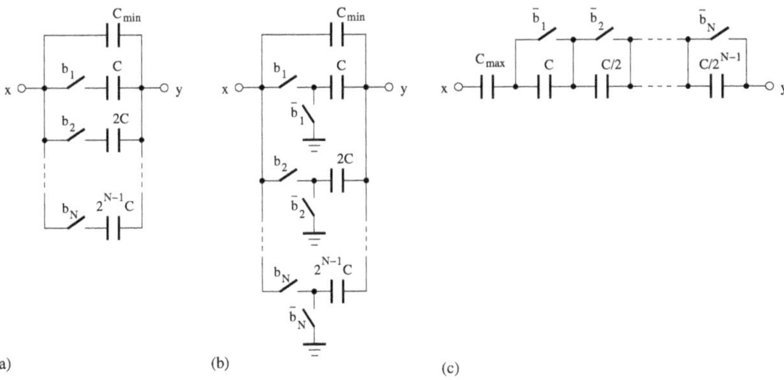

FIGURE 7.11
Programmable capacitor array: (a) Parallel configuration, (b) configuration with a constant node loading, (c) series configuration.

7.4 Programmable capacitor arrays

One approach to achieve the tunability of a circuit characteristic is to use programmable capacitor arrays (PCAs) [15], which are designed to provide capacitance values selectable through a digital interface and ranging from C_{min} to C_{max} in $\triangle C$ increments. PCAs can be composed of binary-weighted capacitors and an un-switched capacitor that defines either C_{min} or C_{max}.

Let the digital word used to program each array be $b_1 b_2 \cdots b_N$, where N is the number of bits. The total capacitance of the parallel capacitor array shown in Figure 7.11(a) is given by

$$C_T = C_{min} + \sum_{j=1}^{N} 2^{j-1} b_j C \tag{7.26}$$

The capacitive load at the node y depends on the digital control word, because a capacitor C_j will either be connected between the nodes x and y if b_j is at the logic high or be floating if b_j is at the logic low. In the PCA of Figure 7.11(b), the capacitive load is maintained constant by switching the capacitor C_j between the node x and ground such that a capacitor never floats. For the series capacitor array depicted in Figure 7.11(c), the total capacitance can be obtained as

$$C_T = \left(\frac{1}{C_{max}} + \sum_{j=1}^{N} \frac{2^{j-1} b_j}{C} \right)^{-1} \tag{7.27}$$

In this case, a given switch will be either open if b_j is at the logic high or closed if b_j is at the logic low.

In practice, the node y is preferably connected to a virtual ground or ground in order to reduce the PCA sensitivity to the digital switching noise. Note also that the high-frequency performance of PCAs can be affected by the parasitic components related to switches.

7.5 Operational amplifiers

An operational amplifier is one of the important active components required in the implementation of SC circuits. In the ideal case, the virtual ground induced on the inverting input node by connecting the noninverting input node to the ground is exploited to cancel the effect of parasitic capacitance on the circuit operation. Two amplifier structures can generally be used in the implementation of SC circuits:

- Operational amplifier (OA), which is equivalent to a voltage-controlled voltage source
- Operational transconductance amplifier (OTA), which is represented as a voltage-controlled current source

Because the load of the amplifiers in this kind of circuit is purely capacitive, both types of devices perform the same operation. However, it appears that designs based on OTAs are more efficient with respect to area and power consumption. The performance parameters of SC circuits, such as speed, dynamic range, and accuracy, are closely linked to the amplifier characteristics. As such, it is essential that the amplifier design reflects a number of considerations, the most important of which are summarized as follows.

Finite gain and bandwidth: The signal gain is frequency dependent. It is approximately constant at low frequencies and is typically in the range of 30 to 90 dB. The frequency at which the gain (which usually decreases in the frequency range of interest with a slope of −20 dB/decade) reaches unity or 0 dB is called unity-gain bandwidth. Generally, unity-gain bandwidths of 1 to 10 MHz are easily achieved in a simple CMOS stage, while it can range up to 100 MHz for amplifiers with a cascode structure.

Transient response time: Because an amplifier in SC circuits is used in a clocked mode, its response time is of considerable importance. If the amplifier output signal is sampled prior to reaching the steady state, the network response will deviate from its specified value and may become distorted. The slew rate and the settling time can be used to determine the amplifier transient behavior.

The *slew rate* can be defined as the maximum rate of change of the amplifier output voltage for a step applied to the input. Its value is about 1 to

15 V/µs for classical amplifier structures and can be increased by one or even two orders of magnitude using slew-enhancement circuit techniques.

The *settling time* is the time required for the amplifier output signal to approximate its final value to within a specified error (usually 0.1 or 1%) when the input signal changes. Typically, it varies from 0.05 to 5 µs.

Linear output range: The transfer characteristic of an amplifier is linear only for a limited range of the input voltage. The maximum output signal is restricted by the supply voltages. If the amplitude of the incoming signal exceeds the maximum amplitude value, the output signal of the network will be clipped to either the minimum level or the maximum level associated with the output signal excursion of the amplifier. This can result in excessive harmonic distortions. In SC circuits, the signal amplitude for which undistorted operation is possible can be maximized by capacitance scaling.

Offset voltage: A practical amplifier can produce an output voltage even if both inputs are grounded. The dc offset voltage is random and can drift, for instance, with temperature. It can be represented by a dc voltage source which is connected in series with the noninverting input of the offset-free amplifier model and can be considered a very low-frequency noise source. The dc offset voltage may occur because of design mask errors and transistor mismatches, and is usually in the 1 to 15 mV range. It can then be minimized using good layout generations or appropriate circuit structures (e.g., correlated double sampling circuits).

For circuits with the amplifier noninverting node connected to the ground, the effect of the dc gain, A_0, and offset voltage, V_{off}, can be estimated by exploiting the following equation

$$V_0 = A_0(V^+ - V^-) \tag{7.28}$$

where $V^+ = V_{off}$. The voltage at the amplifier inverting node can be obtained as

$$V^- = -\mu V_0 + V_{off} \tag{7.29}$$

where $\mu = 1/A_0$.

Other nonideal effects that are present in real amplifiers include, among others, noise, nonzero output resistance, imperfect common-mode signal, and power-supply rejection.

7.6 Track-and-hold (T/H) and sample-and-hold (S/H) circuits

A sample-and-hold circuit is commonly used at the interface between analog and digital systems to hold a sample of the time-varying signal for a period of time so that the high-frequency operation is facilitated. It is commonly realized by combining switching devices, such as MOS transistors, and capacitors.

Ideally, a sample-and-hold circuit takes a sample of the input signal in zero time and holds the sample value during a period T. The output signal obtained in response to a continuous-time unit impulse signal, $\delta(t)$, is known as the impulse response, which can be represented in terms of shifted unit step functions. Thus,

$$h(t) = u_s(t) - u_s(t-T) \tag{7.30}$$

where $u_s(t)$ is the unit step signal. By computing the s-transform of the impulse response, the transfer function, H_{id}, of an ideal sample-and-hold circuit can be obtained as

$$\begin{aligned} H_{id}(s) &= \mathcal{L}[h(t)] \\ &= \mathcal{L}[u_s(t)] - \mathcal{L}[u_s(t-T)] \\ &= (1-\mathrm{e}^{-Ts})\mathcal{L}[u_s(t)] \\ &= \frac{1-\mathrm{e}^{-Ts}}{s} \end{aligned} \tag{7.31}$$

where $\mathcal{L}$ denotes the s-transform. Using $s = j\omega$ and the relationship, $\sin(\omega T/2) = (\mathrm{e}^{j\omega T/2} - \mathrm{e}^{-j\omega T/2})/2j$, it can be shown that

$$\begin{aligned} H_{id}(j\omega) &= \frac{1-\mathrm{e}^{-j\omega T}}{j\omega} \\ &= T\frac{\sin(\omega T/2)}{\omega T/2}\mathrm{e}^{-j\omega T/2} \end{aligned} \tag{7.32}$$

Assuming that $T = 2\pi/\omega_s$, we have

$$H_{id}(j\omega) = \frac{2\pi}{\omega_s}\frac{\sin(\pi\omega/\omega_s)}{\pi\omega/\omega_s}\mathrm{e}^{-j\pi\omega/\omega_s} \tag{7.33}$$

Hence, the magnitude and phase of the transfer function, H_{id}, are given by

$$|H_{id}(j\omega)| = \frac{2\pi}{\omega_s}\left|\frac{\sin(\pi\omega/\omega_s)}{\pi\omega/\omega_s}\right| \tag{7.34}$$

and

$$\angle H_{id}(j\omega) = \begin{cases} -\dfrac{\pi\omega}{\omega_s} & \text{if} \quad \sin\left(\dfrac{\pi\omega}{\omega_s}\right) > 0 \\ -\dfrac{\pi\omega}{\omega_s} + \pi & \text{if} \quad \sin\left(\dfrac{\pi\omega}{\omega_s}\right) < 0 \end{cases} \tag{7.35}$$

Figure 7.12 illustrates the frequency responses of an ideal

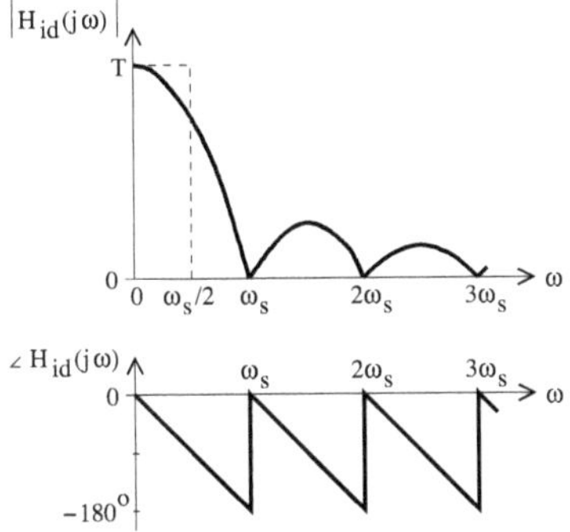

FIGURE 7.12
Frequency responses of an ideal sample-and-hold circuit.

sample-and-hold circuit. In addition to a gain droop, a phase delay is introduced in the passband. By not exhibiting a sharp cutoff frequency response characteristic, the sample-and-hold circuit also transfers the aliased frequency components above one-half of the sampling rate to the output. In general, these image frequencies, whose amplitudes are not sufficiently attenuated, must be removed or attenuated by an appropriate filter.

In practice, the sampling of a signal can be achieved using either a track-and-hold (T/H) circuit or sample-and-hold (S/H) circuit. Because the T/H circuit generally introduces a half-period delay or realizes the transfer function, $z^{-1/2}$, on the signal path, the direct implementation of the S/H or the unit-period delay operator, z^{-1}, requires a cascade of two T/H structures driven by opposite phases of a periodic clock signal, as shown in Figure 7.13(a). The T/H and S/H output signals are depicted in Figure 7.13(b) based on ideal components. The sampling instants are determined by the phases (ϕ_k, $k = 1, 2,$) of the clock signal assumed to exhibit a period T.

A classical T/H circuit, as shown in Figure 7.14(a), consists of a switch, a capacitor, and unity buffer amplifiers, which are necessary to isolate the holding capacitor from the input and output load impedances. An attempt to optimize this T/H architecture using a simple structure for the active buffer and keeping the capacitors involved as small as possible tends to be limited

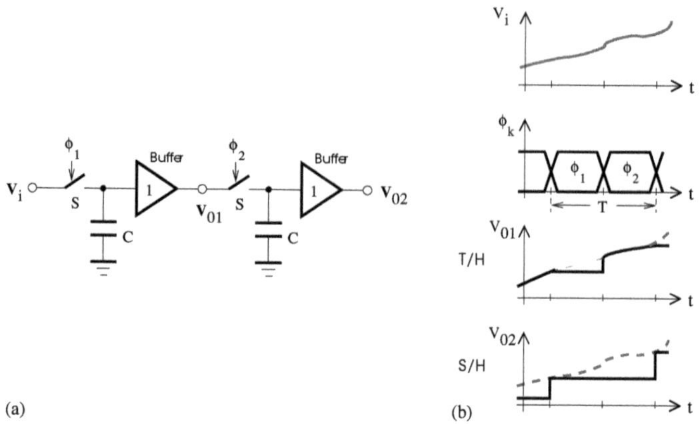

FIGURE 7.13
(a) Circuit diagram of a sampling circuit; (b) ideal representation of the sample-and-hold and track-and-hold output signals.

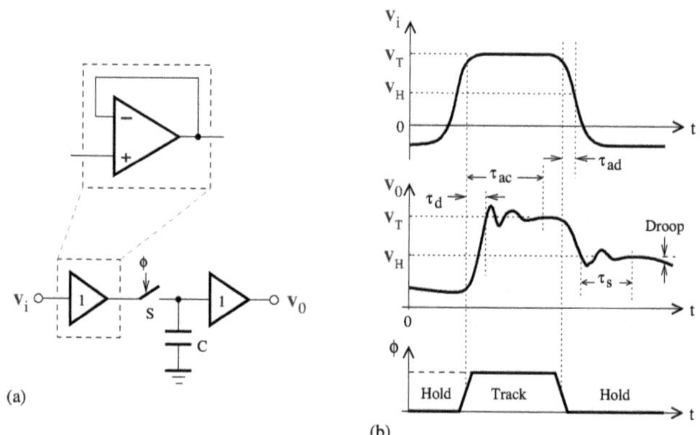

FIGURE 7.14
(a) Circuit diagram of a T/H circuit; (b) transient response of a T/H circuit.

by the exponential relationship existing between the required capacitance and resolution in number of bits. In order for the second buffer amplifier to accurately transmit the sample voltage level to the output, it should exhibit a low input bias current so as not to cause a significant variation in the capacitor charge during the hold mode and a high slew rate to rapidly react to large signal changes.

As the frequency of the input signal increases, the operation of the T/H circuit is affected by various error sources as illustrated in the transient response of Figure 7.14(b). The effects of these errors can be characterized using the following specification parameters.

Hold-to-track delay, τ_d, is the time elapsed from the initiation of the signal sample acquisition to the instant when the output starts to change in response to the input signal.

Acquisition time, τ_{ac}, is the maximum time required to acquire the input signal sample to within a specified error band (e.g., $\pm 0.1\%$ or 1/2 least-significant bit for data converter applications) around the final value of the output.

Effective aperture delay, τ_{ad}, can be defined as the track-to-hold switching delay. It is the time difference between the propagation delays of the input signal and control signal up to the switching instant.

Track-to-hold settling time, τ_s, is the time necessary for the track-to-hold switching transient to settle to within a given error band around its final value.

Droop rate is the variation of the output as a function of the time due to leakage from the hold capacitor. It is generally specified in the hold mode with the input held at a constant dc value.

Feedthrough attenuation ratio is the fraction of the input signal that can appear at the output during the hold mode. It is a measure of the achieved isolation to prevent undesirable coupling of the input signal to the output.

In addition to drift errors, the accuracy of the T/H circuit can be affected by amplifier offset voltages and various noise sources.

In the case where amplifiers are assumed to be ideal, the equivalent model of the T/H circuit in the track mode can be derived as shown in Figure 7.15(a). The switch is modeled by a thermal noise source with the spectral density, $\overline{v_n^2}$, associated with the on-resistance, R_{on}.

• Based on the assumption that the noise signal is filtered by the transfer function, $H(j2\pi f)$, the noise spectral density at the T/H output is given by

$$\overline{v_0^2} = \int_0^{+\infty} |H(j2\pi f)|^2 \overline{v_n^2}\, \mathrm{d}f \tag{7.36}$$

where $\overline{v_n^2} = 4kTR_{on}$, k is Boltzmann's constant, and T is the absolute temperature in Kelvin. The filter transfer function can be expressed as

$$H(j2\pi f) = \frac{1}{1 + j2\pi f R_{on} C} \tag{7.37}$$

Hence,

$$\begin{aligned}\overline{v_0^2} &= 4kTR_{on}\int_0^{+\infty}\frac{1}{1+(2\pi f R_{on}C)^2}\mathrm{d}f \\ &= \frac{2kT}{\pi C}\arctan(2\pi f R_{on}C)\Big|_0^{+\infty} = \frac{kT}{C}\end{aligned} \tag{7.38}$$

Note that $\arctan(0) = 0$ and $\lim_{x\to+\infty}\arctan(x) = \pi/2$. By making the output noise spectral density equal to the quantization noise, the capacitor value can be related to a given resolution. That is,

$$\frac{kT}{C} = \frac{\Delta^2}{12} \tag{7.39}$$

where the least-significant bit (LSB) value can be represented by $\Delta = FSR/2^N$, FSR is the full-scale range, and N is the number of bits. Thus,

$$C = 12kT\left(\frac{2^N}{FSR}\right)^2 \tag{7.40}$$

For applications requiring a high resolution, the capacitor can become impractical to integrate due to the exponential increase of its value with the number of bits.
It should also be noted that the charge droop rate is reduced while the acquisition time tends to become greater as the hold capacitor value is increased. Consequently, the use of larger capacitors is not always adequate.

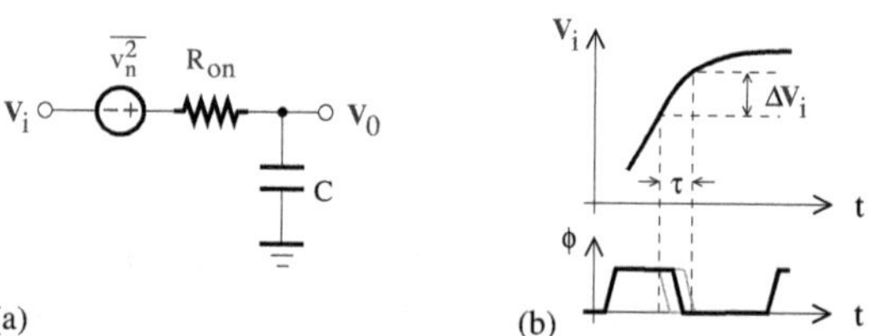

FIGURE 7.15
(a) Equivalent model of the T/H circuit; (b) illustration of the aperture uncertainty.

• Applying Kirchhoff's voltage law around the loop of the T/H equivalent model with the noise source short-circuited yields

$$v_i = R_{on}i(t) + v_0(t) \tag{7.41}$$

where $i(t) = C(dv_O(t)/dt)$. Hence,

$$R_{on}C\frac{dv_O(t)}{dt} + v_O = v_i \tag{7.42}$$

With the assumption that the capacitor is initially discharged, that is, $v_O = 0$ at $t = 0$, it can be shown that

$$v_O(t) = v_i[1 - \exp(-t/R_{on}C)] \tag{7.43}$$

For data converter applications, where the T/H circuit should settle to within the error band of ± 0.5 LSB in $t_s = \epsilon T$, it is required that

$$v_i - v_O(t_s) \ll \triangle/2 \tag{7.44}$$

where ϵ denotes a percentage (usually 50%) of the clock signal period, T. Provided the worst-case condition occurs when the input signal is set to full scale, $v_i = FSR = 2^N\triangle$, and we can obtain

$$2^N \exp(-\epsilon T/R_{on}C) \ll 1/2 \tag{7.45}$$

The switch on-resistance should be sized such that

$$R_{on} \ll \frac{\epsilon T}{2C\ln(2^{N+1})} \tag{7.46}$$

where $T = 1/f_c$ and f_c is the frequency of the clock signal.

• The aperture uncertainty, which is also known as aperture jitter, is the random variations of the occurrence instants of the clock edge (see Figure 7.15). The amplitude of the output error due to this timing deviation can be approximated by

$$\triangle V_i \simeq \frac{\mathrm{d}v_i}{\mathrm{d}t}\tau \tag{7.47}$$

Assuming the random variables are independent, the variance or power associated with the aperture uncertainty is given by

$$P_{\eta_\tau} = \mathrm{E}\left[\left(\frac{\mathrm{d}v_i}{\mathrm{d}t}\right)^2 \tau^2\right] = \mathrm{E}\left[\left(\frac{\mathrm{d}v_i}{\mathrm{d}t}\right)^2\right]\sigma_\tau^2 \tag{7.48}$$

where $\sigma_\tau^2 = \mathrm{E}[\tau^2]$.

Let us consider a sine wave input signal

$$v_i(t) = A\sin(2\pi f\, t) \tag{7.49}$$

where A is the amplitude. The rms value of the rate-of-change of this signal is obtained as

$$\begin{aligned} \mathrm{E}\left[\left(\frac{dv_i}{dt}\right)^2\right] &= \frac{1}{T}\int_0^T (2\pi f\, A)^2 \cos^2(2\pi f\, t)dt \\ &= \frac{(2\pi f\, A)^2}{T}\int_0^T \frac{1+\cos(4\pi f\, t)}{2}dt = \frac{(2\pi f\, A)^2}{2} \end{aligned} \tag{7.50}$$

and the power of the aperture noise becomes

$$P_{\eta_\tau} = 2(\pi f\, A)^2\, \sigma_\tau^2 \tag{7.51}$$

The signal-to-noise ratio (SNR) due to the aperture uncertainty can be computed as

$$\mathrm{SNR}_a = 10\log_{10}\left(\frac{P_{v_i}}{P_{\eta_\tau}}\right), \quad \text{in dB,} \tag{7.52}$$

where $P_{v_i} = \sigma_{v_i}^2 = \mathrm{E}[v_i^2(t)] = A^2/2$ is the input signal power or variance. Hence,

$$\mathrm{SNR}_a = 10\log_{10}\left(\frac{A^2/2}{2(\pi f\, A)^2\, \sigma_\tau^2}\right) = -10\log[(2\pi f)^2\sigma_\tau^2] \tag{7.53}$$

The effect of the aperture uncertainty is considered negligible provided the associated noise contribution remains a few decibels below the thermal noise power.

7.6.1 Open-loop T/H circuit

An open-loop T/H circuit, that is implemented as shown in Figure 7.16(a), can be considered as a linear time-variant system. Its equivalent model is represented in Figure 7.16(b), where R is the total input resistance (or the switch ON-resistance R_{on} in series with the input signal source resistance). Waveforms illustrating the operation of the T/H circuit are depicted in Figure 7.16(c).

In the time domain, assuming an ideal buffer, the T/H circuit can be described by the following equations

$$\frac{dv_O(t)}{dt} = -\frac{1}{RC}v_O(t) + \frac{1}{RC}v_i(t) \quad \text{for} \quad nT \le t \le nT + \tau \tag{7.54}$$

and

$$\frac{dv_O(t)}{dt} = 0 \quad \text{for} \quad nT + \tau \le t < (n+1)T \tag{7.55}$$

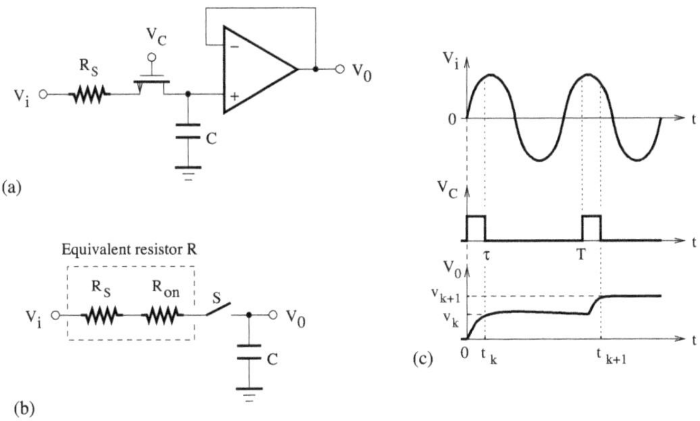

FIGURE 7.16
(a) T/H circuit implementation; (b) equivalent model; (c) waveform representation.

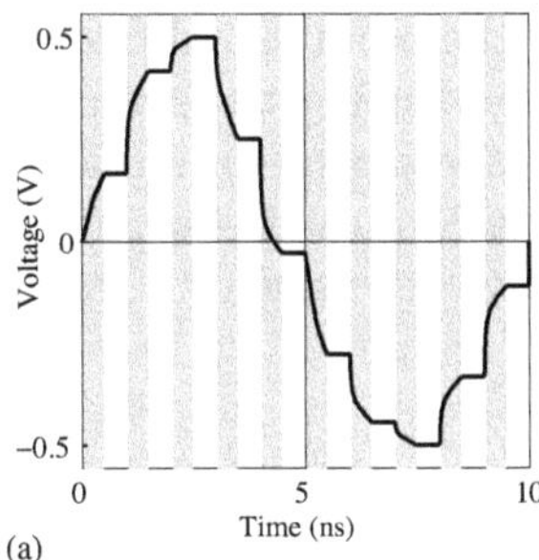

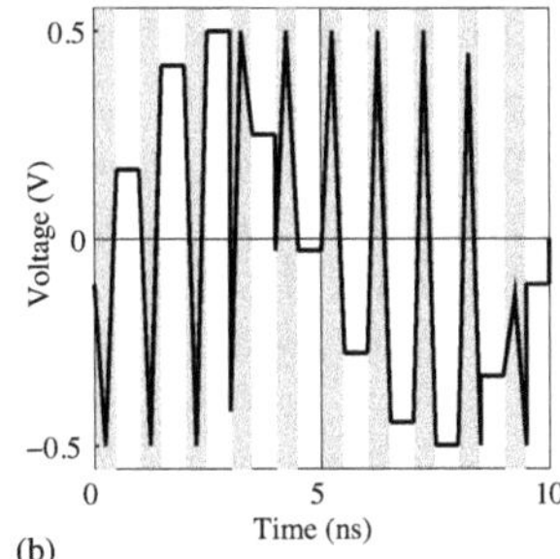

FIGURE 7.17
T/H output waveform representation for a ratio of the input signal frequency to the sampling frequency of 0.1 (a) and 1.1 (b).

where $T = 2\pi/\omega_s$ is the period of the sampling signal. Solving the system of equations (7.54) and (7.55) and using Fourier transform, we arrive at (see Appendix C)

$$V_0(\omega) = \sum_{n=0}^{+\infty} H_n(\omega) V_i(\omega - n\omega_s) \tag{7.56}$$

where

$$H_n(\omega) = \frac{1}{1 + j\omega/\omega_c} \left[\frac{1 - e^{-jn\omega_s \tau}}{jn\omega_s T} - \frac{1 - e^{-j\omega(T-\tau)}}{j\omega T} G(\omega - n\omega_s) \right] \tag{7.57}$$

and

$$G(\omega) = \frac{1}{1 + j\omega/\omega_c} \frac{1 - e^{-\omega_c \tau} e^{-j\omega\tau}}{1 - e^{-\omega_c \tau} e^{-j\omega T}} e^{-j\omega(T-\tau)} \tag{7.58}$$

with $\omega_c = 1/RC$ being the cutoff frequency of the equivalent RC circuit. The output contribution due to the fundamental frequency of the input signal can be estimated using the zeroth-order transfer function given by

$$H_0(\omega) = \frac{1}{1 + j\omega/\omega_c}\left[\frac{\tau}{T} - \frac{1 - e^{-j\omega(T-\tau)}}{j\omega T}G(\omega)\right] \tag{7.59}$$

When $\omega \ll \omega_c$, $1/(1 + j\omega/\omega_c) \simeq 1$, and using $1 - e^{-jx} = 2j\sin(x/2)\mathrm{e}^{-jx/2}$, we obtain

$$H_0(\omega) \simeq \frac{\tau}{T} - \left(1 - \frac{\tau}{T}\right)\frac{\sin\left(\pi\frac{\omega}{\omega_s}\left(1 - \frac{\tau}{T}\right)\right)}{\pi\frac{\omega}{\omega_s}\left(1 - \frac{\tau}{T}\right)}\mathrm{e}^{-j\pi(\omega/\omega_s)(1-\tau/T)}G(\omega) \tag{7.60}$$

where τ/T represents the duty cycle. Note that $\sin(\pi x)/(\pi x)$ is also known as $\mathrm{sinc}(x)$.

Assuming a sinusoidal input signal, Figures 7.17(a) and (b) show the T/H output waveforms for two different ratios of the input signal frequency to the sampling frequency, 0.1 and 1.1, respectively. During the sampling interval, the signal undergoes more rapid changes in the second case than in the first one, because its frequency is higher than the Nyquist frequency (or half the sampling frequency). However, the input sample levels held by the T/H circuit are identical in both cases.

7.6.2 Closed-loop T/H circuits

The T/H circuit accuracy can be improved using closed-loop configurations [16], which are generally known to exhibit a lower speed than open-loop structures due to amplifier stability constraints.

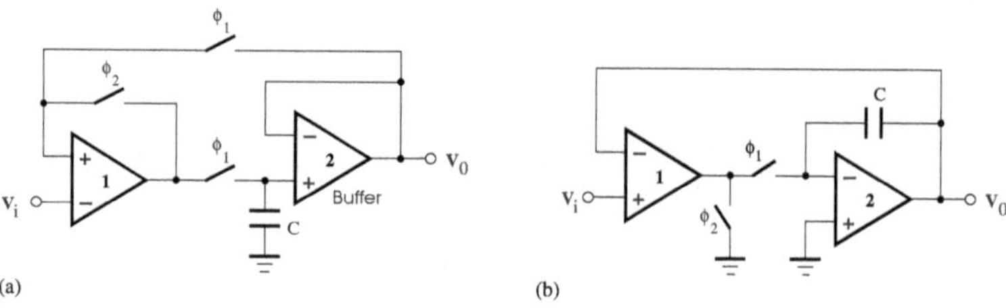

FIGURE 7.18
Circuit diagrams of closed-loop T/H circuits.

In the closed-loop T/H circuit of Figure 7.18(a), the signal is sampled during the clock phase ϕ_1. During the hold phase, ϕ_2, the feedback path between the amplifiers is open, the input amplifier is configured as a buffer to prevent any instability due to an abrupt change in the output level, and the charge stored on the capacitor C is maintained. One disadvantage of this T/H circuit is the undesirable variation in the charge stored on the capacitor with time due to the capacitor leakage current and amplifier input bias current.

A reduction in the output voltage droop can be achieved using the alternative T/H structure depicted in Figure 7.18(b). The capacitor used for storage of the input signal sample is connected between the virtual node and output of the second amplifier. In this way, its charge remains insensitive to the effect of grounded parasitic capacitances. The connection of the first amplifier output to the ground during the hold mode helps reduce the signal feedthrough.

7.7 Switched-capacitor (SC) circuit principle

Switched-capacitor (SC) circuits provide a high-performance solution for the resistor implementations in the analog discrete-time domain. The idea is to use a capacitor C periodically switched between two circuit nodes as shown in Figure 7.19 [12]. Each switch[1] is controlled by one of the clock waveforms,

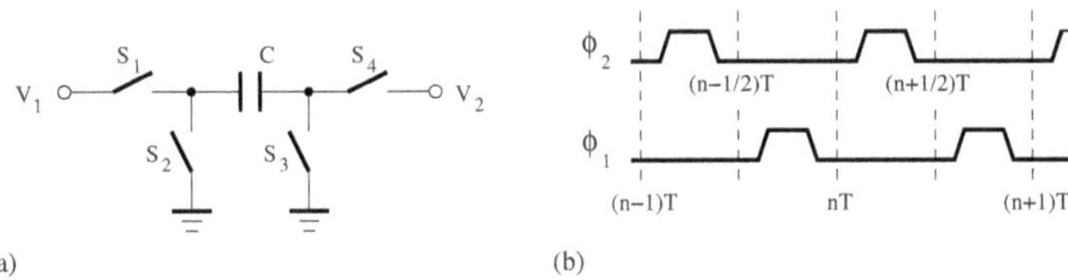

FIGURE 7.19
SC implementation of a resistor: (a) SC branch equivalent to a resistor, (b) switch clock waveforms.

ϕ_1 (or simply 1) or ϕ_2 (or 2), which are the two phases of a nonoverlapping clock signal with the frequency f_c and period T. The circuit operation is as follows: When S_2 and S_3 are closed, C is discharged; when afterwards S_1 and S_4 close, C charges to the voltage $v_1(nT) - v_2(nT)$ and the charge $\Delta q(nT) = C[v_1(nT) - v_2(nT)]$ flows from the input to the output node of the branch. With the assumption that the voltage signals are sampled at a sufficiently high rate, v_1 and v_2 can be considered constant over the sampling period. Thus, the average current transferred during one period is given by

$$\bar{i}(nT) \simeq \frac{\Delta q(nT)}{T} = \frac{v_1(nT) - v_2(nT)}{T/C} \tag{7.61}$$

[1]The simplest realization of a switch in MOS technology is a single transistor. Its drain and source are connected to the nodes to be periodically opened and closed, and the clock signal is applied to the gate.

By analogy to Ohm's law for resistors, it can be concluded that a resistor of value $R \simeq T/C$ has been simulated.

The circuit of Figure 7.19, as described above, operates with a direct path between the input and output nodes. But, if the pair of switches controlled by one clock phase is now S_1 and S_3 or S_2 and S_4, the charge transfer follows an indirect path. In this way, a negative resistor can be implemented.

The direct implementation of a 10-MΩ resistor using an integrated circuit (IC) process with a sheet resistivity of polysilicon lines about 50 Ω/square needs 10^6 μm^2 chip area. In an SC design, only an area of approximately 2.10^3 μm^2 associated with a 1 pF capacitor is necessary. It has been assumed that the switches are minimum-size devices and operate with a clock frequency of 100 kHz. However, the equivalence between a switched-capacitor branch and a resistor relies on an approximation. The errors due to the use of this principle in the design can become dominant for some circuits [13,14]. For this reason, the accurate analysis of SC circuits should be done in the discrete-time domain.

The circuit diagram of an inverting SC integrator is shown in Figure 7.20, while its continuous-time equivalent models during the phases ϕ_1 and ϕ_2 are respectively depicted in Figures 7.21(a) and (b). The input voltage is piecewise

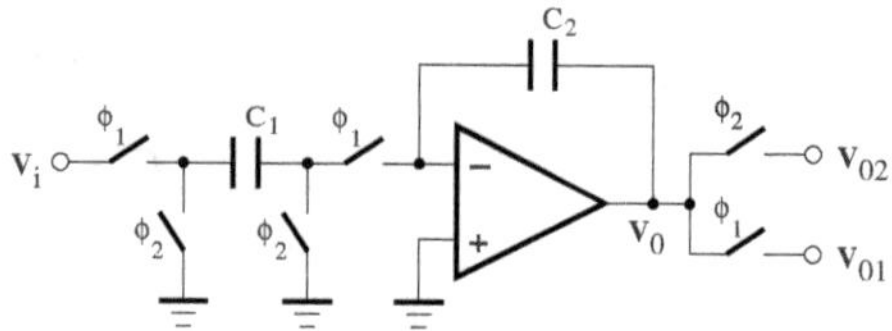

FIGURE 7.20
Circuit diagram of an SC inverting integrator.

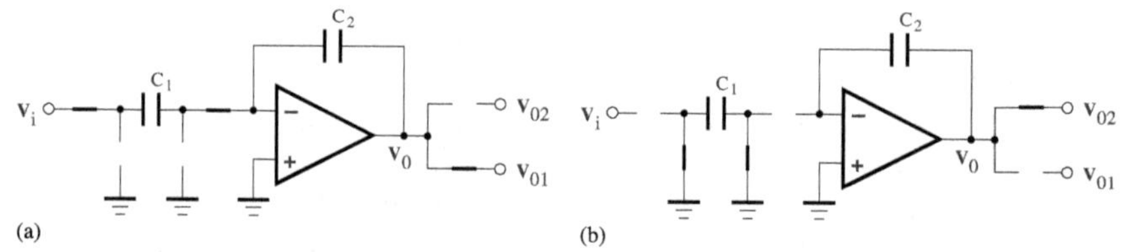

FIGURE 7.21
Continuous-time equivalent models of the SC integrator during phase (a) ϕ_1 and (b) ϕ_2.

constant and is allowed to change at the beginning of every clock phase. The amplifier is assumed to be ideal.

During the second clock phase, that is, for $(n-1)T \leq t < (n-1/2)T$, the capacitor C_1 is discharged to the ground and $V_i[(n-1/2)T] = 0$. The amplifier and the capacitor C_2 have been isolated since the time $(n-1)T$; thus the voltage V_0 has maintained its value,

$$C_2V_0[(n-1/2)T] = C_2V_0[(n-1)T] \tag{7.62}$$

During the clock phase 1, that is, for $(n-1/2)T \leq t < nT$, the application of the charge conservation principle at the inverting node of the amplifier results in

$$C_1\{V_i(nT) - V_i[(n-1/2)T]\} + C_2\{V_0(nT) - V_0[(n-1/2)T]\} = 0 \tag{7.63}$$

By substituting Equation (7.62) into (7.63), the following difference equation can be obtained:

$$-C_1V_i(nT) = C_2\{V_0(nT) - V_0[(n-1)T]\} \tag{7.64}$$

Assuming that the first clock phase is used for the output sampling, we have $V_{01}(nT) = V_0(nT)$, and the transfer function in the z-domain reads

$$H_1(z) = \frac{V_{01}(z)}{V_i(z)} = -\frac{C_1}{C_2}\frac{1}{1-z^{-1}} \tag{7.65}$$

On the other hand, if the output signal is sampled using the second clock phase, that is, $V_{02}(nT) = V_0[(n-1/2)T]$, the integrator transfer function will then become

$$H_2(z) = \frac{V_{02}(z)}{V_i(z)} = -\frac{C_1}{C_2}\frac{z^{-1/2}}{1-z^{-1}} \tag{7.66}$$

The $z^{-1/2}$ term in the transfer function numerator corresponds to the half-period delay introduced between the sampling instants of the input signal and the observation instants of the output signal.

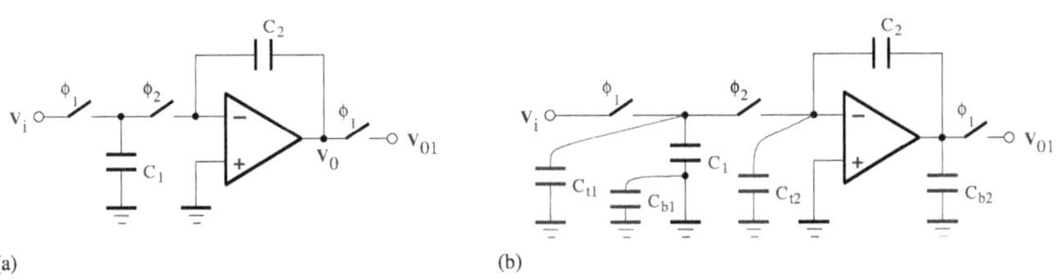

FIGURE 7.22
(a) Stray-sensitive SC integrator; (b) stray-sensitive SC integrator with parasitic capacitances.

An SC integrator can also be implemented as shown in Figure 7.22(a) [21]. Ideally, the capacitor C_1 is first charged to the input signal and then connected to the inverting node of the amplifier. Its switching rate is fixed by a clock signal with two nonoverlapping phases. The charge conservation equation at the end of the second clock phase, ϕ_2, can be written as

$$C_2\{V_0(nT) - V_0[(n-1/2)T]\} = -C_1 V_i[(n-1/2)T] \tag{7.67}$$

During the first clock phase, ϕ_1, we have

$$C_2 V_0[(n-1/2)T] = C_2 V_0[(n-1)T] \tag{7.68}$$

Combining Equations (7.67) and (7.68) yields

$$C_2\{V_0(nT) - V_0[(n-1)T]\} = -C_1 V_i[(n-1/2)T] \tag{7.69}$$

Assuming that the output signal is sampled at the beginning of the phase ϕ_1, $V_0(nT) = V_{01}[(n+1/2)T]$, or $V_0(z) = z^{1/2}V_{01}(z)$, the z-domain transfer function is obtained as

$$H(z) = \frac{V_{01}(z)}{V_i(z)} = -\frac{C_1}{C_2}\frac{z^{-1}}{1-z^{-1}} \tag{7.70}$$

In practice, a parasitic capacitance on the order of 0.1 to 1% of the desired capacitor value exists between the top plate and the substrate, while the one associated with the bottom plate and the substrate may be as high as 20%. These parasitic capacitances are illustrated in the SC integrator circuit diagram of Figure 7.22(b). The effect of parasitic capacitors, which are connected between the ground and either the virtual ground or the amplifier output, can be neglected. The accuracy of the SC integrator is then only dependent on parasitic capacitors, which are in parallel with C_1. With the equivalent input capacitor being $C_1 + C_p$, where the equivalent top plate parasitic capacitor is reduced to a single, lumped capacitor $C_p = C_{t1}$, the transfer function can take the form

$$H(z) = \frac{V_{01}(z)}{V_i(z)} = -\frac{C_1}{C_2}\left(1 + \frac{C_p}{C_1}\right)\frac{z^{-1}}{1-z^{-1}} \tag{7.71}$$

Thus, the capacitance ratio or integrator time constant is affected by a relative error on the order of a few percent.

Now consider the double-sampling SC integrator of Figure 7.23. In contrast to the previous one, the sampling of the input signal along with the integration is realized during both clock phases. The charge conservation law can be written as

$$-C_1' v_i[(n-1/2)T] = C_2\{v_0[(n-1/2)T] - v_0[(n-1)T]\} \tag{7.72}$$

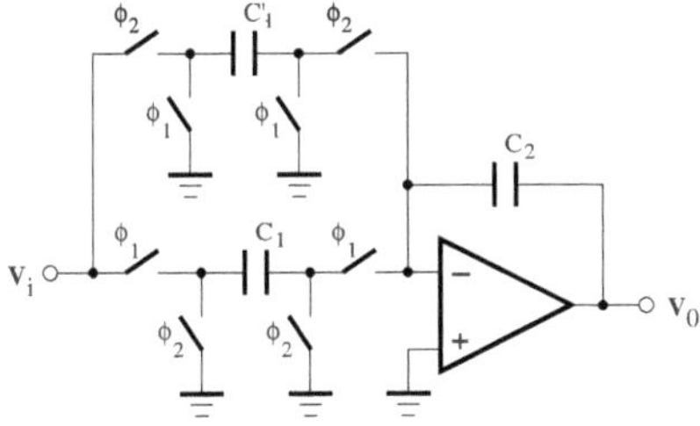

FIGURE 7.23
Double-sampling inverting SC integrator.

during the clock phase 2 and

$$-C_1 v_i(nT) = C_2\{v_O(nT) - v_O[(n-1/2)T]\} \tag{7.73}$$

during the clock phase 1. From the substitution of Equation (7.72) into (7.73), we have

$$-C_1 v_i(nT) - C_1' v_i[(n-1/2)T] = C_2\{v_O(nT) - v_O[(n-1)T]\} \tag{7.74}$$

and therefore, the z-domain transfer function is given by

$$H(z) = \frac{V_O(z)}{V_i(z)} = -\frac{C_1 + C_1' z^{-1/2}}{C_2(1-z^{-1})} \tag{7.75}$$

Note that if $C_1 = C_1'$, the function $H(z)$ can be deduced from $H_1(z)$ using the transformation $z \mapsto z^{-1/2}$.

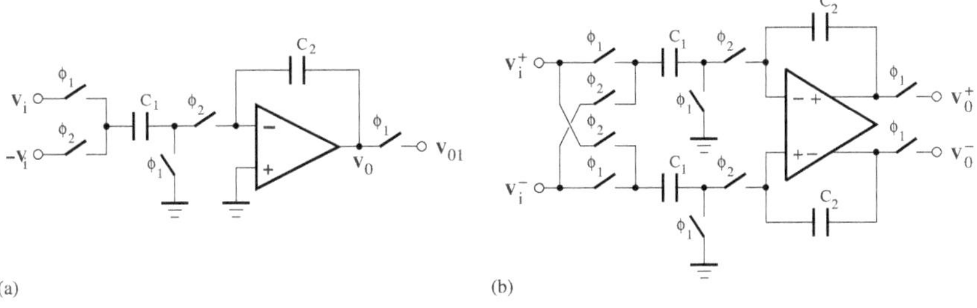

FIGURE 7.24
SC integrator: (a) Single-ended and (b) differential architectures.

Consider the SC integrator structure shown in Figure 7.24(a). During the first clock phase, ϕ_1, or for $(n-1)T \le t < (n-1/2)T$, the capacitor C_1 is connected between the input voltage and ground, while the feedback capacitor C_2 maintains the output voltage constant provided no other capacitor is connected to the amplifier. Hence,

$$C_2 v_O[(n-1/2)T] = C_2 v_O[(n-1)T] \tag{7.76}$$

During the second clock phase, ϕ_2, or for $(n-1/2)T < t \leq nT$, the charges due to $-v_i$ in addition to the one previously stored on C_1 are transferred to C_2. By applying the charge conservation rule, we obtain

$$\begin{aligned} C_1\{-v_i(nT) - v_i[(n-1/2)T]\} \\ = -C_2\{v_O(nT) - v_O[(n-1/2)T]\} \end{aligned} \tag{7.77}$$

Combining Equations (7.76) and (7.77) gives

$$\begin{aligned} C_1\{v_i(nT) + v_i[(n-1/2)T]\} \\ = C_2\{v_O(nT) - v_O[(n-1)T]\} \end{aligned} \tag{7.78}$$

With the assumption that $v_{O1}[(n+1/2)T] = v_O(nT)$, the transfer function in the z-domain is then given by

$$H(z) = \frac{V_{O1}(z)}{V_i(z)} = \frac{C_1}{C_2}\frac{z^{-1/2} + z^{-1}}{1 - z^{-1}} \tag{7.79}$$

For slow-moving signals, it can be assumed that the value of the input sample is updated only at the beginning of the first clock phase and remains constant during the second clock phase. In this case, $v_i[(n-1/2)T] = v_i[(n-1)T]$ and the integrator transfer function becomes

$$H(z) = \frac{V_{O1}(z)}{V_i(z)} = \frac{C_1}{C_2} z^{-1/2}\frac{1 + z^{-1}}{1 - z^{-1}} \tag{7.80}$$

Except for the $z^{-1/2}$ term, this last transfer function can directly be transformed to the one of the continuous-time counterpart using the bilinear transform. Figure 7.24(a) shows the differential implementation of the SC integrator.

A variation up to ±20% can be observed in absolute capacitance values. But, with appropriate layout techniques (e.g., common-centroid layout), the matching accuracy achievable in a capacitor ratio can be as low as 0.1%. Generally, the clock signal is generated by a quartz-crystal oscillator, which is known to have excellent stability and accuracy. SC techniques result in high-precision circuits because their transfer function coefficients are ideally determined by the clock frequency and capacitance ratios.

7.8 SC filter design

The design of SC filters is generally based on the stray-insensitive building blocks shown in Figure 7.25(a), and the transfer functions from the different inputs to the output are equivalent to various types of discrete-time integration and summation, as depicted in Figure 7.25(b).

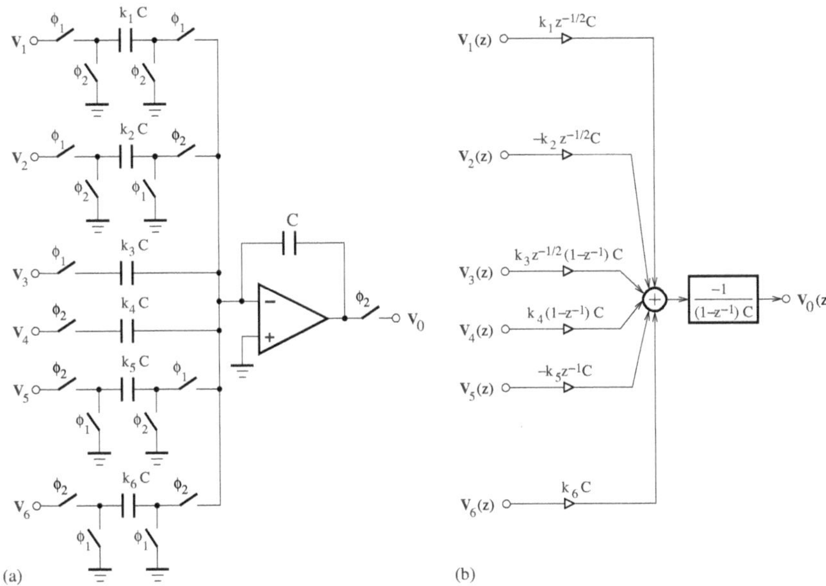

FIGURE 7.25
(a) Circuit diagram of SC building blocks and (b) the corresponding block diagrams.

The filter synthesis should be achieved in terms of discrete-time transfer functions to avoid the approximations related to the equivalence of SC circuits to continuous-time networks. Given a filter specification in the s-domain, the corresponding transfer function, $H(s)$, is mapped into the z-domain, resulting in $H(z)$, which is generally implemented using two basically different structures, the cascade of low-order sections (for biquad design, see [22–26]) and the LC ladder. The s-z mapping can be achieved using the bilinear transform, which preserves the stability and is defined as

$$s = \frac{2}{T}\frac{1 - z^{-1}}{1 + z^{-1}} \tag{7.81}$$

where T is the period of the clock signal. With $s = j\omega$ and $z = e^{j\theta}$, we have

$$\omega = \frac{2}{T}\tan\frac{\theta}{2} \tag{7.82}$$

The frequency of the analog prototype is compressed into an interval of the form $-\pi/T \leq \omega \leq +\pi/T$. The frequency mapping provided by the bilinear transform is nonlinear and the resulting frequency can be distorted, especially at high frequencies.

In general, the SC filter design should consist of the synthesis, the ordering of low-order filter sections in a cascade realization for the maximum dynamic range, and the capacitance assignment steps. The synthesis results in a signal-flow graph that realizes the filter transfer function, and the objective of the capacitance assignment, which can be carried out by an optimization scheme, is to improve the signal swing and reduce the sensitivity to the component nonidealities.

7.8.1 First-order filter

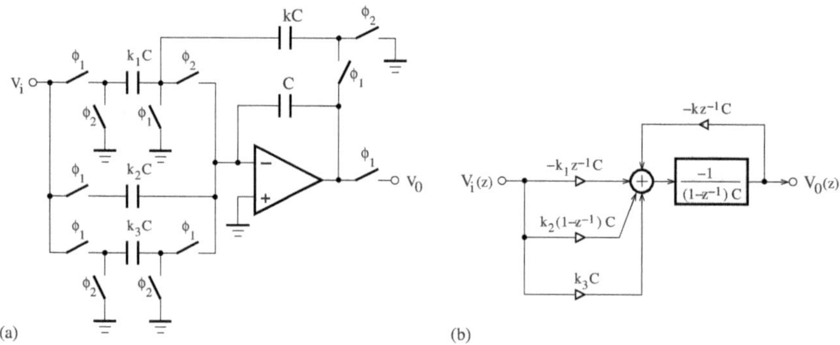

FIGURE 7.26
(a) Circuit diagram of a first-order SC filter; (b) first-order SC filter SFG.

First-order filter sections are required in the synthesis of transfer functions relying on the cascade design method. A generic structure for the first-order SC filter section is depicted in Fig 7.26(a). Based on the associated signal-flow graph (SFG) shown in Fig 7.26(b), its transfer function is given by

$$H(z) = \frac{V_0(z)}{V_i(z)} = \frac{k_1 z^{-1} - k_2(1 - z^{-1}) - k_3}{1 - (1 + k)z^{-1}} \tag{7.83}$$

The type of the resulting filter response is determined by the choice of capacitor values as follows:

- Lowpass filter: $k_2 = k_3 = 0$

$$H_{LP}(z) = \frac{V_0(z)}{V_i(z)} = \frac{k_1 z^{-1}}{1 - (1 + k)z^{-1}} \tag{7.84}$$

- Highpass filter: $k_1 = k_3 = 0$

$$H_{HP}(z) = \frac{V_0(z)}{V_i(z)} = \frac{-k_2(1 - z^{-1})}{1 - (1 + k)z^{-1}} \tag{7.85}$$

- Allpass filter: $k_1 = 0$, $k_2 = 1$ and $k_3 = k$

$$H_{AP}(z) = \frac{V_0(z)}{V_i(z)} = \frac{z^{-1} - (1 + k)}{1 - (1 + k)z^{-1}} \tag{7.86}$$

Thus, the aforementioned circuit is capable of realizing any first-order filter function.

7.8.2 Biquad filter

Let Q and ω_0 be the pole frequency and quality factor, respectively. The s-domain transfer function of a second-order filter is given by

$$H(s) = \frac{N(s)}{D(s)} = \frac{N(s)}{1 + \frac{1}{Q}\frac{s}{\omega_0} + \left(\frac{s}{\omega_0}\right)^2} \tag{7.87}$$

The type of the frequency response depends on the numerator polynomial, $N(s)$. Using the bilinear transform, the equivalent discrete-time transfer function can be written as

$$H(z) = \frac{N(z)}{D(z)} = \frac{N(z)}{1 - (2 - a - b)z^{-1} + (1 - b)z^{-2}} \tag{7.88}$$

where

$$a = \frac{4(\omega_0 T)^2}{4 + 2(\omega_0 T/Q) + (\omega_0 T)^2} \tag{7.89}$$

and

$$b = \frac{4(\omega_0 T/Q)}{4 + 2(\omega_0 T/Q) + (\omega_0 T)^2} \tag{7.90}$$

The poles in the z-domain can be expressed in polar form as

$$z_{p1,p2} = re^{\pm j\theta} \tag{7.91}$$

where $0 \le \theta \le \pi$. Note that

$$a = 1 - 2r\cos\theta + r^2 \tag{7.92}$$

TABLE 7.1
Numerators of Common Second-Order Filter Sections

Filter Type	$N(s)$	$N(z)$
Lowpass	$\left(1+\frac{s}{\omega_z}\right)^2$	a
	$\left(1-\frac{s}{\omega_z}\right)\left(1+\frac{s}{\omega_z}\right)$	az^{-1}
	$\left(1-\frac{s}{\omega_z}\right)^2$	az^{-2}
	$1+\frac{s}{\omega_z}$	$\frac{a}{2}(1+z^{-1})$
	$1-\frac{s}{\omega_z}$	$\frac{a}{2}z^{-1}(1+z^{-1})$
	1	$\frac{a}{4}(1+z^{-1})^2$
Bandpass	$\frac{s}{Q\omega_0}\left(1+\frac{s}{\omega_z}\right)$	$b(1-z^{-1})$
	$\frac{s}{Q\omega_0}\left(1-\frac{s}{\omega_z}\right)$	$bz^{-1}(1-z^{-1})$
	$\frac{s}{Q\omega_0}$	$\frac{b}{2}(1-z^{-1})(1+z^{-1})$
Highpass	$\left(\frac{s}{\omega_0}\right)^2$	$\frac{a}{\omega_0^2T^2}(1-z^{-1})^2$
Notch	$\left(\frac{\omega_1}{\omega_0}\right)^2\left[1+\left(\frac{s}{\omega_1}\right)^2\right]$	$\left(\frac{\omega_1}{\omega_0}\right)^2\frac{a}{a'}[1-(2-a')z^{-1}+z^{-2}]$
Allpass	$1-\frac{s}{Q\omega_0}+\left(\frac{s}{\omega_0}\right)^2$	$(1-b)-(2-a-b)z^{-1}+z^{-2}$

and

$$b = 1 - r^2 \tag{7.93}$$

and the stability requirement is met for $0 \le r < 1$. With

$$\omega_z = \frac{2}{T} \tag{7.94}$$

and

$$a' = \frac{4(\omega_1 T)^2}{1 + (\omega_1 T)^2} \tag{7.95}$$

where T is the clock period. Table 7.1 gives a summary of the different filter types and the corresponding numerators. Three types of notch filters can be distinguished: a lowpass notch for $\omega_1 > \omega_0$, a symmetrical notch for $\omega_1 = \omega_0$, and a highpass notch for $\omega_1 < \omega_0$.

The general circuit diagram of the SC biquad is depicted in Figure 7.27. With reference to the corresponding SFG shown in Figure 7.28, the SC biquad can be described by the following set of equations:

$$\begin{aligned} V_0(z) = &-\frac{[k_5 + k_4 - (k_4 + k_6)z^{-1}]V_i(z)}{D_2(z)} \\ &+ \frac{[k_1 + k_2 - (k_1 + k_3)z^{-1}][k_{13} + k_{12} - (k_{12} + k_{14})z^{-1}]V_i(z)}{D_1(z)D_2(z)} \\ &+ \frac{[k_9 + k_{10} - (k_9 + k_{11})z^{-1}][k_{13} + k_{12} - (k_{12} + k_{14})z^{-1}]V_0(z)}{D_1(z)D_2(z)} \end{aligned} \tag{7.96}$$

and

$$\begin{aligned} V_1(z) = &-\frac{[k_1 + k_2 - (k_1 + k_3)z^{-1}]V_i(z)}{D_1(z)} \\ &+ \frac{[k_5 + k_4 - (k_4 + k_6)z^{-1}][k_9 + k_{10} - (k_9 + k_{11})z^{-1}]V_i(z)}{D_1(z)D_2(z)} \\ &+ \frac{[k_{13} + k_{12} - (k_{12} + k_{14})z^{-1}][k_9 + k_{10} - (k_9 + k_{11})z^{-1}]V_1(z)}{D_1(z)D_2(z)} \end{aligned} \tag{7.97}$$

where

$$D_1(z) = 1 + k_7 - (1 + k_8)z^{-1} \tag{7.98}$$

and

$$D_2(z) = 1 + k_{15} - (1 + k_{16})z^{-1} \tag{7.99}$$

We can then find the next transfer functions

$$H_0(z) = \frac{V_0(z)}{V_i(z)} = \frac{\alpha_0 + \alpha_1 z^{-1} + \alpha_2 z^{-2}}{\gamma_0 + \gamma_1 z^{-1} + \gamma_2 z^{-2}} \tag{7.100}$$

and

$$H_1(z) = \frac{V_1(z)}{V_i(z)} = \frac{\beta_0 + \beta_1 z^{-1} + \beta_2 z^{-2}}{\gamma_0 + \gamma_1 z^{-1} + \gamma_2 z^{-2}} \tag{7.101}$$

where

$$\alpha_0 = (k_1 + k_2)(k_{12} + k_{13}) - (1 + k_7)(k_4 + k_5) \tag{7.102}$$

$$\alpha_1 = (1 + k_8)(k_4 + k_5) + (1 + k_7)(k_4 + k_6) - (k_1 + k_2)(k_{12} + k_{14}) - (k_1 + k_3)(k_{12} + k_{13}) \tag{7.103}$$

$$\alpha_2 = (k_1 + k_3)(k_{12} + k_{14}) - (1 + k_8)(k_4 + k_6) \tag{7.104}$$

$$\beta_0 = (k_4 + k_5)(k_9 + k_{10}) - (1 + k_{15})(k_1 + k_2) \tag{7.105}$$

$$\beta_1 = (1 + k_{16})(k_1 + k_2) + (1 + k_{15})(k_1 + k_3) - (k_4 + k_5)(k_9 + k_{11}) - (k_4 + k_6)(k_9 + k_{10}) \tag{7.106}$$

$$\beta_2 = (k_4 + k_6)(k_9 + k_{11}) - (1 + k_{16})(k_1 + k_3) \tag{7.107}$$

$$\gamma_0 = (1 + k_7)(1 + k_{15}) - (k_9 + k_{10})(k_{12} + k_{13}) \tag{7.108}$$

$$\gamma_1 = (k_9 + k_{11})(k_{12} + k_{13}) + (k_9 + k_{10})(k_{12} + k_{14}) - (1 + k_8)(1 + k_{15}) - (1 + k_7)(1 + k_{16}) \tag{7.109}$$

and

$$\gamma_2 = (1 + k_8)(1 + k_{16}) - (k_9 + k_{11})(k_{12} + k_{14}) \tag{7.110}$$

The denominator coefficients of the z-domain transfer functions can be related to the pole frequency, ω_0, and Q factor of an analog filter according to

$$\omega_0 = \frac{2}{T}\sqrt{\frac{\gamma_0 + \gamma_1 + \gamma_2}{\gamma_0 - \gamma_1 + \gamma_2}} \tag{7.111}$$

and

$$Q = \frac{\sqrt{(\gamma_0 + \gamma_1 + \gamma_2)(\gamma_0 - \gamma_1 + \gamma_2)}}{2(\gamma_0 - \gamma_2)} \tag{7.112}$$

A specific filter type can be realized by selecting the appropriate capacitors to define the numerator polynomial. Note that, with reference to the SFG of Figure 7.25(b), other transfer functions can be obtained by interchanging the clock phases of some switches. Furthermore, to reduce the transfer function sensitivity to some capacitance ratios, it may be necessary to realize a biquad with more than two amplifiers [27].

For a given design, the signal levels at integrator outputs must be set by scaling the capacitances such that the amplifiers can operate with an optimal dynamic range. The maximum output signal is limited by the saturation voltage of the amplifier, while the minimum useful voltage is bounded by the noise level.

Power-efficient SC biquads are designed by cascading inverting and noninverting integrators. In this way, during one clock phase, the feedback loop around the amplifiers is not fully coupled, and the settling requirement of each integrator is relaxed.

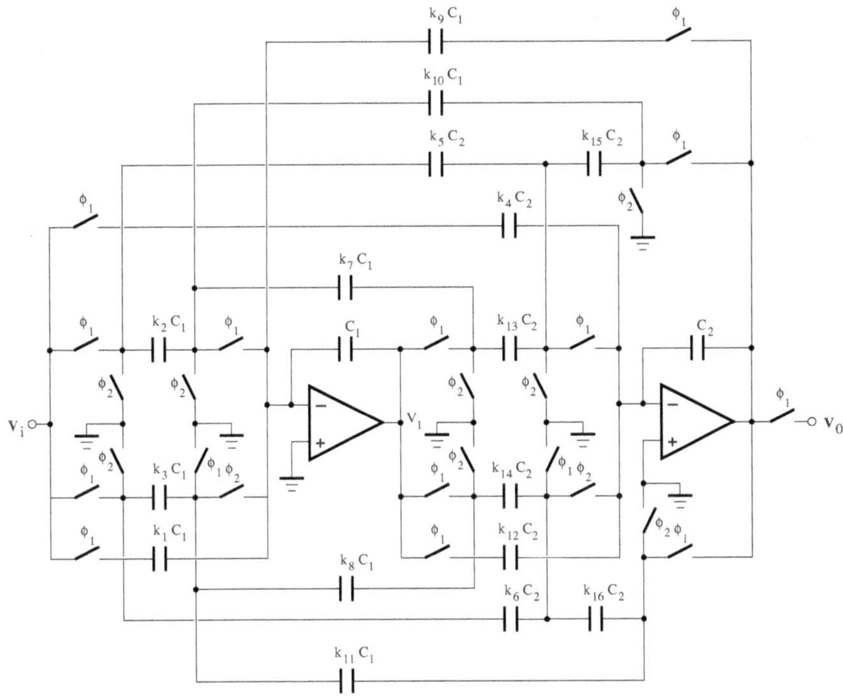

FIGURE 7.27
Circuit diagram of an SC biquad.

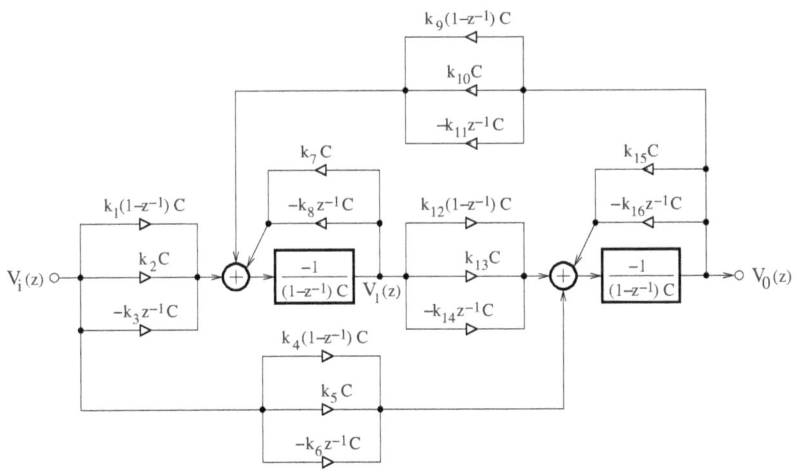

FIGURE 7.28
SFG of the SC biquad.

Principle of capacitance scaling for the optimal dynamic range

Given a second-order transfer function,

$$H(z) = \frac{\delta_0 + \delta_1 z^{-1} + \delta_2 z^{-2}}{\gamma_0 + \gamma_1 z^{-1} + \gamma_2 z^{-2}} \tag{7.113}$$

where $z = e^{j\omega T}$, the maximum value of $H(z)$ is required for the capacitance assignment that maximizes the dynamic range. That is,

$$\left.\frac{d|H(e^{\omega T})|}{d(\omega T)}\right|_{\omega=\omega_m} = 0. \tag{7.114}$$

The resulting equation is given by

$$\cos^2(\omega_m T) + \frac{\kappa}{\eta}\cos(\omega_m T) + \frac{\lambda}{2\eta} = 0 \tag{7.115}$$

where

$$\begin{aligned}\lambda = {} & 2\delta_0\delta_2\gamma_1(\gamma_0 + \gamma_2) - 2\delta_1\gamma_0\gamma_2(\delta_0 + \delta_2) \\ & + \delta_1(\delta_0 + \delta_2)(\gamma_0^2 + \gamma_1^2 + \gamma_2^2) - \gamma_1(\gamma_0 + \gamma_2)(\delta_0^2 + \delta_1^2 + \delta_2^2)\end{aligned} \tag{7.116}$$

$$\eta = 2\delta_0\delta_2\gamma_1(\gamma_0 + \gamma_2) - 2\delta_1\gamma_0\gamma_2(\delta_0 + \delta_2) \tag{7.117}$$

and

$$\kappa = 2\delta_0\delta_2(\gamma_0^2 + \gamma_1^2 + \gamma_2^2) - 2\gamma_0\gamma_2(\delta_0^2 + \delta_1^2 + \delta_2^2) \tag{7.118}$$

Provided $(\kappa/2\eta)^2 - \lambda/2\eta \geq 0$, Equation (7.115) can be solved to

$$\omega_m = \frac{1}{T}\arccos\left(-\frac{\kappa}{2\eta} \pm \sqrt{\left(\frac{\kappa}{2\eta}\right)^2 - \frac{\lambda}{2\eta}}\right) \tag{7.119}$$

and the maximum value of transfer function, $|H(\omega_m)|$, is obtained.

Application to a bandpass SC biquad

Let us consider the bandpass SC biquad shown in Figure 7.29, with the assumption that $k_5 = k_2 k_{14}$. The output and internal transfer functions are, respectively, given by

$$H_0(z) = \frac{V_0(z)}{V_i(z)} = \frac{\alpha_0 + \alpha_1 z^{-1} + \alpha_2 z^{-2}}{\gamma_0 + \gamma_1 z^{-1} + \gamma_2 z^{-2}} \tag{7.120}$$

and

$$H_1(z) = \frac{V_1(z)}{V_i(z)} = \frac{\beta_0 + \beta_1 z^{-1} + \beta_2 z^{-2}}{\gamma_0 + \gamma_1 z^{-1} + \gamma_2 z^{-2}} \tag{7.121}$$

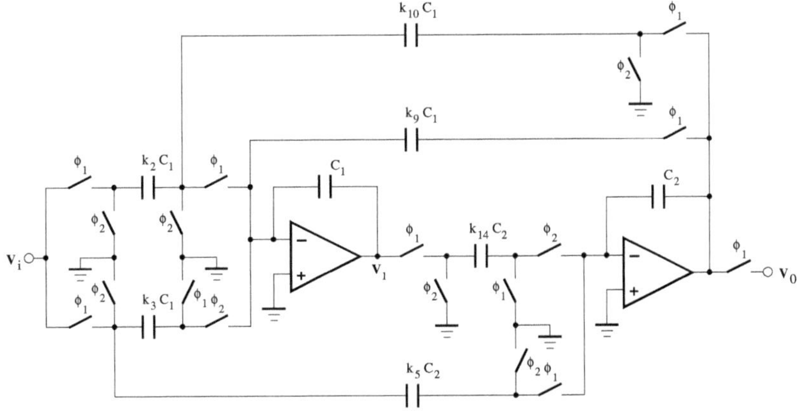

FIGURE 7.29
Circuit diagram of a bandpass SC biquad.

where

$$\alpha_0 = -k_5 \tag{7.122}$$

$$\alpha_1 = k_5 - k_2 k_{14} = 0 \tag{7.123}$$

$$\alpha_2 = k_2 k_{14} = -\alpha_0 \tag{7.124}$$

$$\beta_0 = -k_2 + k_5(k_9 + k_{10}) \tag{7.125}$$

$$\beta_1 = k_2 + k_3 - k_5 k_9 \tag{7.126}$$

$$\beta_2 = -k_3 \tag{7.127}$$

$$\gamma_0 = 1 \tag{7.128}$$

$$\gamma_1 = k_9 k_{14} + k_{10} k_{14} - 2 \tag{7.129}$$

and

$$\gamma_2 = 1 - k_9 k_{14} \tag{7.130}$$

Setting $k_2 = k_3 = k_5 = k$ and $k_{14} = 1$, we obtain

$$\beta_0 = k(1 + \gamma_1) \tag{7.131}$$

$$\beta_1 = k(1 + \gamma_2) \tag{7.132}$$

and

$$\beta_2 = -k \tag{7.133}$$

It then follows that

$$\frac{\lambda}{2\eta} = 1 \tag{7.134}$$

and

$$\frac{\kappa}{2\eta} = \frac{\gamma_1^2 + (1+\gamma_2)^2}{2\gamma_1(1+\gamma_2)} \tag{7.135}$$

where

$$\lambda = \begin{cases} -4\alpha_0^2\gamma_1(1+\gamma_2) & \text{for} \quad H_0(z) \\ -4k^2\gamma_1(1+\gamma_1+\gamma_2)(1+\gamma_2) & \text{for} \quad H_1(z) \end{cases} \tag{7.136}$$

$$\eta = \begin{cases} -2\alpha_0^2\gamma_1(1+\gamma_2) & \text{for} \quad H_0(z) \\ -2k^2\gamma_1(1+\gamma_1+\gamma_2)(1+\gamma_2) & \text{for} \quad H_1(z) \end{cases} \tag{7.137}$$

and

$$\kappa = \begin{cases} -2\alpha_0^2[\gamma_1^2 + (1+\gamma_2)^2] & \text{for} \quad H_0(z) \\ -2k^2(1+\gamma_1+\gamma_2)[\gamma_1^2 + (1+\gamma_2)^2] & \text{for} \quad H_1(z) \end{cases} \tag{7.138}$$

The relation between the transfer function magnitudes can be expressed as

$$|H_1(z)| = \frac{1}{k}|H_0(z)| \left| \frac{\beta_0 + \beta_1 z^{-1} + \beta_2 z^{-2}}{1 - z^{-2}} \right| \tag{7.139}$$

Recalling the definition $z = \cos(\omega T) + j\sin(\omega T)$, it can be shown that for the frequency, ω_m, at which the transfer functions attain their maximum values, we have[2]

$$|H_1(z)| = \frac{1}{k}|H_0(z)| \sqrt{\frac{P}{4[1 - \cos^2(\omega_m T)]}} \tag{7.140}$$

where

$$P = \beta_0^2 + \beta_1^2 + \beta_2^2 + 2\beta_1(\beta_0 + \beta_2)\cos(\omega_m T) \\ + 2\beta_0\beta_2[2\cos^2(\omega_m T) - 1] \tag{7.141}$$

Because

$$\cos(\omega_m T) = -\frac{\gamma_1}{1+\gamma_2} \tag{7.142}$$

we arrive successively at

$$\begin{aligned} |H_1(z)| &= \frac{|H_0(z)|}{2} \sqrt{\frac{Q}{(1+\gamma_2)^2 - \gamma_1^2}} \\ &= \frac{|H_0(z)|}{2} \sqrt{(1+\gamma_2)^2 + 4(1+\gamma_1)} \end{aligned} \tag{7.143}$$

[2]Note that for the derivation of Equation (7.140), it was necessary to use the following trigonometric identities: $\cos(x-y) = \cos(x)\cos(y) + \sin(x)\sin(y)$, $\cos^2(x) + \sin^2(x) = 1$, and $\cos(2x) = 2\cos^2(x) - 1$.

where

$$Q = (1+\gamma_2)^2[(1+\gamma_1)^2 + (1+\gamma_2)^2 + 1 - 2\gamma_1^2] - 2(1+\gamma_1)[2\gamma_1^2 - (1+\gamma_2)^2] \tag{7.144}$$

Hence, the maximum values of the output and internal transfer functions are equal provided,

$$\frac{1}{2}\sqrt{(1+\gamma_2)^2 + 4(1+\gamma_1)} = 1 \tag{7.145}$$

where $\gamma_1 = 1 - k_9$ and $\gamma_2 = k_9 + k_{10} - 2$.

In SC circuits, the high Q factor can result in a large spread of the capacitance ratios. To reduce the sensitivity of the capacitance ratio to the IC process, identical unit capacitors are generally connected in parallel to implement larger ones [28]. The implementation of large capacitance ratios would then require very small unit capacitors, whose reproduction can be affected by significant matching errors due to area variations. Furthermore, an increase in the chip area may be related to the spacing required between the units.

In practice, SC circuits exhibit a frequency response that is limited by the nonideal characteristics of components. They only operate well for input signals at much lower frequencies than the clock frequency. With a biquad consisting of a combination of an inverting and a noninverting integrator, the signals will experience one sampling period delay around the loop. As a result, this topology is less affected by the amplifier gain-bandwidth product.

The determination of capacitances should be achieved to simultaneously maximize the dynamic range and minimize the total chip area required by capacitors and the capacitance spread. The smallest capacitance can be determined based on the output noise requirement and the other capacitors are then scaled appropriately under various design trade-offs (total capacitance, sensitivity to component imperfections, etc.). Capacitance scaling techniques based on analytical models can be applied to only a limited number of biquad structures. To explore all possible solutions without any restrictions, a systematic scaling approach must be numerical and rely on constrained optimization tools.

7.8.3 Ladder filter

In the cascade design of high-order filters, some of the pole Q-factors can become too high, resulting in an increase in the component-value sensitivity and a large chip area. Due to the inherent low sensitivity in the passband, SC ladders [29–33] can then be preferred over the cascade of low-order sections.

SC ladder filters are designed using either the lossless discrete-time integrator (LDI) transform or the bilinear transform. The prototype networks or

SFGs, which are used in the exact LDI design procedure adopted here, include half-unit delay terminations in order to allow the realization of any filter type. The resulting SC ladder filters are canonical, that is, the required number of amplifiers is equal to the filter order. In the case of design techniques based on the bilinear transform, the filter prototype is exactly transformed from the continuous-time domain to the discrete-time domain without any assumption on the network terminations, leading to the use of an extra T/H amplifier in the input stage.

7.9 SC ladder filter based on the LDI transform

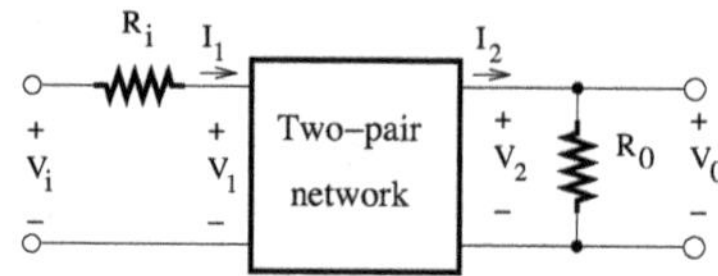

FIGURE 7.30
A doubly terminated discrete-time network.

A doubly terminated discrete-time network is depicted in Figure 7.30 [34]. The input and output terminations,[3] which are assumed to be frequency dependent, are characterized by $R_i(z) = z^{-l_i/2}R_i$, $l_i = \pm 1$, and $R_O(z) = z^{l_O/2}R_O$, $l_O = \pm 1$, respectively. The network consists of discrete-time impedances $L(z) = (z^{1/2} - z^{-1/2})L$ and $C(z) = 1/(z^{1/2} - z^{-1/2})C$. The parameters R_i, R_O, L, and C are constants. The voltages and currents are related by

$$\begin{bmatrix} V_1(z) \\ I_1(z) \end{bmatrix} = \mathbf{T} \begin{bmatrix} V_2(z) \\ I_2(z) \end{bmatrix} \tag{7.146}$$

where $\mathbf{T}$ denotes the chain matrix of the two-pair network given by

$$\mathbf{T} = \begin{bmatrix} A(z) & B(z) \\ C(z) & D(z) \end{bmatrix} \tag{7.147}$$

With $I_1(z) = [V_i(z) - V_1(z)]/R_i(z)$ and $I_2(z) = V_2(z)/R_O(z)$, the overall transfer function is of the form

$$H(z) = \frac{V_O(z)}{V_i(z)} = \frac{1}{A(z) + \dfrac{B(z)}{R_O(z)} + R_i(z)\left[C(z) + \dfrac{D(z)}{R_O(z)}\right]} \tag{7.148}$$

[3]Other forms of the terminations, but which seem to be unsuitable for the design of highpass filters, are $R_i(z) = (z^{1/2} + z^{-1/2})R_i$ and $R_O(z) = (z^{1/2} + z^{-1/2})R_O$.

The reciprocal of the transfer function is $G(z) = 1/H(z)$ and the auxiliary function, $K(z)$, is defined by

$$K(z) = A(z) + \frac{B(z)}{R_0(z)} - R_i(z^{-1})\left[C(z) + \frac{D(z)}{R_0(z)}\right] \tag{7.149}$$

The function $G(z)$ is related to the filter specifications. Provided $K(z)$ is known, the elements of the chain matrix $\mathbf{T}$ are to be determined from the following relations:

$$A(z) + \frac{B(z)}{R_0(z)} = \frac{R_i(z)K(z) + R_i(z^{-1})H(z)}{R_i(z) + R_i(z^{-1})} \tag{7.150}$$

$$C(z) + \frac{D(z)}{R_0(z)} = \frac{K(z) - H(z)}{R_i(z) + R_i(z^{-1})} \tag{7.151}$$

This can only be achieved by assuming that the following requirements are met:

- Either (i) the polynomials $A(z)$ and $D(z)$ are symmetric, and $B(z)$ and $D(z)$ are anti-symmetric rational polynomials, or (ii) $A(z)$ and $D(z)$ are anti-symmetric, and $B(z)$ and $D(z)$ are symmetric rational polynomials.
- The determinant of the chain matrix takes only the values ± 1.

It follows that

$$K(z)K(z^{-1}) - G(z)G(z^{-1}) = \pm[R_i(z) + R_i(z^{-1})]\left[\frac{1}{R_0(z)} + \frac{1}{R_0(z^{-1})}\right] \tag{7.152}$$

This last equation can then be solved numerically for the function $K(z)$. The roots located in the unit circle are to be assigned to $K(z)$ and those outside to $K(z^{-1})$.

With reference to a polynomial defined by

$$P(z) = p_0 z^n + p_1 z^{n-1} + \cdots + p_n \tag{7.153}$$

where p_i $(i = 0, 1, \cdots, n)$ denotes the coefficients, a balanced polynomial can be written as

$$\hat{P}(z) = z^{-n/2}P(z) = p_0 z^{n/2} + p_1 z^{n/2-1} + \cdots + p_n z^{-n/2} \tag{7.154}$$

The polynomial $\hat{P}(z)$ can equivalently be represented by the equation

$$\hat{P}(z) = P^{+}(z) + P^{-}(z) \tag{7.155}$$

where the symmetric polynomial, $P^+(z)$, and the anti-symmetric polynomial, $P^-(z)$, are respectively given by

$$P^+(z) = \frac{\hat{P}(z) + \hat{P}(z^{-1})}{2} = \sum_{i=0}^{(n-1)/2} \hat{p}_i \gamma^{n-2i} \tag{7.156}$$

and

$$P^-(z) = \frac{\hat{P}(z) - \hat{P}(z^{-1})}{2} = \sum_{i=0}^{n/2} \hat{p}_i \gamma^{n-2i} \tag{7.157}$$

The LDI variable, γ, is given by $\gamma = z^{1/2} - z^{-1/2}$, and the coefficients $\hat{p}_i$ can be computed as

$$\hat{p}_i = \begin{cases} p_0 & \text{for } i = 0 \\ p_i - \sum_{j=0}^{i-1} (-1)^{i-j} \binom{n-j}{i-j} \hat{p}_j & \text{otherwise.} \end{cases} \tag{7.158}$$

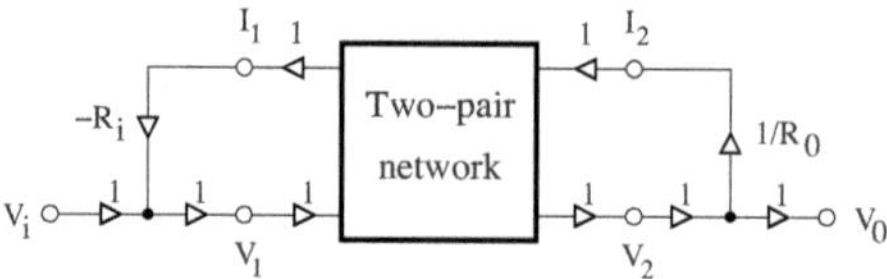

FIGURE 7.31
Signal-flow graph of a doubly terminated discrete-time network.

In the specific case of LC networks, $A(z) = A(z^{-1})$, $B(z) = -B(z^{-1})$, $C(z) = -C(z^{-1})$, $D(z) = D(z^{-1})$, and $\det(\mathbf{T}) = 1$. The chain matrix elements can then be written in terms of the variable γ. The rational polynomials $B(\gamma)/A(\gamma)$, $C(\gamma)/A(\gamma)$, $B(\gamma)/D(\gamma)$, and $C(\gamma)/D(\gamma)$ are reactive driving-point impedance functions. The SFG of a doubly terminated discrete-time network is illustrated in Figure 7.31. The discrete-time network, whose elements are to be determined, is decomposed into a product of low-order sections with the chain matrices $\mathbf{T}_i$, $i = 1, 2, \cdots, N$, such that

$$\mathbf{T} = \prod_{i=1}^{N} \mathbf{T}_i \tag{7.159}$$

The decomposition of the matrix $\mathbf{T}$ from nodes 1 can be achieved recursively using the input function defined as

$$Z_1(\gamma) = \left. \frac{V_1(\gamma)}{I_1(\gamma)} \right|_{I_2=0} = \frac{C(\gamma)}{A(\gamma)} \tag{7.160}$$

After the extraction of the elements $A_i(\gamma)$, $B_i(\gamma)$, $C_i(\gamma)$, and $D_i(\gamma)$, corresponding to $\mathbf{T}_i$, the input function of the remaining network is given by

$$Z_1^r(\gamma) = \frac{C_i(\gamma) - A_i(\gamma)Z_1(\gamma)}{B_i(\gamma)Z_1(\gamma) - D_i(\gamma)} \tag{7.161}$$

which should be zero at the end of the procedure. Carrying on the extraction operation from nodes 2, the above relations become

$$Z_2(\gamma) = -\left.\frac{V_2(\gamma)}{I_2(\gamma)}\right|_{I_1=0} = \frac{B(\gamma)}{A(\gamma)} \tag{7.162}$$

and

$$Z_2^i(\gamma) = \frac{B_i(\gamma) - A_i(\gamma)Z_2(\gamma)}{C_i(\gamma)Z_2(\gamma) - D_i(\gamma)} \tag{7.163}$$

Based on the input functions, the decomposition of the initial chain matrix can be achieved up to a constant reciprocal two-pair characterized by

$$\mathbf{T}_0 = \begin{bmatrix} \frac{1}{k} & 0 \\ 0 & k \end{bmatrix} \tag{7.164}$$

where k is a constant. The possibility of obtaining a network, which is incomplete, should be avoided by using a full-order input function for the chain-matrix extraction. Note that a filter of degree N is to be decomposed into N first-order network sections. The partial fraction expansion of an input function includes terms of the form k_0/γ, $2k_i\gamma/(\gamma^2 + \omega_i^2)$ and $k_\infty\gamma$, where $k_0 \geq 0$, $k_i > 0$, $k_\infty \geq 0$, and ω_i is related to the finite poles.

The signal-flow graphs of ladder filter network sections are summarized in the table illustrated in Figure 7.32 [31, 33]. Let $\mathbf{T}$ be the chain matrix of the original discrete-time filter prototype. Provided that the chain matrix $\mathbf{T}_i$ of the network section to be extracted is known, the characteristic $\mathbf{T}^r$ of the remaining network can be obtained by the following factorization,

$$\mathbf{T}^r = \mathbf{T}_i^{-1}\mathbf{T} \tag{7.165}$$

for the extraction from input nodes, or

$$\mathbf{T}^r = \mathbf{T}\mathbf{T}_i^{-1} \tag{7.166}$$

for the extraction from output nodes. The structures of type I and III realize transmission zeros at infinity, as required for lowpass filters, while the ones of type II and IV implement zeros at the origin, as in the case of highpass filters. The type-I network sections correspond to the next chain matrix.[4]

$$\begin{bmatrix} \alpha\gamma & 1 \\ 1 & 0 \end{bmatrix} \tag{7.167}$$

[4]With $1/Z_1(\gamma) = \tilde{k}_0/\gamma + \sum_{i=1}^{I} 2\tilde{k}_i\gamma/(\gamma^2 + \tilde{\omega}_i^2) + \tilde{k}_\infty\gamma$, the parameter α can be obtained as $\alpha = \tilde{k}_\infty = \lim_{\gamma\to\infty}(1/\gamma Z_1(\gamma))$.

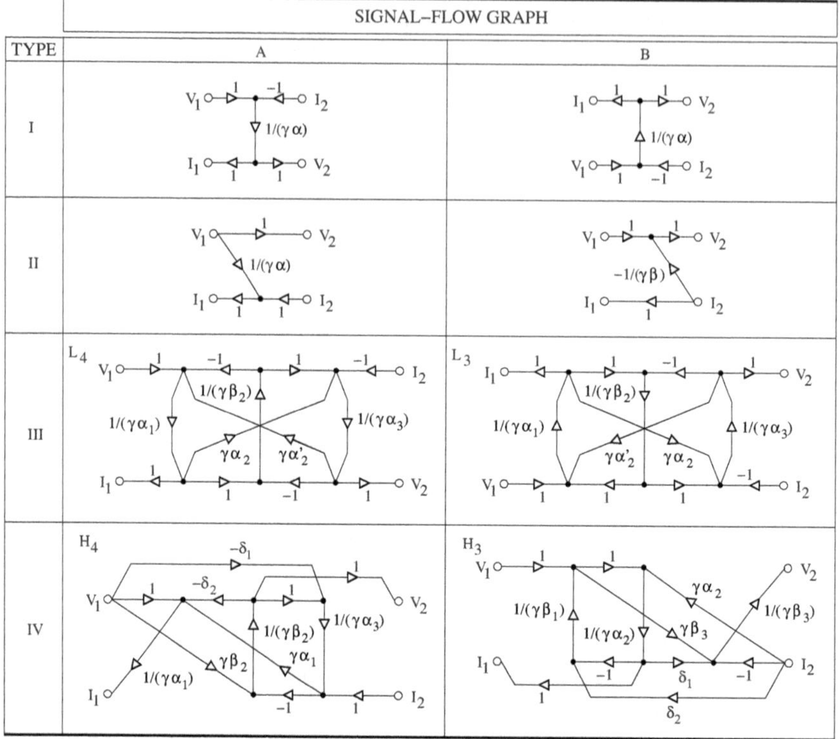

FIGURE 7.32
Signal-flow graphs of ladder filter network sections.

where $\alpha = \lim_{\gamma\to\infty}(1/\gamma Z_1(\gamma))$. We have

$$\begin{bmatrix} 1 & 0 \\ \dfrac{1}{\alpha\gamma} & 1 \end{bmatrix} \quad \text{for the type IIA} \tag{7.168}$$

and

$$\begin{bmatrix} 1 & \dfrac{1}{\beta\gamma} \\ 0 & 1 \end{bmatrix} \quad \text{for the type IIB} \tag{7.169}$$

where $1/\alpha = \lim_{\gamma\to 0}(\gamma Z_1(\gamma))$ and $1/\beta = \lim_{\gamma\to 0}(\gamma Z_2(\gamma))$ are two positive constants.[5]

The networks related to the above chain matrices contain only one integrator and can be used for the realization of all-pole filters. Note that the

[5]Provided $Z_1(\gamma) = k_0/\gamma + \sum_{i=1}^{I} 2k_i\gamma/(\gamma^2+\omega_i^2) + k_\infty\gamma$, the parameter α is given by $1/\alpha = k_0 = \lim_{\gamma\to 0}(\gamma Z_1(\gamma))$. Assuming that the partial fraction expansion of $Z_2(\gamma)$ takes on a form similar to the one of $Z_1(\gamma)$, the following relation can be written as well:
$1/\beta = k_0 = \lim_{\gamma\to 0}(\gamma Z_2(\gamma))$.

network sections should not be extracted in arbitrary sequence. Using the input function Z_1, the first network section will be either of type IA or IIA if the order N of the original discrete-time network is odd, or type IB or IIB if N is even. The following extractions are achieved according to the diagram in Figure 7.33(a). With Z_2, the first network section will be either of type IA or IIA if the order N of the original discrete-time network is even or of the type IB or IIB if N is odd. The subsequent extractions are performed based on the diagram in Figure 7.33(b). From a given network section type, the arrows are directed toward the other type, which can later be extracted.

FIGURE 7.33
Signal-flow graph extraction sequences for network sections of type I and II.

The discrete-time prototype and SFG of a third-order all-pole lowpass filter are shown in Figures 7.34 and 7.35, respectively. It is assumed that $l_i = l_0 = 1$ for the terminations R_i and R_0. The resulting SC filter is shown in Figure 7.36.

For the type-III network sections, we have

$$\begin{bmatrix} \dfrac{(\alpha_1\alpha_3 - \alpha_2\alpha_2')\beta_2\gamma^2 + \alpha_1 + \alpha_3 - \alpha_2 - \alpha_2')\gamma}{\alpha_2\beta_2\gamma^2 + 1} & \dfrac{\alpha_1\beta_2\gamma^2 + 1}{\alpha_2\beta_2\gamma^2 + 1} \\ \dfrac{\alpha_3\beta_2\gamma^2 + 1}{\alpha_2\beta_2\gamma^2 + 1} & \dfrac{\beta_2\gamma}{\alpha_2\beta_2\gamma^2 + 1} \end{bmatrix} \tag{7.170}$$

With $\alpha_2 = \alpha_2'$, a finite zero is realized at $\Omega = 1/\sqrt{\alpha_2\beta_2}$. As shown in the table in Figure 7.32, the network section of type IIIA includes two type IA structures, while the one of type IIIB contains two type IB structures.

The next chain matrices can be written

$$\begin{bmatrix} \dfrac{\alpha_2\beta_3\gamma^2 + 1}{\alpha_2\beta_3\gamma^2 + \delta_1} & \dfrac{[\beta_1(1-\delta_1) + \beta_3(1-\delta_2)]\alpha_2\gamma^2 - \delta_1\delta_2 + 1}{(\alpha_2\beta_3\gamma^2 + \delta_1)\beta_1\gamma} \\ \dfrac{\beta_3\gamma}{\alpha_2\beta_3\gamma^2 + \delta_1} & \dfrac{\beta_3(\alpha_2\beta_1\gamma^2 + 1)}{\beta_1(\alpha_2\beta_3\gamma^2 + \delta_1)} \end{bmatrix} \tag{7.171}$$

for type IVA, and

$$\begin{bmatrix} \dfrac{\alpha_3\beta_2\gamma^2 + 1}{\alpha_3\beta_2\gamma^2 + \delta_1} & \dfrac{\alpha_3\gamma}{\alpha_3\beta_2\gamma^2 + \delta_1} \\ \dfrac{[\alpha_1(1-\delta_1) + \alpha_3(1-\delta_2)]\beta_2\gamma^2 - \delta_1\delta_2 + 1}{\alpha_1(\alpha_3\beta_2\gamma^2 + \delta_1)\gamma} & \dfrac{(1 + \alpha_1\beta_2)\alpha_3\gamma^2}{\alpha_1(\alpha_3\beta_2\gamma^2 + \delta_1)} \end{bmatrix} \tag{7.172}$$

for type IVB. Here, a finite zero is realized at $\Omega = \sqrt{\delta_1/\alpha_3\beta_2} = \sqrt{\delta_2/\alpha_1\beta_2}$.

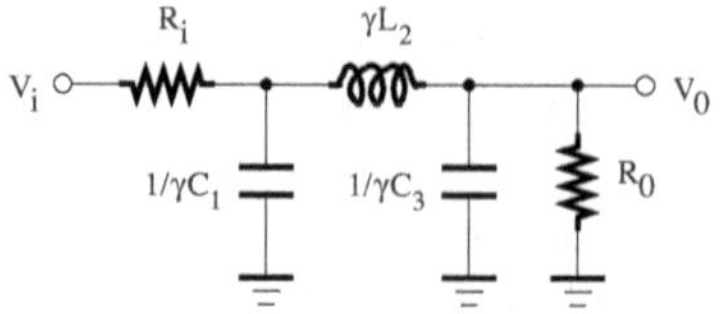

FIGURE 7.34
Circuit diagram of a third-order discrete-time lowpass ladder network.

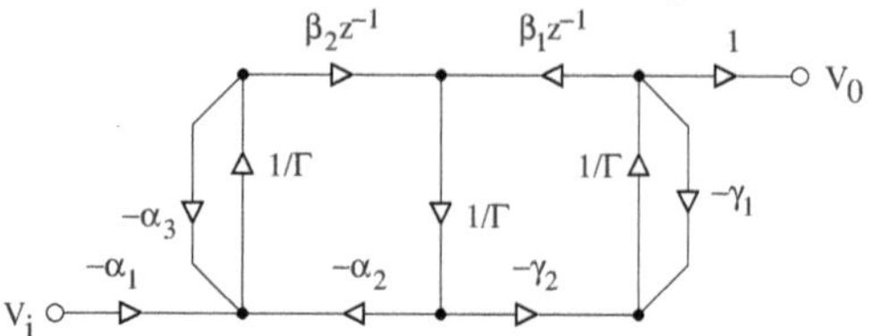

FIGURE 7.35
Signal-flow graph of a third-order lowpass ladder network ($\Gamma = 1 - z^{-1}$).

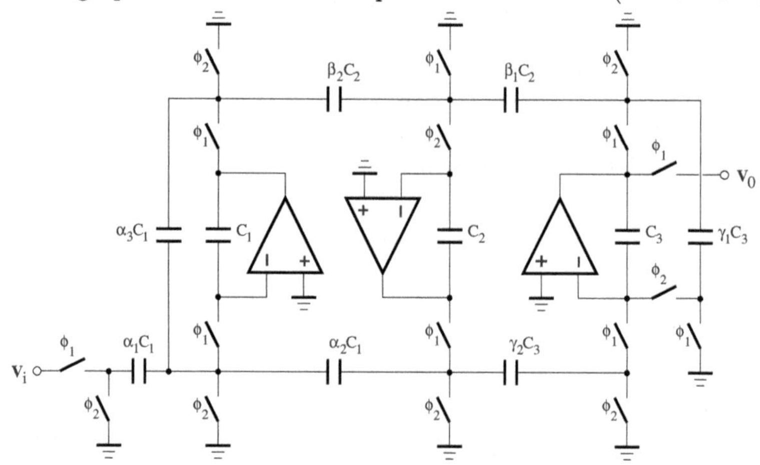

FIGURE 7.36
Circuit diagram of a third-order lowpass ladder SC filter.

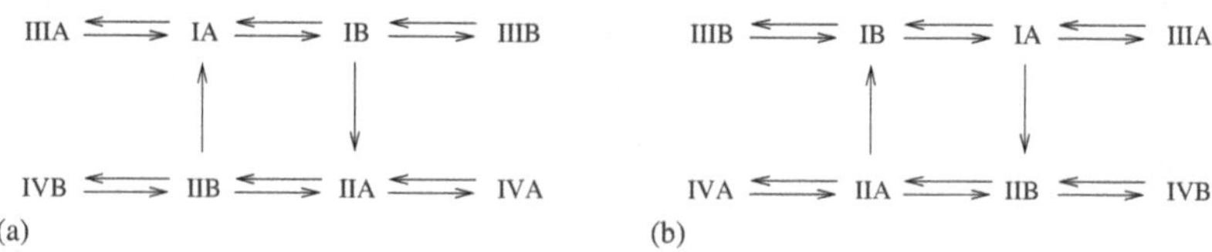

FIGURE 7.37
Signal-flow graph extraction sequences.

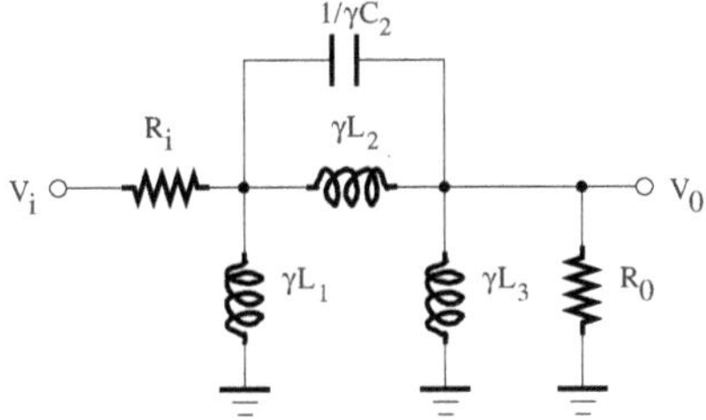

FIGURE 7.38
Circuit diagram of a third-order discrete-time highpass ladder network.

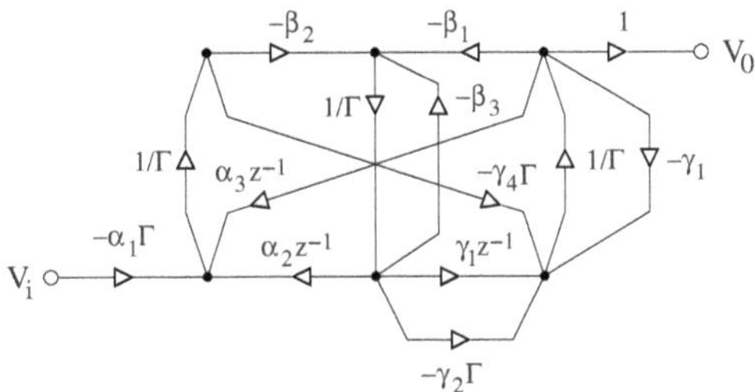

FIGURE 7.39
Signal-flow graph of a third-order highpass ladder network.

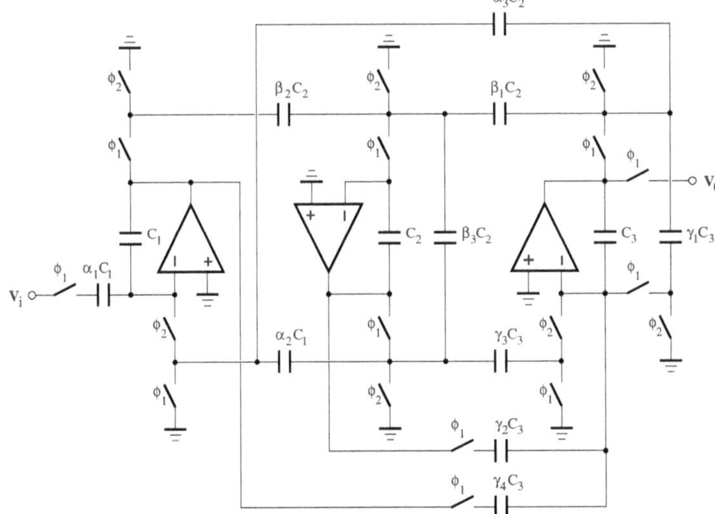

FIGURE 7.40
Circuit diagram of a third-order highpass ladder SC filter.

Note that two type IIA (type IIB) network sections can be extracted from the structure of type IVA (type IVB).

The network sections of type III and IV can be used for the implementation of transfer functions with finite transmission zeros and at least one zero at the origin or at infinity. They should be inserted between the network sections of type I and II, which are also taken out at the first and last steps. Depending on the choice of the input function, Z_1 or Z_2, the extraction of the remaining network sections will be performed according to the diagram shown in Figure 7.37(a) or (b), respectively.

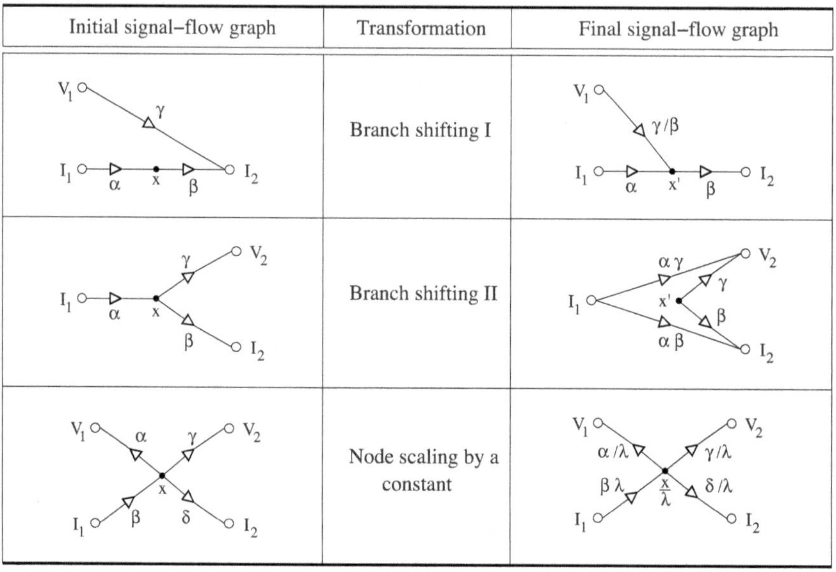

Initial signal–flow graph	Transformation	Final signal–flow graph
	Branch shifting I	
	Branch shifting II	
	Node scaling by a constant	

FIGURE 7.41
Signal-flow graph transformations.

The circuit diagram of a third-order, discrete-time highpass ladder network is shown in Figure 7.38. The terminations R_i and R_0 of the discrete-time network must be chosen such that $l_i = -l_0 = 1$. The resulting SFG is depicted in Figure 7.39 and the corresponding SC filter is shown in Figure 7.40.

Given a discrete-time transfer function obtained from filter specifications, the chain parameters of the network are computed. The filter coefficients are obtained through the factorization of the two-port matrix. The SFG can then be derived and transformed into a form, which is suitable for the SC realization using parasitic-insensitive building blocks. The basic SFG transformations are included in the table of Figure 7.41.

The delays of the input and output terminations must be chosen so that a ladder filter of the order N can be realized with N amplifiers. Note that the last step of the design consists of scaling the amplifier voltage levels for the dynamic range maximization. It should be noted that depending on the SFG transformations, different filter structures realizing the same specification can be derived.

Although SC filters obtained using the design method based on the LDI transform have the advantage of being simple, they can be limited by the achievable attenuation level in the stopband. The design of high-frequency SC filters then preferably relies on the bilinear transform and building blocks that embody LDI integrators.

7.10 SC ladder filter based on the bilinear transform

Using the bilinear transform, SC ladder filters can be designed from RLC filter prototypes or transfer functions that are related to the target response specifications.

7.10.1 RLC filter prototype-based design

In order to preserve a low sensitivity to fluctuations of component values in the passband, the basic design method of high-order SC filters is to model a doubly terminated lossless passive network using SC building blocks [35]. The bilinear-transformed transfer function is exactly realized by properly arranging the phasing between adjacent SC integrators.

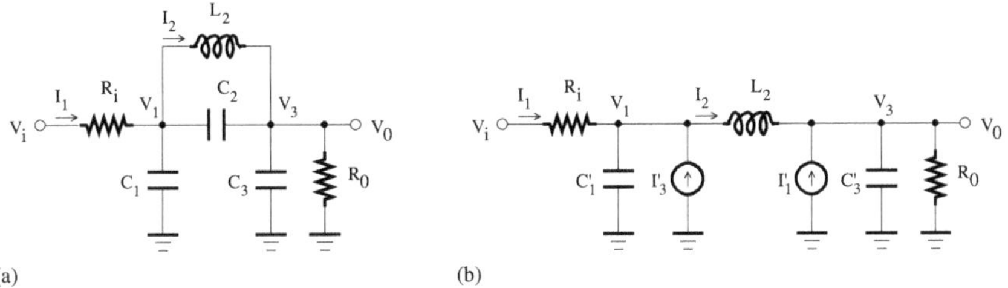

FIGURE 7.42
(a) Third-order RLC lowpass filter; (b) RLC lowpass filter after circuit transformation.

Starting from the prototype of a third-order RLC lowpass ladder filter shown in Figure 7.42(a), the nodal and loop equations can be written as

$$I_1 = \frac{V_i - V_1}{R_i} \tag{7.173}$$

$$V_1 = \frac{I_1 - [I_2 + sC_2(V_1 - V_3)]}{sC_1} \tag{7.174}$$

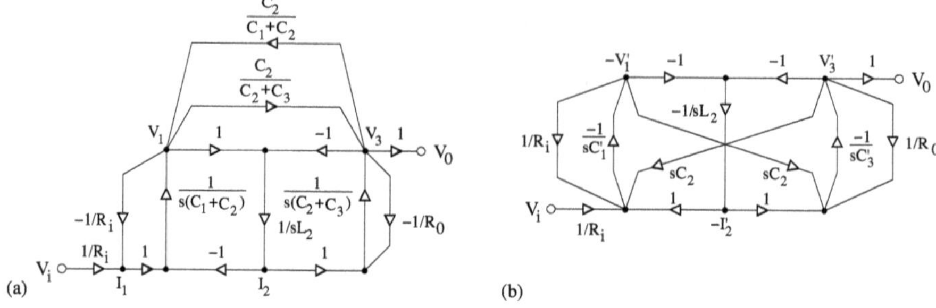

FIGURE 7.43
(a) Third-order LC lowpass filter SFG; (b) third-order LC lowpass filter SFG with the minimum number of integrator stages.

$$I_2 = \frac{V_1 - V_3}{sL_2} \tag{7.175}$$

$$V_3 = \frac{I_2 + sC_2(V_1 - V_3)}{sC_3 + 1/R_0} \tag{7.176}$$

or, equivalently,

$$I_1 = \frac{V_i - V_1}{R_i} \tag{7.177}$$

$$V_1 = \frac{I_1 - I_2}{sC_1'} + \frac{C_2}{C_1'}V_3 \tag{7.178}$$

$$I_2 = \frac{V_1 - V_3}{sL_2} \tag{7.179}$$

$$V_3 = \frac{I_2 - \dfrac{V_3}{R_0}}{sC_3'} + \frac{C_2}{C_3'}V_1 \tag{7.180}$$

where $C_1' = C_1 + C_2$ and $C_3' = C_2 + C_3$. The set of Equations (7.177)–(7.180) can directly be derived from the equivalent circuit of Figure 7.42(b), which is the outcome of the transformation of the initial network using Norton's theorem and has the advantage of requiring the minimum number of reactance elements. It can be represented by the SFG depicted in Figure 7.43(a), where each node represents either a voltage drop across a component or a current flowing through a component. To meet the implementation constraint of using a minimum number of inverting integrator stages, the terms of the previous equations are rearranged such that each state variable is represented by the negative integral of a weighted sum of the state variables and the input voltage. This is achieved by eliminating the current variable I_1 from the equations and scaling V_1 and I_2 by -1, resulting in

$$-V_1 = \frac{-1}{sC_1'}\left[\frac{V_i - V_1}{R_i} - I_2 + sC_2V_3\right] \tag{7.181}$$

$$-I_2 = -\frac{V_1 - V_3}{sL_2} \tag{7.182}$$

$$V_3 = \frac{-1}{sC_3'}\left[-sC_2V_1 - I_2 + \frac{V_3}{R_0}\right] \tag{7.183}$$

The corresponding SFG is shown in Figure 7.43(b). It can also be derived by exploiting the proprieties related to SFG operations such as gain scaling and node elimination.

The bilinear s-to-z transformation is given by

$$s = \frac{2}{T}\frac{1 - z^{-1}}{1 + z^{-1}} = \frac{2}{T}\frac{z^{1/2} - z^{-1/2}}{z^{1/2} + z^{-1/2}} \tag{7.184}$$

where T denotes the clock signal period. By introducing the variable λ as follows

$$\lambda = sT/2 \tag{7.185}$$

the next expression can be derived

$$\lambda = \frac{z^{1/2} - z^{-1/2}}{z^{1/2} + z^{-1/2}} = \frac{(e^{sT/2} - e^{-sT/2})/2}{(e^{sT/2} + e^{-sT/2})/2} = \frac{\sinh(sT/2)}{\cosh(sT/2)} = \tanh(sT/2) \tag{7.186}$$

With the assumption that

$$\rho = \frac{1}{2}(z^{1/2} - z^{-1/2}) = \frac{1}{2}(e^{sT/2} - e^{-sT/2}) = \sinh(sT/2) \tag{7.187}$$

$$\nu = \frac{1}{2}(z^{1/2} + z^{-1/2}) = \frac{1}{2}(e^{sT/2} + e^{-sT/2}) = \cosh(sT/2) \tag{7.188}$$

we have

$$\lambda = \rho/\nu, \qquad \nu^2 = 1 + \rho^2 \qquad \text{and} \qquad \nu + \rho = z^{1/2} \tag{7.189}$$

Using the s-to-z substitution, we derive the filter SFG in the λ-domain, where the capacitor C_k and inductor L_k exhibit the impedances $1/\lambda C_k$ and λL_k, respectively. The direct SC implementation should require bilinear integrators, which are more complicated than the commonly used LDI integrator. To realize the filter SFG using LDI integrators [36], it is required to divide all impedances by the complex variable ν.

Let Z_C, Z_L, and Z_R be the impedances of the capacitor C_k, inductor

L_k, and resistor R_k, respectively, in the initial filter SFG. The division of all impedances by ν results in $\hat{Z}_C$, $\hat{Z}_L$, and $\hat{Z}_R$ given by

$$\hat{Z}_C = \frac{Z_C}{\nu} = \frac{1}{\lambda \nu C_k} = \frac{1}{\rho C_k} \tag{7.190}$$

$$\hat{Z}_L = \frac{Z_L}{\nu} = \frac{\lambda L_k}{\nu} = \frac{\rho L_k}{1+\rho^2} = \left(\frac{1}{\rho L_k} + \frac{\rho}{L_k}\right)^{-1} \tag{7.191}$$

and

$$\hat{Z}_R = \frac{Z_R}{\nu} = \frac{R_k}{\nu} \tag{7.192}$$

respectively. After the scaling, the value C_k of the capacitor is maintained, while the inductor is transformed to a parallel combination of an inductor L_k and a capacitor $1/L_k$, and the resistor impedance becomes frequency dependent. Because the capacitor introduced by the transformation of the inductor should be taken into account in the derivation of the equivalent minimal-reactance network, the bilinear transformation and impedance scaling should first be performed.

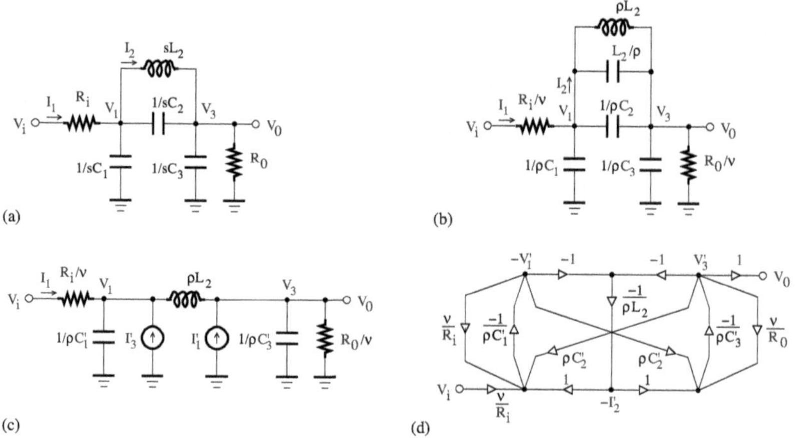

FIGURE 7.44
Design steps of the SC ladder filter based on the third-order RLC lowpass filter prototype: (a) Third-order doubly terminated RLC filter, (b) bilinear-transformed discrete-time equivalent network, (c) modified network derived using Norton's theorem, (d) filter SFG.

The steps required for the derivation of the SFG from the third-order RLC ladder filter are illustrated in Figure 7.44, where $C_1' = C_1 + C_2'$, $C_3' = C_2' + C_3$, $C_2' = C_2 + C_{L_2}$, and $C_{L_2} = 1/L_2$. The set of equations characterizing the filter

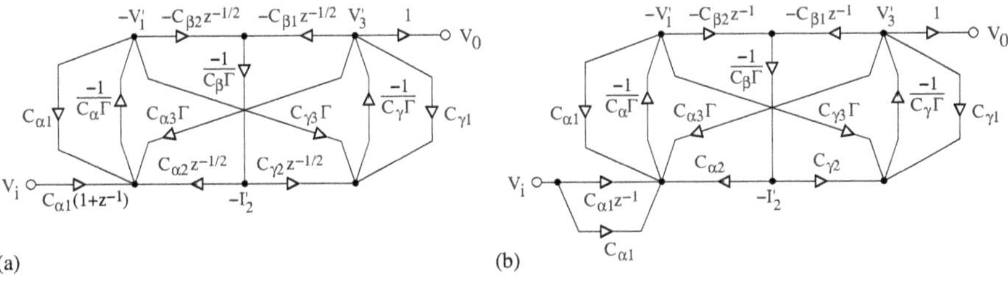

FIGURE 7.45
SFG for the SC implementation of the third-order RLC lowpass filter.

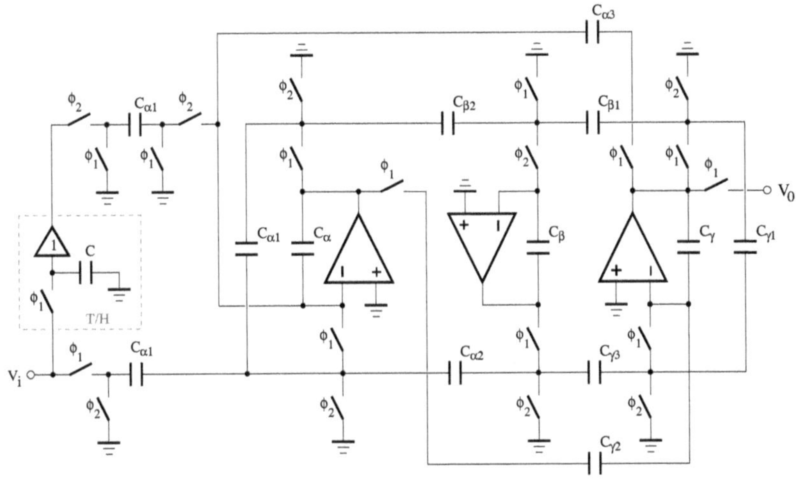

FIGURE 7.46
SC implementation of the third-order LC lowpass filter.

SFG of Figure 7.44(d) can be expressed as

$$-V_1' = \frac{-1}{\rho C_1'}\left[\frac{V_i - V_1'}{R_i/\nu} - I_2' + \rho C_2 V_3'\right], \tag{7.193}$$

$$-I_2' = -\frac{V_1' - V_3'}{\rho L_2} \tag{7.194}$$

$$V_3' = \frac{-1}{\rho C_3'}\left[-\rho C_2 V_1' - I_2' + \frac{V_3'}{R_0/\nu}\right] \tag{7.195}$$

Setting

$$\rho = \frac{z^{1/2}}{2}\Gamma \qquad \text{and} \qquad \nu = \frac{z^{1/2}}{2}(1 + z^{-1}) \tag{7.196}$$

where $\Gamma = 1 - z^{-1}$, we can obtain

$$\frac{1}{2}[(C_1' - 1/R_i)\Gamma + 2/R_i]V_1' = \frac{1}{2}(1 + z^{-1})V_i - z^{-1/2}I_2' + \frac{1}{2}C_2'\Gamma V_3' \quad (7.197)$$

$$I_2' = z^{-1/2}\frac{2}{L_2}\frac{V_1' - V_3'}{\Gamma} \quad (7.198)$$

$$\frac{1}{2}[(C_3' - 1/R_0)\Gamma + 2/R_0]\, V_3' = z^{-1/2}I_2' + \frac{1}{2}C_2'\Gamma V_1' \quad (7.199)$$

On the other hand, a filter SFG based on SC building blocks is illustrated in Figure 7.45. It can be described by the following set of equations:

$$-V_1' = \frac{-1}{C_\alpha \Gamma}\left[C_{\alpha 1}(1 + z^{-1})V_i - C_{\alpha 1}V_1' - C_{\alpha 2}z^{-1/2}I_2' + C_{\alpha 3}\Gamma V_3'\right] \quad (7.200)$$

$$-I_2' = \frac{-1}{C_\beta \Gamma}\left[C_{\beta 2}z^{-1/2}V_1' - C_{\beta 1}z^{-1/2}V_3'\right] \quad (7.201)$$

$$V_3' = \frac{-1}{C_\gamma \Gamma}\left[-C_{\gamma 3}\Gamma V_1' - C_{\gamma 2}z^{-1/2}I_2' + C_{\gamma 1}V_3'\right] \quad (7.202)$$

Rewriting these equations gives

$$(C_\alpha \Gamma + C_{\alpha_1})V_1' = C_{\alpha 1}(1 + z^{-1})V_i - C_{\alpha_2}z^{-1/2}I_2' + C_{\alpha_3}\Gamma V_3' \quad (7.203)$$

$$I_2' = z^{-1/2}\frac{C_{\beta_2}V_1' - C_{\beta_1}V_3'}{C_\beta \Gamma} \quad (7.204)$$

$$(C_\gamma \Gamma + C_{\gamma_1})V_3' = C_{\gamma_2}z^{-1/2}I_2' + C_{\gamma_3}\Gamma V_1' \quad (7.205)$$

Comparing the last sets of equations obtained from the SFGs of Figures 7.44(d) and 7.45, the initial values for the capacitors of the SC ladder filter are derived as

$$\begin{gathered} C_\alpha = (C_1' - C_{R_i})/2 \quad C_{\alpha_1} = C_{R_i} \quad C_{\alpha_2} = 1 \quad C_{\alpha_3} = C_2'/2 \\ C_{\beta_1}/C_\beta = C_{\beta_2}/C_\beta = 2C_{L_2} \\ C_\gamma = (C_3' - C_{R_0})/2 \quad C_{\gamma_1} = C_{R_0} \quad C_{\gamma_2} = 1 \quad C_{\gamma_3} = C_2'/2 \end{gathered} \quad (7.206)$$

where $C_{R_i} = 1/R_i$, $C_{R_0} = 1/R_0$, and $C_{L_2} = 1/L_2$. Taking into account the normalization factor $T/2 = \lambda/s$ of the bilinear transform, these last values take the form $C_{R_i} = T/(2R_i)$, $C_{R_0} = T/(2R_0)$, and $C_{L_2} = T^2/(4L_2)$. The circuit diagram of the SC filter is shown in Figure 7.46, where a T/H circuit is required for the realization of the input branch.

Note that the capacitor values can be further scaled for the optimal dynamic range of amplifiers and the minimum total capacitance. Because the transfer function is only determined by the capacitor ratios, the capacitor values for each integrator can be normalized relative to the corresponding un-switched feedback capacitor.

In general, starting from the RLC filter prototype, the SC ladder filter can be designed by following these steps:

1. Perform the bilinear transformation.
2. Scale all impedances.
3. Construct the minimal-reactance network.
4. Derive the filter SFG.
5. Finally, convert the filter SFG into an SC implementation.

7.10.2 Transfer function-based design of allpass filters

In the continuous-time domain, the transfer function of an N-th-order allpass filter can generally be written as

$$H(s) = \pm \frac{D(-s)}{D(s)} \tag{7.207}$$

where

$$D(s) = \sum_{k=0}^{N} a_k s^k$$

and $a_N = 1$. To meet the stability requirements, the roots of $D(s)$ must be in the left-half s-plane, or equivalently, all of the coefficients a_k should be positive, as is the case for Hurwitz polynomials. By performing the bilinear transformation, the equivalent transfer function in the z-domain is given by

$$H(z) = \pm \frac{z^{-N} D(z^{-1})}{D(z)} \tag{7.208}$$

where

$$D(z) = \sum_{l=0}^{N} a_l z^{-l}$$

and $a_0 = 1$. With the numerator being the mirror-image polynomial of the denominator, it can be shown that the magnitude of the transfer function, $|H(\mathrm{e}^{j\omega})|$, is equal to unity for any frequency, and the phase response is

$$\theta(\omega) = \arg[H(\mathrm{e}^{j\omega})] = \begin{cases} -N\omega - 2\mathrm{Arg}[D(\mathrm{e}^{j\omega})] & \text{if} \quad H(j\omega) \geq 0 \\ \pi - N\omega - 2\mathrm{Arg}[D(\mathrm{e}^{j\omega})] & \text{if} \quad H(j\omega) < 0 \end{cases} \tag{7.209}$$

where it is assumed that the phase of a positive real number is zero while the one of a negative real number is π.

Let us consider a third-order allpass filter with the transfer function $H(z)$ given by

$$H(z) = -\frac{z^{3} D(z^{-1})}{D(z)} \tag{7.210}$$

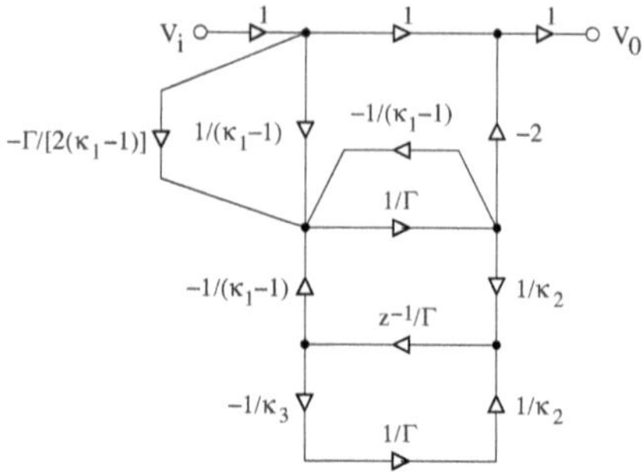

FIGURE 7.47
Signal-flow graph of a third-order allpass ladder network.

where $D(z)$ is a third-order polynomial with real coefficients. The design of the corresponding SC circuit is based on the next decomposition [37]

$$H(z) = 1 - 2\frac{z+1}{2}\frac{1}{1+\tilde{Y}(z)} \tag{7.211}$$

where

$$\tilde{Y}(z) = \frac{1}{2}(z-1) + \frac{1}{2}(z+1)Y(z) \tag{7.212}$$

and the admittance function[6], $Y(z)$, is obtained as

$$Y(z) = \frac{D(z) - z^{-3}D(z^{-1})}{D(z) + z^{-3}D(z^{-1})} \tag{7.213}$$

To proceed further, $Y(z)$ can be expressed in a continued fraction expansion [38] of the form

$$Y(z) = z^{1/2}Y(\gamma) \tag{7.214}$$

where

$$Y(\gamma) = \kappa_1\gamma + \cfrac{1}{\kappa_2\gamma + \cfrac{1}{\kappa_3\gamma}} \tag{7.215}$$

and $\gamma = z^{1/2} - z^{-1/2}$. Note that κ_1 is assumed to be greater than one, and κ_2 and κ_3 are positive constants. The third-order allpass transfer function can be realized by the SFG depicted in Figure 7.47. The resulting SC circuit is shown in Figure 7.48, where the coefficients α_i $(i = 1, 2, 3)$ can be obtained as

$$\alpha_i = \begin{cases} \dfrac{1}{\kappa_1 - 1} & \text{for } i = 1 \\ \dfrac{1}{\kappa_i} & \text{otherwise.} \end{cases} \tag{7.216}$$

[6]Note that, depending on the sign of the transfer function, the ladder network synthesis can be achieved either with $Y(z)$ or $1/Y(z)$.

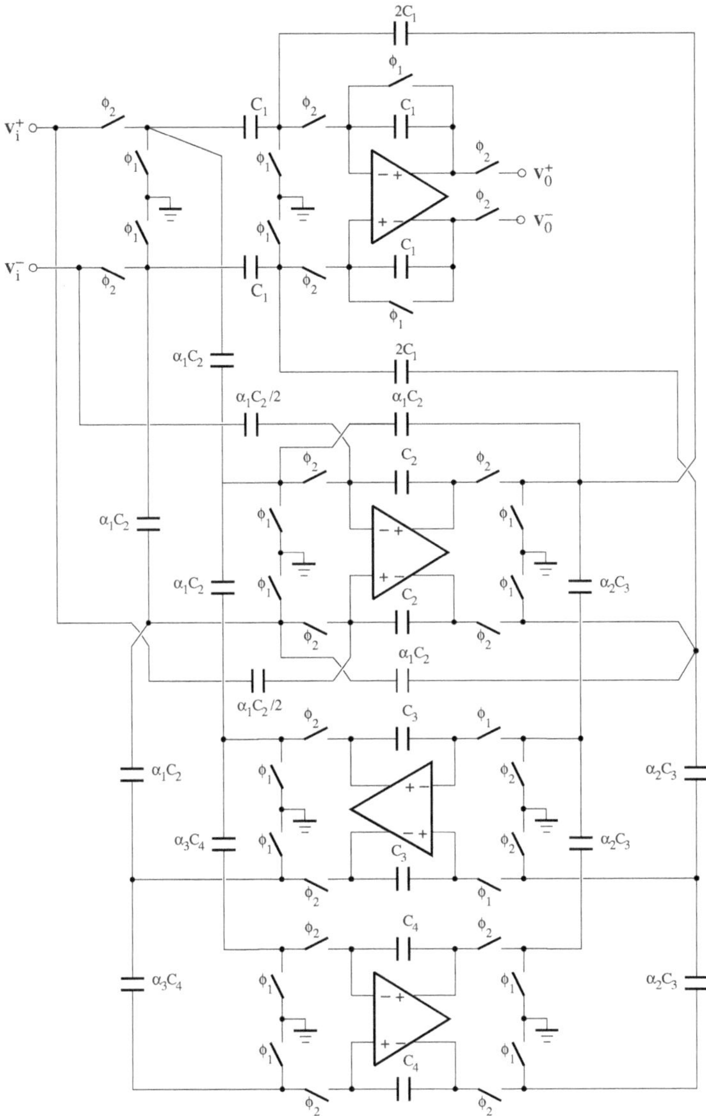

FIGURE 7.48
Circuit diagram of a third-order allpass ladder SC filter.

Here, the SC implementation exploits the availability of signals with both polarities in fully differential structures.

In general for $N > 3$, the admittance function, $Y(z)$, can be decomposed into either a continued fraction expansion or a partial fraction expansion, and the design of an N-th order allpass filter results in two different SC implementations, which require $N + 1$ amplifiers.

7.11 Effects of the amplifier finite gain and bandwidth

Integrators are commonly used in filter design. In this analysis, the stray-insensitive circuit of Figure 7.49 is considered. It can operate as an inverting SC integrator when the switch timings are $S_1(\phi_2)$ and $S_2(\phi_1)$, or a noninverting SC integrator when switch timings are $S_1(\phi_1)$ and $S_2(\phi_2)$. The amplifier

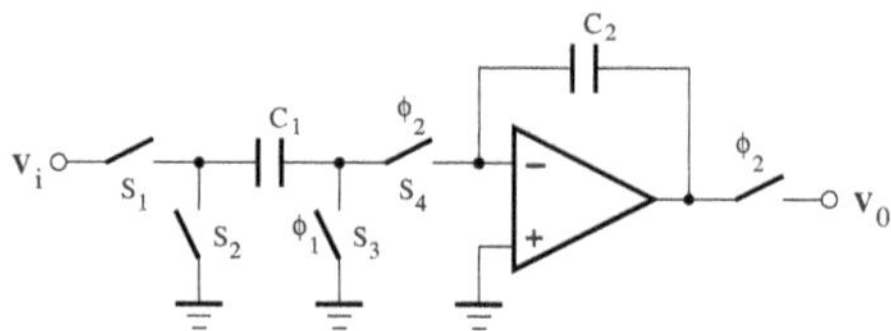

FIGURE 7.49
Lossless discrete-time stray-insensitive integrator.

will have, as in practice, an open loop gain, A, with a finite dc gain,[7] A_0, and a finite transition frequency (or unity-gain frequency), f_t. Thus, the inverting input terminal, v^-, of the amplifier will not be a true virtual ground but will be dependent on the output signal, as $v_0 = A(v^+ - v^-)$.

The equivalent models of the amplifiers, which can be used in the implementation of SC circuits, are shown in Figure 7.50. With the amplifier including an ideal output buffer (see Figure 7.50(a)), the charge transfer is independent of the capacitive load, but the additional pole, which is related to the output voltage follower, can affect the stability of the overall circuit. Generally, the speed performance is then improved for IC designs based on operational transconductance amplifiers (see Figure 7.50(b)).

The amplifier is first described as an ideal voltage-controlled voltage source with a finite dc gain and infinite bandwidth. This model is useful when the distortions due to the finite dc gain must be evaluated. However, in the context of high-frequency applications, the influence of the finite bandwidth on the settling speed becomes dominant. In this case, the analysis is done by modeling the amplifier as an ideal voltage-controlled voltage source with a finite unity-gain frequency and an infinite dc gain [39–41].

For simplicity, the switches are assumed to have a negligible on-resistance and the input signal is considered constant during the charge transfer between

[7] An OTA is generally characterized by a finite transconductance gain, g_m. For a finite output impedance, Z_0, the OTA can be modeled as a voltage-controlled voltage source with the finite dc gain $A_0 = g_m Z_0$.

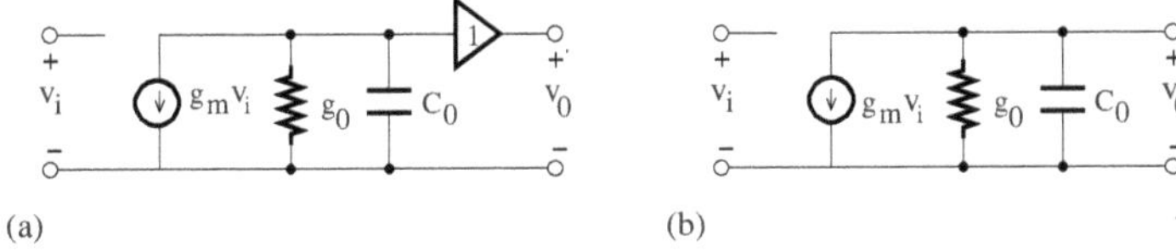

FIGURE 7.50
Equivalent models of (a) an operational amplifier with an ideal output buffer and (b) an operational transconductance amplifier.

the capacitors. The values of all necessary voltages are evaluated at the end of each clock phase, resulting in a set of difference equations that can be readily transformed into the z-domain. The error function, $E(z)$, is then extracted from the circuit transfer function according to the next equations:

$$H(z) = \frac{V_0(z)}{V_i(z)} = H_{id}(z)E(z) \tag{7.217}$$

and

$$E(z)|_{z=\exp(j\omega T)} = [1 + m(\omega)]\exp[j\theta(\omega)] \tag{7.218}$$

where T is the clock period, and $m(\omega)$ and $\theta(\omega)$ are the error magnitude and phase, respectively. For small error values, that is, $m(\omega)$ and $\theta(\omega) \ll 1$, the next approximation can be made

$$E(z)|_{z=\exp(j\omega T)} \simeq 1 + m(\omega) + j\theta(\omega) \tag{7.219}$$

which can be conveniently used for estimating the error magnitude and phase on a complete network. In the independent analysis of the errors due to A_0 and f_t, the error function can be expressed as

$$E(z) = E_{A_0}(z)E_{f_t}(z) \tag{7.220}$$

corresponding to the following expressions,

$$m(\omega) = m_{A_0}(\omega) + m_{f_t}(\omega) \tag{7.221}$$

and

$$\theta(\omega) = \theta_{A_0}(\omega) + \theta_{f_t}(\omega) \tag{7.222}$$

which give the magnitude and phase errors, respectively.

7.11.1 Amplifier dc gain

The gain error is independent of the integrator switch phasing and a dual analysis of the inverting and noninverting structure is not useful.

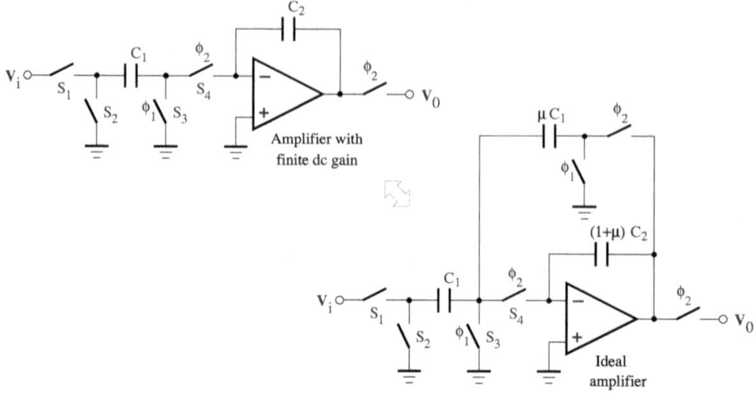

FIGURE 7.51
Equivalent circuit model for the analysis of the finite dc gain effect on the lossless discrete-time stray-insensitive integrator ($\mu = 1/A_0$).

The application of the charge conservation law at the inverting node of the amplifier leads to the following difference equations:

$$C_2\left\{v^-[(n-1/2)T] - v_0[(n-1/2)T]\right\} = C_2\left\{v^-[(n-1)T] - v_0[(n-1)T]\right\} \tag{7.223}$$

at the end of the second clock phase, ϕ_2, or say for $(n-1)T \le t < (n-1/2)T$, and

$$C_1[v_i(nT) - v^-(nT)] \\ = C_2\Big(v^-(nT) - v_0(nT) - \left\{v^-[(n-1/2)T] - v_0[(n-1/2)T]\right\}\Big) \tag{7.224}$$

at the end of the first clock phase, ϕ_1, that is, for $(n-1/2)T \le t < nT$, where T is the clock signal period. Taking into account the fact that we have $v^- = -\mu v_0 + V_{off}$, and combining Equations (7.223) and (7.224) to eliminate the term $v_0[(n-1/2)T]$, we obtain

$$C_1[v_i(nT) + \mu v_0(nT)] + C_2(1+\mu)\left\{v_0(nT) - v_0[(n-1)T]\right\} = 0 \tag{7.225}$$

From the above equation, the z-domain transfer function can be found as

$$H(z) = \frac{V_0(z)}{V_i(z)} = \frac{-C_1}{C_2(1-z^{-1})\left[1+\mu+\mu\dfrac{C_1}{C_2}\dfrac{1}{1-z^{-1}}\right]} \tag{7.226}$$

where $\mu = 1/A_0$. As illustrated in Figure 7.51, this transfer function is equivalent to the one of an ideal integrator (i.e., designed with the ideal amplifier)

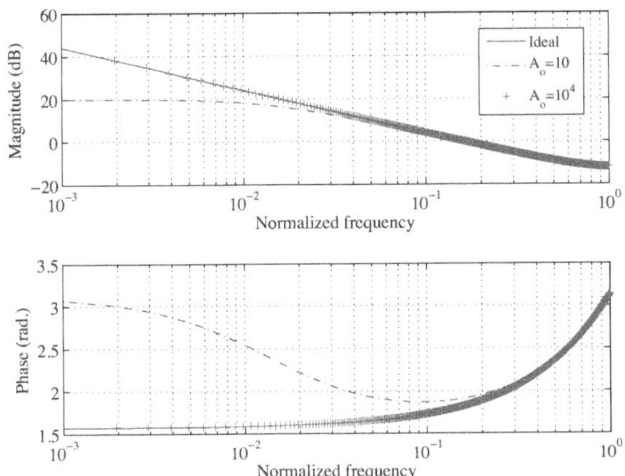

FIGURE 7.52
Finite dc gain effects on the integrator frequency responses.

but with a feedback capacitor of the form $(1+\mu)C_2$ and an extra feedback path providing a damping term equal to μC_1.

In the frequency domain, we have

$$H(j\omega) = H_{id}(j\omega)\left[1+\mu\left(1+\frac{C_1}{2C_2}\right) - j\mu\frac{C_1}{2C_2\tan(\omega T/2)}\right]^{-1} \tag{7.227}$$

where

$$H_{id}(j\omega) = -\frac{C_1}{C_2}\frac{e^{j(\omega T/2)}}{j\sin(\omega T/2)} \tag{7.228}$$

The term in brackets is the same for the two types of integrator. Then, one can identify the next gain and phase errors

$$m_{A_0}(\omega) = m = -\mu\left(1+\frac{C_1}{2C_2}\right) \tag{7.229}$$

and

$$\theta_{A_0}(\omega) = \mu\frac{C_1}{2C_2\tan(\omega T/2)} \tag{7.230}$$

The gain error is found to be frequency independent and can therefore be modeled as an element variation of the integration capacitor. This suggests the use of pre-distortion methods in order to reduce this error. When the frequency increases, the phase error will approach zero.

As confirmed by the test results of Figure 7.52 (see also [39]), the dc gain should be greater than 60 dB in order to reduce the integrator deviation to an acceptable level (i.e., approximately 0.1%).

7.11.2 Amplifier finite bandwidth

In order to analyze the influence of the finite bandwidth on the transfer function of the OTA-based integrator, the amplifier is modeled as an ideal voltage-controlled current source with the short-circuit transconductance, g_m, and an infinite output impedance. Its unity-gain frequency, f_t, is given by $f_t = g_m/(2\pi C_T)$, where C_T is the total load capacitance. Thus, the bandwidth of the OTA depends on the capacitive loading of the circuit; as such, it varies from phase to phase.

The inverting integrator has no delay between the sample instants of the input and output signals. The noninverting one, in contrast, is realized with a half-clock period delay between these sample instants for a 50% duty clock pattern. This delay influences the phase error due to the amplifier finite bandwidth.

7.11.2.1 Inverting integrator

No charge will be transferred by the input branch during the clock phase 1. The charge transfer from the input signal will take place during the clock phase 2. Thus, the amplifier input voltage $v_x(t)$ will be discontinuous at the end of the clock phase 1.

During the clock phase 1, that is, $(n-1)T \leq t < (n-1/2)T$, the amplifier is disconnected from the input signal and we have

$$v_x(t) - v_0(t) = v_x[(n-1)T] - v_0[(n-1)T] \tag{7.231}$$

or in a differential form

$$\frac{dv_0(t)}{dt} = \frac{dv_x(t)}{dt} \tag{7.232}$$

In the time domain, the OTA can be described as

$$g_m v_x(t) + C_L \frac{dv_0(t)}{dt} = 0 \tag{7.233}$$

Substituting Equation (7.232) into (7.233) and solving for the value of $v_x(t)$ at $t = (n-1/2)T$ as

$$v_x[(n-1/2)T]^- = v_x[(n-1)T]e^{-k_1T/2} \tag{7.234}$$

where $k_1 = g_m/C_L$ and the superscript $(-)$ denotes "the time instant just before $t = (n-1/2)T$." At the end of the clock phase 1, $v_x[(n-1/2)T]^-$ will almost be zero. For $t = (n-1/2)T$, Equation (7.231) can then be written as

$$v_0[(n-1/2)T] = v_0[(n-1)T] - v_x[(n-1)T] \tag{7.235}$$

During the clock phase 2, that is, for $(n-1/2)T \leq t < nT$, the charge conservation at the inverting terminal reduces to

$$v_0(t) - v_0[(n-1/2)T] = \frac{C_1 + C_2}{C_2} v_x(t) - \frac{C_1}{C_2} v_i(t) \tag{7.236}$$

Assuming that the input voltage remains constant during the clock phase, the next differential equation is obtained:

$$\frac{dv_0(t)}{dt} = \left(1 + \frac{C_1}{C_2}\right) \frac{dv_x(t)}{dt} \tag{7.237}$$

In this phase, the OTA is described by

$$g_m v_x(t) + C_1 \frac{dv_x(t)}{dt} + C_L \frac{dv_0(t)}{dt} = 0 \tag{7.238}$$

Solving the differential equation obtained by combining Equations (7.237) and (7.238), the value of $v_x(t)$ at $t = nT$ can be written as

$$v_x(nT) = v_x[(n-1/2)T]^+ e^{-k_2T/2} \tag{7.239}$$

where

$$k_2 = \frac{g_m}{C_1 + C_L\left(1 + \frac{C_1}{C_2}\right)} \tag{7.240}$$

and

$$v_0(nT) = v_0[(n-1/2)T] + \left(1 + \frac{C_1}{C_2}\right) v_x[(n-1/2)T]^+ e^{-k_2T/2} - \frac{C_1}{C_2} v_i(nT) \tag{7.241}$$

Note that $v_x[(n-1/2)T]^+$ can be determined from the initial conditions as

$$v_x[(n-1/2)T]^+ = k_0 v_i(nT) \tag{7.242}$$

where $k_0 = C_1(C_2+C_L)/[C_1C_2+C_L(C_1+C_2)]$. To proceed further, Equations (7.235) and (7.242) are substituted into Equation (7.241) and the result is subsequently transformed into the z-domain. Then,

$$H(z) = \frac{V_0(z)}{V_i(z)} = H_{id}(z)E(z) \tag{7.243}$$

where

$$H_{id}(z) = -\frac{C_1}{C_2} \frac{1}{1 - z^{-1}} \tag{7.244}$$

and

$$E(z) = 1 - (1 + \frac{C_2}{C_1})k_0 e^{-k_2T/2} + z^{-1}\frac{C_2}{C_1}k_0 e^{-k_2T/2} \tag{7.245}$$

From Equation (7.245), the gain and phase errors can be computed according to Equation (7.218) as

$$m(\omega) \simeq -\left(1 + \frac{C_2}{C_1}\right)\left(1 - \frac{1}{1 + \frac{C_1}{C_2}} \cos(\omega T)\right) k_0 e^{-k_2T/2} \tag{7.246}$$

and

$$\theta(\omega) \simeq -\frac{C_2}{C_1} k_0 \sin(\omega T) e^{-k_2 T/2} \omega T \tag{7.247}$$

Figure 7.53 shows the frequency responses of the integrator for different values

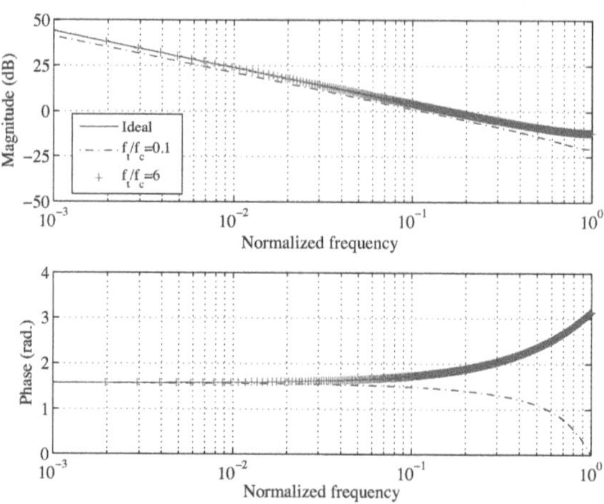

FIGURE 7.53
Finite bandwidth effects on the integrator frequency responses.

of the ratio f_t/f_c.

7.11.2.2 Noninverting integrator

For the noninverting integrator, the charge transfer between the capacitors will take place at the end of the clock phase 2. The analysis of the circuit during the clock phase 1 is very similar to the situation of the inverting integrator.

From Equations (7.231) and (7.232), we have

$$v_x[(n-1/2)T] = v_x[(n-1)T] e^{-k_1 T/2} \simeq 0 \tag{7.248}$$

and

$$v_0[(n-1/2)T] = v_0[(n-1)T] \tag{7.249}$$

During the clock phase 2, that is, $(n-1/2)T \leq t < nT$, the charge conservation law at the inverting input node of the OTA reads

$$v_0(t) - v_0[(n-1/2)T] = \frac{C_1 + C_2}{C_2} v_x(t) + \frac{C_1}{C_2} v_i[(n-1/2)T] \tag{7.250}$$

In the differential form, the above equation can be reduced to Equation (7.237). Because Equation (7.238) is still valid for the description of the OTA, we can obtain

$$v_x(nT) = v_x[(n-1/2)T]^+ e^{-k_2 T/2} \tag{7.251}$$

and

$$v_0(nT) = v_0[(n-1/2)T] + \frac{C_1 + C_2}{C_2} v_x[(n-1/2)T]^+ e^{-k_2 T/2} + \frac{C_1}{C_2} v_i[(n-1/2)T] \tag{7.252}$$

Initially,

$$v_x[(n-1/2)T]^+ = -k_0 v_i[(n-1/2)T] \tag{7.253}$$

The analysis results in the next transfer function given by

$$H(z) = \frac{V_0(z)}{V_i(z)} = H_{id}(z)E(z) \tag{7.254}$$

where

$$H_{id}(z) = \frac{C_1}{C_2} \frac{z^{-1/2}}{1 - z^{-1}} \tag{7.255}$$

and

$$E(z) = 1 - \left(1 + \frac{C_2}{C_1}\right) k_0 e^{-k_2 T/2} \tag{7.256}$$

The associated magnitude and phase errors are

$$m(\omega) = -\left(1 + \frac{C_1}{C_2}\right) k_0 e^{-k_2 T/2} \tag{7.257}$$

and

$$\theta(\omega) = 0 \tag{7.258}$$

In this case, the phase error is eliminated. This is due to an increase in the settling time of the OTA resulting from the additional half-clock period available between the sampling time of the input signal and the beginning of the charge transfer between the capacitors.

In practice, the errors due to the finite bandwidth become negligible by choosing the amplifier unity-gain frequency to be about five times as large as the clock frequency.

7.12 Settling time in the integrator

The circuit diagram of an amplification stage with its capacitive load is depicted in Figure 7.54(a). The amplifier is assumed to have the transfer characteristic shown in Figure 7.54(b). The output current is given by

$$i_0 = \begin{cases} g_m v & \text{for } |v| < V_m \\ \text{sign}(v) I_m & \text{otherwise,} \end{cases} \tag{7.259}$$

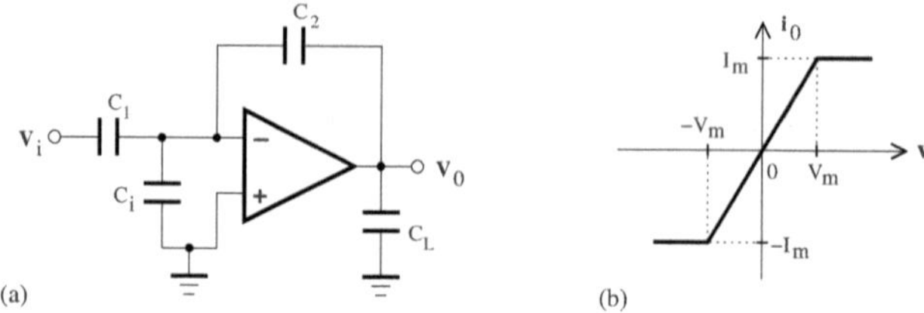

FIGURE 7.54
(a) Amplifier with capacitive loads; (b) voltage-to-current characteristic of the amplifier.

where v denotes the input voltage of the amplifier. The amplifier operates in the linear region with the transconductance g_m provided $|v| < V_m$; otherwise, the output current will be limited by $\pm I_m$ due to the nonlinear effect of the slew rate.

Let us consider the case of the noninverting integrator connected to a step input voltage. Initially,

$$-C_1 v_i = (C_1 + C_i + C_2 + C_L')v(0) \tag{7.260}$$

and

$$v(0) = -\frac{C_1 v_i}{C_1 + C_i + C_2 + C_L'} \tag{7.261}$$

where $C_L' = C_0 || C_L$ and C_i denotes the amplifier input capacitance. The output voltage can be written as

$$v_0(0) = \frac{C_2}{C_2 + C_L'} v(0) \tag{7.262}$$

From the next charge conservation equation,

$$-(C_1 + C_i)[v - v(0)] = C_2\{[v - v(0)] - [v_0 - v_0(0)]\} \tag{7.263}$$

we have

$$v = \beta v_0 - \beta \alpha v_i \tag{7.264}$$

where

$$\alpha = \frac{C_1}{C_2} \tag{7.265}$$

and

$$\beta = \frac{C_2}{C_1 + C_i + C_2} \tag{7.266}$$

With a single-pole model of the amplifier, the differential equation of the circuit is

$$i_0 + \frac{v_0}{r_0} + C_2\frac{\mathrm{d}}{\mathrm{d}t}(v - v_0) + C_L'\frac{\mathrm{d}v_0}{\mathrm{d}t} = 0 \tag{7.267}$$

where $r_0 = 1/g_0$. It can be reduced to

$$\tau\frac{\mathrm{d}v_0}{\mathrm{d}t} + v_0 = -r_0 i_0 \tag{7.268}$$

where

$$\tau = r_0[(1-\beta)C_2 + C_L'] \tag{7.269}$$

For $v \geq V_m$, the output current takes the value I_m and

$$v_0(t) = r_0 I_m\left[1 - \exp\left(-\frac{t}{\tau}\right)\right] + v_0(0)\exp\left(-\frac{t}{\tau}\right) \tag{7.270}$$

At the end of the slewing period, $t = T_{SR}$ and $v(t) = V_m$. Assuming that $\tau \gg T_{SR}$, we can obtain

$$T_{SR} = \left[1 - \frac{V_m + \beta(\alpha v_i - r_0 I_m)}{k(v_0(0) - r_0 I_m)}\right]\tau \tag{7.271}$$

For $v < V_m$, the amplifier operates linearly. With the initial condition at

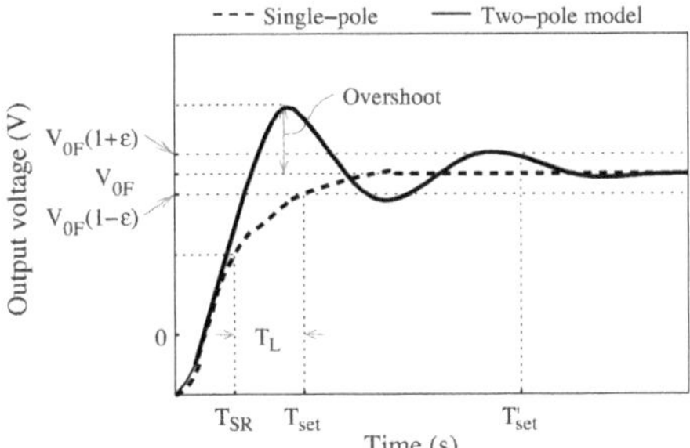

FIGURE 7.55
Amplifier step responses.

$t = T_{SR}$, the output voltage is given by

$$v_0(t) = \frac{\alpha\delta}{1+\delta}v_i\left[1 - \exp\left(-\frac{1+\delta}{\tau}(t - T_{SR})\right)\right] + v_0(T_{SR})\exp\left(-\frac{1+\delta}{\tau}(t - T_{SR})\right) \tag{7.272}$$

where $\delta = \beta g_m r_0$. The settling period, T_{set}, is defined as the time required by the output response to remain within ϵ (generally, about 0.1%) of its final value, V_{0F} (see Figure 7.55). It can be written as

$$T_{set} = T_{SR} + T_L \tag{7.273}$$

where

$$T_L = \frac{\tau}{1+\delta} \ln \left(\frac{v_0(T_{SR}) - V_{0F}}{\epsilon V_{0F}} \right) \tag{7.274}$$

and

$$V_{0F} = \frac{\alpha\delta}{1+\delta} v_i \tag{7.275}$$

The slew rate is approximately given by $SR \simeq (V_{0F} + v_0(0))/T_{SR}$. The amplifier sometimes exhibits a second pole, which is nondominant. In this case, an overshoot can appear in the output transient response, as shown in Figure 7.55.

7.13 Amplifier dc offset voltage limitations

The dc offset voltage is one of the limiting factors in the context of low-voltage applications. It consists of a deterministic and a random component. Because the former is caused by improper bias conditions, it can be reduced to a negligible value by careful circuit design. The latter contributes substantially to the offset voltage and is determined by the level of matching achievable between identical transistors.

To analyze the effect of the offset voltage on the integrator, a voltage source, V_{off}, of unknown polarity is connected to one of the inputs of the amplifier supposed to be ideal. The integrator output voltage can be expressed as

$$V_0(z) = H_{id}(z)V_i(z) + H_0(z)V_{off} \tag{7.276}$$

where $H_{id}(z)$ is the transfer function of the integrator based on the ideal amplifier characteristics and $H_0(z) = C_1/[C_2(1 - z^{-1})]$.

Hence, the offset voltage has essentially a scaling effect on the output voltage. This effect will be practically negligible if an amplifier with low offset voltage (that is, in the range of microvolt) is used.

7.14 Computer-aided analysis of SC circuits

Generally, large-scale integrated circuits are partitioned into small building blocks that can be more easily analyzed. However, because the interactions between the different blocks and parasitic effects may play a significant role, they must be estimated and included in the analysis procedure. This requires the use of an accurate and efficient computer-aided design (CAD) program [42, 43] to simulate the entire circuit. Moreover, even for small-sized networks, the analysis of nonideal effects requires a CAD tool.

The most widely used circuit simulation program is SPICE (Simulation Program with Integrated Circuit Emphasis). It offers a more advanced circuit analysis at the transistor level. However, the SPICE analysis of switched-capacitor circuits, particularly in the frequency domain, is limited by the required amount of computer memory. Hence, we first review the spectral analysis of SC circuits.

bullet **Frequency response of a double-sampling lossless discrete-time integrator**

Let us find the continuous-time frequency response of the circuit in Figure 7.56. Because this circuit does not have a continuous feedthrough path,

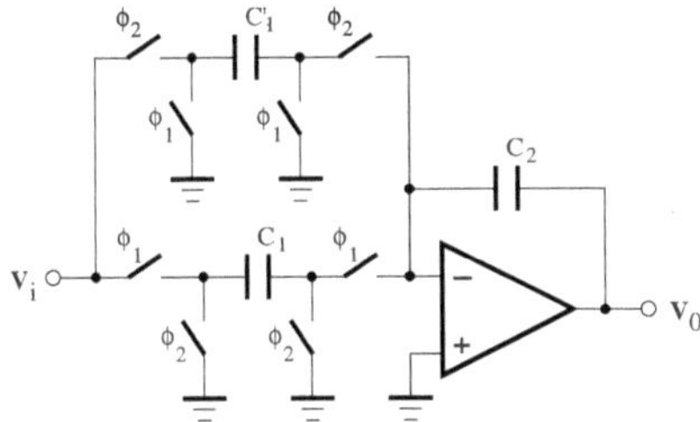

FIGURE 7.56
Double-sampling lossless discrete-time integrator.

its output is always piecewise-constant. Applying Kirchhoff's charge conservation law, for the clock phases 1 and 2, we obtain

$$-C_1 z^{-1/2} V_i^2(z) + C_2 V_0^1(z) - C_2 z^{-1} V_0^2(z) = 0 \tag{7.277}$$

and

$$-C_1 z^{-1/2} V_i^1(z) + C_2 V_0^2(z) - C_2 z^{-1} V_0^1(z) = 0 \tag{7.278}$$

respectively. Solving the above equations with respect to the output voltages, we get

$$V_0^1(z) = \frac{C_1}{C_2} \frac{z^{-1}}{1 - z^{-1}} V_i^1(z) + \frac{C_1}{C_2} \frac{z^{-1/2}}{1 - z^{-1}} V_i^2(z) \tag{7.279}$$

$$V_0^2(z) = \frac{C_1}{C_2} \frac{z^{-1/2}}{1 - z^{-1}} V_i^1(z) + \frac{C_1}{C_2} \frac{z^{-1}}{1 - z^{-1}} V_i^2(z) \tag{7.280}$$

or in matrix form

$$\begin{bmatrix} V_0^1(z) \\ V_0^2(z) \end{bmatrix} = H(z) \begin{bmatrix} 1 & z^{1/2} \\ z^{1/2} & 1 \end{bmatrix} \begin{bmatrix} V_i^1(z) \\ V_i^2(z) \end{bmatrix} \tag{7.281}$$

where

$$H(z) = \frac{C_1}{C_2} \frac{z^{-1}}{1 - z^{-1}} \tag{7.282}$$

The output voltage is observed during both subintervals. The uniform sam-

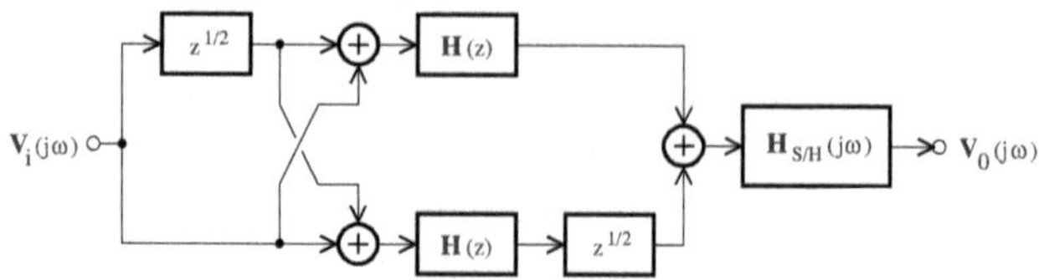

FIGURE 7.57
Block diagram representation of the double-sampling lossless discrete-time integrator.

pling can be modeled mathematically using the following discrete-time input signals, $v_i^1(t)$ and $v_i^2(t)$:

$$v_i^1(t) = v_i(t) \sum_{n=-\infty}^{+\infty} \delta(t - nT) \tag{7.283}$$

$$v_i^2(t) = v_i(t) \sum_{n=-\infty}^{+\infty} \delta(t - nT - T/2) \tag{7.284}$$

The Fourier transform of the sampled input signals is expressed as

$$V_i^1(\omega) = \frac{1}{T} \sum_{n=-\infty}^{+\infty} V_i(\omega - n\omega_s) \tag{7.285}$$

$$V_i^2(\omega) = \frac{1}{T} \sum_{n=-\infty}^{+\infty} V_i(\omega - n\omega_s) e^{j(\omega - n\omega_s)T/2} \tag{7.286}$$

where $\omega_s = 2\pi/T$ and $V_i(\omega)$ is the Fourier transform of the input signal, $v_i(t)$. From the equivalent continuous-time system shown in Figure 7.57, the S/H output spectra of the integrator can be determined as

$$\begin{aligned} V_0(\omega) = H_{SH}(j\omega) \Big[(1 + z^{1/2})H(z)\Big]_{z=e^{j\omega T}} \sum_{n=-\infty}^{+\infty} V_i(\omega - n\omega_s) e^{j(\omega - n\omega_s)T/2} \\ + H_{SH}(j\omega) \Big[(1 + z^{1/2})H(z)\Big]_{z=e^{j\omega T}} \sum_{n=-\infty}^{+\infty} V_i(\omega - n\omega_s) \end{aligned} \tag{7.287}$$

where

$$H_{SH}(j\omega) = \frac{1 - e^{-j\omega T/2}}{j\omega} \tag{7.288}$$

After some simplification, the next expression can be deduced.

$$V_0(\omega) = \frac{C_1}{C_2}\frac{1}{j\omega T}\sum_{n=-\infty}^{+\infty} V_i(\omega - n\omega_s)\left(1 - e^{-jn\omega_s T/2}\right) \tag{7.289}$$

For the baseband, that is, $n = 0$, we obtain

$$V_0(\omega) = \frac{C_1}{C_2}\frac{2}{j\omega T}V_i(\omega) \tag{7.290}$$

This equation indicates that if the clock frequency is greater than the maximum signal frequency, the double sampling integrator will behave exactly like an analog integrator.

bullet **Frequency response of SC circuits**
The analytical spectral analysis of SC circuits can be complicated for large structures. The use of the circuit model of Figure 7.58 allows us to obtain the frequency response directly from the z-domain transfer function based on the direct substitution $z = e^{j\omega T}$. The impulse sampling stage transforms

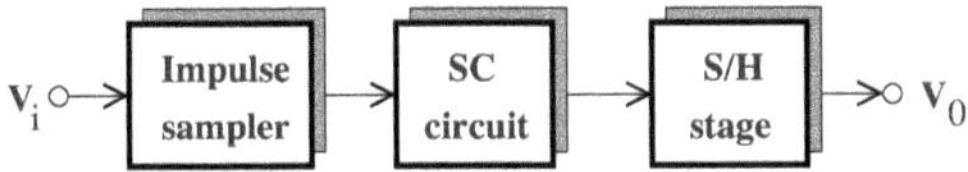

FIGURE 7.58
Frequency domain representation of an SC circuit.

the analog signal into an analog discrete-time signal. It should be emphasized here that, in accordance with the sampling theorem,[8] the maximum frequency of the incoming signal is limited to less than the Nyquist rate, that is, $\omega_s/2 = \pi/T$. The SC circuit can then be represented as a discrete-time signal processor. Its output signal is received and transformed into a staircase signal by the S/H stage.

In general, the frequency domain transfer function can be written as

$$H(j\omega) = \sum_{l=1}^{M}\left[\sum_{k=1}^{M} H^l_{SH}(j\omega)\, H^{(k,l)}(z)\Big|_{z\,=\,e^{j\omega T}}\right] \tag{7.291}$$

where

$$H^l_{SH}(j\omega) = e^{-j\omega(\tau_l/2)}\frac{\tau_l}{T}\frac{\sin(\omega\tau_l/2)}{\omega\tau_l/2} \tag{7.292}$$

Here, $H^{(k,l)}$ is the transfer function of the analog discrete-time circuit in the

[8]The sampling theorem can be stated as follows: A band-limited lowpass signal, $x(t)$, with spectrum $X(j\omega) = 0$ for all $|\omega| > \omega_m$ is uniquely and completely described by a set of sample values $x(nT)$ taken at uniformly spaced time instants separated by $T = \pi/\omega_m$ seconds or less.

z-domain, the superscripts k and l mean that the input voltage is sampled at the beginning of the k-th clock phase and the output voltage is observed during the l-th clock phase. It should be emphasized that the input signal is sampled M times per period of the clock signal and that the output signal is maintained constant during the subinterval of observation, τ_l.

The simulation of SC circuits can be computationally intensive due to their time-varying nature. A program such as SPICE or HSPICE [44] can only provide the transient analysis of SC circuits. In order to obtain the frequency response, it is necessary to use specialized programs [45, 46] supporting only behavioral-level descriptions of the circuit, or SpectreRF [47,48]. In contrast to HSPICE, SpectreRF relies on performing analysis about a periodic operating point. It can then be applied to predict SC circuit characteristics in the time and frequency domains using nonideal components at the transistor level.

7.15 T/H and S/H circuits based on the SC circuit principle

T/H and S/H circuits based on a clock waveform with more than two clock phases can be realized using SC circuit techniques. From the analysis of their associated continuous-time subcircuits, it appears that the circuit settling time is limited by the clock period. As a consequence, the number of clock phases must be limited to the minimum (i.e., two) for high-frequency applications.

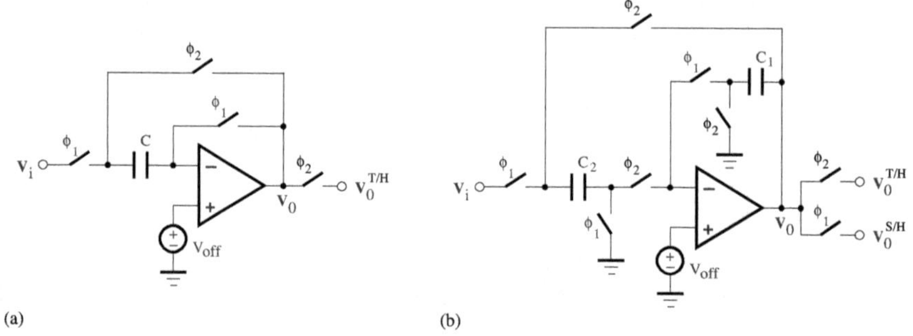

FIGURE 7.59
(a) Offset-compensated T/H circuit; (b) S/H circuit with an improved speed performance.

The T/H circuit depicted in Figure 7.59(a) [17] operates with a two-phase clock signal. Initially, the switches controlled by the first clock phase, ϕ_1, are closed and the capacitor C is connected between the input voltage and the inverting node of the amplifier, which is configured to operate as a unity-gain buffer. During the second clock phase, ϕ_2, the capacitor C is included in the amplifier feedback path, and the amplifier output is equal to a delayed version of the input signal. Because the capacitor C always remains connected to the inverting node of the amplifier, the offset voltage contribution stored on C during ϕ_1, that is, for $(n-1)T \leq t < (n-1/2)T$, is cancelled by the one produced during ϕ_2, that is, for $(n-1/2)T \leq t < nT$. Taking into account the amplifier finite gain, the application of the charge conservation rule at the amplifier inverting node during ϕ_2 yields

$$V^-(nT) - V_0(nT) = V^-[(n-1/2)T] - V_i[(n-1/2)T] \tag{7.293}$$

In general, we have

$$V^- = -\mu V_0 + V_{off} \tag{7.294}$$

where $\mu = 1/A_0$ and A_0 is the amplifier dc gain. We can then combine (7.294) and the expression, $V^-[(n-1/2)T] = V_0[(n-1/2)T]$, obtained at the end of ϕ_1, to write

$$V_0[(n-1/2)T] = V_{off}/(1+\mu) \tag{7.295}$$

Finally, by substituting Equations (7.294) and (7.295) into Equation (7.293), the output voltage is expressed as

$$V_0(nT) = \frac{V_i[(n-1/2)T]}{1+\mu} + \mu\frac{V_{off}}{(1+\mu)^2} \tag{7.296}$$

Ideally, the dc gain is infinite and the offset voltage is negligible, resulting in $V_0(nT) = V_i[(n-1/2)T]$. Due to the fact that the output is reset during the first clock phase, the resulting speed appears to be critically limited by the required value of the amplifier slew rate and settling time.

An S/H circuit, as shown in Figure 7.59(b), does not require a high-speed amplifier [18]. During the clock phase ϕ_1, the capacitor C_2 is charged up to the input voltage, while C_1 is connected between the inverting node and output of the amplifier. During the clock phase ϕ_2, that is, for $(n-1/2)T \leq t < nT$, the capacitor C_2 is placed in the amplifier feedback loop and C_1 is connected as an output load. Hence,

$$C_2[V^-(nT) - V_0(nT)] = -C_2 V_i[(n-1/2)T] \tag{7.297}$$

At the end of the next clock phase ϕ_1 occurring for $nT \leq t < (n+1/2)T$, the charge conservation equation can be written as

$$C_1\left\{V^-[(n+1/2)T] - V_0[(n+1/2)T]\right\} = -C_1 V_0[(n)T] \tag{7.298}$$

Because the voltage at the amplifier inverting node is $V^- = -\mu V_0 + V_{off}$, where $\mu = 1/A_0$ and A_0 is the amplifier dc gain, we can obtain

$$V_0(nT) = \frac{V_i[(n-1/2)T] + V_{off}}{1+\mu} \tag{7.299}$$

and

$$V_O[(n+1/2)T] = \frac{V_O(nT) + V_{off}}{1+\mu} \tag{7.300}$$

In the case where the amplifier is assumed to be ideal, it can be observed that, once per clock period, a sample of the input signal is transferred to the output, whose level is maintained constant up to the next update.

The S/H circuit shown in Figure 7.60(a) consists of capacitors, an amplifier, an inverting voltage buffer, and switches, which are controlled by a clock signal with the two nonoverlapping phases, ϕ_1 and ϕ_2. During the

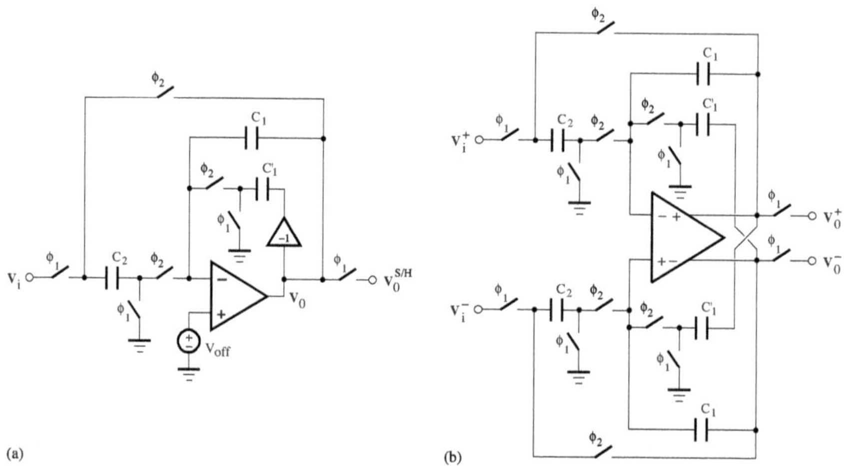

FIGURE 7.60
Circuit diagram of the unity gain S/H circuit ($C_1' = C_1$): (a) single-ended and (b) differential implementations.

first half of the clock signal period (or the first clock phase, ϕ_1), that is, $(n-1)T \leq t < (n-1/2)T$, the capacitor C_2 is charged to V_i, while the capacitors C_1 and C_1' are charged to V_O and $-V_O$, respectively. Next, the amplifier and the feedback capacitors C_1 are isolated; thus the amplifier output voltage of the previous phase is maintained. From the charge conservation equations, we can obtain

$$V_i[(n-1/2)T] = V_i[(n-1)T] \tag{7.301}$$

and

$$V_O[(n-1/2)T] = V_O[(n-1)T] \tag{7.302}$$

During the second clock phase, ϕ_2, that is, $(n-1/2)T \leq t < nT$, all capacitors are connected to the amplifier. Because the charges due to the output voltages and stored on the capacitors C_1 and C_1' have opposite signs, they ideally cancel each other and the new charge redistribution is only determined by the charges stored on the capacitors C_2. The application of the charge conservation rule at the amplifier inverting node gives

$$\begin{aligned} &C_2\{V_0(nT) - V^-(nT) - V_i[(n-1/2)T]\} \\ &\quad + C_1\{-V_0(nT) - V^-(nT) - (-V_0[(n-1/2)T])\} \\ &\quad = C_1\{V^-(nT) - V_0(nT) - (V^-[(n-1/2)T] - V_0[(n-1/2)T])\} \end{aligned} \tag{7.303}$$

Taking into account the fact that $V^- = -\mu V_0 + V_{off}$, and substituting Equations (7.301) and (7.302) into Equation (7.303), we arrive at

$$[(1+\mu)C_2 + 2\mu C_1]V_0(nT) - \mu C_1 V_0[(n-1)T] = C_2 V_i[(n-1)T] + (C_1 + C_2)V_{off} \tag{7.304}$$

Using the z-domain transform, it can be shown that

$$V_0(z) = H_i(z)V_i(z) + H_0(z)V_{off} \tag{7.305}$$

where

$$H_i(z) = \pm\frac{z^{-1}}{1 + \mu + \mu\dfrac{C_1}{C_2}\left(2 - z^{-1}\right)} \tag{7.306}$$

and

$$H_0(z) = \frac{1 + \dfrac{C_1}{C_2}}{1 + \mu + \mu\dfrac{C_1}{C_2}\left(2 - z^{-1}\right)} \tag{7.307}$$

Here, V_{off} represents the amplifier offset voltage and $A_0 = 1/\mu$ is the amplifier dc gain. The sign of the transfer function, H_i, is determined by the signal polarity. Ideally, V_{off} is negligible and $\mu \ll 1$, such that

$$V_0(z) = \pm z^{-1}V_i(z) \tag{7.308}$$

Hence, the input signal is sampled and held for a full-clock period. Figure 7.60(b) [19] shows the differential implementation of the S/H circuit.

The aforementioned S/H structure has the advantage of preventing large signal variations at the amplifier output during the sampling phase, thus reducing the effect of delays caused by the transient response. However, it can only be used to implement a delay with unity gain. Generally, an additional gain stage is required to perform the amplification or attenuation function.

An alternative architecture is shown in Figure 7.61(a) [20]. It features a variable gain and operates as follows. During the first clock phase, ϕ_1, that is, $(n-1)T \leq t < (n-1/2)T$, the capacitor C_1 is connected and charged to the voltages V_i. A similar situation takes place between the capacitor C_2' and the voltage V_0. The feedback capacitor C_2 remains connected between the output and inverting input of the amplifier. Hence,

$$V_i[(n-1/2)T] = V_i[(n-1)T] \tag{7.309}$$

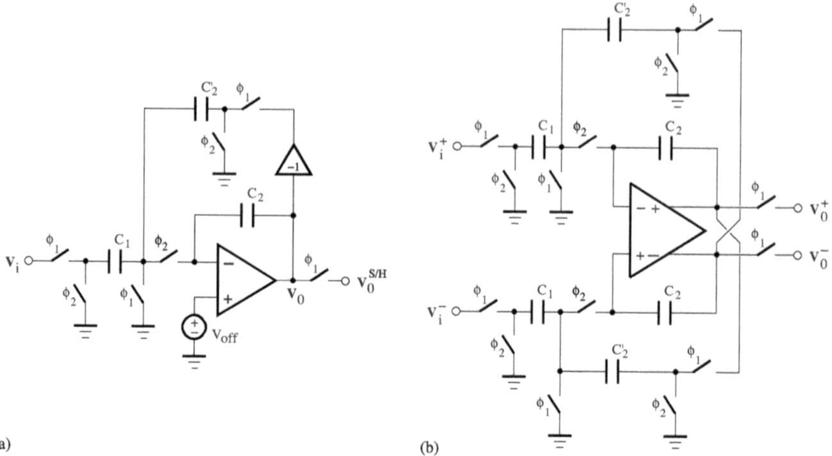

FIGURE 7.61
Circuit diagram of the S/H circuit with a variable gain ($C_2' = C_2$): (a) Single-ended and (b) differential implementations.

and

$$V_0[(n-1/2)T] = V_0[(n-1)T] \tag{7.310}$$

During the second clock phase, ϕ_2, that is, $(n-1/2)T \leq t < nT$, all capacitors are connected to the amplifier inverting input and a cancellation between the same charges stored, respectively, on C_2 and C_2' takes place. The charge conservation equation can be written as

$$\begin{aligned} -C_1\{V^-(nT) + V_i[(n-1/2)T]\} + C_2\{-V^-(nT) + V_0[(n-1/2)T]\} \\ = C_2\{V^-(nT) - V_0(nT) - (V^-[(n-1/2)T] - V_0[(n-1/2)T])\} \end{aligned} \tag{7.311}$$

Using the relationship, $V^- = -\mu V_0 + V_{off}$, in addition to Equations (7.309) and (7.310), we obtain

$$[(1+2\mu)C_2 + \mu C_1]V_0(nT) - \mu C_2 V_0[(n-1)T] = C_1 V_i[(n-1)T] + (C_1+C_2)V_{off} \tag{7.312}$$

The output-input relationship can then be described in the z-domain by

$$V_0(z) = H_i(z)V_i(z) + H_0(z)V_{off} \tag{7.313}$$

where

$$H_i(z) = \pm\frac{C_1}{C_2}\frac{z^{-1}}{1+\mu\left(2+\dfrac{C_1}{C_2}\right) - \mu z^{-1}} \tag{7.314}$$

and

$$H_0(z) = \frac{1 + \dfrac{C_1}{C_2}}{1 + \mu\left(2 + \dfrac{C_1}{C_2}\right) - \mu z^{-1}} \tag{7.315}$$

Here, V_{off} represents the amplifier offset voltage and $A_0 = 1/\mu$ is the amplifier dc gain. In the ideal case, that is, when V_{off} is negligible and $\mu \ll 1$, we obtain

$$V_0(z) = \pm\frac{C_1}{C_2}z^{-1}V_i(z) \tag{7.316}$$

In addition to being sampled and held for a full clock period, the input signal is scaled by a factor set by the capacitor ratio. The differential implementation of the S/H circuit shown in Figure 7.61(b) offers the advantage of generating the inverting version of the output voltage without the need for an inverting gain stage.

7.16 Circuit structures with low sensitivity to nonidealities

The operation of SC circuits is strongly influenced by the amplifier characteristics. The purpose of the circuit techniques discussed in this section is to relax the constraints that must be imposed on the amplifier structure.

Two techniques are described in this section in order to improve the performance of SC circuits. They are respectively based on the use of additional active components or the correlated double sampling (CDS) scheme. In a CDS-based structure, the error signal is sampled and stored on a capacitor. It is then subtracted from the samples of the input signal in order to achieve the compensation. The dc offset voltage, which is constant over the time, can be cancelled in this way. This can also be the case of any noise source that varies very slowly.

In the following analysis, the circuit output voltage, V_0, is written as

$$V_0(z) = H_i(z)V_i(z) + H_0(z)V_{off} \tag{7.317}$$

Note that V_i is the input voltage; V_{off} and H_0 denote the offset voltage and its associated transfer function, respectively; and $H_i = H_{id} \cdot E$, where

$$H_{id}(z) = \begin{cases} -\dfrac{C_1}{C_2}\dfrac{1}{1 - z^{-1}} & \text{for the integrator} \\ -\dfrac{C_1}{C_2} & \text{for the gain stage} \end{cases} \tag{7.318}$$

and the error function, E, is due to the amplifier dc gain. For convenience, a 50% duty cycle clock pattern is assumed and the amplifier dc gain, A_0, will be indicated by $\mu = 1/A_0$.

7.16.1 Integrators

For applications requiring a high precision, SC integrators can be designed to exhibit a low sensitivity to the dc gain of the amplifier.

• The circuit technique presented in [49] for the realization of an integrator with a low sensitivity to the amplifier dc gain uses a unity-gain buffer. If this latter is ideal, a phase-error-free integrator will be obtained. Figure 7.62 shows the compensated integrator circuit. By inspection of the circuit, it can be

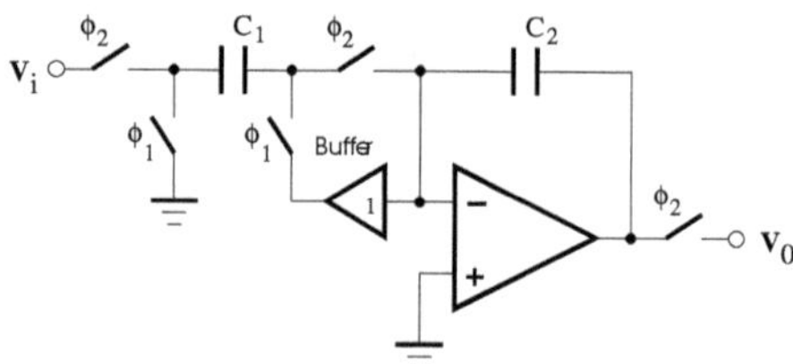

FIGURE 7.62
Phase-error free integrator.

found that during the clock phase 1, the output voltage will not change and therefore the error signal will be stored on the input capacitor C_1. The charge transfer takes place during the clock phase 2. Then, we have

$$v_0[(n-1/2)T] = v_0[(n-1)T] \tag{7.319}$$

during the clock phase 1 and

$$\begin{aligned} C_1\left\{v_i(nT) + \mu v_0(nT) - (1-\alpha)\mu v_0[(n-1/2)T] - \alpha V_{off}\right\} \\ + (1+\mu)C_2\left\{v_0(nT) - v_0[(n-1)T]\right\} = 0 \end{aligned} \tag{7.320}$$

during the clock phase 2. The buffer gain has the value $1-\alpha$, where α is the gain error. The functions E and H_0 are given by

$$E(z) = \frac{1}{1 + \mu + \mu\dfrac{C_1}{C_2} + \alpha\mu\dfrac{C_1}{C_2}\dfrac{z^{-1}}{1-z^{-1}}} \tag{7.321}$$

and

$$H_0(z) = \alpha\frac{C_1}{C_2(1-z^{-1})}E(z) \tag{7.322}$$

respectively. This technique will be efficient if α is on the order of a few percent and A_0 is greater than 100.

• The schematic diagram of an integrator based on the CDS scheme [50] is depicted in Figure 7.63. The charge conservation principle can be used to

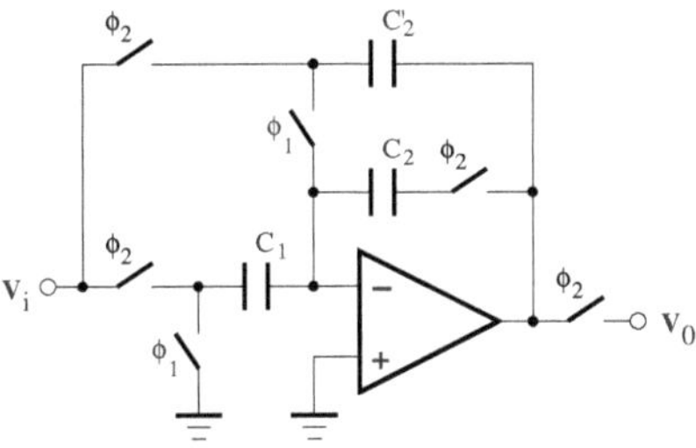

FIGURE 7.63
CDS compensated integrator I.

derive the following equations

$$C_1 \{-v_i[(n-1)T] + \mu v_0[(n-1/2)T] - \mu v_0[(n-1)T]\} \\ + C_2' \{-v_i[(n-1)T] - (1+\mu)v_0[(n-1/2)T] + v_0[(n-1)T] + V_{off}\} = 0 \tag{7.323}$$

during the clock phase 1 and

$$C_1 \{v_i(nT) + \mu v_0(nT) - \mu v_0[(n-1/2)T]\} \\ + (1+\mu)C_2 \{v_0(nT) - v_0[(n-1)T]\} = 0 \tag{7.324}$$

during the clock phase 2. The capacitor C_2' is included in the feedback path during the clock phase 1 and enables C_1 to discharge. The charge previously stored on C_2' is then used for the purpose of compensation. The resulting error function can be written as

$$E(z) = \frac{1 - \dfrac{\mu C_1}{\mu C_1 + (1+\mu)C_2'}\left(1 - \dfrac{C_2'}{C_1}\right) z^{-1}}{1 + \dfrac{\mu C_1}{(1+\mu)C_2} \dfrac{1}{1-z^{-1}} \left(1 - \dfrac{C_2' + \mu C_1}{\mu C_1 + (1+\mu)C_2'} z^{-1}\right)} \tag{7.325}$$

The contribution of the dc offset voltage to the output signal can be deduced from the transfer function

$$H_0(z) = \frac{\dfrac{-\mu C_2'}{\mu C_1 + (1+\mu)C_2'}}{1 + \dfrac{\mu C_1}{(1+\mu)C_2} \dfrac{1}{1-z^{-1}} \left(1 - \dfrac{C_2' + \mu C_1}{\mu C_1 + (1+\mu)C_2'} z^{-1}\right)} H_{id}(z) \tag{7.326}$$

The compensation strategy is effective when $C_2' = C_1$. Furthermore, the compensated integrator in a configuration with N inputs will require N capacitors C_2'.

• The circuit of Figure 7.64 [51, 52] also relies on CDS switching for the amplifier nonideality reduction. If $v_O[(n-1/2)T]$ is made equal to $v_O[(n-1)T]$, the transfer function, H_i, will be free of a phase error and the gain error will be $-\mu(1 + C_1/C_2)$. This ideal situation is only approximated by the next difference equations valid at the end of the clock phase 1:

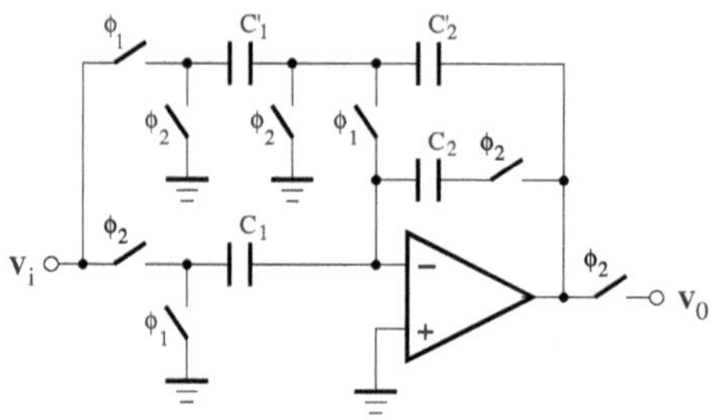

FIGURE 7.64
CDS compensated integrator II.

$$\begin{aligned} C_1\{-v_i[(n-1)T] + \mu v_O[(n-1/2)T] - \mu v_O[(n-1)T]\} \\ + C_1'\{v_i[(n-1/2)T] + \mu v_O[(n-1/2)T] - V_{off}\} \\ + C_2'\{(1+\mu)v_O[(n-1/2)T] - v_O[(n-1)T] - V_{off}\} = 0 \end{aligned} \tag{7.327}$$

By using the charge conservation laws at the end of the clock phase 2, we arrive at

$$\begin{aligned} C_1\{v_i(nT) + \mu v_O(nT) - \mu v_O[(n-1/2)T]\} \\ + C_2(1+\mu)\{v_O(nT) - v_O[(n-1)T]\} = 0 \end{aligned} \tag{7.328}$$

The functions E and H_0 can then be derived from the z-domain output-input relation as

$$E(z) = \frac{1 + \dfrac{\mu C_1}{\mu(C_1 + C_1') + (1+\mu)C_2}(C_1' z^{-1/2} - C_1 z^{-1})}{1 + \dfrac{\mu C_1}{(1+\mu)C_2}\dfrac{1}{1-z^{-1}}\left(1 - \dfrac{\mu C_1 + C_2'}{\mu(C_1 + C_1') + (1+\mu)C_2'}z^{-1}\right)} \tag{7.329}$$

and

$$H_0(z) = \frac{\dfrac{-\mu(C_1' + C_2')}{\mu(C_1 + C_1') + (1+\mu)C_2}}{1 + \dfrac{\mu C_1}{(1+\mu)C_2}\dfrac{1}{1-z^{-1}}\left(1 - \dfrac{\mu C_1 + C_2'}{\mu(C_1 + C_1') + (1+\mu)C_2'}z^{-1}\right)} H_{id}(z) \tag{7.330}$$

respectively. The choice of capacitors C_1' and C_2' is dictated by the minimization of the transfer function deviations. If the input voltage varies very slowly, that is, $v_i[(n-1/2)T] = v_i[(n-1)T]$, $C_1' = 2C_1$ and $C_2' = C_2$, the integrator dc gain will be reduced to

$$H_i(1) = H_{id}(z)E(z)|_{z=1} = \mu^{-2}\frac{1 + \mu\left(1 + 4\dfrac{C_1}{C_2}\right)}{1 + 2\dfrac{C_1}{C_2}} \tag{7.331}$$

Thus, the effective gain is now a function of μ^{-2} (or A_0^2) rather than A_0, as in the case of the uncompensated structure.

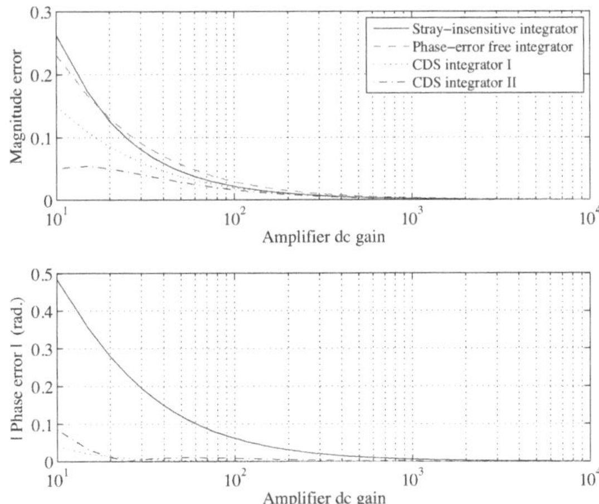

FIGURE 7.65
Plots of the integrator gain and phase errors ($C_1/C_2 = 2$ and $f/f_c = 0.05$).

The effects of the amplifier dc gain on the different integrator structures are shown in Figure 7.65. The CDS compensated integrators seem to offer better performance than other structures, but they also require extra switches and capacitors.

• A goal in designing an SC integrator that is insensitive to amplifier nonidealities is to maintain the input and feedback capacitors connected at the summing nodes during both clock phases, so that the charge transfer between the capacitors is made without an error. Based on this design principle, an implementation of the CDS technique was proposed in [52], but the output of the resulting circuit (see Figure 7.64) is valid only during one clock phase. Furthermore, this circuit works well only under the assumption that the input voltage is approximately constant over two consecutive half-clock cycles.

An integrator circuit that operates without constraints on the input voltage is depicted in Figure 7.66 [58]. Note that the second input path capacitor of

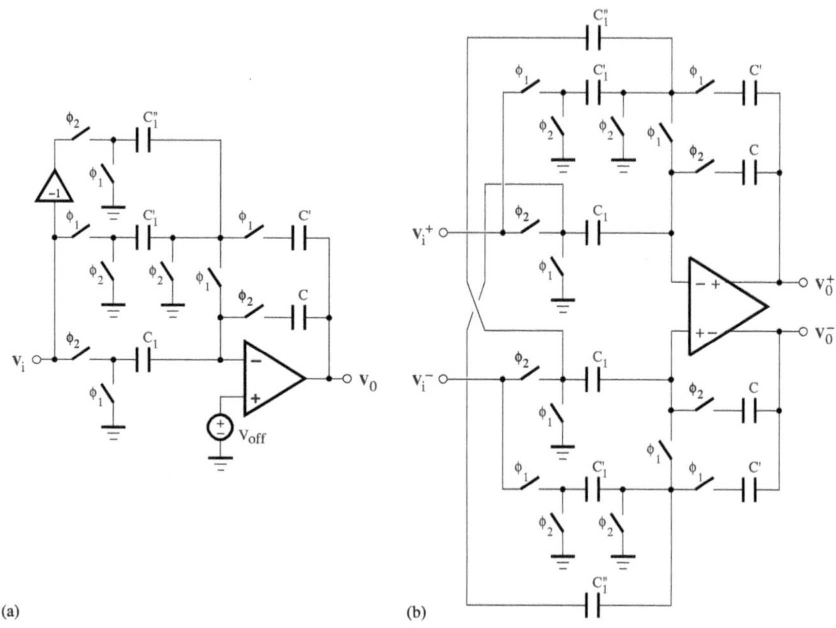

FIGURE 7.66
Circuit diagram of a double-sampling, low-sensitivity integrator ($C = C'$ and $C_1 = C'_1 = C''_1$): (a) Single-ended and (b) differential versions.

the integrator of [52], which is two times the first input path capacitor value, is divided into two equal capacitors: one of these is connected to the input voltage and the other is connected to a negative version of the input voltage. Ideally, the application of the charge conservation law at the amplifier input nodes leads to the following difference equation,

$$Cv_0(nT) = Cv_0[(n-1)T] - C_1 v_i(nT) \tag{7.332}$$

for the clock phase 2. In this modified SC integrator, the capacitor C''_1 is charged during the clock phase 1 by a negative version of the input voltage. In this way, its charge is used in the next clock phase (clock phase 2) to cancel the charge generated by the capacitor C_1, which is always connected at the summing node. Then, the next relation can be written as

$$Cv_O[(n-1/2)T] = Cv_O[(n-3/2)T] - C_1 v_i[(n-1/2)T] \tag{7.333}$$

for the clock phase 1. It appears that the output $v_O[(n-1/2)T]$ in the clock phase 1 and the output $v_O(nT)$ in the clock phase 2 represent a discrete-time integration of a given input signal sample.

Taking into account the effect of the finite dc gain, $A_0 = 1/\mu$, and offset voltage, V_{off}, of the operational amplifier, the application of the charge conservation law at the end of the clock phase ϕ_1, which is characterized by $(n-1)T \leq t < (n-1/2)T$, gives

$$\begin{aligned} C_1\{\mu v_O[(n-1/2)T] - v_i[(n-1)T] - \mu\{v_O[(n-1)T]\}\\ + C_1'\{v_i[(n-1/2)T] + \mu\{v_O[(n-1/2)T] - V_{off}\}\\ + C"_1\{\mu v_O[(n-1/2)T] + v_i[(n-1)T] - V_{off}\}\\ + C_2(1+\mu)\{v_O[(n-1/2)T] - v_O[(n-3/2)T]\} = 0 \end{aligned} \tag{7.334}$$

For the clock phase ϕ_2, or the time instants defined by $(n-1/2)T \leq t < nT$, the charge conservation law can be written as

$$C_1\{v_i(nT) + \mu v_O(nT) - \mu v_O[(n-1/2)T]\} + C_2(1+\mu)\{v_O(nT) - v_O[(n-1)T]\} = 0 \tag{7.335}$$

Combining Equations (7.334) and (7.334) and taking the z-transform of the resulting difference equation, we can obtain

$$V_O(z) = H_i(z)V_i(z) + H_O(z)V_{off} \tag{7.336}$$

where $H_i = H_{id} \cdot E$. Here,

$$H_{id}(z) = \pm\frac{C_1}{C}\frac{1}{1-z^{-1}} \tag{7.337}$$

$$E(z) = \frac{1 + \mu\dfrac{C_1}{C}x^{-1}z^{-1/2}}{1 + \mu + \dfrac{\mu\dfrac{C_1}{C}}{1-z^{-1}}\left[1 - \mu\dfrac{C_1}{C}x^{-1}z^{-1}\left(1 + \dfrac{1+\mu}{\mu\dfrac{C_1}{C}}z^{-1/2}\right)\right]} \tag{7.338}$$

$$H_O(z) = H_{id}(z) \cdot \frac{2\mu\dfrac{C_1}{C}x^{-1}}{1 + \mu + \dfrac{\mu\dfrac{C_1}{C}}{1-z^{-1}}\left[1 - \mu\dfrac{C_1}{C}x^{-1}z^{-1}\left(1 + \dfrac{1+\mu}{\mu\dfrac{C_1}{C}}z^{-1/2}\right)\right]} \tag{7.339}$$

and

$$x = 1 + \mu + 3\mu \frac{C_1}{C} \tag{7.340}$$

where H_{id} represents the integrator ideal transfer function and E gives an estimation of the transfer function deviation caused by the finite amplifier gain. At dc, that is, for $z = 1$, the transfer function, H_i, is reduced to

$$|H_i(1)| = \mu^{-2} \frac{1 + \mu + 4\mu \frac{C_1}{C}}{2\frac{C_1}{C}} \tag{7.341}$$

Therefore, the effective gain of the amplifier is squared ($\mu^{-2} = A_0^2$).

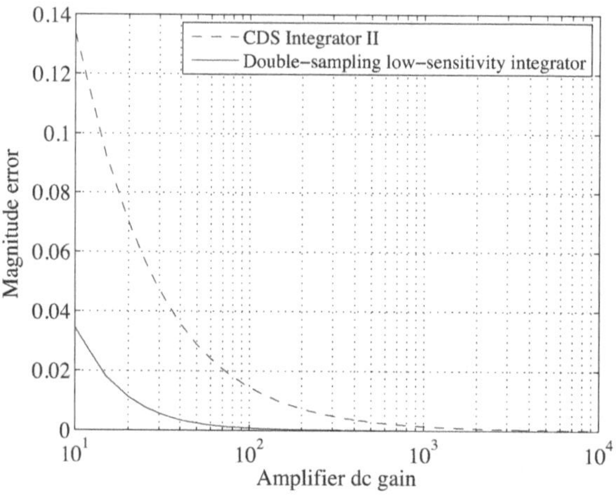

FIGURE 7.67
Gain errors of low-sensitivity integrators ($f = 100$ kHz, $f_c = 2$ MHz, and $C_1 = C$).

The error function magnitudes were computed as $1 - |E|$, for the structures of Figures 7.64 and 7.66. As Figure 7.67 demonstrates, the double-sampling circuit topology is far superior in magnitude accuracy. The action of the SC integrator on the offset voltage is characterized by the transfer function H_0. Due to the CDS switching, the offset voltage contribution is primarily determined by the size of only two capacitors (here, C_1' and C_1''), as is the case for the integrator shown in Figure 7.64.

7.16.2 Gain stages

A conventional gain stage structure is shown in Figure 7.68 [53]. Ideally, its output voltage in the z-domain can be written as

$$V_0(z) = -(C_1/C_2)V_i(z), \tag{7.342}$$

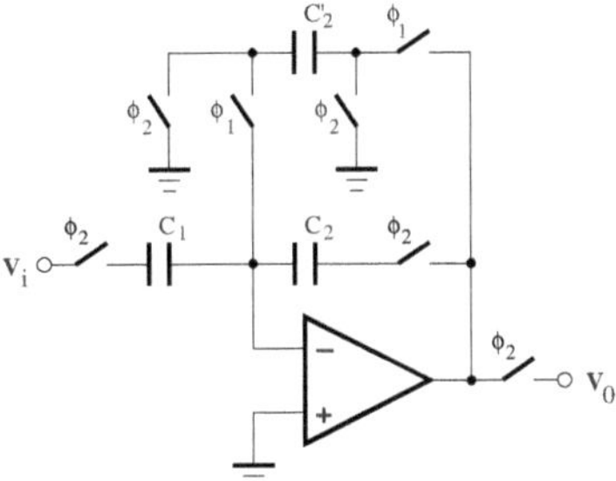

FIGURE 7.68
Uncompensated gain stage.

Taking into account the finite dc gain, we obtain

$$V_0(z) = -\frac{C_1}{C_2} \frac{1}{1 + \mu\left(1 + \dfrac{C_1}{C_2}\right)} V_i(z) \tag{7.343}$$

The resulting gain is then affected by an error term dependent on μ. For typical component values, a gain error on the order of 1% is to be expected.

In order to attenuate the above-mentioned deviation, the CDS technique can be used for the design of gain stage circuits.

- The circuit shown in Figure 7.69 was proposed in [51]. During the clock

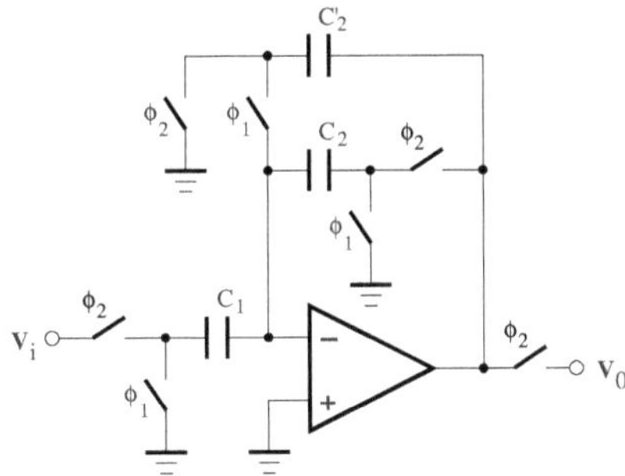

FIGURE 7.69
CDS compensated gain stage I.

phase 1, C_2' plays the role of the feedback capacitor and the other capacitors are connected between the amplifier inputs and the ground. The charge conservation equation is then

$$\begin{aligned}C_1 \{-v_i[(n-1)T] + \mu v_0[(n-1/2)T] - \mu v_0[(n-1)T]\} \\ + C_2 \{\mu v_0[(n-1/2)T] - (1+\mu) v_0[(n-1)T]\} \\ + C_2' \{(1+\mu) v_0[(n-1/2)T] - v_0[(n-1)T] - V_{off}\} = 0\end{aligned} \tag{7.344}$$

During the clock phase 2, the capacitor C_1 is connected to the input voltage and a charge transfer can take place between C_1 and C_2. This results in the following equation,

$$\begin{aligned}C_1 \{v_i(nT) + \mu v_0(nT) - \mu v_0[(n-1/2)T]\} \\ + C_2 \{(1+\mu) v_0(nT) - \mu v_0[(n-1/2)T]\} = 0\end{aligned} \tag{7.345}$$

From the above equations, we can obtain

$$E(z) = \frac{1 - \dfrac{\mu(C_1+C_2)}{C_2' + \mu(C_1+C_2+C_2')} z^{-1}}{1 + \mu\left(1 + \dfrac{C_1}{C_2}\right) - \dfrac{\mu(C_1+C_2)}{C_2' + \mu(C_1+C_2+C_2')}\left[1 + \dfrac{C_2'}{C_2} + \mu\left(1 + \dfrac{C_1}{C_2}\right)\right] z^{-1}} \tag{7.346}$$

and

$$H_0(z) = \frac{C_2'}{C_2} \times \frac{\dfrac{\mu(C_1+C_2)}{C_2' + \mu(C_1+C_2+C_2')}}{1 + \mu\left(1 + \dfrac{C_1}{C_2}\right) - \dfrac{\mu(C_1+C_2)}{C_2' + \mu(C_1+C_2+C_2')}\left[1 + \dfrac{C_2'}{C_2} + \mu\left(1 + \dfrac{C_1}{C_2}\right)\right] z^{-1}} \tag{7.347}$$

It can be observed that the gain is frequency dependent due to the CDS high-pass filter effect.

- The circuit diagram of another gain stage that relaxes the amplifier specifications is shown in Figure 7.70 [51]. Here, the input and output branches are duplicated in order to provide an anticipatory amplification during the clock phase 1. The circuit operation can be described by

$$\begin{aligned}C_1 \{-\, v_i[(n-1)T] + \mu v_0[(n-1)T] - \mu v_0[(n-1)T]\} \\ + C_2 \{\mu v_0[(n-1/2)T] - (1+\mu) v_0[(n-1)T]\} \\ + C_1' \{v_i[(n-1/2)T] + \mu v_0[(n-1/2)T] - V_{off}\} \\ + C_2' \{(1+\mu) v_0[(n-1/2)T] + V_{off}\} = 0\end{aligned} \tag{7.348}$$

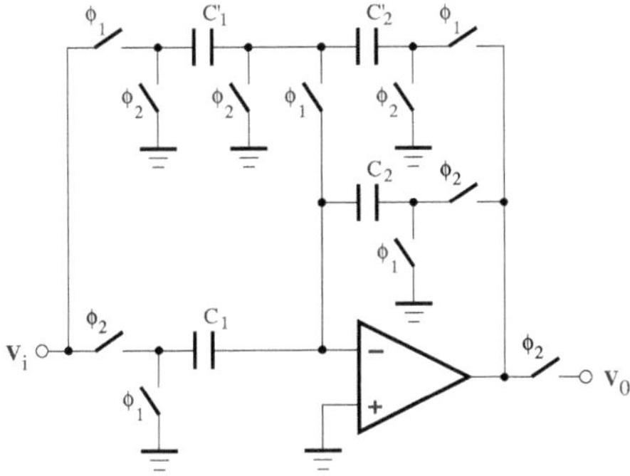

FIGURE 7.70
CDS compensated gain stage II.

During the clock phase 2, C_1 and C_2 form the signal path around the amplifier and the appropriate output signal is generated according to the following relation:

$$\begin{aligned} C_1 \{v_i(nT) + \mu v_0(nT) - \mu v_0[(n-1/2)T]\} \\ + C_2 \{(1+\mu)v_0(nT) - \mu v_0[(n-1/2)T]\} = 0 \end{aligned} \tag{7.349}$$

In this case, we have

$$E(z) = \frac{1 + \dfrac{\mu(C_1+C_2)}{C'_2 + \mu(C_1 + C'_1 + C_2 + C'_2)}\left(\dfrac{C'_1}{C_1} z^{-1/2} - z^{-1}\right)}{\left(1 + \mu + \mu\dfrac{C_1}{C_2}\right)\left(1 - \dfrac{\mu(C_1+C_2)}{C'_2 + \mu(C_1 + C'_1 + C_2 + C'_2)} z^{-1}\right)} \tag{7.350}$$

and

$$H_0(z) = \frac{C'_2}{C_2}\left(1 + \frac{C'_1}{C'_2}\right) \frac{\dfrac{\mu(C_1+C_2)}{C'_2 + \mu(C_1 + C'_1 + C_2 + C'_2)}}{\left(1 + \mu + \mu\dfrac{C_1}{C_2}\right)\left(1 - \dfrac{\mu(C_1+C_2)}{C'_2 + \mu(C_1 + C'_1 + C_2 + C'_2)} z^{-1}\right)} \tag{7.351}$$

The capacitors C'_1 and C'_2 can be chosen to satisfy the relations $C'_1 = C_1$ and $C'_2 = C_2$.

The error functions of the aforementioned gain stages are plotted in Figure 7.71. The performance provided by CDS-compensated gain stages appears to be superior to the ones of the other circuit structures.

In high-frequency applications, the amplifier is designed to have a moderate dc gain and a slew rate of about hundreds of millivolts per nanosecond (mV/ns). Because some improved building blocks operate with a nonoverlapping switching scheme, the amplifier may saturate during the brief intervals,

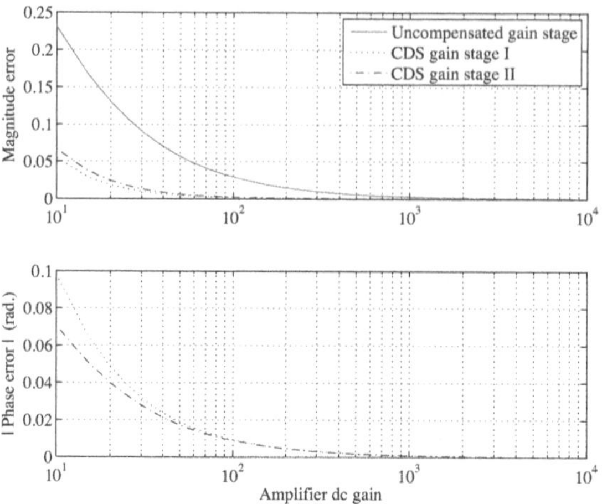

FIGURE 7.71
Plots of the gain stage errors due to the amplifier finite dc gain ($C_1/C_2 = 2$ and $f/f_c = 0.05$).

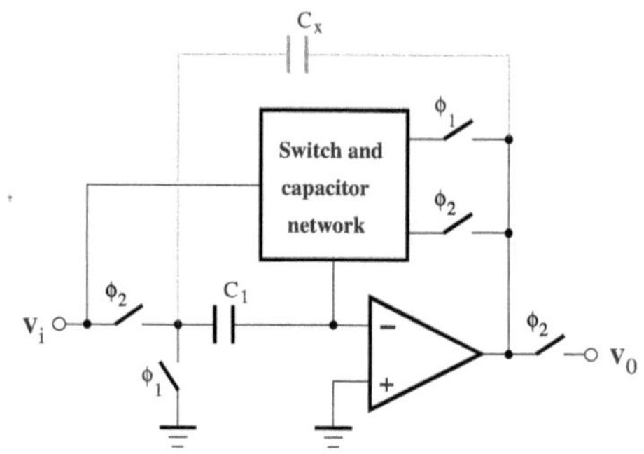

FIGURE 7.72
SC circuit with a transient spike compensation based on the use of small feedback capacitors.

when the two clock phases take their low level, and in which the amplifier does not have a negative feedback path. One solution to reduce the effect of the resulting transient spike is to use a small capacitor C_x [54], as shown in Figure 7.72, in order to maintain a closed loop around the amplifier.

Other circuit configurations for low-sensitivity structures were proposed in [55–57] (see Circuit design assessment 7 at the chapter end). The principle was to perform a preliminary charge transfer before the desired one in order to obtain a close approximation of the amplifier error signal. This latter was then stored on an auxiliary capacitor and subsequently used for compensation.

However, the presence of a stray-sensitive node can substantially limit the precision in practical realizations.

7.17 Low supply voltage SC circuits

Due to the increasing importance of portable systems for data processing in instrumentation and multimedia communication applications, the analog circuitry of modern mixed-signal integrated circuits has to operate with low supply voltage. The use of switched capacitor techniques results in circuits having a high accuracy and a good dynamic range. However, the low supply voltage does not allow a suitable control of the switches whose overdrive is signal dependent (e.g., input switch and switches at the amplifier output). Figure 7.73 shows the on-conductances of a CMOS switch operating with the

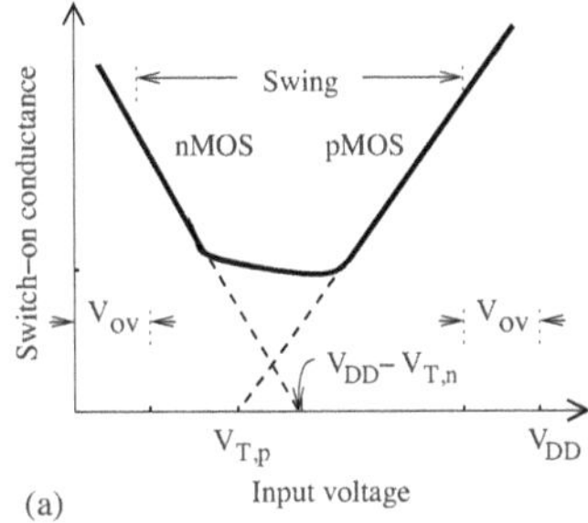

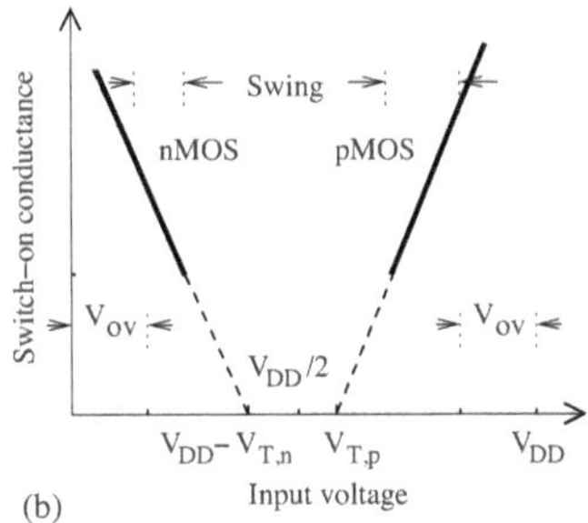

FIGURE 7.73
On-conductances of a CMOS switch operating (a) normally and (b) with a low supply voltage.

appropriate and low supply voltages, respectively. By decreasing the supply voltage to about $V_{T,n} + |V_{T,p}|$, there is a range of the input voltage around $V_{DD}/2$ for which the nMOS and pMOS transistors will not conduct (see Figure 7.73(b)). In the low-voltage circuit design, it is preferable to use either the nMOS or pMOS transistor as a switch. A supply voltage of at least, $V_T + V_{ov}$, where V_T is the transistor threshold voltage and V_{ov} is the highest voltage level of the signal to be switched, is then required [59].

Three approaches have been proposed in the literature for the design of low-voltage switched-capacitor circuits:

1. The first approach is to use lower threshold voltage transistors [60]. But this technique suffers from the high cost associated with the required process technology. Furthermore, the switch-off leakage is much higher than in the classical case.

2. The second one consists of using voltage multipliers to generate the clock voltages that can drive the switches [61].
3. The third alternative is the switched-amplifier (SA) technique [62, 63]. In this case, the critical switches are eliminated and their functions are realized by the amplifier, which can be turned off and on by a clock signal. This can be realized, for instance, by a switch introduced between the amplifier core and the power supply line.

The circuit diagram of a bootstrapped switch [64] is shown in Figure 7.74. It is equivalent to the simplified structure of Figure 7.75(a). The states of the switch are determined by the phases, ϕ_1 and ϕ_2, of the clock signal represented in Figure 7.75(b). The voltage V_{DD} is periodically connected between the input and the gate of the transistor switch so that the gate voltage required to turn on the switch can now be higher than the supply voltage V_{DD}. Transistors

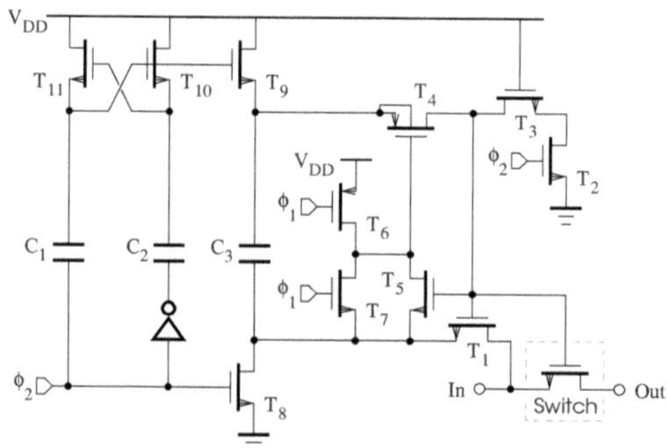

FIGURE 7.74
Circuit diagram of a bootstrapped switch.

T_3 and T_5 are used to allow a design within the reliability limits of the IC process. The voltages V_{DS} and V_{GD} sustained by T_2 during its on state are decreased by T_3. Due to the regulation achieved by T_5, the voltage V_{GS} of T_4 should not exceed the supply voltage, V_{DD}.

The switch is off when ϕ_1 is low and ϕ_2 is high. The capacitor C_3, which is isolated from the switch by T_1 and T_4, is charged by V_{DD}. It should be sufficiently large to mitigate the effect of parasitic capacitances on the boosted clock signal. To reduce the error due to the subthreshold charge leakage, the switch implemented by the nMOS transistor T_9 is controlled by a level-shifted signal with the levels V_{DD} and $2V_{DD}$. The basic idea for the signal generation is to use the charge pump circuit consisting of capacitors C_1 and C_2, which are connected to V_{DD} via the cross-coupled transistors T_{10} and T_{11}, and an inverter.

During the on state of the switch, that is, when ϕ_1 is high and ϕ_2 is low, a connection is established by T_7 between the capacitor C_3 and the gate of T_4, and the gate of the switch is then bootstrapped to $V_{DD} + V_i$, where V_i is the

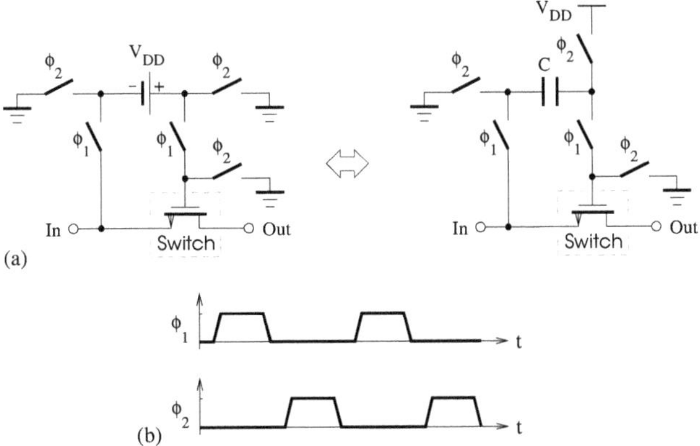

FIGURE 7.75
(a) Principle of a bootstrapped switch; (b) timing diagram.

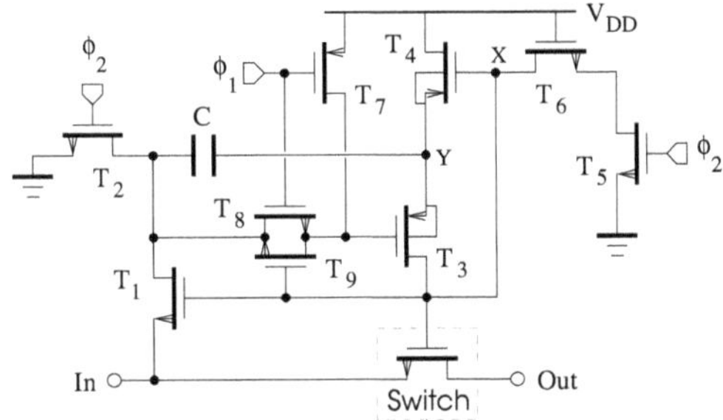

FIGURE 7.76
Alternative circuit diagram of a bootstrapped switch.

input voltage. The device T_1 maintains constant the voltage V_{GS} across the switch.

The on-resistance of the switch is given by

$$R_{on} = \frac{1}{K(V_{DD} - V_T)} \tag{7.352}$$

where $K = \mu_n C_{ox}(W/L)$ is the transconductance parameter and V_T is the threshold voltage. Due to the V_T dependence, it is sensitive to the body effect of the transistor.

An alternative circuit diagram of a bootstrapped switch [65] is depicted in Figure 7.76. In addition to transistors, $T_1 - T_5$, that operate as switch, other transistors are used to allow a rail-to-rail operation of switches and to avoid any oxide overstress by limiting the gate-source/drain voltage to V_{DD}. The bulk terminals of transistors T_3 and T_4 are connected to the node Y to prevent the source junctions from being forward biased. The gate connections of T_3

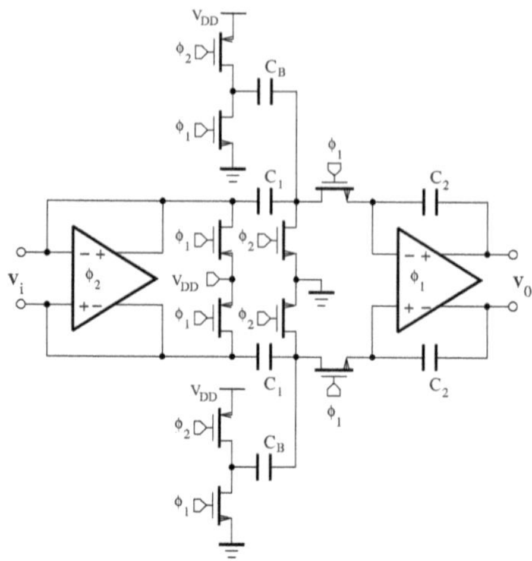

FIGURE 7.77
Circuit diagram of an integrator based on the switched-amplifier technique.

and T_4 help limit voltage differences across the transistors as the voltage at the node Y can become higher than V_{DD}. The transistor T_3 can be turned on by T_8 at the beginning of the clock phase ϕ_1 or by T_9 when the capacitor C is charged by the input voltage. To maintain the drain-source voltage of T_6 less than V_{DD}, the voltage at the node X should not exceed $2V_{DD} - V_T$ because the worst-case value of the source voltage of T_6 obtained during the off-state (phase ϕ_1) is about $V_{DD} - V_T$.

A fully differential SA integrator is depicted in Figure 7.77. The nMOS switches are connected to ground, while the pMOS switches are related to the supply voltage, V_{DD}. The branch including the capacitor C_B is used to optimize the dynamic range. In the steady state, it can be shown that

$$C_1 V_{0,dc} + C_B V_{DD} = C_1 V_{DD} \tag{7.353}$$

where $V_{0,dc} = V_{DD}/2$ provided that $C_B = C_1/2$. It was assumed that the input dc level is set to the ground. Note that the amplifier architecture should be chosen with the objective of minimizing the turn-on time, which can limit the achievable sampling frequency. Furthermore, the set of transfer functions that can be realized with the SA approach is limited due to the fact that the amplifier output is defined and can be used only during one clock phase.

7.18 Summary

SC circuits used for the implementation of S/H and T/H circuits, gain stages, and integrators should be designed to be less sensitive to component nonidealities. This can be achieved by optimizing the circuit performance based on the analysis of the limitations due to the practical characteristics of components (amplifiers, switches and capacitors), and by introducing some refinements at the circuit level. Furthermore, in the specific case of filter design, the circuit accuracy can also depend on the synthesis method or filter architecture. Hence, SC ladder filters are less sensitive to capacitance mismatches or fluctuations in the passband than cascaded biquad structures.

7.19 Circuit design assessment

1. **Offset-free tunable gain stage**
   Consider the offset-free tunable gain stage shown in Figure 7.78, where ϕ is a 50% duty cycle clock signal.

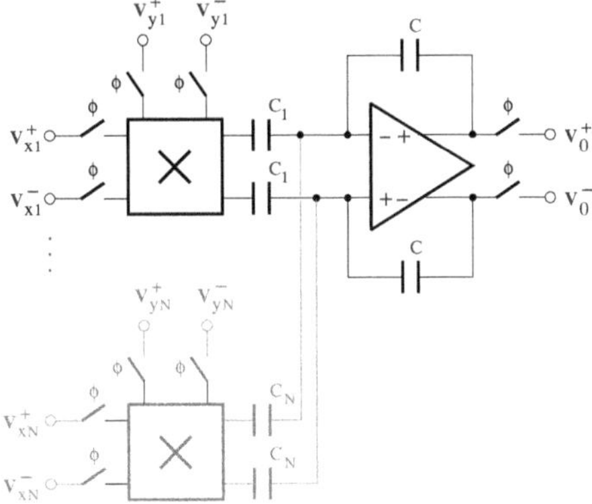

FIGURE 7.78
Offset-free tunable gain stage.

   Analyze the charge transfer taking place between the capacitors C_j and C to prove that the output voltage is not affected by the offset voltage contributions due to the multipliers and amplifier.

   Assuming that the amplifier gain is $A = 1/\mu$, show that the output voltage can be expressed as

$$V_0(z) = \pm \frac{1}{C(1+\mu)} \sum_{j=1}^{N} C_j \cdot \triangle V_j(z) \tag{7.354}$$

where $v_0 = v_0^+ - v_0^-$, and $\triangle v_j$ is the voltage across each capacitor C_j.

2. **Integrator with input parasitic capacitors**
The circuit diagram of an SC inverting integrator is depicted in Figure 7.79.

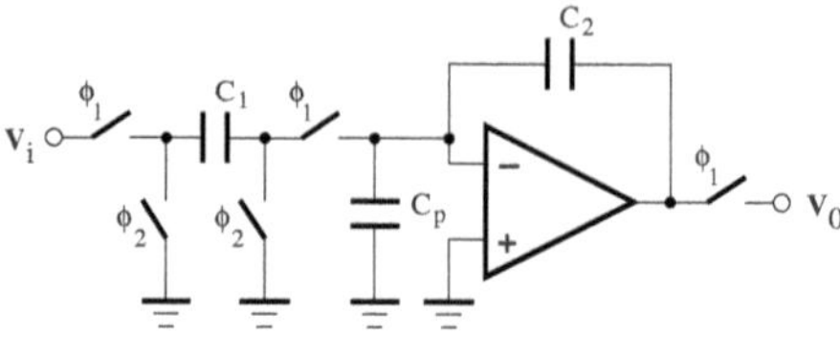

FIGURE 7.79
SC inverting integrator.

Taking into account the effect of the finite *dc* gain, $A_0 = 1/\mu$, and parasitic capacitor, C_p, show that the transfer function can be derived as

$$\begin{aligned} H(j\omega) &= \frac{V_0(j\omega)}{V_i(j\omega)} \\ &= H_{id}(j\omega)\left[1 + \mu\left(1 + \frac{C_p + C_1/2}{C_2}\right) - j\mu \frac{C_1}{2C_2\tan(\omega T/2)}\right]^{-1} \end{aligned} \tag{7.355}$$

where

$$H_{id}(j\omega) = -\frac{C_1}{C_2}\frac{e^{j(\omega T/2)}}{j\sin(\omega T/2)} \tag{7.356}$$

3. **Track-and-hold circuit**
The circuit of Figure 7.80 is designed with $C_1 = C_2 = C$ and $C_1' = 2C$.

Let v^- be the voltage at the inverting node of the amplifier. Verify that

$$C\big(v^-[(n-1/2)T] - v_0[(n-1/2)T] - \{v^-[(n-1)T] - v_i[(n-1)T]\}\big) = 0 \tag{7.357}$$

during the clock phase, ϕ_2, that is, for $(n-1)T \le t < (n-1/2)T$,

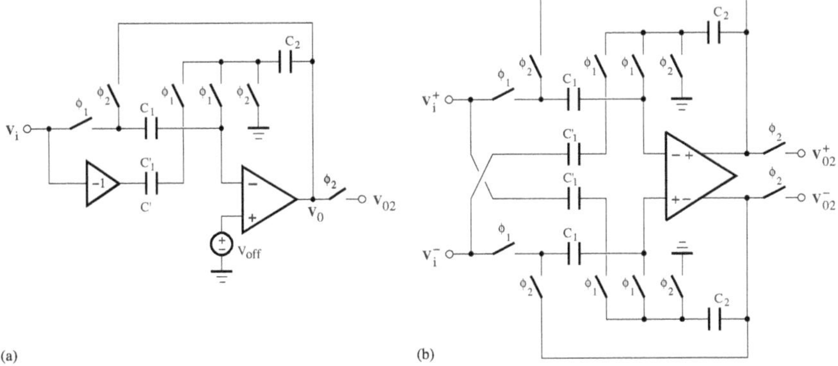

FIGURE 7.80
Circuit diagram of a low-sensitivity T/H ($C_1 = C_2 = C$ and $C_1' = 2C$): (a) Single-ended and (b) differential versions.

and

$$\begin{aligned} C\left(v_i(nT) - v^-(nT) - \{v_0[(n-1/2)T] - v^-[(n-1/2)T]\}\right) \\ + 2C\left(-v_i(nT) - v^-(nT) - \{-v_i[(n-1)T] - v^-[(n-1)T]\}\right) \\ = C\{v^-(nT) - v_0(nT) + v_0[(n-1/2)T]\} \end{aligned} \tag{7.358}$$

during the clock phase, ϕ_1, that is, for $(n-1/2)T \leq t < nT$.

With the assumption that $v_{02}(nT) = v_0[(n-1/2)T]$, show that

$$V_{02}(z) = z^{-1/2}\frac{(1+\mu-\mu z^{-1})V_i(z) + (1+\mu)V_{off}}{(1+\mu)(1+4\mu) - \mu(4+3\mu)z^{-1}} \tag{7.359}$$

where $\mu = 1/A_0$, A_0 and V_{off} are the dc gain and offset voltage of the amplifier, respectively.

In the case of an amplifier with ideal characteristics (dc gain, bandwidth, offset voltage), deduce that $V_0(z) = z^{-1/2}V_i(z)$.

4. **Analysis of a sample-and-hold circuit**
Consider the circuit structure shown in Figure 7.81, where $C_1 = C_1'$. During the first clock phase, the capacitor C_1 charges up to V_i, and the capacitor C_1' charges up to $-V_i$. The charges acquired by C_1 and due to the input voltages cancel those acquired by C_1'. During the second clock phase, the capacitor C_1 is included in the amplifier feedback path while the capacitors C_1' and C_2 are discharged.

Assuming that v^- is the voltage at the inverting node of the am-

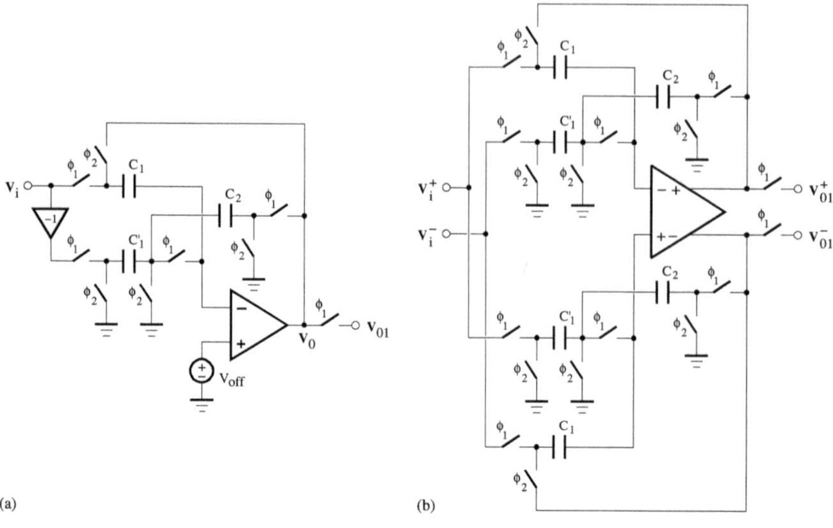

FIGURE 7.81
Circuit diagram of an S/H structure ($C_1 = C_1'$): (a) Single-ended and (b) differential versions.

plifier, verify that

$$v^-[(n-1/2)T] - v_O[(n-1/2)T] - \{v^-[(n-1)T] - v_i[(n-1)T]\} = 0 \tag{7.360}$$

during the clock phase, ϕ_2, that is, for $(n-1)T \leq t < (n-1/2)T$, and

$$\begin{aligned} C_1\left(v_i(nT) - v^-(nT) - \{v_O[(n-1/2)T] - v^-[(n-1/2)T]\}\right) \\ + C_1'\left\{-v_i(nT) - v^-[(n-1/2)T]\right\} = C_2[v^-(nT) - v_O(nT)] \end{aligned} \tag{7.361}$$

during the clock phase, ϕ_1, that is, for $(n-1/2)T \leq t < nT$.

Taking into account the dc gain $A_0 = 1/\mu$ and offset voltage V_{off} of the amplifier, determine the circuit transfer function, $H(z)$.

In the ideal case, deduce that $v_O(nT) = v_i[(n-1)T]$, or equivalently $V_O(z) = z^{-1}V_i(z)$.

5. **Improved bootstrapped switch technique**
The bootstrapped switch of Figure 7.82 [66] was designed to be less sensitive to the body effect of the transistor. During the on state, the voltage $V_{DD} + V_i + V_T$ is applied to the gate of the switch. This latter is grounded during the off state.

Show that the on-resistance of the switch is given by

$$R_{on} = \frac{1}{K(V_{DD} - V_B)} \tag{7.362}$$

where $V_B = \sqrt{2I_B/K_1}$, $K_1 = \mu_n C_{ox}(W/L)_1$, and K is the transconductance parameter of the switch.

What is the effect of the mismatch between the switch and T_1 on the circuit operation?

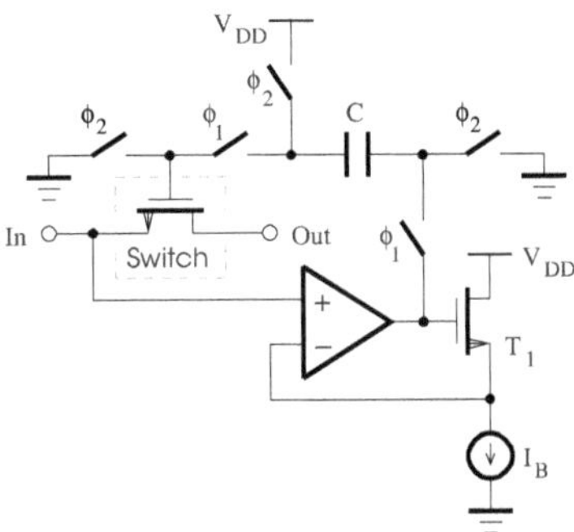

FIGURE 7.82
Circuit diagram of a bootstrapped switch.

6. **Analysis of first-order filter sections**
   With the assumption that the input signal is slow moving, that is, $V_i(n - 1/2) = V_i(n)$, show that the transfer function of the filter circuit of Figure 7.83 can be written as

$$H(z) = \frac{V_O(z)}{V_i(z)} = \frac{\alpha - \beta + \beta z^{-1}}{1 + \gamma - z^{-1}} \tag{7.363}$$

   Determine the type of filter corresponding to $\alpha = 1$, $\beta = 2$, and $\gamma = 1$.

   Consider the circuit diagram of a first-order allpass filter shown in Figure 7.84.
   Show that the filter transfer function is given by

$$H(z) = \frac{V_O(z)}{V_i(z)} = \frac{\alpha z^{-1} - \beta}{\alpha - \beta z^{-1}} \tag{7.364}$$

   What is the advantage of this circuit structure with respect to the mismatch between the capacitor ratios?

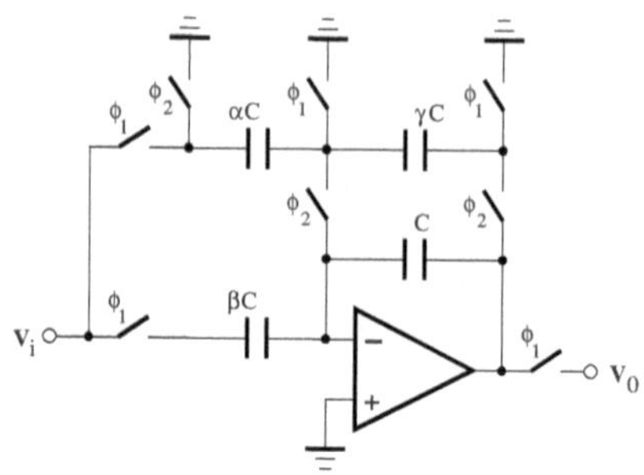

FIGURE 7.83
Circuit diagram of a first-order allpass filter.

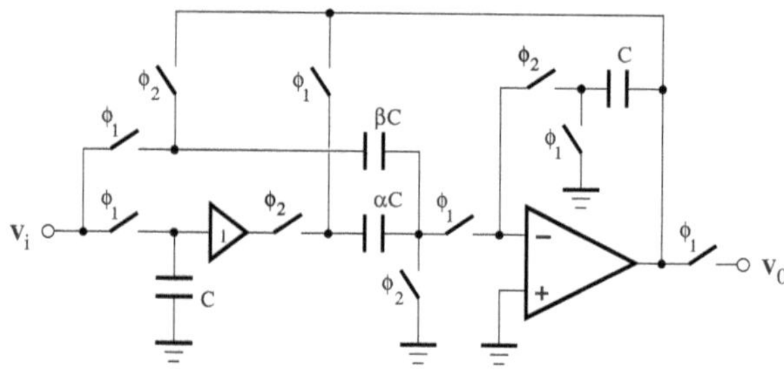

FIGURE 7.84
Circuit diagram of a first-order structurally allpass filter.

7. **Low-sensitivity single-sampling integrator**
Given $\mu = 1/A_0$, where A_0 is the amplifier dc gain, and $C_3 = C_2$, verify the following relations for the single-ended version of the low-sensitivity integrator shown in Figure 7.85 [56].
During the clock phase 1, that is, for $(n-1)T \leq t < (n-1/2)T$,

$$C_2\{(1+\mu)v_0[(n-1/2)T] - v_0[(n-1)T] + v_x[(n-1)T] - V_{off}\} + C_3\{\mu v_0[(n-1/2)T] - \mu v_0[(n-1)T] - v_x[(n-1)T]\} = 0 \tag{7.365}$$

During the clock phase 2, that is, for $(n-1/2)T \leq t < nT$,

$$C_1[v_i(nT) - v_x(nT)] + C_2\{v_0(nT) - (1+\mu)v_0[(n-1/2)T] - v_x(nT) + V_{off}\} = 0 \tag{7.366}$$

and

$$v_x(nT) - \mu\{v_0[(n-1/2)T] - v_0(nT)\} = 0 \tag{7.367}$$

because there is no current flowing through C_3.

Show that the output voltage can be written as

$$V_0(z) = H_i(z)V_i(z) + H_0(z)V_{off} \tag{7.368}$$

where

$$H_0(z) = -\frac{\mu}{1+2\mu}H_i(z) \tag{7.369}$$

$$H_i(z) = H_{id}(z)E(z) \tag{7.370}$$

$$H_{id}(z) = -\frac{C_1}{C_2}\frac{1}{1-z^{-1}} \tag{7.371}$$

and

$$E(z) = \frac{1}{1+\mu+\mu\frac{C_1}{C_2}\frac{1}{1-z^{-1}} - \mu\frac{1+\mu}{1+2\mu}\frac{C_1}{C_2}\frac{z^{-1}}{1-z^{-1}}} \tag{7.372}$$

Repeat the above question assuming that $v_0[(n-1/2)T] = v_0(nT)$. What is the effect of a parasitic capacitor, C_p, connected to the stray-sensitive node x, on the error function $E(z)$?

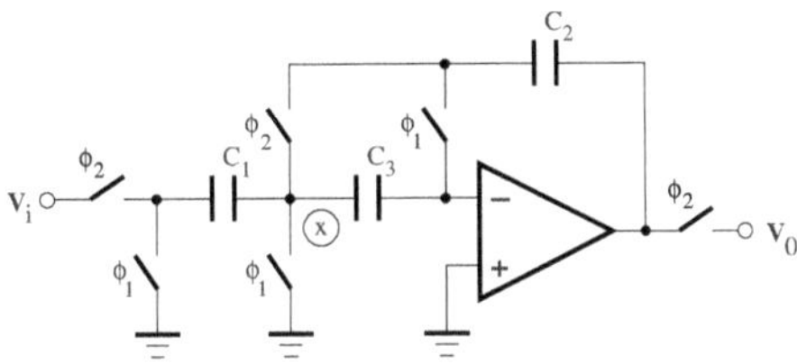

FIGURE 7.85
Circuit diagram of a low-sensitivity integrator.

8. **Design of an anti-aliasing filter**
In DSP applications, aliasing can occur whenever the input signal contains spectral components at frequencies greater than one half of the sampling frequency.

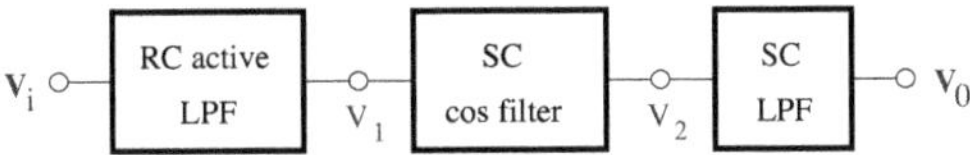

FIGURE 7.86
Block diagram of an anti-aliasing filter.

Suppose that we are required to design an anti-aliasing filter to restrict the bandwidth of a 200-kHz input signal to be sampled at $f_s = 800$ kHz. Because the sampling frequency is only four times the signal passband, it may be difficult to attenuate high-frequency

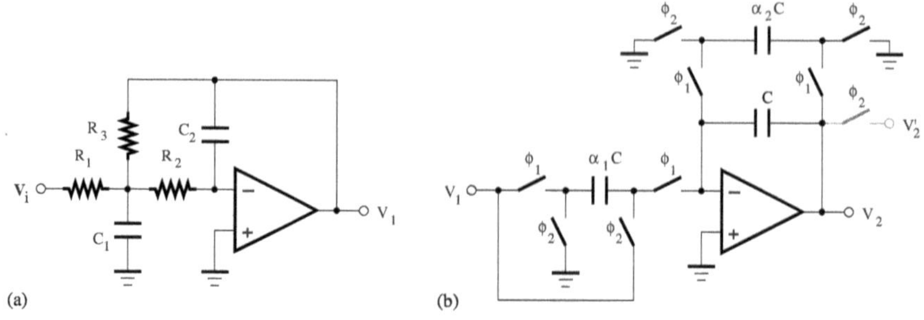

FIGURE 7.87
(a) Circuit diagram of a second-order RC active lowpass filter (LPF); (b) circuit diagram of a first-order cos SC filter.

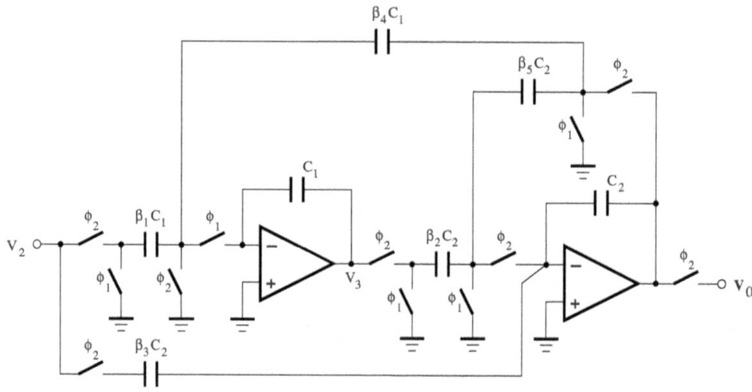

FIGURE 7.88
Circuit diagram of a second-order SC lowpass filter.

components that can be aliased into the passband using a conventional pre-filter based only on RC circuits. Figure 7.86 shows the block diagram of the anti-aliasing filter. The switched-capacitor section is assumed to operate with a sampling frequency of 1.6 MHz.

Show that the transfer function of the RC biquad depicted in Figure 7.87(a) is of the form

$$H_1(s) = \frac{V_1(s)}{V_i(s)} = \frac{-k\omega_0^2}{s^2 + \left(\dfrac{\omega_0}{Q}\right)s + \omega_0^2} \tag{7.373}$$

where

$$k = \frac{R_3}{R_1} \tag{7.374}$$

$$\omega_0 = \sqrt{\frac{1}{R_2 R_3 C_1 C_2}} \tag{7.375}$$

and

$$Q = \frac{\sqrt{\dfrac{C_1}{R_2 R_3 C_2}}}{\dfrac{1}{R_1} + \dfrac{1}{R_2} + \dfrac{1}{R_3}} \tag{7.376}$$

The RC filter section is designed to provide a 40-dB attenuation for frequency components at 1.4 MHz. Using the Butterworth approximation, the normalized transfer function can be obtained as

$$H_1(s') = \frac{-k'\omega_0'^2}{s'^2 + \left(\dfrac{\omega_0'}{Q'}\right)s' + \omega_0'^2} = \frac{-1}{s'^2 + \sqrt{2}s' + 1} \tag{7.377}$$

where

$$s' = \frac{s}{\omega_p} \tag{7.378}$$

$$k' = \frac{R_3'}{R_1'} \tag{7.379}$$

$$\omega_0' = \sqrt{\frac{1}{R_2' R_3' C_1' C_2'}} \tag{7.380}$$

and

$$Q' = \frac{\sqrt{\dfrac{C_1'}{R_2' R_3' C_2'}}}{\dfrac{1}{R_1'} + \dfrac{1}{R_2'} + \dfrac{1}{R_3'}} \tag{7.381}$$

Assuming that $R_1' = R_2' = R_3'$ and $C_1' = C_2'$, determine k', ω_0', Q', and the normalized values, R_j' $(j = 1, 2, 3)$ and C_j' $(j = 1, 2)$, of the filter components.

Find the values of the filter components using $C_j = C_j'/\omega_p \tilde{R}$ and $R_j = R_j'\tilde{R}$, where $\tilde{R} = 1$ kΩ.

Consider the first-order cos SC filter depicted in Figure 7.87(b). Assuming that the amplifier is ideal, write the charge conservation equations during each of both clock phases.

Deduce that the transfer function can be put into the form

$$H_2(z) = \frac{V_2'(z)}{V_1(z)} = -\frac{\alpha_1 z^{-1/2}(1 + z^{-1/2})}{1 + \alpha_2 - z^{-1}} \tag{7.382}$$

Determine the ratio α_1/α_2 by setting the dc gain, $|H_2(z)|_{z=1}$, equal to unity.

For the second-order SC lowpass filter illustrated in Figure 7.88, use the signal-flow graph to show that the transfer function can be written as

$$H_3(z) = \frac{V_O(z)}{V_2(z)} = -\frac{\beta_3 + (\beta_1\beta_2 - 2\beta_3)z^{-1} + \beta_3 z^{-2}}{1 + \beta_5 + (\beta_2\beta_4 - \beta_5 - 2)z^{-1} + z^{-2}} \tag{7.383}$$

Verify that $\beta_1/\beta_4 = 1$ if the dc gain, $|H_3(z)|_{z=1}$, is equal to unity.

Using the bilinear transform and the Butterworth approximation function, determine the coefficients β_j ($j = 1, 2, 3, 4, 5$) based on the requirement that the overall anti-aliasing filter should provide more than 40-dB attenuation for frequencies above 600 kHz.

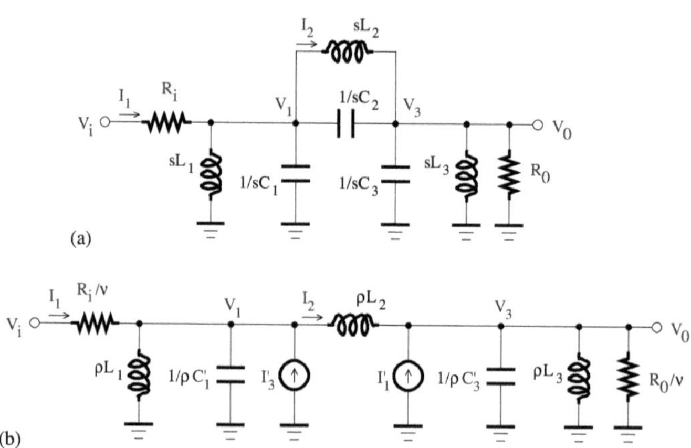

FIGURE 7.89
(a) Continuous-time and (b) discrete-time bandpass filter prototypes.

9. **Design of a bandpass SC ladder filter**
After prewarping the passband and stopband specifications from the discrete-time domain to the continuous-time domain, the bandpass RLC ladder filter prototype of Figure 7.89(a) was found to satisfy the target filtering requirements.

Given $\rho = (z^{1/2} - z^{-1/2})/2$, verify that the next design step should result in a discrete-time filter prototype, as shown in Figure 7.89(b), where

$$C_1' = C_1 + \frac{1}{L_1} + C_2 + \frac{1}{L_2} \tag{7.384}$$

$$C_2' = C_2 + \frac{1}{L_2} \tag{7.385}$$

$$C_3' = C_3 + \frac{1}{L_3} + C_2 + \frac{1}{L_2} \tag{7.386}$$

$$I_1' = \rho C_2' V_1 \tag{7.387}$$

and

$$I_3' = \rho C_2' V_3 \tag{7.388}$$

Derive the corresponding z-domain SFG and convert it into an SC filter.

Bibliography

[1] T. Ndjountche, *Dynamic Analog Circuit Techniques for Real-Time Adaptive Networks,* Aachen, Germany: Shaker Verlag, 2000.

[2] L. Larson and G. C. Temes, "Signal conditioning and interface circuits," In S. K. Mitra and J. F. Kaiser, Editors, *Handbook for Digital Signal Processing,* Chap. 10, pp. 677–720, New York, NY: John Wiley & Sons, 1993.

[3] S. Hirano and E. Hayara, "Pre- and postfiltering effects on SCF stopband attenuation," *IEEE Trans. on Circuits and Systems*, vol. 38, pp. 547–551, May. 1991.

[4] D. J. Allstot and W. C. Black Jr., "Technological design considerations for monolithic MOS switched-capacitor filtering systems," *Proc. of the IEEE*, vol. 71, pp. 967–986, Aug. 1983.

[5] R. Unbehauen and A. Cichocki, *MOS Switched-Capacitor and Continuous-Time Integrated Circuits and Systems*, Berlin, Germany: Springer Verlag, 1989.

[6] A. Robertini, J. Goette, and W. Guggenbühl, "Nonlinear distortions in SC-integrators due to nonideal switches and amplifiers," *Proc. of 1989 IEEE Int. Symp. on Circuits and Systems*, pp. 1692–1695, May 1989.

[7] B. Sheu and C. Hu, "Switch-induced error voltage on switched capacitor," *IEEE J. of Solid-State Circuits*, vol. 19, pp. 519–525, Aug. 1984.

[8] G. Wegmann, E. A. Vittoz, and F. Rahali, "Charge injection in analog MOS switches," *IEEE J. of Solid-State Circuits*, vol. 22, pp. 1091–1097, Dec. 1987.

[9] D. G. Haigh and B. Singh, "A switching scheme for SC filters which reduces the effect of parasitic capacitances," in *Proc. 1983 IEEE Int. Symp. on Circuits and Systems*, pp. 586–589, April 1983.

[10] E. Suarez, P. R. Gray, and D. A. Hodges, "All-MOS charge redistribution analog-to-digital conversion techniques–Part II," *IEEE J. of Solid-State Circuits*, vol. 10, pp. 379–385, Dec. 1975.

[11] C. Eichenberger and W. Guggenbühl, "On charge injection in analog MOS switches and dummy switch compensation techniques," *IEEE Trans. on Circuits and Systems*, vol. 37, pp. 256–264, Feb. 1990.

[12] K. Martin and A. S. Sedra, "Strays-insensitive switched-capacitor filters based on the bilinear z-transform," *Electronics Letters*, vol. 19, pp. 365–366, June 1979.

[13] Y. Tsividis, "Analytical and experimental evaluation of the switched-capacitor filter and remarks on the resistor/switched capacitor correspondence," *IEEE Trans. on Circuits and Systems*, vol. 26, pp. 140–144, Feb. 1979.

[14] K. Martin and A. S. Sedra, "Transfer function deviations due to resistor-SC equivalence assumption in switched-capacitor simulation of LC ladders," *Electronics Letters*, vol. 16, pp. 387–389, May 1980.

[15] A. M. Durham, J. B. Hughes, and W. Redman-White, "Circuit architectures for high linearity monolithic continuous-time filtering," *IEEE Trans. on Circuits and Systems–II*, vol. 39, pp. 651–657, Sept. 1992.

[16] D. A. Johns and K. Martin, *Analog Integrated Circuit Design*, New York, NY: John Wiley & Sons, 1997.

[17] Y. A. Haque, R. Gregorian, R. W. Blasco, R. A. Mao, and W. E. Nicholson, "A two chip PCM voice CODEC with filters," *IEEE J. of Solid-State Circuits*, vol. 14, pp. 961–969, Dec. 1979.

[18] J. J. F. Rijns and H. Wallinga, "Stray-insensitive switched-capacitor sample-delay-hold buffers for video frequency applications," *Electronics Letters*, vol. 27, no. 8, pp. 639–640, April 1991.

[19] G. Nicollini, P. Confalonieri and D. Senderowicz, "A fully differential sample-and-hold circuit for high-speed applications," *IEEE J. of Solid-State Circuits*, vol. 24, pp. 1461–1465, Oct. 1989.

[20] T. Ndjountche and R. Unbehauen, "Analog discrete-time basic structures for adaptive IIR filters," *IEE Proc.-Circuits Devices and Systems*, vol. 147, pp. 250–256, Aug. 2000.

[21] B. J. Hosticka, R. W. Brodersen, and P. R. Gray, "MOS sampled data recursive filters using switched capacitor integrators," *IEEE J. of Solid-State Circuits*, vol. 12, pp. 600–608, Dec. 1977.

[22] P. E. Fleischer and K. R. Laker, "A family of active switched-capacitor biquad building blocks," *Bell Systems Tech. J.*, vol. 58, pp. 2235–2269, Dec. 1979.

[23] E. I. El-Masry, "Strays-insensitive state-space switched-capacitor filters," *IEEE Trans. on Circuits and Systems*, vol. 30, pp. 474–488, July 1983.

[24] S. Signell, "On selectivity properties of discrete-time linear networks," *IEEE Trans. on Circuits and Systems*, vol. 31, pp. 275–280, March 1984.

[25] U. Weder and A. Moeschwitzer, "Comments on 'Design techniques for improved capacitor area efficiency in switched-capacitor biquads,"' *IEEE Trans. on Circuits and Systems*, vol. 37, pp. 666–668, May 1990.

[26] W.-H. Ki and G. C. Temes, "Optimal capacitance assignment of switched-capacitor biquads," *IEEE Trans. on Circuits and Systems*, vol. 42, pp. 334–342, June 1995.

[27] A. Petraglia and S. K. Mitra, "Switched-capacitor equalizers with digitally programmable tuning characteristics," *IEEE Trans. on Circuits and Systems*, vol. 38, pp. 1322–1331, Nov. 1991.

[28] J. L. McCreary, "Matching properties, and voltage and temperature dependence of MOS capacitors," *IEEE J. of Solid-State Circuits*, vol. 16, pp. 608–616, Dec. 1981.

[29] D. A. Vaughan-Pope and L. T. Burton, "Transfer function synthesis using generalized doubly terminated two-pair networks," *IEEE Trans. on Circuits and Systems*, vol. 24, pp. 79–88, Feb. 1977.

[30] M. Kaneko and M. Onoda, "Z-domain exact design for LDI leapfrog switched-capacitor filters," *Proc. of 1984 IEEE Int. Symp. on Circuits and Systems*, pp. 304–307, May 1984.

[31] A. Kaelin and G. S. Moschytz, "Exact design of arbitrary parasitic-insensitive elliptic SC-ladder filter in the z-domain," *Proc. of 1988 IEEE Int. Symp. on Circuits and Systems*, pp. 2485–2488, June 1988.

[32] S. O. Scanlan, "Analysis and synthesis of switched-capacitor state-variable filters," *IEEE Trans. on Circuits and Systems*, vol. 28, pp. 85–93, Feb. 1981.

[33] A. Muralt, "The design of switched-capacitor filters based on doubly-terminated two-pair signal-flow graphs," PhD thesis, Swiss Federal Institute of Technology, ETH-Center, Zurich, Switzerland, 1993.

[34] E. S. K. Liu, L. E. Turner, and L. T. Bruton, "Exact synthesis of LDI and LDD ladder filters," *IEEE Trans. on Circuits and Systems*, vol. 31, pp. 369–381, April 1984.

[35] R. B. Datar and A. S. Sedra, "Exact design of strays-insensitive switched-capacitor ladder filters," *IEEE Trans. on Circuits and Systems*, vol. 30, pp. 888–898, Dec. 1983.

[36] M. S. Lee and C. Chang, "Switched-capacitor filters using the LDI and bilinear transformations," *IEEE Trans. on Circuits and Systems*, vol. 28, pp. 265–270, April 1981.

[37] B. Nowrouzian, "A new synthesis technique for the exact design of switched-capacitor LDI allpass networks," *Proc. of the 1990 IEEE Int. Symp. on Circuits and Systems*, pp. 2185–2188, 1990.

[38] A. M. Davis, "A new z-domain continued fraction expansion," *IEEE Trans. on Circuits and Systems*, vol. 29, pp. 658–662, Oct. 1982.

[39] K. Martin and A. S. Sedra, "Effects of the op amp finite gain and bandwidth on the performance of switched-capacitor filters," *IEEE Trans. on Circuits and Systems*, vol. 28, pp. 822–829, Aug. 1981.

[40] D. B. Ribner and M. A. Copeland, "Biquad alternatives for high-frequency switched-capacitor filters," *IEEE J. of Solid-State Circuits*, vol. 20, pp. 1085–1095, Dec. 1985.

[41] W. M. C. Sansen, H. Qiuting, and K. A. I. Halonen,"Transient analysis of charge transfer in SC filters-gain error and distortion," *IEEE J. of Solid-State Circuits*, vol. 22, pp. 268–276, April 1987.

[42] J. Vlach, K. Singhal, and M. Vlach, "Computer oriented formulation of equations and analysis of switched-capacitor networks," *IEEE Trans. on Circuits and Systems*, vol. 31, pp. 753–765, Sept. 1984.

[43] N. Fröhlich, B. M. Riess, U. A. Wever, and Q. Zheng, "A new approach for parallel simulation of VLSI circuit on a transistor level," *IEEE Trans. on Circuits and Systems*, vol. 45, pp. 601–613, June 1998.

[44] *HSPICE Simulation and Analysis User Guide*, Release U-2003.03-PA, Synopsys, March 2003.

[45] S. C. Fang, Y. Tsividis, and O. Wing, "SWITCAP: A switched-capacitor network analysis program, part I: Basic features," *IEEE Circuits and Dev. Mag.*, vol. 5, pp. 4–10, Dec. 1983.

[46] S. C. Fang, Y. Tsividis, and O. Wing, "SWITCAP: A switched-capacitor network analysis program, part II: Advanced applications," *IEEE circuits and Dev. Mag.*, vol. 5, pp. 41–46, Dec. 1983.

[47] *Affirma RF Simulator (SpectreRF) Theory (v446)*, Cadence Design Systems, June 2000.

[48] *Affirma RF Simulator (SpectreRF) User Guide (v446)*, Cadence Design Systems, April 2001.

[49] G. Fischer and G. S. Moschytz, "SC Filters for high frequencies with compensation for finite-gain amplifiers," *IEEE Trans. on Circuits and Systems*, vol. 32, pp. 1050–1056, Oct. 1985.

[50] G. C. Temes and K. Haug, "Improved offset-compensation schemes for the switched-capacitor circuits," *Electronics Letters*, vol. 20, pp. 508–509, June 1984.

[51] L. E. Larson, K. W. Martin, and G. C. Temes, "GaAs switched-capacitor circuits for high-speed signal processing," *IEEE J. of Solid-State Circuits*, vol. 22, pp. 971–981, Dec. 1987.

[52] L. E. Larson and G. C. Temes, "Switched-capacitor building blocks with reduced sensitivity to finite amplifier gain, bandwidth, and offset voltage," *Proc. of 1987 IEEE Int. Symp. on Circuits and Systems*, pp 334–338, May 1987.

[53] R. Gregorian, K. W. Martin, and G. C. Temes, "Switched-capacitor circuit design," *Proc. of the IEEE*, vol. 71, pp. 941–966, Aug. 1983.

[54] H. Matsumoto and K. Watanabe, "Spike-free SC circuits," *Electronics Letters*, vol. 8, pp. 428–429, 1987.

[55] K. Nagaraj, K. Singhal, T. R. Viswanathan, and J. Vlach, "Reduction of finite-gain effect in switched-capacitor filters," *Electronics Letters*, vol. 21, pp. 664–665, June 1985.

[56] K. Nagaraj, J. Vlach, T. R. Viswanathan, and K. Singhal, "Switched-capacitor integrator with reduced sensitivity to amplifier gain," *Electronics Letters*, vol. 22, pp. 1103–1105, Oct. 1986.

[57] K. Nagaraj, T. R. Viswanathan, K. Singhal, and J. Vlach, "Switched-capacitor circuits with reduced sensitivity to amplifier gain," *IEEE Trans. on Circuits and Systems*, vol. 34, pp. 571–574, May 1987.

[58] T. Ndjountche and R. Unbehauen, "Improved structures for programmable filters: Application in a switched-capacitor adaptive filter design," *IEEE Trans. on Circuits and Systems*, vol. 46, pp. 1137–1147, Sept. 1999.

[59] E. A. Vittoz, "The design of high-performance analog circuits on digital CMOS chips," *IEEE J. of Solid-State Circuits*, vol. 20, pp. 657–665, June 1985.

[60] Y. Matsuya and J. Tamada, "1 V power supply low-power consumption A/D conversion technique with swing suppression noise shaping," *IEEE J. of Solid-State Circuits*, vol. 29, pp. 1524–1530, Dec. 1994.

[61] G. Nicollini, A. Nagari, P. Confalonieri, and C. Crippa, "A −80 dB THD 4 V_{pp} switched-capacitor filter for 1.5V battery-operated systems," *IEEE J. of Solid-State Circuits*, vol. 31, pp. 1214–1219, Aug. 1996.

[62] J. Crols and M. Steyaert, "Switched-opamp: An approach to realize full CMOS switched-capacitor circuits at very low power supply voltages," *IEEE J. of Solid-State Circuits*, vol. 29, pp. 936–942, Aug. 1994.

[63] A. Baschirotto and R. Castello, "A 1-V 1.8 MHz CMOS switched-opamp SC filter with rail-to-rail output swing," *IEEE J. of Solid-State Circuits*, vol. 32, pp. 1979–1987, Dec. 1997.

[64] A. M. Abo and P. Gray, "A 1.5-V, 10-bit, 14.3-Ms/s CMOS pipeline analo-to-digital converter," *IEEE J. of Solid-State Circuits*, vol. 34, pp. 599–606, May 1999.

[65] M. Dessouky and A. Kaiser, "Very low-voltage digital audio $\Delta\Sigma$ modulator with 88 dB dynamic range using local switch bootstrapping," *IEEE J. of Solid-State Circuits*, vol. 36, pp. 349–355, March 2001.

[66] A. K. Ong, V. I. Prodanov, and M. Tarsia, "A method for reducing the variation in on resistance of a MOS sampling switch," *Proc. of 2000 IEEE Int. Symp. on Circuits and Systems*, vol. V, pp. 437–440, May 2000.

A

Transistor Sizing in Building Blocks

CONTENTS

Analog circuits, such as amplifiers and comparators, must satisfy many requirements, often conflicting ones. The same structure can be designed to meet different sets of specifications for a variety of applications.

Transistor sizing affects the performance characteristics (speed, power consumption, chip area, etc.) of integrated circuits. It is an iterative process that is generally achieved using SPICE-based simulation tools. Because the design parameters such as transistor sizes and bias currents depend on the performance metrics that are nonlinear functions, their optimal values may be derived only using computation-intensive programs. The effects of manufacturing yields are minimized by taking into account the process and operating-condition variations.

The design variables are the aspect ratios (width over length) of transistors, the value of the passive components (capacitors and resistors), and the value of bias currents and bias voltages. The optimization of the design variables for a component can only improve the operation characteristics (speed, precision, power dissipation, and area) to an extent allowed by the chosen topology.

For any given IC process, the determination of transistor sizes may require transistor matching data extracted from previous wafer fabrications. The minimum transistor area, WL, is set by the matching requirements, while parasitic capacitances are increased by scaling up transistor dimensions. The W/L ratio is determined by the drain current level and the operating region.

Depending on the design specifications, the transistor channel length, L, can be selected intuitively. In general, the use of long channel lengths for transistors can provide a high output resistance (or a high voltage gain), and a low unity-gain frequency. In the other hand, the choice of short channel length (or minimum length) transistors comes with a high unity-gain frequency, and a low output resistance (or a low voltage gain), while the resulting small gate areas is associated with an increase of the input-referred noise voltage and offset voltage.

For low power consumption, the bias currents should be chosen to be as small as possible while satisfying the noise and speed requirements.

A.1 MOS transistor

Two different types of MOS transistors are generally available for circuit design. In nMOS transistors, electrons are majority carriers in the source, which should operate at a voltage lower than the one of the drain, while in pMOS transistors, holes are generated in the source, which should be biased at a voltage higher than that of the drain. The operation of MOS transistors can be described as follows:

nMOS	pMOS
Cutoff region	
$V_{GS} < V_{T_n}$	$V_{SG} < -V_{T_p}$
$I_D = 0$	$I_D = 0$
Linear region	
$V_{GS} > V_{T_n}$, $V_{DS} < V_{DS(sat)}$	$V_{SG} > -V_{T_p}$, $V_{SD} < V_{SD(sat)}$
$I_D = K_n[2(V_{GS} - V_{T_n})V_{DS} - V_{DS}^2]$	$I_D = K_p[2(V_{SG} + V_{T_p})V_{SD} - V_{SD}^2]$
Saturation region	
$V_{GS} > V_{T_n}$, $V_{DS} \geq V_{DS(sat)}$	$V_{SG} > -V_{T_p}$, $V_{SD} \geq V_{SD(sat)}$
$I_D = K_n(V_{GS} - V_{T_n})^2(1 + \lambda_n V_{DS})$	$I_D = K_p(V_{SG} + V_{T_p})^2(1 + \lambda_p V_{SD})$

Note that, for the nMOS transistor,

$$K_n = \mu_n(C_{ox}/2)(W/L), \qquad V_{DS(sat)} = V_{GS} - V_{T_n}, \quad \text{and} \quad V_{T_n} > 0$$

and for the pMOS transistor,

$$K_p = \mu_p(C_{ox}/2)(W/L), \qquad V_{SD(sat)} = V_{SG} + V_{T_p}, \quad \text{and} \quad V_{T_p} < 0$$

The saturation voltage, $V_{DS(sat)}$ or $V_{SD(sat)}$, is also referred to as the overdrive. The linear region is also known as the ohmic or triode region.

In the saturation region, the transconductance and conductance can be derived as

$$\begin{aligned} g_m = \frac{\partial I_D}{\partial V_{GS}} &= 2K_n(V_{GS} - V_{T_n})(1 + \lambda_n V_{DS}) \\ &= 2\sqrt{K_n I_D(1 + \lambda_n V_{DS})} = \frac{2I_D}{V_{GS} - V_{T_n}} \end{aligned} \tag{A.1}$$

and

$$g_{ds} = \frac{\partial I_D}{\partial V_{DS}} = \frac{\lambda_n I_D}{1 + \lambda_n V_{DS}} \tag{A.2}$$

for nMOS transistors, and

$$\begin{aligned} g_m = \frac{\partial I_D}{\partial V_{SG}} &= 2K_p(V_{SG} + V_{T_p})(1 + \lambda_p V_{SD}) \\ &= 2\sqrt{K_p I_D (1 + \lambda_p V_{SD})} = \frac{2I_D}{V_{SG} + V_{T_p}} \end{aligned} \tag{A.3}$$

and

$$g_{sd} = \frac{\partial I_D}{\partial V_{SD}} = \frac{\lambda_p I_D}{1 + \lambda_p V_{SD}} \tag{A.4}$$

for pMOS transistors. To simplify the circuit analysis, the effect of λ can at first be neglected in g_m expressions.

In submicron CMOS technology, transistors are characterized using the Berkeley short-channel IGFET model (BSIM), which is supposed to better take into account short-channel effects. Let us consider an nMOS transistor with a known (W/L) aspect ratio. Given a value of the gate-source voltage, the values of the drain current can be obtained by running SPICE simulations with BSIM parameters for at least two values of the drain-source voltage. These values can then be inserted into an $I - V$ characteristic of the form

$$I_D = K_n' \frac{W_{eff}}{L_{eff}} (V_{GS} - V_{T_n})^2 (1 + \lambda_n V_{DS}) \tag{A.5}$$

to determine the parameters K_n' and λ_n required for hand calculations. For a better fit, this must be achieved in a region around the drain-source saturation voltage, which defines the operating point of the transistor. Here, the effective channel width and length,[1] W_{eff} and L_{eff}, are used due to the fact that the mask-defined sizes (W and L) are reduced by the amount of lateral diffusions, which are specified in the BSIM model card. Similarly, the values of K_p' and λ_p can be determined using the expression

$$I_D = K_p' \frac{W_{eff}}{L_{eff}} (V_{SG} + V_{T_p})^2 (1 + \lambda_p V_{SD}) \tag{A.6}$$

Parameter extraction example

Using BSIM models for a 0.13-µm CMOS process, the simulated drain currents for both nMOS and pMOS transistors with the effective aspect ratio of 20 µm/0.12 µm can be obtained as shown in Table A.1.

[1]Based on BSIM3v3 parameters, the effective channel width and length are of the form $W_{eff} = W - 2 \cdot WINT$ and $L_{eff} = L - 2 \cdot LINT$, where $WINT$ and $LINT$ are specified in the BSIM model card.

BSIM SPICE Model Card Example

```
*
* SPICE BSIM3 VERSION 3.1 PARAMETERS
* SPICE 3f5 Level 8, Star-HSPICE Level 49, UTMOST Level 8
* 0.13 μm CMOS process
*
* Temperature_ parameters=Default
*
.model CMOS NMOS (
```

+VERSION = 3.1	TNOM=27	TOX= 3.2E-9
+XJ=1E-7	NCH=2.3549E17	VTH0=0.0458681
+K1=0.3661767	K2=-0.0334177	K3=1E-3
+K3B=4.0506568	W0=1E-7	NLX=1E-6
+DVT0W=0	DVT1W=0	DVT2W=0
+DVT0=1.4508861	DVT1=0.1491907	DVT2=0.2337763
+U0=436.7862785	UA=-3.86228E-10	UB=3.278288E-18
+UC=4.781785E-10	VSAT=1.929894E5	A0=1.9927058
+AGS=0.751416	B0=1.840348E-6	B1=5E-6
+KETA=0.05	A1=7.776166E-4	A2=0.3
+RDSW=150	PRWG=0.3498753	PRWB=0.1103551
+WR=1	WINT=4.847999E-9	LINT=1.039837E-8
+DWG=1.179843E-8	DWB=8.997945E-9	VOFF=-0.0270176
+NFACTOR=2.5	CIT=0	CDSC=2.4E-4
+CDSCD=0	CDSCB=0	ETA0=2.751524E-6
+ETAB=-0.0111499	DSUB=4.060052E-6	PCLM=1.9774199
+PDIBLC1=0.9702431	PDIBLC2=0.01	PDIBLCB=0.1
+DROUT=0.9994828	PSCBE1=7.965102E10	PSCBE2=5.021019E-10
+PVAG=0.5368546	DELTA=0.01	RSH=7
+MOBMOD=1	PRT=0	UTE=-1.5
+KT1=-0.11	KT1L=0	KT2=0.022
+UA1=4.31E-9	UB1=-7.61E-18	UC1=-5.6E-11
+AT=3.3E4	WL=0	WLN=1
+WW=0	WWN=1	WWL=0
+LL=0	LLN=1	LW=0
+LWN=1	LWL=0	CAPMOD=2
+XPART=0.5	CGDO=3.74E-10	CGSO=3.74E-10
+CGBO=1E-12	CJ=9.581273E-4	PB=0.9758836
+MJ=0.4044874	CJSW=1E-10	PBSW=0.8002027
+MJSW=0.6	CJSWG=3.3E-10	PBSWG=0.8002027
+MJSWG=0.6	CF=0	PVTH0=2.009264E-4
+PRDSW=0	PK2=1.30501E-3	WKETA=0.013236
+LKETA=0.0327523	PU0=4.4729531	PUA=1.66833E-11
+PUB=0	PVSAT=653.2294237	PETA0=1E-4
+PKETA=-9.655097E-3	)	

```
*
.MODEL CMOS PMOS (
```

+VERSION=3.1	TNOM=27	TOX=3.2E-9
+XJ=1E-7	NCH=4.1589E17	VTH0=-0.2219851
+K1=0.2770146	K2=5.044386E-3	K3=0.0971898
+K3B=6.5020562	W0=1E-6	NLX=2.628685E-7
+DVT0W=0	DVT1W=0	DVT2W=0
+DVT0=9.146632E-3	DVT1=1	DVT2=0.1
+U0=111.7597102	UA=1.237083E-9	UB=1.90335E-21
+UC=-1.69849E-11	VSAT=1.22678E5	A0=2
+AGS=0.4944995	B0=5.266819E-6	B1=5E-6
+KETA=0.0118456	A1=0.4157385	A2=0.9596542
+RDSW=109.2955948	PRWG=-0.4797803	PRWB=0.5
+WR=1	WINT=0	LINT=7.536533E-9
+DWG=5.119137E-9	DWB=-1.84021E-8	VOFF=-0.1022829
+NFACTOR=1.5332272	CIT=0	CDSC=2.4E-4
+CDSCD=0	CDSCB=0	ETA0=0.0125544
+ETAB=-6.066043E-3	DSUB=2.751452E-3	PCLM=0.3090456
+PDIBLC1=0	PDIBLC2=-1.27526E-13	PDIBLCB=0.1
+DROUT=1.003724E-3	PSCBE1=2.606086E9	PSCBE2=8.052708E-10
+PVAG=0.0181389	DELTA=0.01	RSH=7
+MOBMOD=1	PRT=0	UTE=-1.5
+KT1=-0.11	KT1L=0	KT2=0.022
+UA1=4.31E-9	UB1=-7.61E-18	UC1=-5.6E-11
+AT=3.3E4	WL=0	WLN=1
+WW=0	WWN=1	WWL=0
+LL=0	LLN=1	LW=0
+LWN=1	LWL=0	CAPMOD=2
+XPART=0.5	CGDO=3.42E-10	CGSO=3.42E-10
+CGBO=1E-12	CJ=1.156238E-3	PB=0.8
+MJ=0.4407762	CJSW=1.125225E-10	PBSW=0.8
+MJSW=0.1152909	CJSWG=4.22E-10	PBSWG=0.8
+MJSWG=0.1152909	CF=0	PVTH0=4.284016E-4
+PRDSW=60.4471984	PK2=2.405903E-3	WKETA=0.0352518
+LKETA=0.0207754	PU0=-1.4797175	PUA=-5.65562E-11
+PUB =7.212046E-25	PVSAT=-50	PETA0 = 1.069996E-5
+PKETA=-4.427073E-3	)	

Use the values of the threshold voltages, $V_{T_n} = 0.46$ V and $V_{T_p} = -0.44$ V, to determine the transistor parameters K'_n, λ_n, K'_p, and λ_p.

Solution

Knowing the values of the threshold voltages, V_{T_n} and V_{T_p}, we can solve Equations (A.5) and (A.6) for

$$K'_n = 288.4\ \mu\text{A}/V^2 \quad \lambda_n = 0.072\ V^{-1}$$
$$K'_p = 48.2\ \mu\text{A}/V^2 \quad \lambda_p = 0.229\ V^{-1}$$

TABLE A.1
Simulated Drain Currents

nMOS Transistor, $V_{GS} = 0.95$ V			pMOS Transistor, $V_{SG} = 0.95$ V		
V_{DS} (V)	0.50	1.10	V_{SD} (V)	0.50	1.10
I_D (mA)	11.96	12.45	I_D (mA)	23.29	26.15

Note that due to a further reduction of the transistor channel length, the value of the channel-length modulation parameter is increased, yielding a diminution of the small-signal drain-source resistance.

A.2 Amplifier

Basically, there are two structure types for the design of amplifiers: single-stage and multistage structures. Using a given IC process, an amplifier can be designed to meet the following specifications:

Supply voltage	V_{DD} (V)
Capacitive load	C_L (pF)
Slew rate	SR (V/µs)
Settling time	t_s (µs)
DC voltage gain	A_0 (dB)
Gain-bandwidth product	GBW (MHz)
Phase margin	ϕ_M (°)
Power consumption	P (mW)
Input-referred noise	$\overline{v_{n,i}^2}$ (nV/$\sqrt{\text{Hz}}$ at f Hz)

For a given amplifier, the same dc common-mode voltage is applied to each input node. Its value can vary from the maximum $V_{ICM,max}$ to the minimum $V_{ICM,min}$. Similarly, the output dc voltage is assumed to be bounded by $V_{0,max}$ and $V_{0,min}$.

One of the simplest single-stage differential amplifiers is depicted in Figure A.1. It consists of a differential transistor pair, a current mirror, and a biasing circuit. The slew rate can be expressed as

$$SR = I_5/C_L \tag{A.7}$$

where I_5 is the drain current of the transistor T_5. Hence,

$$I_5 = SR \cdot C_L \tag{A.8}$$

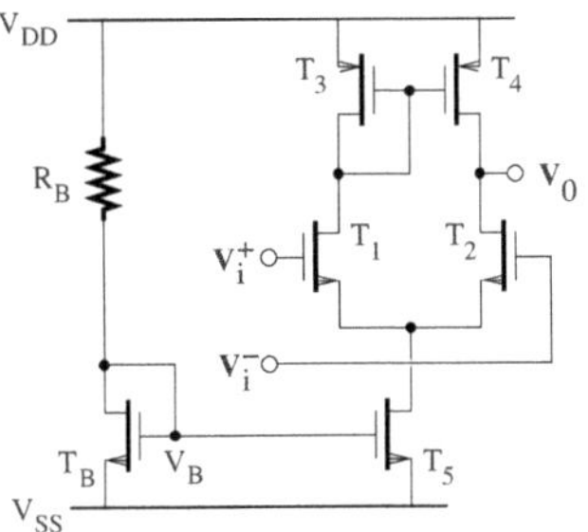

FIGURE A.1
Circuit diagram of a single-stage differential amplifier.

Ideally, the current I_5 should be equal to the bias current I_B. Because the accuracy achieved for the value of I_5 can be limited by the biasing circuit, it is good practice to set the bias current value with a safe margin, that is, $I_B = I_5(1 + x)$, where x is a fractional number denoting the worst-case variation of the current.

The gain-bandwidth product or unity-gain frequency is of the form

$$GBW = \frac{g_{m1}}{C_L} \tag{A.9}$$

The transconductance of the transistor T_1 can be derived as

$$g_{m1} = 2K_n(V_{GS_1} - V_{T_n}) = GBW \cdot C_L \tag{A.10}$$

The aspect ratio of the transistor T_1 is then given by

$$\frac{W_1}{L_1} = \frac{GBW \cdot C_L}{\mu_n C_{ox}(V_{GS_1} - V_{T_n})} = \frac{GBW \cdot C_L}{\mu_n C_{ox} V_{DS_1(sat)}} \tag{A.11}$$

where

$$V_{DS_1(sat)} = \frac{2I_{D_1}}{g_{m1}} = \frac{I_B}{GBW \cdot C_L} \tag{A.12}$$

Note that the transistors T_1 and T_2 are matched, that is, $W_1/L_1 = W_2/L_2$.

Because the dc current flowing through the transistors T_3 and T_4 is $I_B/2$, we can obtain

$$\frac{W_3}{L_3} = \frac{2I_{D_3}}{\mu_p C_{ox}(V_{SG_3} + V_{T_p})^2} = \frac{I_B}{\mu_p C_{ox}(V_{SG_3} + V_{T_p})^2} \tag{A.13}$$

where

$$V_{SG_3} = V_{DD} - V_{ICM,max} + V_{T_n} \tag{A.14}$$

and

$$\frac{W_4}{L_4} = \frac{W_3}{L_3} \tag{A.15}$$

The bias current flows through the transistors T_5 and T_6 and it can be found that

$$\frac{W_5}{L_5} = \frac{2I_{D_5}}{\mu_n C_{ox}(V_{GS_5} - V_{T_n})^2} = \frac{2I_B}{\mu_n C_{ox}(V_{GS_5} - V_{T_n})^2} \tag{A.16}$$

where

$$V_{GS_5} - V_{T_n} = V_{DS_5(sat)} = V_{ICM,min} - V_{SS} - V_{GS_1} \tag{A.17}$$

and

$$\frac{W_6}{L_6} = \frac{W_5}{L_5} \tag{A.18}$$

The value of the resistor R_B is given by

$$R_B = \frac{V_{DD} - V_{GS_6} - V_{SS}}{I_B} \tag{A.19}$$

where $V_{GS_6} = V_{DS_6(sat)} + V_{T_n}$.

The power consumption of the differential stage is estimated to be

$$P = (V_{DD} - V_{SS})I_B \tag{A.20}$$

For the above differential stage, the dc gain is typically about 35 dB, and the phase margin should be at least 60° to meet the stability requirement.

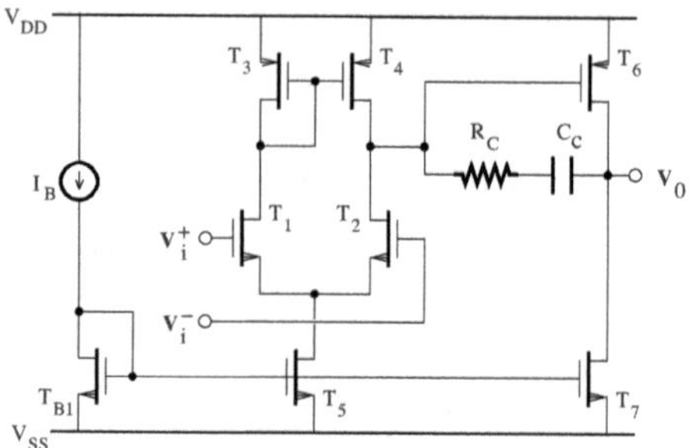

FIGURE A.2
Circuit diagram of a two-stage amplifier.

Let us consider the two-stage amplifier [1,2] shown in Figure A.2. The input stage consists of a differential transistor pair loaded by a current mirror, while the output stage is a common source amplifier with an active load. The frequency stabilization is achieved using a compensation network.

All transistors are biased to operate in the saturation region. Applying Kirchhoff's voltage law between the positive input node and the positive supply voltage, we can obtain

$$V_{DD} - V_i^+ = V_{SG_3} + V_{DS_1} - V_{GS_1} \tag{A.21}$$

where $V_i^+ = V_{ICM,max}$ and $V_{DS_1} = V_{GS_1} - V_{T_n}$. The transistor T_1 operates in the saturation region provided

$$\sqrt{\frac{2I_{D_3}}{\mu_n C_{ox}(W_3/L_3)}} \leq V_{DD} - V_{ICM,max} + V_{T_p} + V_{T_n} \tag{A.22}$$

On the other hand, the application of Kirchhoff's voltage law between the positive input node and the negative supply voltage gives

$$V_i^+ - V_{SS} = V_{GS_1} + V_{DS_5} \tag{A.23}$$

where $V_i^+ = V_{ICM,min}$ and $V_{GS_1} = V_{DS_1} + V_{T_n}$. The operation of the transistor T_5 in the saturation region then requires that

$$\sqrt{\frac{2I_{D_1}}{\mu_n C_{ox}(W_1/L_1)}} + \sqrt{\frac{2I_{D_5}}{\mu_n C_{ox}(W_5/L_5)}} \leq V_{ICM,min} - V_{T_n} - V_{SS} \tag{A.24}$$

The source-drain voltage of the transistor T_6 is given by

$$V_{SD_6} = V_{DD} - V_0 \tag{A.25}$$

To bias T_6 in the saturation region, the following condition must be fulfilled:

$$\sqrt{\frac{2I_{D_6}}{\mu_n C_{ox}(W_6/L_6)}} \leq V_{DD} - V_{0,max} \tag{A.26}$$

In the case of the transistor T_7, the drain-source voltage is of the form

$$V_{DS_7} = V_0 - V_{SS} \tag{A.27}$$

The saturation condition corresponds to

$$\sqrt{\frac{2I_{D_7}}{\mu_n C_{ox}(W_7/L_7)}} \leq V_{0,min} - V_{SS} \tag{A.28}$$

The spectral density of the output noise can be computed as

$$\overline{v_{n,o}^2} = g_{m6}^2 R_2^2 [\overline{v_{n,6}^2} + \overline{v_{n,7}^2} + R_1^2 (g_{m1}^2 \overline{v_{n,1}^2} + g_{m2}^2 \overline{v_{n,2}^2} + g_{m3}^2 \overline{v_{n,3}^2} + g_{m4}^2 \overline{v_{n,4}^2})] \tag{A.29}$$

where R_1 and R_2 are the output resistances of the first and second stages, respectively. Assuming that $g_{m1} = g_{m2}$, $g_{m3} = g_{m4}$ $\overline{v_{n,1}^2} = \overline{v_{n,2}^2}$, $\overline{v_{n,3}^2} = \overline{v_{n,4}^2}$, and $\overline{v_{n,6}^2} = \overline{v_{n,7}^2}$, the spectral density of the input-referred noise can be derived as

$$\overline{v_{n,i}^2} = \frac{\overline{v_{n,o}^2}}{g_{m1}^2 g_{m6}^2 R_1^2 R_2^2} = 2\frac{\overline{v_{n,6}^2}}{g_{m1}^2 R_1^2} + 2\overline{v_{n,1}^2}\left[1 + \left(\frac{g_{m3}}{g_{m1}}\right)^2 \frac{\overline{v_{n,3}^2}}{\overline{v_{n,1}^2}}\right] \tag{A.30}$$

Because $g_{m1}R_1 \gg 1$, $\overline{v_{n,1}^2} = 8kT/(3g_{m1})$, and $\overline{v_{n,3}^2} = 8kT/(3g_{m3})$, we obtain

$$\overline{v_{n,i}^2} \simeq 2\overline{v_{n,1}^2}\left[1 + \left(\frac{g_{m3}}{g_{m1}}\right)^2 \frac{\overline{v_{n,3}^2}}{\overline{v_{n,1}^2}}\right] = \frac{16kT}{3g_{m1}}\left(1 + \frac{g_{m3}}{g_{m1}}\right) \tag{A.31}$$

To reduce the noise contribution due to the transistors acting as loads or current sources, the transconductance g_{m1} (or equivalently g_{m2}) should be made as large as possible. The gain-bandwidth product is given by

$$GBW = \frac{g_{m1}}{C_C} \tag{A.32}$$

where C_C is the compensation capacitor. The slew rate can be written as

$$SR = \frac{I_{D_5}}{C_C} = \frac{I_{D_7} - I_{D_5}}{C_L} \tag{A.33}$$

where C_L is the total load capacitor at the output node. Hence,

$$I_{D_7} = SR(C_C + C_L) \tag{A.34}$$

Because the positive input common-mode range is of the form

$$ICMR^+ = V_{DD} - V_{ICM,max} = V_{SD_3(sat)} - V_{T_n} \tag{A.35}$$

and $I_{D_3} = I_{D_1} = I_{D_5}/2$, it can be found that

$$g_{m3} = \frac{2I_{D_3}}{V_{SD_3(sat)}} = \frac{SR \cdot C_C}{ICMR^+ + V_{T_n}} \tag{A.36}$$

Substituting Equations (A.32) and (A.36) into (A.31) gives

$$C_C = \frac{16kT}{3 \cdot GBW \cdot \overline{v_{n,i}^2}}\left[1 + \frac{SR}{GBW(ICMR^+ + V_{T_n})}\right] \tag{A.37}$$

From Equation (A.32),

$$g_{m1} = \sqrt{2\mu_n C_{ox}(W_1/L_1)I_{D_1}} = GBW \cdot C_C \tag{A.38}$$

Assuming that $I_{D_1} = I_{D_5}/2 = SR \cdot C_C/2$, we have

$$\frac{W_1}{L_1} = \frac{GBW^2 C_C}{\mu_n C_{ox} SR} \tag{A.39}$$

Note that the aspect ratios of T_1 and T_2 are identical. The headroom voltage for the negative input common-mode range input can be expressed as

$$ICMR^- = V_{ICM,min} - V_{SS} = V_{DS_5(sat)} + V_{DS_1(sat)} + V_{T_n} \tag{A.40}$$

where $V_{DS_1(sat)} = SR/GBW$. The current I_{D_5} is given by

$$I_{D_5} = \mu_n C_{ox}(W_5/L_5)V^2_{DS_5(sat)}/2 = SR \cdot C_C \tag{A.41}$$

Combining Equations (A.40) and (A.41), it can be shown that

$$\frac{W_5}{L_5} = \frac{2 \cdot SR \cdot C_C}{\mu_n C_{ox}(ICMR^- - V_{T_n} - SR/GBW)^2} \tag{A.42}$$

Note that the transistors T_5 and T_{B_1} are designed with the same aspect ratio. The transistors T_5 and T_7 are biased in the saturation region and $V_{GS_5} = V_{GS_7}$. Hence,

$$\frac{I_{D_7}}{I_{D_5}} = \frac{W_7/L_7}{W_5/L_5} \tag{A.43}$$

From Equation (A.33), we find that

$$\frac{I_{D_7}}{I_{D_5}} = 1 + \frac{C_L}{C_C} \tag{A.44}$$

and therefore

$$\frac{W_7}{L_7} = \left(1 + \frac{C_L}{C_C}\right)\frac{W_5}{L_5} \tag{A.45}$$

Based on the small-signal analysis of the amplifier, the zero of the transfer function is rejected at infinity by choosing the compensation resistor such that

$$R_C = \frac{1}{g_{m6}} \tag{A.46}$$

The second pole of the transfer function is approximately located at

$$\omega_{p_2} \simeq -\frac{g_{m6}}{C_L} \tag{A.47}$$

where C_L is the load capacitor. The amplifier is designed to exhibit the behavior of a two-pole system, that is, $\omega_{p_3} \geq 10 \cdot GBW$. Hence, its phase margin is of the form

$$\phi_M \simeq 90^\circ - \arctan\frac{GBW}{|\omega_{p_2}|} \tag{A.48}$$

Combining Equations (A.47) and (A.48) gives

$$g_{m6} = GBW \cdot C_L \cdot \tan(\phi_M). \tag{A.49}$$

Note that, for a given angle x, $\tan(90^\circ - x) = 1/\tan(x)$. Because we have $g_{m6} = \sqrt{2\mu_n C_{ox}(W_6/L_6)I_{D_6}}$ and $I_{D_6} = I_{D_7}$, it can be shown that

$$\frac{W_6}{L_6} = \frac{[GBW \cdot C_L \tan(\phi_M)]^2}{2\mu_n C_{ox} I_{D_7}} \tag{A.50}$$

To reduce the offset voltage due to the difference, which can exist between the values of currents I_{D_6} and I_{D_7} when the amplifier input nodes are grounded, we should have $V_{SD_4} = V_{SD_3} = V_{SG_3}$ and $V_{SD_4} = V_{SG_6}$. It can also be deduced that $I_{D_3} = I_{D_4} = I_{D_5}/2$ and $I_{D_6} = I_{D_7}$. The aspect ratios of transistors T_3, T_4, T_5, T_6, and T_7 satisfy the following relation,

$$\frac{W_3/L_3}{W_6/L_6} = \frac{W_4/L_4}{W_6/L_6} = \frac{1}{2}\frac{W_5/L_5}{W_7/L_7} \tag{A.51}$$

Equation (A.51) can be used for the determination of the aspect ratio of T_3 and T_4.

For the resulting amplifier, the dc gain is approximately given by

$$A_0 = -\frac{g_{m1}}{g_2 + g_4} \cdot \frac{g_{m6}}{g_6 + g_7} \tag{A.52}$$

where g_2, g_4, g_6, and g_7 denote the drain-source conductance of transistors T_2, T_4, T_6, and T_7, respectively. The power dissipation can be computed as

$$P = (V_{DD} - V_{SS})(2I_{D_5} + I_{D_7}) \tag{A.53}$$

The results of the initial transistor sizing for the two-stage amplifier may be optimized using appropriate CAD programs [3]. The maximum and minimum sizes of transistors are primarily set by process-related variations and limitations of the IC fabrication technique.

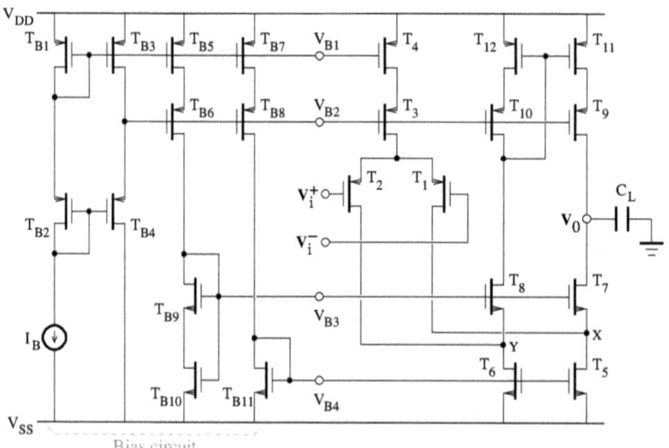

FIGURE A.3
Circuit diagram of a folded-cascode amplifier.

In cases where a large gain is required with a high gain-bandwidth product and a limited power budget, a folded-cascode amplifier [4, 5], as shown in

Figure A.3, is the architecture of choice. Due to symmetry, the transistors T_1 and T_2, T_5 and T_6, T_7 and T_8, T_9 and T_{10}, and T_{11} and T_{12} are matched. As a result of the small-signal analysis, the *dc* gain of the folded-cascode amplifier can be written as

$$A_0 = g_{m1} r_0 \tag{A.54}$$

where

$$r_0 = \frac{1}{\dfrac{(g_1 + g_5)g_7}{g_{m7} + g_{mb7}} + \dfrac{g_{11} g_9}{g_{m9} + g_{mb9}}} \tag{A.55}$$

and g_k $(k = 1, 5, 9.11)$ is the drain-source conductance of transistor T_k. The slew rate is given by

$$SR = \frac{I_{D_3}}{C_L} \tag{A.56}$$

where $I_{D_3} = I_{D_4} = I_1$ and I_1 is the bias current of the differential transistor pair. The gain-bandwidth product is of the form

$$GBW = \frac{g_{m1}}{C_L} \tag{A.57}$$

Because $g_{m1} = \sqrt{2\mu_p C_{ox}(W_1/L_1) I_{D_1}}$ and $I_{D_1} = I_{D_3}/2$, we obtain

$$\frac{W_1}{L_1} = \frac{GBW^2 C_L}{\mu_p C_{ox} SR} \tag{A.58}$$

Using Equation (A.56), the aspect ratios of transistors T_3 and T_4 can be computed as

$$\frac{W_3}{L_3} = \frac{2I_{D_3}}{\mu_p C_{ox} V^2_{SD_3(sat)}} = \frac{2 \cdot SR \cdot C_L}{\mu_p C_{ox} V^2_{SD_3(sat)}} \tag{A.59}$$

and

$$\frac{W_4}{L_4} = \frac{2I_{D_4}}{\mu_p C_{ox} V^2_{SD_4(sat)}} = \frac{2 \cdot SR \cdot C_L}{\mu_p C_{ox} V^2_{SD_4(sat)}}, \tag{A.60}$$

where $V_{SD_3(sat)} = V_{SD_4(sat)}$, $V_{SD_3(sat)} = (V_{DD} - V_{SG_1} - V_{ICM,max})/2$, and $V_{SG_1} = V_{SD_1(sat)} - V_{T_p}$. The power consumption of the amplifier is of the form

$$P = (V_{DD} - V_{SS})(I_1 + 2I_2) \tag{A.61}$$

where $I_1 = I_{D_3} = I_{D_4}$ and $I_2 = I_{D_7} = I_{D_8}$. Using Equations (A.56) and (A.61), the aspect ratio of the transistor T_{11} is derived as

$$\frac{W_5}{L_5} = \frac{2I_{D_5}}{\mu_n C_{ox} V^2_{DS_5(sat)}} \tag{A.62}$$

where

$$I_{D_5} = I_2 + I_1/2 = \frac{P}{2(V_{DD} - V_{SS})} \tag{A.63}$$

Because

$$V_i^- = V_{SS} + V_{DS_5(sat)} + V_{SD_1(sat)} - V_{SG_1} \tag{A.64}$$

it can be shown that

$$V_{DS_5(sat)} = V_{ICM,min} - V_{SS} - V_{T_p} \tag{A.65}$$

where $V_i^- = V_{ICM,min}$. For the size of the transistor T_7, we find

$$\frac{W_7}{L_7} = \frac{2I_{D_7}}{\mu_n C_{ox} V_{DS_7(sat)}^2} \tag{A.66}$$

where

$$I_{D_7} = I_2 = \frac{1}{2}\left(\frac{P}{V_{DD} - V_{SS}} - SR \cdot C_L\right) \tag{A.67}$$

and

$$V_{DS_7(sat)} = V_{0,min} - V_{SS} - V_{DS_5(sat)} \tag{A.68}$$

The small-signal analysis indicates that the transfer function of the folded-cascode amplifier has four poles and two zeros [6]. Because two of the poles are approximately compensated by the zeros, the amplifier phase margin can be expressed as

$$\phi_M \simeq 180^\circ - \arctan\frac{GBW}{|\omega_{p_1}|} - \arctan\frac{GBW}{|\omega_{p_2}|} \tag{A.69}$$

where $\omega_{p_1} = -GBW/A_0$ and $\omega_{p_2} = -g_{m7}/C_p$. Note that A_0 is the amplifier dc gain and C_p is the total parasitic capacitance at the source of the transistor T_7 ($C_p \simeq C_{gs_7} \simeq 2W_7L_7C_{ox}/3$). To proceed further, we can simplify Equation (A.69) to

$$\phi_M \simeq 90^o - \arctan\frac{GBW}{|\omega_{p_2}|} \tag{A.70}$$

or equivalently

$$g_{m7} = GBW \cdot C_p \cdot \tan\phi_M \tag{A.71}$$

Here, g_{m7} and C_p are functions of the width and length of the transistor T_7, making difficult the determination of the width-to-length ratio from Equation (A.71) alone. However, using the transconductance expression in the saturation region given by

$$g_{m_7} = \mu_n C_{ox}\left(\frac{W_7}{L_7}\right)(V_{GS_7} - V_{T_n}) \tag{A.72}$$

the length of the transistor T_7 can be obtained from Equation (A.71) as,

$$L_7 = \sqrt{\frac{3}{2}\frac{\mu_n V_{DS_7(sat)}}{GBW \cdot \tan\phi_M}} \tag{A.73}$$

where $V_{DS_7(sat)} = V_{GS_7} - V_{T_n}$

The aspect ratios of transistors T_9 and T_{11} are, respectively, given by

$$\frac{W_9}{L_9} = \frac{2I_{D_9}}{\mu_p C_{ox} V_{SD_9(sat)}^2} \tag{A.74}$$

and

$$\frac{W_{11}}{L_{11}} = \frac{2I_{D_{11}}}{\mu_p C_{ox} V_{SD_{11}(sat)}^2} \tag{A.75}$$

where

$$I_{D_9} = I_{D_{11}} = I_{D_7} = \frac{1}{2}\left(\frac{P}{V_{DD} - V_{SS}} - SR \cdot C_L\right) \tag{A.76}$$

and

$$V_{SD_9(sat)} = V_{SD_{11}(sat)} = (V_{DD} - V_{0,max})/2 \tag{A.77}$$

Given the current I_B, the biasing circuit should be sized such that it can generate the required voltages V_{B_1}, V_{B_2}, V_{B_3}, and V_{B_4}. All transistors, except T_{B10}, are biased in the saturation region. Transistors $T_{B1} - T_{B8}$ operate as a current mirror delivering the bias current to each of the loads $T_{B9} - T_{B10}$ and T_{B11}. Hence,

$$\begin{aligned} W_3/L_3 &= W_{B1}/L_{B1} = 4(W_{B2}/L_{B2}) = W_{B3}/L_{B3} = W_{B4}/L_{B4} \\ &= W_{B5}/L_{B5} = W_{B6}/L_{B6} = W_{B7}/L_{B7} = W_{B8}/L_{B8} \end{aligned} \tag{A.78}$$

The currents flowing through transistors T_{B9} and T_{B10}, that are supposed to operate in the saturation and triode regions, respectively, can be expressed as,

$$I_1 = \frac{1}{2}\mu_n C_{ox}\left(\frac{W_{B9}}{L_{B9}}\right)(V_{GS_{B9}} - V_T)^2 \tag{A.79}$$

and

$$I_1 = \frac{1}{2}\mu_n C_{ox}\left(\frac{W_{B10}}{L_{B10}}\right)\left[2(V_{GS_{B10}} - V_T)V_{DS_{B10}} - V_{DS_{B10}}^2\right] \tag{A.80}$$

where

$$V_{GS_{B9}} = V_{DS(sat)} + V_T \tag{A.81}$$

$$V_{DS_{B10}} = V_{DS(sat)} \tag{A.82}$$

and

$$V_{GS_{B10}} = V_{GS_{B9}} + V_{DS_{B10}} = 2V_{DS(sat)} + V_T \quad \text{(A.83)}$$

Combining Equations (A.79) and (A.80), we obtain

$$\frac{\mu_n C_{ox}}{2}\left(\frac{W_{B9}}{L_{B9}}\right)V_{DS(sat)}^2 = \frac{\mu_n C_{ox}}{2}\left(\frac{W_{B10}}{L_{B10}}\right)\left[2(2V_{DS(sat)})V_{DS(sat)} - V_{DS(sat)}^2\right] \quad \text{(A.84)}$$

Hence,

$$\frac{W_{B9}}{L_{B9}} = 3\left(\frac{W_{B10}}{L_{B10}}\right) \quad \text{(A.85)}$$

The gate-source voltage of the transistor T_{B10} can be written as

$$V_{GS_{B10}} = V_{GS_7} + V_{DS_5} \quad \text{(A.86)}$$

where $V_{DS_5} = V_{DS(sat)}$. Using the I-V equation of transistors T_{B10}, T_7, and T_5, we arrive at

$$\sqrt{\frac{I_1}{\frac{\mu_n C_{ox}}{2}\left(\frac{W_{B10}}{L_{B10}}\right)}} = \sqrt{\frac{I_2}{\frac{\mu_n C_{ox}}{2}\left(\frac{W_7}{L_7}\right)}} + \sqrt{\frac{I_2 + I_1/2}{\frac{\mu_n C_{ox}}{2}\left(\frac{W_5}{L_5}\right)}} \quad \text{(A.87)}$$

The aspect ratio of the transistor T_{B10} can then be computed as,

$$\frac{W_{B10}}{L_{B10}} = \left(\sqrt{\frac{I_2/I_1}{\frac{W_7}{L_7}}} + \sqrt{\frac{I_2/I_1 + 1/2}{\frac{W_5}{L_5}}}\right)^{-2} \quad \text{(A.88)}$$

For the transistors T_{B11} and T_5, we have

$$V_{GS_{B11}} = V_{GS_5} \quad \text{(A.89)}$$

Writing each gate-source voltage as a function of the current flowing through the transistor and the threshold voltage gives

$$V_T + \sqrt{\frac{I_1}{\frac{\mu_n C_{ox}}{2}\left(\frac{W_{B11}}{L_{B11}}\right)}} = V_T + \sqrt{\frac{I_2 + I_1/2}{\frac{\mu_n C_{ox}}{2}\left(\frac{W_5}{L_5}\right)}} \quad \text{(A.90)}$$

Finally, we obtain

$$\frac{W_{B11}}{L_{B11}} = \frac{1}{I_2/I_1 + 1/2}\left(\frac{W_5}{L_5}\right) \quad \text{(A.91)}$$

The amplifier sizing is based on the approximate value of the parasitic capacitance C_p. If the phase margin specification is not met, the actual value of C_p should be extracted from the resulting amplifier and used to redo the calculations of transistor sizes.

In practice, due to IC process variations, a small safety margin is often added to $V_{DS(sat)}$ or $V_{SD(sat)}$ to ensure that the transistors remain in the saturation region. That is,

$$V_{DS} - (V_{GS} - V_{T_n}) \geq V_{DS(sat)} \tag{A.92}$$

and

$$V_{SD} - (V_{SG} + V_{T_p}) \geq V_{SD(sat)} \tag{A.93}$$

For submicrometer circuit designs, the saturation voltage is generally in the range from 0.2 V to 0.6 V.

The total power consumption of the amplifier is given by

$$P_T = (V_{DD} - V_{SS})(I_1 + 2I_2) + P_B \tag{A.94}$$

where $P_B = 4(V_{DD} - V_{SS})I_B$ is the power consumption of the bias circuit.

Note that the design equations of the folded-cascode amplifier can be further simplified in the case where $I_1 = I_B$ and $I_2 = I_B/2$.

A.3 Comparator and latch

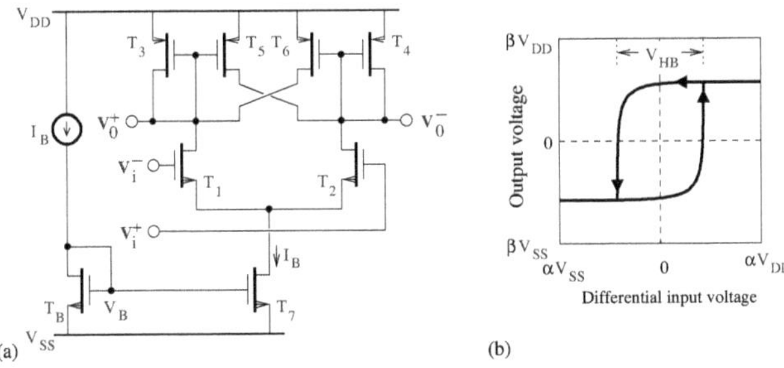

FIGURE A.4
(a) Comparator circuit with hysteresis; (b) comparator characteristics in the noninverting configuration.

One of the structures that can be used for the design of a comparator is depicted in Figure A.4(a). It consists of an input differential stage loaded by cross-coupled transistors. Due to the positive feedback provided by the loading transistors, the comparator transfer characteristic exhibits hysteresis as shown in Figure A.4(b).

Given the values of some specifications (slew rate, gain-bandwidth product, input common-mode voltage, input-referred noise, power dissipation), the sizes of transistors composing the input stage can be determined in the same way as in the case of an amplifier.

The difference between the positive and negative trigger points is the hysteresis band, which is given by

$$V_{HB} = V_{trig^+} - V_{trig^-} \tag{A.95}$$

Assuming that the transistors T_1 and T_2 are matched, that is, $W_1/L_1 = W_2/L_2$, we obtain

$$V_{trig^+} = \sqrt{\frac{I_B}{2K'(W_1/L_1)}} \frac{\sqrt{(W_5/L_5)/(W_3/L_3)} - 1}{\sqrt{1 + (W_5/L_5)/(W_3/L_3)}} \tag{A.96}$$

and

$$V_{trig^-} = \sqrt{\frac{I_B}{2K'(W_1/L_1)}} \frac{1 - \sqrt{(W_6/L_6)/(W_4/L_4)}}{\sqrt{1 + (W_6/L_6)/(W_4/L_4)}} \tag{A.97}$$

For a symmetrical structure, it is required that $V_{trig^+} = -V_{trig^-}$. Hence, $W_3/L_3 = W_4/L_4$ and $W_5/L_5 = W_6/L_6$. Given the values of I_B, $2K'(W_1/L_1)$ and the specification V_{trig^+}, the ratio $x = (W_5/L_5)/(W_3/L_3)$ can be determined by solving the next equation, which is derived from Equation (A.96) and can be written as

$$x + \frac{2}{\Delta^2 - 1}\sqrt{x} + 1 = 0 \tag{A.98}$$

where $\Delta = V_{trig^+}\sqrt{2K'(W_1/L_1)/I_B}$. Solving Equation (A.98) gives

$$\sqrt{x} = -\frac{1}{\Delta^2 - 1} \pm \sqrt{\frac{1}{(\Delta^2 - 1)^2} - 1} \tag{A.99}$$

where $|\Delta| < \sqrt{2}$ and $\Delta \neq 1$. To choose between the plus and minus sign in the valid expression of x, we note that the condition $x > 1$ should be realized in order for the comparator to operate with a positive feedback.

A D latch [7] can be implemented as shown in Figure A.5(a). The input data are applied differentially to a pair of transistors connected through clocked transistors to two cross-coupled inverters.

Ideally, the transition between the low state and high state should occur at the mid-point of the logic swing or $V_{DD}/2$ (see Figure A.5(b)). This is achieved when both transistors are in the saturation region. Hence,

$$\mu_n(W_5/L_5) = \mu_p(W_7/L_7) \tag{A.100}$$

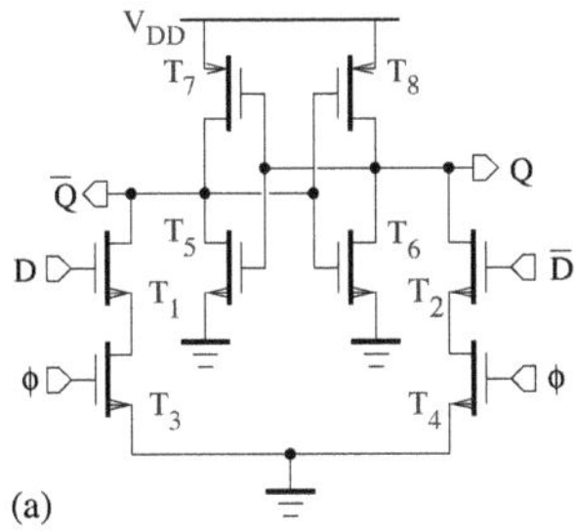

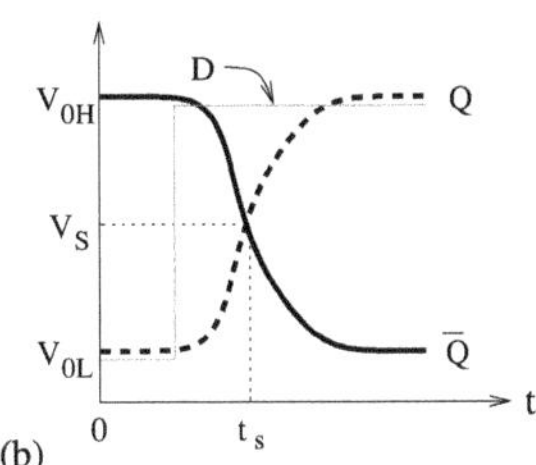

FIGURE A.5
(a) Circuit diagram of a D latch; (b) output waveforms.

and

$$\mu_n(W_6/L_6) = \mu_p(W_8/L_8) \tag{A.101}$$

Provided $\mu_n/\mu_p = r$, the difference in charge mobilities is compensated in an inverter by choosing the aspect ratio (W/L) of the pMOS transistor to be r times the one of the nMOS transistor.

To switch the latch from the low state to the high state, the low level of the circuit section $T_3 - T_1 - T_7$ should be below the switching threshold of the inverter $T_5 - T_7$. Assuming that the transistors are biased in the triode region, and transistors T_1 and T_3 can be reduced to a single transistor with the aspect ratio W_k/L_k, it can be shown that

$$K_{p_7}[2(V_{SG_7} + V_{T_p})V_{SD_7} - V_{SD_7}^2] = K_{n_k}[2(V_{GS_k} - V_{T_n})V_{DS_k} - V_{DS_k}^2] \tag{A.102}$$

where $V_{SG_7} = V_{GS_k} = V_{DD}$ and $V_{SD_7} = V_{DS_k} = V_{DD}/2$. With $V_{T_n} \leq |V_{T_p}|$, we arrive at the following condition:

$$K_{n_k} \geq K_{p_7} \tag{A.103}$$

Substituting $K_{n_k} = \mu_n C_{ox}(W_k/L_k)$ and $K_{p_7} = \mu_p C_{ox}(W_7/L_7)$ into Equation (A.103) gives

$$(W_k/L_k) \geq (\mu_p/\mu_n)(W_7/L_7) \tag{A.104}$$

Because $(W_1/L_1) = (W_3/L_3) = 2(W_k/L_k)$, we can obtain

$$(W_1/L_1) = (W_3/L_3) \geq 2(\mu_p/\mu_n)(W_7/L_7) = 2(W_5/L_5) \tag{A.105}$$

Similarly, due to the circuit symmetry, it can be found that

$$(W_2/L_2) = (W_4/L_4) \geq 2(\mu_p/\mu_n)(W_8/L_8) = 2(W_6/L_6) \tag{A.106}$$

Because the above derivation of transistor sizes is based on first-order transistor models and simplified circuit operation, it may still be necessary to use simulation results to adequately adjust the transistor widths and lengths.

A.4 Transistor sizing based on the g_m/I_D methodology

The g_m/I_D design method is valid for all operation regions of the transistor. It makes use of either SPICE or analytical transistor models to relate transistor characteristics to the transconductance efficiency, g_m/I_D. For each transistor, the channel width can then be calculated from the selected drain current, inversion coefficient (or operation region), and channel length.

TABLE A.2
Transistor Parameters in 0.18-µm CMOS Technology

Parameter	nMOS	pMOS
I_T (μA)	0.63	0.12
n	1.23	1.21
μ ($\mu A/V^2$)	490.8	98.6
C_{ox} ($fF/\mu m^2$)	7.79	7.53

Table A.2 gives some useful transistor parameters in 0.18-µm CMOS technology.

Consider the circuit diagram of a common-source amplifier shown in Figure A.6(a), where the p-channel transistor is loaded by an ideal current source and a load capacitance. To derive the small-signal performance characteristics, the transistor equivalent model of Figure A.6(b) is used. Due to the fact that $v_{sb} = 0$, the small-signal operation of the common-source amplifier is not affected by g_{mb} and C_{SB}.

With reference to the amplifier small-signal equivalent circuit shown in Figure A.6(c), the output node equations can be written as follows:

$$I_O(s) = -g_m(-V_i(s)) + sC_{GD}(V_O(s) - V_i(s)) + (g_{ds} + sC_{DB})V_O(s) \quad \text{(A.107)}$$

and

$$I_O(s) = -(G_L + sC_L)V_O(s) \quad \text{(A.108)}$$

where $G_L = 1/R_L$. Solving for the voltage gain gives

$$A_v = \frac{V_O(s)}{V_i(s)} = -\frac{g_m - sC_{GD}}{g_{ds} + G_L + s(C_{GD} + C_{DB} + C_L)} \quad \text{(A.109)}$$

or equivalently,

$$A_v = \frac{V_O(s)}{V_i(s)} = -A_{v0}\frac{1 - s/\omega_z}{1 - s/\omega_p} \quad \text{(A.110)}$$

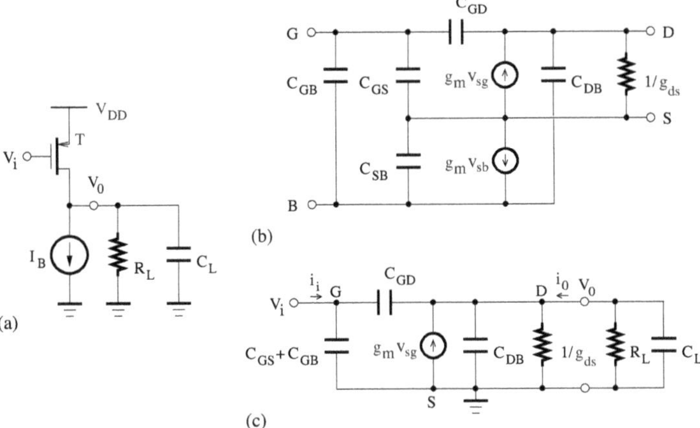

FIGURE A.6
(a) Common-source amplifier; (b) transistor small-signal equivalent model; (c) amplifier small-signal equivalent circuit.

where

$$A_{v0} = g_m/(g_{ds} + G_L) \tag{A.111}$$
$$\omega_z = g_m/C_{GD} \tag{A.112}$$

and

$$\omega_p = -(g_{ds} + G_L)/(C_{GD} + C_{DB} + C_L) \tag{A.113}$$

Generally, $C_L \gg C_{GD} + C_{DB}$ and the -3 dB bandwidth is reduced to

$$BW = |\omega_p| \simeq (g_{ds} + G_L)/C_L \tag{A.114}$$

The gain-bandwidth product is then given by

$$GBW = A_{v0} \times BW \simeq \frac{g_m}{C_L} = \left(\frac{g_m}{I_D}\right)\left(\frac{I_D}{C_L}\right) = \frac{g_m}{I_D} \cdot SR \tag{A.115}$$

where SR is the slew rate. Note that $GBW = 2\pi f_T$, where f_T is the transition frequency (or the frequency at which the voltage gain is reduced to unity or 0 dB).

In the saturation region and with the assumption that $g_{ds} \gg G_L$, we have

$$A_{v0} \simeq \frac{g_m}{g_{ds}} = \left(\frac{g_m}{I_D}\right)\left(\frac{I_D}{g_{ds}}\right) \simeq \frac{g_m}{I_D} \cdot V_A \tag{A.116}$$

Because g_m/I_D can be related to the inversion coefficient, IC, it can also be deduced that the *dc* gain, A_{v0}, and the transition frequency, f_T, are functions of the IC.

The drain current that is valid in all operation regions of the transistor can be expressed as

$$I_D = 2n\mu C_{ox}U_T^2\left(\frac{W}{L}\right)\ln^2\left(1+\mathrm{e}^{\dfrac{V_{GS}-V_T}{2nU_T}}\right) \tag{A.117}$$

where U_T is the thermal voltage ($U_T = kT/q = 25.8$ mV at the room temperature, $T = 300$ K), and n represents the substrate factor and ranges from 1.2 to 1.5. The current I_D can be rewritten as

$$I_D = 2n\mu C_{ox}U_T^2\left(\frac{W}{L}\right)IC \tag{A.118}$$

where the inversion coefficient, IC, is given by

$$IC = \ln^2\left(1+\mathrm{e}^{\dfrac{V_{GS}-V_T}{2nU_T}}\right) \tag{A.119}$$

The voltage overdrive can then be obtained as

$$V_{ov} = V_{GS} - V_T = 2nU_T\ln\left(\mathrm{e}^{\sqrt{IC}}-1\right) \tag{A.120}$$

In the weak inversion region, the assumption $IC \ll 1$ leads to $\mathrm{e}^{\sqrt{IC}} \simeq 1+\sqrt{IC}$, and it can be shown that

$$V_{ov} = V_{GS} - V_T \simeq nU_T\ln(IC) \tag{A.121}$$

In the strong inversion region, the voltage overdrive is equal to the drain-source saturation voltage, $V_{DS(sat)}$. A simplification is obtained by using the fact that $IC \gg 1$ and $\mathrm{e}^{\sqrt{IC}} \gg 1$. Hence,

$$V_{ov} = V_{DS(sat)} = V_{GS} - V_T \simeq 2nU_T\sqrt{IC} \tag{A.122}$$

Here, the voltage overdrive changes as the square root of the inversion coefficient.

Let g_m be the transistor transconductance that is defined as, $g_m = \partial I_D/\partial V_{GS}$. It can be shown that the transconductance efficiency is given by

$$\frac{g_m}{I_D} = \frac{1-\mathrm{e}^{-\sqrt{IC}}}{nU_T\sqrt{IC}} \tag{A.123}$$

In the weak inversion region, $IC \leq 0.1$ and $g_m/I_D \simeq 1/nU_T$, while in the strong inversion region, $IC > 10$ and we have $g_m/I_D \simeq 1/(V_{GS} - V_T)$. The transconductance efficiency can be appropriately interpolated [8] using a simple expression of the form

$$\frac{g_m}{I_D} = \frac{1}{nU_T(\sqrt{IC + 0.5\sqrt{IC} + 1})} \quad \text{(A.124)}$$

The transconductance efficiency is essentially dependent on the inversion coefficient, but it can be affected by short-geometry effects, such as the velocity saturation.

The intrinsic gain, g_m/g_{ds}, can be expressed as a function of the transconductance efficiency. Hence,

$$\frac{g_m}{g_{ds}} = \left(\frac{g_m}{I_D}\right)\left(\frac{I_D}{g_{ds}}\right) \quad \text{(A.125)}$$

where g_{ds} is the drain-source conductance. For long-channel transistors, $I_D/g_{ds} = V_A + V_{DS} \simeq V_A$, where $V_A = 1/\lambda$ represents the Early voltage and λ is the channel length modulation parameter. The intrinsic gain is a function of the Early voltage, which is generally the result of physical effects such as the channel-length modulation and drain-induced barrier lowering.

A practical approach to transistor sizing for CMOS circuits then consists of using plots of transistor parameters (transconductance efficiency, intrinsic gain, etc.) as functions of either the inversion coefficient or the overdrive.

Analog design based on the g_m/I_D methodology is attractive because it allows the transistor sizing for any given biasing point of the entire operating range extending from the weak inversion region to the strong inversion region.

The transconductance efficiency, g_m/I_D, is independent of the transistor size, and can be derived from experimental data and fitted with a SPICE model or an analytical model that provides a continuous representation of the drain-source current and small-signal characteristics in all regions of operations.

The design flow for the g_m/I_D sizing methodology is as follows:

1. Determine the transconductance, g_m, from circuit specifications such as gain, bandwidth, and power consumption.
2. Choose L depending on the IC process and circuit requirements (short channel for high transition frequency or say bandwidth and long channel for a high output resistance that translates into a high gain).

3. Depending on the transistor operating (strong, moderate or weak) region, obtain the ratio g_m/I_D (large g_m/I_D for low power and large signal swing, and small g_m/I_D for high speed).
4. Determine the drain-source current I_D from the values of g_m and g_m/I_D.
5. Compute the drain-source conductance, g_{ds}, from the values of g_m and g_m/g_{ds} if necessary, as in the case of a transistor operating in the moderate inversion (triode or linear) region.
6. Use the current density or I_D/W plot to find W.
7. Size the devices of the bias circuit.

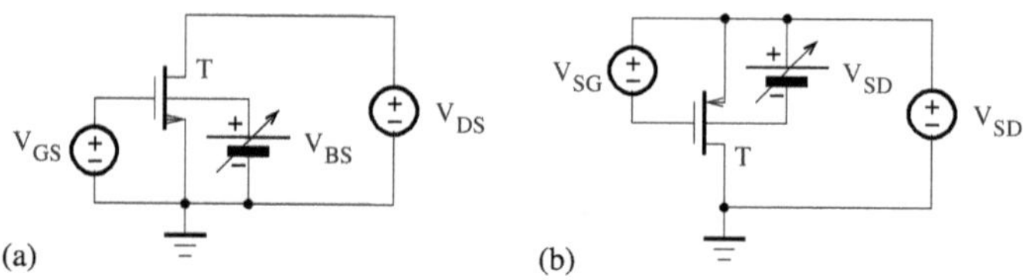

FIGURE A.7
Transistor characterization circuits: (a) nMOSFET, (b) pMOSFET.

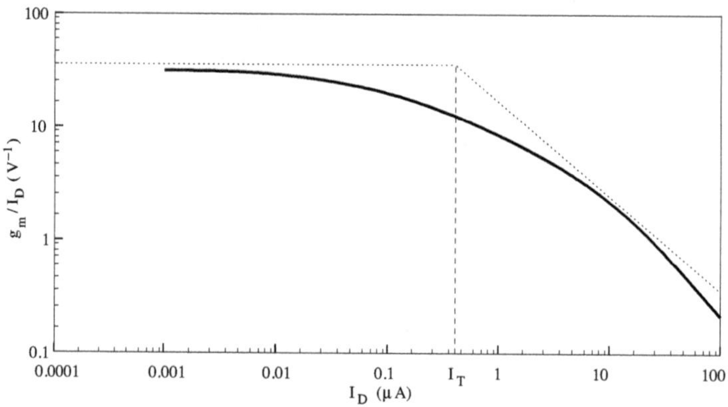

FIGURE A.8
Plot of g_m/I_D versus I_D.

The g_m/I_D design method can be implemented using a graphical lookup approach, and transistor sizing is performed by exploiting the plots of some transistor characteristics obtained from SPICE simulations.

Figure A.7 shows circuits whose SPICE input files can be used to plot various characteristics of n-channel and p-channel transistors.

The transconductance is found by estimating the ratio of the incremental

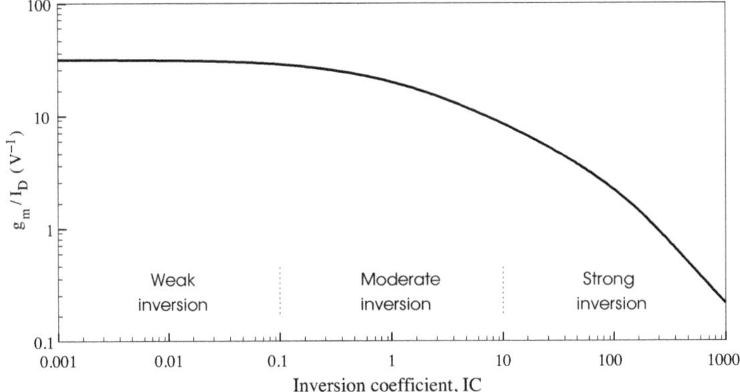

FIGURE A.9
Plot of g_m/I_D versus IC.

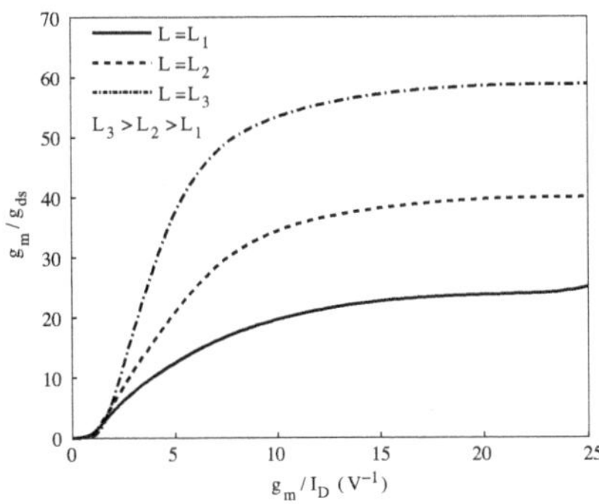

FIGURE A.10
Plots of g_m/g_{ds} versus g_m/I_D.

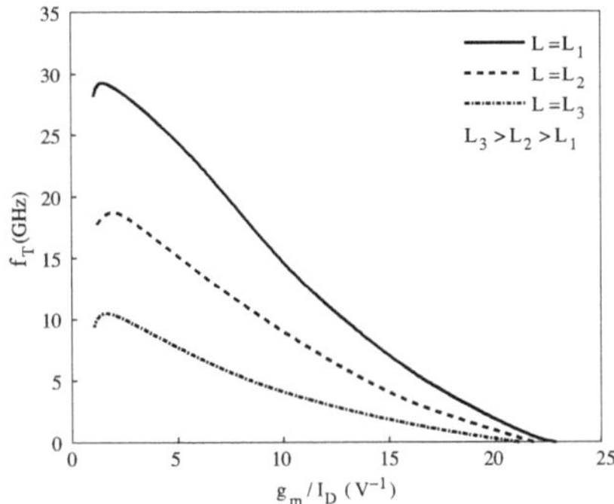

FIGURE A.11
Plots of f_T versus g_m/I_D.

changes in the drain-source current and the gate-source voltage by slightly varying the gate voltage from the operating point. It may be required to vary the supply voltage so that the drain-source voltage can remain constant during the whole process.

The conductance is obtained by computing the ratio of the incremental changes in the drain-source current and the drain-source voltage by slightly varying the supply voltage, while keeping the gate-source voltage constant.

The technology current, I_T, can be determined from the plot of g_m/I_D versus I_D, as shown in Figure A.8. The center of the moderate inversion region corresponds to the intersection between the asymptotes related to the constant function in the weak inversion region and the square law in the strong inversion region. The plot of g_m/I_D versus IC is depicted in Figure A.9. At the center of the moderate inversion region, where the drain current is equal to the technology current, $IC = 1$. Note that the technology current is a function

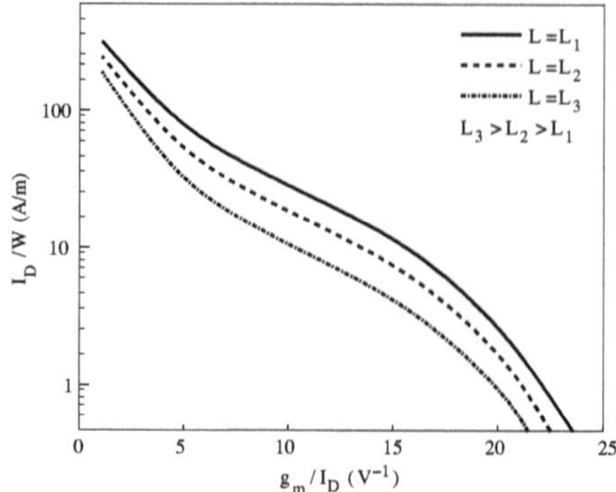

FIGURE A.12
Plots of I_D/W versus g_m/I_D.

of the surface carrier mobility, μ, and the voltage division ratio, n, between the gate oxide capacitor and the substrate depletion MOS capacitor. It is then not constant, because μ and n are dependent on the biasing condition and IC technology.

The sizing procedure starts by the selection of the transistor operating region or equivalently the value of the IC, and then the determination of the corresponding transconductance efficiency, g_m/I_D, or vice versa. The next step consists of estimating the transistor size by using the plots shown in Figures A.10, A.11, and A.12, respectively, for the intrinsic gain, g_m/g_{ds}, the transition frequency, f_T, and the current density, I_D/W, versus the transconductance efficiency, g_m/I_D, and for several channel lengths.

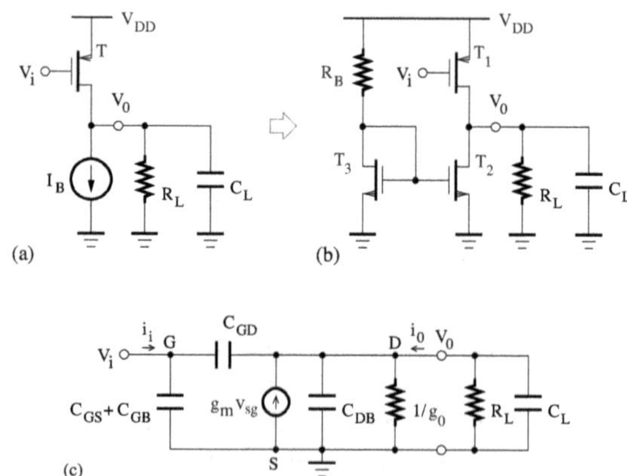

FIGURE A.13
(a) Common-source amplifier biased by a current source; (b) practical implementation of the common-source amplifier and (c) its small-signal equivalent circuit.

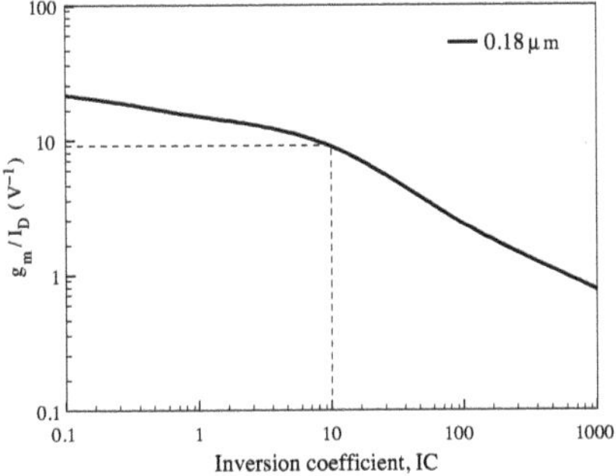

FIGURE A.14
Plot of g_m/I_D versus IC.

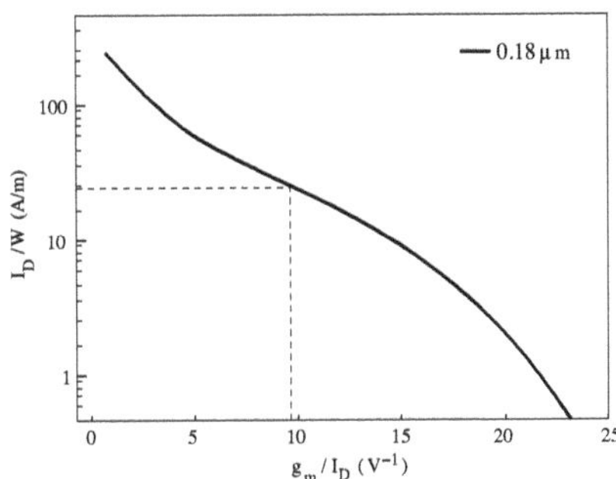

FIGURE A.15
Plot of I_D/W versus g_m/I_D.

In practice, the common-source amplifier biased by a current source (see Figure A.13(a)) can be implemented by the circuit of Figure A.13(b), where the transistors T_2 and T_3 are identical. Taking into account the drain-source resistance of the transistor T_2, that plays the role of the current source, the small-signal equivalent circuit is now represented, as shown in Figure A.13(c), where $g_m = g_{m1}$ and $g_0 = g_{ds1} + g_{ds2}$.

Using the 0.18-μm CMOS technology, determine the sizes of the transistor T_1 required to meet the following specifications:

- Transition frequency, $f_T = GBW/(2\pi) = 400$ MHz
- Load capacitor, $C_L = 2$ pF

The transistors are biased to maximize the output voltage swing.

Assuming that both transistors operate in the strong inversion (saturation) region, the range of the output voltage is defined by

$$V_{DS2(sat)} \geq V_0 \geq V_{DD} - |V_{SD1(sat)}| \tag{A.126}$$

In the strong inversion region, the voltage $V_{DS(sat)}$ is of the form,

$$V_{DS(sat)} = V_{GS} - V_T \simeq 2nU_T\sqrt{IC} \tag{A.127}$$

where $IC \geq 10$. The output voltage swing is maximized by biasing the transistors slightly above the saturation region. Hence, the value of the inversion coefficient should be selected in the order of 10.5.

The requirement of a high transition frequency is easily achieved by using a transistor with the minimum channel length. From the plot of g_m/I_D versus IC, shown in Figure A.14, where $L = 0.18$ μm, we have $(g_m/I_D)_{IC=10.5} = 8.9$.

The transition frequency equation gives

$$g_m = 2\pi f_T C_L = 2\pi(400 \times 10^6)(2 \times 10^{-12}) = 5.024\ mS \tag{A.128}$$

The drain-source current is obtained as

$$I_D = \frac{g_{m1}}{(g_m/I_D)_{IC=10.5}} = \frac{5.024 \times 10^{-3}}{8.9} = 564\ \mu A \tag{A.129}$$

The plot of I_D/W versus g_m/I_D is depicted in Figure A.15, where it can be deduced that $(I_D/W)_{g_m/I_D=8.9} = 28.1$. The transistor width is then given by

$$W = \frac{I_D}{(I_D/W)_{g_m/I_D=8.9}} = \frac{564 \times 10^{-6}}{28.1} = 20\ \mu m \tag{A.130}$$

Remarks

The aspect ratio of transistors T_2 and T_3 can be determined in two ways.

- In the strong inversion region, it can be shown that

$$V_{DS(sat)} = V_{GS} - V_T = 2\frac{I_D}{g_m} \tag{A.131}$$

or equivalently,

$$\frac{g_m}{I_D} = \frac{2}{V_{DS(sat)}} \tag{A.132}$$

Using the plot of the current density, I_D/W, versus the transconductance efficiency, the width of each of the transistors T_2 and T_3 can be computed as,

$$W = \frac{I_D}{(I_D/W)_{g_m/I_D=2/V_{DS(sat)}}} \tag{A.133}$$

where $I_D = I_B$.

- In the strong inversion region, we have $V_{ov} = V_{DS(sat)} = V_{GS} - V_T$ and the inversion coefficient is given by

$$IC = \left[\ln\left(1 + \mathrm{e}^{\frac{V_{DS(sat)}}{2nU_T}}\right)\right]^2 \tag{A.134}$$

The aspect ratios of each of the transistors T_2 and T_3 can be obtained from the inversion coefficient definition as,

$$\frac{W}{L} = \frac{I_D}{I_T IC} \tag{A.135}$$

where $I_D = I_B$ and $I_T = 2n\mu_n C_{ox} U_T^2$ is the technology current.

The resistance R_B is given by

$$R_B = \frac{V_{DD} - V_{GS}}{I_B} \tag{A.136}$$

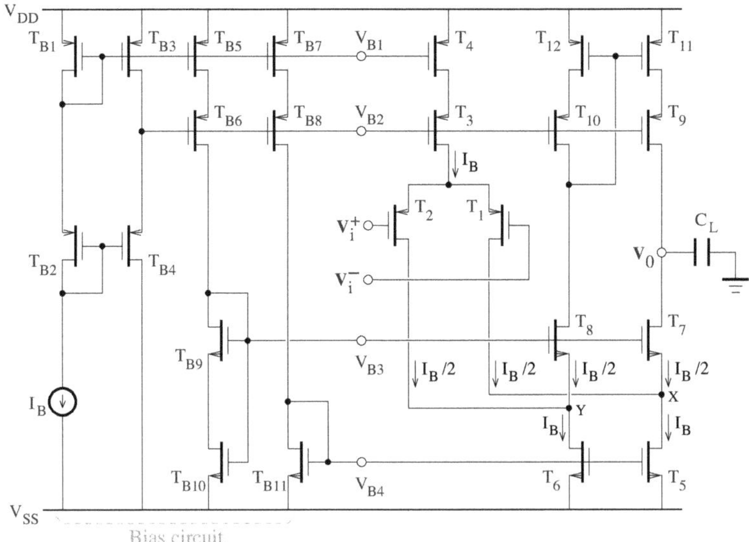

FIGURE A.16
Folded-cascode amplifier.

where $V_{GS} = V_{DS(sat)} + V_T$, and V_T is the threshold voltage.

□ Folded-cascode amplifier

Let us consider the folded-cascode amplifier shown in Figure A.16. The folded-cascode amplifier is less affected by the frequency degradation due to the power supply rejection ratio than the two-stage amplifier.

The gain-bandwidth product is of the form

$$GBW = 2\pi f_T = \frac{g_{m_1}}{C_L} \tag{A.137}$$

where f_T is the transition frequency, and the slew rate is defined as,

$$SR = \frac{I_B}{C_L} \tag{A.138}$$

where I_B is the bias current of the differential transistor pair or the drain current of transistors T_3 and T_4. Assuming that $I_{D_1} = I_B/2$, we have

$$\frac{g_{m_1}}{I_{D_1}} = \frac{4\pi f_T}{SR} \tag{A.139}$$

Assuming that the value of the transconductance efficiency is v_1, and using the plot of I_{D_1}/W_1 versus g_{m_1}/I_{D_1}, the transistor width is computed as

$$W_1 = W_2 = \frac{I_{D_1}}{(I_{D_1}/W_1)_{g_{m_1}/I_{D_1}=v_1}} \tag{A.140}$$

In practice, to meet the requirement for low noise, the widths of the input differential transistor pair should be greater than 100 μm.

In a well-designed amplifier, a phase margin greater than or equal to 60° requires that the first pole be dominant and also that the frequency of the second pole be higher than the gain-bandwidth product. From the small-signal analysis, we have

$$g_{m_7} = GBW \cdot C_p \cdot \tan \phi_M \tag{A.141}$$

where ϕ_M is the phase margin, and C_p is the total parasitic capacitance at the source of the transistor T_7. For initial calculations, $C_p \simeq C_{gs_7}$, and the capacitor C_{gs_7} is a linear function of $W_7 L_7$. Given the transconductance efficiency g_{m_7}/I_{D_7} and the fact that

$$W_7 L_7 = \frac{L_7^2}{IC_7} \cdot \frac{I_{D_7}}{I_S} \tag{A.142}$$

Equation (A.141) can be used to determine the length of transistor T_7.

TABLE A.3
Specifications of Transistors $T_3 - T_{12}$

	Source-drain saturation voltage	Inversion coefficient
$T_3 - T_4$	$V_{SD_3(sat)} = (V_{DD} - V_{SG_1} - V_{ICM,max})/2$	IC_3
$T_5 - T_6$	$V_{DS_5(sat)} = V_{ICM,min} - V_{SS} - V_{T_p}$	IC_5
$T_7 - T_8$	$V_{DS_7(sat)} = V_{0,min} - V_{SS} - V_{DS_5(sat)}$	IC_7
$T_9 - T_{10}$	$V_{SD_9(sat)} = V_{SD_{11}(sat)} = (V_{DD} - V_{0,max})/2$	IC_9
$T_{11} - T_{12}$	$V_{SD_{11}(sat)} = (V_{DD} - V_{0,max})/2$	IC_{11}

From the equations of the source-drain or drain-source saturation voltages given in Table A.3, where

$$V_{SG_1} = V_{SD_1(sat)} - V_{T_p} \tag{A.143}$$

we can obtain the inversion coefficients of transistors $T_3 - T_{12}$. Using the plot of g_{m_j}/I_{D_j} versus IC_j, the corresponding value, say v_j, of the transconductance efficiency is obtained.

Assuming a given value of the drain current and using the plot of I_{D_j}/W_j versus g_{m_j}/I_{D_j}, the transistor width is then estimated to be

$$W_j = \frac{I_{D_j}}{(I_{D_j}/W_j)_{g_{m_j}/I_{D_j}=v_j}} \tag{A.144}$$

For the bias circuit, the aspect ratios of transistors $T_{B1} - T_{B8}$ can be computed as follows:

$$\begin{aligned} W_3/L_3 &= W_{B1}/L_{B1} = 4(W_{B2}/L_{B2}) = W_{B3}/L_{B3} = W_{B4}/L_{B4} \\ &= W_{B5}/L_{B5} = W_{B6}/L_{B6} = W_{B7}/L_{B7} = W_{B8}/L_{B8} \end{aligned} \tag{A.145}$$

It can be shown that the aspect ratios of transistors $T_{B9} - T_{B10}$ are given by

$$W_{B10}/L_{B10} = (1/3)(W_{B9}/L_{B9}) \tag{A.146}$$

$$= \left(\frac{1}{\sqrt{2(W_3/L_3)}} + \frac{1}{\sqrt{(W_5/L_5)}}\right)^{-2} \tag{A.147}$$

Finally, the aspect ratios of transistors T_{B11} can be found from

$$W_{B11}/L_{B11} = W_5/L_5 \tag{A.148}$$

The *dc* gain of the designed amplifier is given by

$$A_0 = \frac{g_{m1}}{\dfrac{(g_1 + g_5)g_7}{g_{m7} + g_{mb7}} + \dfrac{g_{11}g_9}{g_{m9} + g_{mb9}}} \tag{A.149}$$

It can be rewritten as,

$$A_0 = \frac{g_{m1}}{I_{D_1}} V_{A_1} \left(\frac{1 + \dfrac{I_{D_5}}{V_{A_5}} \dfrac{V_{A_1}}{I_{D_1}}}{n_7 \dfrac{g_{m7}}{I_{D_7}} V_{A_7}} + \frac{\dfrac{I_{D_{11}}}{V_{A_{11}}} \dfrac{V_{A_1}}{I_{D_1}}}{n_9 \dfrac{g_{m9}}{I_{D_9}} V_{A_9}} \right)^{-1} \tag{A.150}$$

where, for the transistor T_k, $V_{A_k} = I_{D_k}/g_k$ is the Early voltage and $g_{mb_k}/g_{m_k} = n_k - 1$.

The total power consumption of the amplifier is

$$P_T = 2(V_{DD} - V_{SS})I_B + P_B \tag{A.151}$$

where $P_B = 4(V_{DD} - V_{SS})I_B$ is the power consumption of the bias circuit.

□ Two-stage amplifier

A two-stage amplifier is shown in Figure A.17, where the compensation resistor is implemented by the transistor T_C which operates in the moderate inversion (or triode) region.

For each transistor, T_j, assuming that the value of the transconductance

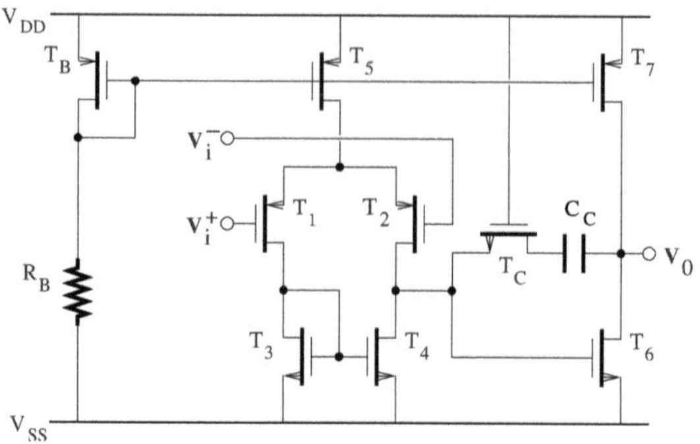

FIGURE A.17
Two-stage amplifier.

efficiency, g_{m_j}/I_{D_j}, is υ_j, and using the plot of I_{D_j}/W_j versus g_{m_j}/I_{D_j}, the transistor width is computed as

$$W_j = \frac{I_{D_j}}{(I_{D_j}/W_j)_{g_{m_j}/I_{D_j}=\upsilon_j}} \tag{A.152}$$

Initially, it can be assumed that the minimum value is chosen for the length of all transistors.

The gain-bandwidth product can be written as,

$$GBW = 2\pi f_T = \frac{g_{m_1}}{C_C} \tag{A.153}$$

where f_T is the transition frequency, and the slew rate is defined as,

$$SR = \frac{I_B}{C_C} \tag{A.154}$$

where I_B is the bias current of the differential transistor pair or the drain current of the transistor T_5. Assuming that $I_{D_1} = I_B/2$, we have

$$\frac{g_{m_1}}{I_{D_1}} = \frac{4\pi f_T}{SR} \tag{A.155}$$

Note that the sizes of transistors T_1 and T_2 are identical.

The transconductance efficiency of the transistor T_5 is given by

$$\frac{g_{m_5}}{I_{D_5}} = \frac{2}{V_{DS_5(sat)}} \tag{A.156}$$

where $V_{DS_5(sat)} = ICMR^- - |V_{T_p}| - V_{SD_1(sat)}$, $V_{SD_1(sat)} = SR/GBW$, and

$I_{D_5} = I_B$.

For transistors T_7 and T_5, we have

$$V_{SG_7} = V_{SG_5} \tag{A.157}$$

or equivalently,

$$\frac{g_{m_7}}{I_{D_7}} = \frac{g_{m_5}}{I_{D_5}} \tag{A.158}$$

where

$$I_{D_7} = (1 + C_L/C_C)I_{D_5} = (1 + C_L/C_C)I_B \tag{A.159}$$

To achieve a phase margin of at least 60°, the second pole should be sufficiently higher than the gain-bandwidth product. Hence,

$$\omega_{p_2} \geq 2.2 \cdot GBW \tag{A.160}$$

where $\omega_{p_2} \simeq g_{m_6}/C_L$. The transconductance efficiency of the transistor T_6 can be computed as,

$$\frac{g_{m_6}}{I_{D_6}} = 2.2 \cdot \frac{C_L}{C_C} \cdot \frac{I_{D_1}}{I_{D_6}} \cdot \frac{g_{m_1}}{I_{D_1}} \tag{A.161}$$

where $I_{D_1} = I_B/2$ and $I_{D_6} = I_{D_7} = (1 + C_L/C_C)I_B$.

The cancellation of the systematic offset that can be observed when the same voltage is applied to both differential inputs is realized by assuring that

$$V_{DS_4} = V_{DS_3} = V_{GS_3} \quad \text{and} \quad V_{DS_4} = V_{GS_6} \tag{A.162}$$

Hence,

$$\frac{g_{m_3}}{I_{D_3}} = \frac{g_{m_4}}{I_{D_4}} = \frac{g_{m_6}}{I_{D_6}} \tag{A.163}$$

where $I_{D_3} = I_{D_4} = I_{D_5}/2 = I_B/2$.

The compensation resistor is implemented by the transistor T_C, that operates in the triode region, where

$$I_D = \frac{1}{2}\mu_n C_{ox}\left(\frac{W_C}{L_C}\right)[2(V_{GS} - V_{T_n})V_{DS} - V_{DS}^2] \tag{A.164}$$

If $V_{DS} \ll 2(V_{GS} - V_{T_n})$, we can obtain

$$I_D \simeq \mu_n C_{ox}\left(\frac{W_C}{L_C}\right)(V_{GS} - V_{T_n})V_{DS} \tag{A.165}$$

The compensation resistance is then given by

$$R_C = \frac{I_D}{V_{DS}} = \frac{1}{\mu_n C_{ox}(W_C/L_C)(V_{GS} - V_{T_n})} \tag{A.166}$$

For the transistor T_C, it can also be shown that

$$\frac{g_m}{I_D} = \frac{1}{V_{GS} - V_T} \tag{A.167}$$

where $V_{GS} = V_{DD} - V_{DS_4(sat)} - V_{SS}$. Due to the capacitor C_C, the current flowing through the transistor T_C and some voltages existing between its nodes are reduced to *ac* components.

The compensation resistor will be chosen either as $R_C = 1/g_{m_6}$ if the zero is to be placed at infinity or as $R_C = (1 + C_L/C_C)/g_{m_6}$ if the second pole ω_{p_2} should be cancelled by the zero ω_z.

Due to the fact that the parasitic capacitances at the output nodes of the first and second stages, C_1 and C_2, are dependent on the transistor sizes, the amplifier is designed using an approximate value of the compensation capacitance C_C. Based on the small-signal analysis, the actual value of the capacitor C_C can be obtained as,

$$C_C = \frac{1}{2}\frac{\omega_{p_2}/GBW}{g_{m_6}/g_{m_1}}\left[C_1 + C_2 + \sqrt{(C_1 + C_2)^2 + 4\frac{\omega_{p_2}/GBW}{g_{m_6}/g_{m_1}}C_1 C_2}\right] \tag{A.168}$$

Some design iterations may then be necessary to update the transistor sizes accordingly to the exact value of C_C.

Assuming that the transistors operate in the strong inversion region, the input common-mode range (ICMR) is defined as follows,

$$V_{SS} + V_{DS_3(sat)} + V_{T_n} - |V_{T_p}| \geq ICMR \geq V_{DD} - V_{SD_5(sat)} - V_{SD_1(sat)} - |V_{T_p}| \tag{A.169}$$

It should be verified that the size of transistors T_3 and T_4 also satisfy the specification of the input common-mode range.

The output swing is given by

$$V_{SS} + V_{DS_6(sat)} \geq V_0 \geq V_{DD} - V_{DS_7(sat)} \tag{A.170}$$

The transistors T_6 and T_7 should be sized such that the output swing specification is also met.

The amplifier *dc* gain can be written as

$$A_0 = \frac{g_{m_1}}{|I_{D_1}|}\left(\frac{1}{1/|V_{A_1}| + 1/V_{A_3}}\right)\frac{g_{m_6}}{I_{D_6}}\left(\frac{1}{1/V_{A_6} + 1/|V_{A_7}|}\right) \tag{A.171}$$

where $I_{D_1} = I_{D_3}$ and $I_{D_6} = I_{D_7}$.
The power dissipation can be computed as

$$P = (V_{DD} - V_{SS})(2I_{D_5} + I_{D_7}) \tag{A.172}$$

A.5 Bibliography

[1] P. R. Gray and R. G. Meyer, "MOS operational amplifier design — A tutorial overview," *IEEE J. of Solid-State Circuits*, vol. 17, pp. 969–982, Dec. 1982.

[2] G. Palmisano, G. Palumbo, and S. Pennisi, "Design procedure for two-stage transconductance operational amplifiers: a tutorial," *Analog Integrated Circuits and Signal Processing*, vol. 27, pp. 179–189, June 2001.

[3] M. del M. Hershenson, S. P. Boyd, and T. H. Lee, "Optimal design of a CMOS op-amp via geometric programming," *IEEE Trans. on CAD of Integrated Circuits and Systems*, vol. 20, pp. 1–21, Jan. 2001.

[4] C.-C. Shih, and P. E. Gray, "Reference refreshing cyclic analog-to-digital and digital-to-analog converters," *IEEE J. of Solid-State Circuits*, vol. 21, pp. 544–554, Aug. 1986.

[5] S. M. Mallya and J. H. Nevin, "Design procedures for a fully differential folded-cascode amplifier," *IEEE J. of Solid-State Circuits*, vol. 24, pp. 1737–1740, Dec. 1989.

[6] H. C. Yang, M. A. Abu-Dayeh, and D. J. Allstot, "Small-signal analysis and minimum setling time design of a one-stage folded-cascode CMOS operational amplifier," *IEEE Trans. on Circuits and Systems*, vol. 38, pp. 804–807, July 1991.

[7] J. Yuan and C. Svensson, "New single-clock CMOS latches and flipflops with improved speed and power savings," *IEEE J. of Solid-State Circuits*, vol. 32, pp. 62–69, Jan. 1997.

[8] C. Enz, F. Krummenacher, and E. A. Vittoz, "An analytical MOS transistor model valid in all regions of operation and dedicated to low-voltage and low-current applications," *Analog Integrated Circuits and Signal Processing*, vol. 8, pp. 83–114, July 1995.

B

Signal Flow Graph

CONTENTS

Signal flow graphs (SFGs) [1] are suitable for the representation of continuous-time and discrete-time systems in the s-domain and z-domain, respectively. They can be considered a network diagram in which nodes are connected by unidirectional branches. Each node represents a system variable, and a scaling factor is indicated along each branch. The analysis and design of small and medium-size systems is often facilitated by the use of signal flow graph techniques.

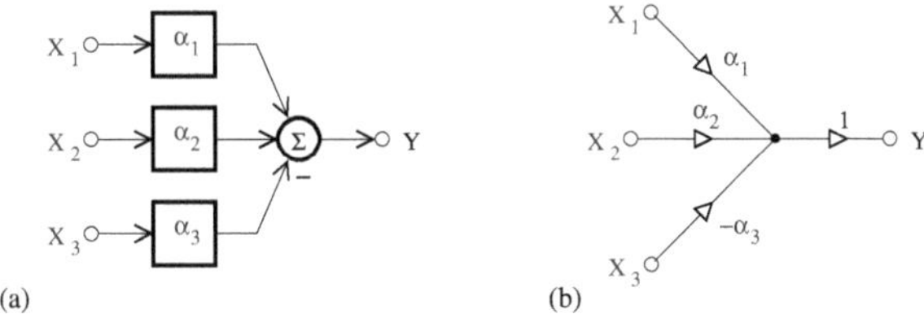

FIGURE B.1
(a) Block diagram and (b) signal flow graph representations of a system.

A block diagram and an SFG contain the same network information. They can represent a system of equations or an equation. The SFG representation is then not unique for a network, which can be characterized by equations in different forms. In the special case of Figure B.1, where the input nodes are labeled X_1, X_2, and X_3, the variable Y at the output node can be expressed as

$$Y = \alpha_1 X_1 + \alpha_2 X_2 - \alpha_3 X_3 \tag{B.1}$$

For the SFG, the summer is implemented by a node. Hence, a mixed node, that is, a node that receives both incoming and outgoing branches, sums the weighted signals of all incoming branches and the result is transmitted to all outgoing branches.

B.1 SFG reduction rules

Initial signal–flow graph		Final signal–flow graph
	Parallel rule	
	Series rule	
	Recursion rule	
	Node–splitting rule	

FIGURE B.2
Basic SFG reduction rules.

SFGs are often used in system analysis. Generally, the objective is to determine the relationship or transfer function between various variables. This can be achieved by the application of a set of rules such that nodes are absorbed and branches are combined to form new branches with the equivalent scaling factors. The reduction process is repeated until an SFG with the desired degree of complexity, or a residual graph in which only one path exists between the input node and the selected output node, is obtained. Basic SFG reduction rules are summarized in Figure B.2. However, the reduction procedure can become complicated as the system complexity increases. An alternative method for finding the relationship between system variables can be based on Mason's gain formula [2].

B.2 Mason's gain formula

To proceed further, it is necessary to give the following definitions:

- A path is a series connection of branches following the direction of the signal flow from one node to a given node.
- A forward path is a path from the input node to the output node that does not go through the same node twice.
- A loop is a path that starts and ends at the same node, without passing through any other node more than once.
- A path gain is the product of the gains or scaling factors of all its branches.
- Two loops are *nontouching* if they do not have any node in common.

Mason's gain formula gives the transfer function T from an input (or source) node to an output (or sink) node in the form

$$T = \frac{\sum_{k}^{P} T_k \triangle_k}{\triangle} \tag{B.2}$$

where

$\triangle$ = 1 – (Sum of all individual loop gains) + (Sum of the product of the loop gains of all possible combinations of nontouching loops taken two at a time) – (Sum of the product of the loop gains of all possible combinations of nontouching loops taken three at a time) + (Sum of the product of the loop gains of all possible combinations of nontouching loops taken four at a time) $- \cdots + (-1)^r$(Sum of the product of the loop gains of all possible combinations of nontouching loops taken r at a time),

P is the total number of forward paths between the input and output nodes,

T_k is the gain of the k-th forward path between the input and output nodes,

$\triangle_k$ is the value of $\triangle$ for the part of the graph not touching the k-th forward path.

The term $\triangle$ can be considered the determinant of the graph, while $\triangle_k$ represents the cofactor of the k-th forward path.

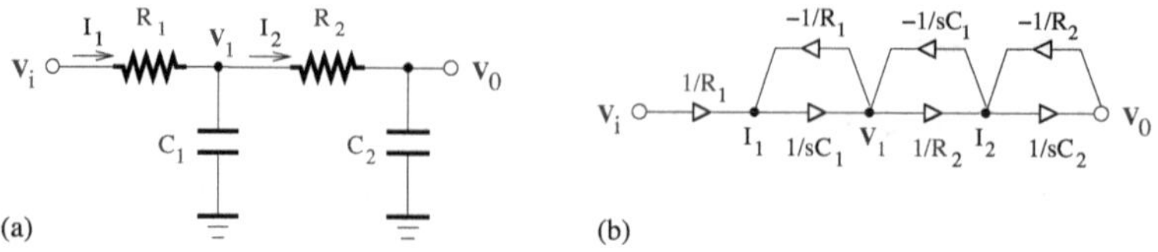

FIGURE B.3
(a) RC network; (b) equivalent SFG representation.

Consider the circuit diagram of the RC network depicted in Figure B.3(a). Using Kirchhoff's current law and Kirchhoff's voltage law gives

$$I_1 = \frac{V_i - V_1}{R_1} \tag{B.3}$$

$$V_1 = \frac{I_1 - I_2}{sC_1} \tag{B.4}$$

$$I_2 = \frac{V_1 - V_0}{R_2} \tag{B.5}$$

and

$$V_0 = \frac{I_2}{sC_2} \tag{B.6}$$

An SFG of the RC network can then be derived as shown in Figure B.3(b), where there are three individual loops with the gains $P_{11} = -1/sC_1R_1$, $P_{21} = -1/sC_1R_2$, and $P_{31} = -1/sC_2R_2$. Because the first and third loops are nontouching, we have $P_{12} = 1/s^2C_1C_2R_1R_2$. Hence, the determinant $\triangle$ is given by

$$\begin{aligned} \triangle &= 1 - (P_{11} + P_{21} + P_{31}) + P_{12} && \text{(B.7)} \\ &= 1 + 1/sC_1R_1 + 1/sC_1R_2 + 1/sC_2R_2 + 1/s^2C_1C_2R_1R_2 && \text{(B.8)} \end{aligned}$$

The gain of the forward path joining the input node and output node is

$$T_1 = 1/s^2C_1C_2R_1R_2 \tag{B.9}$$

Because the forward path touches all three loops, it can be shown that

$$\triangle_1 = 1 \tag{B.10}$$

Note that $\triangle_1$ is obtained from $\triangle$ by removing the contributions due to the loops that touch the forward path. The circuit transfer function is then of the form

$$T = \frac{V_0}{V_i} = \frac{T_1\triangle_1}{\triangle} \tag{B.11}$$

$$= \frac{1}{s^2C_1C_2R_1R_2 + s(C_1R_1 + C_2R_1 + C_2R_2) + 1} \tag{B.12}$$

It can be deduced that the aforementioned RC circuit is a second-order low-pass filter, whose transfer function only exhibits finite poles.

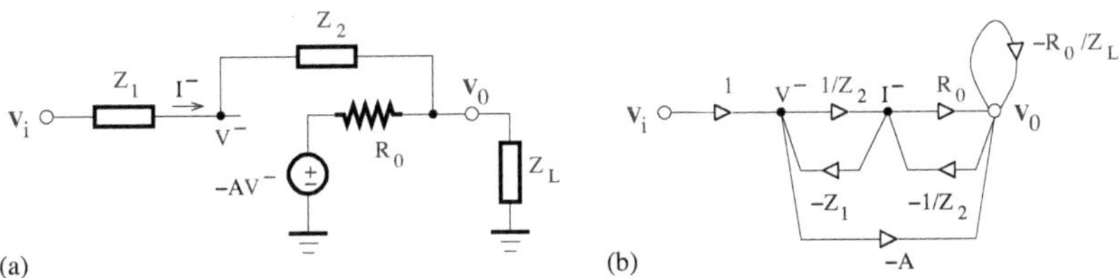

FIGURE B.4
(a) Equivalent model of an amplifier and (b) its SFG representation.

As a second example of circuit analysis using Mason's gain formula, consider the equivalent model of an amplifier shown in Figure B.4(a). Using Kirchhoff's laws, the following equations can be derived:

$$V^- = V_i - Z_1 I^- \tag{B.13}$$

$$I^- = \frac{V^- - V_0}{Z_2} \tag{B.14}$$

$$V_0 = -AV^- + R_0\left(I^- - \frac{V_0}{Z_L}\right) \tag{B.15}$$

The corresponding SFG representation is obtained as illustrated in Figure B.4(b). Four individual loops can be identified. Their gains are of the form $P_{11} = -Z_1/Z_2$, $P_{21} = -R_0/Z_2$, $P_{31} = -AZ_1/Z_2$, and $P_{41} = -R_0/Z_L$. We also have $P_{12} = Z_1R_0/Z_2Z_L$ because the loops P_{11} and P_{41} are nontouching. The term $\triangle$ is then given by

$$\triangle = 1 - (P_{11} + P_{21} + P_{31} + P_{41}) + P_{12} \tag{B.16}$$

$$= 1 + Z_1/Z_2 + R_0/Z_2 + AZ_1/Z_2 + R_0/Z_L + Z_1R_0/Z_2Z_L \tag{B.17}$$

The SFG exhibits two forward paths, whose gains can be written as, $T_1 = -A$ and $T_2 = R_0/Z_2$. Each forward path has a common node with all loops, and it can then be found that $\triangle_1 = 1$ and $\triangle_2 = 1$. Therefore, the transfer function between the input and output nodes reads

$$T = \frac{V_0}{V_i} = \frac{T_1\triangle_1 + T_2\triangle_2}{\triangle} \tag{B.18}$$

$$= -\frac{Z_2}{Z_1} \cdot \frac{1 - \frac{1}{A} \cdot \frac{R_0}{Z_2}}{1 + \frac{1}{A}\left(1 + \frac{Z_2}{Z_1} + \frac{R_0}{Z_1} + \frac{R_0}{Z_L} + \frac{Z_2R_0}{Z_1Z_L}\right)} \tag{B.19}$$

For a high gain, A, the transfer function is reduced to the ratio $-Z_2/Z_1$ and an inverting amplifier is realized.

B.3 Bibliography

[1] S. J. Mason, "Feedback theory — Some properties of signal flow graphs," *Proc. of the IRE*, vol. 41, pp. 1144–1156, Sept. 1953.

[2] S. J. Mason, "Feedback theory — Further properties of signal flow graph," *Proc. of the IRE*, vol. 44, pp. 920–926, July 1956.

C

Notes on Track-and-Hold Circuit Analysis

CONTENTS

A track-and-hold (T/H) circuit samples a continuous-time signal by periodically capturing and holding signal levels. Its transfer function can be derived as follows.

C.1 T/H transfer function

Let $v_i(t)$ and $v_O(t)$ be the input and output voltages, respectively, of the linear time-variant network characterized by the next equation

$$\frac{dv_O(t)}{dt} + A_k v_O(t) = B_k v_i(t) \quad \text{for} \quad nT + \sigma_{k-1} \leq t \leq nT + \sigma_k \tag{C.1}$$

where A_k and B_k are constant coefficients. To proceed further, the output voltage can be expressed as

$$v_O(t) = \sum_{k=1}^{K} v_{O,k}(t) \tag{C.2}$$

where

$$v_{O,k}(t) = v_O(t) w_k(t) \tag{C.3}$$

and

$$w_k(t) = \begin{cases} 1 & nT + \sigma_{k-1} \leq t \leq nT + \sigma_k \\ 0 & \text{otherwise} \end{cases} \tag{C.4}$$

For a T/H circuit controlled by a sampling signal with the period T, Figure C.1 shows waveform plots, where the duration of the sampling and hold intervals are τ and $T - \tau$, respectively. For $k = 2$, $\sigma_1 = \tau$ and $\sigma_2 = T$.

The response of the linear time-variant network can be constrained to take the value zero outside the definition interval by disconnecting the input

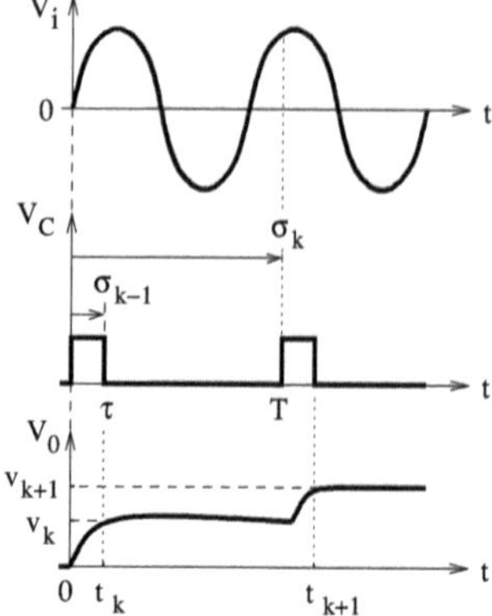

FIGURE C.1
T/H waveform representation.

source and subtracting the final conditions at the interval end [1,2,3]. Hence, for $-\infty \leq t \leq \infty$, we have

$$\frac{dv_{0,k}(t)}{dt} = A_k v_{0,k}(t) + B_k v_{i,k}(t) + \sum_{n=-\infty}^{\infty} [v_0(t)\delta(t - nT - \sigma_{k-1}) - v_0(t)\delta(t - nT - \sigma_k)] \tag{C.5}$$

where

$$v_{i,k}(t) = v_i(t)w_k(t) \tag{C.6}$$

and $\delta(t)$ is known as the delta or unit impulse function. Equation (C.5) can be solved using the Fourier transform because it is valid for all t. Hence,

$$(j\omega - A_k)V_{0,k}(\omega) = B_k \mathcal{F}[v_i(t)w_k(t)] + \sum_{n=-\infty}^{\infty} (\mathcal{F}[v_0(t)\delta(t - nT - \sigma_{k-1})] - \mathcal{F}[v_0(t)\delta(t - nT - \sigma_k)]) \tag{C.7}$$

and

$$V_0(\omega) = \sum_{k=1}^{K} V_{0,k}(\omega) \tag{C.8}$$

The multiplication of $v_i(t)$ and $w_k(t)$ in the time domain is transformed into a convolution product in the frequency domain. That is,

$$\mathcal{F}[v_i(t)w_k(t)] = \sum_{n=-\infty}^{\infty} \frac{1 - e^{-j\omega_s n\tau_k}}{jn\omega_s T} e^{-j\omega_s n\sigma_{k-1}} V_i(\omega - n\omega_s) \tag{C.9}$$

Assuming that the sampled output voltage and the input voltage are related by $G_k(\omega)$ or the transfer function at the switching instant $nT + \sigma_k$, it can be shown that

$$\sum_{n=-\infty}^{\infty} \mathcal{F}[v_0(t)\delta(t - nT - \sigma_k)] = \sum_{n=-\infty}^{\infty} (G_k(\omega)\mathcal{F}[v_i(t)]) * \delta(\omega - n\omega_s)\frac{e^{-j\omega_s n\sigma_k}}{T} \tag{C.10}$$

where $*$ is the convolution operation. Or equivalently,

$$\sum_{n=-\infty}^{\infty} \mathcal{F}[v_0(t)\delta(t-nT-\sigma_k)] = \sum_{n=-\infty}^{\infty} G_k(\omega-n\omega_s)\frac{e^{-j\omega_s n\sigma_k}}{T}V_i(\omega-n\omega_s) \tag{C.11}$$

Combining (C.7), (C.9), and (C.11), we obtain

$$V_{0,k}(\omega) = \sum_{n=-\infty}^{\infty} H_{n,k}(\omega)V_i(\omega - n\omega_s) \tag{C.12}$$

where

$$\begin{aligned} H_{n,k}(\omega) = \frac{1}{j\omega - A_k}\Bigg[&B_k\frac{1 - e^{-j\omega_s n\tau_k}}{jn\omega_s T}e^{-j\omega_s n\sigma_{k-1}} \\ &+ G_{k-1}(\omega - n\omega_s)\frac{e^{-j\omega_s n\sigma_{k-1}}}{T} - G_k(\omega - n\omega_s)\frac{e^{-j\omega_s n\sigma_k}}{T}\Bigg] \end{aligned} \tag{C.13}$$

The transfer function of the linear time-variant network is then of the form,

$$H_n(\omega) = \sum_{k=1}^{K} H_{n,k}(\omega) \tag{C.14}$$

Note that K can be taken equal to 2 for the T/H circuit.

Consider the next first-order differential equation

$$\frac{dv_{0,k}(t)}{dt} + A_k v_{0,k}(t) = B_k v_i(t) \tag{C.15}$$

The output response can be obtained as

$$v_{0,k}(t) = \phi_k(t - t_0)v_{0,k}(t_0) + B_k\int_{t_0}^{t} \phi_k(t - \tau)v_i(\tau)\mathrm{d}\tau \tag{C.16}$$

$$\phi_k(t) = e^{A_k t} \tag{C.17}$$

where the initial condition is specified at the instant t_0. In the case of a sinusoidal input voltage, $v_i(t) = e^{j\omega t}$, we have

$$v_{0,1}(t) = e^{-\omega_c(t-t_0)}v_{0,1}(t_0) + \frac{1}{1 + j\omega/\omega_c}[e^{j\omega(t-t_0)} - e^{-\omega_c(t-t_0)}]e^{j\omega t_0} \tag{C.18}$$

during the sampling phase, where $-A_1 = B_1 = 1/RC$, and

$$v_{0,2}(t) = v_{0,2}(t_0) \tag{C.19}$$

during the hold phase, where $A_2 = B_2 = 0$. Due to the fact that the output voltage is continuous, the initial value of an interval is equal to the final value of the previous interval. Assuming that $t_0 = nT$, we arrive at

$$v_0((n+1)T) = e^{-\omega_c \tau} v_0(nT) + \frac{1}{1 + j\omega/\omega_c}[e^{j\omega\tau} - e^{-\omega_c \tau}]e^{j\omega nT} \tag{C.20}$$

Equation (C.20) can be considered as a first-order difference equation, whose solution is composed of steady-state and transient voltage terms. After the transient voltage contribution has vanished, the frequency response remains only dependent on the steady-state voltage. Using $v_0((n+1)T) = v_0(nT)e^{j\omega T}$, the steady-state output voltage is obtained as

$$v_0(nT) = G(\omega)e^{j\omega nT} \tag{C.21}$$

where

$$G(\omega) = \frac{1}{1 + j\omega/\omega_c} \frac{e^{j\omega\tau} - e^{-\omega_c \tau}}{e^{j\omega T} - e^{-\omega_c \tau}} \tag{C.22}$$

$$= \frac{1}{1 + j\omega/\omega_c} \frac{1 - e^{-\omega_c \tau} e^{-j\omega\tau}}{1 - e^{-\omega_c \tau} e^{-j\omega T}} e^{-j\omega(T-\tau)} \tag{C.23}$$

The transfer function $G(\omega)$ can be modeled as a cascade of a first-order RC filter whose output is subtracted from its delayed and scaled version, and a network with the transfer function $e^{-j\omega(T-\tau)}/(1 - e^{-\omega_c \tau} e^{-j\omega T})$.

C.2 Bibliography

[1] T. Ström and S. Signell, "Analysis of periodically switched linear circuits," *IEEE Trans. Circuits Syst. I*, vol. 24, no. 10, pp. 531–541, Oct. 1977.

[2] A. Opal and J. Vlach, "Analysis and sensitivity of periodically switched linear networks," *IEEE Trans. Circuits Syst. I*, vol. 36, no. 4, pp. 522–532, Apr. 1989.

[3] M. C. M. Soer, E. A. M. Klumperink, P.-T. de Boer, F. E. van Vliet, and B. Nauta, "Unified frequency-domain analysis of switched-series-RC passive mixers and samplers," *IEEE Trans. Circuits Syst. I*, vol. 57, no. 10, pp. 2618–2631, Oct. 2010.

Index

G

H

I

J

K

L

O

P

R

S

T

Milton Keynes UK
Ingram Content Group UK Ltd.
UKHW021934071024
449327UK00022B/1798